APPLICATIONS *and* COMPUTATIONAL ELEMENTS *of* INDUSTRIAL HYGIENE

T0199389

APPLICATIONS *and* COMPUTATIONAL ELEMENTS *of* INDUSTRIAL HYGIENE

Martin B. Stern
S.Z. Mansdorf

CRC Press
Taylor & Francis Group
Boca Raton London New York

CRC Press is an imprint of the
Taylor & Francis Group, an **informa** business
A TAYLOR & FRANCIS BOOK

CRC Press
Taylor & Francis Group
6000 Broken Sound Parkway NW, Suite 300
Boca Raton, FL 33487-2742

First issued in paperback 2019

© 1999 by Taylor & Francis Group, LLC
CRC Press is an imprint of Taylor & Francis Group, an Informa business

No claim to original U.S. Government works

ISBN-13: 978-1-56670-197-6 (hbk)
ISBN-13: 978-0-367-40040-8 (pbk)

Library of Congress Card Number 98-6443

Library of Congress Cataloging-in-Publication Data

Applications and computational elements of industrial hygiene / edited by
Martin B. Stern, S. Z. Mansdorf.
p.   cm.
Includes bibliographical references and index.
ISBN 1-56670-197-X (alk. paper)
1. Industrial hygiene.   I. Stern, Martin B.   II. Mansdorf, S. Z. (Seymour Zack) ,
1947-.
RC967.A67   1998
616.9 ' 803—dc21   98-6443

**Visit the Taylor & Francis Web site at
http://www.taylorandfrancis.com**

**and the CRC Press Web site at
http://www.crcpress.com**

PREFACE

Martin B. Stern, MPH, CIH

Industrial hygiene has emerged as a respected technical profession that links subject areas that range from biology, chemistry, and toxicology, to engineering, mathematics, and management. An industrial hygienist has a need to problem-solve in a manner comparable to detective work: an unknown exposure event occurs and the industrial hygienist must piece together all of the critical events that lead up to the employee's exposure and attempt to determine what (and how) the employee was exposed, and develop recommendations to protect the worker from future exposure events or episodes. At times the information available is sufficient to determine action, however at other times the industrial hygienist is remanded to invest significant amounts of time, energy, and resources into solving the exposure event.

The reason *Applications of Industrial Hygiene* was written was to provide a tool for individuals preparing for one of the certification examinations, for use as textbook, and for use as a reference source for professionals in the field. The *Application's* work is a compilation of contributions from 14 seasoned industrial hygiene professionals, coming from academia, the private sector, and combinations of academia and the workplace. The individuals who were selected to contribute chapters were selected based on their technical expertise, experience, and competency. It was felt by the editors that it would not be effective for one or two authors to create an industrial hygiene book, as invariably the work would lack sufficient depth and breadth. To avoid this, each contributor brought specialized knowledge and experience to the table for the chapters that they authored.

The information contained in the work is designed to be straightforward and comprehensible, thus useable and extractable from an application standpoint. The editors desired to create a textbook which presents users with a "soup-to-nuts approach" in each chapter by including detailed technical coverage for the industrial hygiene rubrics. Each chapter was designed to provide an extensive overview of the given industrial hygiene rubric, industrial/workplace applications, monitoring/sampling techniques (where applicable), personal protective equipment needs (where applicable), and key computational needs. An additional value-added component is that the chapters are equipped with case studies so that textbook users can further relate the technical information contained in each chapter to a real-world application. At the culmination of each chapter, there are questions and answers to enable users to demonstrate their understanding of the chapter's content. By employing this approach for each chapter, the editors feel that users of the *Application's* work will gain critical technical knowledge, as well as computational skills, for each of the topics covered.

The *Application's* book contains over 250 calculations. Unlike other textbooks that have multiple contributors, the equations contained within the book follow a common format in each chapter: listing the equation's variables, describing what each variable represents, and where applicable, a sample calculation is provided to demonstrate how to use the calculation in a "practical situation." The equations/calculations are presented in such a manner so that the equations are readily evident in each chapter. Each equation has been assigned an independent number for easy reference. It is hoped that this approach will minimize users from having to obtain multiple industrial hygiene titles when studying for one of the certification examinations, preparing for a homework assignment/test in a graduate or undergraduate course, or referencing information to solve a problem in the workplace.

The editors feel that this work will be a welcome addition to your industrial hygiene collection, and are certain that this work will be the source that you will most frequently go to for information, because it will be easy to use, yet contain the detail needed to solve the more complex issues. This work is efficiently prepared, as you will not find a lot of *filler* information, the material typically included to increase the page count and subsequently the price. The filler information usually does not help you solve problems, but helps the author and editors make their publishing commitments. The editors purposely avoided this type of information. The *Application's* text contains the necessary information that you will need.

We hope you enjoy the *Application's* work, and it assists you in the solution of your industrial hygiene problem needs.

CONTRIBUTORS

Editors

S.Z. (Zack) Mansdorf, PhD, CIH, CSP
A. D. Little
Cambridge, MA

Martin B. Stern, MPH, CIH
AT&T
Warren, NJ

Contributing Authors

Peter Bellin, PhD, CIH
California State University
Northridge, CA

Anthony Joseph, PhD
Indiana University
Indiana, PA

D. Jeff Burton, BS, MS, PE, CIH, CSP
IVE, Inc.
Bountiful, UT

James P. Kohn, PhD, CIH, CSP
East Carolina University
Greenville, NC

Jack Daugherty, CIH, CHMM
Vickers, Inc.
Jackson, MS

Howard K. Pelton, PE
Pelton Marsh Kinsella
Dallas, TX

Robert Hague, PhD, CIH
University of Medicine & Dentistry of NJ
Piscataway, NJ

Ilse Stoll, MS
Lucent Technology Inc.
Allentown, PA

Stephen K. Hall, PhD, CIH
Oakland University
Oakland, MI

Wesley R. Van Pelt, PhD, CIH, CHP
Wesley R. Van Pelt Associates, Inc.
Paramus, NJ

R. Timothy Hitchcock, MSPH, CIH
IBM Corporation
Research Triangle Park, NC

Celeste Winterberger, PhD, CIT
East Carolina University
Greenville, NC

ACKNOWLEDGMENTS

I would like to thank the following individuals for their support, patience, and consideration during the preparation of the *Application's* book:

To my wife, Heather for her patience, support, and love, during the *Application's* book preparation, as I routinely burned the "midnight oil" and was a borderline zombie while working on the book. She routinely pushed me during times I wanted to say the hell with the book (as well as other things). She was the epitome of consideration during the book writing process and I love her;

To my children, Chad, Marty, and Christina, for not getting too frustrated when Daddy had work to do and couldn't play, even though the reality is I would have rather been out playing with them;

To my parents, for pushing me in my youth, and encouraging me to take the paths of *most* resistance. Their parenting has been influential in helping me to be more than someone who watches life pass by and takes the easy train;

To Ken McCombs, CRC Acquisitions Editor, for having the guts to move on a project idea and provide blind-faith support throughout the project. He's the kind of person I would want on any writing project. Also from CRC, Susan Alfieri, who provided support through many of the project's more "challenging" periods.

Special thanks to: Dave Nargis (AT&T), Kelly Stefanik (AT&T), and Paul Kim (Johnson & Johnson) for taking personal time to review selected chapters.

MBS

I would like to acknowledge the unfailing encouragement and hard work of two key persons responsible for the successful completion of this book. First, Ken McCombs of CRC Press. Ken was an invaluable aid and resource in the shepherding of the book. Second, my coeditor, Marty Stern. Marty was the person really responsible for the genesis of the book. He put up with my delays in a kind spirit but with an unending dedication to accomplishing a valuable work that all of the authors could be proud of achieving. The reader should know that my part in this worthwhile venture was small with Marty being the person really responsible for its success.

SZM

TABLE OF CONTENTS

CHAPTER 5   ANALYTICAL CHEMISTRY FOR INDUSTRIAL HYGIENISTS 241

Ilse Stoll

Howard K. Pelton

Chapter 1

INTRODUCTION

S. Zack Mansdorf, PhD, CIH, CSP, PE

I. OVERVIEW

Industrial hygiene is a profession dedicated to the anticipation, recognition, evaluation, and control of environmental factors that could result in the injury, illness, impairment, or adversely affect the well-being or efficiency of workers and members of the community. It is a profession that has its roots in many of the traditional sciences (e.g., chemistry, engineering, mathematics, medicine, physics, etc.) as well as the social sciences. Like medicine, it is commonly thought to be both a science and art. A discipline based on established scientific laws and principles which also requires deductive skills and imagination (hence the art).

Industrial hygiene, as a profession, has experienced a phenomenal growth in the last nearly three decades since the passage of the Occupational Safety and Health Act of 1970 in the U.S. While this growth rate is likely to moderate, it is clear that the profession is here to stay and will prosper not only in the U.S. but worldwide.

The remainder of this chapter is organized into seven sections. The first part contains a brief history of industrial hygiene. This is followed by an overview of the structure and function of two important government agencies important to occupational health and safety. These are the Occupational Health and Safety Administration (OSHA) and the National Institute for Occupational Safety and Health (NIOSH). The next section describes the two major professional membership organizations for industrial hygienists in the U.S. These are the American Industrial Hygiene Association (AIHA) and the American Conference of Governmental Industrial Hygienists (ACGIH). This section is followed by a description of the professional certification body for industrial hygienists–the American Board of Industrial Hygienists (ABIH). Finally, a review of the basic tenets of the art and science of industrial hygiene is presented: namely, the anticipation, recognition, evaluation, and control of occupational and community hazards.

II. A BRIEF HISTORY OF INDUSTRIAL HYGIENE

Industrial hygiene, as we know it today, has a relatively recent history as a distinct professional practice. Nevertheless, it is rooted in the ancient history of medicine and the sciences. Since it is related to the "industrial"

1-56670-197-X/99/$0.00=$.50
© 1999 by CRC Press LLC

activities of the human species, the first historical reference might be made as much as one million years ago (B.C.) when Australoptihecus walked the earth and suffered the consequences of making stone tools and hunting. Needless to say, little is thought to have occurred in relation to the recognition or control of these "occupational" hazards of ancient times. Even through the Copper and Bronze Ages (circa 5000 B.C.) with their related mining and smelting activities, it was not until about 2000 B.C. that an anonymous Egyptian wrote in *Papyrus Selier II*:

> I have seen...the metal worker at his work; he is grilled at the mouth of the furnace. The mason, exposed to all weathers and all risks, builds without clothing. His arms are worn out with work, his food is mixed up with dirt and rubbish....

From that brief description of toil, it was not until the time of Hippocrates (460-370 B.C.) that occupational disease was first thought to have been recorded. *Hippocrates*, the Greek physician often called the "Father of Medicine," described the devastating effects of lead poisoning to miners and metallurgists. He also wrote of fullers, cleaners, and dyers of cloth who used their own urine (improved with wine) as a cleaning agent and subsequently would develop fever, cough, and swelling of the groin and neck.. While he was of the belief that good health was the result of the balance of "humors" of the body, he did little to suggest causation or prevention of the adverse health effects of mining and smelting lead.

The next major historical figure with a connection to industrial hygiene was Caius Plinius Secundus, known as *"Pliney, the Elder."* Pliney, the Elder (23-79 A.D.) was a Roman scholar who described the use of crude respirators made from animal bladders, "...lest they should inhale the pernicious dust." used by certain mining trades especially in the mining of cinnabar (red mercuric sulfate). It is also interesting to note that his scientific curiosity resulted in his death from watching the eruption of Mount Vesuvius. Pliney, the Elder, might then be thought of as the first "industrial hygienist" to suggest a control measure to lessen the adverse effects of exposure to mineral dusts. The next historical figure of note was *Galen* (130-200). Galen was a Greek physician who attended the Roman emperor Marcus Aurelius. He is remembered by us for his work in describing the adverse effects of acid mists to copper miners. As an example, he had written of a visit to a cave in Cyprus where workers (and Galen) suffered from the "fumes" generated by the greenish waters resulting from copper mining. This waste byproduct was commonly used to make vitriol (a greenish fluid of cooper and other sulfates used in medicine and for dyeing). It was common for miners working in copper mines to have their teeth completely eroded from exposure to acid mists.

From the era of Galen, it is not until the middle centuries that there are any further writings of any significance. An industrial hygienist and student of English, George Krafcisin, in his Masters thesis has suggested that *Chau-*

cer described the ill effects of the alchemists trade in his *Canterbury Tales* written around 1390. In this medieval tale, Chaucer describes the ill effects of "fumes diverse of metals" to a Yeoman. The Yeoman, who is described as sweating and of a dull leaden hue, when asked by his host (one of the Canterbury Pilgrims) why his face is so pale, replies-"I am so used in the fyr to blowe/That it halth chaunged my colour, I trowe." Krafcisin attributes this and other inferences in this early literature to the alchemist's trade of working with mercury and arsenic in their quest to turn lead to gold. In the same era (1472), *Ulrich Ellenbog*, of the mining town Augsburg, wrote an eight-page booklet which discussed the toxic effects of working with mercury and lead. However, it was not until the time of *Philippus Paracelsus* (1493-1541) that a full treatise was written on the hazards of mining. His book, *Von der Bergsucht und andersen Bergkrankheiten* (On the Miners' Sickness and Other Diseases of Miners) was published after his death in 1567. Paracelsus, also known by the name Theophrastus Bombastus von Hohenheim, was a Swiss physician, alchemist, and scientist. He was town physician and lectured in Basel on the value of observational evidence. Since this was the age of humoral healing and bleeding, of which Paracelsus was not a proponent, he was considered by many of his contemporaries to be a "quack." While many of his beliefs were contrary to practices of today, he is credited with the early use of medicinals. His contributions to industrial hygiene were related to his writings about chronic lung and stomach diseases from the effects of mining. He wisely attributed these ill effects to the vapors and emanations from smelting. He also accurately described the symptoms of mercurialism and wrote of the effects of "choke damp," which is carbon dioxide. More importantly, he suggested that avoiding smelting emanations would reduce the chance of disease. Paracelsus also contributed to the field of toxicology and our concept of thresholds with his now-famous statement, "All substances are poisons; there is none that is not a poison. The right dose differentiates a poison and a remedy." This basic tenant lead to the concept of our current *Threshold Limit Values*.

At about the same time as that of Paracelsus, Georg Bauer completed a scholarly work titled, *De Re Metallica,* comprising 12 books about mining technology and its associated hazards. Bauer is best known by his Latin name of *Georgius Agricola.* While his books were published a full 11 years before (1556) those of Paracelsus, I have listed him after Paracelsus since he was born and died slightly before his contemporary. He was born at Glauchau in Saxony at the beginning of the Renaissance in 1494. Thirty-three years later from his position of town physician for Jachimsthalm (in a Bohemian mining district), he carefully studied the mining techniques and practices of the era. This period of experience and his position as town physician would provide the material for his famous literary work on mining. Of interest to us, Agricola described asthma and the ulceration the lungs caused by inhalation of certain dusts. He described what is known today as silicosis, as well as TB and other diseases of the lung. In one section of his famous works, he graphically describes the effects of working in mines and the short life expectancy of miners in the Carpathian Mountains:

> ...some mines are so dry that they are entirely devoid of water and this dryness causes the workmen even greater harm, for the dust which is stirred up by digging, penetrates into the windpipe and lungs, and produces difficulty in breathing and the disease which the Greeks called asthma. If the dust has corrosive qualities, it eats away the lungs, and implants consumption in the body. In the mines of the Carpathian Mountains women are found who have married seven husbands, all of whom this terrible consumption has carried to a premature death.

The work of Agricola was principally devoted to mining technology and not health. Nevertheless, he does include information on ventilation design and health issues. It should also be noted that Agricola, like Pliney the Elder, suggested covering of the mouth and nose (with loose veils) to reduce the effects of the dusty conditions. Many of the practices of this period can be seen in the woodcuts which accompany *De Re Metallica* published by Dover Press from a translation done by H.C. Hoover and L.H. Hoover (the famous American President Herbert Hoover and his wife).

It was over a century later that the next significant work was published. *Bernardino Ramazzini*, (1633-1714) commonly referred to as the "Father of Industrial Medicine," published *De Morbis Arificum Diatiba* (Diseases of Workers) originally in 1700, and then in an expanded version just a year before his death in 1713. Ramazzini was professor of medicine at the University of Modena, and later chair of medicine at the University of Padua in Italy. While not as well traveled as others before his time, he did visit and treat a large number of workers of all types in his local area. He also relied on the scholarly works by Galen, Agricola, and others before him to classify and comment on occupational disease. In one instance early in his career, he describes the incident that he says "...first gave me the idea of writing this treatise on diseases of workers." In this anecdote, he describes observing a man cleaning a cesspit [ccsspool]. He writes:

> I pitied him at the filthy work and asked him why he was working so strenuously and why he did not take it more quietly so as to avoid the fatigue that follows overexertion...[The worker] said, no one who has not tried it can imagine what it cost to stay more than four hours in this place; it is the same thing as being struck blind...

It should be noted at this point that Ramazzini states that blindness is common to cleaners of privies and cesspits just as the worker had described in the story.

Ramazzini was also the first physician to carefully describe the common occupations of his era and the ill effects of exposures from work. As our final example from his work, he describes the problem of lead intoxication of potters:

> In almost all cities there are other workers who habitually incur the serious maladies from the deadly fumes of metals. Among these are the potters...when they need roasted or calcined lead for glazing their pots, they make the lead in marble vessels, and in order to do this they hang a wooden pole from the roof, fasten a square stone to its end, and then turn it round and round. During this process or again when they use tongs to daub the pots with molten lead before putting them into the furnace, their mouths, nostrils, and the whole body take in the lead poison that has been melted and dissolved in water; hence they are soon attacked by grievous maladies. First their hands become palsied, then they become paralytic, splenetic, lethargic, cachectic, and toothless, so that one rarely sees a potter whose face is not cadaverous and the color of lead.

He is also credited with suggesting preventive measures which led to such changes as the wet grinding of flint. Finally, he is best remembered for suggesting that the patient history include the question: "What is your occupation?" The first instance of formally suggesting that doctors consider a person's occupation as the potential cause of disease. This, of course, is a standard part of any medical history today.

Further literature contributions on the hazards of working men and women were not made until the 1900s. Nevertheless, others also contributed during this span of time and the start of the industrial revolution. *Sir Percival Pott*, an English physician (one of the early founders of the science of epidemiology), deduced the first connection to occupationally related disease in chimney sweeps in 1775. He noted an unusual incidence of scrotal cancer in young boys working for chimney sweeps. Scrotal cancer would normally be a very rare disease but was common to boys in this profession. He concluded it was related to their work exposure to soot. The young boys used to climb up and down chimneys, which was indeed sad and very hazardous. Pott described it as follows:

> ...The fate of these people seems singularly hard; in their infancy they are most frequently treated with great brutality, and almost starved with cold and hunger; they are thrust up narrow, and sometimes hot chimneys, where they are buried, burned and almost suffocated; and when they get to puberty, become liable to a most noisome, painful, and fatal disease.

This pioneering work and the work of others led to the Chimney Sweeps Act of 1840 which prohibited the use of young boys to climb up chimneys.

About the time of the work of Sir Percival Pott, a famous American statesman also wrote of the dangers of work and certain lifestyles. *Benjamin Franklin* is credited with an interest in the effects of exposure to lead. He wrote of the "West Indian Gripes," which is stomach pain resulting from the drinking of West Indies rum made from lead stills. This was a common problem in the "Colonies." He also was familiar with the occupational problems of working with lead. He wrote to his friend Benjamin Vaughan in 1786 concerning the plight of those making and using lead type for printing. Franklin had earlier experienced a case of lead poisoning himself when working with hot lead type. He wrote to his friend about a discussion he had with a fellow "letter-founder":

> ...his People, who work'd over the little Furnances of melted Metal, were not subject to that Disorder; he made light of any Danger from the Effluvai, but ascrib'd it to Particles of the Metal swallow'd with their Food by slovenly Workmen, who went to their Meals after handling the Metal, without well-washing their Fingers, so that some of the metalalline Particles were taken off by their Bread and eaten with it. This appear'd to have some Reason in it. But the Pain I had experienc'd made me still afraid of those Effluvia.

Later in the same letter, Franklin describes a visit to a French hospital, La Charite, which was famous for the treatment of lead poisoning. He describes a pamphlet listing the occupations treated for lead intoxication which included Plumbers, Glasiers [sic], Painters, Stonecutters, and Soldiers. Franklin, like a modern-day industrial hygienist, did not understand the connection of stonecutters and soldiers and writes:

> ...These I could not reconcile to my Notion that Lead was the Cause of that Disorder. But on mentioning this Difficulty to a Physician of that Hospital, he inform'd me that the Stonecutters are continually using melted lead to fix the End of Iron Balustrades in Stonc; and that the Soldiers had been employ'd by Painters-Labourers in Grinding of Colours.

Franklin was clearly a man of many skills and interests and could have been an excellent member of our profession.

About the time of the Chimney Sweeps Act (1840), *Charles Thackrah*, wrote *The Effects of Arts, Trades and Professions and All Civic States and Habits of Living Life and Longevity*. Thackrah was a physician and epidemiologist who treated workers in the manufacturing town of Leeds. Among his studies, he noted that workers employed on dry grinding died between the ages of 28-32 while those employing a wet grinding method typically lived to 40 to 50 years. This is an early application of an incidence rate approach leading to control of occupational hazards without knowing the specific causation of disease.

The rapid industrialization in England during the 1800s brought with it an equal growth in the injuries and illnesses of workers toiling under difficult working conditions. The first attempt to correct this problem occurred in 1802 with the passage of the Health and Morals of Apprentices Act. This Act restricted the working hours for apprentices to 12 hours per day in cotton and wool mills, required religious instruction, and some sanitary measures. Following this, the English Factory Acts of 1833 were enacted. The English established the principle of compensating injured workers. This was followed by U.S. legislation in Massachusetts that established employer's liability for worker injuries in 1877. Prior to this time, a worker would have to prove the employer was at fault and the employer had a number of defenses that he could employ. These included the notion that by accepting employment in a hazardous occupation or job, the worker accepted the risk.

In 1878, a revised English Factory Act created a central inspection bureau to assure minimal standards for factories. Also of note, the physician *Thomas Legge* became the first medical inspector of factories in 1898. However, the government's unwillingness to support the ILO White Lead Convention led to his becoming the Medical Advisor to the Trades Union Congress, where he published *Industrial Maladies*, a definitive work for this period. This was followed by the publication of *Dangerous Trades* by *Sir Thomas Oliver* in 1908, which also contributed greatly to the understanding of occupational lead poisoning. By 1913, the first state industrial hygiene programs in the U.S. were established in New York and Ohio. There also were other contributions from European scientists and doctors during this era; however, the next major figure to emerge was an American, *Dr. Alice Hamilton*. Although two other American doctors published works (Charles Brigham of the Michigan Board of Health published a brief essay on worker diseases in 1875 followed by a later study by Benjamin McCready on diseases of trade workers), it was Alice Hamilton that completed and published the most definitive and alarming studies in the early 1900s.

Alice Hamilton received her medical degree from the University of Michigan in 1893 with continuing studies at John Hopkins among other schools. She began her long career in occupational medicine through her experience supervising a one-year study of lead poisoning for the Illinois Commission on Occupational Disease in 1910. This was followed by employment as a special investigator for the Bureau of Labor Statistics over the next 10 years. She took a teaching position in occupational medicine at Harvard University in 1919 (very unusual for a woman to occupy such a position), which she held until 1935. During this time, she published *Industrial Poisons in the United States* in 1925. This was a definitive work by a true pioneer in the profession. Two aspects of her work are of particular interest. First, she effectively documented the clear and devastating effects resulting from overexposure to lead, a common occupational hazard. Even though these effects were widely acknowledged, this problem has continued into modern times. Second, she documented the effects of overexposure to white phosphorus used in the manufacture of matches. In this case, workers developed what was called "Phossy jaw", which was a disfiguring disease of the jaw and could lead to death. In her autobiography, *Exploring the Dangerous Trades*, Dr. Hamilton described this occupational disease as follows:

> Phossy jaw is a very distressing form of industrial disease. It comes from breathing the fumes of white or yellow phosphorus, which gives off fumes at room temperature, or from putting into the mouth food or gum with fingers smeared with phosphorus. Even drinking from a glass which has stood on the workbench is dangerous. The phosphorus penetrates into a defective tooth and down through the roots to the jawbone, killing the tissue cells which then become the prey of suppurative germs from the mouth, and abscesses form. The jaw swells and the pain is intense...sometimes the abscess forms in the upper jaw and works up into the orbit, causing the loss of an eye. In severe cases one lower jawbone may have to be removed, or an upper jawbone–perhaps both.

In her autobiography, she goes on to state that she had been told by medical doctors there were no cases of Phossy jaw in the United States, "...because American match factories were so scrupulously clean." She later learned of a survey that found more than 150 cases in southern match factories. This finding led to the heavy taxation of white phosphorus and a complete ban on import or export to discourage its use in the U.S. This also is a classic example of the industrial hygiene principle of substitution. In this case, the less toxic and less hazardous red phosphorus for white phosphorus was already in use in England in 1891 (origin of the name "safety match"), while U.S. workers were still suffering the effects of phosphorus exposure at least 20 years later.

Following the contributions of Alice Hamilton, the industrial hygiene profession progressed rapidly. This progress was induced by the World Wars, with major advances made during World War II. While one would like to think this was due to an increased interest in human welfare and public health, it was most likely due to the recognized advantage the U.S. gained through the increased production of war goods. In fact, there was a slogan at the time from the U.S. Public Health Service which was "...keep him healthy; keep him at work." This would appear to be the first instance where health and safety of the workforce proved to provide a major economic and strategic advantage.

In the period of the early 1900s through the end of World War II there were some major events which also should be noted. In 1913, The New York State Department of Labor established a division of industrial hygiene.

The following year and on a federal level, a section on industrial hygiene was organized by the U.S. Public Health Service. Alice Hamilton headed the first academic program in industrial hygiene (Department of Applied Physiology) at Harvard University in 1918. The National Safety Council began accident record keeping by 1921.

During the late 1920s though the 1960s, most of the progress in industrial hygiene was directly or indirectly sponsored by the government, although many of the major industries had industrial hygiene activities usually under their medical departments. Government efforts included funding of industrial hygiene at the state level, and the establishment of government agencies or programs such as the Tennessee Valley Authority, the Works Projects Administration, the Social Security Administration, in the 1930s and 1940s and the Public Health Service at the federal level. In particular, the Public Health Service had responsibility for much of the industrial hygiene effort directed at assuring adequate war goods production, and later for the investigation of significant occupational disease outbreaks, such as asbestosis. Other major players in this time period included the Bureau of Mines which was very involved with dust control, since silicosis and other dust-related diseases were prevalent during this period.

In closing this very brief historical overview, it should be noted that more has been learned about occupational disease control in the last 50 years than the entire span of history described before that period in this chapter. It should also be noted that this is true of other sciences as well, such as medicine.

Table 1.1 summarizes the major legislative history of regulations, state actions, and other significant events over approximately the last decade related to safety, health, and the environment.

III. OCCUPATIONAL SAFETY AND HEALTH ADMINISTRATION

No action, event, or regulation has had a more profound effect on the safety and health profession than the Occupational Safety and Health Act. Since the promulgation of the Act in 1970, industrial hygienists and safety professionals have become an essential ingredient for commerce and government in the United States, its territories, and possessions. While the influence of OSHA on business and government seems to have diminished with less enforcement activities lately, there is little doubt that it is still the most influential government agency for most industrial hygienists.

The Occupational Safety and Health Administration (OSHA) was established by the Occupational Safety and Health Act of 1970 (also known as the Williams-Steiger Occupational Safety and Health Act) signed on December 29, 1970, by President Nixon. This Act also created a number of other entities, including the National Institute for Occupational Safety and Health (NIOSH) and the OSHA Review Commission. OSHA was placed under the Department of Labor as an enforcement agency (administrative power to make rules and issue penalties). The Agency is run by an Assistant Secretary reporting to the Cabinet-level Secretary for Labor.

Table 1.1
A Century of Legislation and Significant Events Related to Safety, Industrial Hygiene, and the Environment

1878	Revised English Factory Act
1908	Limited compensation benefits for U.S. Civil Service employees
1911	First state compensation law in New Jersey
1912	U.S. Public Health Service and U.S. Bureau of Mines begin joint health surveys
1913	First state industrial hygiene programs (New York and Ohio)
1918	First academic program in industrial hygiene (Harvard)
1936	Walsh-Healy Public Contracts Act requires government contractors to meet workplace standards
1938	American Conference of Governmental Industrial Hygienists (called the National Conference of Governmental Industrial Hygienists)
1939	American Industrial Hygiene Association established
1966	Metallic and Nonmetallic Mine Safety Act
1969	Federal Coal Mine Health and Safety Act
1970	Occupational Safety and Health Act (Williams-Steiger Act) established OSHA and NIOSH
1970	Clean Air Act (amended several times)
1972	Noise Control Act (environmental noise)
1972	Clean Water Act (amended several times)
1976	Toxic Substances Control Act
1976	Resource Conservation and Recovery Act
1977	Federal Mine Safety and Health Act
1980	Comprehensive Environmental Response, Compensation, and Liability Act (amended several times)
1986	Superfund Amendments and Reauthorization Act

OSHA has as its stated purpose, "...to assure, so far as possible, every working man and woman in the nation safe and healthful working conditions." In order to achieve this stated purpose, the Act authorized promulgation and enforcement of safety and health standards; provides for assistance and encouragement of states in their efforts to assure safe and healthful working conditions; and, provides for research, information, education, and training in the field of occupational safety and health. Under the guise of the Act, OSHA has responsibility and authority for most commercial business (and the federal government through an Executive Order) in the United States, its territories, and possessions. Mining is regulated separately through the Mine Safety and Health Act, and the federal agency (Mine Safety and Health Administration) created by that Act. Figure 1.1 depicts early miners circa late 1800s using a canary to determine if atmospheric conditions in the mine were "safe."

Figure 1.1   Early miners use a canary to determine if air in a mine is hazardous (Courtesy: *Mine Safety Appliances Company*, Pittsburgh, PA).

As noted above, an important part of the Occupational Safety and Health Act was the partnership established with the states. Under Section 18 of the Act, the states can elect to write and enforce their own rules related to safety and health that are under the jurisdiction of federal OSHA provided they meet the minimum criteria established by OSHA. If these criteria are met (basically safety and health standards and enforcement capabilities that are at least as effective as the federal system), the federal OSHA program will cost share with the states. It should also be noted that the states may elect to cover only a portion of the programs that the federal OSHA covers, with federal OSHA covering all remaining aspects (e.g., Connecticut's plan only covers public employees). Even with the incentive of permitting the states to establish and enforce their own programs, the cost sharing feature has been attractive to less than half the states. Therefore, most of the states have federal enforcement of OSHA through regional, district, and local federal OSHA offices. Some of the state programs, such as the one in California, can be more comprehensive and stringent than the federal program. It should also be noted that essentially all states offer some type of free consultation service for business. This is usually administered through the states.

In OSHA's early years many consensus standards were adopted as developed by: the American National Standards Institute (ANSI), the National Electrical Code (NEC), the National Fire Protection Association (NFPA), the American Conference of Governmental Industrial Hygienists (ACGIH) Threshold Limit Values (TLVs) for Chemical Substances, and others. These adopted standards applied to general industry, maritime industries, and others. In addition to these adopted standards, the Assistant Secretary has the authority to issue, revise, modify, and revoke health and safety standards. Since the inception of OSHA, a large number of standards have been promulgated. The standards issued by OSHA are generally of four types: (1) *Design standards* are those which specify the design of the controls or equipment used such as the ventilation standards. (2) *Performance standards* are those which state what the objective is that must be met but not the method by which to obtain the result. Most of the Permissible Exposure Limits for chemical substances are performance standards (they do not specify the methods that must be used to achieve acceptable exposure limits). (3) *Vertical standards* are those that apply to a particular industry such as diving. (4) *Horizontal standards* are those that apply across all workplaces such as the requirements for sanitation or the standards for portable and fixed ladders.

While it is beyond the scope of this chapter to detail all of the OSHA regulations, those of most likely interest to industrial hygienists include:

- Permissible Exposure Limits—numerical limits for airborne contaminant exposure in the workplace
- Specific substances standards—requirements for working with specific substances which usually include monitoring, medical surveillance, training, written programs, and exposure limits (e.g., asbestos, cadmium, formaldehyde, inorganic arsenic, lead, etc.)
- Ventilation—requirements for ventilation for many processes or operations
- Noise—requirements for working in hazardous noise areas and establishment of a hearing conservation program for those exposed to noise above 85 decibels (A-weighted scale)
- Radiation—requirements for those exposed to both ionizing and non-ionizing radiation
- Hazardous Waste Operations and Emergency Response—requirements for those working with hazardous wastes and those providing emergency response to hazardous chemical spills and releases
- Respiratory Protection Program—requirements for those workers using respirators
- Personal Protective Equipment—requirement for hazard assessment, selection, worker training, and other requirements for workers using protective clothing and equipment
- Hazard Communication—requirements for informing workers of the hazards of the chemicals with which they work or for which there might be exposure
- Laboratory Safety—requirements for laboratories where workers are potentially exposed to hazardous chemicals

- Process Safety Management–requirements for conducting a process hazard analysis and worker training for certain highly hazardous materials
- Blood-borne Pathogens–requirements for workers with potential exposure to blood-borne pathogens such as AIDS and hepatitis

In addition to the standards noted above, OSHA has a *General Duty Clause*, which requires that the workplace be free of recognized hazards that are causing, or are likely to cause, death or serious physical harm. While this requirement would seem to cover a vast array of potential hazards that are not specifically regulated, the courts have found that the hazard must be generally recognized by industry and a feasible means available for control of the hazard. This has significantly curtailed the effective use of this aspect of the Act.

Enforcement of the Act is carried out through the network of regional, area, and local OSHA offices. The criteria for inspections under OSHA are based on the following criteria: **(1)** notice of imminent danger situations, **(2)** fatalities or serious injuries, **(3)** programmed high-hazard inspections, and **(4)** employee complaints of a general nature. The OSHA compliance officer has a right of entry which has been upheld by the courts (you can require they get a court order). Additionally, the compliance officer has a right to speak to the employer or employees in private, inspect any and all equipment, materials, and processes, collect samples, and perform other duties related to the investigation. Based upon the findings of the investigation, the compliance officer may issue citations consisting of: **(1)** repeat violations (can be considered egregious), **(2)** willful violations (can also be considered egregious), **(3)** serious violations, **(4)** other than serious violations, and **(5)** *de minimis* violations. Based upon the number and type of citations and compliance history of the company, fines can be assessed ranging upwards of $70,000 for each violation. Employers have several routes of appeal ranging from an informal meeting with the area director, through the Occupational Safety and Health Review Commission and the federal courts.

IV. NATIONAL INSTITUTE FOR OCCUPATIONAL SAFETY AND HEALTH

The National Institute for Occupational Safety and Health (NIOSH) was also established by the Occupational Safety and Health Act of 1970. It is currently within the Department of Health and Human Services under the Centers for Disease Control. NIOSH is responsible for research on the safety and health effects of exposures in the workplace and for providing training and information to workers and employers. As part of this responsibility, they provide training grants to colleges and universities through their Educational Resource Center program.

Major programs at NIOSH include the development of recommendations for regulations by OSHA through their criteria document program,
conducting health hazard evaluations at the request of employees or employers, the preparation of the *Registry of Toxic Effects of Chemical Substances*, the conduct of industry-wide studies on health effects of work exposures, control measures for those exposures, and training.

One major initiative at NIOSH is their National Occupational Research
Agenda (NORA). NORA is an effort to list and establish cooperative efforts (with private and public sectors) on the national priorities for research
in occupational health and safety. The NORA priority research areas are
shown in Table 1.2.

NIOSH is not a regulatory agency; however, they do certify respirators
and conduct research in mining under authority of the Mine Safety and
Health Act.

The next three sections of this chapter describe the major membership
organizations for industrial hygienists, followed by the certifying body for
the credential *CIH*.

Table 1.2
NORA Priority Research Areas

Category	Priority Research Areas
Disease and injury	Allergic and irritant dermatitis
	Asthma and chronic obstructive pulmonary disease
	Fertility and pregnancy abnormalities
	Hearing loss
	Infectious disease
	Low back disorders
	Musculoskeletal disorders of the upper extremities
	Traumatic injuries
Work environment and workforce	Emerging technologies
	Indoor environment
	Mixed exposures
	Organization of work
	Special populations at risk
Research tools and approaches	Cancer research methods
	Control technology and personal protective equipment
	Exposure assessment methods
	Health services research
	Intervention effectiveness research
	Risk assessment methods
	Social and economic consequences of work-place illness and injury
	Surveillance research methods

V. AMERICAN INDUSTRIAL HYGIENE ASSOCIATION

The American Industrial Hygiene Association (AIHA) is a nonprofit membership organization for industrial hygienists founded in 1939. It is the largest single professional organization of industrial hygienists in the world with an approximate membership of 13,000. The AIHA is located at 2700 Prosperity Ave., Suite 250, Fairfax, VA 22031 [Telephone (703) 849-8888]. The mission statement of the AIHA is as follows:

> The American Industrial Hygiene Association is an organization of professionals dedicated to the anticipation, recognition, evaluation, and control of environmental factors arising in or from the workplace that may result in injury, illness, impairment, or affect the well-being of workers and members of the community...[it] is devoted to promoting, protecting, and enhancing the profession to affect positively the health and well-being of workers and members of the community.

Membership is open to anyone with an interest in industrial hygiene. There are eight classes of membership which comprise categories of Full, Retired, Associate, Affiliate, Student, Fellow, Honorary, and Organizational. Major programs or activities of the AIHA include technical committees, special interest groups, service committees, a government affairs department, laboratory accreditation programs, continuing education training and seminars, an employment service, publications (including the *American Industrial Hygiene Association Journal*), local sections organized geographically and internationally, and an annual technical conference and exposition sponsored with the American Conference of Governmental Industrial Hygienists.

VI. AMERICAN CONFERENCE OF GOVERNMENTAL INDUSTRIAL HYGIENISTS

The American Conference of Governmental Industrial Hygienists (ACGIH) was founded in 1938 with a change to its present name in 1946. Membership is open to anyone interested in industrial hygiene; however, voting rights are limited to those with a governmental or university affiliation (e.g., federal, state, and local governments and government contractors as well as universities and colleges). Their are seven categories of membership which include Full, Associate, Technical, Student, Emeritus, Honorary, and Affiliate. Their membership today is approximately 6000. ACGIH is located at 1330 Kemper Meadow Drive, Cincinnati, OH 45240 [Telephone number (513) 742-2020]. The mission and purposes of ACGIH are:

> To be an indispensable resource for industrial hygienists and related professionals worldwide. To promote excellence in environ-

mental and occupational health. To provide technical information of the highest quality. To benefit the occupational health and well-being of people worldwide. To serve the membership and continually improve the organization, including its financial and human resources.

Major programs or activities of ACGIH include technical committees, service committees, continuing education programs, publications (including their journal, *Applied Occupational and Environmental Hygiene*), symposia, and sponsorship with AIHA of an annual technical conference and exposition. ACGIH is probably best known for its publication of the *Threshold Limit Values and Biological Exposure Indices* booklet, which is published annually and used by industrial hygienists worldwide.

VII. AMERICAN BOARD OF INDUSTRIAL HYGIENE

The American Board of Industrial Hygiene (ABIH) is the certifying body for the *Certified Industrial Hygienist* or *CIH* credential. ABIH was established in 1960, with the first exams given in 1963. This voluntary credential is the hallmark of the practice of industrial hygiene. It is widely recognized in the United States and in many other countries of the world. There are approximately 6000 certified industrial hygienists. The requirements for certification include the successful completion of an exam (consisting of two parts); 5 years of experience in industrial hygiene; and appropriate academic credentials acceptable to the Board (generally a degree with "hard science" minimum course requirements) and references. Once all criteria are met, the CIH credential is awarded with a requirement for continuing education and professional activities. Persons completing their certification automatically become members of the American Academy of Industrial Hygienists. Specific details on the requirements for certification may be obtained from the American Board of Industrial Hygiene, 4600 West Saginaw Street, Suite 101, Lansing, MI 48917-2737. The telephone number for ABIH is (517) 321-2638.

The last section of this chapter briefly describes the basic tenets of the practice of industrial hygiene. These tenets are the anticipation, recognition, evaluation, and control of occupational hazards or stresses.

VIII. THE TENETS OF INDUSTRIAL HYGIENE

Anticipation is the recognition of potential hazards or stresses that might impact a worker through the review of industrial processes, equipment, and operations prior to their installation. This work generally requires the greatest level of experience and expertise in industrial hygiene, since the actual process or activity cannot be viewed or surveyed. *Recognition* is the ability to identify environmental hazards and stresses in the occupational setting and the ability to understand their effect on the worker and their well-being. This facet also requires a high degree of knowledge and skill and is usually accomplished by a visual survey of the workplace and interviews. This phase of industrial hygiene work can be facilitated by the researching of the process or activity and materials used prior to the site survey. *Evaluation* is the qualitative and quantitative measurement of the magnitude of the hazards or stresses to the worker. This phase is usually accomplished by on-site exposure measurement using various techniques such as direct reading instruments, collection devices, and other means. As depicted in Figure 1.2, an industrial hygienist uses a direct-reading instrument to evaluate workplace contaminants levels. Once the numerical values of the magnitude of exposures are known, they can usually be compared to requirements or recommendations for exposures. From this, a risk assessment can be prepared and appropriate measures taken to control the exposures as necessary.

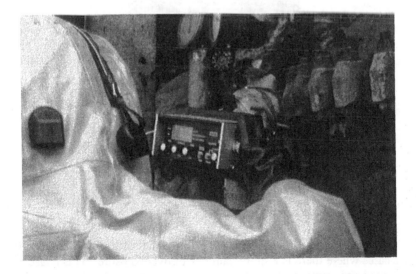

Figure 1.2   Industrial hygienist conducting air montoring to evaluate workplace atmospheric conditions (Courtesy: *Mine Safety Appliances Company*, Pittsburgh, PA).

Figure 1.3   Personal protective equipment (PPE) employed to control worker exposure (Courtesy: *Kappler Safety Group*, Guntersville, AL).

Figure 1.4   Controlling worker exposure through ventilation/engineering control (Courtesy: *United Air Specialists, Inc.*, Cincinnati, OH).

Control is usually the final phase of most industrial hygiene work for those cases where there is a significant risk or overexposure. This phase can require the prescription of various control measures to eliminate or reduce the hazards or stresses in order to alleviate any adverse effects on the worker. Control measures can range from engineering controls such as ventilation to work practice controls such as rotation of workers. Once the control measures are put in place, another round of evaluation is usually necessary to assure that the control measures are effective in reducing risk to the worker. Figure 1.3 depicts personal protective equipment used to control worker exposure, while Figure 1.4 shows ventilation to remove contaminants from worker breathing zones and control exposure.

REVIEW QUESTIONS

1. What single historical event resulted in the greatest advance in the art and science of industrial hygiene?

    a.   WWII
    b.   WWI
    c.   The Industrial Revolution
    d.   The invention of respirators

2. Scrotal cancer was an occupational hazard of what trade?
    a.   Sanitation workers
    b.   Coal miners
    c.   Chimney sweeps
    d.   Vitriol makers

3. What was the causative agent for the occurrence of scrotal cancer?

    a.   Mineral acids
    b.   Fumes
    c.   White phosphorus
    d.   Soot from coal

4. Vitriol was a common byproduct of what industry?

    a.   Tanning
    b.   Flax making
    c.   Copper mining
    d.   Silver mining

5. What was the occupational disease noted by Ramazzini for workers cleaning out a cesspit?

a.   Dysentery
    b.   Blindness
    c.   Loss of teeth
    d.   Lung disease

6. Agricola described the short life expectancy of miners in what graphic way?

    a.   Employer strikes
    b.   Carpathian Mountain women
    c.   Saxony festivals
    d.   Clothes handed down

7. What was the name of the famous book that Agricola wrote?

    a.   *Hunters Diseases of Occupations*
    b.   *De Re Metallica*
    c.   *Diseases of Workers*
    d.   *Dangerous Trades*

8. The book by Agricola focused on what subject?

    a. Mining
    b. Farming
    c. Guilds
    d. Woodworking

9. Who is commonly referred to as the "Father of Industrial Medicine"?

    a. Hippocrates
    b. Pliney, the Elder
    c. Galen
    d. Ramazzini

10. Who is first credited with suggesting the preventive measures that led to the wet grinding of flint?

    a. Ellenbog
    b. Paracelsus
    c. Ramazzini
    d. Agricola

11. What other historical figure actually did a study to compare the onset of death between wet grinding and dry grinding?

    a. Thackrah
    b. Ramazzini
    c. Chaucer
    d. Ellenbog

12. Who wrote the book, *Industrial Poisons in the United States*?

    a. Hamilton
    b. Paracelsus
    c. Pott
    d. Franklin

13. What was the cause of the occupational disease Phossy Jaw?

    a. Red phosphorus
    b. Blue phosphorus
    c. White phosphorus
    d. Vitriol

14. Who is commonly called the "Father of Medicine"?

    a. Hippocrates
    b. Ramazzini
    c. Legge
    d. Hamilton

15. Who was the person credited with first recommending the use of animal parts as respirators?

    a.   Thackrah
    b.   Paracelsus
    c.   Agricola
    d.   Pliney, the Elder

16. Galen was famous for describing occupational disease among what workers?

    a.   Masons
    b.   Chimney sweeps
    c.   Miners
    d.   Plumbers

17. For what principle of toxicology is Paracelsus famous?

    a.   Dose-response
    b.   LD50's
    c.   Immune response
    d.   Exposure limits

18. Describe at least two of the effects of lead poisoning as Ramazzini observed and wrote of them.

    a.   Weakness and constipation
    b.   Loss of teeth and paralysis
    c.   Loss of hair and blindness
    d.   Increase in facial color and melancholy

19. Benjamin Franklin is known to us for writing of what occupational disease?

    a.   Lead poisoning
    b.   Phossy jaw
    c.   Blindness
    d.   Coal miners lung

20. What did the Health and Morals of Apprentices Act of 1802 do for working conditions?

    a.   Reduced work hours to 8
    b.   Reduced work hours to 4
    c.   Reduced work hours to 12
    d.   Eliminated young workers

21. When was the Occupational Safety and Health Act promulgated into law?

    a.   1972
    b.   1976
    c.   1980
    d.   1970

22. What does the General Duty Clause require?

    a. Workplace free from recognized hazards
    b. Workplace with hazard controls
    c. Workplace free of hazards
    d. A general duty to perform work

23. What is a "horizontal" standard?

    a. Applies to certain workplaces
    b. Applies to all workplaces
    c. Applies to certain industries
    d. Requires level work surfaces

24. The NIOSH NORA program is:

    a. National Occupational Research Agenda
    b. National Occupational Reference Assistance
    c. National On-site Research Assistance
    d. Not Occupationally Reasonably or Achievable

25. ACGIH is well known for publishing:

    a. PELs
    b. TLVs
    c. Local Section Guidelines
    d. RTECS

ANSWERS

1.	a.	14.	a.
2.	c.	15.	d.
3.	d.	16.	c.
4.	c.	17.	a.
5.	b.	18.	b.
6.	b.	19.	a.
7.	b.	20.	c.
8.	a.	21.	d.
9.	d.	22.	c.
10.	c.	23.	b.
11.	a.	24.	a.
12.	a.	25.	b.
13.	c.		

References

ABIH, *American Board of Industrial Hygiene* (Roster), ABIH, Lansing, MI, 1997.

ACGIH, *Membership Directory and Information Guide*, ACGIH, Cincinnati, OH, 1997.

Agricola, G., *De Re Metallica*, (Translated by H.C. and L.H. Hoover), Dover Publications, New York, 1950.

AIHA, Chap. 13, *The American Industrial Hygiene Association-Its History and Personalities, 1939-1990*, Clayton, G. & Clayton, F., Eds., AIHA and the Federal Government, AIHA, Fairfax, VA, 1994.

AIHA, Fundamental of Industrial Hygiene Course Notes (historical overview), an American Industrial Hygiene Association course developed by S.Z. Mansdorf & Associates, Inc., Fairfax, VA, 1991.

AIHA, *Who's Who in Industrial Hygiene*, AIHA, Fairfax, VA, 1997.

ECON, Lead concerns date back to the days of Ben Franklin, *ECON*, March issue, 54-56, 1992.

Hamilton, A., *Exploring the Dangerous Trades* (an autobiography of Alice Hamilton, M.D.), AIHA, Fairfax, VA, 1995.

Hunter, D., *The Diseases of Occupations*, 6th ed., Little, Brown & Company, Boston, 1987, 14-15.

Kaufman, S., Chap. 1, in *Protecting People At Work*, McLaury, J., Eds., OSHA, U.S. Government Printing Office, Washington, 1980.

Krafcisin, G., The Poisoning of Chaucer's Canon and His Yeoman, Masters Thesis, Northwestern University, Chicago, 1995.

Mansdorf, S. Z., Chap. 1, *The Complete Manual of Industrial Safety*, Complying with safety legislation and regulations, Prentice Hall, Englewood Cliffs, NJ, 1993.

Murray, R., Section 1, History, in *Hunter's Diseases of Occupations*, Raffle, P. A., Lee, W. R., McCallum, R. I., and Murray, R., Eds., Little Brown & Co., Boston, 1987. NIOSH, National Occupational Research Agenda (Update), National Institute for Occupational Safety and Health, NIOSH Publication 97-138, 1997.

Ottonboni, M. A., *The Dose Makes the Poison*, Vincente, Berkeley, CA, 1984.

Perkins, J. L., Chap. 2, *Modern Industrial Hygiene*, Vol. 1, Industrial Hygiene–Historical Perspective, Van Nostrand Reinhold, New York, 1997.

Ramazzini, B., *Diseases of Workers* (Translated by W.C. Wright), University of Chicago Press, Chicago, 1940.

Rose, V., Chap. 1, History and philosophy of industrial hygiene, *The Occupational Environment: Its Evaluation and Control*, DiNardi, S. R, Ed., American Industrial Hygiene Association, Fairfax, VA, 1997.

Sherwood, R. J., Review of occupational hygiene, in *Proc. Industrial Hygiene and Medicine Symp.*, Kaohsiung Medical College, Taiwan, 1997, 149-165.

Stannard, J. N., Breathing Is An Old Habit, Banquet Address, Dept. of Community Medicine and Dept. of Radiology, University of California, San Diego, CA.

Waldron, H. A., A brief history of scrotal cancer, *British Journal of Industrial Medicine*, 40: 390-401, 1983.

Chapter 2

GENERAL PRINCIPLES

Celeste Winterberger, PhD, CIT

I. OVERVIEW

Chemical properties deal with the characteristics of a material such as its structure and formation from the elements. This section of the chapter will discuss the areas of inorganic and organic chemistry and their importance to the industrial hygiene professional.

A. Inorganic Chemistry

Inorganic chemistry is the study of compounds derived from minerals and other non-carbon compounds. These materials compose the largest portion of the periodic table. In the next paragraphs, chemistry terms of most importance to the industrial hygiene professional will be discussed.

1. *Heat of reaction*

The *heat of reaction* or ΔH is the change in potential energy that occurs during a chemical reaction. It is usually expressed in kilojoules (kJ). If a reaction is *exothermic* (releases heat), the heat of reaction will be expressed as a negative number. *Endothermic reactions* (absorbs heat) will have the energy given off so the heat of reaction is expressed as a positive number. In the following reaction, oxygen is produced from the decomposition of potassium chlorate, as such:

$$2KClO_3 \rightarrow 2KCl + 3O_2 \qquad (2.1)$$

Sample Calculation: If 45 kJ of energy are required for each potassium molecule (mole) used in the reaction, then the heat of reaction could be calculated using the formula:

ΔH = (moles of reactant)(heat energy of the reactant/mole) (2.2)
so:
ΔH = (2 moles of $KClO_3$)(45 kJ/1 mole of $KClO_3$)
Since the moles cancel each other out, the equation can be reduced to:
$\qquad \Delta H$ = (2)(45 kJ)
$\qquad \Delta H$ = 90 kJ (an endothermic reaction)

2. pH

pH is used to measure the acidity or alkalinity of a solution. It uses a scale of from 1 to 14, with 7 being neutral. Acids (H_3O^+) have pH values of from 0 to 6, with the lowest values being more acidic. Litmus paper turns red when exposed to acids. Alkalis (bases, OH^-) have pH values ranging from 8 to 14, with the highest numbers indicating the greatest alkalinity. Litmus paper will turn blue in the presence of alkalis.

3. *Ionization potential*

Ionization potential deals with the ease with which atoms can give up their electrons. The amount of energy needed for ionization to occur decreases to the left and down in the periodic chart. Typically, those materials that are most likely to ionize are found on the left side of the periodic chart in Groups IA (except hydrogen) and IIA.

Table 2.1
Activity Series of Metals

Compound	Reacts with Water	Displaces H_2 from Acids	Combines Directly with Oxygen	Oxidizes Easily Reduced by H at Elevated Temperatures to Free Metals and H_2O	Oxides decompose when heated
Rubidium	Violently	Yes	Yes	No	No
Potassium	Violently	Yes	Yes	No	No
Calcium	Slowly	Yes	Yes	No	No
Magnesium	Steam	Yes	Yes	No	No
Aluminum	Steam	Yes	Yes	No	No
Manganese	Steam	Yes	Yes	No	No
Zinc	Steam	Yes	Yes	No	No
Chromium	Steam	Yes	Yes	No	No
Iron	No	Yes	Yes	Yes	No
Nickel	No	Yes	Yes	Yes	No
Tin	No	Yes	Yes	Yes	No
Lead	No	Yes	Yes	Yes	No
Hydrogen	No	No	Yes	Yes	No
Copper	No	No	Yes	Yes	No
Bismuth	No	No	Yes	Yes	No
Antimony	No	No	Yes	Yes	No
Mercury	No	No	yes	Yes	Yes
Silver	No	No	No	Yes	Yes
Platinum	No	No	No	Yes	Yes
Gold	No	No	No	Yes	Yes

4. *Electrochemical potential*

Electrochemical potential describes how active metals will react when exposed to certain materials such as water. The electrochemical series shows which metals precipitate out faster when they are combined with certain substances. Table 2.1 shows which metals are most active and under what conditions.

5. *Hydrolysis*

A *hydrolysis reaction* occurs when a negative ion (anion) combines with water to produce a weak base (OH⁻) or when a positive ion (cation) combines with water to produce a weak acid (H_3O^+). For the industrial hygiene professional, this is of importance because it can be used as a disposal method for hazardous chemicals, which can be neutralized by water.

6. *Oxidation/reduction*

An *oxidation reaction* occurs when a material loses electrons during a chemical reaction. For example, take the following reaction:

$$Na \rightarrow Na^+ + e^- \qquad (2.3)$$

The sodium atom has lost an electron, which creates a positive ion (cation) of sodium and one electron.

A *reduction reaction* occurs when a material gains electrons during a chemical reaction. Note the following reaction.

$$2e^- + Cl_2 \rightarrow 2Cl^- \qquad (2.4)$$

Here, the chlorine molecule has gained two electrons which now makes it a negative ion (anion).

Many reactions involve both oxidation and reduction reactions. These reactions are called *redox reactions*. Using the reactions described above, we can create the following redox reaction.

$$2Na + Cl_2 \rightarrow 2NaCl \qquad (2.5)$$

Notice that the sodium gives up two electrons by oxidation to the gas chlorine, which takes these electrons by the process of reduction. This reaction of sodium and chloride produces the substance called sodium chloride, which is more commonly known as table salt.

7. *Reactivity/stability*

Reactivity is the ability of a substance to undergo rapid, sometimes violent reactions under certain conditions. For example, when water and sodium come into contact, the heat generated by the reaction can cause both the hydrogen and sodium to ignite. Explosive materials, such as TNT, are considered highly reactive. For the industrial hygiene professional, it is important to note that, in addition to water combining with a substance, an acid, a base, or air can cause an uncontrolled reaction to occur. Unfortunately, some materials need no additional substance to react and are considered self-reactive. This means they contain both an oxidant and a reducing agent in the same compound. Nitroglycerin is a good example of a self-reactive substance. Table 2.2 contains a list of reactive compounds and structures that may be of importance to the industrial hygiene professional.

Table 2.2
Reactive Compounds and Structures of
Importance to Industrial Hygienists

Inorganic	Organic
Nitrous oxide (N_2O)	Allenes (C = C = C)
Nitrogen halides (NCl_2, NI_2)	Dienes (C = C-C = C)
Interhalogen compounds (BrCl)	Azo compounds (C-N = N-C)
Halogen oxides (ClO_2)	Triazenes (C-N = N-N)
Halogen azides (ClN_2)	Hydroperoxides (R-OOH)
Hypohalites (NaClO)	Peroxides (R-OO-R´)
	Alkyl nitrates ($R-O-NO_2$)
	Nitro compounds (R- NO_2)

Stability is the exact opposite of reactivity. If a substance is not very reactive, it is known as stable. Some of the most stable elements are the noble gases (Group VIII in the periodic chart). Since these gases have the maximum number of electrons in their outermost shell, they have no need to gain or lose electrons to fill their outer shell.

8. *Activation energy*

Activation energy is the amount of energy needed for two colliding molecules to form an activated complex. If the molecules do not have enough energy to form an activated complex at collision, they would just bounce off each other. Different molecules have different activation energies. For example, if white phosphorus comes into contact with air, the white phosphorus will immediately react with the oxygen to produce a very hot flame. However, the violent reaction between oxygen and hydrogen, which is used to power spacecraft must occur at a temperature of 400°C. Therefore, the activation energy is higher for the oxygen/hydrogen reaction than the phosphorus/oxygen reaction. For the industrial hygiene professional, this means that certain compounds must be segregated from each other if they have low activation energies.

B. Organic Chemistry

The basis for *organic chemistry* is carbon. Carbon can bond in a number of diverse ways. It can form single bonds, double bonds, and triple bonds with other compounds. In addition, carbon bonding can occur as straight chains, branched chains, or rings. Organic compounds can be found in industrial compounds, synthetic polymers, agricultural chemicals, and biological compounds. There are four categories of organic materials: hydrocarbons, oxygen-containing compounds, nitrogen-containing compounds, sulfur-containing compounds, organohalides, phosphorus-containing compounds, or a combination of these compounds. The next section will discuss some of the organic chemistry concepts of importance to the industrial hygiene professional.

1. *Consecutive reactions*

Consecutive reactions occur when two reactions happen in such a way that they can be combined into a single reaction without an intermediate step. An example of the notation for consecutive reactions is:

$$\text{organic reactant} \xrightarrow[\substack{\textbf{reagent and solvent used} \\ \textbf{in second reaction}}]{\substack{\textbf{reagent and solvent} \\ \textbf{used in the first reaction}}} \text{organic product} \quad (2.6)$$

The reaction of ethylene to produce ethanol is considered a consecutive reaction and is written as follows:

$$CH_2 = CH_2 \xrightarrow[\textbf{(2) H}_2\textbf{O/NaOH/H}_2\textbf{O}]{\textbf{(1) BH}_3\textbf{/THF}} CH_3CH_2OH \qquad (2.7)$$

2. *Reverse reactions*

Reverse reactions are those chemical reactions where the products of forward reaction can reverse to form the original compound. Reactions that are reversible are called two opposing chemical reactions. The decomposition of ammonia to form hydrogen and nitrogen is an example of a reversible reaction:

$$2NH_3 \; \underset{\longleftarrow}{\overset{\longrightarrow}{}} \; 3H_2 + N_2 \qquad (2.8)$$

3. *Competing reactions*

There are some reactions that occur which can produce two chemically different products. However, only one product is ever observed. This is because of the difference in transition-state energies of the two competing reactions. For example, the addition of hydrogen bromide to propene could have both the following results:

$$CH_3\text{-}CH=CH_2 + HBr \rightarrow CH_3\text{-}\overset{\overset{\textbf{Br}}{|}}{C}\text{-}\overset{\overset{\textbf{H}}{|}}{C}H_2 \qquad (2.9a)$$
$$\textbf{2-bromopropane}$$

or:

$$\text{CH}_3\text{-CH=CH}_2 + \text{HBr} \rightarrow \overset{\overset{\displaystyle H}{|}}{\text{CH}_3}\text{-}\overset{\overset{\displaystyle Br}{|}}{\text{C}}\text{-CH}_2 \qquad (2.9b)$$

1-bromopropane
(not observed)

In this case, the first reaction is the observable one because its transition-state energy is closer to that of the original intermediate, $\text{CH}_3\text{CH=CH}_2 + \text{H}^+$.

4. Reaction kinetics

Reaction kinetics deals with determining the speed or rate of reactions. Using the principles of reaction kinetics, one can uncover the mechanics of reactions and describe the sequence of bond breaking and bond forming. In addition, the energy associated with each step can be calculated. However, ordering reactions is of most importance to the industrial hygiene professional.

a. First order reactions

First order reactions take place without an intermediate step. Enough energy is generated to completely change the reactants into the product. The decomposition of dinitrogen pentoxide is a good example of a first order reaction.

$$2\text{N}_2\text{O}_5 \rightarrow 4\text{NO}_2 + \text{O}_2 \qquad (2.10)$$

b. Second order reactions

Second order reactions have an intermediate step which occurs before the product is finished. It is written using the following format.

$$\text{reactants} \rightleftarrows \text{ intermediates } \rightleftarrows \text{ products} \qquad (2.11)$$

An example of a second order reaction is combining nitrogen monoxide and ozone, which produces nitrogen oxide and oxygen. This reaction has been linked to the depletion of the ozone layer.

$$\text{NO} + \text{O}_3 \rightarrow \text{NO}_2 + \text{O}_2 \qquad (2.12)$$

In order to determine the order of this reaction, one must look at the reactants. The break-down of nitrogen monoxide is a first order reaction. It is also a first order reaction to break down the oxygen. Since the sum of the orders, $1 + 1$, equals two, the reaction is considered a second order reaction.

c. *Third and higher order reactions*

In *third* and higher order reactions, you must also add the sum of the orders to determine the order of the reaction. For example, a reaction of nitrogen monoxide and chlorine proceeds as follows:

$$2NO + Cl_2 \rightarrow 2NOCl \qquad (2.13)$$

To determine the order of this reaction, one should first notice that there are 2 molecules of nitrogen monoxide. If there was only one, it would be a first order reaction, but because there are two, one must multiply by two which causes this portion of the reaction to be second order. The chlorine molecule is a first order reaction. Therefore, this reaction would be considered a third order $(2 + 1)$ reaction.

In order to determine the order for any reaction, the industrial hygiene professional should first note the number of molecules involved in the reaction and the number of bonds between atoms in the molecule. The number of molecules will determine the order for that molecule (see example above). When you have double or triple bonds between molecules, each bond that needs to be broken takes two steps, one to break the bond and make it an ion, and a second to combine and create the product. Then, one can simply sum the orders to determine the order for that reaction.

d. *Zero order reactions*

According to Loudan, a *zero order reaction* occurs when changing the concentration of the second reactant in a reaction has no effect. This makes the second reactant a zero order reaction. An example of this type of reaction is:

$$D + E \rightarrow F \qquad (2.14)$$

where the rate only doubles if the concentration of D is doubled. In this equation, D is considered first order while E is considered zero order. Therefore, the entire equation is only considered first order even though there are two reactants $(1 + 0)$.

5. *Catalysis*

A reaction that involves the use of a catalyst is called *catalysis*. Catalysts are used to speed up reactions by reducing the activation energy of the reactants. It is important to note that a catalyst is not used in the reaction. This means that the catalyst's mass and form remain the same.

a. *Contact catalysts*

Contact catalysts are used to provide a surface on which a reaction can occur. A catalytic converter on an automobile is a good example of a contact catalyst. Inside the converter, fine particles of platinum or palladium give the gases of carbon monoxide and unburned fuel a surface that lowers the activation energy so these combustion byproducts can be quickly changed to less noxious materials.

b. *Enzyme catalysts*

Enzyme catalysts are big protein molecules which catalyze a reaction only at one specific site on the enzyme, called the active site. Increases of a million times the reaction rate are very common. In fact, the reaction between water and carbon dioxide to form carbonic acid occurs 10 million times faster in the presence of the enzyme carbonic anhydrase.

6. *Chemistries of typical workplace contaminants*

In terms of organic chemicals, there are six types of materials of importance to the industrial hygiene professional. Each will be discussed in detail in the paragraphs below.

The first group of organic chemicals is the hydrocarbons. Hydrocarbons are compounds that contain only carbon and hydrogen. It is important to note that these materials are all gases in this form and most are highly flammable. The first group is called the alkanes. Alkanes are also known as paraffins (for their waxy feel) or aliphatic hydrocarbons. Molecules in this family are joined by single bonds. Methane, or CH_4, is the first member of the alkanes. Some other commonly found alkanes include propane (C_3H_8, a fuel), butane (C_4H_{10}, lighter fuel), and octane (C_8H_{18}, gasoline). Alkenes, also known as olefins, are the second type of hydrocarbon. In these substances, the carbons are held by double bonds. Ethene or ethylene (C_2H_4) is the simplest of the alkenes and when the double bond is broken, the ethylene monomer can be used to produce one of the most popular plastics, polyethylene. Alkynes have triple bonds between the carbons which makes them more reactive. The final hydrocarbon family are the aromatic or aryl compounds. These compounds are composed of six single- and double-bonded carbon atoms arranged in a ring. Benzene, a highly toxic material, is the primary member of the aromatics and is composed of only one ring. Naphthalene is the simplest member of the aromatics having two or more rings.

Organooxygen compounds have oxygen-containing functional groups. These include epoxides, alcohols, phenols, ethers, aldehydes, ketones, and carboxylic acids. Ethylene oxide (an epoxide) is used as a chemical intermediate, sterilant, and fumigant. It is a toxic, colorless, flammable, and explosive gas, which is considered both a mutagen and a carcinogen. Methanol or wood alcohol is a flammable, volatile, and colorless liquid. It is used as both a solvent and a fuel. If methanol is ingested, large doses can cause death or, in lower dosages, blindness. Phenols are toxic materials which are used for chemical synthesis or polymer manufacture. Acrolein (an aldehyde) is a very dangerous, highly reactive chemical. Exposure to oxygen causes acrolein to form explosive peroxides. It is toxic by all routes of exposure. Acetone (a ketone) is typically used as a solvent and is highly flammable. Finally, propionic acid is an example of a carboxylic acid.

Organonitrogen compounds contain nitrogen in the form of amines, nitrosamines, and nitro compounds. A typical toxic amine compound would be methylamine. It is flammable, as well as being an eye, skin, and mucous membrane irritant. Dimethylnitrosamine is used as an industrial solvent. It is known to cause liver damage and jaundice. In addition, these compounds can be carcinogenic. Solid 2,4,6-trinitrotoluene (TNT) has been used as an explosive for a long time. Hepatitis and aplastic anemia have been found in workers exposed to this compound.

Organohalide compounds have halogen-substituted hydrocarbon molecules. This means that each compound has fluorine, chloride, bromine, or iodine atoms in its structure. Alkyl halides in this group include dichloromethane (found in paint strippers), carbon tetrachloride (refrigerants), and 1,2-dibromoethane (an insecticide). The alkenyl or olefinic organohalides include: vinyl chloride (used to produce polyvinyl chloride, PVC), a known carcinogen, trichlorethylene (used for degreasing and as a drycleaning solvent), tetrachloroethylene, and hexachlorobutadiene (used as a hydraulic fluid). Aryl halides are used in chemical synthesis and as pesticides and solvents. They are derivatives of benzene and toluene. Polychlorinated biphenyls (PCBs), highly toxic materials, are an example of a halogenated biphenyl. Chlorofluorocarbons (CFCs), halons, and hydrogen-containing chlorofluorocarbons are of significant importance to the environment. CFCs, once used primarily as refrigerants and aerosol propellants, are believed to have caused the breakdown of the ozone layer and have been banned from production. Halogens used in fire extinguishers as halon have also been implicated in the depletion of the ozone layer and are being phased out. Hydrogen containing chlorofluorocarbons (HFCs) are being touted as the substitute for CFCs as refrigerants and plastic foam blowing agents. Chlorinated phenyls such as pentachlorophenol, are used to treat wood against fungi and insect infestation. The byproduct of that process causes hazardous waste, which has been known to cause liver damage and dermatitis.

Organosulfur compounds are very often noted for their "rotten egg" or garlic odors. Methanethiol, an example of thiols and thioethers, is used as a leak-detecting additive for methane, propane, and butane. Two examples of nitrogen-containing organosulfur compounds are thiourea and phenylthiourea. These compounds are used as rodenticides but are also highly toxic to humans. Dimethylsulfoxide (DSMO) and sulfolane are examples of sulfoxides and sulfones. DSMO has a number of uses including as a paint and varnish remover and as a biological anti-inflammatory. Sulfolane is used as part of the BTX process where it helps in the extraction of benzene, toluene, and xylene from alkanes. Organic esters of sulfuric acids include methylsulfuric and ethylsulfuric acids, which are strong irritants.

Organophosphorus compounds consist of alkyl and aryl phosphines, organophosphate esters, and phosphorothionate esters. Methyl phosphine, a reactive gas, and triphenylphosphine, a toxic material, are examples of alkyl and aryl phosphines. Examples of organophosphate esters include tri-*o*-cresyl-phosphate (TOCP), a poison, and tetraethylpryophosphate (TEPP), an insecticide, which kills everything, including human beings. Parathion, a potent insecticide, is an example of a phosphorothionate ester.

II. GENERAL REVIEW OF PHYSICAL PROPERTIES

Physical properties refer to the interaction of materials with various forms of energy. These properties can be measured without destroying the material. In the following section, some of the physical properties of importance to the industrial hygiene professional will be discussed.

A. Absolute Zero

Absolute zero is the temperature at which translational motion (motion from point to point) ceases. The exact numerical value is -273.15°C or 0K. In order to convert temperature values from Celsius to Kelvin, one would use the following formula:

$$T(K) = [t(C) + 273] \qquad (2.15)$$

B. Change of State

A change of state occurs when a solid transforms to a liquid or when a liquid changes to a gas. These are the three states of matter. A solid is a material, which has a specific shape and volume. Liquids have a definite volume but not a definite shape. Gases are materials that have no definite shape and no definite volume.

C. Phase Change/Phase Equilibria

When matter changes from one state to another, a phase change occur. For example, if one watches snow melt, there is a point where the snow will begin changing from solid water to liquid water. This in-between state is called slush. Slush contains both snow and water, and the heat energy being released by the phase change, allows it to continue. Phase equilibria occurs when the snow has changed completely to water.

D. Boiling Point/Condensation Point

The *boiling point* occurs when a liquid reaches a temperature where a phase change to a gaseous state occurs. Boiling points of materials are affected by atmospheric pressure. Therefore, water's boiling point decreases at higher altitudes with lower pressures.

The *condensation point* is the opposite of the boiling point. It occurs when a gas is cooled to a temperature where a phase change to a liquid results. When materials such as liquid oxygen are produced, it involves the cooling and increasing of pressure of gaseous oxygen to the point where it condenses to form a liquid.

E. Melting Point/Freezing Point

The *melting point* is the temperature at which a solid becomes a liquid. Freezing points occur when a liquid material transforms to a solid material. Pressure can affect both the melting point and freezing point of materials.

F. Viscosity

Viscosity measures the flow resistance of a liquid. Different liquids have different viscosities. Syrup and molasses have a high viscosity while water and gasoline have low viscosities. Temperature also affects viscosity. Higher temperatures create less viscous liquids and lower temperatures make liquids more viscous.

G. Surface Tension

Surface tension is the tendency of a liquid's surface to contract. This is what causes water to form drops and allows some insects to walk on water. The surface tension of a liquid can be modified by certain agents. For example, laundry detergents reduce the surface tension of water so that it will penetrate the clothing. For small hazardous materials spills of oil on water, sometimes a material is used to breakdown the surface tension of the oil and dissolve it into the water.

H. Latent Heat

When molecules are undergoing a phase change, the heat energy being generated is used to either to break the bonds or initiate bonding within the material. This energy cannot be measured by a thermometer, therefore it is *latent* (unmeasurable). When water boils, the thermometer will tell you when the temperature reaches 100°F. However, the measurable temperature does not increase for a short period of time until water vapor begins to appear. Then, the temperature will begin to increase again.

I. Sensible Heat

Sensible heat is the measure of the heat energy given off to the environment. It is measured using a thermometer.

J. Vapor Pressure

Vapor pressure is "the pressure exerted by the vapor with the container on the sides of the container." Vapor pressure will vary with temperature. The higher the temperature, the higher the vapor pressure.

K. Density/Specific Gravity

Density is the ratio of mass to volume. If a material has a high density, such as lead, it will feel heavier than the same volume of aluminum which has a lower density.

Specific gravity is the mass of a substance as compared to the mass of an equal volume of water. If a material has a specific gravity above 1, it will sink. Substances with specific gravities of 1 or less will float.

L. Solubility

Solubility deals with the maximum amount of a compound that will dissolve in a solvent at a given temperature. At 20°C, 205 grams of sugar will be completely soluble in 100 grams of water. However, there are some materials that are insoluble, such as oil and vinegar.

III. GENERAL REVIEW OF COMBUSTION PROPERTIES

Combustion is a complex process. One must understand the components that go into causing a fire to begin and propagate. These components are described by the *fire tetrahedron*. The different portions of this tetrahedron are: fuel, oxygen, heat of ignition, and chain reaction. These four items are all necessary for combustion to be maintained.

Fuel is basically any material that will burn. Fuels can be classified into four main categories.

1. Carbon-based and other readily oxidizable nonmetals
2. Hydrocarbons
3. Compounds containing carbon, hydrogen, and oxygen
4. Metals

Oxygen is the second part of the fire tetrahedron. It provides the atmosphere that allows the fire to continue burning. Oxygen is the hardest element of the fire tetrahedron to control. In our atmosphere, oxygen composes 21% of the air we breathe. This is often enough for most materials to burn. However, air is not the only source of oxygen. Materials called oxidizers can also add oxygen to support combustion reactions. An example of how the amount of oxygen can affect the speed of combustion occurred in the Apollo 1 Capsule. Because of the pure oxygen atmosphere, the fire spread rapidly and killed the three astronauts inside.

The *heat of reaction* is another element of the fire tetrahedron. This heat causes a material to begin burning. Different materials have different heats of reaction. For example, paper has a heat of ignition of 451°F. However, in order for solids to ignite, they must first decompose and start to emit flammable vapors. This decomposition is called *pyrolysis*. In liquids, the decomposition process is called *vaporization*. Flammable gases do not need to decompose since they are already in a form, which can combine with oxygen. Once these vapors and gases have been released and the heat of ignition is reached, flames can be seen.

The final part of the fire tetrahedron is the *chain reaction*. As the chemical bonds are broken, energy is released. The energy from this exothermic reaction is given off as heat. As the temperature rises, the breakdown of fuel continues, which allows the combustion process to continue.

In the following paragraphs, some important fire terms will be defined. It is hoped that by understanding this terminology the industrial hygiene professional will better be able to deal with the area of fire prevention.

A. Flash Point

The *flash point* is the lowest temperature at which a material gives off enough vapors to form an ignitable mixture. At this point, the fire will quickly flash across the surface of the material. It is the flash point temperature that is most often found in fire protection literature.

B. Fire Point

The *fire point* is defined as "the temperature, usually about 5 to 10 degrees above the flash point, at which the ignitable mixture will continue to burn." While most references mention only the flash point, the fire point is important because it is the temperature at which combustion will be sustained.

C. Autoignition Temperature

The *autoignition temperature* is the temperature at which a material will ignite without an external source of ignition. It is also known as the ignition temperature or autogenous ignition temperature. If a liquid is heated and any ignition sources isolated, there will be a point where the liquid will ignite. Most of these temperatures are determined under laboratory conditions, but in real-world situations this temperature can be affected by other variables. Therefore, it is important to note that this temperature should not be considered as a constant.

D. Flammable/Explosive Limits

Flammable limits are used to determine when the vapor concentration is either too low or too high to allow the combustion reaction to occur. For example, if there is too little gasoline vapor per volume of air in the combustion chamber of a car engine, the engine will not start. In this situation, the gasoline vapor to air ratio is considered too lean to ignite. At the other end of the spectrum, if the gasoline's vapor volume is too high compared to that of the air, ignition will also not occur. This mixture is often referred to as too rich. When a car engine floods, one can often smell gasoline because the concentration of those particular vapors is very high. The point at which a vapor/air mixture is too lean for combustion is called the *lower flammability limit (LFL)*, while the point at which a vapor/air mixture is too rich for combustion is called the *upper flammability limit (UFL)*. In oxygen-rich environments, this flammability range will increase. However, under oxygen-deficient conditions, it will contract. Industrial hygiene professionals are usually most concerned about the LFL since any measurements used to determine atmospheric suitability for employees are based on this ratio. In most cases, one would not want employees working in vapor concentrations anywhere near the LFL. However, the percentage below the LFL is usually dictated by the volatility of the vapor, temperature, and oxygen content.

Explosive limits are very similar to flammable limits. The *lower explosive limit (LEL)* is used to determine when the concentration of material is too small for an explosion to occur. Similarly, *the upper explosive limit (UEL)* defines the level at which there is too much material present for an explosion to occur. Once again, the industrial hygiene profession is usually more concerned with the LEL. They must determine what concentration below the LEL is acceptable for a given material (refer to Figure 2.1). One factor of importance in this determination that is often ignored is the effect of pressure. Increasing the pressure will raise the flash point, while lowering the pressure will decrease the flash point. Other factors that should also be considered include: oxygen content, chemical composition, turbulence, and confinement.

E. Exotherm

An *exotherm* is any material which is capable of releasing energy when it burns. This energy is typically given off as heat, which allows the combustion reaction to continue.

F. Spontaneous Combustion

Spontaneous combustion happens when combustible solids are heated to the ignition point by an internal source of heating. It most commonly occurs in agricultural products that are stored wet. This allows bacterial oxidation to occur, which releases heat. Spontaneous combustion can also occur with coal or charcoal, as well as minerals of vegetable origin. In addition, there have been cases of spontaneous combustion caused by violent chemical reactions with either air, water, or other compounds. When phosphorous is exposed to air, it will immediately ignite.

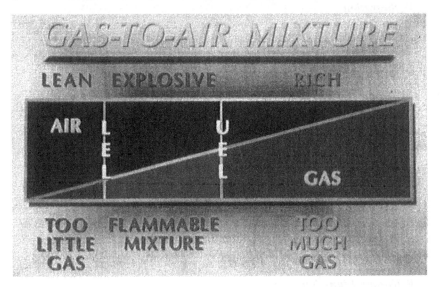

Figure 2.1   Relationship of the Lower Explosive Limit (LEL) and the Upper Explosive Limit (UEL) to explosions and combustion (Courtesy: *Mine Safety Appliances Company*, Pittsburgh, PA).

G. Explosibility

An *explosion* can be defined as "the effect produced by a sudden violent expansion of gases which may or may not be accompanied by shock waves, and/or disruption of enclosing materials." For the industrial hygiene professional, explosions occur due to high concentrations of materials such as vapors, gases, or dusts. In addition, one should also be aware of *boiling liquid expanding vapor explosions (BLEVEs)*. These occur when a liquefied flammable gas container (such as liquefied propane) is heated up causing the liquid to boil and change to a gas, which expands and can cause the container to explode. Therefore, knowing the LEL and UEL of potentially explosive materials is of extreme importance.

1. Detonation

A *detonation* can be defined as "an explosive chemical reaction with a release rate of less than 1/100 of a second." It proceeds faster than the speed of sound and produces a shock wave. However, very special conditions are needed for detonation to occur. Since the range between the LEL and UEL is smaller than the flammability range, it is less likely that an explosion will happen. Another important thing to consider is the size of pressure vessels and pipes. A length to diameter ratio greater than 10 is more favorable for detonation.

2. *Deflagration*

Deflagration occurs at a speed slower than the speed of sound and causes no shock wave. Most explosions are defined as deflagrations.

IV. GENERAL REVIEW OF MATHEMATICS

In order to help you better understand some of the calculations found in subsequent chapters, there will a review of some mathematical principles. There will be three areas covered: basic algebra, logarithms, and trigonometry.

A. Basic Algebra

Algebra is a mathematical science that uses symbols (typically letters) to represent numbers or members of a specified set of numbers and are related to operations that hold for all numbers in the set. These algebraic rules help individuals solve equations that involve unknown quantities.

1. *Polynomial functions*

Polynomials are defined using the following formula:

$$f(x) = a_n x^n + a_{n-1} x^{n-1} + \ldots + a_1 x + a_0 \qquad (2.16)$$

where a is an element of ($\in$) R. This is an expression of a formal polynomial. Of importance for the industrial hygiene professional is the addition of polynomials and the multiplication of polynomials, which are discussed in the next section.

The addition of polynomials can be described by the following equations:

$$f(x) = a_0 + a_1 x + a_1 x^2 \ldots, \qquad (2.17)$$

where the coefficients of a_i are all from ring R and only finitely many of the coefficients are different from zero. Therefore, the coefficients for $f(x)$ are:

$$a_i = (a_0, a_1, \ldots) \qquad (2.18)$$

where $a_i \in$ R and all but a finite number of a_i are zero. Another polynomial, $g(x)$, from the same ring R can been seen below:

$$g(x) = b_0 + b_1 x + b_1 x^2 \ldots, \qquad (2.19a)$$

where $b_i = (b_0, b_1, ...)$. The sum of these two polynomials can be written as:

$$f(x) + g(x) = (a_0 + b_0) + (a_1 + b_1)x + (a_2 + b_2)x^2 + ... \quad (2.19b)$$

or:

$$f(x) + g(x) = \sum_k (a_k + b_k)x^k \quad (2.19c)$$

This corresponds to the vector addition of $a + b = (a_0 + b_0, a_1 + b_1, ...)$.

The product of two polynomials, f and g, can be calculated using the following formula.

$$f(x)g(x) = p_0 + p_1x + p_2x^2 + ... \quad (2.20a)$$

where:

$$p_k = a_0b_k + a_0b_k + ... + a_kb_0 = \sum_{i+j=k} = a_ib_j. \quad (2.20b)$$

2. Exponential functions
When a quantity is raised to some power, it is known as an exponential function. For example:

$$5^2 = 25 \quad (2.21)$$

However, for algebraic operations, there are a number of principles, which can be used to solve equations with exponentials.

• Multiplication of exponentials
The rule used for multiplying exponential numbers of the same base is:

$$x^n x^m = x^{n+m} \quad (2.22)$$

There is really no way to simplify an expression if the bases are different ($x^n y^m = x^n y^m$). However, if the exponents are similar but the bases are different, the rule is:

$$x^n y^n = (xy)^n \quad (2.23)$$

In addition, the exponential can be raised to a higher power. The rule for this function is:

$$(x^n)^m = x^{nm} \quad (2.24)$$

- *Negative exponents*

    If the sign of an exponent of an exponential number in the nu-
    merator of an expression is changed, it is written to the denomi-
    nator or vice versa. The rule can be written as:

$$x^{-n} = 1/x^n \qquad (2.25a)$$

or:

$$1/x^{-n} = x^n \qquad (2.25b)$$

- *Division*

    Division in exponential terms is accomplished by using negative
    exponents. The rule for division can be written as:

$$x^n/x^m = x^{n-m} \qquad (2.26)$$

B. Logarithms

There are two types of logarithms that are of importance for industrial
hygienists. They are the common logarithm and natural logarithm. Common logarithms use a base of 10 and are denoted by $\log x$. Noise calculations use common logarithms. Natural logarithms use base e and are used
most often in ventilation calculations. To denote natural logarithms, the
term $\ln x$ is used.

There are three basic rules that are used with logarithms. These rules
are:

- The logarithm of the product of two numbers equals the sum of
    their logarithms.

$$\log xy = \log x + \log y \qquad (2.27)$$

- The logarithm of the ratio of two numbers is equal to the differ-
    ence of the logarithm of the numerator and the logarithm of the
    denominator.

$$\log x/y = \log x - \log y \qquad (2.28)$$

- The logarithm of an exponential number equals the product of the
    exponent and the logarithm of the base.

$$\log x^n = n \log x \qquad (2.29)$$

A corollary to the last principle is that a logarithm of the nth root of a
number is equal to the ratio of the logarithm of the number to n. This is
written as follows:

$$\log x^{1/n} = (\log x)/n \qquad (2.30)$$

Sometimes it could be necessary to change the base of a logarithm, for example, changing a common logarithm to a natural logarithm. Taking a logarithm to the base b of $y = a^x$ gives the following result.

$$\log_b y = \log_b a^x \qquad\qquad (2.31a)$$

so:

$$\log_b y = x \log_b a \qquad\qquad (2.31b)$$

Since it is known that $x = \log_a y$, it can be substituted into the equation above to give:

$$\log_b y = \log_a y \log_b a \qquad\qquad (2.32a)$$

This equation can be rewritten in the following form:

$$\log_a y = \log_b y / \log_b a \qquad\qquad (2.32b)$$

Therefore, the base conversion formula is:

$$\log y = \ln y / \ln 10 \text{ where } (\ln 10)^{-1} = 0.43429 \qquad (2.33a)$$

$$\ln y = \log y / \log e \text{ where } (\log e)^{-1} = 2.3026 \qquad (2.33b)$$

C. Trigonometry

Trigonometry is the study of the relationship of angles and sides of triangles. In this section, the use of sine, cosine, and tangent functions will be discussed.

The *Pythagorean theorem*, $c^2 = a^2 + b^2$, is used to describe the sides of a right triangle where c indicates the hypotenuse and a and b are the sides. In addition, the sum of the interior angles must equal 180° ($\alpha + \beta + \gamma = \pi$). Since we know that $\gamma = 90°$, then:

$$\alpha + \beta = \pi/2 \qquad\qquad (2.34)$$

This formula shows us that angles α and β are complementary angles since their sum is equal to 90°. Formulae for the sine, cosine, and tangent functions of complimentary angles can be written as follows:

$\sin \alpha = a/c$	$\sin \beta = b/c$
$\cos \alpha = b/c$	$\cos \beta = a/c$
$\tan \alpha = a/b$	$\tan \beta = b/a$

By knowing these relationships, other basic relationships can be postulated. The first of these is that the sine of any angle θ is equal to the cosine of the complimentary angle of θ, therefore:

$$\sin \theta = \cos (90° - \theta) \qquad (2.35a)$$

and:

$$\cos \theta = \sin (90° - \theta) \qquad (2.35b)$$

Further, it can be suggested that since $\sin \alpha = \cos \beta$ and $\cos \alpha = \sin \beta$, the tangent of α will be:

$$\tan \alpha = \sin \alpha/\cos \alpha \qquad (2.36a)$$

and:

$$\tan \beta = \sin \beta/\cos \beta \qquad (2.36b)$$

therefore:

$$\tan \theta = \sin \theta/\cos \theta \qquad (2.36c)$$

Besides these basic formulae, there are three other relationships, which are of interest. These formulae are called the addition formulae and are as follows:

$$\sin (\alpha \pm \beta) = \sin \alpha \cos \beta \pm \sin \beta \cos \alpha \qquad (2.37a)$$

$$\cos (\alpha \pm \beta) = \cos \alpha \cos \beta \pm \sin \alpha \sin \beta \qquad (2.37b)$$

$$\tan (\alpha \pm \beta) = (\tan \alpha \pm \tan \beta)/(1 \pm \tan \alpha \tan \beta) \qquad (2.37c)$$

There are also two laws, which are important to the trigonometry of right triangles. The first is the law of cosines, which is defined as:

$$c^2 = a^2 + b^2 - 2ab \cos \gamma \qquad (2.38)$$

The second law, the law of sines, is defined as:

$$a/\sin \alpha = b/\sin \beta = c/\sin \gamma \qquad (2.39)$$

V. GENERAL REVIEW OF PHYSICS

Physics is "the science of matter and energy and of the interactions of the two." In the following paragraphs, some of the terms of importance to the industrial hygiene professional will be discussed.

A. Mechanics

In *classical mechanics*, it is important to understand the kinematics of the motion and path of a particle in reference to a point in space. This point in space is described using a reference frame or coordinate system. Using the common Cartesian coordinate system, the position of a particle in time t can be described as its displacement vector (r):

$$r = xi + yj + zk \qquad (2.40)$$

If these spatial coordinates are seen as a function of time, then:

$$x = x(t) \qquad (2.41a)$$
$$y = y(t) \qquad (2.41b)$$
$$z = z(t) \qquad (2.41c)$$

which leads to the definition of instantaneous translational velocity. It is defined by the following equation:

$$v \equiv dr/dt \qquad (2.42)$$

whose functional units are expressed in meters/second or inches/second. A three-dimensional vector can then be expressed in terms of its rectangular components.

$$v = v_x i + v_y j + v_z k \qquad (2.43)$$

where the components of velocity are described by:

$$v_x \equiv dx/dt \qquad (2.44a)$$
$$v_y \equiv dy/dt \qquad (2.44b)$$
$$v_z \equiv dx/dt \qquad (2.44c)$$

The equations above deal with problems involving Einsteinian relativity. By using derivations of classical relativity, the equations for average translational velocity can be developed as follows:

$$\overline{v} \equiv \Delta r/\Delta t \qquad (2.45a)$$
$$\overline{v}_x \equiv \Delta x/\Delta t \qquad (2.45b)$$
$$\overline{v}_y \equiv \Delta y/\Delta t \qquad (2.45c)$$
$$\overline{v}_z \equiv \Delta z/\Delta t \qquad (2.45d)$$

where Δ means the change in a parameter. For example, if the first z coordinate was 5 and the next is 10, the change (Δz) would be equal to 10 minus 5 so Δz would be equal to 5.

Acceleration is the time rate of change of velocity. Mathematically, the equation for instantaneous translational acceleration is:

$$a \equiv dv/dt = d^2 r/dt^2 \qquad (2.46a)$$
$$a = a_x i + a_y j + a_z k \qquad (2.46b)$$

The components of this equation are:

$$a_x \equiv dv_x/dt = d^2x/dt^2 \tag{2.47a}$$
$$a_y \equiv dv_y/dt = d^2y/dt^2 \tag{2.47b}$$
$$a_z \equiv dv_z/dt = d^2z/dt^2 \tag{2.47c}$$

In order to compute the average translational acceleration, the following equation should be used:

$$\bar{a} \equiv \Delta v/\Delta t = \bar{a}_x i + \bar{a}_y j + \bar{a}_z k \tag{2.48}$$

The Cartesian components of this equation are:

$$\bar{a}_x \equiv \Delta v_x/\Delta t \tag{2.49a}$$
$$\bar{a}_y \equiv \Delta v_y/\Delta t \tag{2.49b}$$
$$\bar{a}_z \equiv \Delta v_z/\Delta t \tag{2.49c}$$

The basic units of acceleration are m/s^2 or in/s^2.

Up to this point, only kinematics have been discussed. In order to find how external factors affect the state of motion of a particle or system of particles, the science of dynamics must be explored. Newton developed three laws of motion which deal with dynamics and are stated as:

- A body in a state of rest or uniform motion will continue in that state unless acted upon by an external unbalanced force.
- The net external force acting on a body is equal to the time rate of change of the body's momentum.
- For every force acting on a body there exits a reaction force, equal in magnitude and oppositely directed, acting on another body.

Newton's second law can be described mathematically with the following equation:

$$F = ma \tag{2.50}$$

Essentially, this equation states that the force on an object is equal to its mass times acceleration.

B. Wave Motion and Sound

Sound waves use a medium of air while water waves have a medium of water. Electromagnetic waves use electric and magnetic fields as the medium.

Sound waves are defined by their frequency, wavelength, and velocity. The frequency of a sound wave is the number of times per second that a vibrating body traces one complete cycle. It is measured in Hertz (Hz). Frequency is frequently thought of as pitch. Humans can hear sounds in the range of 20 Hz to 20,000 Hz. The distance a wave travels in one cycle is called its wavelength. Wavelength is measured in feet or meters and is expressed by the letter lambda (λ). Velocity, called the speed of sound, can be calculated using the following formula:

$$c = f\lambda \qquad (2.51)$$

Where:

c = Speed of sound, ft/s or m/s
f = Frequency of the sound, Hz
λ = Wavelength, ft or m

C. Electricity and Magnetism

Electricity is transmitted in the form of current through a conductive medium. One of the important terms used in the description of electrical current is based on Coulomb's law.

$$F_e = q_1 q_2 / R^2 \qquad (2.52)$$

Where:

F_e = Electrostatic force, dynes
q_1 = Electrical charge, particle 1, statcoulomb or electrostatic unit of charge
q_2 = Electrical charge, particle 2, statcoulomb or electrostatic unit of charge
R = Distance between two charges, cm

In this equation, F_e, the electrostatic force (measured in dynes) is equal to the total electrical charge (indicated by q) of two bodies divided by the square of the distance between the two charges, R. The electrical charge of each body is determined using the following equation:

$$q = (n_+ - n_-)e \qquad (2.53)$$

Where:

q = Electrical charge, statcoulomb or electrostatic unit of charge
n_- = Negative particle charge
n_+ = Positive particle charge
e = Charge of an electron, 4.802×10^{-10} statcoulomb

The unit for q is the statcoulomb or electrostatic unit of charge. Fundamental particles have either a negative (n_-) or a positive charge (n_+). When the difference between numbers of each charge is determined, it is multiplied by e, the charge of an electron. This value is 4.802×10^{-10} statcoulomb. Then the total charge for each body must be calculated. An example calculation to find the electrical force on a system follows:

Sample Calculation: Given that $q_1 = +50$ statcoulomb, $q_2 = -5$ statcoulomb, and R = 10 cm, determine the electrostatic force for the system:

$$F_e = q_1 q_2 / R^2$$
$$F_e = (50)(5)/10^2$$
$$F_e = 250/100$$
$$F_e = 2.5 \text{ dynes directed to the right toward } q_1$$

Knowing Coulomb's law helps us to determine the electric current (*I*) flowing through a wire, using the following equation.

$$I = (n_r - n_l)e/t \qquad (2.54)$$

Where:

I   = Electric current, amperes
$n_r - n_l$ = Indicates the net flow of electrons
e   = Charge of an electron, 4.802×10^{-10} stacoulomb
t   = Time it takes for current to go from right to left

The term $n_r - n_l$ is used to indicate the net flow of electrons. Flow time (t) is the amount of time it takes for the current to go from right to left. The charge of an electron is indicated by the term *e* and is measured in statcoulombs.

While it is important to determine the amount of electrical current, one intuitively knows that this current cannot remain constant throughout its transmission because of resistance. This reality is known as Ohm's law which states that current is directly proportional to the potential difference between the ends of the wire and inversely proportional to the resistance. It can written mathematically as:

$$I = V/R \qquad (2.55)$$

Where:

I = Current, amperes
V = Voltage, volts
R = Resistance, ohms

Magnetism occurs when an electrical field is generated around an object. The magnet has opposing charges (+ or -) at its poles. These poles will attract opposite charges and repel similar charges. The electrical field generated around a magnetic object is dependent on the size of the object and the strength of the magnet.

D. Heat Transfer

Heat transfer deals with how quickly heat energy can be passed from one object to another object. There are three methods of heat transfer, conduction, convection, and radiation.

1. *Conduction*

Conduction occurs when two objects come into contact. The hotter item will pass the heat to the cooler item. When a person touches a hot item, conduction occurs and if it is hot enough, the person will quickly jump back to keep from getting burned.

2. *Convection*

When heat is transferred by air currents, it is called *convection*. In humans, if the skin temperature is warmer than the air temperature, the air next to the skin is warmed and carried away from the body by wind currents. This is also the principle of operation used in convection ovens. Air is circulated around the baking item, which causes more even cooking to occur since hot spots are avoided.

3. *Radiation*

Radiation is the transfer of heat between two objects using electromagnetic radiation. When a person feels warmth on a cold sunny day, the skin is being warmed by the radiation from by sun.

E. Electromagnetic Spectrum

Different *electromagnetic radiation* sources generate discrete energy frequencies. The electromagnetic spectrum covers the gamut of frequencies of different radiation sources.

Table 2.3
The Electromagnetic Spectrum

Type of Radiation	Wavelength Interval (in meters)
Cosmic rays	10^{-16} to 10^{-14}
Gamma rays	10^{-14} to 10^{-12}
X-rays	10^{-12} to 10^{-9}
Ultraviolet	10^{-7} to 10^{-8}
Visible	10^{-7} to 10^{-6}
Infrared	10^{-6} to 10^{-3}
Microwaves	10^{-2} to 10
Broadcast Waves	10^{-1} to 10^{5}

F. Radiation

Radiation is a form of electromagnetic energy. There are two forms of radiation: ionizing and nonionizing.

1. *Nonionizing radiation*

Nonionizing radiation begins in the far ultraviolet range of the electromagnetic spectrum and goes through the far infrared portion of the electromagnetic spectrum. Included in this range are the problems associated with the use of ultraviolet light, visible light, microwaves, radio waves, radar, and lasers.

2. *Ionizing radiation*

As referenced from the ionizing radiation chapter in *Fundamentals of Industrial Hygiene, 3rd ed.*, *ionizing radiation* is "electromagnetic or particulate radiation capable of producing ions, directly or indirectly, by interaction with matter. In biological systems, such radiation must have a photon energy greater than 10 electron volts. This excludes most ultraviolet bands and all longer wavelengths." The following are types of ionizing radiation:

- Alpha particles
- Beta particles
- Neutrons
- X-radiation
- Gamma radiation

It is important for the industrial hygiene professional to remember that ionizing radiation causes more damage to biological systems than does nonionizing radiation.

REVIEW QUESTIONS

1. A material loses electrons:

    a.   In a reduction reaction
    b.   When it is stable
    c.   Because it has a low heat of reaction
    d.   In an oxidation reaction

2. Which of following materials is most active when combined with water?

    a. Lead   b. Potassium   c. Iron   d. Tin

3. The difference between consecutive and competing reactions is that competing reactions:

    a.   Have an intermediate step
    b.   Can produce two different chemical products
    c.   Do not have an intermediate step
    d.   Have a special notation

4. A catalyst is used:

    a.   To add to the chemical product
    b.   To lower the activation energy of a chemical reaction
    c.   To increase the activation energy of a chemical reaction
    d.   Both a and c

5. Enzymes can increase reactions over:

    a. 10 times   b. 100 times   c. 10,000 times   d. 1 million times

6. The group of hydrocarbons that contains CFCs is:

    a. Hydrocarbons   b. Organonitrogens   c. Organohalides   d. Organosulfurs

7. Which of the following is most viscous at room temperature?

    a. Motor oil   b. Syrup   c. Water   d. Gasoline

8. If heat energy can be measured, it is called:

    a. The heat of reaction   b. Specific heat   c. Latent heat   d. Sensible heat

9. The least controllable element of the fire tetrahedron is:

    a. The oxygen   b. The fuel   c. The chain reaction   d. The heat of reaction

10. The temperature at which combustion can be maintained is:

    a.   The flash point
    b.   The ignition temperature
    c.   The fire point
    d.   Both a and c

11. Which of the following terms is of *most* importance to the industrial hygiene professional?

    a.   Upper flammable limit
    b.   Lower flammable limit
    c.   Lower explosive limit
    d.   Both a and c

12. If a car floods, the mixture:

    a.   Has too much gasoline vapor
    b.   Is too rich
    c.   Is above the upper flammability limit
    d.   All of the above

13. Spontaneous combustion occurs in:

    a.   Stacked papers
    b.   Agricultural products stored wet
    c.   Propane gas
    d.   Gasoline

14. Common logarithms:

    a.   Are used in noise computations
    b.   Have a base of e
    c.   Are used in ventilation calculations
    d.   None of the above

15. The Pythagorean theory states:

    a.   That sines, cosines, and tangents can be added
    b.   That the sum of the interior angles of a right triangle is 360°
    c.   That the interior angles are all equal
    d.   That sum of the squares of the sides will equal the square of the hypotenuse

16. In classical mechanics, the term used to describe the motion and path of a particle in reference to a point in space is:

    a   Cartesian coordinates
    b.   Kinematics
    c.   Dynamics
    d.   Instantaneous translational velocity

17. In terms of sound waves, the frequency is:

    a.   The number of times per second that a vibrating body traces in one cycle
    b.   The speed of sound
    c.   Expressed by the letter λ
    d.   The distance a wave travels in one cycle

18. Heat transfer caused by sun is called:

    a. Conduction   b. Convection   c. Radiation   d. None of the above

19. In the electromagnetic spectrum, the visible spectrum is located at frequencies of:

    a. 10^{-16} to 10^{-14} meters
    b. 10^{-7} to 10^{-6} meters
    c. 10^{-6} to 10^{-3} meters
    d. 10^{-2} to 10 meters

20. Which of the following radiation sources causes the most problems in biological systems?

    a. Microwaves   b. Lasers   c. Radar   d. Gamma radiation

21. Determine the order of the following equations:

    a. $CH_3CH_2Cl \rightarrow C_2H_4 + HCl$
    b. $H_2 + I_2 \rightarrow 2HI$
    c. $2NO_2 \rightarrow 2NO + O_2$
    d. $N_2O_4 \rightarrow 4NO_2 + O_2$

22. If the temperature is 32° C, what is the temperature in °K? For 15° C?

23. Find the solution to log 25 times log 55.

24. Find the sine of 30° plus 45°.

25. Find the speed of sound (c), if f is equal to 726 Hz and λ is equal to 0.445 m.

ANSWERS

1	d	11.	b.
2.	b.	12.	a.
3.	b.	13.	b.
4.	b.	14.	d.
5.	d.	15.	d.
6.	c.	16.	b.
7.	b.	17.	a.
8.	d.	18.	c.
9.	a.	19.	b.
10.	c.	20.	d.

21. a 1st Order
    b. 2nd Order
    c. 2nd Order
    d. 2nd Order

22. a.   °K = °C + 273K
         °K = 32°C + 273K
         °K = 305K
    b.   °K = °C + 273K
         °K = 15°C + 273K
         °K = 288K

23.   log (25 x 55) = log 25 + log 55
         log (25 x 55) = 1.3979 + 1.7404
         log (25 x 55) = 3.1383

24.   sine(30 + 45) = [(sine 30)(cos 45)] + [(sine 45)(cos 30)]
       sine(30 + 45) =[(0.5) x (0.707)] + [(0.707) x (0.866)]
       sine(30 + 45) = 0.3536 + 0.6122
       sine(30 + 45) = 0.9658
25. $c = f\lambda$
       c = 726 Hz x 0.445 m
       c = 323 m/s

References

Artin, M., *Algebra*, Prentice Hall, Englewood Cliffs, NJ, pg. 350-351, 1991.

Atkins, K. R., *Part D, Physics, 2nd ed.*, John Wiley & Sons, New York, 1970.

Budinski, K. G., Chap. 2, *Engineering Materials: Properties and Selection, 5th ed.*, Prentice Hall, Englewood Cliffs, NJ, 1996.

Burns, M. L., Appendix A, *Modern Physics for Science and Engineering*, Harcourt Brace Jovanovich, San Diego, 1988.

Burns, M. L., Chap. 1, *Modern Physics for Science and Engineering*, Harcourt Brace Jovanovich, San Diego, 1988.

Burns, M. L., *Modern Physics for Science and Engineering*, Harcourt Brace Jovanovich, San Diego, pg. 181, 1988.

Cheever, C. L., Chap. 10, Ionizing radiation, *Fundamentals of Industrial Hygiene, 3rd ed.*, Plog, B. A. Ed., National Safety Council, Itasca, IL, 1988.

Cheever, C. L., Ionizing radiation, *Fundamentals of Industrial Hygiene, 3rd ed.*, Plog, B. A., Ed., National Safety Council, Itasca, IL, pg 207, 1988.

Keenan, C. W. and Wood, J. H., *General College Chemistry, 4th ed.*, Harper & Row, New York, pg. 140, 1971.

Keenan, C. W. and Wood, J. H., Chap. 12, *General College Chemistry, 4th ed.*, Harper & Row, New York, 1971.

Ladwig, T. H., Chap. 2, *Industrial Fire Prevention and Protection*, Van Nostrand Reinhold, New York, 1991.

Ladwig, T. H., *Industrial Fire Prevention and Protection*, Van Nostrand Reinhold, New York, pg. 27, 1991.

Ladwig, T. H., Chap. 4, *Industrial Fire Prevention and Protection*, Van Nostrand Reinhold, New York, 1991.

Ladwig, T. H., *Industrial Fire Prevention and Protection*, Van Nostrand Reinhold, New York, pg. 67, 1991.

Ladwig, T. H., Chap. 10, *Industrial Fire Prevention and Protection*, Van Nostrand Reinhold, New York, 1991.

Ladwig, T. H., *Industrial Fire Prevention and Protection*, Van Nostrand Reinhold, New York, pg. 224, 1991.

Largent, E. J., Olishifski, J. B. and Anderson, L. E., Nonionizing radiation, in *Fundamentals of Industrial Hygiene, 3rd ed.*, Plog, B. A. Ed., National Safety Council, Itasca, IL, pg 228, 1988.

Largent, E. J., Olishifski, J. B. and Anderson, L. E., Chap 11, Nonionizing radiation, in *Fundamentals of Industrial Hygiene, 3rd ed.*, Plog, B. A., Ed., National Safety Council, Itasca, IL, 1988.

Loudon, G. M., *Organic Chemistry 3rd ed.*, The Benjamin/Cummings Publishing Company, Redwood City, CA, pg. 391-392, 1995.

Malone, L. J., *Basic Concepts of Chemistry, 4th ed.*, John Wiley & Sons, New York, pg. 532, 1994.

Malone, L. J., Chap. 8, *Basic Concepts of Chemistry, 4th ed.*, John Wiley & Sons, New York, 1994.

Malone, L. J., Chap. 5, *Basic Concepts of Chemistry, 4th ed.*, John Wiley & Sons, New York, 1994.

Malone, L. J., Chap. 12, *Basic Concepts of Chemistry, 4th ed.*, John Wiley & Sons, New York, 1994.

Malone, L. J., Chap. 13, *Basic Concepts of Chemistry, 4th ed.*, John Wiley & Sons, New York, 1994.

Malone, L. J., Chap. 14, *Basic Concepts of Chemistry, 4th ed.*, John Wiley & Sons, New York, 1994.

Malone, L. J., Chap. 2, *Basic Concepts of Chemistry, 4th ed.*, John Wiley & Sons, New York, 1994.

Malone, L. J., *Basic Concepts of Chemistry, 4th ed.*, John Wiley & Sons, New York, pg. 286, 1994.

Malone, L. J., Chap. 10, *Basic Concepts of Chemistry, 4th ed.*, John Wiley & Sons, New York, 1994.

Malone, L. J., Chap. 11, *Basic Concepts of Chemistry, 4th ed.*, John Wiley & Sons, New York, 1994.

Manahan, S. E., Chap. 20, *Environmental Chemistry, 6th ed.*, CRC Press/Lewis Publishers, Boca Raton, FL, 1994.

Manahan, S. E., *Environmental Chemistry, 6th ed.*, CRC Press/Lewis Publishers, Boca Raton, FL, pg. 542, 1994.

Manahan, S. E., Chap 26, *Environmental Chemistry, 6th ed.*, CRC Press/Lewis Publishers, Boca Raton, FL, 1994.

Noll, G. G., Hildebrand, M. S., and Yvorra, J. G. *Hazardous Materials: Managing the Incident, 2nd ed.*, Fire Protection Publications, Stillwater, OK, pg. 231, 1995.

Noll, G. G., Hildebrand, M. S., and Yvorra, J. G. *Hazardous Materials: Managing the Incident, 2nd ed.*, Fire Protection Publications, Stillwater, OK, pg. 267, 1995.

Ohanian, H. C., *Modern Physics, 2nd ed.*, Prentice Hall, Englewood Cliffs, NJ, pg. 17, 1995.

Olishifski, J. B. and Standard, J. J., Chap 9, Industrial noise, in *Fundamentals of Industrial Hygiene, 3rd ed.*, Plog, B. A., Ed., National Safety Council, Itasca, IL, 1988.

Ouellett, R. J. and Rawn, J. D., Chap. 7, *Organic Chemistry*, Prentice Hall, Upper Saddle River, NJ, 1996.

Ouellett, R. J. and Rawn, J. D., Chap. 3, *Organic Chemistry*, Prentice Hall, Upper Saddle River, NJ, 1996.

Ouellett, R. J. and Rawn, J. D., *Organic Chemistry*, Prentice Hall, Upper Saddle River, NJ, pg. 130, 1996.

Plog, B. A., Ed., *Fundamentals of Industrial Hygiene 3rd ed.*, National Safety Council, Itasca, IL, pg. 880 1988.

Sanders, M. S. and McCormick, E. J., *Human Factors in Engineering and Design, 7th ed.*, McGraw-Hill, pg. 552-553, 1993.

Schmid, G. H., *Organic Chemistry*, Mosby, St. Louis, pg. 293-294, 1996.

Schmid, G. H., Chap. 3, *Organic Chemistry*, Mosby, St. Louis, 1996.

Soukhanov, A. H., Ed., *The American Heritage Dictionary of the English Language, 3rd ed.*, Miflin, Boston, pg. 1367, 1992.

Chapter 3

TOXICOLOGY

Stephen K. Hall, Ph.D., CIH

I. OVERVIEW

Toxicology is the study of the adverse effects of chemicals on living organisms. Like medicine, toxicology is both a science and an art; the science portion of toxicology deals with the observational or data-gathering and the art portion deals with prediction. In most cases, these two portions of the discipline are linked since the "facts" generated by the science portion are used to develop the "prediction" of the adverse effects of toxic chemicals.

Earliest man was well aware of the toxic effects of animal venoms and poisonous plants. His knowledge was used for hunting and for waging more effective warfare. In the mythology and literature of classic Greece, one finds many references to poisons and their use. The Romans also made considerable use of poisons to remove their political adversaries.

A significant figure in the history of toxicology was probably the renaissance man of the sixteenth century, Philippus Audreolus Theophrastus Bombastus von Hohenhein, more commonly known as Paracelsus, who formulated many revolutionary views which remain an integral part of modern toxicology. He realized that a toxic substance consisted of a chemical entity. Paracelsus is best remembered by toxicologists for the following views: "All substances are poisons. There is none which is not a poison. The right dose differentiates a poison from a remedy." As a result of Paracelsus establishing the foundation of toxicology, the dose-response relationship, many toxicologists consider Paracelsus "The Father of Toxicology."

II. GENERAL TOXICOLOLGY PRINCIPLES

One of the objectives of studies in toxicology is to produce harmful effects (i.e., *toxicity*), measure and analyze the doses at which toxicity occurs (i.e., the *dose-response relationship*), and assess the probability that injury or illness will occur under specified conditions of use (i.e., *hazard and risk assessment*).

A distinction must be made between toxicity and hazard. An extremely toxic chemical that is in a sealed container on a shelf has inherent toxicity but presents little or no hazard. When the chemical is removed from the shelf and used by a worker in a closed space and without appropriate protection, the hazard becomes great. Thus, the manner of use affects how hazardous the chemical will be in the workplace. Again, two chemicals may possess the same degree of toxicity but present different degrees of hazard. Carbon monoxide, for example, is odorless, colorless, and nonirritating to the eye, nose, and throat while ammonia has a pungent odor and is an eye, nose, and throat irritant. By comparison, ammonia, with the warning properties, presents a lesser degree of hazard. It should be emphasized that the toxicity of a chemical is not a physical constant, such as melting point or vapor pressure, and usually only a general statement can be made concerning the toxic nature of some of the chemicals.

A. Classification of Toxic Chemicals

Toxic substances are classified in a variety of ways, depending on the interests and needs of the classifier. Toxic chemicals may be classified in general terms such as physical state (gas, vapor, fiber, dust, fumes, etc.), their labeling requirements (irritant, corrosive, flammable, explosive, etiologic agent, etc.), their chemistry (aromatic amine, halogenated hydrocarbon, polycyclic aromatic hydrocarbon, etc.), or in accordance with their poisoning potential (extremely toxic, highly toxic, moderately toxic, slightly toxic, etc.).

Classification of toxic chemicals in terms of their site of injury (liver, kidney, bone marrow, etc.), their use (pesticide, solvent, food additive, etc.), their source (coke oven emissions, plant toxin, etc.), their effects (carcinogenic, mutagenic, etc.), and their mechanism of action (asphyxiation, enzyme inhibition, methemoglobin producer, etc.) are usually more informative than classification by general terms, but the more general terms such as air pollutants, occupation-related agents, and acute and chronic hazards can provide a useful focus on a specific problem.

It is evident from the above that no single classification will be applicable for the entire spectrum of toxic chemicals and that combinations of classification systems or classification based on other factors may be needed to provide the best rating system for a special purpose. Nevertheless, classification systems that take into consideration both the chemical and biologic properties of the substance and the exposure characteristics are most likely to be useful for legislative or control purposes and for toxicology in general.

Adverse or toxic effects in a biologic system are not produced by a chemical unless that chemical or its metabolites reach appropriate sites in the body at a concentration and for a length of time sufficient to produce the toxic manifestation. Whether or not a toxic response occurs is dependent, therefore, on the chemical and physical properties of the toxic substance, the exposure situation, and the susceptibility of the biologic system. Thus, to characterize fully the potential hazard or toxicity of a specific chemical we need to know not only what type of effect it produces and the dose required to produce the effect, but also information about the chemical, the exposure, and the subject. The major factors that influence toxicity as it relates to the exposure situation for a specific chemical are the *route* and *site of administration* and the *duration* and *frequency* of exposure.

B. Route and Site of Exposure

In considering the health hazards of chemicals, it is necessary to describe how a chemical gains entrance into the body and then into the bloodstream. The *rate of absorption* is dependent upon the concentration and solubility of the chemical. Chemicals in aqueous solution are absorbed more rapidly than those in oily suspension. Absorption is also enhanced at sites that have increased blood flow or large absorptive surfaces, e.g., the adult lung and gastrointestinal tract, whose surfaces are the size of a tennis court and a football field, respectively. Once absorbed into the bloodstream, a chemical may elicit general effects or more than likely, the toxic effects will be localized in specific tissues or target organs Occupational exposure to toxic chemicals most frequently is the result of inhalation and skin absorption, whereas accidental and suicidal poisoning occurs most frequently by oral ingestion.

1. *Inhalation*
The most common route of occupational exposure is inhalation or pulmonary absorption. Gaseous and volatile toxic chemicals may be inhaled and absorbed through the pulmonary epithelium and mucous membranes in the respiratory tract. Access to the circulation is rapid because the surface area of the lungs is large and the blood flow is great. The nasal hair, the cough reflex, and the mucociliary barrier help prevent dust particles and fumes from reaching the lung. The solubility of chemicals also affects their absorption. Highly water-soluble gases such as ammonia and chlorine are absorbed in the upper airways and cause marked irritation there. This serves as a warning and limits the injury to the lung. Noxious gases of low water solubility such as nitrogen dioxide and phosgene, which have few early warning properties, reach the lungs and cause delayed injury there.

2. *Skin and eye absorption*

Many toxic chemicals pass through the skin, intact or broken. The amount of skin absorption is generally proportional to the surface area of contact and to the lipid solubility of the toxic chemical. The *epidermis*, or outer layer, acts as a lipid barrier, and the *stratum corneum*, or middle layer, provides a protective barrier against noxious agents. The *dermis*, or inner layer, is freely permeable to many toxic chemicals. Absorption is enhanced by toxic chemicals that increase the blood flow to the skin. It is also enhanced by use of occlusive skin coverings (e.g., permeable clothes and industrial gloves) and topical application of fat-solubilizing vehicles. Hydrated skin is more permeable than dry skin. The thick skin on the palms of the hands and the soles of the feet is more resistant to absorption than is the thin skin on the face, neck, and scrotum. Burns, abrasions, dermatitis, and other injuries to the skin may alter its protective properties and allow absorption of larger quantities of the toxic substance. The eye is also a ready site of absorption in the occupational environment. When chemicals enter the body through the conjunctiva, they bypass hepatic elimination and may cause severe systemic toxicity. This may occur when organophosphate pesticides are splashed into the eyes.

3. *Ingestion*

The problem of ingesting chemicals is not widespread but accidental swallowing does occur. Ingestion of inhaled chemicals can also occur because chemicals deposited in the respiratory tract can be carried out to the throat by the action of the mucociliary escalator of the respiratory tract. These chemicals are then swallowed and significant absorption from the gastrointestinal tract may occur. The amount of gastrointestinal absorption is usually proportional to the gastrointestinal surface area and its blood flow, and also depends on the physical state of the substance. Most toxic chemicals are absorbed in the small intestine. Therefore, chemicals that accelerate gastric emptying will increase the absorption rate, while factors that delay gastric emptying will decrease it. Some toxic chemicals may be affected by gastric juice, e.g., the acidity of the stomach may release cyanide products and form hydrogen cyanide gas, which is even more toxic than the cyanide salt.

The vehicle and other formulation factors can markedly alter the absorption following inhalation, skin absorption, or ingestion. In addition, the route of administration can influence the toxicity of chemicals, e.g., a chemical that is detoxified in the liver would be expected to be less toxic when ingested than when inhaled. Toxic effects by any route of exposure can also be influenced by the concentration of the chemical in its vehicle, the total volume of the vehicle and the properties of the vehicle to which the biologic system is exposed, and the rate at which exposure occurs.

C. Duration and Frequency of Exposure

Toxicologists usually divide the exposure to chemicals into two major categories: acute and chronic. *Acute* exposure is defined as exposure to a chemical/toxin for less than 24 hours and it usually refers to a single and continuous exposure to a high concentration of a chemical. Typically, acute exposures are characterized by rapid absorption of the offending chemical, and are usually related to an accident, release, or unplanned event. In contrast to acute exposure, *chronic* exposure refers to repeated exposure to a toxin for more than 3 months. For many chemicals, the toxic effects following an acute exposure are quite different from those produced by chronic exposure, e.g., the primary acute toxic manifestation of benzene is central nervous system depression or may even result in death, but chronic exposures to benzene can affect the blood cell production capability of the bone marrow and result in leukemia.

Acute exposure to chemicals that are rapidly absorbed is likely to produce immediate toxic effects, but acute exposure can also produce delayed toxicity that may or may not be similar to the toxic effects of chronic exposure. Conversely, chronic exposure to toxic chemicals may produce some acute effects after each exposure, in addition to the long-term, low-level, or chronic effects of the agent. In characterizing the toxicity of a specific chemical, it is evident that information is needed not only for the single-dose (acute) and long-term (chronic) effects, but also for exposures of intermediate duration, i.e., subacute and subchronic. *Subacute* exposure refers to repeated exposure to a chemical/toxin for one month or less while *subchronic* exposure is for one to three months.

The other time-related factor that is important in the temporal characterization of exposure is the frequency of exposure. In general, fractionation of the dose reduces the effect. A single dose of a chemical that produces an immediate severe effect might produce less than half the effect when given in two divided doses and no effect when given in ten doses over a period of several hours or days. Such fractionation effects occur when metabolism or excretion occurs between successive doses or when the injury produced by each administration is partially or fully reversed prior to the next exposure. Chronic toxic effects occur if the chemical accumulates in the biologic system, e.g., absorption exceeds metabolism and/or excretion, if it produces irreversible effects, or if there is insufficient time for the system to recover from the toxic damage within the exposure frequency interval.

D. Dose-Response Relationship

A *dose-response relationship* exists when changes in dose are followed by consistent changes in response, as shown in dose-response curves. A variety of toxicological phenomena can be demonstrated by these curves. Figures 3.1A and 3.1B show the intensity of the response to various doses in an individual. Because Figure 3.1B is in a logarithmic scale, the shape of the curve is more linear. This makes it easier to determine values for specific points on the curve. The frequency of a response in a population can be related to dose as a frequency distribution as in Figure 3.2, or as a cumulative frequency as in Figures 3.3 and 3.4.

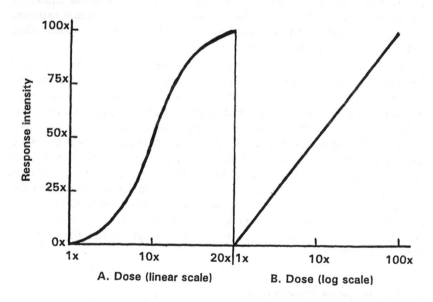

Figure 3.1　　Dose-response relationship (A: linear scale; B: log scale).

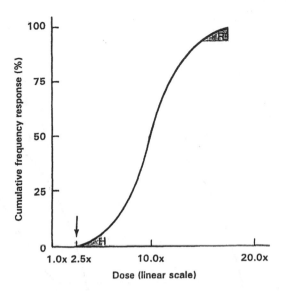

Figure 3.2 Dose-response relationship indicating threshold (arrow); hypersusceptible group (H); and resistant group (R).

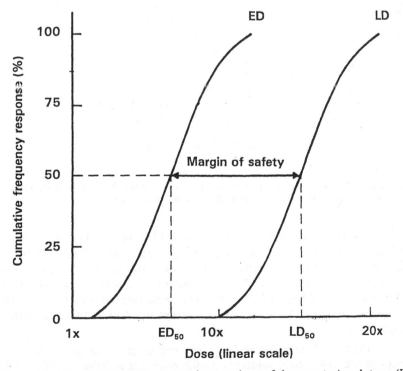

Figure 3.3 Dose-response curves comparing two doses of the same toxic substance (ED: effective dose; LD: lethal dose).

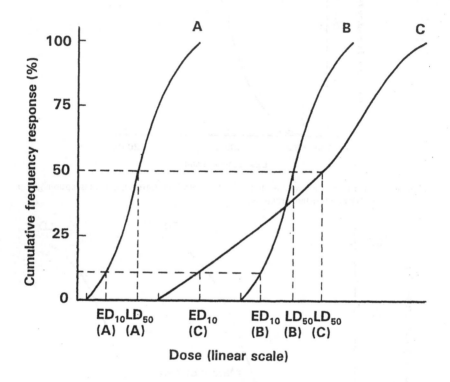

Figure 3.4   Dose-response curves comparing the doses at which the same toxic effect is elicited by three different toxic substances A, B, and C.

In Figure 3.2, the existence of a threshold is indicated by the arrow at the point where the curve intersects the dose coordinate. Doses below this point do not produce a response. Individuals who exhibit the response at doses well below the average or the mean are considered *hypersusceptible* (H in lower left of Figure 3.2), while those who respond only to doses well above the average or the mean are considered *resistant* (R in upper right of Figure 3.2).

In Figure 3.3, cumulative frequency curves are used to compare two doses of the same toxic substance to the dose that is lethal to 50% of the population (LD_{50}) and the dose that has an effect on 10% (ED_{10}). The ED_{10} may, for example, represent an effect that is not harmful, such as odor. The ratio between comparable points on the curves, i.e., the ratio of LD_{50} to ED_{10}, will then represent the margin of safety for odor as a warning against the lethal effect.

In Figure 3.4, cumulative frequency curves are used to compare the doses at which the same toxic effect is elicited by three different toxic chemicals (A, B, and C). Chemical A is clearly the most toxic because at every dose level a greater percentage of the population exhibits the response to A than to B or C. The LD_{50}, the ED_{10}, and the threshold for A are all lower than the corresponding values for B and C. The comparison between B and C is less clear and demonstrates the need to consider the entire dose-response curve rather than individual points when comparing toxicities. Because the LD_{50} of B is lower than that of C, at this dose B is more toxic than C. However, because the ED_{10} of C is lower than that of B, at the lower dose C is more toxic than B. The shape of a dose-response curve is important for assessing the hazard of a toxic chemical. A chemical that has a low threshold and shallow dose-response curve, such as C, may be more hazardous at low doses, while a chemical that has a steep dose-response curve, such as B, may be more hazardous as the dose increases. Adequate assessment of the hazard of a toxic chemical requires evaluation of dose-response data over a wide range of doses.

E. Distribution and Storage

Toxic chemicals are transported via the blood to various portions of the body. Some are removed by the lymph, and some insoluble compounds are transported through tissues such as the lung via cells such as macrophages. Most toxic chemicals enter the bloodstream and are distributed into interstitial and cellular fluids. The pattern of distribution depends on the physiologic and physicochemical properties of the chemical. The initial phase of distribution usually reflects the cardiac output and regional blood flow. Lipid soluble chemicals that penetrate membranes poorly are restricted in their distribution, and their potential sites of action are therefore limited. Distribution may also be limited by the binding of toxic chemicals to plasma proteins. Toxic chemicals can accumulate in higher concentration in some tissues as a result of pH gradients, binding to special cellular proteins, or partitioning into lipids.

Toxic chemicals are often concentrated in a specific tissue. Some toxic chemicals achieve their highest concentration at their site of toxic action, such as carbon monoxide, which has a very high affinity for hemoglobin, and paraquat, which accumulates in the lung. Other toxic chemicals concentrate at sites other than the site of toxic action. For example, lead is stored in bone, whereas the symptoms of lead poisoning are due to lead in the soft tissues. The compartment where the toxic chemical is concentrated can be thought of as a storage depot. While stored, the toxic chemical seldom harms the organism. *Storage depots*, therefore, could be considered as protective mechanisms, preventing the accumulation of high concentrations of the toxic chemicals at the site of toxic action. The toxic chemicals in these depots are always in equilibrium with the free form in plasma, and as the chemical is metabolized or excreted from the body, more is released from the storage site. As a result, the biologic half-life of stored compounds can be very long.

Liver and kidney have a high capacity to bind chemicals, and these two organs probably concentrate more toxic chemicals than other organs. This might be related to the fact that liver and kidney are very important in the elimination of toxicants from the body. Kidney and liver have a capacity to excrete many toxicants and the liver has a high capacity to metabolize them.

Because many organic chemicals entering the environment are highly *lipophilic*, a characteristic that permits rapid penetration of cell membranes and uptake by tissues, it is not surprising that these highly lipophilic chemicals distribute and concentrate in body fat. This has been demonstrated for a number of chlorine-containing chemicals such as DDT, PCB, chlordane, and many others.

Bone can also serve as a reservoir for chemicals such as fluoride, lead, and strontium. It is well known that bone is a major storage site for lead. For example, 90% of lead in the body is found in the skeleton. The deposition and storage of toxic chemicals in bone may or may not be detrimental. Lead is not toxic to bone, but the chronic effects of fluoride deposition in bone is skeletal fluorosis, or osteosclerosis; and radioactive strontium can result in osteosarcoma and other neoplasms.

F. Metabolism and Excretion

Toxic substances that are lipid soluble may go through a series of metabolic conversions, commonly known as *biotransformation*, to produce more polar, i.e., water-soluble, products and thereby enhance removal by urinary excretion. The most common site for biotransformation is the liver, but it can also occur in plasma, lung, or other tissue. Biotransformation may result in either a decrease (*detoxification* or *inactivation*) or an increase (*activation*) in the toxicity of a chemical. Differences in the metabolism of toxic chemicals account for much of the observed differences between individuals and between animal species.

Biotransformation occurs in the liver by oxidation, reduction, hydrolysis, and conjugation. Microsomal enzymes play a key role in the process, and the activity of the microsomal enzyme system can be induced by many environmental factors as well as pharmacologic agents. Both normal individual differences in microsomal enzyme activity and susceptibility to induction are genetically determined and account for the marked variability in bioavailability of many toxic substances. Other factors that regulate key liver enzyme systems are hormones, which account for some sex-dependent differences, and disease states, e.g., the presence of hepatitis, cirrhosis, or heart failure. Because the activity of many hepatic metabolizing systems is low in neonates, particularly premature neonates, they may be much more susceptible to toxic substances that are inactivated by liver metabolism.

Toxic chemicals are eliminated from the body by various routes. The kidney is a very important organ for excretion of toxicants, and more toxicants are eliminated from the body by this route than any other routes. Other routes can be major pathways for excretion of specific chemicals. For example, many toxic chemicals metabolized by the liver are excreted in the bile and later eliminated in the stool, whereas lungs are important for excretion of gaseous molecules such as carbon monoxide. All body secretions appear to have the ability to excrete foreign chemicals and toxic chemicals or their metabolites can be excreted in sweat, saliva, tears, and breast milk, and there may be minor excretion in hair or skin.

G. Tests for Toxic Effects

Much of the information about the toxic effects of different chemicals comes from studying various strains and species of animals. Although tests in animals are the most common methods of identifying chemicals that cause toxicity, the results are difficult to extrapolate to humans, given the disparity among life spans (18-24 months for rodents versus 75 years for humans). In addition, different strains and species of animals may show both qualitative and quantitative differences in the pattern or intensity of response to a toxic chemical. Even with the best statistical approaches and the best evidence of toxic responses in animals, there is no certain way of estimating the incidence of toxicity or determining the type of response to a toxic chemical in a human population. Furthermore, there is no absolute certainty that safety factors for exposure to a toxic chemical based on studies in animals would be valid for humans.

Tests for acute effects are usually performed when there are no data available on the potential toxicity of a single exposure or a few exposures to a specific chemical. An appropriate route of administration is chosen, and a specific end point, e.g., death of the laboratory animal, is selected. The signs and symptoms before death are observed, and the animal is later examined for gross and histologic damage to tissues. In some cases, topical application of a chemical is used to test for skin or eye injury.

Tests for chronic effects are performed in animals when long-term human exposure to a specific chemical is anticipated or a long latency period between exposure and toxicity is expected. Rats and mice are usually exposed from a few weeks of age until their premature death or their sacrifice at the end of the expected life span.

Teratologic tests involve exposing pregnant female animals to a specific chemical at a critical time during pregnancy and then examining their offspring for malformations. Usually two or three species are used for comparison and controls. In reproductive studies, male and female animals are exposed to a chemical and subsequently observed for reproductive failure or success. In cases of successful reproduction, the first- and second-generation offspring are also observed for their ability to reproduce. In cases of unsuccessful reproduction, male animals are often tested for sperm motility, count, and morphology.

III. METAL EXPOSURE AND TOXIC RESPONSES

Increasing technologic use of metals is one measure of progress of civilization since early man's emergence from the Stone Age. This has posed potential hazard to man from the time metals were fashioned into spears to present-day exposures to space-age alloys. Today, metals are used extensively in the workplace and employee exposure can result from numerous industrial operations. Of the 80 or so elements that are classified as metals, more than half of them are of industrial and economic significance. As a group, metals exhibit a very wide range of biological, chemical, physical, and toxicological properties.

While a few metals have found some use in medicine, e.g., inert metals such as tantalum, platinum, and gold used as surgical implants, and lithium in the treatment of manic depression, all metals are potentially toxic at high concentrations, even though some of them are also essential. An *essential metal* is one for which a deficiency results in impairment of function that is relieved only by administration of that metal. At excess concentrations deleterious effects begin to set in, the biological response becomes unfavorable, and the substance becomes toxic. Some of these essential metals and their daily requirements are listed in Table 3.1.

Animals maintain a concentration of essential nutrients within the optimal range by a complex set of physiological reactions termed *homeostasis*. Concentrations of all essential metal ions are under homeostatic control. Toxicity is usually unrelated to essentiality. Almost any substance in excess ultimately becomes toxic.

Metabolism of metals refers to all of the processes by which metals are handled by the body. Absorption, storage, and elimination are the most important processes in the metabolism of inorganic metal compounds. Organometallic compounds are metabolized by different biochemical pathways.

Common routes of metal absorption are through inhalation and the gastrointestinal tract. Metals in the air may be inhaled as a vapor or as an aerosol, i.e., fumes or dust particulate. Fumes and vapors of cadmium and mercury are readily absorbed from the alveolar space, as are many organometallic compounds such as tetraethyl lead. Large particles of metal aerosol (>10 mm) are trapped by the upper respiratory tract. They are then cleared by the mucociliary escalator, swallowed, and absorbed into the blood through the gastrointestinal tract. Small particles (< 5 mm) may reach the alveolar space or gas exchange portion of the lungs. If they are water soluble, they are then rapidly absorbed from the alveoli into the blood. Organometallic compounds may also be absorbed through the skin.

Table 3.1

Essential Metals and Daily Requirements

Metal	Daily Requirement (mg)
Calcium	0.8
Chromium	0.1
Cobalt	3
Copper	3
Iron	10 - 20
Magnesium	700
Manganese	4
Molybdenum	0.3
Potassium	2000 - 5000
Selenium	100
Sodium	1000 - 2000
Zinc	15

Gastrointestinal absorption of metals and their compounds varies widely and depends on many factors such as the solubility, the chemical form, the presence of other materials, the competition for binding sites, as well as the physiological state of the gastrointestinal tract. For example, the salts of cadmium, lead, and tin are poorly absorbed (<10%) while the salts of arsenic and thallium are almost completely absorbed (>90%).

After metal ions enter the plasma, they are available for distribution throughout the body. Distribution usually occurs rapidly and the rate of distribution to the tissues of each organ is determined by the blood flow through the organ. The eventual distribution of a metal compound is largely dependent on the ability of the chemical to pass through the cell membrane and its affinity for the various binding sites.

Toxicants are often concentrated in a specific tissue. The compartment where the toxicant is concentrated can be thought of as a storage depot, e.g., lead is stored in bone. While stored, the toxicant seldom harms the hosts. Storage depots, therefore, could be considered as protective mechanisms, preventing the accumulation of high concentrations of toxicants at the site of toxic action. The toxicants in these depots are always in equilibrium with free toxicant in plasma, and as the chemical is excreted from the body, more is released from the storage site.

Most metal ions have a high affinity for functionally essential amino acid side chains such as sulfhydryl, histidyl, or carboxyl groups, and can react directly with proteins to alter enzymatic function. They may also bind to cofactors, vitamins, and substrates, thereby altering the availability of these cell constituents for biological function. Since particular metals play essential roles in enzyme catalysis, such biological processes are sensitive to alteration by toxic metals that are chemically similar to the essential metal. Finally, the presence of an excess of any metal may lead to depletion of essential metals.

Chelation is the formation of a metal ion complex in which the metal ion is associated with one or more electron donors referred to as a ligand. The ligand may be *monodentate, bidentate,* or *multidentate,* i.e., it may attach or coordinate using one or two or more donor atoms. Bidentate ligands form ring structures that include the metal ion and the two ligand atoms are attached to the metal. The donor molecule is properly referred to as a chelating agent. This term is derived from the Greek *chela,* for claw.

Chelating agents are generally nonspecific in regard to their affinity for metals. To varying degrees they will mobilize and enhance the excretion of a rather wide range of metals, including essential metals in the body. Therefore, the use of chelating agents as a prophylactic without close medical supervision is condemned by the medical profession. Among the common chelating agents in use today are diethylenetriaminepentaacetic acid and dimercaptosuccinic acid.

The following are the summaries of toxic effects of metals arranged in alphabetical order. These metals are selected for discussion because of their economic value, industrial use, and, more importantly, documentation of their toxic effects. Each summary contains the clinical manifestation of signs and symptoms of toxicity, target organs or systems, metabolism, carcinogenicity, and detoxification.

A. Aluminum

Aluminum is presently considered a nonessential element in humans. A variety of biochemical interactions with aluminum have been investigated. Many of these effects pertain to aluminum neurotoxicity. Acute neurological disorders termed *dialysis dementia* or *dialysis encephalopathy* have been reported in chronic hemodialysis patients. Patients developing such neurological disorders have been shown to have high levels of aluminum in their serum and brain tissue. Perhaps the most intriguing and potentially significant aspect of aluminum toxicity is the evidence implicating aluminum in the pathogenesis of Alzheimer's disease, an insidious neurodegenerative disorder of unknown etiology for which there is no effective treatment or cure.

Pulmonary effects of aluminum occur following inhalation of bauxite fumes. The resulting pulmonary fibrosis producing both restrictive and obstructive pulmonary function impairment has been described by Schaver. Interestingly, inhalation of aluminum mists was used in the 1930s to serve as prophylaxis of pulmonary fibrosis due to inhalation of silica particles.

Aluminum compounds can affect absorption of other elements in the gastrointestinal tract and alter intestinal function. Aluminum inhibits fluoride absorption and may decrease the absorption of calcium and iron. The binding of phosphorus or the interference with phosphate absorption can lead to *osteomalacia*, a condition marked by the softening of the bones.

Absorbed aluminum is highly bound in protein, making it difficult for removal. In addition, there are large stores in bone and nervous tissue that are difficult to access. Chelation of aluminum using deferrioxamine alters the dialyzability of aluminum and facilitates its removal by hemodialysis.

B. Antimony

Antimony and its compounds are not essential to humans. Antimony poisonings today are rare in highly developed countries because of the limited use. The toxic effects of antimony compounds are similar to those of the corresponding arsenic compounds. There are two oxidation states for antimony: the trivalent form and the pentavalent form. The trivalent forms are more toxic than pentavalent forms, but only a small fraction of the pentavalent is reduced to the trivalent form in the liver. The disposition of the trivalent and pentavalent forms differ. Trivalent antimony is concentrated in red blood cells and liver, whereas the pentavalent form is mostly in plasma. Both forms are excreted in feces and urine, but more pentavalent antimony is excreted in urine, whereas there is greater gastrointestinal excretion of trivalent antimony.

Chronic effects due to antimony and its compounds are alterations of the electrocardiogram, especially T-wave abnormalities, and myocardial changes. Other chronic effects include liver toxicity, characterized by jaundice and fatty degeneration, pulmonary congestion and edema, and popular skin eruptions have been reported. Antimony miners have developed disabling, but benign forms of pneumoconiosis.

The most toxic antimony compounds are stibine (SbH_3), the tri- and pentachlorides, and the tri- and pentafluorides. Stibine is a powerful hemolytic poison. One of the earliest signs of overexposure is hemoglobinuria. Fatalities caused by antimony pentachloride have been reported. After oral ingestion of toxic antimonial compounds, irrigating of the stomach, administration of 2,3-dimercaptopropanol, and a liver-protecting therapy are recommended.

C. Arsenic

Arsenic is one of the oldest poisons used by man. The effects of arsenic poisoning were described in detail in the pre-Christian era. In general, the naturally occurring arsenic is pentavalent while that added to the environment is trivalent. It is generally true that trivalent arsenic compounds are more toxic than pentavalent compounds and that natural oxidation favors the conversion of trivalent arsenic to the pentavalent form.

There are two forms in acute arsenic poisoning: the paralytic form and the gastrointestinal form. The paralytic form is observed if large doses of arsenic are absorbed quickly. Symptoms develop quickly and are usually characterized by constriction of the throat followed by dysphagia. Death may occur by general paralysis. In the gastrointestinal form, abnormal symptoms such as watery diarrhea caused by paralysis of the central mechanism of the capillary control in the intestinal tract are dominant. This may result in a decrease in blood volume and blood pressure falls to shock levels. This in turn results in disturbed heart action and causes death.

Chronic arsenic poisoning may show many forms. The skin is the major organ of interest since most effects are first seen here. It begins as local erythema with burning and itching, giving the skin a mottled appearance. This is followed by swelling and sometimes vesicular eruptions. Melanosis is first seen on the eyelids, temples, neck, and then spreads through the trunk. This is also often accompanied by hyperkeratosis, hyperhidrosis (excessive sweating on palms and soles), and warts. Mee's lines (horizontal white bands across nails) are also commonly seen and are considered to be a diagnostic accompaniment of chronic arsenic poisoning. Nasal septum ulceration is seen after long industrial exposure.

Arsine (AsH_3) is the most toxic arsenic compound. It is a colorless gas with a slight garlic-like odor generated by side reactions. Once absorbed, arsine is gradually oxidized in the body to arsenic trioxide. During this oxidation the protein of the red blood cells is denatured, resulting in hemolysis. Invariably, the first sign observed is hemoglobinuria, coloring the urine up to port wine hue. Jaundice starts at the second or third day and rapidly spreads over the whole body. As a result of the rapid destruction of the red blood cells, large quantities of free hemoglobin block the renal tubules with hemoglobin crystals and fragments of cells. This is manifested by increasing oliguria, followed by anuria, leading to uremia and death.

While there has been some controversy, the epidemiologic evidence indicates that industrial and agricultural exposure to arsenic is implicated in cancer of the skin and respiratory tract. The individuals at greatest risk are smelter workers. For this reason, the International Agency for Research on Cancer (IARC) has declared that "there is sufficient evidence that inorganic arsenic compounds are skin and lung carcinogens in humans." It is interesting to note that up to now it has only been possible to show that arsenic has a mutagenic as well as teratogenic effect in animal experiments but nobody succeeded in producing malignant tumors in animals.

During the Second World War in Britain, the so-called British Anti-Lewisite (BAL), 2,3-dimercaptopropanol, a bidentate chelating agent, was developed in order to protect against an attack with arsenic-containing chemical warfare agents. The mechanism of action is to compete with enzymes in the reactions for arsenic-containing compounds since by the formation of a stable five-membered ring, arsenic is strongly bound. Other dithiols such as sodium 2,3-dimercaptopropanesulfonate and 2,3-dimercaptosuccinic acid have also been found to be effective chelating agents. In arsine poisoning, however, the condition can only be treated symptomatically. Chelating agents are not effective.

D. Barium

Barium is not an essential element in the body. Barium ions are taken up, retained, and excreted in much the same way as calcium ions. While the insoluble forms of barium, particularly barium sulfate, are not toxic by the oral route because of minimal absorption, the soluble barium compounds are quite toxic. Accidental poisoning from ingestion of soluble barium compounds has resulted in gastroenteritis, muscular paralysis, decreased pulse rate, and ventricular fibrillation and extrasystoles. The toxicity is related to a competitive inhibition of potassium ions and removal of sulfate ions. Baritosis, a benign pneumoconiosis, arises from the inhalation of barium sulfate dust and barium carbonate. It is not incapacitating, but does produce radiologic changes in the lungs. The radiologic changes are reversible with cessation of exposure.

In the rare case of acute barium poisoning, subcutaneous atropine injections to prevent colic and respiration-supporting measures are recommended. In case of electrocardiogram changes or hypokalemia symptoms, a potassium infusion therapy should be of value. Chelating agents are not effective in barium poisoning.

E. Beryllium

Beryllium is not an essential element in the body. Beryllium and its compounds form insoluble precipitates at about pH 7 and are rarely absorbed from the gastrointestinal tract. However, contact with water-soluble beryllium salts may result in acute dermatitis and/or ocular lesions. Cutaneous injuries from beryllium metal, alloys, or oxide may require surgical treatment with excision of the foreign substance. Long-term skin contact with beryllium and its compounds may result in contact dermatitis and skin sensitization. In the bloodstream, a colloidal beryllium phosphate is formed and deposited in the liver, spleen, and bone marrow.

The most dangerous mode of entry into the body is by the respiratory tract, and effects are predominantly pulmonary. Pulmonary beryllium disease has two different forms: an acute chemical tracheo broncho-pneumonia, and a chronic form with granulomatous lesions of the lung. Acute pulmonary beryllium diseases are caused by the inhalation of large doses of beryllium or its soluble compounds in finely dispersed forms. The acute forms correlate with the concentration of beryllium exposure.

Chronic pulmonary beryllium disease, or *berylliosis*, is quite different from that of the acute form, with severe shortness of breath being the leading symptom. Pulmonary X-rays show miliary mottling, a haziness to "snow-flurry" effect. Histopathological examination of lung tissue shows interstitial granulomatosis. A dose-response relationship between extent of exposure and severity of disease is distinctly absent.   As the fibrosis increases, bleb formation is common and pneumothorax occurs.   This progressive form of berylliosis is accompanied by a decreased life expectancy.

Chelating agents for removal of deposited tissue beryllium have been explored.   Among these, aurintricarboxylic acid has been found to be effective in mice and rats if given parenterally within 8 hours after intravenous injection of an otherwise lethal dose of beryllium sulfate. However, chelating agents are not effective in chronic beryllium poisoning cases.

F. Cadmium

Cadmium is not an essential element in the body. The main organs for cadmium accumulation in humans are the liver and kidneys.   After absorption, cadmium is transported to the liver where it stimulates synthesis of metallothionein. Cadmium bound to metallothionein released from the liver moves via the blood to the kidney. The kidney is a critical organ in long-term, low-level exposure. In the kidney, cadmium-metallothionein accumulates in tubular cells. Since cadmium first affects the proximal tubule's reabsorption capabilities, the first effect to be detected is proteinuria.   Later signs include amino aciduria, glucosuria, decreased urine-concentrating ability, and abnormalities in handling uric acid, calcium, and phosphorus. The mineral problems may lead to kidney stones and osteomalacia, softening of the bone due to mineral loss.

Acute toxicity of cadmium in the workplace is usually via the respiratory route and is characterized by lung edema, cell proliferation, and fibrosis. Symptoms of respiratory exposure in the workplace include coughing, shortness of breath, irritation of the upper respiratory tract, and loss of sense of smell. Yellow staining of the teeth in heavy industrial exposure has been reported.

Itai-itai (ouch-ouch) disease was reported in Japan. This cadmium-induced disease was a result of dietary intake of the metal. Symptoms include painful sites spread all over the body, and difficulty in walking. The condition progresses and bone fractures are common. Pathological changes include osteomalacia and osteoporosis, especially prevalent in postmenopausal women with deficient diets.

An increase in carcinoma of the prostate was first noted in a mortality study of battery workers in England but this was not found in a study of a large worker population. The National Institute for Occupational Safety and Health (NIOSH) has reviewed a number of epidemiological studies and recommended that cadmium and its compounds be regarded as potential occupational carcinogens.

Chelation therapy is not available for cadmium toxicity in humans. Experimental studies have shown that the action of chelating agents on the pharmacokinetics of cadmium depends on the time of administration of the agent after cadmium exposure. When chelating agents are given shortly after cadmium exposure and before new metallothionein has been synthesized, the dithiols such as 2,3-dimercaptopropanol and penicillamine increase the biliary excretion of cadmium while ethylenediaminetetraacetic acid, diethylenetriaminepentaacetic acid, and related chelating agents increase urinary excretion. For chronic exposure, when cadmium is bound to metallothionein, there is little effect from chelation therapy. In the case of itai-itai, large doses of vitamin D given over a period of months are effective in relieving painful symptoms.

G. Chromium

Chromium exists in several valence states. Only the trivalent and hexavalent are biologically significant. Trivalent chromium is an essential element in the body. It plays a role in glucose and lipid metabolism. Trivalent chromium is poorly absorbed by the body regardless of the route of administration, while hexavalent chromium is more readily absorbed. Hexavalent chromium can be reduced to trivalent by the gastrointestinal tract, thereby reducing uptake.

In occupational exposures, the lungs are the primary route while skin is considered a minor route of entry for both trivalent and hexavalent chromium compounds. In general, chromium compounds in the trivalent state are of a low order of toxicity. The toxic effects of occupational exposure to high levels of chromium have been recognized for over 200 years. The overt signs of chromium toxicity, e.g., perforation of the nasal septum, skin ulcers or chrome holes, and liver and kidney damage, are rarely seen today. A common human toxicity associated with chromium exposure is allergic contact dermatitis.

Lung cancer is considered to be an occupational hazard for workers exposed to chromium in a wide variety of industrial and commercial occupations. The majority of information on chromium-induced carcinogenesis comes from human epidemiology studies of occupationally exposed workers. There is a good correlation between the dose of chromium and the relative risk of developing lung cancer. Hexavalent chromium compounds induce cancer in experimental animals at the site of exposure while trivalent chromium compounds are inactive. However, there has been little success in inducing cancer in animals by topical application to the skin, and in humans there is no evidence for an increased risk of skin cancer, even at sites of severe skin ulceration.

The acute clinical manifestations of chromium toxicity are rarely seen today and there has been only limited investigation of systematic chromium detoxification.

H. Cobalt

Cobalt is essential as a component of vitamin B_{12} required for the population of red blood cells and prevention of pernicious anemia. The effect that cobalt enhances the formation of red blood cells and hemoglobin has been used for therapeutic purposes. This therapy, however, led to hypothyroidism and thyroid hyperplasia. Acute intoxications caused by repeated ingestion of soluble cobalt salts occurred in Canada and the United States. In these cases, cobalt salts had been added to beer as a foam stabilizer. Excessive beer drinkers showed cardiopathy. Up to 50% of these cases were fatal. Extreme cardiomegaly was accompanied by lower systolic blood pressure. Drinkers of cobalt-containing beer also suffered from hypothyroidism and thyroid hyperplasia. NIOSH has recommended thyroid palpation as part of the medical surveillance of workers exposed to cobalt.

Inhalation of cobalt dust may result in pulmonary fibrosis. This so-called hard-metal disease is a disease of the compliance of the lung caused by fibrotic alteration of the tissue. At the workplace of hard-metal industry, cobalt occurs together with the carbides of tungsten or titanium. In spite of this, it is commonly believed today that cobalt itself is the etiologic factor of hard-metal disease. Occupational exposure to cobalt also gives rise to respiratory sensitization, bronchitis, asthma, as well as emphysema. Dependent on the concentration of cobalt in the air, increases and decreases of red blood cell count and the content of hemoglobin have been observed.

Contact of cobalt with the skin may lead to dermatitis or eczema, especially if there are skin lesions. Allergic dermatitis of an erythematous papular type may also occur, and affected persons may show positive skin tests.

Little is known about therapy for acute or chronic human exposure to cobalt. It has been postulated that cobalt complexes are formed *in vitro*. The formation of these chelates might prevent cobalt from binding with biochemically important compounds containing sulfhydryl groups and thus might reduce the toxic response of reabsorbed cobalt ions.

I. Copper

Copper is an essential element for all vertebrates. It is part of several of the most vital enzymes, including tyrosinase, cytochrome oxidase, and amine oxidases. It is essential in the incorporation of iron into hemoglobin. The primary target organ for accumulation of copper is the liver; the metabolism of copper involves the turnover of the copper-containing enzymes.

There are two genetically inherited inborn errors of copper metabolism that are, in a sense, a form of copper toxicity. *Wilson's disease* is characterized by excessive accumulation of copper in liver, brain, kidneys, and cornea. Affected individuals fall largely into two broad categories: one with symptoms involving primarily the nervous system; the other involving the liver. When the accumulated copper is removed by effective chelation therapy, the symptoms of the disease regress. Wilson's disease is apparently synonymous with copper toxicity. *Menke's disease*, apparently an inability to absorb copper, is a copper deficiency. This disease is characterized by rapidly progressive cerebral degeneration and the presence of spirally twisted hair. It appears that in Menke's disease, copper is available but it is not metabolized in the normal manner.

Copper in dust, fumes, or sprays may produce *brass chills* which is a form of *metal fume fever*. Chronic exposure may result in nasal ulceration and bleeding. Various damage to lung tissue has been reported in cases of severe intoxication after exposure to copper sulfate sprays. Exposure to copper dust may also cause discoloration of the skin. Although contact allergy to copper is very rare, there are reports of isolated incidences due to exposure to copper metals or salts. Oral ingestion may cause nausea, vomiting, diarrhea, and intestinal cramps. Liver and kidney functions may be disrupted giving rise to jaundice and cirrhosis. In severe cases of intoxication, hemolytic anemia and death may follow.

The treatment of choice in copper poisoning or Wilson's disease is oral administration of penicillamine. It can be administered orally in capsules. The immediate and most dramatic effect of the administration of penicillamine is a marked increase in urinary copper excretion. However, there are some undesirable side effects. With the use of triethylenetetramine dihydrochloride, marked cupriuria and clinical improvement in individuals with Wilson's disease have been achieved. The modes of action for these two chelating agents are apparently different.

J. Iron

Iron is an essential component of several cofactors, including hemoglobin and the cytochromes. There is an active, complicated homeostatic mechanism for maintaining proper iron levels in the body. In this homeostatic mechanism, the divalent form is absorbed into the gastrointestinal mucosa. The adequacy of iron stores in the body, however, seems to be the major controlling factor in the absorption of iron by the gastrointestinal tract. Absorption increases when iron stores are low and decreases when body stores are sufficient.

Acute pulmonary exposures to iron compounds are associated with kidney and liver damage. Altered respiratory rates and convulsions are among the neurological effects. Roentgenological changes in the lungs, referred to as *siderosis*, iron or hematite pneumoconiosis, or arc welders lung occur following inhalation of iron oxide fumes. These changes are similar to those seen in silicosis or miliary tuberculosis.

Chronic exposure to or excessive intake of iron may lead to hemosiderosis or hemochromatosis. *Hemosiderosis* refers to a condition in which there is generalized increased iron content in the body tissues, particularly the liver and reticuloendothelial system. *Hemochromatosis*, on the other hand, indicates demonstrable histologic hemosiderosis and diffused fibrotic changes of the affected organ.

Deferrioxamine is the chelating agent of choice in the treatment of iron poisoning. It complexes with iron ions to form ferrioxamine, which is then excreted via the kidney; 1 gram of deferrioxamine binds 85 mg of iron.

K. Lead

The essentiality of lead has been proposed at different times but it has never been fully demonstrated to the satisfaction of the scientific community. The symptoms of lead poisoning were noted by Greek, Roman, and Arabian physicians long before they were ascribed to lead. Lead *colic* was described by Hippocrates in a man who extracted metals, and the relationship of constipation colic, pallor, and paralysis was noted by Nicander in the second century. By the early part of this century, the clinical signs and symptoms of lead toxicity were documented in considerable detail. Nonspecific signs and symptoms include loss of appetite, metallic taste in the mouth, constipation and obstipation, pallor, malaise, weakness, insomnia, headache, irritability, muscle and joint pains, fine tremors, and colic.

The central nervous system is probably the most clinically significant. Symptoms vary from ataxia to stupor, coma, and convulsions. *Lead encephalopathy* is the term commonly used to describe the damage to the brain by lead. *Peripheral neuropathy*, characterized by wrist drop and foot drop, is the classical manifestation of lead toxicity.

Lead has multiple hematopoietic effects. In lead-induced anemia, the red blood cells are microcytic and hypochromic as in iron deficiency and usually there are increased numbers of reticulocytes with basophilic stippling. Two basic defects of lead-induced anemia are shortened life span for red blood cells and impairment of heme synthesis. As a consequence of the latter, there is a marked increase in circulating blood levels and urinary excretion of d-aminolevulinic acid. Lead also decreased ferrochelatase activity. This enzyme catalyzes the incorporation of the ferrous ion into the porphyrin ring structure. Failure to insert iron into protoporphyrin results in depressed heme formation. The excess protoporphyrin takes the place of heme in the hemoglobin molecule and zinc is chelated at the center of the molecule at the site usually occupied by iron. Red blood cells containing zinc protoporphyrin are intensely fluorescent and may be used to diagnose lead toxicity.

In workers with years of exposure to lead, toxicological effects on the kidney are most often observed. In the early stages, morphological and functional changes in the kidney are confined to the renal tubules and are most pronounced in proximal tubular cells. In advanced stages, irreversible chronic interstitial nephropathy characterized by vascular sclerosis, tubular cell atrophy, interstitial fibrosis, and glomerular sclerosis may occur. Gout occurs in about 50% of persons with chronic lead nephropathy.

Severe lead toxicity has long been known to cause sterility, abortion, and neonatal mortality and morbidity. It has been found that female lead workers are abortifacient and that male lead workers have a high incidence of sterile marriages.

The possibility of carcinogenic effects of lead has been receiving increased attention. There is evidence that lead can induce cancer in kidneys of rodents when fed high doses of lead. However, there is currently no evidence that lead is carcinogenic to man.

The most commonly used chelating agent in the removal of lead from the body is the calcium salt of disodium ethylenediaminetetraacetate. The lead chelate formed by exchange of calcium for lead is promptly excreted in the urine. For treatment of children with lead poisoning, the statement issued by the Centers for Disease Control should be consulted.

L. Manganese

Manganese is an essential element. It is a cofactor in a number of enzymatic reactions, particularly those involved in phosphorylation, cholesterol, and fatty acid synthesis. Manganese toxicity has been observed primarily among workers associated with the mining, refining, and manufacturing of manganese. In these individuals, overt signs of toxicity normally occur as the result of chronic inhalation of massive amounts of airborne manganese. The initial expression of manganese toxicity is often characterized by a severe psychiatric disorder. If the individual is removed from the high-manganese environment, some improvement of the psychiatric symptoms of the toxicity may occur. With continued exposure, however, the symptoms may progress remarkably similar to those noted in Parkinson's disease. Individuals at this point tend to show mask-like faces, difficulty in walking, and exaggerated reflexes. Removal of the individual from the high-manganese environment at this point will not result in a remission of the disorder even tough tissue manganese levels may decrease to normal values. An increase in respiratory diseases such as pneumonitis and pneumonia is frequently observed with milder forms of manganism.

In addition to the extensive neural tissue damage which can occur with chronic manganese toxicity, an iron-responsive anemia is a common finding with orally induced manganese toxicity. This anemia is presumably the result of an inhibitory effect of manganese or gastrointestinal uptake of iron. Reproduction and immune system dysfunction, nephritis, testicular damage, pancreatitis, and hepatic damage have all been reported.

Chelation theory of patients suffering from chronic manganese poisoning with penicillamine or ethylenediaminetetraacetate showed some improvement in patients. However, the improvements observed were transient. Currently, there is debate with regard to the efficiency of chelation therapy in the treatment of manganism as it has been reported that it is of little value in the treatment of this disorder. Support for this view is given by the observation that removal of an individual from a high-manganese environment results in a rapid loss of excess manganese from the body. Although body loss of manganese is reflected by lower concentrations of the element in the brain, lesions that occurred during exposure to the high-manganese environment are not reversible.

M. Mercury

Mercury is not an essential element in the body. Mercury and its compounds can be classified according to their dominant toxic characteristics into the following groups: **(1)** elemental mercury and those ionic compounds that can decompose to the vapor form in the environment; **(2)** inorganic mercury, e.g., both the mercury(I) and mercury(II) ions; **(3)** short-chain alkylmercurials such as methyl- and ethylmercury; and **(4)** organomercurials with more than two carbon atoms in the liquid, e.g., phenylmercury.

Exposure to high concentrations of mercury vapor can produce acute pneumonitis. Chronic low-level exposure can increase the incidence of weight loss caused by loss of appetite. Higher levels can produce chronic mercurialism identified by the four classical signs: **(1)** gingivitis; **(2)** salivation; **(3)** increased irritability; and **(4)** muscular tremors. Proteinuria may also occur. The primary target organ of elemental mercury is the central nervous system. It is oxidized to the mercury(II) ion. This reaction limits, but does not prevent, the accumulation of mercury by the brain.

The first signs of the ingestion of inorganic mercury(II) salts are caused by a corrosive effect on the alimentary tract, followed by oliguria and acute renal failure. The prolonged ingestion of mercury(I) salts can cause an idiosyncratic disease characterized by irritability, loss of body weight, acrodynia (painful extremities), "pink" disease (rash), and photophobia. Neither the occurrence of the disease nor its severity is dose related. The disposition of organomercurials with more than two carbon atoms in the molecule is essentially the same as for inorganic mercury. This is because the carbon-mercury bond is rapidly cleaved *in vitro*, yielding the mercury(II) ion and an organic moiety.

Short-chain alkylmercurials such as methylmercury primarily damage the central nervous system. The first symptom is paresthesia and the first clinical sign is ataxia, followed in severe cases by the constriction of the visual field and deafness. Ethylmercury has the same effect on the central nervous system but signs of renal damage such as proteinuria are also present. All organomercurials can cause redness of skin, burning sensation, blisters, and hypersensitivity reactions after repeated exposure. Transplacental transport is significant for alkylmercury as well as for elemental mercury.

Chelating agents are used to remove mercury from the body. The first thiol chelating agent, dimercaptopropanol, successfully decreased the mortality of acute mercury(II) intoxication, but the majority of the patients had adverse reactions to therapeutic doses. Dimercaptopropanol has two other drawbacks. It is ineffective when given orally and ineffective as a chelating agent against alkylmercury. The next major advance was the discovery that penicillamine was an effective chelating agent even when administered orally. More recently, N-acetyl-*dl*-penicillamine, 2,3-dimercaptopropanesulfonate, and 2,3-dimercaptosuccinic acid have been discovered to be even more effective and less toxic than penicillamine. They are all water soluble and can be given orally.

N. Nickel

Nickel is an essential trace element in chicks, rats, and swine but not in humans. Occupational health hazards from exposure to nickel compounds fall into three major categories: hypersensitivity, cancer, and respiratory disorders.

Hypersensitivity to nickel, or *nickel itch*, is a common cause of allergic dermatitis. The cutaneous lesions of nickel itch begin as a *papular erythema* (redness) of the hands or other areas of skin that contact nickel. The lesions gradually become *eczematous* (weeping) and undergo *lichenification* (scaling) in the chronic state.

Nickel refining workers have had increased mortality rates from cancers of the lungs and nasal cavities, attributed to exposure to nickel compounds with low aqueous solubilities, such as nickel sulfide and nickel oxide. Increased risks of other malignant tumors, including cancers of the larynx, kidney, prostate, or stomach have been noted in some refinery workers.

Owing to its volatility, lack of strong odor, and propensity for inadvertent formation, nickel carbonyl is generally considered the most hazardous compound of nickel. Symptoms of exposure to nickel carbonyl are of two distinct types. The immediate symptoms of toxicity consist of headache and vomiting. These are relieved by fresh air. Delayed and severe symptoms may then develop insidiously hours or even days after exposure. These symptoms include dyspnea, tachycardia, cyanosis, headache, dizziness, and profound weakness. Adrenal, hepatic, and renal damage may also develop. Diffuse interstitial pneumonitis and cerebral hemorrhage or edema are the usual causes of death.

The chelating agent of choice for treatment of acute nickel poisoning is penicillamine since it greatly promotes the renal excretion of nickel. In nickel carbonyl poisoning, diethyldithiocarbamate and tetraethylene-pentamine are the most effective chelating agents if administered immediately after inhalation. The therapeutic efficacy of these chelating agents reflects their ability to form an intracellular nickel complex with gradual release of the nickel complex into extracellular fluids. Penicillamine and triethylenetetramine are effective in the extracellular complexation of nickel and thus enhance its renal clearance. Thus, if treatment of nickel carbonyl poisoning is administered several hours after exposure, penicillamine or triethylenetetramine would be the chelating agents of choice.

O. Selenium

Selenium is an essential nutrient. Its best known biochemical function is as a component of the enzyme glutathione peroxidase. The enzyme prevents oxidative damage of important cell constituents.

The first and most characteristic sign of selenium absorption is a "garlic odor" of the breath due to small amounts of dimethyl selenide. In selenium intoxication, the main symptoms are brittle hair with intact follicles, and new hair has no pigment as well as brittle nails with spots of longitudinal streaks on the surface. Damaged nails are replaced by new, thickened nails, rough on the surface. In more severe cases, fluids effuse from around the nail bed. Another common symptom is lesions on the skin, mainly on the backs of hands and feet, the outer side of the legs and forearms, and the neck. Affected skin becomes red and swollen. Allergy to selenium may cause a pink discoloration of the eyelids and palpcbral conjunctivitis ("rose-eye").

In addition to the apparent protective effect against some carcinogenic agents, selenium is an antidote to the toxic effects of certain metals. At appropriate concentrations, mutual detoxification of selenium and arsenic, selenium and cadmium, selenium and copper, selenium and mercury, and selenium and thallium has been demonstrated. However, the mechanisms underlying these interactions are unknown.

In treatment of selenium intoxication, a combination of dimercaptopropanol and vitamin C was found to be an effective antidote. Symptomatic treatment with oxygen has been recommended as a treatment for selenium oxide and hydrogen selenide inhalation. Painful skin, nail, and eye disorders caused by selenium have been successfully treated with solutions or ointments of thiosulfate.

P. Thallium

Thallium is not an essential element in the body. Whether acute or chronic, the most characteristic features in *thallotoxicosis* are those involving the nervous system, skin, and cardiovascular tract. The usual patterns of damage in the peripheral nervous system are those of the "dying-back" type, with some involvement of the central nervous system. Hair loss generally occurs during thallium poisoning. Cutaneous effects may also include dry, scaly skin and impairment of nail growth often resulting in the appearance of Mee's lines. Thallium poisoning may also produce a complex pattern of cardiovascular responses. Small doses of thallium have been reported to produce increases in heart rate and blood pressure but large doses may produce hypotension and bradycardia. Cardiovascular symptoms are often accompanied by retrosternal pain or electrocardiogram abnormalities such as flattening or inversion of the T-wave.

Current treatment of thallium poisoning is primarily directed to the elimination of the metal from the body. Treatment with Prussian blue, potassium ferric hexacyanoferrate(II), can accelerate the fecal elimination of thallium by forming a stable, unabsorbable compound in the intestinal tract. Other antidotes such as diethyldithiocarbamate and diphenyl-thiocarbazone are contraindicated because they form lipophilic chelates with thallium which are redistributed to central nervous system structures and may cause more neurological damage.

Q. Tin

Tin is not an essential element in the body. In general, metallic tin and inorganic tin compounds have low toxicity. Inhalation of tin dioxide over a number of years may lead to *stannosis*, a pneumoconiosis with benign characteristics. Tin hydride, however, is highly toxic. Its effect is comparable to that of arsine.

In contrast to the inorganic tin compounds, some of the organic compounds are strongly toxic. Those with short alkyl groups such as methyl and ethyl groups are particularly dangerous. They are derived from tetravalent tin. Their toxicity increases with the number of alkyl groups. Trimethyl tin chloride and triethyl tin chloride deserve particular attention since they are highly neurotoxic and cause neurological and psychiatric disorders.

Inorganic tin compounds are very poorly absorbed when taken orally. Organotin compounds are decomposed by hydrolysis, but this reaction proceeds only slowly. Currently, there is no known antidote for the treatment of tin poisoning.

R. Vanadium

Vanadium is an essential element for chickens, rats, and some plants but is currently not known to be essential in man. In occupational exposure, the upper respiratory tract is the main target. Vanadium compounds, especially vanadium pentoxide, are strong irritants of the eyes and the respiratory tract. Other respiratory effects of vanadium include allergic contact dermatitis, asthma, and a green discoloration of the tongue.

Eighteen chelating agents as antidotes for acute vanadium intoxication were investigated and most showed some activity. The most promising antidote in vanadium poisoning is ascorbic acid.

S. Zinc

Zinc is an essential element in the body. Zinc compounds are not very toxic to humans. Zinc oxide has posed perhaps the greatest toxic risk and then only from inhalation of zinc oxide fumes in an industrial setting. Reports of zinc toxicity due to occupational exposure are very rare. In fact, lack of adequate zinc intake appears to be a far greater threat to human health than overexposure.

The exceedingly low toxicity and tightly regulated absorption and metabolism of zinc have made active measures directed toward detoxification unnecessary. The nausea, vomiting, and diarrhea that accompany acute exposure are self-limiting, and symptoms associated with metal fume fever characteristically abate within hours.

IV. SOLVENT EXPOSURE AND TOXIC RESPONSES

A *solvent* is any relatively nonreactive substance, usually a liquid at room temperature, that dissolves another substance resulting in a solution. Since most of the substances that solvents are used to dissolve in industry are organic, most industrial solvents are organic chemicals. Solvents are employed in a wide variety of applications, including as dry-cleaning agents, chemical intermediates, degreasers, and liquid extracts. There is a wide range in the ability of solvents to dissolve a given substance, and a similar range in their toxicities and relative hazards to potentially exposed workers. Exposure to solvents occur primarily through inhalation of vapors and through skin contact.

Solvents may cause toxic effects in an exposed individual. They affect several organ systems. The central and peripheral nervous systems are particularly susceptible, with effects ranging from slight decreases in nerve conduction velocity, to narcosis and death. The acute neurological effects are related to the anesthetic property of organic solvents, manifesting as transient symptoms such as lightheadedness and dizziness. Chronic neurological effects may include loss of intellectual function and memory. The blood, lungs, liver, kidneys, and skin also may be adversely affected by exposure to a particular solvent. Classically, the halogenated hydrocarbons are capable of inducing fatty changes and cirrhosis of the liver. *Dermatitis*, a common result of prolonged or repeated contact with solvents, is due primarily to defatting of the skin tissues. Selected solvents have been linked to the destruction of the bone marrow. Other solvents are known to cause cardiovascular effects, renal effects, as well as other health effects in the exposed individual. Some solvents are known human carcinogens; others are animal carcinogens suspected of possessing carcinogenic activity in humans.

Organic solvents can be divided into families according to chemical structure and the attached functional groups. Toxicological properties tend to be similar within a group, such as liver toxicity from chlorinated hydrocarbons and respiratory tract irritation from ketones. The basic structures are aliphatic, alicyclic, and aromatic hydrocarbons. The functional groups include halogens, alcohols, ketones, ethers, esters, amines, and others.

Since organic solvents are generally volatile liquids, inhalation is the primary route for occupational exposure. The pulmonary retention or uptake for most organic solvents ranges from 40% to 80% at rest. Because physical labor increases pulmonary ventilation and blood flow, the amount of organic solvent delivered to the alveoli and the amount absorbed are likewise increased.

Upon direct contact with the skin, the lipid solubility of organic solvents results in most being absorbed through the skin. However, skin absorption is also determined by water solubility and volatility. Solvents that are soluble in both water and lipid are most readily absorbed through the skin. Highly volatile substances are less well absorbed. For a number of solvents, dermal absorption contributes to overall exposure sufficient to result in a *skin* notation in the Threshold Limit Values (TLVs) of the American Conference of Governmental Industrial Hygienists (ACGIH). For a few solvents, significant absorption of vapors through the skin can also occur.

After absorption, organic solvents tend to be distributed to fatty tissues. In addition to adipose tissue, this includes the nervous system and liver. Since distribution occurs via the blood and since the blood-tissue membrane barriers are usually rich in lipids, solvents are also distributed to organs with large blood flows, such as cardiac and skeletal muscle. Persons with greater amounts of adipose tissue will accumulate greater amounts of an organic solvent over time. Most organic solvents will cross the placenta and also enter breast milk.

Some organic solvents are extensively metabolized and some not at all. The metabolism of a number of organic solvents plays a key role in their toxicity and in some cases the treatment of intoxication. The role of toxic metabolites will be discussed in their respective sections in this chapter. Biological monitoring can provide a more accurate measure of exposure than environmental monitoring for some solvents and this important topic will be discussed and summarized in a later chapter.

Excretion of organic solvents occurs primarily through exhalation of unchanged compound, elimination of metabolites in urine, or a combination of both. Solvents that are poorly metabolized are excreted primarily through exhalation.

The following are the summaries of toxic effects of organic solvents according to their chemical family and functional group. These solvents are selected for discussion because of their economic value, industrial use, and documentation of their toxic effects on exposed individuals. Each summary contains the clinical manifestation of signs and symptoms of toxicity, target organs or systems, metabolism, and carcinogenicity. Wherever appropriate, specific solvents within a chemical family are discussed.

A. Aliphatic Hydrocarbons

Aliphatic hydrocarbons are saturated or unsaturated, branched or unbranched, open carbon chains. They are further divided into alkanes (saturated hydrocarbons with carbon-to-carbon single bonds), alkenes (unsaturated hydrocarbons with one or more carbon-to-carbon double bonds), and alkynes (unsaturated hydrocarbons with one or more carbon-to-carbon triple bonds). Synonyms are paraffins, olefins, and acetylenes, respectively. Compounds of lower molecular weight containing fewer than four carbons are gases at room temperature, whereas larger molecules containing from 5 to 16 carbons, are liquids, and those having more than 16 carbons are usually solids.

A number of liquid aliphatic hydrocarbons are used in relatively pure form as solvents and also are the major constituents of a number of petroleum distillate solvents. The liquid alkanes are important ingredients in gasoline, which accounts for most of the pentane and hexane used worldwide. Hexane is an inexpensive general-use solvent in solvent glues, quick-drying rubber cements, varnishes, inks, and extraction of oils from seeds.

The alkanes are generally of low toxicity. The first three alkanes (methane, ethane, and propane) are simple inert asphyxiants whose toxicity is related only to the amount of available oxygen remaining in the environment. The vapors of the lighter, more volatile liquids, pentane through nonane, are respiratory tract irritants and anesthetics, while the heavier liquids, known as liquid paraffins, are primarily defatting agents. Hexane and heptane are most commonly used as general-purpose solvents. They cause anesthesia, respiratory tract irritation, and dermatitis and are associated with neurobehavioral dysfunction.

n-Hexane is produced during the cracking and fractional distillation of crude oil and is used in such applications as printing of laminated products; vegetable oil extraction; as a solvent in glues, paints, varnishes, and inks; as a diluent in the production of plastics and rubber; and as a minor component of gasoline. Vapor concentrations of many hundreds of parts per million are tolerated for several minutes without causing discomfort among workers. Peripheral neuropathy was reported in workers involved in laminating polyethylene products. The proximate neurotoxin is the metabolite 2,5-hexanedione, which is excreted in the urine. Methyl ethyl ketone (MEK) and possibly methyl isobutyl ketone (MiBK) potentiate the neurotoxicity of n-hexane.

The liquid alkenes are not widely used as solvents but are common chemical intermediates. They are more reactive than alkanes, a property that leads to their use as monomers in the production of polymers, such as polyethylenes from ethylene, polypropylene from propylene, and synthetic rubber and resin copolymers from 1,3-butadiene. The alkenes are similar in toxicity to the alkanes. The double bonds increase lipid solubility and therefore irritant and anesthetic potencies, compared to corresponding alkanes. n-Hexene does not cause peripheral neuropathy, as does n-hexane. Dienes are more reactive than alkenes. This reactivity is utilized in the production of polymers but may in some cases result also in additional health hazards. 1,3-Butadiene was found to be carcinogenic in animals, while propylene and ethylene were not.

B. Alicyclic Hydrocarbons

Alicyclic hydrocarbons are saturated or unsaturated molecules in which three or more carbon atoms are joined to form a ring structure. The saturated compounds are called cycloalkanes, cycloparaffins, or naphthenes. The cyclic hydrocarbons with one or more double bonds are called cycloalkenes or cyclo-olefins. Cyclohexane is the only alicyclic hydrocarbon that is widely used as an industrial solvent for fats, oils, waxes, resins, and certain synthetic rubbers, and as an extractant of essential oils in the perfume industry. However, most of the cyclohexane produced is used in the manufacture of nylon. Cyclohexene is used in the manufacture of adipic, maleic, and cyclohexane carboxylic acids. Methylcyclohexane is used as a solvent for cellulose ethers and in the production of organic synthetics.

The aliphatic hydrocarbons are similar in toxicity to their alkane or alkene counterparts in causing respiratory tract irritation and central nervous system depression, although their acute toxicity is low. The danger of chronic poisoning is relatively slight because these compounds are almost completely eliminated from the body. Alicyclic hydrocarbons are excreted in the urine as sulfates or glucuronides, the particular content of each varying. Small quantities of these compounds are not metabolized and may be found in blood, urine, and expired breath.

C. Aromatic Hydrocarbons

Aromatic hydrocarbons are characterized by the presence of the aromatic nucleus. The basic aromatic nucleus is the benzene ring. Aromatic hydrocarbons have enjoyed wide usage as solvents and as chemical intermediates. They are produced chiefly from crude petroleum and to a lesser extent from coal tar. Aromatic hydrocarbons used as solvents include benzene, toluene (methyl benzene), xylene (dimethyl benzene), ethyl benzene, cumene (isopropyl benzene), and styrene (vinyl benzene). They have a characteristic aromatic (sweet) odor.

Benzene is still currently widely used in manufacturing, for extraction in chemical analyses, and as a specialty solvent. Approximately half the benzene produced is used to synthesize ethyl benzene for the production of styrene. Toluene and the xylenes are two of the most widely used industrial solvents used in paints, adhesives, and the formulation of pesticides. Ethyl benzene is used chiefly as an intermediate in the manufacture of plastics and rubber. Cumene is used to manufacture phenol and acetone.

All the aromatic hydrocarbons are extensively metabolized. The metabolites vary with the substituents on the benzene ring. Benzene is metabolized mainly to phenol and excreted in urine as conjugated phenol and dihydroxyphenols. Toluene is primarily metabolized to benzoic acid and excreted in urine as the glycine conjugate, hippuric acid. Xylenes are metabolized to methylbenzoic acids and excreted in urine as the glycine conjugates, methylhippuric acid. Ethyl benzene is metabolized and excreted in urine as mandelic acid. Styrene is metabolized and excreted in urine primarily as mandelic acid and, to a lesser extent, phenylglyoxylic acid.

The aromatic hydrocarbons are generally stronger respiratory tract irritants and anesthetics than the aliphatics containing the same number of carbon atoms in the molecule. Substitution on benzene (toluene, xylene, ethyl benzene, cumene, and styrene) increases lipid solubility and toxicities slightly. As a family, aromatic hydrocarbons cause acute anesthetic effects, dermatitis, and respiratory tract irritation, and are associated with neurobehavioral dysfunction. Benzene is noted for its effects on the bone marrow: aplastic anemia that may itself be fatal or progress to leukemia. There is no evidence that any of the alkyl-substituted benzenes have any of the myelotoxic effects. Earlier reports of effects of the substituted benzenes on the bone marrow were probably due to their contamination with benzene.

D. Petroleum Distillates

Petroleum distillate solvents are mixtures of petroleum derivatives distilled from crude petroleum at a particular range of boiling points. Each is a mixture of aliphatic, alicyclic, and aromatic hydrocarbons, the relative concentration of each depending on the particular petroleum distillate fraction. They have a "hydrocarbon" or "aromatic" odor depending on the relative concentrations of aliphatic or aromatic hydrocarbons. "Petroleum ether" represents more than half of the total industrial solvent usage. Kerosene (stove oil) is used as a fuel as well as a cleaning and thinning agent. In industry, petroleum distillates may be referred to by any of the following names: naphtha, coal tar naphtha, petroleum naphtha, mineral spirits, Stoddard solvent, and others.

The hazard of a particular petroleum distillate fraction is related to concentrations of the various classes of hydrocarbons it contains. Most of the aliphatic fractions are alkanes, including *n*-hexane. Therefore, the risk of peripheral neuropathy must be considered, particularly with exposure to petroleum ether, which may contain a significant percentage of *n*-hexane. As the fraction becomes heavier (higher boiling point, increasing number of carbons), the percentage of aromatic hydrocarbons and therefore the toxicity increases. However, this increase in toxicity is offset by a decrease in volatility. Petroleum distillate solvents cause anesthetic effects, respiratory tract irritation, and dermatitis, and have been associated with neurobehavioral dysfunction.

E. Chlorinated Hydrocarbons

The addition of chlorine to a hydrocarbon increases the stability and decreases the flammability of the resulting compounds. Chlorinated hydrocarbons are typically colorless, volatile liquids with excellent solvent properties. Chemically, they consist of saturated or unsaturated carbon chains in which one hydrogen atom or more have been replaced by one or more chlorine atoms. Hydrocarbons having only one or two chlorine atoms are usually flammable and less toxic than similar hydrocarbons with complete chlorine substitution. Chlorinated hydrocarbon solvents are moderately well absorbed by inhalation. Skin absorption of vapor is usually insignificant but skin absorption following prolonged contact of the skin with liquid can be significant.

Chlorinated hydrocarbons have found wide use as solvents in degreasing, dewaxing, dry-cleaning, and extracting processes. Six chlorinated aliphatic hydrocarbons are used as general industrial solvents: methylene chloride, chloroform, carbon tetrachloride, methyl chloroform, trichloroethylene, and perchloroethylene. Methylene chloride is used as a paint stripper and extraction agent. Chloroform is used for extraction and spot cleaning. Carbon tetrachloride is used primarily as a chemical intermediate and in small quantities as a spot cleaning agent. Methyl chloroform is used in vapor degreasers and increasingly as a general cleaning and thinning agent. Historically, trichloroethylene was the principal solvent used in vapor degreasers, and it is being replaced by methyl chloroform, which is less toxic. Perchloro-ethylene has replaced mineral spirits and carbon tetrachloride as the primary dry-cleaning solvent because of the flammability of the former and the toxicity of the latter.

Biological monitoring of chlorinated hydrocarbons is based on their pattern of metabolism and excretion, which varies with their structure. Methylene chloride is both excreted unchanged in exhaled air and metabolized to carbon monoxide. Chloroform and carbon tetrachloride are each excreted unchanged in exhaled air and metabolized. However, little information is available on biologic monitoring for either. Methyl chloroform, trichloroethylene, and perchloroethylene are excreted unchanged in exhaled air and metabolized and excreted as trichloroethanol and trichloroacetic acid.

F. Alcohols

Alcohols are hydrocarbons substituted with a single hydroxyl group. They are widely used as cleaning agents, thinners, and diluents; as vehicles for paints, pesticides, and pharmaceuticals; as extracting agents; and as chemical intermediates. Methyl alcohol is widely used as an industrial solvent and as an adulterant to denature ethanol to prevent its abuse when used as an industrial solvent. Another important industrial use of methyl alcohol is in the production of formaldehyde. Isopropyl alcohol is used as rubbing alcohol and in the manufacture of acetone. Cyclohexanol is used to produce adipic acid for the production of nylon. In general, the aliphatic alcohols with more than five carbon atoms are divided into the plasticizer range (6-11 carbons) and the detergent range (>12 carbons).

Low-molecular weight alcohols are volatile and up to 50% of the inhaled vapor of these low-molecular weight alcohols can be absorbed into the body. In addition, vapors of methyl alcohol, isopropyl alcohol, n-butyl alcohol, and iso-octyl alcohol can be absorbed through the skin and they carry the ACGIH TLV *skin* notation.

The primary alcohols are metabolized by hepatic alcohol dehydrogenase to aldehydes, and by aldehyde dehydrogenase to carboxylic acids. The optic neuropathy and metabolic acidosis caused by methyl alcohol have been attributed to its metabolism to formaldehyde and then to formic acid. Metabolic interactions of ethanol with chlorinated hydrocarbon solvents, such as *degreaser's flush* in workers exposed to trichloroethylene, are due to competition for alcohol and aldehyde dehydrogenases, with subsequent accumulation of the alcohol and aldehyde and resulting reaction. Secondary alcohols are primarily metabolized to ketones, e.g., isopropyl alcohol is metabolized to acetone.

The aliphatic alcohols are more potent central nervous system depressants and irritants than the corresponding aliphatic hydrocarbons, but they are weaker skin and respiratory tract irritants than aldehydes or ketones. Respiratory tract and eye irritation usually occurs at lower concentrations than central nervous system depression and thus serves as a useful warning property.

Methyl alcohol is toxicologically distinct owing to its toxicity to the optic nerve, which could result in blindness. Inhalation exposure to ethanol and propanols result in simple irritation and central nervous system depression. Auditory and vestibular nerve injury in workers exposed to n-butyl alcohol has been reported.

G. Glycols

Glycols are hydrocarbons with two hydroxyl groups attached to two adjacent carbon atoms in an aliphatic chain.   Glycols are used as antifreezing agents and as solvent carriers and vehicles in a variety of chemical formulations. Only ethylene glycol is in common general industrial use as a solvent, but large volumes of the other glycols are used as vehicles and chemical intermediates.

The glycols have such low vapor pressures that inhalation is only of moderate concern unless heated or aerosolized.   Ethylene glycol and diethylene glycol are metabolized to glycol aldehyde, glycolic acid, glyoxylic acid, oxalic acid, formic acid, glycine, and carbon dioxide. Oxalic acid is the cause of acute renal failure and metabolic acidosis that occur following ingestion of ethylene glycol. Urinalysis for oxalic acid may be useful in biological monitoring of ethylene glycol exposure.   The metabolic pathways of methyl alcohol and ethylene glycol may be competitively blocked by the administration of ethyl alcohol.

H. Phenols

Phenols are aromatic hydrocarbons with one or more hydroxyl groups attached to the benzene ring. The simplest of the compounds is phenol, which contains only one hydroxyl group on a benzene ring. Other examples include cresol (methyl phenol), catechol (1,2-benzenediol), resorcinol (1,3-benzenediol), and hydroquinone (1,4-benzenediol). Phenol is used as a cleaning agent and disinfectant, but its primary use is as a chemical intermediate for resins and pharmaceuticals. Cresol is used primarily as a disinfectant. Catechol is used in photography, fur dying, leather tanning, and as a chemical intermediate.   Resorcinol is used as a chemical intermediate for adhesives, dyes, and pharmaceuticals. Hydroquinone is used in photography, as a polymerization inhibitor, and as an antioxidant.

Phenol is well absorbed both by inhalation of vapors and by dermal penetration of vapors and liquids. Phenol and cresol have the ACGIH TLV *skin* notations. Phenols are potent irritants that can be corrosive at high concentrations. As a result of their ability to complex with, denature, and precipitate proteins, they can be cytotoxic to all cells at sufficient concentrations. Direct contact with concentrated phenol can result in burns, local tissue necrosis, systemic absorption, and tissue necrosis in the liver, kidneys, urinary tract, and heart.   As it does with aliphatic alcohols and other volatile organic solvents, central nervous system depression also occurs.

I. Ketones

Ketones are hydrocarbons with a carbonyl group attached to a secondary carbon atom. Ketones are widely used as solvents for surface coatings with natural and synthetic resins; in the formulation of inks, adhesives, and dyes; in chemical extraction and manufacture; and as cleaning agents. Acetone, methyl ethyl ketone (MEK), and cyclohexanone are in most common use as industrial solvents. Consumer exposure to acetone is common in the form of nail polish remover and general-use solvent. Acetone is also used in the manufacture of methacrylates while cyclohexanone is used to make caprolactam for nylon.

Ketones are well absorbed by inhalation of vapors and to some extent after skin contact with liquid. Cyclohexanone has the ACGIH TLV *skin* notation. Most ketones are rapidly eliminated unchanged in urine and exhaled air and by reduction to their respective secondary alcohols, which are conjugated and excreted or further metabolized to a variety of compounds, including carbon monoxide.

Ketones are colorless liquids with good warning properties: strong odors or irritation to the skin, eyes, and respiratory tract at levels below those that cause central nervous system depression. Headaches and nausea as a result of the odor have been mistaken for central nervous system depression.

Methyl *n*-butyl ketone (M*n*BK), or 2-hexanone, is another hexacarbon solvent such as *n*-hexane that was thought to have little potential for health hazard. It is used as a paint thinner, cleaning agent, solvent for dye printing, and in the lacquer industry. Methyl *n*-butyl ketone causes the same type of peripheral neuropathy as *n*-hexane. It is also metabolized to the neurotoxic diketone, 2,5-hexanedione, to an even greater extent than *n*-hexane, and therefore poses an even greater hazard.

J. Ethers

Ethers consist of two hydrocarbon groups joined by an oxygen linkage. The two commonly used ethers as industrial solvents are ethyl ether and dioxane. Ethyl ether, the simplest ether, has been used extensively in the past as a general anesthetic, and has been historically known as *ether*. It has now been replaced by agents less flammable and with less aftereffects. Ethyl ether is used as a solvent for waxes, fats, oils, and gums. Dioxane is used as a solvent for a wide range of organic products, including cellulose esters, rubber, and coatings; in the preparation of histologic slides; and as a stabilizer in chlorinated solvents.

Ethyl ether is well absorbed by the inhalation of vapors. Its volatility limits skin absorption. Absorbed ethyl ether is excreted unchanged in exhaled air. The rest may be metabolized by enzymatic cleavage of the oxygen link to acetaldehyde and acetic acid. Ethyl ether is a potent anesthetic. Higher ethers are relatively more potent irritants. Dioxane is both an anesthetic and irritant but has also caused acute kidney and liver necrosis in workers. It is well absorbed by inhalation of vapors and through skin contact with liquid and has the ACGIH TLV *skin* notation. Dioxane is metabolized almost entirely to -hydroxyethoxyacetic acid and excreted in urine.

Occupationally, exposure to chlorinated ethers, bis(chloromethyl) ether (BCME) and chloromethyl methyl ether (CMME), is much more significant. Bis(chloromethyl) ether is used as an alkylating agent in the manufacture of polymers, as a solvent for polymerization reactions, in the preparation of ion exchange resins, and as an intermediate for organic synthesis. Bis(chloromethyl) ether has an extremely suffocating odor even in minimal concentrations. Bis(chloromethyl) ether is a known human carcinogen. There have been several reports of increased incidence of human lung carcinomas among workers exposed to bis(chloromethyl) ether as an impurity. The latency period is relatively short—10 to 15 years. Smokers, as well as nonsmokers, may be affected.

Chloromethyl methyl ether is a highly reactive methylating agent and is used in the chemical industry for synthesis of organic chemicals. Commercial grade chloromethyl methyl ether contains from 1 to 7% bis(chloromethyl) ether. Several studies of workers with chloromethyl methyl ether manufacturing exposure have shown an excess of bronchogenic carcinoma. It is not known whether or not the carcinogenic activity of chloromethyl methyl ether is due to bis(chloromethyl) ether contamination, but this may be a moot question inasmuch as two of the hydrolysis products of chloromethyl methyl ether can combine to form bis(chloromethyl) ether.

K. Esters

Esters are hydrocarbons that are derivatives of an organic acid and an alcohol. They are named after their parent alcohols and organic acids respectively, e.g., ethyl acetate for the ester of ethyl alcohol and acetic acid. The organic acid may be aliphatic or aromatic and may contain other substituents. Esters are an industrially important group of compounds. They are used in plastics and resins, as plasticizers, in lacquer solvents, in flavors and perfumes, in pharmaceuticals, and in industries such as automotive, aircraft, food processing, chemical pharmaceutical, soap, cosmetic, surface coating, textile, and leather.

Many esters have extremely low odor thresholds. Their distinctive sweet smells serve as good warning properties. Because of this property, *n*-amyl acetate (banana oil) is used as an odorant for qualitative fit testing of respirators. Esters are very rapidly metabolized by plasma esterases to their parent organic alcohols and acids. In general, esters are more potent anesthetics than corresponding alcohols, aldehydes, or ketones but are also strong respiratory tract irritants. Odor and irritation usually occur at levels below central nervous system depression. Their systemic toxicity is determined to a large extent by the toxicity of the corresponding alcohol.

There are four basic types of physiological effects of esters, and these can generally be related to structure: (1) anesthesia and primary irritation are characteristic of most simple aliphatic esters; (2) lacrimation, vesication, and lung irritation are due to the halogen atom in halogenated esters; (3) cumulative organic damage to the nervous system or neuropathy can be caused by some, but not all, phosphate esters; and (4) most aliphatic and aromatic esters used as plasticizers are physiologically inert.

L. Glycol Ethers

Glycol ethers are alkyl ether derivatives of ethylene, diethylene, triethylene, and propylene glycol. The acetate derivatives of glycol ethers are included and are considered toxicologically identical to their precursors. They are known by formal chemical names, e.g., ethylene glycol monomethyl ether; common chemical names, e.g., 2-methoxyethanol; and trade names, e.g., methyl cellosolve.

The glycol ethers are widely used solvents because of their solubility or *miscibility* in water and most organic liquids. They are used as diluents in paints, lacquers, enamels, inks, and dyes; as cleaning agents in liquid soaps, dry-cleaning fluids, and glass cleaners; as surfactants, fixatives, desiccants, antifreeze agents, and deicers; and in extraction and chemical synthesis. Since the first two members of this family, methyl cellosolve and ethyl cellosolve, were found to be potent reproductive toxins in laboratory animals, there has been a shift in use to butyl cellosolve and to diethylene and propylene glycol ethers.

The glycol ethers are well absorbed by all routes of exposure owing to their universal solubility. The acetate derivatives are rapidly hydrolyzed by plasma esterases to monoalkyl ethers. The ethylene glycol monoalkyl ethers maintain their ether linkages and are metabolized by hepatic alcohol and aldehyde dehydrogenases to their respective aldehyde and acid metabolites. The acid metabolites of 2-methoxyacetic acid and 2-ethoxyacetic acid are responsible for the reproductive toxicities of 2-methoxyethanol and 2-ethoxyethanol. These metabolites are excreted in urine unchanged or conjugated to glycine and may be used as biologic exposure indices.

Cases of encephalopathy have been reported in workers exposed to 2-methoxyethanol over a long period of weeks to months. Manifestations have included personality changes, memory loss, difficulty in concentrating, lethargy, fatigue, loss of appetite, weight loss, tremor, gait disturbances, and slurred speech. Bone marrow toxicity manifested as pancytopenia has been reported in workers exposed to 2-methoxyethanol and 2-ethoxyethanol.

Male reproductive toxicity has been observed in laboratory animals for 2-methoxyethanol and 2-ethoxyethanol in reduction in sperm count, impaired sperm motility, increased numbers of abnormal forms, and infertility. The same glycol ethers that are testicular toxins have been shown to be teratogenic in the same species of laboratory animals.

M. Glycidyl Ethers

The glycidyl ethers consist of a 2,3-epoxypropyl group with an oxygen linkage to another hydrocarbon group. They are synthesized from epichlorohydrin and an alcohol. Only the monoglycidyl ethers are in common use. The epoxide ring of glycidyl ethers makes these compounds very reactive, so their use is confined to processes that utilize this property such as reactive diluents in epoxy resin systems.

The glycidyl ethers have low vapor pressures, so that inhalation at normal air temperatures is not usually a concern. However, the curing of epoxy resins often generates heat, which may vaporize some glycidyl ether. Reported effects of glycidyl ether exposure have been confined to dermatitis of both the primary irritant and allergic contact type. Dermatitis can be severe and may result in second-degree burns. Asthma in workers exposed to epoxy resins may be due to exposure to glycidyl ethers. Glycidyl ethers are testicular toxins in laboratory animals.

N. Aliphatic Amines

The aliphatic amines are derivatives of ammonia in which one or more hydrogen atoms are replaced by a hydrocarbon group. They can be classified as primary, secondary, or tertiary according to the number of substitutions on the nitrogen atom. They are used to some extent as solvents but to a greater degree as chemical intermediates. They are also used as catalysts for polymerization reactions, preservatives (bactericides), corrosion inhibitors, drugs, and herbicides.

Amines are basic compounds and may form strongly alkaline solutions which can be highly irritating and cause damage on contact with eyes and skin. Amines are well absorbed by inhalation, and some have an ACGIH TLV *skin* notation. Skin absorption may be significant as many are capable of cutaneous sensitization. The vapors of the volatile amines cause eye irritation and a characteristic corneal edema, with visual changes of halos around lights, that is reversible. Irritation occurs wherever contact with the vapors occurs, including the respiratory tract and skin. Direct contact with the liquid can produce serious eye or skin burns. Allergic contact dermatitis has been reported from exposures to ethyleneamines.

O. Aromatic Amines

The aromatic amines are aromatic hydrocarbons in which at least one hydrogen atom has been replaced by an amino group. The hydrogen atoms in the amino group may be replaced by aryl or alkyl groups, giving rise to secondary and tertiary amino compounds. Their most important uses are as intermediates in the manufacture of dyestuffs and pigments; however, they are also used in the chemical, textile, rubber, dyeing, paper, and other industries.

Most of the aromatic amines in the free base form are readily absorbed through the skin in addition to the respiratory route. The two major toxic effects are methemoglobinemia and cancer of the urinary tract. Other effects may be hematuria, cystitis, anemia, and skin sensitization. Several of the aromatic amines have been shown to be carcinogenic in humans or animals or both. The most common site of cancer is the bladder, but cancer of the pelvis, ureter, kidney, and urethra do occur.

Aniline is a clear, colorless, oily liquid with a characteristic odor. It is widely used as an intermediate in the synthesis of dyestuffs. It is also used in the manufacture of rubber accelerators and antioxidants, pharmaceuticals, marking inks, tetryl, optical whitening agents, photographic developers, resins, varnishes, perfumes, shoe polishes, and many organic chemicals. Absorption of aniline, whether from inhalation of the vapor or from skin absorption of the liquid, causes anoxia due to the formation of methemoglobin. Moderate exposure may cause only cyanosis. As oxygen deficiency increases, the cyanosis may be associated with headache, weakness, irritability, drowsiness, dyspnea, and unconsciousness. Methemoglobin levels, and/or urinary excretion of p-aminophenols, can be used for biologic monitoring for aniline exposure.

P. Miscellaneous Solvents

1. Carbon disulfide

Carbon disulfide is a highly refractive, flammable liquid which in pure form has a sweet odor, and in commercial and reagent grades, has a foul smell. Its odor can be detected at about 1 ppm, but the sense of smell fatigues rapidly, and therefore odor does not serve as a good warning property. Carbon disulfide is used as a solvent for phosphorus, sulfur, selenium, bromine, iodine, alkali cellulose, fats, waxes, lacquers, camphor, resins, and cold vulcanized rubber. It is also used in degreasing, chemical analysis, electroplating, grain fumigation, oil extraction, and dry-cleaning.

Carbon disulfide vapor in sufficient quantities is severely irritating to eyes, skin, and mucous membranes.   Inhalation of vapor may be compounded by percutaneous absorption of liquid or vapor.   Contact with liquid may cause blistering with second- and third-degree burns.   Skin sensitization may occur.   Skin absorption may result in localized degeneration of peripheral nerves which is most often noted in the hands. Respiratory irritation may result in bronchitis and emphysema, though these effects may be overshadowed by systemic effects. Intoxication from carbon disulfide is primarily manifested by psychological, neurological, and cardiovascular disorders. Acute exposures may result in extreme irritability, uncontrollable anger, suicidal tendencies, and a toxic manic depressive psychosis.   Chronic exposures have resulted in insomnia, nightmares, defective memory, and impotency.   Less dramatic changes include headache, dizziness, and diminished mental and motor ability, with staggered gait and loss of coordination.

Atherosclerosis and coronary heart disease have been significantly linked to exposure to carbon disulfide. Atherosclerosis develops notably in the blood vessels of the brain, glomeruli, and myocardium. A significant increase in coronary heart disease mortality has been observed in carbon disulfide workers. Abnormal electrocardiograms may occur.

Carbon disulfide can be determined in expired air and blood. It is metabolized and excreted in the urine as 2-thiothiazolidine-4-carboxylic acid.

2. Dimethylformamide

Dimethylformamide is a colorless liquid which is soluble in both water and organic solvents. It has a fishy, unpleasant odor at relatively low concentrations, but the odor has no warning property. Dimethylformamide has powerful solvent properties for a wide range of organic compounds. It finds particular usage in the manufacture of polyacrylic fibers, butadiene, purified acetylene, pharmaceuticals, dyes, petroleum products, and other organic chemicals. However, these properties also result in its being well absorbed by all routes of exposure.

Dimethylformamide is a potent hepatotoxin and has been associated with both hepatitis and pancreatitis following occupational exposure. It has produced kidney damage in animals.   Recent studies have associated dimethylformamide exposure with testicular cancer.   Exposure can be monitored biologically by measuring monomethylformamide and related metabolites in urine.

3. Turpentine

Turpentine is a mixture of substances called terpenes, primarily pinene. Gum turpentine is extracted from pine pitch, wood turpentine from wood chips. Inhalation of vapor and percutaneous absorption of liquid are the usual paths of occupational exposure. High vapor concentrations are irritating to the eyes, nose, and bronchi. Aspiration of liquid may cause direct lung irritation resulting in pulmonary edema and hemorrhage. Turpentine liquid may produce allergic contact dermatitis. The incidence of sensitization varies with the type of pine, being generally higher with European than American pines. Limonene is a terpene used as a solvent for art paints that also causes allergic contact dermatitis. Eczema from turpentine is quite common and has been attributed to the auto-oxidation products of the terpenes (formic acid, formaldehyde, and phenols). Liquid turpentine splashed in the eyes may cause corneal burns and demands emergency treatment.

Turpentine vapor in acute concentrations may cause central nervous system depression. It also produces kidney and bladder damage. Chronic nephritis with albuminuria and hematuria has been reported as a result of repeated exposures to high concentrations.

V. TOXIC RESPONSES OF THE LUNG

The lung is a common site of occupational disease. Because the lung has a limited number of ways in which to respond to injury, a wide variety of chemicals cause only a few familiar patterns of disease. Acute responses include upper airway obstruction, bronchoconstriction, alveolitis, and pulmonary edema; and chronic responses include asthma, fibrosis, and cancer. The specific type of response is dictated by the site of deposition of the noxious chemical, the dose and duration of exposure, the susceptibility of lung cells, and the interaction between the chemical and local host defense mechanisms.

Deposition of inhaled materials is primarily dependent on water solubility for gases and particle size for solids. Because water-soluble gases, such as ammonia and sulfur dioxide, are almost entirely removed from inspired air by the aqueous layer lining the nose, oropharynx, and upper airways, they are most likely to cause injury to the upper respiratory tract. Water-insoluble gases, such as nitrogen dioxide and phosgene, are able to bypass the upper airways and cause injury to the distal airways and alveoli.

Though most particles are not perfect spheres, the deposition pattern of inhaled particles is similar to that of spheres and is best described in terms of *aerodynamic diameter*. During periods of quiet breathing through the nose, virtually all particles with aerodynamic diameters in excess of 10 mm are deposited on the nasal mucosa. During strenuous exercise, however, because of increased airflow and mouth breathing, up to 20% of particles between 10 and 20 mm in diameter are deposited within the airways. Particles between 3 and 10 mm in diameter can be deposited throughout the tracheobronchial tree. More central deposition is favored by high rates of inspiratory flow, by airway obstruction, and by the presence of increased quantities of mucus. Particles between 0.1 and 3 mm in diameter are preferentially deposited within the alveoli. Smaller particles tend to remain in the airstream and are exhaled. A fiber with an *aspect ratio* of 3 to 1 or higher (i.e., length exceeds its width by at least three-fold) is deposited on the basis of aerodynamic diameter rather than length, which explains why long fibers up to 25 mm in length are often deposited in alveoli.

A. Pneumoconiosis

Pneumoconiosis is an occupational disease of the lungs caused by the accumulation of certain inorganic and organic dusts and the reaction of pulmonary tissue to the dusts. Originally, pneumoconiosis included only particulate matter in the solid phase. In recent years, however, the definition has been broadened to include aerosols other than dusts. The lesions of pneumoconiosis are generally divided into two major categories. The first, simple pneumoconiosis, is a disorder in which there are discrete small lesions scattered throughout the lung in various profusion and with varying degrees of fibrosis. An example of simple pneumoconiosis is silicosis or coal worker's pneumoconiosis. The second, complicated pneumoconiosis or progressive massive fibrosis, affects a small percentage of persons with pneumoconiosis. In this case, fibrotic nodules coalesce and encompass blood vessels and airways. In the past, tuberculosis was a common accompaniment of this condition.

A type of pneumoconiosis produced by dusts that are inert is known as benign pneumoconiosis. There is no pulmonary tissue reaction to deposition of the dusts, even when there are large accumulations in the lungs. There also is no functional impairment despite somewhat dramatic X-ray findings. Barium sulfate and other barium, iron, and tin dusts are inert.

B. Silicosis

Silicosis is a pathological condition of the lungs resulting from the inhalation of particulate matter containing free silica (SO_2). It is important to distinguish between silica in the free state as silicon dioxide and in the combined state as the various silicates. Silicosis is prevalent in many industries, and of all the pneumoconioses, it claims the largest number of victims. Silicosis is generally divided into three stages: slight, moderate, and severe. The first stage, so-called simple silicosis, supervenes in a worker who has been exposed to free silica dust for many years. The onset of symptoms is marked by dyspnea on exertion, slight at first and later increasing in severity. Cough may be present and is usually unproductive. Impairment of working capacity is slight or absent. In the second stage, dyspnea and cough become established and further physical signs appear. Some degree of impairment of working capacity begins to develop. In the third stage, dyspnea progresses to total incapacity. Right heart hypertrophy and then failure may supervene. The concentrations of dust that can be inhaled without danger vary according to the nature of the dust and also to the length of time during which it is breathed. Intermittent exposure to high concentrations of dust may be more dangerous than exposure to lower concentrations of dust over a longer time period. The harder the work, the more deeply a worker will have to breathe, and consequently will inhale more dust. Individuals also vary greatly in their capacity to respond to inhaled dusts.

Acute silicosis is a form of silicosis that can occur in workers exposed to high levels of respirable silica over a relatively short period of time. The occupations in which the disease has been described are sandblasting, rock drilling, lens grinding, and the manufacture of abrasive soaps. The history is typically of progressive dyspnea, fever, cough, and weight loss. The disease is rapidly progressive, death occurring in respiratory failure.

Nodular silicosis is a respiratory condition associated with the inhalation of silicon oxide particles ranging from 0.1 to 3 mm in size stored in excessive amounts in the lungs. The particles characteristically produce fibrous nodules that are identified as *dust granulomas*. The nodules develop by a peculiar fibrotic process in which fibrous tissue is laid down in concentric rings about the central core of foci of silica particles, enveloping the particles by repeated layers of fibrous tissue until they resemble the layers of an onion. The final macroscopic, round, hard, discrete fibrous nodules are usually 2 to 4 mm in size. The nodules cause many individual alveoli to be compressed and collapse. Distortion and some breakdown of alveolar septa develops, causing minute foci of emphysema to develop about the nodules. No demonstrable impairment of the lung function results because of the great functional reserve capacity of the lung.

C. Silicate Pneumoconiosis

Silicates have a crystal structure containing the SiO_4^{2-} tetrahedron arranged as isolated units, as single or double chains, as sheets, and as a three-dimensional network. There are also fibrous "rock-forming silicates," variously described as asbestiform, elongate, fibrous, bladed, prismatic, and columnar. At present, there is no concurrence on the effects of silicates on humans by minerals not generally regarded as asbestos. Because of the diversity of silicates, the health effects of these minerals are discussed according to their mineralogical classification.

The *olivine group* of igneous rocks are iron and magnesium silicates with island structures of SiO_4^{2-} units. Olivine is used principally in foundries, primarily as a special sand for mold-making in brass, aluminum, and magnesium foundries. It is used as a refractory (bricks) material, in mixes for furnace linings, and as a source of magnesium in fertilizer. No disease or pneumoconiosis from olivine minerals has been observed in humans.

The *kyanite group* of minerals are aluminum silicates with island structures of SiO_4^{2-} units. This group of minerals has been in great demand for making high-grade refractories, such as in spark plugs, thermocouple tubing, and refractory bricks in electric and forging furnaces and cement kilns. Mild fibrosis or pneumoconiosis may result from exposure to kyanite. Silicosis is possible if cristobalite is present.

The *pyroxene group* forms chains of $(Si_2O_4)^{4-}$ units bound together by covalent oxygen bonds. The commercially mined pyroxenes are spodumen and wallastonite. Spodumen, a comparatively rare mineral, is one of the commercial sources of lithium. Wallastonite is used in the ceramic industry, as a replacement for nonfibrous materials in brake linings, as an insulation for electronic equipment and thermal insulation such as mineral wool, and many other uses. There is no known disease associated with exposure.

The most common clay minerals belong to the kaolinite. The structural unit of this group is the siloxane sheet, $(Si_2O)^{2-}$, which exhibits excellent cleavage, easy gliding, and a greasy feel. The leading consumer of kaolin is the paper industry where it is used to fill and coat the surface. It is also used as a filler in both natural and synthetic rubber, as a paint extender, a filler in plastics, and in the manufacture of ceramics. *Kaolinosis* is a pneumoconiosis produced by kaolin. The pneumoconiosis is mainly nodular or massive fibrosis of the lung. Symptoms are dyspnea on exertion and productive cough. Pulmonary tuberculosis and emphysema are often found. A benign pneumoconiosis is more commonly seen on chest radiographs. Other silicates with sheet structures are talc, bentonite, and Fuller's earth. The effects of exposure to these silicates are discussed below.

1. Talc

Talc is an extremely versatile mineral that has found many uses. The principal uses of talc include extender and filler pigment in the paint industry, coating and filling of paper, ceramic products, filler material for plastics, and roofing products. The character of pneumoconiosis associated with talc exposure depends on the composition of the talc dust inhaled. When asbestos is the dominant mineral, the disease is characteristic of asbestos-induced disorders. When talc is associated with quartz, the reaction of the lung to quartz is modified, giving rise to localized fibrocellular lesions, and is called talco-silicosis. Talcosis is pneumoconiosis caused by deposition of pure talc, i.e., asbestos-free and quartz-free. Granulomas are observed in chest radiographs.

2. Bentonite

Bentonite is used as a foundry sand bond, drilling mud where penetrated rocks contain only fresh water, bleaching clay, and pelletizing of talconite ore. Silicosis has been reported in bentonite workers. No disease has been associated with bentonite alone.

3. *Fuller's earth*

Fuller's earth is a porous colloidal aluminum silicate. The term is a catch-all for clay or other fine-grained earthy material suitable for use as an absorbent and bleach. The disease is pneumoconiosis, often resembling silicosis. There is no recognized disease specific for Fuller's earth.

The framework structure group of silicates is probably the most important because nearly three-fourths of the rocky crust of the earth is composed of these minerals which are stable and strongly bonded. Quartz, tridymite, and cristobalite are the three principal crystalline polymorphs and can be transformed from one to the other under different conditions of temperature and pressure. In the mixed dust pneumoconioses, the pathology depends to a large extent upon the relative proportion of quartz present in the airborne dust. Those with a quartz content of less than about 0.1% tend to develop small nodular areas in the lungs in almost direct proportion to the total amount of dust deposited. On the other hand, dusts in which the quartz content ranges from about 2% to about 18-20% of the total dust tend to produce lesions that more nearly resemble those seen in classical silicosis.

4. *Diatomaceous earth*

A respiratory disorder caused by inhalation of amorphous silicon dioxide particles is known as diatomaceous earth pneumoconiosis. In sufficient quantities, this amorphous diatomaceous earth may produce a mild linear reticulation without clinical symptoms. After calcining, 14-60% of the amorphous silicon dioxide is transformed into crystalline silicon dioxide, mostly in the form of cristobalite. When stored in the lung in sufficient quantity, it produces a diffuse fibrosis. A delicate mesh of fibrous tissue produced by the thickened alveolar wall is present.

D. Coal Workers' Pneumoconiosis

Coal workers' pneumoconiosis is a condition in which coal dust and similar forms of carbon are stored in the lungs, forming an often nondisabling anthracosis. In addition to heavy exposure to coal dust, many coal miners, particularly those mining anthracite coal, are exposed to high concentrations of silica dust. In such exposed individuals, the accumulation in the lungs of two types of dust—silica and coal—results in the development of *anthracosilicosis*. Many hard coal miners develop that condition with all the characteristics of silicosis, plus the results of lung deposits of coal dust. Initially, in coal workers' pneumoconiosis, the disease is manifested characteristically by the coal macule and later by coal micronodules and nodules resulting in simple coal workers' pneumoconiosis. In some cases, large lesions of 1-3 cm in diameter, or even massive consolidated lesions develop, resulting in progressive massive fibrosis.

1. *Black lung disease*

Black lung is a legislatively defined term which encompasses the classical medical definition of coal workers' pneumoconiosis, but is defined by the Federal Mine Safety and Health Act of 1977, (P.L. 91-173 as amended by P.L. 95-164), as "a chronic dust disease of the lung arising out of employment in an underground coal mine." The definition is used to cover disability primarily from chronic airways obstruction that is associated with coal mine dust exposure. Coal miners who meet any of the qualifications criteria in the Act, and in the judgment of the examining physician and administrative law judge, that they have developed their condition in association with coal mine employment, may be compensated for total disability.

2. *Caplan's syndrome*

Caplan's syndrome is a form of pneumoconiosis that may develop in a person afflicted with or susceptible to a rheumatoid disease and who has been exposed to a dust hazard. It was first described in coal miners with rheumatoid arthritis but subsequently found in other mining operations. The symptoms result from an interaction between the inhaled mineral dust in the lung and rheumatoid factor. Its most characteristic feature is larger, more rapidly developing nodules, than those seen in simple silicosis. An increased prevalence of progressive massive fibrosis is seen in these individuals.

E. Asbestosis

Asbestos is a group of impure magnesium silicate mineral fibers with a theoretical formula of $Mg_6Si_4O_{10}(OH)_8$. There are two major types of asbestos fibers: *amphibole* (short and straight) and *serpentine* (long, flexible, and curly). Categories defined on the basis of commercial use are *chrysotile* (serpentine), *amosite, actinolite, anthophyllite, crocidolite,* and *tremolite* (amphiboles). Only amosite, chrysotile, and crocidolite are of economic importance. Chrysotile is basically a sheet silicate mineral rolled into itself to form a hollow tube. This tube constitutes the basic fibril of chrysotile. All amphibole asbestos types are similar in crystal structure: they consist of double chains of linked silicon oxygen tetrahedra between which metallic ions are sandwiched. More than 90% of all asbestos used in the United States is of the chrysotile variety. Asbestos cement products constitute the major use of asbestos, followed closely by floor products or materials used in the construction industry. Materials containing asbestos have been extensively used in construction and shipbuilding for purposes of fireproofing and for decoration.

Asbestos fibers vary considerably in size. Amosite fibers are the largest. The smallest fibers are too fine to be seen with an optical microscope and must be viewed through an electron microscope. The small fibers may be inhaled from airborne dust to form residues in lung tissue. Inhalation of asbestos fibers is associated with a number of health problems, including asbestosis, lung cancer, mesothelioma, and pleural plaques. In the lung of asbestos-exposed persons may be found the presence of asbestos bodies, or ferruginous bodies, which are inhaled asbestos fibers that have become encased in a deposit of protein, calcium, and iron salts in the lung tissue. It is believed that the coating is produced by pulmonary macrophages in order to isolate the fibers and prevent them from becoming the source of a fibrosing action in the lung tissue.

Asbestosis is the name of the pneumoconiosis produced by the inhalation of asbestos fibers. It is characterized by diffuse interstitial fibrosis of the lung parenchyma, often accompanied by thickening of the visceral pleura and sometimes calcification of the pleura. The risk for asbestosis seems to vary with the amount of asbestos fiber regularly inhaled, other lifestyle factors such as cigarette smoking, and the type of asbestos to which the worker was exposed. One study indicates that crocidolite fibers present the greatest risk and chrysotile fibers the least risk.

F. Fiberglass

Glass fibers are manufactured from a melt of a batch containing silica, limestone, aluminum hydroxide, soda ash, and borax. Fiberglass is produced by drawing or blowing the molten glass into fine fibers that are flexible but retain the tensile strength of glass. The risk from inhalation of respirable fibers is not fully evaluated. Among the hazards to workers are those due to abrasions. Some workers may develop dermatitis from the mechanical action of the fibers on the skin, and a small percentage of them may develop an allergic dermatitis from the binder. Fiberglass is capable of producing a mechanical, transitory skin irritation characterized by a maculopapular eruption. It usually is noted at pressure points, such as around the waist, collar, and wrists. This temporary irritation usually begins to decrease within 3 to 5 days after beginning work with the material. Workers experience no lasting adverse effects once they are removed from exposure to the material.

G. Man-made Mineral Fibers

Man-made mineral fibers are those made from glass, natural rock or any readily fusible slag. They differ from naturally occurring fibers, such as asbestos, which are crystalline in structure and differ chemically. Man-made mineral fibers are glassy cylinders and can never split longitudinally. They only break across. As they are destroyed, they form fragments that no longer have the character of fibers. The structure of asbestos fiber is totally different. It is always present as bundles, never as a single fiber. Individual fibrils may be as small as 2 to 30 mm in diameter. Because of the size of individual asbestos fibers, a bundle of asbestos fibers of the same diameter as mineral fibers would contain almost 1 million fibers. The health effects of man-made mineral fibers may be different, depending on the length and diameter of the individual fibers. Some studies indicate that no significant changes are observed in chest radiographs or pulmonary function tests of workers involved in the production and use of man-made mineral fibers. However, there is some evidence that pre-existing respiratory conditions, such as asthma and bronchitis, may be aggravated by exposure to man-made mineral fibers. Fibrosis is not a hazard of man-made mineral fibers and post-mortem examinations of the lungs of persons who have worked with man-made mineral fibers show no significant difference from those of the lungs of city dwellers who have had no special exposure to man-made mineral fibers.

H. Natural Fibers

Natural fibers is a term applied to any flexible filamentous substance with a length that is many times that of the diameter. Natural animal fibers include those of wool and silk. The wool category is sometimes broadened to include alpaca, camel, goat, and mohair fibers. Silk is the only natural fiber that occurs as a continuous filament. The more important vegetable fibers are cotton, flax, hemp, jute, ramie, and sisal. Vegetable fibers are sometimes subdivided according to whether they are derived from the stem, leaf, fruit, or seed of the plant. Flax, hemp, jute, and ramie are classified as bast fibers, obtained from plant stems. Sisal is a leaf fiber. Fruit or seed fibers include cotton, kapok, and the coir fibers of coconut husks.

Exposure to vegetable fibers is a major occupational hazard. Of the three vegetable fibers associated with respirable dust exposure during processing, cotton is the predominant textile fiber, followed by flax which is woven into linen, and soft hemp, traditionally used for rope and net making, but now largely replaced by synthetic fibers.

1. *Byssinosis*

Byssinosis is the generic name applied to acute and chronic airway disease among those who process cotton, flax, and hemp fibers. In the textile industry, it is also known as *brown lung disease*. It is complicated by the presence of foreign materials and microorganisms, such as molds and fungi, that collect on fabrics. The acute response is characterized by a sensation of chest tightness upon return to exposure following a holiday or weekend break. This symptom is often accompanied by a cough, which becomes productive with time, and occasionally by shortness of breath. Measurement of lung function upon return to exposure often reveals modest decreases in expiratory flow rates over the working shift. For most affected individuals, these findings will diminish or disappear on the second day of work. With prolonged exposure, both the symptoms and functional changes become more severe. Dyspnea becomes the prominent complaint while decrements in expiratory flow rates over a work shift are often marked, and clear clinical and physiological evidence of chronic obstructive lung disease emerges.

I. Metal Dusts

A group of occupational lung disorders is due to the inhalation of a variety of different inorganic materials, mainly metals of one sort or another. While the total number of exposed workers at risk may be large, the actual number of individuals employed in any single industry or exposed to any specific agent may not be large. Consequently, occupational health information is not as readily available as it is in larger industries where health and safety programs are better recorded. Another consideration is that many of the mineral dust pneumoconioses cause only chest radiographic alterations but are not associated with medical disability. Because of this, there is less interest or impetus to monitor these pneumoconioses.

An *aerosol* is an airborne solid or liquid particle produced by grinding, cutting, hot operations of forging or welding, painting and spraying, extraction and crushing, shaping, and chemical reactions. Aerosols can occur in the form of dusts, fumes, smokes, or mists and fogs, according to their physical nature, their particle size, and their method of generation. *Dusts* are usually produced by mining or ore reduction, resulting in disintegration of solid materials. *Fumes* are produced by such processes as combustion, distillation, calcination, condensation, sublimation, and chemical reactions. They form true colloidal systems in air. *Smokes* are colloidal systems produced by incomplete combustion. *Mists* and *fogs* cover a wide range of particle sizes and are considered to be primarily liquid.

1. *Aluminum*

The metal of aluminum is never found in the elemental state. Its primary sources are the ores cryolite and bauxite. Most hazardous exposures to aluminum occur in smelting and refining processes. Aluminum is mostly produced by electrolysis of bauxite, Al_2O_3, dissolved in molten cryolite, Na_3AlF_6. Aluminum's effect on the lungs differs depending upon the composition of the inhaled aluminum-containing dust. Schaver described a series of cases of pulmonary disease in a group of furnace workers processing bauxite. *Schaver's disease*, as this occupational disease has been named, is a special form of silicosis involving ultramicroscopic silica particles, fume in character, that produce an immediate, intense irritant reaction in the bronchioles and alveoli when they are inhaled. The condition leads to acute bronchiolitis and diffuse fibrosis. Although silica is present in crude bauxite only as a 5-7% impurity, crystalline free silica comprises 30-40% of the furnace fumes.

2. *Barium*

Baritosis is a benign pneumoconiosis that results from the inhalation of dusts of barium sulfate or barium ores. The main ore of barium is baryte. Baryte is used principally as a constituent of lithopone, a white pigment employed in the manufacture of paints. It is also used as a filler in textiles, rubber, soaps, cements, and plasters. Barium is highly insoluble and radiopaque, which allows it to be used safely as a radiographic contrast medium. The inert dust of barium compounds in nonfibrogenic and baritosis is not associated with any respiratory symptoms or functional impairment. However, the radiographic appearances are quite striking. The deposits of barium appear as multiple, dense, small, rounded opacities.

3. *Beryllium*

Inhalation of any beryllium compound is potentially hazardous. It can produce either acute or chronic beryllium disease. Acute beryllium disease, the acute response to inhaling toxic beryllium compounds, is defined as disease which lasts less than 1 year, occurs during exposure to beryllium, and includes any of the following: nasopharyngitis, tracheitis, bronchitis, pneumonitis, dermatitis, and conjunctivitis. Ulceration may also be present, and nasal septum perforation can occur. For acute disease, the more soluble beryllium compounds, including beryllium fluoride, beryllium sulfate, and ammonium beryllium fluoride, have been implicated as the cause of both upper and lower respiratory abnormalities. In addition, acute pneumonitis has been associated with beryllium oxide, carbide, oxyfluoride, hydroxide, and zinc beryllium silicate. Chronic beryllium disease is caused by inhalation of the fumes of the metal or an alloy containing beryllium, lasts longer than 1 year, and usually causes both systemic and pulmonary abnormalities. Skin lesions may develop and the lesions are reddish, papulovesicular, and pruritic. Radiological pattern is nonspecific, showing an image that resembles sarcoidosis, tuberculosis, mycosis, or other lung disease. Reactions to beryllium are believed to involve the immune system through formation of an antigen by a beryllium ion combining with a protein or other natural body substance. The term berylliosis should not be used as it implies two false conclusions: (1) beryl ore itself causes disease, and (2) beryllium disease is similar to pneumoconiosis.

4. Cadmium

Several forms of cadmium are hazardous to workers. All occur as respirable dusts, and the metal also vaporizes if heated. On an equal weight basis, cadmium vapor is considered more toxic than the dust. Most cases of acute cadmium intoxication have been associated with welding, soldering, or silver brazing. The manifestations of toxicity are chiefly respiratory. The onset of symptoms may be delayed several hours or until the worker has left the scene of exposure. The clinical picture may simulate that of an acute infection, or be mistaken for metal fume fever, particularly among welders. Slight exposure is attended by drying and irritation of the upper respiratory tract. Pulmonary edema may occur within hours of severe exposure and persist for days or weeks. Fatality following pulmonary edema is well documented. While recovery from edema generally appears to be complete within weeks, shortness of breath and impaired pulmonary function persist for years in some instances. With chronic exposure to cadmium, nasal passages become inflamed, and there is loss of the sense of smell owing to damage to the olfactory nerve. The teeth show yellow discoloration. Chronic exposure to cadmium, however, does not seem to result in any major hazard for the lung.

5. Hard metal disease

Hard metal disease is an occupational lung disease in which the respiratory system is adversely affected by exposure to tungsten carbide. Tungsten carbide is produced by blending and heating tungsten and carbon in an electric furnace, and then mixing in a ball mill with cobalt to form a matrix for tungsten carbide crystals and other metals such as nickel, chromium, and titanium. Because tungsten carbide is extremely hard and resistant to heat, it is used in the manufacture of metal-cutting tools, dental drills, and bearings. There are two major types of pulmonary reactions that occur among workers exposed to tungsten carbide: an asthma syndrome and diffuse interstitial pulmonary fibrosis. Early symptoms are cough and scanty mucoid sputum. Later, the worker complains of shortness of breath, which worsens progressively. Pulmonary function measurements reveal reduced lung volume. The pattern is that of classical restrictive disease without significant airway obstruction. Chest radiographs may show a fine reticular nodular pattern, and the heart outline is often blurred, as in asbestosis.   Fine honeycombing may develop, as in other pulmonary fibrosis.   The disease is seldom seen in workers with less than 10 years exposure to tungsten carbide materials.

6. Iron

Inhalation of metallic iron or iron compounds causes *siderosis*. Siderosis is a relatively benign pneumoconiosis, characterized by large accumulations of inorganic iron-containing macrophages in the lungs with minimal reactive fibrosis.   In its pure form, the condition probably does not progress to true nodulation as seen with silicosis and is usually asymptomatic.   It is known chiefly for the abnormal changes produced on chest radiographs.   Siderosis is seen in its purest form in arc welders, oxyacetylene cutters, and silver finishers.   During arc welding and oxyacetylene cutting, iron is melted and boiled by the heat of the arc or torch.   The iron is emitted as particles of ferrous oxide which are immediately oxidized to ferric oxide and appear as blue-gray fumes. Prolonged inhalation of these fumes can lead to siderosis.   Silver finishers use what is known as jeweler's rouge to polish their unfinished wares.   The rouge is composed of iron oxide and is often applied with a buffer that generates a cloud of small iron and silver particles.   When iron is inhaled in conjunction with other fibrogenic mineral dusts, pulmonary fibrosis results. This is referred to as mixed-dust pneumoconiosis.   Most of the affected miners have worked in the mines for more than 20 years.   The symptoms and signs are relatively nonspecific.   The miners often complain of shortness of breath, cough, and reddish-brown sputum.   The shortness of breath is worse in miners who have massive fibrosis.   The amount of fibrosis is in general related to the free silica content.

7. Manganese

A respiratory condition associated with the inhalation of particles of manganese compounds is known as manganese pneumonitis. The effects of manganese on the nervous system are well established. However, a high incidence of bronchitis and pneumonia has been reported in a group of workers involved in the manufacture of potassium permanganate. Manganese pneumonitis is slow to respond to treatment but apparently leaves no permanent damage.

8. Metal fumes

Metal fume fever is a common occupational disease in environments where workers are exposed to the fumes of certain metals, as in foundries, rolling mills, welding operations, galvanizing operations, and molten metal processing. It is characterized by a feverish reaction to the inhalation of finely divided particles of metallic oxides.   While zinc, copper, and magnesium are the chief offenders, cadmium, iron, manganese, nickel, selenium, tin, and antimony are responsible in some instances. The disease has an acute onset, and although there is no chronic form of metal fume fever, repeated episodes occur. The symptoms may develop in a new worker on his first day on the job and also in experienced workers on reporting to work after a weekend break, hence the popular term of "Monday morning fever." Metal fume fever symptoms include thirst and a metallic taste sensation. There is usually a time lag of several hours between exposure and the onset of symptoms. Later, the worker has rigors, high fever, muscular aches and pains, headache, and a generalized feeling of weakness. There may be nausea, vomiting, and mental disturbances, usually marked by agitation. The worker sweats profusely, and the condition is often mistaken for influenza. The diagnosis of metal fume fever is dependent upon the worker's occupational history. There is no recognized treatment of the disease.

9. Mercury

Three chemical forms of mercury pose occupational hazards: elemental mercury, inorganic salts, and organic salts. Liquid elemental mercury vaporizes readily at ambient temperatures. Exposure by inhalation occurs with both the vapor and the inorganic salts as dusts. Most cases of acute intoxication are accidental, e.g., following rupture of a large mercury-containing receptacle in a confined space. If the vapor has been inhaled, the clinical picture will generally reflect injury to the lung (chest pain, cough, shortness of breath) plus general toxemia (fever, chills, profound weakness, anorexia, and joint pain). In nonfatal cases, recovery is rapid and may be complete within 24 hours. Chronic exposure to mercury vapor will affect the central nervous system. The clinical picture is termed "erethism." Headache and various personality changes are described, including increased irritability, depression, paranoia, insomnia, and loss of memory and mental acuity. Motor disturbances also occur. Tremors of the limbs, particularly of the hands, are often an early sign of chronic intoxication. Muscular coordination can become impaired.

10. Tin

Exposure to dust or fumes of inorganic tin may cause a benign pneumoconiosis known as *stannosis*. Stannosis produces distinctive progressive changes in the lungs as recorded on chest radiographs as long as exposure continues, but there is no fibrosis and no evidence of disability. Because tin is radiopaque, early diagnosis is possible.

11. Titanium

Titanium pneumoconiosis is a respiratory disorder caused by the inhalation of dust of a titanium compound. Titanium oxide is used as white pigment in the manufacture of paint. Titanium carbide finds extensive use in the manufacture of tools. There is some evidence that titanium oxide may produce radiographic abnormalities similar to those seen following the inhalation of iron and tin dusts. However, the condition is relatively benign, and there is no associated pulmonary impairment.

J. Organic Dusts

The inhalation of organic dusts may lead to two distinct pulmonary responses. First, and more common, is occupational asthma. It occurs most commonly in *atopic subjects*, i.e., susceptible individuals, and is characterized by changes in the airflow resistance in the conducting system of the lungs. The second and less common type of reaction is known as *hypersensitivity pneumonitis*. It affects the lung parenchyma, namely the respiratory bronchioles and alveoli, and does not appear to be related to atopy.

K. Occupational Asthma

Asthma is a disease characterized by an increased responsiveness of the airways to various stimulants and manifested by slowing of forced expiration which changes in severity either spontaneously or with treatment. Occupational asthma is a disorder with generalized obstruction of the airways, usually reversible, and caused by the inhalation of substances or materials that a worker manufactures or uses directly, or that are incidentally present at the work site. Individuals with occupational asthma complain of tightness of chest, nocturnal cough, wheeziness, and shortness of breath. Initially, these symptoms occur only while the individual is at work but later they may persist at home and on weekends. Workers who are atopic are more prone to develop occupational asthma and may do so with a relatively short exposure. Nevertheless, normal individuals may be affected until they have been sensitized.

1. *Immunologic mechanism*

Immunologic mechanisms seem operative in occupational asthma, often affecting atopic individuals. Atopic individuals have a unique response to intranasal immunizations with certain protein and carbohydrate antigens by producing high concentrations of skin-sensitizing antibodies. This reaction is rarely observed in nonatopic (normal) subjects immunized by the same schedule. Atopic individuals develop symptoms rapidly and are forced to leave their job at an early stage because they develop diseases, while nonatopic individuals represent a survivor population.

Immunologic reactions are classified into four types. *Type I* reactions are immediate and are mediated by a specific immunoglobulin, IgE. When an atopic individual is exposed to an antigen, there is an increase in the IgE specific to the antigen. The specific IgE reacts with the cell-bound antibody to form bivalent complexes and these, in turn, trigger a series of enzymatic reactions that ultimately result in the release of mediators, such as histamine, serotonin, slow-reacting substances of anaphylaxis (SRS-A), and eosinophil chemotactic factor of anaphylaxis (ECF-A), causing the asthmatic reaction. *Type II*, or *cytotoxic-type reactions*, are mediated by reaction of antibody with surface antigen of the cell and formation of an immune complex. There is no evidence to suggest that this type of immunologic reaction takes place in occupational asthma. The *Type III* responses, which are related to the *Arthus phenomenon* and are associated with the presence of precipitins in the blood, occur several hours after the challenge. They are due to immunoglobulin, IgG. *Type IV*, delayed (24-48 hours) response, or *tuberculin reaction*, are cell-mediated immunities. This type of immunologic mechanism may play some role in certain types of occupationally induced asthma.

Different patterns of asthmatic reactions can be observed following bronchial provocation with specific antigens. The reactions fall into three main patterns: immediate, delayed, or combined reactions in which both immediate and delayed reactions occur. Immediate reactions occur within minutes and are relatively short in duration. The immunologic mechanism responsible for the immediate asthmatic reactions is Type I. Delayed reactions occur hours after bronchial provocation and may last for several hours or even a couple of days. There is evidence of a Type III precipitating antibody immune-complex allergic reaction. In a Type I reaction, the decline in ventilatory capacity is usually evident within 10-15 minutes. If it is a type III response, the decline is often delayed for several hours. Skin tests in which the reaction is immediate have been beneficial in evaluating workers with occupational asthma associated with exposure to certain agents.

Many substances in the work environment can cause occupational asthma. The list of known agents continues to expand and any current list will, of necessity, be incomplete. It has been recognized that an individual with an atopic background may become sensitized to virtually any natural product of appropriate antigenicity and particle size. With respect to synthetic agents, atopy is probably of less importance although bronchial hyperactivity in atopics probably makes them more susceptible to agents causing occupational asthma, regardless of mechanism.

2. Substances of vegetable origin

Substances of vegetable origin are probably the most commonly reported causes of occupational asthma. Carpenters, joiners, and sawmill workers become sensitized to sawmill dust, fungal spores, and substances used to treat wood. Wood dusts most often implicated include cedar, oak, mahogany, keejat, African zebra wood, and Western red cedar. Occupational asthma due to grain allergy is found principally in millers and bakers, although it may occur in farm workers handling grain. Outbreaks of asthma have occurred in people exposed to a prevailing wind carrying dust from neighboring mills.

3. Substances of animal origin

Substances of animal origin include animal hair, skin cells, mites, small insects, molds, dander, and bacterial and protein dust. Shepherds, farmers, jockeys, laboratory technicians, animal handlers, veterinarians, and others who are in regular contact with animals may develop occupational asthma associated with IgE antibodies. Allergic pulmonary and skin symptoms are most likely to occur in animal handlers. Occupational asthma is also associated with the use of a protein enzyme manufactured by fermentation of *Bacillus subtilis*. Workers exposed to relatively high concentrations of the dust have developed occupational asthma. Immediate skin reactions to extracts of *Bacillus subtilis* have been reported. Both immediate and late bronchoconstrictor responses were found in some individuals on bronchial challenge.

4. Substances of chemical origin

Many chemicals, both simple and complex, are associated with occupational asthma. Among inorganic chemicals, the complex salts of platinum, when given sufficiently long exposure, will result in virtually 100% sensitization. Other inorganic chemicals known to cause occupational asthma include nickel salts, chromium salts, and sodium and potassium persulfates. Organic chemicals that can cause occupational asthma include the amines, such as ethylenediamine, *p*-phenylenediamine, etc.; anhydrides, such as phthalic anhydride, trimellitic anhydride, etc.; pharmaceuticals, such as ampicillin, penicillin, etc.; miscellaneous chemicals such as formaldehyde, organophosphate insecticides, etc.; and of course, diisocyanates. Diisocyanates are used to manufacture polyurethane foams. Two compounds have been incriminated as causes of occupational asthma: toluene diisocyanate and diphenylmethane diisocyanate.

L. Hypersensitivity Pneumonitis

Hypersensitivity pneumonitis is a respiratory disorder that is primarily a pulmonary inflammation of the interstitial tissues. It results from sensitization and recurrent exposure to inhaled foreign substances. The disease is diffuse, predominantly mononuclear inflammation of the lung parenchyma, particularly the terminal bronchioles and alveoli. The inflammation often organizes into granulomas and may progress to fibrosis.

Most individuals who develop hypersensitivity pneumonitis are exposed through their occupation.

Although a large number of organic dusts have been identified as causes of hypersensitivity pneumonitis, the pathophysiological effects are similar no matter which dust is responsible. Not everybody who is repeatedly exposed to the antigenic organic dust develops hypersensitivity pneumonitis, though a small percentage does. Similarly, while there is good evidence that a substantial proportion of the subjects who are exposed to the antigen develop antibodies, the presence of antibodies alone is not necessarily an indication that the individual has, or is likely to suffer from, hypersensitivity pneumonitis.

1. Acute form

The clinical presentation of hypersensitivity pneumonitis depends on the immunologic response of the individual, the antigenicity and particle size of the inhaled dust, and the intensity and frequency of exposure. The most common form of clinical presentation of hypersensitivity pneumonitis is an acute, explosive episode of respiratory and systemic symptoms occurring in temporal relationship to inhalation of the offending antigen. The symptoms occur 4 to 6 hours after exposure and include sudden onset fever, chills, shortness of breath, and a dry cough. Pulmonary function tests may show a decrease in forced vital capacity and a decrease in forced expiratory volume in 1 second, with minimal alteration of the ratio of the two variables. The radiographic appearances are those of a diffuse acinous filling process predominantly affecting the mid and lower zones. Because the episode mimics acute influenza or other viral episodes, broad-spectrum antibiotics are often used for therapy of suspected bronchitis or pneumonia. The spontaneous recovery suggests effective antibiotic therapy, but the episode recurs with reexposure. If not treated, symptoms and signs gradually regress over a period of a week to 10 days.

2. Chronic form

Besides the acute form of hypersensitivity pneumonitis, a chronic form exists. The chronic form occurs with repeated low-dose exposures, and although on the first few occasions there may be mild fever and chills, the continued low-dose insults are not so obviously related to occupational exposure. The afflicted individual notices the onset of dyspnea and this is accompanied by cough, malaise, weakness, and weight loss. Pulmonary function tests show a restrictive ventilatory pattern with a decrease in diffusion capacity. Chest radiographs may show diffuse interstitial fibrosis or honeycombing. Granulomas may be found dispersed in the interstitial fibrosis. Bronchiolar walls are thickened by collagen and contain lymphocyte infiltrations and their lumens are obstructed by granulation tissue. The obstruction may lead to peripheral destruction of alveoli with honeycombing.

Hypersensitivity pneumonitis may occur following the inhalation and subsequent sensitization of antigens in a wide variety of organic materials. Offending agents may be bacterial, fungal, animal proteins, chemicals, or yet undefined. Each class of these offending agents is discussed below.

3. Substances of bacterial origin

The major occupations and industries associated with hypersensitivity pneumonitis are those in which moldy vegetable compost is handled—which by its very nature is contaminated with thermophilic actinomycetes. Thus, farmers, sugarcane workers, and mushroom compost handlers are exposed.

a. Bagassosis

Bagassosis is a type of hypersensitivity pneumonitis found in moldy sugarcane waste, or bagasse. Bagasse is a fibrous substance that remains after the juice containing the melted sugar is removed from the cane.

b. Farmer's lung

Farmer's lung is caused by inhalation of organic dusts containing fungal spores. It is often cited as the classic example of an organic dust allergy involving the alveoli. The source of the fungal spores is found in moldy hay, grain silage or straw.

c. Mushroom worker's lung

Mushroom worker's lung is a condition similar to bagassosis and farmer's lung. It apparently is caused by inhalation of spores that thrive in the same environment as the common mushroom, which is grown on compost whose main components are straw and horse manure.

4. Substances of fungal origin

A variety of saprophytic fungi have been found to cause hypersensitivity pneumonitis. Industries in which raw wood products are handled are prone to the development of hypersensitivity pneumonitis.

a. Maple bark stripper's disease

Maple bark stripper's disease affects lumber workers employed in stripping the bark from maple logs. The fungus grows beneath the bark of the tree. Removal of the bark during the stripping process liberates the spores into the air.

b. *Malt worker's lung*

Malt worker's lung is similar to farmer's lung. During the germination of barley, malt frequently is turned on open floors, resulting in the generation of a heavy green dust containing a high concentration of fungal spores.

c. *Woodworker's lung*

Woodworker's lung is a progressive, irreversible, interstitial hypersensitivity pneumonitis that develops in workers manufacturing wood pulp. Exposure to oak dust and fine mahogany dust can produce an immediate sensitivity reaction in wood workers.

d. *Sequoiosis*

Exposure to redwood dust can cause a hypersensitivity known as *sequoiosis*.

e. *Suberosis*

Suberosis develops in workers exposed to cork dust. In chronic suberosis, fibrotic nodules and arteriolitis are present, while the acute form of the disease is characterized by a granulomatous pneumonitis similar to that of farmer's lung.

f. *Cheese worker's lung*

Cheese worker's lung is caused by exposure to clouds of *Penicillium casei* that are produced when mold is washed from the surface of cheeses.

M. Miscellaneous Inhaled Agents

Some inhaled chemical agents are hazardous primarily to the respiratory system, while others also affect distant organs and tissues. Examples of the former are ammonia and chlorine, while examples of the latter are nitrogen and ozone. How extensively the respiratory system is involved, particularly following acute or accidental exposure, is determined largely by the concentration of the agent and duration of exposure. Other factors that may modify the individual's response include pre-existing heart or lung disease, prior long-term exposure to the same agent, level of activity during exposure, and the individual's age. Occupational agents with relatively low solubility in aqueous solution tend to shift their primary effect to the periphery of the respiratory system. Thus, nitrogen dioxide and ozone are notable for the bronchiolar and parenchymal injury that they produce at relatively low concentrations.

1. *Arsine*

Arsine is a colorless, very highly toxic gas with a faint, garlic-like odor. Although arsine as such is not used in industry, it is apt to be encountered wherever arsenic, even in scrap metals, becomes moist under reducing conditions with free hydrogen. In addition, arsine is involved in the smelting and refining of various arsenic-containing ores. Upon inhalation, arsine damages lungs and passes into the bloodstream where it causes hemolytic destruction of the red blood cells. Invariably, the sign observed in arsine poisoning is *hemoglobinemia*, appearing with discoloration of the urine up to port wine hue. Jaundice sets in on the second or third day and may be intense. Coincident with these effects is a severe hemolytic-type anemia.   Severe renal damage may occur with oliguria or complete suppression of urinary function, leading to uremia and death. Where death does not occur, recovery is prolonged.

2. *Hydrogen sulfide*

Hydrogen sulfide is a colorless, flammable, and explosive gas with a powerful nauseating odor. The smell is generally perceptible at 0.77 ppm; at 4.6 ppm, it is quite noticeable; while at 27 ppm, it is strong and unpleasant. Hydrogen sulfide is not used directly in industry but is formed frequently as a byproduct of certain processes. Hydrogen sulfide is heavier than air and therefore tends to accumulate under stagnant conditions in deep cavities, such as tunnels, vats, and cellars. It occurs in many natural gas and petroleum products and in some mining and refining operations. It is also found in sewer gas that results from bacterial fermentation. The toxicity of hydrogen sulfide is attributable to both biochemical and direct initiative actions. As a surface irritant, hydrogen sulfide primarily affects the eyes and respiratory system. In tissue liquids, the gas dissociates into hydrosulfide (HS^-) and sulfide (S^{2-}) ions, which by inactivating a number of respiratory enzymes, interferes with the cellular metabolism of oxygen.   As a consequence, the respiratory center in the brain may cease to function, causing apnea and sudden death. Hydrogen sulfide is deadly and, like the cyanide, exceedingly rapid in its action. However, it can also be insidious, as the warning by the sense of smell is lost early owing to fatigue or paralysis of the olfactory nerve from small amounts of the gas. If death does not occur because of asphyxiation, recovery is usually complete.

3. *Osmium tetroxide*

Osmium has only limited commercial use. Its principal forms of production are as metallic osmium and osmium tetroxide, also called osmic acid. It is used principally in histology laboratories to fix and stain tissue. The second major use is in the drug industry as a catalyst in the production of steroid hormones. The properties that distinguish osmium tetroxide as a fixative are responsible for its toxicity. It reacts with lipids, nucleic acids, and proteins. The structure and function of proteins are thereby altered. As a consequence, a variety of enzymes may be destroyed. Osmium tetroxide vapors irritate the surfaces of the skin, eyes, and respiratory tract. The subject may have smarting of the eyes, lacrimation, and see halos around lights. Corneal ulcers may occur. Among all respiratory irritants, osmium vapors appear to strike with the most dramatic intensity. Depending on the degree of exposure, all strata of the respiratory system may be involved. Coughing has been the most frequent symptom. More severe exposure may cause a sense of chest constriction coupled with difficulty in breathing. Headache behind or above the eyes is relatively common.

4. *Oxides of nitrogen*

The term "oxides of nitrogen" is reserved for nitric oxide and nitrogen dioxide. Both gases are byproducts of a variety of combustive processes associated with high temperature. Typically they co-exist together although their relative concentrations vary widely, depending on the nature of the combustive process. Nitrogen dioxide is the more toxic of the two gases. Oxides of nitrogen may also be encountered on the farm in silos, often with resulting serious disabling effects in "silo-filler's disease." Nitrogen dioxide unites with water to form nitric and nitrous acids. If the dose is overwhelming, pulmonary edema, or even death, may occur.

5. *Ozone*

Ozone, an allotropic form of oxygen, is colorless, highly reactive, and unstable. Ozone is used as an oxidizing agent in the organic chemical industry; as a disinfectant for food in cold storage rooms and for water; for bleaching textiles, waxes, flour, mineral oils and their derivatives, paper pulp, starch, and sugar; for aging liquor and wood; for processing certain perfumes, vanillin, and camphor; in treating industrial wastes; in the rapid drying of varnishes and printing inks; and in the deodorizing of feathers. Ozone is irritating to the eyes and all mucous membranes. It aggravates chronic respiratory diseases, such as asthma and bronchitis, and can cause structural and chemical changes in the lungs.

6. *Phosgene*

Phosgene, or carbonyl chloride, is a highly toxic, colorless, and noncombustible gas with an odor resembling that of musty hay in low concentrations. In high concentrations, it is irritating and pungent. Along with chlorine, phosgene was used as a poisonous gas in World War I. Industrially, it is used to make acid chlorides and anhydrides and for preparing numerous other organic chemicals. It can be formed spontaneously during welding, degreasing, and other operations where chlorinated hydrocarbon solvents come in contact with high-energy sources. Accidental exposure to phosgene may pass unnoticed before the onset of symptoms. After exposure, there may be a feeling of tightness in the chest, cough, nausea, and vomiting. There is irritation of the eyes, nose, and throat, with respiration becoming more rapid. In addition to respiratory symptoms, there may be evidence of nervous system involvement: dizziness, headache, blurred vision, mental confusion, and muscular twitching. Death may result from respiratory or cardiac failure.

7. Polymer fumes

Polymer fume fever begins several hours after exposure to the heat-degraded polymer, polytetrafluoroethylene, also known as teflon. Polytetrafluoroethylene breaks down at a temperature of 250 to 300°C. At that point it liberates a collection of aliphatic and cyclic fluorocarbon compounds. Many are powerful irritants. The disorder is characterized by brief but acute attacks of chest tightness, choking, a dry cough, and occasional rigors.

8. Vanadium pentoxide

Vanadium pentoxide is an industrial catalyst in oxidation reactions, used in glass and ceramic glazes, a steel additive, and used in welding electrode coatings. Of all the vanadium compounds, the pentoxide is probably the most hazardous to health. Absorption of vanadium pentoxide through the lungs results in chronic toxicity, irritation of the respiratory tract, pneumonitis, and anemia. Initial symptoms are profuse tearing and a burning sensation in the eyes, rhinitis that may be accompanied by nosebleeds, a productive cough, sore throat, and chest pain. A green-black discoloration of the tongue has been described.

9. Animal proteins

Serum protein has been implicated as a causative agent of hypersensitivity pneumonitis. Avian proteins may be inhaled in the dispersed dried excreta of caged pigeons, chickens, parakeets, or budgerigars. The clinical conditions have been variously described as bird breeder's lung (avian droppings), snuff lung (bovine and porcine pituitary snuff), technician's lung (rat urine), and furriers lung (animal hair).

VI. TOXIC RESPONSES OF THE BLOOD

Chemical-induced toxic responses of the blood have been historically associated with the development of the chemical industry. Common factors contributing to epidemics of toxicity have been the rapid introduction of many new chemicals and the exposure of large numbers of workers without adequate education or protection. As new chemicals are introduced and new products become available, it is important to be aware of potential mechanisms of toxicity so that the epidemic poisonings of the past will not be repeated.

A. Bone Marrow Suppression

Some chemicals can exert a toxic effect upon the hematopoietic system, resulting in abnormalities in the blood cells themselves, or on their production in the bone marrow. Classification of such chemicals is based on these two effects. Bone marrow effects are mostly *hypoplasia*, or underproduction of blood cells, or sometimes *hyperplasia*, over-production of blood cells. In extreme cases, the bone marrow effects may be aplasia, or no production. Some of the conditions associated with hypoplasia are: *erythropenia* (deficiency of red blood cell production), *leukopenia* (deficiency of platelet production), *pancytopenia* (deficiency of all cell production), or any combination of these.

Bone marrow suppression is characterized by a deficiency of all or some cellular elements in peripheral blood. This condition results from either a decrease in production of cells or an inability of bone marrow to manufacture adequate numbers of these cells. Bone marrow suppressors are of two groups: those that regularly produce suppression upon exposure if a sufficient dose is given (e.g., arsine, benzene, radioactive chemicals, as well as ionizing radiation), and those occasionally producing suppression that may be related to an unknown biochemical idiosyncracy of the worker (e.g., organic arsenical, gold compounds, sulfonamide, and trinitrotoluene).

Signs and symptoms of bone marrow suppression are variable and often insidious, depending upon the degree of pancytopenia. The general conditions of anemia, weakness, fatigue, irritability, and gastrointestinal distress may be present. Hemorrhage and blood loss from mucous surfaces are common and infection can become a serious problem.

B. Hemolytic Anemia

Hemolytic anemia is due to an increased rate of red blood cell destruction. Frequently, there is shortened life span or an increased fragility of these red blood cells. Chemicals cause hemolytic anemia by at least three mechanisms: those that produce hemolysis directly in all persons if a sufficient dose is given (e.g., arsine, benzene, lead, methyl chloride, phenylhydrazine, and trinitrotoluene); those that produce hemolysis by an immune mechanism (e.g., quinine and quinidine); and those that affect people with certain genetic defects such as glucose-6-phosphate dehydrogenase deficiency (e.g., acetanilide, naphthalene, phenylhydrazine, potassium perchlorate, and sulfanilamide). Clinical tests that are available for the determination of hemolytic anemia are blood smears, cell counts, cell morphology, Heinz body accumulation, and urinary levels of erythrocyte breakdown products.

C. Polycythemia

Polycythemia is the presence of an abnormally high red blood cell count, an increased hemoglobin concentration, or increased hematocrit. A number of chemicals have been reported to cause polycythemia. The best-known agent is cobalt. Besides the polycythemic effect, cobalt produces *reticulocytosis*, i.e., an increase in the number of precursor red blood cells, bone marrow hyperplasia, and increased erythropoietic activity in the spleen and liver.

D. Carboxyhemoglobin

Hemoglobin is the oxygen-carrying protein of the red blood cells. The globin, or protein chains, has irregularly folded conformations that enclose the heme group in a hydrophobic pocket that forms the oxygen binding site. The active site of the heme group is an Fe^{2+} ion situated in a porphyrin ring. The reversible binding of oxygen by hemoglobin is called oxygenation. When the hemoglobin molecule is fully saturated, all four oxygen molecules are thought to be equivalent, and any one of them may be the first to be released. The release of the first oxygen, however, will greatly facilitate the release of the second oxygen molecule. In the same manner, the release of the second oxygen facilitates the release of the third oxygen. Release of the fourth oxygen does not occur under normal physiologic conditions.

Carbon monoxide is perhaps the best-known example of a chemical agent that decreases the oxygen transport of the blood and produces hypoxia, a condition in which there is an inadequate supply of oxygen to the tissues. The mechanism whereby carbon monoxide elicits this toxic effect results from the fact that carbon monoxide is a stronger ligand for hemoglobin than is oxygen, and therefore has a stronger binding affinity. This means that carbon monoxide molecules compete more successfully than oxygen molecules for the hemoglobin binding sites that normally carry oxygen. In humans this carbon monoxide binding affinity has a value of 220 at pH 7.4 for human blood. This means that carbon monoxide binds the hemoglobin 220 times more tightly than oxygen. Thus, at equal concentrations of the two gases, the blood would contain 220 times more carboxyhemoglobin than oxyhemoglobin.

The carbon monoxide binding affinity differs among species. In the canary, for example, the carbon monoxide binding affinity is only about 110. At steady state, the canary contains less carboxyhemoglobin than a human for any given carbon monoxide concentration. In the old days, miners took caged canaries into the shafts with them as an early warning device of carbon monoxide accumulation or oxygen depletion. Canaries were useful indicators because their respiration rates and metabolic rates are much higher than those of humans. As a result, they achieve any change in equilibrium much more quickly than miners do. At high carbon monoxide concentrations, i.e., those toxic to the canary are well above those toxic for the miners, the canary dies first and warns the miner to leave. At low carbon monoxide concentrations, however, a steady state is attained in the bird first but the amount of hemoglobin deoxygenated in the canary is not as high as that deoxygenated in humans and the miner expires first, warning the canary to leave. The decisive dividing line is about 0.2% of carbon monoxide in air. Below this concentration, the miner probably dies first and above this concentration, the canary dies first.

One of the most important concerns of occupational toxicology has been the relationship of atmospheric carbon monoxide level, carboxyhemoglobin level in blood, and physiologic effects. The relationship between carboxyhemoglobin percent saturated in blood and the respective toxic effects is summarized in Table 3.2. It is important to note that normal carboxyhemoglobin levels in blood are not zero, even in instances with no detectable carbon monoxide level. This small but detectable level of carboxyhemoglobin is a natural component of the blood formed from the normal breakdown of hemoglobin. It has been determined that the carboxyhemoglobin level of non-smoking adults breathing carbon monoxide-free air was between 0.3 and 0.5% in blood, and that of tobacco smokers was between 3 and 10% depending on the number of cigarettes smoked and the manner of smoking, inhaling or not inhaling. Exposure to dichloromethane can also produce increased carboxyhemoglobin levels.

Table 3.2
Carboxyhemoglobin Levels and Toxic Effects

% COHb in Blood	Toxic Effects
0 - 10	No symptoms.
10 - 20	Tightness across forehead, possible slight headache, dilation of cutaneous blood vessels.
20 - 40	Headache and throbbing in temples. Severe headache, weakness, dizziness, dimness of vision, nausea, vomiting, collapse.
40 - 50	Same as previous item with more possibility of collapse & syncope. Increased respiration and pulse.
50 - 60	Syncope, increased respiration & pulse, coma with intermittent convulsions & Chyne-Stokes respiration.
60 - 70	Coma with intermittent convulsions. Depressed heart action & respiration. Possible death.
70 - 80	Weak pulse & slow respiration. Respiratory failure & death.

The obvious and specific antagonistic to carbon monoxide poisoning is oxygen. After termination of the exposure, respirations must be supported artificially if necessary. In advanced cases of carbon monoxide poisoning, the use of oxygen hyperbaric chambers can increase significantly the rate of conversion of carboxyhemoglobin to oxyhemoglobin *in vitro*. Exchange transfusion has been used for moribund victims.

E. Methemoglobin

The iron in the heme of hemoglobin is normally in the ferrous state (Fe^{2+}), but it can be oxidized by certain chemicals to the ferric state (Fe^{3+}). The resulting pigment is known as *methemoglobin*. *Methemoglobinemia* can be hereditary, owing to rare genetic defects in the hemoglobin molecule, or lack of the red blood cell enzyme that converts methemoglobin back to hemoglobin after the red blood cells take up oxygen in the alveoli. A more common form of the disorder, however, is secondary or toxic methemoglobinemia, caused by exposure to a chemical agent. Substances that cause toxic methemoglobinemia may be strong oxidizers, others may interfere with the enzymatic function of the red blood cells, and still others may block the return of methemoglobin molecules to hemoglobin. In any event, a hemoglobin molecule that has been oxidized to methemoglobin is no longer a viable oxygen transporter in the bloodstream. The result is a form of anemia. Another effect is that hemoglobin molecules still able to transport oxygen to the body tissues bind the oxygen molecules so firmly, because of the effect of methemoglobin on the oxygen dissociation curve, that tissue cells have difficulty in obtaining oxygen. Thus, the presence of methemoglobin in the bloodstream has a double adverse effect and the severity of the condition varies with the concentration of methemoglobin.

In the normal person, the natural process termed "auto-oxidation" is believed to account for the steady-state small amounts of 1% methemoglobin in blood. Generally, cyanosis becomes apparent when the methemoglobin concentration exceeds 15%, but most people do not exhibit any symptoms until about 20%. The industrial terms of "blue lip" or "huckleberry pie face" refer to the cyanotic complexion as a result of methemoglobinemia. At levels of about 20-70%, depending on the individual, methemoglobin weakness accompanied by dizziness, headaches, tachycardia, or dyspnea may ensue.

Chemicals can convert hemoglobin to methemoglobin either directly or indirectly. Some of these chemicals are listed in Table 3.3. Direct action is associated with nitrites, chlorates, hydrogen peroxide, hydroxylamine, quinone, and methylene blue. Nitrites and nitrates, which are reduced to nitrite by gastrointestinal bacteria, act by destabilizing the oxygen-hemoglobin complex, allowing oxygen itself to oxidize the iron to the Fe^{3+} state. Methylene blue accomplishes oxidation by acting as a hydrogen donor in the presence of molecular oxygen. In contrast, indirect action occurs with aniline, nitrobenzene, and other amino-, nitro-, and aryl-compounds. These chemicals are active *in vivo* in the indirect formation of methemoglobin.

Table 3.3

Chemicals That Can Cause Methemoglobinemia

Acetanilide	Naphthylamine
Aminophenol	Nitrates
Ammonium nitrate	Nitrobenzene
Amyl nitrite	Nitroglycerin
Aniline	Nitrophenol
Arsine	p-Aminophenol
Bismuth subnitrate	p-Bromoaniline
Bromates	p-Nitroaniline
Chloronitrobenzene	Phenylenediamine
Dimethylamine	Phenylhydrazine
Dinitrophenol	Potassium chlorate
Dinitrotoluene	Resorcinol
Hydroquinone	Sulfonamide
Hydroxylamine	Toluidine
Methylene blue	Trinitrotoluene

Almost all methemoglobin-generating chemicals have additional toxic side effects. For example, inorganic nitrites and organic aliphatic nitrite or nitrate chemicals not only cause methemoglobin formation, but also vasodilation. However, it is generally agreed that a reduction in the circulating titer of abnormal pigment is a desirable therapeutic goal. If the methemoglobin is contained within intact and functional erythrocytes, the intravenous administration of methylene blue usually evokes a dramatic response. Ascorbic acid also appears to be capable of reducing methemoglobin to hemoglobin and is useful to those persons genetically deficient in glucose-6-phosphate dehydrogenase. Glucose-6-phosphate dehydrogenase deficiency is a biochemical genetic condition involving the red blood cells. This genetic deficiency is a sex-linked trait, occurring homozygously only in males and present to some degree in virtually all racial groups. The prevalence rates for this trait are shown in Table 3.4.

F. Porphyria

Porphyrias are a group of disorders characterized by abnormalities in the heme biosynthetic pathway that result in the abnormal accumulation of heme precursors. Although many porphyrias are genetic disorders of enzymatic activity, acquired porphyria has been observed following exposure to various chemicals. In addition to aluminum and lead, other industrial chemicals have been identified to induce toxic porphyrias. These chemicals include chlorophenols, dioxin, hexachlorobenzene, and vinyl chloride.

Table 3.4
Prevalence of Glucose-6-Phosphate Dehydrogenase Deficiency

Racial Group	Prevalence (%)
African Americans	16.0
Filipinos	12.0-13.0
Greeks	2.0-32.0
Chinese	2.0-5.0
Jews	1.0-11.0
Scandinavians	1.0-8.0
Asian Indians	0.3
White Americans	0.1
British	0.1

Source: Office of Technology Assessment, United States Congress (1983).

G. Cyanide Poisoning

Cyanide inhibits cytochrome oxidase. As a result, oxidative metabolism and phosphorylation are compromised. Electron transfer from cytochrome oxidase to molecular oxygen is blocked, peripheral tissue oxygen tensions rise, and the dissociation gradient for oxyhemoglobin decreases.

The treatment of acute cyanide poisoning involves, as the initial step, the generation of a safe level of methemoglobin. This can be accomplished by the administration of amyl nitrite by inhalation and the intravenous injection of sodium nitrite. Free cyanide then combines with methemoglobin to form *cyanmethemoglobin*, which cannot transport oxygen. Although very stable, cyanmethemoglobin will eventually dissociate to yield free cyanide. Therefore, the second step of treatment involves the intravenous administration of sodium thiosulfate, which mediates the conversion of cyanide to the much less toxic thiocyanate that is readily excreted in urine. Methemoglobin is restored endogenously to functional blood pigment by the intracellular reductase system.

H. Hydrosulfide Poisoning

The hydrosulfide anion (HS⁻) is as potent an inhibitor of cytochrome oxidase as the cyanide anion, and the signs and symptoms of hydrosulfide poisoning are similar to cyanide poisoning. The treatment for hydrosulfide poisoning is the same as that used for cyanide poisoning. The hydrosulfide anion also forms a complex with methemoglobin known as sulfmethemoglobin, except that no further treatment is needed to degrade the sulfmethemoglobin. Because of its ability to react with disulfide bonds under physiologic conditions, the hydrosulfide anion can be inactivated by oxidized glutathione and other simple sulfides. The sulfide, so generated *in vivo*, may be metabolized to sulfates.

VII. TOXIC RESPONSES OF THE LIVER

The liver is the second largest organ of the body after the skin and is the most important organ for the detoxification of foreign chemicals. It is the first organ to be exposed to potentially toxic chemicals absorbed via the stomach and intestines, and is the major organ for the detoxification of such chemicals. The liver is also important in the detoxification of chemicals absorbed into the bloodstream by other routes such as the skin and lungs.

Despite its ability to detoxify many chemical agents, toxic injury to the liver (*hepatotoxicity*) is one of the most common manifestations of industrial/occupational toxicity. Many solvents, degreasing agents, heavy metals, and dyes can induce liver damage in exposed workers. A variety of pharmaceutical agents and dietary foodstuffs are also hepatotoxic and can enhance the toxicity of occupational agents. Hepatotoxicity is a broad term as chemical injury to the liver occurs in many forms and by many mechanisms. Distinct anatomical and clinical forms of hepatotoxicity occur with different agents. Several forms of hepatic injury are discussed below and may occur upon exposure to certain industrial chemicals (Table 3.5).

A. Fatty Liver

Fatty liver (*steatosis*) is an excess accumulation of lipid in the parenchymal tissue. Microscopically, the hepatocytes in fatty liver appear to be filled with numerous lipid droplets that push aside other cellular components. Fatty liver can occur through excess fat synthesis, decreased fat degradation, or impaired fat secretion. It is a common and reversible acute toxic response to many hepatotoxicants such as carbon tetrachloride and ethanol.

B. Hepatic Death

Death of hepatocytes is the end result of excessive or prolonged exposure to hepatotoxic agents. It is often preceded by fatty liver. The mechanisms of hepatocyte death vary with the toxic agent, but a common factor appears to be the reduction of cellular energy (ATP) levels. This leads to the inactivation of many ATP-dependent functions of the cell needed to maintain viability, including ion and water balance. This in turn results in the accumulation of excess calcium ions and water in the cells, inactivation of cellular enzymes, and swelling and bursting of cellular components and the other cell membranes. When the latter process occurs, the cell is dead.

Table 3.5
Chemicals and Liver Injury

Liver Injury	Chemical
Fatty liver	Bromobenzene Carbon tetrachloride Ethanol Valproic acid Yellow phosphorus
Hepatocyte death	Bromobenzene Carbon tetrachloride Ethanol Trichloroethylene Trinitrotoluene
Hepatitis	Isoniazid Nitrofurantoin
Cholestasis	Dichloroethylene Estrogens Ethanol Manganese Methylene dianiline Organic arsenicals
Cirrhosis	Ethanol Methotrexate
Blood vessel disorders	Arsenic Decarbazine Microcystin
Liver cancer	Aflatoxin Arsenic Thorium dioxide Vinyl chloride

C. Hepatitis

Hepatitis, or liver inflammation, can occur in response to chemically induced hepatic cell injury, or as a result of viral infection. It is characterized by the activation of Kupffer cells and the presence of other inflammatory and immune system cells such as neutrophils, macrophages, lymphocytes, and plasma cells. These cells function to remove cellular debris, destroy foreign agents, produce antibodies, and repair damaged tissues. Many of these cells produce toxic oxygen-free radicals as a means to destroy foreign organisms, but at excessive levels, can also damage parenchymal cells.

D. Cholestasis

Cholestasis is a reduction of bile formation or impaired secretion of specific bile components. It is not necessarily associated with hepatocyte death, but may be the result of hepatocellular injury or death. Cholestasis can also occur following injury to bile duct cells or blockage of bile ducts. A number of toxic agents can induce cholestasis through these mechanisms. Methylene dianiline is one agent that injures bile duct cells and causes cholestasis. Associated with cholestasis is a phenomenon known as jaundice in which the skin and eyes of the affected individual appear yellowish due to the accumulation of bilirubin. Normally, the liver degrades hemoglobin, the oxygen-carrying pigment of red blood cells, to a yellowish compound called bilirubin which is excreted with the bile. When the secretion or flow is reduced, bilirubin accumulates in blood and tissues, resulting in jaundice.

E. Cirrhosis

Cirrhosis, or liver scarring, is the result of chronic toxic injury or inflammation of the liver parenchyma in which the damaged tissue is replaced with fibrous scar tissue. Repeated exposure to hepatotoxic chemicals or chronic inflammation of the liver can lead to cirrhosis. The scar tissues appear as tough fibrous bands and may reduce blood and bile flow and greatly impairs the metabolic capacity of the liver. Cirrhosis is a common occurrence in alcoholics and is irreversible. It often results in liver failure and death of the afflicted individual.

F. Blood Vessel Disorders

Certain industrial chemicals can directly damage liver endothelial cells, impairing blood flow and affecting many hepatic functions. Dacarbazine, a cancer chemotherapeutic drug, is more toxic to the liver endothelial cells than the hepatocytes. Damage to the liver endothelial cells can cause hepatocytes to swell and block blood flow.

G. Liver Cancer

Many industrial chemicals have been found to induce liver cancer. While many of these chemicals do so in experimental conditions in the laboratory, agents such as aflatoxins and alcohol are clearly liver carcinogens in man. The types of cancers that arise in the liver are specific to the carcinogenic agent. Most hepatic cancers develop from hepatocytes and are known as hepatocellular carcinomas. However, some carcinogens cause the formation of liver tumors from other cells. Vinyl chloride and arsenic, for example, induce tumors from endothelial cells. Thorium dioxide, also known as Thorotrast, was used from 1920 to 1950 as a radioactive contrast agent. This agent induces tumors from endothelial cells, hepatocytes, and biliary cells.

VIII. TOXIC RESPONSES OF THE KIDNEY

The kidneys are complex organs which perform many functions vital to maintenance of homeostasis, including excretion of wastes; regulation of extracellular fluid volume and electrolyte concentration; conversion of inactive vitamin D to its active form; and the systemic release of erythropoietin, renin, vasoactive prostaglandins, and kinins. Despite their relatively small size, the kidneys receive 25% of the cardiac output and thus are potentially exposed to large doses of toxic materials. Because of its function, osmotic gradients that develop in the kidney can concentrate toxic substances to levels far in excess of those found in other organs. Since the kidney is capable of acidifying the urine, various solutes may occur in ionic forms. This combination of factors may explain why the kidney is affected by a variety of toxic chemicals. A number of toxic chemicals in the occupational environment can cause acute or chronic renal dysfunction. These chemicals vary in their effects and fall into four principal categories: heavy metals, halogenated hydrocarbons, pesticides, and miscellaneous chemicals. Renal toxicity seldom occurs as an isolated finding in occupational toxicology, but rather in conjunction with other systemic consequences of toxic exposure.

A. Heavy Metals

Exposure to most heavy metals results in renal toxicity. Acute renal dysfunction may be observed within hours to a few days following relatively high-dose exposure to certain divalent heavy metals. A potent renal toxin is mercury, which can be introduced into the body in three forms: elemental mercury, inorganic mercury, and organic mercury. All forms of mercury are nephrotoxic. The mechanism involved in mercury toxicity primarily involves combination with sulfhydryl groups and inhibition of oxidative enzyme systems. Cadmium toxicity results in enhanced synthesis within the liver of the metal binding protein, metallothioncin, which in turn enhances cadmium nephrotoxicity. Renal toxicity has also been observed with chromium, arsenic, gold, iron, antimony, thallium, and lead.

B. Halogenated Hydrocarbons

Generally, halogenated hydrocarbons such as carbon tetrachloride, chloroform, ethylene chlorohydrin, ethylene dichloride, tetrachloroethane, tetrachloroethylene, and vinylidene chloride are both hepatotoxic and nephrotoxic, although considerable variability of the nephrotoxicity is seen due to species, sex, and strain differences. Production of nephrotoxic halogenated hydrocarbon metabolites occurs through a number of mechanisms. They may be produced directly by the kidney, or produced in the liver and transported to the kidney with resultant damage. A non-nephrotoxic metabolite may be produced in the liver and then transported to the kidney, where further biotransformation results in the production of the ultimate nephrotoxin.

Other chemicals that can cause acute renal dysfunction include some miscellaneous hydrocarbons such as dioxane, toluene, phenol, pentachlorophenol, dinitrophenol, dinitro-*o*-cresol, cellosolve, methylcellosolve, and butylcellosolve. Exposure to arsine and ingestion of yellow phosphorus are also known to cause acute renal dysfunction.

IX. TOXIC RESPONSES OF THE SKIN

The skin is the largest organ of the body and acts as an interface between the external and internal environments. The skin functions as a barrier to external chemical agents and in maintaining the body's internal composition. With its dead outer layer, the skin is an excellent permeability barrier to certain chemical agents, but readily allows the entry of others. For many chemicals, the skin is the major route of entry. The rate of absorption through the skin is an important determinant in the toxicity of many hazardous chemicals.

The toxic responses that occur in the skin are dependent not only upon the chemical nature of the agent, but also upon the site of contact and method of application. Work-related skin diseases account for about one-fifth of all cases of occupational disease in the United States. Of all the reported cases, contact dermatitis is the most common skin disorder. Four-fifths of these cases are associated with exposure to irritating chemicals such as solvents, cutting oils, detergents, alkalis, and acids. Ultraviolet light may react with certain chemicals such as coal tar and creosote to cause irritant contact dermatitis at sites of exposure to the chemical and light. The rest of the cases can be related to a specific contact sensitizer such as epoxy resins, chromium, plant resins, and many others.

A. Acute Irritant Contact Dermatitis

A cutaneous irritant is an agent that produces an inflammatory response in the skin at the site of contact by direct action without the involvement of an immunologic mechanism. In man, cutaneous irritation results in *erythema* (reddening), *edema* (swelling), *vesiculation* (blistering), *scaling*, and *thickening* of the epidermis. Acute irritation is a local, reversible inflammatory response of the skin to direct injury by a single application of a corrosive chemical without the involvement of the immune system. Acid burns are characterized by immediate damage and early onset of resolution.

Alkali burns tend to cause injury over a number of hours subsequent to contact and thus the initial extent of the injury may underestimate the final outcome.

B. Chronic Irritant Contact Dermatitis

Chronic irritant contact dermatitis is probably the most common occupational skin disease. Workers appear to experience a change in the irritability of their skin after weeks or even years of exposure to a mild irritant, e.g., detergent, soap, solvent, cutting oil, and others. The disorder often develops gradually. Once the eruption becomes severe enough for the worker to seek treatment, it tends to persist. Though many industrial workers experience desiccation or "chapping" of the skin, the disorder that leads to disability appears to be different. It is more severe and persists to a varying degree after further irritant exposure ends.

C. Allergic Contact Dermatitis

Allergic contact dermatitis is the result of a cell-mediated immune response. This form of contact dermatitis is important because of the specificity of the response and the low amounts of allergic material that may be necessary to elicit a response. Generally, the development of allergic contact dermatitis is characterized by three stages. After the initial exposure to a potentially allergic compound, there is a refractory period of a few days or longer in which sensitization does not take place. After this period, the induction period occurs wherein sensitivity to the antigen develops. This may require 10 to 21 days. After the individual has become sensitive to the antigen, re-exposure will result in an allergic reaction following a characteristic delay of 12 to 48 hours. Once an allergic reaction has occurred, the individual may remain sensitive to the antigen for the remainder of his or her lifetime.

Chemicals that can cause allergic contact dermatitis are generally of low molecular weight. There is a great range in the antigenic potency of such chemicals. Strong allergens are usually lipid-soluble, protein reactive, aromatic compounds with molecular weights less than 500. Some important occupational contact allergens include metals (chromium, cobalt, nickel, and organomercurials), rubber additives (p-phenylenediamine, resorcinol monobenzoate, and thiuram sulfides), epoxy oligomer, methyl methacrylate and other acrylic monomers, phenolic compounds, aliphatic amines, and formaldehyde. There is a great diversity between individuals in the ability of these and other substances to elicit allergic responses. The phenomenon of cross-sensitization may also occur for two or more chemicals. This is the process whereby an individual who is sensitized to one chemical is also sensitive to a second because of its similar chemical structure.

D. Photosensitization

Photosensitization is an abnormal adverse reaction of the skin to ultraviolet (UV) and/or visible light. The two most important forms of photosensitization are *phototoxicity* and *photoallergy*. Phototoxicity is a chemically induced increase in reactivity of a tissue after exposure to UV and/or visible radiation through a nonimmunologic mechanism. Phototoxic responses are dependent on the nature and concentration of the chemical in the skin and the amount of light exposure. They may occur after contact, ingestion, or injection of causal agents. The skin and eyes are the major organs affected. Pathologic changes may include swelling, redness, and blistering. Hyperpigmentation may also occur after a reaction. Phototoxic agents include polyaromatic hydrocarbons (anthracene, acridine, and phenanthrene), tetracyclines, phenothiazides, and furocoumarins such as 8-methoxypsoralen, which is used to treat psoriasis.

Photoallergic reactions result from cell-mediated immune reactions similar to allergic contact dermatitis. They can also spread to areas of the skin not exposed to light. However, the vast majority of these reactions result from topically administered agents interacting with UV light. The role of light in photoallergy is believed to be involved in the conversion of a chemical to an allergic form. Two mechanisms have been proposed. Radiation may chemically change a photosensitizer into an antigenic form. An example of this would be sulfanilamide, which is converted by UV light to the potent allergen *p*-hydroxyaminobenzene sulfonamide. Alternatively, the light-excited photosensitizer may react with proteins, producing an antigenic photosensitizer-protein product. Some reported photoallergens are halogenated salicylanilides, sulfonamides, coumarin derivatives, sunscreen components (glycerol *p*-aminobenzoic acid), and several plant products. In general, photoallergens have little phototoxicity.

E. Occupational Vitiligo

Depigmentation of the skin at the site of contact with a depigmenting chemical usually results from inhibition of melanin biosynthesis. In most instances, cessation of exposure results in slow repigmentation. Allergic contact sensitivity to the depigmenting chemical occasionally occurs, in which event widespread persistent depigmentation may occur outside the areas of original contact. Table 3.6 lists the industrial chemicals known to be associated with work-induced vitiligo. All are antioxidants used in the manufacture of such products as rubber, plastic, and soluble oils.

Table 3.6

Chemicals That Can Cause Occupational Vitiligo

p-t-Amyl phenol
o-Benzyl-p-chlorophenol
4-t-Butyl catechol
4-t-Butyl phenol
p-Cresol
Hydroquinone
Hydroquinone monobenzyl ether
Hydroquinone monomethyl ether
o-Phenyl phenol

F. Chemical Acne and Chloracne

A number of chemicals can produce conditions similar to those seen in acne vulgaris, the typical form of acne. These include greases and oils, coal tar pitch, creosote, and a number of cosmetic preparations. Systemically administered drugs including iodides, bromides, and isoniazid can also cause acne.

Chloracne is a more specific and severe form of acne. It is characterized by small, strawberry-colored cysts and comedones (blackheads). Inflammatory pustules and abscesses may also occur. The most sensitive areas of the skin are below and to the outer side of the eye and behind the ear. The eruption may be limited to these areas or may involve nonexposed areas, especially the scrotum in males. It may continue well after initial exposure to the inciting agent because of the release of the chemical from body stores. Agents that cause chloracne include polyhalogenated dibenzofurans, polychlorinated dibenzodioxins (including 2,3,7,8-tetrachlorodibenzo-p-dioxin or TCDD), polychloronapthalenes, polyhalogenated biphenyls (PCBs and PBBs), and polychlorinated azoxybenzenes. Many of these substances are byproducts during the manufacture of other polychlorinated substances. TCDD is the most potent inducer of chloracne. The development of chloracne following exposure to PCBs and PBBs is probably due to contaminants such as tri- and tetrachlorodibenzofurans (TCDFs) rather than the PCBs or PBBs themselves.

G. Other Toxic Skin Reactions

A wide variety of other toxic skin reactions may occur in addition to those already discussed.

- *Physical dermatitis* is produced by fiberglass and other fibers and results in an intense pruritic (itching) reaction and pinpoint-sized reddened papules. The development of this condition is directly related to the fiber diameter (which must be greater than 4.5 mm) and inversely related to fiber length.

- *Urticarial reactions* (wheal-and-flare reactions) are produced within 30 to 60 minutes of exposure to a variety of agents. They are caused by a variety of chemicals found in plants (nettles), animals (caterpillars and jellyfish), and other sources. These reactions are thought to be due to the release of histamine and other substances from immune cells and are characterized by rapidly occurring intense burning, itching, and redness.

- *Cutaneous granuloams* usually appear as slightly reddened, more or less flesh-colored papules that may be clustered and associated with inflammatory changes. They are generally localized to areas of foreign substance contact. They result from mononuclear cells attempting to "wall off" poorly soluble foreign body and antigenic substances in the dermis. These include talc and silica and certain metals such as beryllium, chromium, and zirconium salts.

- *Hair damage and alopecia* may occur through direct effects on the external hair or through damage to the hair-producing cells of the hair follicle. Alkali, thioglycolates, and oxidizing agents such as peroxides and perborates can dissolve hair keratin and cause softening, matting, and increased fragility of the hair. Growth inhibitory agents such as alkylating agents and antimetabolites may impede the growth of follicular cells and the production of hair keratin. Dyes (indigo), metals (copper and cobalt), and acids (picric acid) may change the color of the hair upon contact. Some chemicals such as thallium, phenyl glycidyl, and dixyrazine can cause hair loss.

X. TOXIC RESPONSES OF THE EYE

Recognition of the toxic effects of chemicals and protection from those that may be splashed into the eyes are vital to prevent visual damage. The ready availability of facilities for cleansing and irrigation of the face and eyes in the workplace is of the utmost importance, since initial steps for treatment of chemical burns, especially those due to strong alkalis and acids, must be carried out immediately by the victim, fellow workers, or anyone else in close proximity.

A. Chemical Burns

Strong alkalis and acids cause the most severe and damaging chemical injuries that can be sustained by the eye. Alkali burns are commonly due to sodium and potassium hydroxide used as cleaning agents, to calcium hydroxide used in mason's mortar and plaster, and to anhydrous ammonia used in fertilizer. Battery acids and the strong acids used to clean metal in the electroplating industry are also common causes of severe eye injury.

Alkalis affect the lipid in cell membranes and thereby reduce the normal barriers to diffusion. This allows the chemical to penetrate rapidly the interior of the eye. Because alkalis are not quickly neutralized by tissue, their destructive action can continue for hours if they are not diluted and removed immediately by irrigation of the eye. In contrast, acids tend to be fixed by protein in tissues, and this neutralizes them in a relatively shorter period of time and keeps them from penetrating as deeply.

B. Cataract

Normal lenses in eyes are transparent permitting, light to pass through and allowing it to be focused on the retina. Any opacity in the lens is called a cataract, which can be caused by a variety of unrelated circumstances. One form of cataract is known as toxic cataract, which is chemically induced by exposure to, or ingestion of, certain chemicals.

In addition to a variety of toxic effects, systemic administration of 2,4-dinitrophenol causes cataracts in some individuals. This occurred in the 1930s when the chemical was introduced as an anti-obesity agent and was sold without prescription. Several hundred human cataracts resulted.

The soluble salts of thallium acetate and thallium sulfate have been used as insecticides, as rodenticides, and, at one time, as a systemic or topical depilatory agent. Ingestion or application causes a variety of toxic symptoms, including in some rare instances, cataracts.

At the present, the most common cause of chemical-induced cataracts seems to be prolonged use of corticosteroids, either topical or systemic. At least two years of moderate to high doses are usually necessary to produce cataract.

Table 3.7

Chemicals and Cardiovascular Injury

Cardiovascular Injury	Chemical
Cardiac arrhythmia	Arsenic
	Carbamate insecticides
	Chlorinated hydrocarbons
	Chlorofluorocarbons
	Organophosphates
Coronary artery disease	Carbon disulfide
	Carbon monoxide
Hypertension	Cadmium
	Carbon disulfide
	Lead
Ischemic heart disease (Nonathero-matous)	Ethylene glycol dinitrate
	Nitroglycerin
Myocardial injury	Antimony
	Arsenic
	Arsine
	Cobalt
	Lead
Peripheral arterial occlusive disease	Arsenic
	Lead

XI. TOXIC RESPONSES OF THE CARDIOVASCULAR SYSTEM

Heart disease and stroke cause the majority of deaths in the United States. The major risk factors for coronary heart disease—family history, hypertension, diabetes, lipid abnormality, and cigarette smoking—explain only a small portion of the cases. Other factors such as stress and exposure to occupational or environmental toxic chemicals are believed to contribute to the development of heart disease, though the magnitude of the risk is unknown. The types and possible toxic causes of cardiovascular disease are shown in Table 3.7. Massive exposure may occur, such as in acute carbon monoxide poisoning, but toxic cardiovascular disease is usually the result of chronic low-level exposures.

A. Carbon Disulfide

Carbon disulfide is a solvent widely used in the manufacture of carbon tetrachloride and ammonium salts, and as a degreasing solvent. It is also the solvent used in the rubber and viscose rayon industries. Chronic exposure to carbon disulfide is one of the best-documented toxic causes for accelerated atherosclerosis and coronary heart disease. Epidemiologic studies have indicated that there is a 2.5- to 5-fold increase in the risk of death from coronary heart disease in workers exposed to carbon disulfide.

B. Carbon Monoxide

Carbon monoxide is the most widely distributed of all industrial toxic chemicals and accounts for the greatest number of intoxications and deaths. It is formed wherever combustion engines or other types of combustion are present. Acute intoxication with carbon monoxide can cause myocardial infarction or sudden death. Chronic high-level carbon monoxide exposure may result in congestive cardiomyopathy. The solvent methylene chloride is metabolized within the body to carbon monoxide.

C. Organic Nitrates

In the 1950s, an epidemic of sudden death in 30 young munitions workers who hand-packed cartridges was observed. It was subsequently discovered that abrupt withdrawal from excessive exposure to organic nitrates, particularly nitroglycerin and ethylene glycol dinitrate, may result in myocardial ischemia even in the absence of coronary artery disease. Occupations in which workers may be exposed to organic nitrates include explosives manufacturing, construction work involving blasting, weapons handling in the armed forces, and pharmaceutical manufacturing of nitrates.

D. Halogenated Hydrocarbons

Exposure to various solvents and propellants may result in cardiac arrhythmia, syncope with resultant accidents at work, or sudden death. Most serious cases of arrhythmia have been associated with abuse of, or industrial exposure to, halogenated hydrocarbon solvents, e.g., 1,1,1-trichloroethane and trichloroethylene, or exposure to chlorofluorocarbon (freon) propellants. Nonhalogenated solvents, and even ethanol, present similar risks.

E. Organophosphate and Carbamate Insecticides

Intoxication with organophosphate and carbamate insecticides can produce diverse cardiovascular disturbances, including tachycardia and hypertension, bradycardia and hypotension, heart block, and ventricular tachycardia. Acute phosphate and carbamate insecticide poisoning affects the circulatory system and may be fatal. Chronic poisoning may cause neuropsychiatric disturbances.

F. Heavy Metals

A number of heavy metals have been associated with disturbances in cardiovascular function. Electrocardiographic abnormalities have been observed in workers exposed to antimony. Although these changes usually resolve after removal from exposure, a few studies have reported increased cardiovascular mortality rates in exposed workers.

Acute arsenic poisoning can cause electrocardiographic abnormalities. Subacute arsenic poisoning caused by ingestion of arsenic-contaminated beer has been associated with cardiomyopathy and cardiac failure. Chronic arsenic poisoning has been reported to produce *blackfoot disease*. Arsine gas causes red blood cell hemolysis. Massive hemolysis produces hyperkalemia, which can result in cardiac arrest. Arsine may also directly affect the myocardium, causing a greater magnitude of cardiac failure than would be expected from the degree of anemia.

In Quebec City, Canada, in 1965 and 1966, an epidemic of cardiomyopathy occurred in heavy drinkers of beer to which cobalt sulfate had been added as a foam stabilizer. Cobalt is known to depress oxygen uptake by mitochondria of the heart and to interfere with energy metabolism in a manner biochemically similar to the effects of thiamine deficiency. Since individuals receiving higher doses of cobalt for therapeutic reasons have not developed cardiomyopathy, perhaps cobalt, excessive alcohol consumption, and nutritional deprivation acted synergistically to produce cardiomyopathy in this epidemic.

XII. TOXIC RESPONSES OF THE NERVOUS SYSTEM

Neurologic dysfunction, neuropathies, and encephalopathies can be caused by exposure to a wide variety of chemicals in the workplace. Among these agents are heavy metals (e.g., arsenic, lead, manganese, mercury, thallium, and others), solvents and other organic chemicals (e.g., acrylamide, carbon disulfide, *n*-hexane, methyl *n*-butyl ketone, perchloroethylene, toluene, trichloroethylene, and vinyl chloride), noxious gases (e.g., carbon monoxide, ethylene oxide, and nitrous oxide), and organophosphates.

A. Heavy Metals

Acute arsenic exposure usually consists of accidental ingestion. If an individual survives the acute syndrome, neuropathy usually appears 1 to 3 weeks later, which is characterized by axonal degeneration with occasional demyelination. Chronic exposure to arsenic may also lead to neuropathy but it evolves more slowly than in acute cases.

Acute exposure to lead, occurring almost exclusively by ingestion in children with pica, causes encephalopathy with cerebral edema. Gastrointestinal, hematopoietic, and renal effects are well described. Chronic exposure, by ingestion or inhalation, results in peripheral motor neuropathy, which may affect the upper limbs more severely than the lower limbs and particularly involves muscles supplied by the radial nerve.

After prolonged inhalation exposure to manganese, from 6 months to over 20 years, encephalopathy (manganese madness) develops, with hallucinations, emotional instability, and bizarre behavior. A few months after onset of these symptoms, generalized weakness, dysarthria, ataxia, tremor, impotence, and headache appear. Later, extrapyramidal features develop that are similar to those in Parkinson's syndrome.

Neurologic disease may follow exposure to both organic and inorganic mercury. Exposure to inorganic mercury is usually by inhalation and possibly by ingestion. Symptoms may develop after many years of exposure, particularly if exposure is infrequent and the dosage is low. The most characteristic feature of inorganic mercurialism is a coarse, rapid tremor that begins in the hands and may slowly spread to involve the eyelids, the tongue, the facial muscles, and the remainder of the extremities. There may also be muscular excitability, weakness, and personality changes such as those that gave rise to the expression "mad as a hatter" among workers in the felt hat industry. Chronic exposure to organic mercury causes mental impairment, tremor, ataxia, dysarthria, deafness, and constriction of visual fields.

Acute ingestion exposure to high doses of thallium causes encephalopathy with prominent ataxia. In severe cases, this may progress to psychosis, convulsions, and coma. Alopecia is a characteristic of thallium poisoning.

B. Organic Chemicals

The acrylamide monomer (not the polymer) is neurotoxic, and most instances of acrylamide intoxication occur by inhalation and skin absorption during its manufacture. Acute exposure causes an encephalopathy that is usually of brief duration. Peripheral neuropathy may develop both as a delayed manifestation of acute exposure (2-3 weeks later) and with chronic exposure. The earliest signs following acute exposure are numbness in the feet, accompanied by profuse sweating of the hands and feet, and occasionally urinary retention. This is accompanied by sensory loss as well as a generalized loss of reflexes. Effects of low-dose acrylamide exposure may be delayed for up to 3 years.

Acute inhalation exposure to carbon disulfide in high concentrations produces narcosis. Subacute inhalation exposure produces encephalopathy manifested mainly as dramatic psychologic disturbance with mania, profound depression, and hallucination. Attacks on relatives and co-workers and attempted suicides have been reported. Chronic inhalation exposure to low concentrations produces sensory and motor dysfunctions.

n-Hexane and methyl n-butyl ketone (MnBK) are metabolized to 2,5-hexanedione, which is responsible for most, if not all, of the neurotoxic effects. Acute inhalation exposure to n-hexane and MnBK produces euphoria, which leads to abuse of these chemicals by inhalation ("glue-sniffing"). Chronic inhalation exposure causes distal axonal neuropathy, which results in sensory and peripheral dysfunction.

Acute inhalation exposure to perchloroethylene in high concentrations produces symptoms of acute respiratory depression which may result in death. Irreversible encephalopathy, personality change, and dementia can result from inhalation exposure at low concentrations.

Toluene, like n-hexane, has become an abused chemical because of the euphoria that results from acute inhalation exposure. Chronic brain dysfunction occurs in toluene-sniffers. Visual loss from optic atrophy has been reported.

Acute inhalation exposure to trichloroethylene in high concentrations produces neuropathy that involves both the sensory and motor functions. Sensory loss due to trichloroethylene occurs in a peculiar circumstantial distribution on the face and involving the optic and facial nerves. Chronic inhalation exposure may cause an encephalopathy characterized by neuropsychologic disturbances and sometimes overt dementia, visual impairment, and tremor.

Acute inhalation exposure to vinyl chloride produces encephalopathy characterized by euphoria and, later, depression of consciousness. With repeated exposure, neuropsychologic effects may occur, such as headache, insomnia, difficulty in concentrating, and memory loss.

C. Noxious Gases

Acute inhalation exposure to carbon monoxide results in encephalopathy characterized by headache, confusion, and somnolence, leading to coma and death if exposure continues. Recovery may be incomplete, leaving the individual with diffuse hypoxic encephalopathy or focal deficits. Occasionally, following apparently good recovery, there is a delayed deterioration.

There are reports about the neurotoxicity of ethylene oxide, used extensively as a sterilizing agent for heat-sensitive biomedical materials. The most prominent effect of inhalation exposure is on peripheral nerves, with distal sensory loss and some weakness. Encephalopathy may accompany the neuropathy or may occur independently.

Acute inhalation exposure of nitrous oxide in high concentrations causes encephalopathy, but recovery is complete even after multiple exposures. Repeated exposure on a daily basis for many months results in a neuropathy resembling that associated with vitamin B_{12} deficiency.

D. Organophosphates

The organophosphates are one of the few classes of neurotoxins in which the target enzymes are known. These compounds inhibit both acetylcholinesterase, which explains the central nervous system and neuromuscular toxicity, and neuropathy target esterase, an enzyme that is inhibited in neuropathy. Acute poisoning by inhalation, ingestion, or skin absorption produces a cholinergic crisis characterized by encephalopathy in which convulsions, muscle twitching, weakness, lacrimation, and involuntary defecation and urination are common. In milder cases, there is nausea, vomiting, and diarrhea lasting from a few hours to days. The severity of these cholinergic features depends on the severity of exposure and the particular organophosphate. With some organophosphates, if the acute episode is survived, a delayed neuropathy develops. There is little relationship between the severity of the acute effects and subsequent development of neuropathy, which in fact may develop in the absence of obvious acute effects.

XIII. TOXIC RESPONSES OF THE IMMUNE SYSTEM

Immune hypersensitivity reactions play an important part in many occupational health disorders in the workplace. In 1700, Benardino Ramazzini, considered the founder of modern occupational medicine, reported that, after repeated exposures to flour dust, bakers often developed respiratory problems, a disease now called *baker's asthma*. Today, many occupational health disorders have been shown to be caused by immune reactions to environmental factors in the workplace. The most common immune hypersensitivity occupational disorders include allergic asthma or rhinitis, hypersensitivity pneumonitis, and allergic contact dermatitis. The reactions are dependent on the host, the duration, the degree and type of sensitization, and the antigen. In addition, there are chemicals that can induce occupational immune hypersensitivity disorders.

A. Allergic Asthma and Allergic Rhinitis

Allergic asthma and allergic rhinitis occur when sensitized workers inhale specific antigens. In general, atopic individuals are predisposed to sensitization to large-molecular-weight inhalants (proteins) such as animal danders, pollens, and house dust. The reactions occur in three patterns: **(1)** the immediate onset of symptoms after the inhalation of antigen; **(2)** a late onset reaction occurring 3 to 6 hours after exposure, often after a worker has returned to the home environment, and lasting in some cases 12 to 36 hours; **(3)** and an immediate, followed by a late onset response.

B. Hypersensitivity Pneumonitis

Hypersensitivity pneumonitis is a pulmonary disease resulting from sensitization and subsequent exposure to a variety of inhalant organic dusts. Sensitization to bacterial products, small amounts of serum present in the excreta of animals, thermophilic actinomycetes, fungi, and vegetable proteins has produced hypersensitivity pneumonitis. Examples include pigeon breeder's disease, farmer's lung, humidifier lung, and bagassosis. The incidence varies with the type and frequency of antigen exposure.

C. Allergic Contact Dermatitis

After a latent period of about 2 weeks following exposure to the sensitizing chemical, which binds to a protein component in the skin to become a complete antigen, the body develops a cell-mediated immune response. Subsequent re-exposure leads to a lymphocyte-mediated inflammatory reaction in the skin associated with development of edema of the epidermis with collection of fluid between epidermal cells and development of vesicles or even bullae. This response is specific for the given allergen, though cross-reactions to chemically similar substances occasionally occur. The immune response, once acquired, tends to be permanent. Much less exposure is required to elicit this allergic response than is usually necessary for initial induction of the immune process. Table 3.8 summarizes the many common causes of occupational allergic contact dermatitis.

Table 3.8

Allergic Contact Dermatitis and Occupational Groups

Chemical	Occupational Group
Acrylic monomers	Acrylic ink manufacturers
	Dental technicians
	Printers
Epoxy resins	Aircraft assembly workers
	Construction workers
	Electrical utility workers
	Electronics workers
	Painters
Formaldehyde	Embalmers
	Insulation workers
	Resin manufacturers
	Woodworkers
Heavy metals (Co, Cr, Ni, Pt)	Cement workers
	Electronics workers
	Hairdressers
	Jewelers
p-Phenylenediamine and related dyes	Hairdressers
	Rubber workers
Rubber accelerators (thiurams, mercaptobenzothiazole)	Tire builders
	Tire repairmen
	Workers wearing rubberized articles

D. Chemical Sensitizers

Workers in industrial plants may be exposed to a wide variety of chemical sensitizers. Two that have been extensively studied are toluene diisocyanates and trimellitic anhydride. Isocyanates are used in the manufacture of pesticides, polyurethane foams, and synthetic varnishes. There are many case reports of obstructive airway problems related to toluene diisocyanate (TDI). These occur with equal frequency in atopic and nonatopic workers. Trimellitic anhydride (TMA) is used in the manufacture of plastics, epoxy resins, and paints. TMA dust or fumes have been associated with four types of syndromes:

- The immediate reaction syndrome: The individual may have rhinitis, conjunctivitis, or asthma.

- The late reacting-systemic syndrome: This is characterized by cough, occasional wheezing, dyspnea, and systemic symptoms of malaise, chills, myalgia, and arthralgia.

- The pulmonary disease-anemia syndrome: This develops after exposure to TMA fumes. It occurs after repeated high-dose exposure to the volatile fumes of TMA sprayed on heated metal surfaces to prevent corrosion. Hemolytic anemia and respiratory failure are evident.

- The irritant respiratory syndrome: This occurs with the first high-dose exposure to TMA powder and fumes. Sensitized individuals develop cough and dyspnea.

XIV. TOXIC RESPONSES OF THE FEMALE REPRODUCTIVE SYSTEM

Occupational exposure to chemicals in human females may be hazardous to a wide range of reproductive processes. Altered fertility, low birth weight, spontaneous abortion, transplacental carcinogenesis, congenital malformations, mutagenesis, and developmental abnormalities are among the effects on reproduction that have been recognized to result from toxic occupational exposure. Over the past several decades there has been a threefold increase in the number of women employed in the work force in the United States, and many of these women have become employed in hazardous occupations, including those traditionally limited to men. In addition, during recent years an increasing number of women have remained in the workplace until near the end of pregnancy. Thus, many more women and their unborn children are being exposed to chemical hazards of the workplace.

Toxic chemicals may directly damage the female germ cells, i.e., sex cells in the ovaries, or induce adverse effects on developmental processes in the embryo and/or fetus. Injury to the germ cells in the ovary may lead to genetic abnormality which may last for generations, while embryonal and fetal damage usually lead to various deformities or mental retardation. There are at least three developmental stages during which the embryo and the fetus are at risk. The first stage begins with conception and includes the first 17 days following conception. During this period, the egg is fertilized to form a one-cell zygote; subsequent diversions form a ball of cells (32 to 256 cells) and these cells implant on the uterine wall. The second stage of embryonic development spans from 18 to 55 days following conception and is characterized by folding of the embryo and organogenesis, e.g., formation of all organ systems. The third stage, which is the fetal period, spans from 56 days to term and principally involves growth and maturation of existing structures.

An acute injury during the first period of development is almost always lethal. During this stage the embryo is composed of a small number of cells and injury to these cells usually results in cell death. If a lethal dose of a chemical is given to the mother, abortion or resorption of the embryo will occur. However, if the dose is not lethal to the embryo, then recovery follows usually with no structural deformity in the fetus. If injury occurs at the second stage of development, i.e., during organogenesis, then various deformities result. During this stage, different organs have their own critical period of development. therefore, this entire period is extremely sensitive. During the third stage, fully formed fetal organs are not prone to injury. Although minor malformations can be induced, the developing brain is prone to injury due to the continuous developmental process. It should be noted that development continues even after birth, particularly in the skeletal, muscular, and nervous systems. Consequently, dangerous chemical exposure during the third stage of development may also lead to fetal growth retardation.

Currently, the list of chemicals which are known to cause reproductive effects in females is quite short in comparison to approximately 1000 chemicals in laboratory animals.   This lack of information is due to the inherent difficulty in working with humans.

In females, about 45% of the reports on reproductive toxicity deal with *embryo toxicity* and *fetotoxicity*. Placental toxicity usually causes spontaneous abortions. Occupational exposure to chemicals may alter the integrated function of the hypothalamus-pituitary-ovarian-uterian axis, suggesting some relationship between the chemical exposures and impaired fertility, ovarian toxicity, abnormal hormone production, menstrual disturbance, and uterine toxicity.

Female reproductive toxicants can be broadly divided into two categories: direct acting and indirect acting. The direct-acting toxicants cause direct damage to the subcellular organelles and macromolecules within the cells (Table 3.10), while the indirect-acting toxicants alter the metabolic activity of a cell or cause hormonal imbalance (Table 3.11).

Occupational exposure to sex steroids and contraceptives have been associated with impaired ovarian function and disturbance in the menstrual cycle. Women employed in agricultural work are exposed to pesticides and herbicides. These chemicals contain organohalide and organophosphorus compounds. Both these compounds have been associated with impaired ovarian function, disturbance in menstrual cycle, and infertility. Occupational exposure to carbon disulfide in rayon fiber manufacturing plants is known to cause menstrual disturbance in female workers. Lead, cadmium, mercury, manganese, and tin have also been reported to cause abnormal menses and spontaneous abortions. Reports suggest that exposure to gasoline vapors in the industrial setting can alter ovarian functions. Menstrual disturbances were also noted in women exposed to benzene, chlorine, hexachlorocyclohexane, and chlorobenzene. Similar observations of altered menses were observed in women exposed to formaldehyde vapors and ethylene oxide. Women employed in plastic industries are exposed to a variety of chemicals including polymerizers, plasticizers, and vinyl chloride. These chemicals have also been reported to cause menstrual disturbances.

Table 3.10

Chemicals with Direct-Acting Female Reproductive Effects

Chemical	Effect	Mechanism
Alkylating agents	Altered menstruation Amenorrhea	Follicle toxicity
Cadmium	Follicular atresia Persistent diestrus	Vascular toxicity Direct toxicity
Halogenated hydrocarbons	Altered menstruation Amenorrhea Infertility Sterility	Estrogen agonists
Lead	Abnormal menstruation Decreased fertility Ovarian atrophy	Decreased FSH Decreased progesterone
Mercury	Abnormal menstruation	Follicle toxicity Granulosa cell proliferation

Table 3.11

Chemicals with Indirect-Acting Female Reproductive Effects

Chemical	Effect	Mechanism
Barbiturates	Increased steroid clearance	Enzyme induction
Cigarette smoke	Altered menstruation Impaired fertility Reduced age at menopause	Follicle destruction Blocked ovulation
Cytoxan	Amenorrhea Premature ovarian failure	Follicle destruction
DDT metabolites	Altered steroid metabolism	Enzyme induction
Halogenated hydrocarbons	Abnormal menstruation	Decreased FSH Decreased Leutinizing Hormone
Polycyclic aromatic hydrocarbons	Impaired fertility	Follicle destruction

Occupational exposure of chemicals may cause ovarian toxicity by directly destroying oocytes or the follicles, or may cause genetic mutations leading to developmental disorders or carcinogenesis. Depending on the site and mechanism of action of an ovarian toxin, a period of subfertility or infertility may occur.

A study was conducted on the relationship between paternal occupational exposures to six commonly used organic solvents, namely, styrene, toluene, xylene, tetrachloroethylene, tricholoethylene, and 1,1,1-trichloroethane, on spontaneous abortions and congenital malformations. Some indications of increased risk of spontaneous abortion was noted as a result of preconceptional paternal exposure to these solvents. However, due to the lack of a significant number of cases, these results were not conclusive.

XV. TOXIC RESPONSES OF THE MALE REPRODUCTIVE SYSTEM

Toxic chemicals can impair the male reproductive system at several sites, including neural, endocrine, gonadal, as well as accessory glands. They can also affect male sexual behavior. In the male reproductive system, a number of separate physiological processes must work in a harmonious fashion in order to maintain normal fertility. These physiological processes range from the central and autonomic nervous system to secretory functions of various glands and the endocrine regulatory system. Disturbances at any level may disrupt the normal reproductive capacity of a man. The impact of toxic chemicals on the testis may cause genetic disorders. It is known that recessive mutations in germ cells may be accumulated undetected for generations before being expressed.

Frequently, in toxicological studies, the male reproductive system is not examined in routine autopsy. Even when there are clear indications of the toxic nature of certain chemicals impairing fertility, studies are conducted mainly on females. This is due to an obvious lack of easily measurable reproductive indicators in the male system. For example, in pregnant women, the occurrence of a menstrual cycle may indicate spontaneous abortion or pregnancy wastage. Chemical agents may also appear in breast milk. Therefore, in many cases, breast milk can be regarded as an indicator of reproductive toxicity. Whereas in the male, semen analysis does not provide a definite clue of infertility. There is no good laboratory test for clinical evaluation of human male reproductive function. Testicular biopsy is not performed to investigate the effect of toxic chemicals. As a result, there is no means to study the effect of environmental pollutants on the testis except testing the end points, i.e., sperm production in semen. In the testis, unlike the ovaries, division of cells is a continuous process. Dividing cells are far more sensitive to any toxicant than cells at rest. Therefore, the testis is at greater risk.

The gonads play a crucial role in developing the reproductive capacity of a man. Adverse interactions of testicular cells with reproductive toxic chemicals may cause germ cell mutation, cell death, gonadal tumor, impaired hormone production, and loss of fertility. A number of chemicals are known to cause damage or have been suspected to cause damage to the male reproductive process (Table 3.12). However, this is a very short list.

Table 3.12

Chemicals with Male Reproductive Effects

Known Effects	Inconclusive Effects	No Effects
Carbon disulfide	Boron	Anesthetic gases
Dibromochloropropane	Cadmium	Epichlorohydrin
Lead	Carbaryl	Ethylene glycol monoethyl ether
	Dinitrotoluene	
	Ethylene dibromide	Formaldehyde
	Kepone	Glycerin
	Methylmercury	Polybrominated bi-phenyls
	Toluenediamine	

In 1977, the discovery of infertility and sterility among male employees exposed to dibromochloropropane (DBCP) is one example of the dangerous consequences of the lack of systematic investigation. DBCP is a liquid nematocidal agent that was widely used to save perennial crops without damaging the plant. In the United States, citrus, grapes, peaches, pineapple, soybeans, and tomatoes used to be treated with DBCP. The reproductive toxic effects of DBCP were not discovered by scientists and physicians, but the workers themselves noted the paucity of children of the men exposed to DBCP. After this problem was reported, a thorough investigation was conducted. In several separate studies, it was shown that occupational exposure to DBCP caused testicular damage. Testicular biopsies of men exposed to DBCP showed spermatogenic damage. It was suggested that DBCP directly damaged the spermatogonia. In some severely affected men, many seminiferous tubules had no germ cells. Some men even became permanently azoospermic.

A controversial example of male reproductive toxicant is 2,3,6,8-tetrachlorodibenzo-*p*-dioxin or TCDD. Exposure to TCDD present in Agent Orange by the Vietnam veterans remained unresolved. Although there are reports of increasing congenital abnormalities among children fathered by the exposed veterans compared to the children fathered by unexposed individuals, according to U.S. government reports, these data are not conclusive.

Dimethylformamide (DMF), a commonly used industrial solvent, is also used in the production of paint, artificial fibers, and drugs. It is estimated that over 100,000 workers in the United States are exposed to this chemical. Men exposed to DMF in the leather tanning industry and the aircraft industry have been reported to have testicular cancer. Testicular cancer is not rare among young adult men in the United States. In fact, the rate of testicular cancer appears to be on the increase in the past several decades. Further work is needed to verify these initial findings of the association of DMF with testicular cancer.

The adverse effect of metal on male reproduction has long been known. In fact, it dated back to ancient Rome, where lead in drinking water vessels of the upper class Romans was associated with the declining fertility. It was shown that lead had a direct effect on spermatogenesis, causing abnormal or low sperm production, thus causing male infertility. Lead also was reported to cause chromosomal abnormalities. Cadmium has also been known specifically for many years to cause testicular damage. Although the effects of cadmium on the human testis have been studied for the last 25 years by a large number of investigators, the results are inconclusive.

XVI. BIOLOGICAL MONITORING OF METAL EXPOSURE

Many metals, especially their ions, are playing a double role in the physiology of organisms: some are indispensable for normal life, while most of them are toxic at elevated concentrations, i.e., they adversely affect the activity and well-being of living organisms. Recent years have brought an increasing concern for the potential toxic effects of metals, metal ions, inorganic compounds, as well as organometallic compounds.

There are three important aspects of metal distribution in the body. The first is that, unlike organic molecules, metals are often not susceptible to metabolic detoxification mechanisms. The second aspect is the relative ease with which metals, their ions, as well as inorganic compounds, can circulate among body tissues and fluids, which allows them to settle among the tissues to which they are most tightly bound. The third aspect is the propensity of metals to partition quickly out of the blood into tissues, making it difficult in the evaluation of body burden.

Blood and urine levels of metal usually reflect recent exposure and correlate best with acute effects. An exception is beryllium, the concentration in blood and urine has been only qualitatively linked to exposure. Beryllium in urine may be detected for years after removal from exposure. Another exception is urine-cadmium where increased metal in urine reflects renal damage related to accumulation of cadmium in the kidney. Speciation of toxic metals in urine may provide diagnostic insights. For example, cadmium metallothionein in urine may be of greater toxicologic significance than cadmium chloride.

It has been known for many years that hair contains metals. This may be considered an excretory mechanism for certain metals. Since metals appear to have no functional role in the hair protein, their content in hair seems to vary with exposure. Hair, however, is not suitable as a dynamic system for evaluating past exposure to metals since the individual hair shafts do not have access throughout their length to any fluid transport system. Since hair grows at the rate of about 1 cm in 30 days, clippings would be expected to reflect a historical exposure at best.

The various aspects of metal toxicology are not clearly understood for all metals. As a matter of fact, the body of knowledge concerning the individual metals is extremely spotty. A fairly complete picture emerges only for a few metals, namely, lead, mercury, and cadmium. These metals have been studied most intensively for one reason or another. They serve as models. The following are biological monitoring profiles for metals of industrial significance.

A. Aluminum

Aluminum does not represent an important health hazard in industry. In a healthy person, aluminum appears to be poorly absorbed by the oral route and probably also through the respiratory tract. While the threat of aluminum exposure to healthy individuals has not been established conclusively, toxicity due to aluminum exposure in exposed individuals with impaired renal function has been widely documented. Observations made in these individuals suggest that the relationship between the concentration of aluminum in blood and urine has not been characterized.

B. Antimony

Antimony and its compounds are not essential to humans. Environment and nutrition influence antimony concentration in human tissues, causing large differences among individuals. Great variations of antimony content were found in blood, urine, and different tissues.

C. Arsenic

In industry, workers are usually exposed to inorganic arsenic compounds. Whatever the arsenic compound to which the workers are exposed, the absorbed arsenic is rapidly eliminated in the urine, either in the unchanged form or in the methylated forms of monomethylarsonic acid and cacodylic acid (dimethylarsonic acid). The biological monitoring of workers should be carried out by measuring the total amount of arsenic present in urine collected at the end of the shift or at the beginning of the next shift. As some marine organisms may contain very high concentrations of organoarsenicals of negligible toxicity that are also rapidly excreted in urine, workers should be instructed to refrain from eating seafood for 2 or 3 days before urine collection. As in urine, the arsenic concentration in blood reflects mainly recent exposure. Arsenic in hair is an indicator of the amount of inorganic arsenic absorbed during the growth period of the hair.

D. Beryllium

While the major toxicological effects of beryllium are on the lungs, after absorption, beryllium does not localize in the lungs but it is widely distributed in the body. Beryllium can be determined in blood and urine but at the present time these analyses can only be used as qualitative tests for confirmation of exposure. It is not known to what extent the concentrations of beryllium in the blood and urine may be influenced by recent exposure and by the amount already stored in the body. Urinary excretion of beryllium may be detected for years after removal from exposure. Blood investigations of workers exposed to beryllium indicate that the beryllium ion is bound to a protein to form a beryllium antigen which induces the cell-mediated beryllium hypersensitivity and can be determined *in vitro* by an increase of lymphoblast transformation and also an inhibition of macrophage migration.

E. Cadmium

Cadmium is a cumulative toxin and mostly bound to red blood cells in blood. Absorbed cadmium accumulates mainly in the kidney and in the liver, with about half of the body burden found in these two organs. In all tissues, cadmium is bound mainly to metallothionein, a protein of high sulfhydryl content with a capacity for binding cadmium, copper, and zinc. The body burden of smokers is about twice that of non-smokers. Cadmium level in blood appears to reflect mainly recent exposure levels. When the total amount of cadmium absorbed has not yet saturated all the available cadmium binding sites, the cadmium concentration in urine reflects mainly the cadmium level in the body. When integrated exposure has been so high as to cause a saturation of binding sites, cadmium levels in urine may be related partly to the body burden and partly to recent exposure. In urine, cadmium is bound mainly to metallothionein. Metallothionein analysis presents an advantage over cadmium metal analysis in that it is not subject to external contamination. A radioimmunoassay has been developed for the determination of metallothionein in urine. The determination of cadmium in hair has been proposed to evaluate past exposure.

F. Chromium

Chromium absorption depends on the oxidation state of the metal. In the trivalent state, chromium is very poorly absorbed. The toxicity of chromium is attributed to the hexavalent state, which is also responsible for the carcinogenicity of the metal. Currently, there is insufficient data to allow evaluation of blood chromium measurements. Except for chromates, all chemical forms of chromium are rapidly cleared from the blood. The major excretory route for absorbed chromium is in the urine. Chromium concentration in urine can be used as an index of recent exposure to hexavalent soluble chromium compounds.

G. Cobalt

Cobalt is an essential metal present in Vitamin B_{12}. Generally, cobalt does not seem to accumulate in a specific target organ. The concentrations of cobalt in blood are highly variable. The high variability may be due to sample contamination. Absorbed cobalt is largely excreted in the urine. Urinary concentrations of cobalt in 24-hour urine collection appear to correlate well with occupational exposure.

H. Copper

Copper is an essential trace element for all vertebrates. Blood concentrations of copper vary widely from city to city. Abnormally high values have been observed from people living in copper smelting vicinities. There appears to be a sex difference. Males typically have higher blood concentrations than females, but females have higher day-to-day variation. The major excretion route of copper is the bile. Urinary copper is reabsorbed.

I. Iron

Iron is found in virtually every food. The rate of absorption of iron is inversely related to the state of the body's iron stores and it is a complicated process. Absorbed iron is transported by transferrin, a globulin which is normally one-third saturated. There is diurnal variation in serum iron as well as day-to-day variation. Excessive absorbed iron is excreted in the urine.

J. Lead

Lead is a cumulative toxin that is absorbed by the lungs and the gastrointestinal tract. In blood, lead is mainly bound to red blood cells. In a steady-state situation, lead in blood is considered to be the best indicator of the concentration of lead in soft tissue and hence of recent exposure. However, it does not necessarily correlate with the total body burden. Lead in urine reflects the amount of lead recently absorbed. In the same individual, lead in urine fluctuates with time more than lead in blood. The biological tests for lead exposure are mainly based on the interference of lead with several stages of the heme synthesis pathway. Because of the inhibition of d-aminolevulinic dehydrase by lead, d-aminolevulinic acid accumulates in tissue and is excreted in urine. The inhibition of ferro-chelatase in young red blood cells leads to an accumulation of protoporphyrin which exists as zinc protoporphyrin. Other enzymatic methods are also available but not commonly used. Determination of lead in hair is another method that could be used for evaluation of past exposure to lead.

K. Manganese

Manganese is absorbed mainly through the lungs. Gastrointestinal absorption appears to be negligible. In blood, manganese is present mainly in red blood cells. Independent studies have failed to find any significant correlation between manganese exposure and the blood manganese level in workers. Concentrations in red blood cells have been reported to be increased in persons with rheumatoid arthritis. Excretion occurs mainly through the bile. Very small amounts are eliminated in the urine and hair.

L. Mercury

Mercury absorption depends on the form of the metal. Elemental mercury is well absorbed by inhalation. Inorganic mercury compounds are well absorbed after ingestion. Organic mercury compounds are well absorbed by all routes. Once absorbed, mercury is bound to the sulfhydryl groups of proteins. In blood, inorganic mercury is equally distributed between plasma and red blood cells. Inorganic mercury is excreted mainly in the urine and feces. If exposure to mercury vapor has lasted for at least 1 year, there is a correlation between the concentrations of mercury in blood or urine and the intensity of recent exposure as well as the occurrence of clinical signs of intoxication. Biological monitoring of organic mercury compounds depends on the organic moiety. Short-chain alkyl compounds, mainly methylmercury, are found mainly in red blood cells and excreted through feces. In this case, urinary determination has no practical value. Aryl and alkoxyalkyl mercury compounds liberate inorganic mercury *in vivo* and the concentration of mercury in blood as well as urine is indicative of the exposure intensity.

M. Nickel

Nickel is not a cumulative toxin. It is absorbed mainly through the respiratory tract but also possibly through the gastrointestinal tract and the skin. Once absorbed, it is rapidly excreted. Excretion occurs predominantly in the urine. Several studies have demonstrated that the concentrations of nickel in plasma and urine are indicators of recent exposure to soluble nickel compounds. While a single blood sample concentration of nickel correlates quite well with air sample concentration, urinary concentration of nickel would only correlate well if a 24-hour urine sample is collected.

N. Selenium

Selenium compounds are well absorbed orally and seem to be well absorbed through the lungs. Under conditions of exposure to high concentrations, a volatile metabolite, dimethylselenide, may be eliminated through the lungs. Blood and urine concentrations of selenium reflect mainly recent exposure. It is important to note that concentration of selenium in blood and urine may vary considerably, depending on the dietary intake.

O. Thallium

Thallium is easily absorbed by any route of exposure. Blood is not a reliable indicator of thallium exposure. Only a small fraction of the body burden of thallium in poisoned subjects is in the plasma. Excretion is predominantly in the urine. Determination of thallium in urine is probably a more reliable indicator of exposure than its determination in blood.

P. Tin

Absorption of tin depends on the form. Ingested inorganic tin is very poorly absorbed, on the order of about 1% Organotin compounds are better absorbed from the gastrointestinal tract. Short-chain alkyltin compounds may be absorbed through the skin. Whereas metallic tin and inorganic tin compounds are relatively nontoxic, a number of organotin compounds are highly poisonous. Most absorbed tin is found in the red blood cells. However, large variations from day-to-day and week-to-week have been observed and such variations have been attributed to dietary intake of tin. Inorganic tin is excreted in the urine, while organotin is excreted in the bile.

Q. Vanadium

Vanadium is absorbed mainly by the pulmonary route. The oral absorption rate appears to be negligible. In blood, absorbed vanadium is bound to plasma transferrin. Currently, the co-relationship between concentrations of vanadium in blood and exposure level has not been discovered. Absorbed vanadium is rapidly excreted in the urine. The determination of vanadium in urine has been proposed for evaluating recent exposure to the metal.

R. Zinc

Zinc is an essential metal and available in food. Absorbed zinc is present in the red blood cells, white blood cells, plasma, and serum. Although zinc values in blood serum are generally reliable, serum zinc is lower in women taking oral contraceptives, in pregnant women, and in persons undergoing certain stresses such as infections. Excretion of zinc is largely in the feces. Part of the zinc excreted is reabsorbed.

XVII. BIOLOGICAL MONITORING OF SOLVENT EXPOSURE

Solvent molecules can gain entry into a worker's body by inhalation, skin absorption, or ingestion. They may also gain entry by more than one route concurrently. Traditionally, inhalation has been the major exposure route of concern for the worker. Consequently, industrial hygienists have concentrated on keeping atmospheric levels of chemical below effect levels known as Threshold Limit Values (TLVs) which are issued annually by the American Conference of Governmental Industrial Hygienists (ACGIH). TLVs refer to airborne concentrations of substances and represent conditions under which it is believed that nearly all workers may be repeatedly exposed day after day without experiencing adverse health effects.

Examination of the chemical substances listed by the ACGIH show that a number of the listed substances also has the *skin* notation, which refers to the potential compound's contribution to the overall exposure by the cutaneous route, including mucous membranes and eye, either by airborne or, more particularly, by direct contact with the substance. Vehicles can also alter skin absorption. While little quantitative data are available describing absorption of vapors and gases through skin, the consensus of industrial hygiene toxicologists is that substances having a skin notation may present an additional problem, particularly if a significant area of the body is exposed to the substance for a long period of time. Hence, measurements of atmospheric concentrations of chemical substances with the *skin* notation will not give an accurate indication of the quantity of chemical that might reach the target organ.

Even if a chemical substance does not have a *skin* notation, and even if there exists a relationship between the airborne concentration and the worker by a specific route, one still cannot expect that the determination of the airborne concentration may allow an estimate of the total amount of chemical substance absorbed by the exposed worker. There are differences in workers' personal hygiene habits, as well as differences in the individual absorption rate of the substance through any portal of entry.

A. Biotransformation of Foreign Compounds

Most foreign compounds entering the body are subject to biotransformation, commonly known as metabolic transformation. It is important to note that while biotransformation of foreign compounds often involves detoxification, the same metabolic pathway may actually lead to an increase in the toxicity of some compounds.

One important aspect of biotransformation of foreign compounds is increased excretability, and thus a decrease in toxicity. A major route of excretion is the renal system. The kidney is built in such a way that it can handle electrolytes better than nonelectrolytes. Thus, the higher the degree of ionization at body pH, the more readily the chemicals are excreted by the kidney. Ionization in turn depends on the degree of polarizability of the chemical. Polarizability of a chemical is related to the geometry and molecular size. Asymmetric and large molecular substances are more polarizable than symmetric and small ones. The general mechanism of detoxification is summarized in Figure 3.5.

Biotransformation occurs mainly in the liver although other tissues such as the kidneys or lungs may also be involved. These enzymatic reactions occurs in two phases: a Phase I reaction involves oxidation, reduction, and hydrolysis; and a Phase II reaction is a conjugation or synthetic reaction. The primary function of Phase I reactions is to add or alter the functional groups on the parent molecule. These functional groups then permit the Phase I product to undergo a Phase II reaction. In Phase II, the Phase I product is covalently bonded to an endogenous molecule, producing a conjugate, which results in an increase in size of the foreign compound and hence its excretability. For example, toluene is oxidized in Phase I to form benzoic acid, which conjugates with glycine in Phase II and is excreted in the urine as hippuric acid. The relationship between Phase I and Phase II reactions is summarized in Figure 3.6.

B. Oxidation Reactions

Oxidation reactions are essential to animal life. In fact, the enzymatic makeup of living organisms is geared toward oxidizing foodstuff. The lower primary aliphatic alcohols are closely related to carbohydrates and are metabolized to carbon dioxide and water. The higher primary alcohols and glycols are oxidized progressively through the corresponding aldehydes and acids, while secondary alcohols are oxidized to ketones. In biological monitoring of solvent exposure, it is necessary for the occupational health professional to be well aware of the metabolic pathway(s) of foreign chemicals in question and in what biological specimen the index chemical can be determined. Biological monitoring of exposure to aliphatic alcohols and glycols is summarized in Table 3.13, aldehydes in Table 3.14, ketones in Table 3.15, and ethers in Table 3.16.

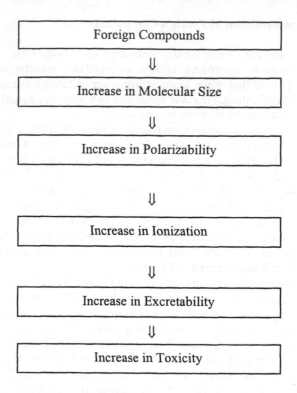

Figure 3.5   General mechanism of detoxification.

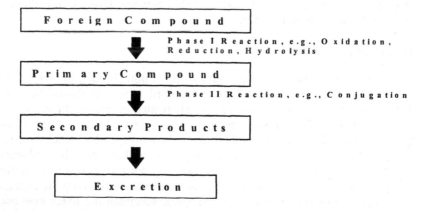

Figure 3.6   Biotransformation of foreign compounds.

Table 3.13

Biological Monitoring of Exposure to Alcohols and Glycols

Alcohol/Glycol	Exposure Index	Biologic Specimen
Allyl alcohol	Ally alcohol	Breath
	Acrolein	Blood
Chloroethanol	Chloroacetic acid	Urine
Diethylene glycol	Oxalic acid	Urine
Ethyl alcohol	Ethyl alcohol	Breath
Ethylene glycol	Ethylene glycol	Serum
	Oxalic acid	Urine
Isopropyl alcohol	Acetone	Blood
	Acetone	Urine
Methyl alcohol	Methyl alcohol	Breath
	Methyl alcohol	Urine
	Formic acid	Blood
	Formic acid	Urine

Table 3.14

Biological Monitoring of Exposure to Aldehydes

Aldehyde	Exposure Index	Biologic Specimen
Acetaldehyde	Acetaldehyde	Blood
Chloraldehyde	Trichloroethanol	Urine
Formaldehyde	Formaldehyde	Blood
	Formic acid	Urine
Furfural	Furoic acid	Urine

Table 3.15

Biological Monitoring of Exposure to Ketones

Ketone	Exposure Index	Biologic Specimen
Acetone	Acetone	Breath
	Acetone	Blood
	Acetone	Urine
	Formic acid	Urine
Methyl ethyl ketone	Methyl ethyl ketone	Breath
	Methyl ethyl ketone	Blood
	Methyl ketone	Urine
Methyl isobutyl ke-tone	4-Methyl-2-pentanone	Urine
	4-Hydroxy-4-methyl-2-pentanone	Urine
Methyl n-butyl ke-tone	2,5-Hexanedione	Urine

Table 3.16
Biological Monitoring of Exposure to Ethers

Ether	Exposure Index	Biologic Specimen
Diethyl ether	Diethyl ether	Breath
	Diethyl ether	Blood
Dioxane	b-Hydroxyethoxy acetic acid	Urine
Ethylene glycol monoethyl ether	Ethoxyacetic acid	Urine
Ethylene glycol monomethyl ether	Methoxyacetic acid	Urine

While the oxidation of oxygen-containing hydrocarbons such as alcohols and aldehydes is quite straightforward, the oxidation of unsubstituted aliphatic and alicyclic hydrocarbons occurs on a secondary carbon atom next to the primary carbon atom (Table 3.17).

Unsubstituted aromatic hydrocarbons are oxidized to a corresponding aromatic alcohol, which is then conjugated with an endogenous chemical and excreted (Table 3.18).

Table 3.17
Biological Monitoring of Exposure to Aliphatic and Alicyclic Hydrocarbons

Hydrocarbon	Exposure Index	Biologic Specimen
Cyclohexane	Cyclohexane	Breath
	Cyclohexanol	Blood
	Cyclohexanol	Urine
n-Heptane	n-Heptane	Breath
	n-Heptane	Blood
	2-Heptanol	Urine
	2,6 Heptanedione	Urine
n-Hexane	n-Hexane	Breath
	n-Hexane	Blood
	2-Hexanol	Urine
	2,6-Heptanedione	Urine
2-Methylpentane	2-Methylpentane	Breath
	2-Methyl-2-pentanol	Urine
3-Methylpentane	3-Methylpentane	Breath
	3-Methyl-2-pentanol	Urine

Table 3.18

Biological Monitoring of Exposure to Aromatic Hydrocarbons

Aromatic Hydrocarbon	Exposure Index	Biologic Specimen
Benzene	Benzene	Breath
	Benzene	Blood
	Phenol	Urine
Biphenyl	4-Hydroxybiphenyl	Urine
Ethylbenzene	Ethylbenzene	Blood
	Ethylphenol	Urine
Isopropylbenzene	Dimethylphenylcarbinol	Urine
Naphthalene	α- and β-Naphthol	Urine
Styrene	Styrene	Breath
	Styrene	Blood
	Mandelic acid	Urine
	Phenylglyoxylic acid	Urine
Toluene	Toluene	Breath
	Toluene	Blood
	Hippuric acid	Urine
	o-Cresol	Urine
Xylene	Xylene	Breath
	Xylene	Blood
	Methylhippuric acid	Urine

Halogenated hydrocarbons may be excreted unchanged, remain in the living organism unaltered, or oxidized to an alcohol and the corresponding carboxylic acid (Table 3.19).

While aliphatic amines are oxidatively deaminated into the corresponding aldehydes by loss of ammonia, oxidation of aromatic amines may occur either on the amino group or on a secondary carbon atom in the ring (Table 3.20). Aliphatic and aromatic nitro compounds are not further oxidized, but many of them produce a physiological effect known as methemoglobinemia, which affects oxygen transport in the blood.

Table 3.19

Biological Monitoring of Exposure to Halogenated Hydrocarbons

Halogenated Hydrocarbon	Exposure Index	Biologic Specimen
Carbon tetrachloride	Carbon tetrachloride	Breath
	Carbon tetrachloride	Blood
Chloroform	Chloroform	Breath
	Chloroform	Blood
p-Dichlorobenzene	2,5-Dichlorophenol	Urine
Hexachlorobenzene	Hexachlorobenzene	Blood
Methyl chloroform	Methyl chloroform	Breath
	Methyl chloroform	Blood
	Trichloroethanol	Urine
	Trichloroacetic acid	Urine
Perchloroethylene	Perchloroethylene	Breath
	Perchloroethylene	Blood
Polychlorinated bi-phenyl	Polychlorinated bi-phenyl	Blood
1,1,2,2-Tetrachloroethane	Dichloroacetic acid	Urine
	Glycolic acid	Urine
	Trichloroethanol	Urine
	Trichloroacetic acid	Urine
1,1,2-Trichloroethane	1,1,2-Trichloroethane	Breath
	1,1,2-Trichloroethane	Blood
	1,1,2-Trichloroethane	Urine
	Chloroacetic acid	
Vinyl chloride	Thioglycolic acid	Urine

Table 3.20

Biological Monitoring of Exposure to Nitrogen-Containing Chemicals

Chemical	Exposure Index	Biologic Specimen
Acetonitrile	Cyanide	Blood
	Thiocyanate	Urine
Acrylonitrile	Cyanide	Blood
	Thiocyanate	Urine
Aniline	Methemoglobin	Blood
	p-Nitrophenol	Urine
Benzidine	Benzidine	Urine
Dimethyl formamide	Dimethyl formamide	Blood
	N-Methylformamide	Urine
Dinitrobenzene	Methemoglobin	Blood
	Nitroaniline	Blood
Ethylene glycol	Ethylene glycol dinitrate	Urine
Nitrobenzene	Methemoglobin	Blood
	p-Nitrophenol	Urine
Nitroglycerine	Nitroglycerine	Blood
Trinitrotoluene	Aminodinitrotoluene	Blood

C. Reduction Reactions

Reduction reactions in the metabolism of organic compounds are much less common than oxidation reactions. This is because they go counter to the general trend of biochemical reactions in living tissue. Yet it must be realized that all enzymatic reactions in a living organism are fundamentally reversible. A well-known example of a reduction reaction is the biotransformation of chloral hydrate, the oldest hypnotic, into trichloroethanol. In the case of aromatic polynitro compounds, typically only one of the nitro groups becomes reduced.

D. Hydrolysis Reactions

Certain organic compounds require cleavage or degradation before they can be further metabolized. The most common reaction of this type is the hydrolysis of an ester to an alcohol and an acid. For example, aspirin is hydrolyzed *in vitro* to acetic acid and salicylic acid, the latter containing an alcohol functional group. Amides, hydrazides, and nitriles may also be hydrolyzed.

E. Conjugation Reactions

Conjugation reactions appear to have developed chiefly for the metabolism of foreign compounds. This is the reaction that is most successful in terms of detoxification. The typical endogenous compounds involved in conjugation are amino acids (such as glycine) and their derivatives. A few simpler endogenous chemical species (such as acetate and sulfate) are also occasionally involved in conjugation. The conjugation of glycine with benzoic acid to form hippuric acid, has been discussed.

F. Biological Monitoring Measurements

Biological monitoring of exposure to industrial chemicals provides occupational health personnel with a tool for assessing a worker's overall exposure to an index chemical, i.e., the chemical of analytical interest through measurement of the appropriate determinant(s) in biological specimens collected from the worker at a specified time. In short, this means evaluation of the internal exposure of the worker to a chemical substance, i.e., the internal dose, by a biological method. The determinant can be the chemical itself or its metabolite(s), or a characteristic biochemical change induced by the chemical. Depending on the chemical and the analyzed biological parameter, the term *internal exposure* or *internal dose* may cover different concepts. It may mean the amount of the chemical recently absorbed, or the amount of the active species bound to the sites of action.

Biological monitoring takes into consideration absorption by routes other than the lungs. Many industrial chemicals enter the body of a worker by multiple routes. The greatest advantage is that the biological parameter of exposure is more directly related to the adverse health effects that are attempted to be prevented, as compared to any atmospheric measurement. Therefore, it offers a better assessment of the potential health risk to the worker than ambient air monitoring.

Biological monitoring measurements can be made in exhaled air, urine, blood, or other biological specimens such as hair, milk, saliva, tears, nail, perspiration, adipose tissue, and others. Based on the determinant, the specimen chosen, and the time of sampling, the measurements can indicate the intensity of a recent exposure, an average daily exposure, a chronic cumulative exposure, or even a past exposure.

G. Exhaled Air Analysis

It has been known for decades that many industrial organic chemicals that are inhaled or absorbed through the skin are later excreted to some degree in exhaled air. Metabolites of some of these chemicals may also be exhaled. Examples of such chemicals include acetone, benzene, carbon tetrachloride, diethyl ether, ethanol, styrene, toluene, xylene, and many others. The concentrations of these chemicals in the exhaled air, however, have been found to decrease exponentially with time. Despite the fact that this is the method of preference for most workers, studies to date indicate that this measurement technique has some shortcomings. First of all, although the body tends to integrate the external exposure, this monitoring technique reflects principally the most recent exposure level. Secondly, the decay curves for many of these chemicals and times of exposure are frequently indistinguishable within 2 hours after exposure, and it is impractical to instruct workers to collect alveolar air samples after work and return for analysis the following day.

H. Urinalysis

Urinalysis is probably the second most preferred method of biological monitoring among workers because it is not invasive. It usually relies on analysis for a metabolite in most organic chemicals. However, there have been great variations reported between authors for the translation of urinary levels of the index chemical to exposure levels of the workplace chemical. Despite the disagreement in translating urinary levels to body burden, there is no question that for many chemicals a relationship does exist. Problems could arise in urinalysis when the index chemical may come from diet, medication, or some other nonoccupational source. For example, workers should not have taken Chloraceptic, which contains sodium phenolate, prior to biological monitoring for benzene exposure.

I. Blood Analysis

Blood analysis is an invasive technique that carries some resultant risk to the worker. Most workers find it objectionable and would object strongly to phlebotomy or blood drawing. Generally, the order of worker acceptance for sampling for biological monitoring would be expected to be exhaled air, urine collection, and last of all, phlebotomy. There is no question that total blood levels of index chemicals reflect total dosage. For example, the correlations of inorganic lead exposure, biological effects, and blood concentrations for inorganic lead have been extensively reviewed.

J. Miscellaneous Analysis

Human breast milk analysis has not been used to monitor exposure to industrial chemicals, even though milk could be considered as an excretory mechanism. Because of the small number of lactating women in the workplace, human breast milk analysis has very restricted utility for monitoring industrial chemical exposure.

Saliva, tears, perspiration, hair, nail, and adipose tissue, although not usually considered excretory fluids, do contain chemicals transferred from blood. To date, very little meaningful work has been reported on the analysis of these tissues as a measure of industrial exposure.

XVIII. CASE STUDY

On three successive days, a migrant farm worker in California had expressed discomfort associated with a pesticide product being used to treat apples. The worker complained about headaches, nausea, blurred vision, and dizziness. On the fourth day using the product, the worker experienced difficulty breathing and was taken to a nearby hospital for treatment by a co-worker.

The co-worker explained to the doctor that they were apple pickers. Fortunately, the doctor had some previous training and experience with industrial hygiene and agricultural applications, as he had spent a summer internship working as an industrial hygiene technician. He served as a technician for a consortium established to evaluate exposures during agricultural work, so he was equipped with an understanding of the exposure pathways, products routinely employed, and their toxicological significance.

He asked the co-worker if *they* had been using any pesticide products, and the co-worker said they had not. The doctor then asked if there had been any pesticide products applied in any of the areas where they had been picking apples over the past week or so. This question elicited an affirmative response from the co-worker. The doctor asked if the worker or co-worker saw a material safety data sheet (MSDS) for the product. Both workers looked puzzled and said no. As well, neither worker knew the name of the product that was applied either.

Without a product name or MSDS, the doctor's task would be somewhat more challenging, as many pesticide products and groups elicit similar effects, though toxicologically, the mechanisms of action can be exceedingly different. The doctor asked the workers for the name of the farm that they worked so he could determine what pesticide had been used. The doctor contacted McCombs Farm & Garden where the workers were employed and learned that the product being used in the apple orchards was a malathion-based product. There was no MSDS available on the farm, but the doctor now could proceed nonetheless. Malathion is an organophosphate pesticide. This pesticide group is notorious for its ability to inhibit cholinesterase. Inactivation of cholinesterase results in elevated acetylcholine during synaptic transmission in the nervous system.

Based on the doctor's diagnosis after tests were performed on the ill worker, it was surmised that the worker did not sustain an "excessive" exposure to malathion. The worker was kept overnight and asked to avoid any potential exposure to malathion. McCombs Farm & Garden was contacted and informed about the toxicity of malathion and asked to allow at least 1½ days after malathion application before workers returned to picking apples. Workers involved in malathion application should be equipped with appropriate personal protective equipment, including proper respiratory protection and chemical-resistant clothing. Additional cases were not identified after the doctor's recommendations were implemented.

REVIEW QUESTIONS

1. A worker has been observed with the following: port wine hue in urine, severe jaundice, and gradual suppression of urinary function. He is diagnosed to be suffering from:

   a. Arsine poisoning
   b. Lead poisoning
   c. Benzene poisoning
   d. Malathion poisoning

2. Chelating agents are frequently used in the rapid removal of heavy metals from workers suffering from heavy metal poisoning. An example of a chelating agent is:

   a. Aurintricarboxylic acid
   b. British Anti-Lewisite
   c. Deferrioxamine
   d. All of these

3. Exposure to methylene chloride may elevate the concentration of which of the following in blood?

   a. Zinc protoporphyrin
   b. Carboxyhemoglobin
   c. d-Aminolevulinic acid
   d. Methemoglobin

4. Urinalysis was conducted on a worker at the end of his 8-hour shift and phenylglyoxylic acid was detected. The worker was most likely exposed to:

   a. Phenol   b. Oxalic acid   c. Lead   d. Styrene

5. Inhalation of bauxite over a number of years may result in:

   a. Itai-itai disease
   b. Metal fume fever
   c. Schaver's disease
   d. Wilson's disease

6. Which of the following biological indices may indicate exposure to carbon disulfide?

   a. 2-Thiothiazolidine-4-carboxylic acid
   b. 4-Thiothiazolidine-4-carboxylic acid
   c. 4-Thiothiazolidine-2-carboxylic acid
   d. 2-Thiothiazolidine-2-carboxylic acid

7. Cobalt sulfate was once used as a foam stabilizer in beer. As a result of this, some heavy drinkers died suddenly of:
   a. Liver failure
   b. Heart failure
   c. Kidney failure
   d. Hemolytic anemia

8. Cataracts are opaque spots in the lens that impair vision. Which of the following chemicals may produce cataracts?

    a. Pentanol
    b. Naphthol
    c. Dinitrophenol
    d. Cresol

9. Exfoliative dermatitis is a skin disorder characterized by erythema with desquamation, scaling, itching, and loss of hair. This can be caused by exposure to:

    a. Osmium tetroxide
    b. Chromium
    c. Vanadium pentoxide
    d. Arsenic

10. Mee's lines are white transverse lines on the fingernails seen in chronic:

    a. Zirconium poisoning
    b. Carbon disulfide poisoning
    c. Methyl *n*-butyl ketone poisoning
    d. Arsenic poisoning

11. 2,5-Hexanedione in urine is used for the biological monitoring of occupational exposure to:

    a. Methyl *iso*-butyl ketone
    b. *n*-Hexane
    c. Methyl ethyl ketone
    d. *n*-Heptane

12. Chronic exposure to inorganic fluorides may result in:

    a. Osteosarcoma
    b. Osteosclerosis
    c. Osteomalacia
    d. Kidney stones

13. Agranulocytosis is a condition characterized by a marked decrease in the number of granulocytes in blood. Which of the following chemicals may cause such a condition?

    a. Nitrobenzene
    b. Benzene
    c. *n*-Hexane
    d. Dimethyl sulfoxide

14. Cutaneous porphyria is characterized by skin fragility with bullous lesions in areas exposed to sunlight, hyperpigmentation, and hypertrichosis. This may be caused by:

    a. Benzene
    b. Polychlorinated biphenyl
    c. Dioxin
    d. Hexachlorobenzene

15. In lead-acid batteries, a lead alloy is used as grids, connectors, and posts. In addition to lead, which other metal would you recommend for biological monitoring?

    a. Antimony   b. Beryllium   c. Cadmium   d. Arsenic

16. Conjugation reactions are developed chiefly for the biotransformation and elimination of foreign compounds. Benzoic acid is eliminated from the body through conjugation with:

    a. Hippuric acid   b. Glycine   c. Alanine   d. Phosphate

17. Anemia is a reduction below normal in the number of red blood cells, in the quantity of hemoglobin, or in the volume of packed cells. The type of anemia caused by exposure to lead is best described as:

    a.   Macrocytic, hypochromic
    b.   Microcytic, hyperchromic
    c.   Microcytic, hypochromic
    d.   Macrocytic, hyperchromic

18. Phosphorus is a nonmetallic element that exists in several allotropic forms. Which of the following allotropes of phosphorus is "relatively nontoxic?"

    a. Yellow   b. Black   c. Red   d. Purple

19. Biological monitoring and analysis for cadmium metal is frequently subject to external contamination. A radioimmunoassay method has been developed for the determination of the following in urine:

    a.   Albumin
    b.   Metallothionein
    c.   Globulin
    d.   High density lipoprotein (HDL)

20. The recommended treatment for methemoglobinemia is intravenous administration of:

    a.   Amyl nitrite
    b.   Sodium thiosulfate
    c.   Calcium disodium ethylenediaminetetraacetic acid
    d.   Methylene blue

21. Which of the following would you recommend as the biological exposure index for ethylene glycol exposure?

    a.   Ethylene in expired breath
    b.   Ethylene glycol in blood
    c.   Ethylene glycol in urine
    d.   Methoxyacetic acid in urine

22. British Anti-Lewisite (BAL) or 2,3-dimercaptopropanol was developed as an antidote for:

    a.   Arsenic
    b.   Chlorine gas
    c.   Organophosphates
    d.   Lead

23.   According to the statement issued by the Centers for Disease Control (CDC), the most effective chelation therapy for lead is the administration of:

    a.   British Anti-Lewisite followed by calcium disodium ethylenediamine tetraacetate
    b.   British Anti-Lewisite only
    c.   Penicillamine
    d.   Calcium disodium ethylenediamine tetraacetate only

24.   Alzheimer's disease is an insidious neurodegenerative disorder for which there is no effective treatment or cure. Which of the following metals is a suspect?

    a. Lead   **b.** Manganese   **c.** Mercury   **d.** Aluminum

25.   Codeine, a narcotic analgesic, is metabolized *in vivo* to morphine.   This biotransformation reaction is a(n):

    a.   Oxidation reaction
    b.   Reduction reaction
    c.   Hydrolysis reaction
    d.   Conjugation reaction

ANSWERS

1.	a.	**10.**	d.	**18.**	c.
2.	d.	**11.**	b.	**19.**	d.
3.	b.	**12.**	b.	**20.**	d.
4.	d.	**13.**	d.	**21.**	d.
5.	c.	**14.**	d.	**22.**	a.
6.	a.	**15.**	a.	**23.**	a.
7.	b.	**16.**	b.	**24.**	d.
8.	c.	**17.**	c.	**25.**	c.
9.	d.				

References

Adams, R. M., Ed., *Occupational Skin Diseases*, Saunders, Philadelphia, PA, 1989.

Baselt, R. C., *Biological Monitoring for Industrial Chemicals*, Biomedical Publications, Davis, CA, 1980.

Beckett, W. S. and Bascom, R., Eds., *Occupational Lung Disease*, Hanley & Belfus, Philadelphia, PA, 1992.

Carson, B. L., Ellis III, H. V., and McCann, J. L., *Toxicology and Biological Monitoring of Metals in Humans*, Lewis Publishers, Boca Raton, FL, 1986.

Clayton, G. D. and Clayton, F. E., Eds., *Patty's Industrial Hygiene and Toxicology*, 4th ed., 6 volumes, John Wiley & Sons, New York, 1993-94.

Cullen, M. R., Ed., *Workers with Multiple Chemical Sensitivities*, Hanley & Belfus, Philadelphia, PA, 1987.

Dean, J. H., Luster, M. I., Munson, A. E., and Kimber, I., *Immunotoxicology and Immunopharmacology*, 2nd ed., Raven Press, New York, 1994.

Ellenhorn, M. J. and Barceloux, D. G., *Medical Toxicology: Diagnosis and Treatment of Human Poisoning*, Elsevier, Amsterdam, The Netherlands, 1988.

Fergusson, J. E., *The Heavy Elements: Chemistry, Environmental Impact and Health Effects*, Pergammon Press, New York, 1990.

Fishbein, L. and Furst, A., Eds., *Biological Effects of Metals*, Plenum Press, New York, 1987.

Gardner, D. E., Crapo, J. D., and McClellan, R. O., Eds., *Toxicology of the Lung*, 2nd ed., Raven Press, New York, 1993.

Hall, S. K., *Chemical Safety in the Laboratory*, CRC Press/Lewis Publishers, Boca Raton, FL, 1994

Hall, S. K., Chakraborty, J., and Ruch, R. J., *Chemical Exposure and Toxic Responses*, CRC Press/Lewis Publishers, Boca Raton, FL, 1997.

Hall, S. K. and Cissik, J. H., *Pulmonary Health Risks from Asbestos Exposure and Smoking*, in Sourcebook on Asbestos Diseases, Volume 5, Peters, G. A. and Peters, B. J., Eds., Butterworth Legal Publishers, Salem, NH, 1991.

Hall, S. K., Markiewicz, D. S., and Sherman, L. D., Biological monitoring of workers exposed to hazardous materials, *Proceedings of the Third Annual Presentation of the Hazardous Materials Management Conference*, Rosemont, IL, 1990.

Hayes, A. W., Ed., *Principles and Methods of Toxicology*, 3rd ed., Raven Press, New York, 1994.

Ho, M. H. and Dilon, H. K., Eds., *Biological Monitoring of Exposure to Chemicals*, John Wiley & Sons, Inc., New York, 1987.

Kamrin, M. A., *Toxicology: A Primer on Toxicology Principles and Applications*, Lewis Publishers, Chelsea, MI, 1988.

Kimber, I. and Maurer, T., *Toxicology of Contact Hypersensitivity*, Taylor & Francis, Bristol, PA, 1996.

Klaassen, C. D., Eds., *Casarett and Doull's Toxicology: The Basic Science of Poisons*, 5th ed., Pergammon Press, New York, 1995.

Kneip, T. J. and Crable, J. V., Eds., *Methods for Biological Monitoring*, American Public Health Association, Washington, D.C., 1988.

Lauwerys, R. R. and Hoet, P., *Industrial Chemical Exposure—Guidelines for Biological Monitoring*, 2nd ed., CRC Press/Lewis Publishers, Boca Raton, FL, 1993.

Malzulli, F. N. and Maibach, H. I., *Dermatology*, 5th ed., Taylor & Francis, Bristol, PA, 1996.

Merchant, J. A., *Occupational Respiratory Diseases*, DHHS (NIOSH) Publication No. 86-102, Washington, D.C., 1986.

Paul, M., *Occupational and Environmental Reproductive Hazards: A Guide for Clinicians*, Williams & Wilkins, Philadelphia, PA, 1993.

Proctor, N. H. and Hughes, J. P., *Chemical Hazards in the Workplace*, Lippincott, Philadelphia, 1978.

Rest, K., *Proceedings of Multiple Chemical Sensitivity Workshop*, Princeton Scientific, Princeton, NJ, 1992.

Seiler, H. G., Sigel, H., and Sigel, A., *Toxicity of Inorganic Compounds*, Marcel Dekker, New York, 1988.

Sheldon, L. et al., *Biological Monitoring Techniques for Human Exposure to Industrial Chemicals*, Noyes Publications, Pard Ridge, NJ, 1986.

Smialowicz, R. and Holsapple, M. P., *Experimental Immunotoxicology*, CRC Press, Boca Raton, FL, 1995.

Snyder, R., *Ethel Browning's Toxicity and Mechanism of Industrial Solvents*, Elsevier, New York, 1992.

Sorsa, M. and Vainio, H., *Mutagens in Our Environment*, Alan R. Liss, New York, 1982.

Sullivan, J. B., Jr. and Krieger, G. R., *Hazardous Materials Toxicology*, Williams & Wilkins, Baltimore, MD, 1992.

Watterson, A., *Pesticide User's Health and Safety Handbook, An International Guide*, Van Nostrand Reinhold, New York, 1988.

Chapter 4

AIR SAMPLING

Stephen K. Hall, PhD

I. OVERVIEW

Air sampling is performed for a variety of reasons. The primary reason for air sampling is to identify and quantify specific chemical, biological, and physical agents that may be present in the occupational environment, especially in a worker's breathing zone. Other important reasons for air sampling include routine surveillance, evaluation of the effectiveness of control methods, and the compliance status with respect to various occupational health standards. Reliable measurements of airborne contaminants are useful in determining appropriate personal protective equipment needs, determining whether (and which) engineering controls can achieve permissible exposure limit levels, delineating areas where protection is needed, assessing the potential health effects of exposure, and determining the need for specific medical monitoring.

Air sampling provides guidance on evaluating potentially harmful exposures to workers from hazardous chemical, biological, and physical agents. *Integrated methods* are the best way to monitor exposures to a specific chemical, and a large number of methods are available. Real-time or direct-reading techniques are increasingly important, since they are the best way to examine the exposure profile and identify sources. However, it should be pointed out that the emphasis of sampling should not be on air sampling alone. All types of sampling, air, bulk, biological, and wipe, are needed in today's complex evaluations of hazardous environments. Therefore, the most sensible solution is some combination of these approaches, depending on the situation.

In summary, this chapter is organized with the following in mind. First, one needs to determine what types of sampling must be done. What chemicals will be sampled? One usually attempts to identify the exposure first. Then, a selection of the sampling methods, equipment, and materials is made. Generally, this selection is done to keep sampling options as broad as possible. Although sometimes it is known exactly what types of chemicals must be sampled, in general, even in the most classic type of industrial operation, these chemicals are going to change either as the process changes, or as new chemicals are introduced as replacements for older ones. Contaminants previously thought harmless may become a concern and sampling will be necessary. The next step is to plan the sampling event by selecting an appropriate methodology, calibration equipment, and ensure that there is adequate quality control so that results are not negated by poor technique or other problems. Finally, evaluation and interpretation of the sampling data are performed to determine how the data can best be applied to assist in minimizing any hazards that might be present.

II. AIR CONTAMINANT CONSIDERATIONS

Generally, air sampling varies in complexity and purpose. It ranges from the relatively straightforward sample and identification of chemicals in a well-characterized industry, such as a foundry, to the often difficult and complex identification associated with airborne contaminants at a hazardous waste site. Different types of surveys often dictate different approaches. For example, an insurance survey usually involves limited sampling, and consultants generally must operate within the contractual scope of the proposal accepted by their client. A health and safety professional assigned to a single industrial facility usually performs repetitious sampling for the same contaminants. In consulting and insurance work, rarely is there opportunity to collect enough samples for high statistical accuracy. Mistakes requiring a survey to be redone can be costly, especially if extensive travel or time expenditures are involved. In general, an acceptable strategy is one that prioritizes needs, optimizes resources, is readily implemented, and is cost effective.

In devising an air sampling program, it is essential to consider the following basic requirements. Any manipulation of sampling equipment in the field should be kept to a minimum and the sampled air must follow the shortest possible route to reach the collection medium. The sampling instrument should provide an acceptable efficiency of collection for the contaminants involved, and this efficiency must be maintained at a rate of air flow that can provide sufficient contaminant for the intended procedure for subsequent sample analysis. The collected air sample should be obtained in a chemical form that is stable during transport to the analytical laboratory. Consequently, any use of unstable or otherwise hazardous sampling media should be avoided.

A. Physical States of Air Contaminants

The physical states of air contaminants may be divided into a few broad categories depending on physical characteristics.

Gases are fluids that occupy the entire space of their enclosure and can be liquified only by the combined effects of increased pressure and decreased temperature. Examples of gases are helium, hydrogen, carbon monoxide, ethylene oxide, formaldehyde, hydrogen sulfide, and radon. *Va-*

pors are the evaporation product of substances that are also liquid at room temperatures, such as benzene, toluene, and styrene. It can also be the *sublimation* (evaporation directly from a solid) product at room temperature, such as iodine. Although gases and vapors behave similarly from a thermodynamic standpoint, the reason for making the distinction is because in many instances they are collected by different devices.

An *aerosol* is a system consisting of airborne solid or liquid particles that are dispersed in a gas stream, usually the atmosphere. Aerosols are generated by fire, erosion, sublimation, condensation, and abrasion of minerals, metallurgical materials, organic and other inorganic substances in construction, manufacturing, mining, agriculture, and transportation among other operations. Aerosol classifications depend on the physical nature, particle size, and method of generation. Dusts, fumes, smoke, soot, mist, and fog are all terms used to describe certain types of aerosols.

Dusts are generated from solid inorganic or organic materials reduced in size by mechanical processes such as chipping, crushing, drilling, grinding, pulverizing, and other abrasive actions occurring in natural and commercial operations. Dusts can be derived from inorganic minerals, such as asbestos, limestone, and silica, as well as various metals. Organic dusts are derived from vegetable, soil, and animal sources. Examples include grain dusts of barley, oats, wheat, and hay, straw, wood, cotton, as well as sewage sludge. Dust can also be derived from animal dander, insects, mites, fungal spores, and pollen. Microbials can often be found mixed with dusts, especially those that are organic based. Most often, dust particles are somewhat spherical in shape. The concern for dusts is for those particles with an aerodynamic equivalent diameter of less than 10 mm because these remain suspended in the atmosphere for a significant period of time and are *respirable*. The aerodynamic equivalent diameter is the diameter of a hypothetical sphere of unit density having the same terminal settling velocity in air as the particle in question, regardless of its geometric size, shape, and true density.

Fumes are produced by chemical reactions and by such processes as combustion, distillation, calcination, condensation, and sublimation. Fumes are solids that are the result of condensation of solids from an evaporated state. Examples are welding fumes, hot asphalt, and volatilized polynuclear aromatic hydrocarbons from coking operations. When a metal or plastic evaporates, the atoms or molecules disperse singly into the air and form a uniform gaseous mixture. In the air, they combine rapidly with oxygen and recondense, forming a very fine particulate ranging in size from 1.0 mm to 0.0001 mm. Automobile "exhaust fumes" and "paint fumes" are inappropriate terms since gases and vapors and airborne mists are not classified as fumes.

Smokes are products of incomplete combustion of carbonaceous materials such as coal, oil, tobacco, and wood. Smoke particles generally range from 0.3 mm to 0.5 mm in diameter and are characterized by optical density.

Mists and *fogs* are suspended liquid droplets but the distinction between these two terms has not been fully defined. They are generated by condensation from the gaseous to the liquid state or by mechanically breaking up a liquid into dispersed state by spraying, splashing, foaming, or atomizing. Examples are oil mists produced from metal-working fluids during parts machining and the solvent mists observed above electroplating tanks. Some mists can have a vapor component as well. For example, volatile solvents are contained in paint spray mists. When there is no need to differentiate between the particle and droplet components of an aerosol, the collective term *particulate matter* is used.

Fibers are particles whose length exceeds their width. They can be generated from minerals, such as asbestos, and man-made sources, including fiberglass, if the composition of the material lends itself to disintegration producing such particles. For the purpose of classification, some fibers are assigned a minimum size criterion, also known as the *aspect ratio*, e.g., asbestos particles must be at least three times longer than they are wide to be considered a fiber for occupational sampling purposes. Fibers are thought to behave differently when in the lung than particles in the shape of a sphere. Organic sources, such as hemp and animal fibers, also exist.

For the purpose of air sampling, several factors are important when considering aerosols. Particle size is often dependent on the means by which the aerosol was generated. When airborne, particles can collide and stick together. If the purpose of sampling is simply to collect the total mass of dust, this factor will not influence the results to a significant degree, since it is likely that these larger particles will still be collected. If size-selective sampling is done, the results may not reflect the actual airborne dust composition, because it will appear as if there are fewer particles and they are larger in size. This consideration can be especially important when sampling near combustion sources because thermal effects can promote coagulation of smaller particles, leading to changes in the particle number and composition of the aerosol.

B. Odor of Air Contaminants

Odor is a human sensory experience. The human nose is a highly sensitive instrument capable of detecting extremely low concentrations of certain chemicals. *Odors* are defined as sensations resulting when volatile chemicals interact with the olfactory system in a body, causing impulses to be transmitted to the brain. The type and amount of odor are both important in fixing the signal sent to the brain. Certain chemicals are virtually nonodorous and cannot be detected at any level. Other substances are odorous but quickly intoxicate (dull) the human odor sensors, resulting in a discontinued sensation of the odors. This sensation is commonly referred to as olfactory fatigue.

Very low concentrations of an odorous substance can produce an odor sensation indicating the presence of the odor vapors. This is the *odor recognition threshold*. At this level, the brain may not be capable of recognizing the specific odor. At higher concentrations, the odor sensation becomes recognizable. This is the *odor recognition threshold*. For example, hydrogen sulfide has one of the lowest odor detection thresholds, 0.0005 parts per million (ppm) by volume or 0.5 parts per billion (ppb). At this level, a sensitive nose can detect the presence of an odor but does not recognize it. In fact, if one had to provide a description at this level, it would usually be described as smelling like chocolate, instead of the commonly described odor of rotten eggs at the odor recognition threshold. The difference between the detection and recognition thresholds varies by concentrations in the range of two- to ten-fold for the majority of odorous chemicals.

Obviously, there are individual differences in levels of odor perception, including differences in the level of odor detected when odor panelists come from a "clean air" background versus if they have spent time in an industrial atmosphere, as well as a decrease in the ability to detect odors with increasing age, or as the result of desensitizing activities such as cigarette smoking. Table 4.1 lists odor recognition thresholds of selected chemicals.

The discussion so far has implied that odors are gases or vapors. However, odors can also be related to particulate matter. A solid material can be volatile and release odorous gases. Small particles (10 mm and smaller) can also be inhaled into the respiratory system and interact with body fluids to create a flavor which could give the apparent sensation of an odor. Many odorous vapors are also absorbed in the body fluids of the respiratory system and we, therefore, "taste" them.

C. Vapor Properties of Air Contaminants

The degree of volatility of an air contaminant can affect the sampling method since it often dictates the physical state of the compound as it is found in the air. *Vapor pressure* is a measure of the ability of a compound to become airborne. The higher the vapor pressure, the higher the airborne concentration probability. Vapor pressures should be corrected to the temperature at which they were measured. Air contaminants that are gases at room temperature will have vapor pressures greater than standard atmosphere, i.e., 760 mmHg or 1 atm. Volatile compounds have vapor pressures greater than 1 mmHg at room temperature and exist entirely in the vapor phase. Semivolatile compounds have vapor pressures of 10^{-7} mmHg to 1 mmHg and can be present in both the vapor and particulate state. Nonvolatile compounds have vapor pressures of less than 10^{-7} mmHg and are found exclusively in the particle-bound state. Some of the classes of semivolatile organic compounds of particular interest to investigators include polynuclear aromatics, organochlorine pesticides, organophosphate pesticides, polychlorinated bipheyls, furans, and dioxins. Table 4.2 lists the vapor pressure of selected chemicals.

Table 4.1
Sample Odor Recognition Thresholds for Select Chemicals

Chemical	Odor Recognition Threshold	Odor Description
Acetaldehyde	0.2	Apple sweet
Acetic acid	24	Vinegar
Acetic anhydride	0.4	Sour acid
Acetone	100	Nail polish remover
Acrylonitrile	0.2	Onion garlic
Ammonia	55	Pungent
Aniline	1	Pungent
Benzene	5	Aromatic
Bromine	0.05	Pungent
Butyl acetate	20	Fruity
Butyl aldehyde	0.04	Sweet rancid
Carbon disulfide	0.2	Sulfide
Carbon tetrachloride	100	Sweet, pungent
Chlorine	5	Bleach
Chloroform	300	Characteristic
Dimethylamine	0.05	Fishy, pungent
Dioxane	6	Sweet, alcohol
Ethanol	10	Alcohol
Ethyl ether	0.3	Hospital
Formaldehyde	1	Pungent
Hydrochloric acid gas	10	Pungent
Hydrogen sulfide	0.8	Rotten egg
Isoamyl acetate	7	Banana
Isopropyl alcohol	200	Rubbing alcohol
Methanol	100	Alcohol
Methyl chloroform	400	Chloroform
Methyl methacrylate	0.3	Pungent
Naphtha	100	Aromatic
Phosgene	1	New-mown hay
Phosphine	0.02	Onion, mustard
Pyridine	0.02	Burnt, pungent
Stoddard solvent	30	Kerosene
Sulfur dioxide	5	Choking
Toluene	40	Airplane glue
Toluene diisocyanate	2	Mediciny, pungent
Turpentine	200	Solvent
Xylene	200	Aromatic

Table 4.2
Vapor Pressures of Selected Chemicals at 25 °C

Chemical	Vapor Pressure in mm Hg
Acetic acid	19
Acetone	363
Acetonitrile	94
Acrylonitrile	119
Benzene	96
Carbon disulfide	473
Carbon tetrachloride	117
Diethyl ether	553
Ethanol	63
Ethyl acetate	93
Ethyl benzene	9.8
Formic acid	43
n-Hexane	182
Hydrocyanic acid	742
Methanol	140
Methyl acetate	253
Methyl chloroform	143
Methylene chloride	420
Nitromethane	37
Pyridine	25
Styrene	8.6
Tetracholorethylene	351
Toluene	32
Trichloromethane	236

Occasionally, a chemical will be described as being heavier than air. In these cases, the property of *vapor density* is the point of reference. Low ambient temperature increases vapor density, as does high humidity. Vapor density can be important in situations where high concentrations are present due to an emergency release, buildup to combustible concentrations, or chemical release in a confined space where there is a lack of air movement. In reality, when comparing molecular weights, most organic compounds are heavier than air and the most important criterion is the ratio of contaminant to available air currents and air volume for mixing. In reality, most organic compounds are going to mix with air and disperse readily throughout an area. Table 4.3 lists vapor densities of some selected chemicals.

Table 4.3
Vapor Densities of Selected Chemicals at 25°C

Chemical	Vapor Density in g/L
Air	1.00
Ammonia	0.59
Butane	2.04
Carbon dioxide	1.52
Carbon monoxide	0.97
Carbon tetrachloride	5.32
Chlorine	2.45
Ethane	1.04
Ethylene	0.97
Gasoline	3.93
Hydrogen	0.07
Hydrogen sulfide	1.17
Isobutane	2.01
Methane	0.55
Methyl chloride	1.78
Methyl chloroform	4.60
Methyl isocyanate	1.97
Nitrogen	0.97
Propane	1.52
Sulfur dioxide	2.20
Tetrachloroethylene	4.53

Sample Calculation: A simple calculation can be done to show the dispersion property of most organic compounds. Trichloroethylene, a commonly used dry cleaning agent and in the extraction of caffeine from coffee, has a vapor density of 4.53 g/L. Assuming there is ample air for dilution, the following calculations indicate the effective vapor density of a 1000 ppm trichloroethylene mixture. (Note the vapor density of air is 1, and the vapor density of trichoroethylene is 4.53):

1000 ppm = 0.1% = 1 part trichloroethylene to 999 parts of air
0.001 x (4.53 g/L) = 0.00453
0.999 x (1.0) = 0.999
0.999 + 0.00453 = 1.00353 or approximately 1.004

The value *1.004* is the effective vapor density of the mixture. Therefore, the trichloroethylene-air mixture compared to clean air would have a ratio of 1004:1000, *not* 4.53:1. Since 1000 ppm is 20 times the safe occupational exposure limit for trichloroethylene, a mixture normally encountered in these situations would contain much less. Thus, the effects of window ventilation, cross currents, wind, traffic, and heat are often more important than molecular weight and vapor density.

III. SAMPLING METHOD CONSIDERATIONS

There are two basic sampling methods for the collection of airborne contaminants. The first method involves the use of an air moving device, or a *pump*, to obtain a definite volume of air at a known temperature and pressure. This method is known as *active* sampling and, depending on the concentration and the analytical method used, the contaminant may be analyzed either with or without further concentration. The second method does not involve any air moving device. Instead, the air monitoring device depends entirely on the phenomenon of diffusion of airborne contaminants to achieve trapping in a collection medium. This method is known as *passive* sampling. Whereas active sampling can be applied to gases, vapors, as well as aerosols and bioaerosols, passive sampling is currently used for gases, vapors, and bioaerosols only. Each of these sampling methods is discussed in greater detail in the following sections.

A. Active Sampling

Active sampling can involve either *grab sampling* or *integrated sampling*. In grab sampling, an actual sample of air is collected in a suitable container and the collected air sample is considered *representative* of the atmospheric conditions at the monitoring site at the time of sampling. Numerous types of grab sampling devices are available. These include evacuated flasks (Figure 4.1), gas or liquid displacement containers (Figure 4.2), flexible plastic containers, and hypodermic syringes.

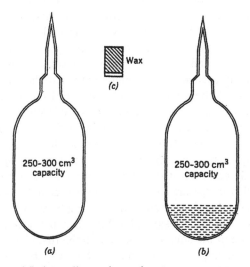

Figure 4.1   Evacuated flask to collect grab samples.

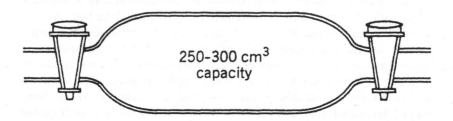

Figure 4.2   Liquid displacement containers used to collect grab samples.

Integrated sampling is used when the concentration of airborne contaminant is low, or when the contaminant concentration fluctuates with time. It is also used when only the time-weighted average exposure value is desired. The contaminant in these cases is extracted from the air and concentrated by an absorbing solution or collected on sorbent media. Absorbers vary in characteristics, depending on the gas or vapor to be collected. Four basic types of absorbers are available: simple bubbling or gas washing bottles, spiral and helical absorbers, fritted bubblers, and glass bead columns. The function of these absorbers is to provide sufficient contact between the contaminant in the air and the absorbing solution.

For nonreactive and insoluble vapors, adsorption is the method of choice. Commonly used sorbents include activated charcoal, silica gel, and porous polymers. Sorbent tubes, which are useful for personnel integrated sampling of most organic vapors, contain two interconnected chambers in series filled with the adsorbent. In an activated charcoal tube, for example, the first chamber contains 100 mg of charcoal; this chamber is separated from the backup section, containing 50 mg of charcoal, by a piece of urethane foam. The contents of the two chambers are analyzed separately to determine whether or not the charcoal in the first chamber has become saturated and lost an excessive amount of the sample to the second chamber. Sampling results are not considered valid when the second chamber contains 20% or more of the amount collected on the first chamber.

An active sampling train may consist of a pump or source of suction, a flow regulator such as an orifice or a nozzle, a flowmeter to indicate the flow rate, a collection device such as an absorber, sorbent, or a filter, a probe or sampling line, and a prefilter to remove any particulate matter that may interfere with sample collection or laboratory analysis.

B. Passive Sampling

One of the more noteworthy developments in air sampling technology has been the availability of passive dosimeters for a broad list of vapors and gases. These sampling devices are small, lightweight, and inexpensive. They have no moving parts to break down and can be conveniently used unattended. They depend solely on permeation or diffusion of gaseous contaminants to achieve trapping in a collection medium. Commercially available units usually require no calibration since this is generally provided by the manufacturer.

Permeation devices utilize a polymeric membrane of silicone rubber as a barrier to ambient atmosphere. Gaseous contaminants penetrate the membrane and are transported through the membrane to a collection medium such as activated charcoal, silica gel, or ion exchange granules. Permeation across the membrane is controlled by the solubility of the vapor or gas in the membrane material and by diffusion of the dissolved molecules across the membrane under a concentration gradient. Diffusion devices are provided with a porous barrier to minimize atmospheric turbulence. Molecules diffuse through the barrier to a stagnant air layer and then are collected on the adsorbent material. Specific information on sampling rates of the airborne contaminants is supplied by the manufacturers.

All the organic passive sampling dosimeters use activated charcoal as the collection medium. Both permeation and diffusion devices are commercially available. They can be used to sample any organic compound that can be actively sampled by the charcoal tube method. It is important to note that the sampling and analysis of these organic passive dosimeters are affected by environmental factors such as atmospheric pressure, ambient temperature, relative humidity, and face velocity. The overall accuracy of passive dosimeters is ± 25% in the range of 0.5 to 2.0 times the environmental standard. It should be stressed that the manufacturer's recommendations on how, and under what conditions the passive samplers may be used, should be strictly followed.

Today, there is a growing market for passive monitors due to their ease of use. However, it is generally accepted that integrating methods involving the use of a pump provide better accuracy and precision. Most passive samplers are tested against active methods prior to use.

IV. SAMPLING STRATEGY CONSIDERATIONS

Air sampling of workplace air is the key element of an occupational control program. The person conducting air sampling must be as familiar as possible with the particular operation. The investigator should be aware of the reactions being run, the reagent chemicals used, and the end-products, by-products, and chemical wastes generated. He should also know what personal protective measures are provided, how engineering controls are being used, and how many workers are exposed. An experienced industrial hygiene professional often can evaluate quite accurately the magnitude of chemical and physical stresses associated with an operation on a qualitative walk through. However, in order to document the actual airborne concentration of contaminants, it would be necessary for the industrial hygienist to use air sampling devices and instrumentation. Regardless of the objectives of the air sampling program, the investigator must take into consideration the following parameters in order to be able to implement the correct air sampling strategy.

A. Sampling Location

There are three primary approaches for air sampling: area, personal, and source. *Area sampling* is most common when the general work area is being monitored for air contaminants. A best approximation would involve placing a number of sampling devices around a given area so that all are monitoring at the same time to create a "map" of the levels of air contaminant present. Obviously, the results of area sampling will have a very different interpretation than those of personnel monitoring.

If sampling is being done to determine compliance with an Occupational Safety and Health Administration (OSHA) standard, personnel must be monitored within the breathing zone unless prohibited by the sampling method or the workplace environment. Breathing zone is defined by OSHA to be a hemisphere forward of the shoulders with a radius of approximately 15 to 23 cm. *Personal sampling* is usually collected with a portable sampling apparatus consisting of a pump and collection media attached to the shirt or jacket collar of a worker to collect a sample in the proximity of the employee's *breathing zone*. Personnel usually fall into one of two categories: those assigned to a particular station, or those who roam throughout an operation in the course of their work. An assembly line worker is an example of the first situation and a forklift operator is an example of the second situation. Sampling data of the latter is more difficult to interpret than on stationary workers.

Studies comparing area sampling with personal sampling have shown that area sampling may significantly underestimate exposures, especially in environments with moderate to great variability in air concentrations during the workday.

Source sampling is conducted by placing the sampling apparatus near a source where the investigator wants to get an estimate of whether or not significant contamination may be released. Area sampling is more appropriate for the measurement of source emissions than for approximating personal exposures.

B. Sampling Duration

Brief period samples are often referred to as *grab* or *instantaneous* samples, whereas longer period samples are termed *integrated* or *continuous* samples. Although there is no sharp dividing line between the two categories, grab samples are generally obtained over a period of less than 5 minutes and are best for determining peak concentrations of contaminant, whereas integrated samples are taken for longer periods, ranging anywhere from a few minutes up to a full shift of 8 hours.

The usual objective in airborne chemical contaminant monitoring is to maintain an occupational environment below a contaminant's applicable OSHA's Permissible Exposure Limit (PEL). The ACGIH's Threshold Limit Values (TLV) refer to airborne concentrations of substances and represent conditions under which it is believed that nearly all workers may be repeatedly exposed day after day without adverse effect. Both the OSHA PELs and the ACGIH TLVs assume an 8-hour workday, 5-day work week, and a "healthy" work population. Brief period samples include the threshold limit value—ceiling (TLV–C), the concentration that should not be exceeded even instantaneously, and the threshold limit value—short term exposure limit (TLV–STEL), the maximal concentration to which workers can be exposed for a period up to 15 minutes continuously. The longest period sample is the threshold limit value—time-weighted average (TLV–TWA), the time-weighted concentration for a normal 8-hour workday for a 40-hour workweek. OSHA PELs are incorporated into current legal standards, *Code of Federal Regulations* 29 CFR 1910.1000, Table Z-1 through Z-3, for evaluating the exposure of workers to airborne contaminants in the workplace. The OSHA PELs are legally enforceable standards. The equations below can be used to convert TLV values from parts per million (ppm) to mg/m^3, and vice versa.

$$TLV_{mg/m^3} = \frac{(TLV_{ppm}) \times (MW)}{24.45} \qquad (4.1)$$

or:

$$TLV_{ppm} = \frac{(TLV_{mg/m^3}) \times (24.45)}{(MW)} \qquad (4.2)$$

Where:

TLV = 8-hour threshold limit value, ppm or mg/m^3

MW = Gram molecular weight of substance

Recommended exposure limits (RELs), prepared by the National Institute for Occupational Safety and Health (NIOSH), typically refer to exposures up to 10 hours per day, and 40 hours per week.

Generally, a full sampling period refers to an 8-hour workday. One method is to collect a single sample to quantify the exposure over the entire period. A preferable approach is to collect multiple consecutive samples of equal or unequal time duration that equal the total sampling period when combined. An example would be collecting 4 samples, each approximately 2 hours in length over a workday. There are several advantages to this type of sampling. First, if a single sample is lost during the sampling period due to equipment failure, gross contamination, or during transport to the analytical laboratory, at least some data will have been collected to evaluate the exposure. Second, the use of multiple samples will reduce the effects of sampling and analytical errors. Lastly, collection of several samples provides the sampling professional with more information about the exposure variations throughout the workday, including those associated with different tasks. The following two equations can be employed to determine airborne contaminant concentrations based on a number of different variables.

1. Determination if a TLV is exceeded based on exposure to a number of compounds with similar toxicologic effects (e.g., acetone, methyl ethyl ketone, and *sec*-butyl acetate). The threshold limit value is exceeded if the calculated TLV is in excess of 1:

$$\frac{C_1}{T_1} + \frac{C_2}{T_2} + \frac{C_3}{T_3} + ... \frac{C_n}{T_n} = 1 \qquad (4.3)$$

Where:

C = Measured airborne concentration of each contaminant

T = 8-hour Threshold Limit Value for each contaminant

2. Cases when the source contaminant is a *liquid* mixture and the airborne concentration is assumed to be similar:

$$TLV_{liquid} = \frac{1}{\frac{f_a}{T_a} + \frac{f_b}{T_b} + \frac{f_c}{T_c} + ... + \frac{f_n}{T_n}} \qquad (4.4)$$

Where:

f = Percent of contaminant in liquid mixture, where 50% = 0.5; 30% = 0.3; 20% = 0.2; etc.

T = Threshold Limit Value, mg/m^3

Brief sampling periods are best for following the phases of a cyclic process and for determining *peak airborne concentrations* of brief duration, but they usually require analytical methods capable of detecting and measuring very low concentrations of the collected airborne contaminants. Short period sampling may also be accomplished by trapping a small portion of the atmosphere in a previously evacuated container. It is also suitable for primary irritants such as sulfur dioxide; whereas integrated sampling methods are best for evaluating cumulative systemic poisons such as benzene and mercury. Each sampling method has social value and it is essential to develop a capability to do both. It is always important that the sample contains sufficient contaminant to be above the minimum quantity that can be measured reliably by the chosen analytical method.

C. Sampling Efficiency

One of the most important factors in the collection of atmospheric contaminants is the efficiency of the monitoring device for the particular contaminant in question. In many cases, the efficiency need not be 100% as long as it is known and is constant over the range of concentrations being evaluated. It should, however, be above 90%. For many types of monitoring devices, the collection efficiency of the concentrating device must be measured.

Several methods for determining the monitoring efficiency of collecting devices are available: (a) by series testing where enough samples are arranged in series so that the last sample does not recover any of the test compound; (b) by monitoring from a gas-tight chamber containing a known gas or vapor concentration; (c) by comparing results obtained with a device known to be accurate; and (d) by introducing a known amount of gas or vapor into a sampling train containing the absorber being tested. The chief advantage of grab sampling methods is that collection efficiency is essentially 100%, provided correction is made for completeness of evacuation of the container and assuming no losses as a result of leakage.

D. Sampling Volume

Prior to conducting air sampling, the sampling professional must determine the minimum and maximum sample volumes allowed by the sampling method. The governing standard must also be taken into consideration. Collection of the minimum volume is necessary in order to collect enough sample to meet the detection limit of the analytical method, while the maximum volume is intended to protect the integrity of the compound that has been collected, as well as to assure that the sample will not be overloaded. When high sensitivity is desired, longer sampling durations should be used, but the sampling professional must stay within the limits of the maximum sampling volume and flow rate of the method. Since final air volumes are used in calculating the concentration of contaminants, accurate pump flow rates are necessary. This involves doing a flow rate calibration. In order to calculate the minimum and maximum sample volumes, the following equations may be used:

$$V_{(min)} = \frac{MDL}{0.2 \text{ x (PEL)}} \qquad (4.5)$$

Where:

$V_{(min)}$ = Minimum sample volume, liters
MDL = Minimum detection limit, g
PEL = OSHA Permissible Exposure Limit, g/L

$$V_{(max)} = \frac{\text{Tube Vapor Capacity, mg}}{2.0 \text{ x (PEL)}} \qquad (4.6)$$

Where:

$V_{(max)}$ = Maximum sample volume, liters
PEL = OSHA Permissible Exposure Limit, mg/m^3

Prior to submitting air samples to the laboratory for analysis, the collection volumes must be calculated. The total minutes a pump ran for each sample are multiplied by the flow rate in liters, as such:

$$V = T \text{ x } Q \qquad (4.7)$$

Where:

V = Sample volume, liters or cm^3
T = Time, minutes
Q = Volumetric flow rate for sample collection, liters per minute (LPM) or cm^3/min

If a different flow rate was measured when doing the post-calibration than was originally set, the sampling professional must make a determination of what flow to use. If working in compliance, the sampling professional would favor the higher flow rate. On the other hand, if the sampling professional works independently, the lower flow rate would be used, thus calculating a smaller volume, which in turn leads to a higher concentration when the analysis is done by the laboratory. In practice, however, most often the two flow rates are averaged.

Evaluating the impact of air contaminants necessitates the accurate determination of the amount of the contaminant present in a unit volume of air. This value defines the concentrations of the contaminant and is determined either directly from the airstream or following collection on a suitable medium. Calibrations are performed to establish the relationship between the instrument's response and the airborne chemical contaminant concentration being measured. The reference standards used must be accurate and precise to produce well-characterized and reproducible calibrations.

Table 4.4
Volume Correction Factors for Elevation

Altitude Difference in 1000 feet	Sampling Location is Higher	Sampling Location is Lower
0.0	1.000	1.000
0.5	1.009	0.991
1.0	1.019	0.982
1.5	1.028	0.973
2.0	1.038	0.964
2.5	1.047	0.955
3.0	1.057	0.955
3.5	1.067	0.946
4.0	1.077	0.937
4.5	1.087	0.920
5.0	1.090	0.912
5.5	1.107	0.903
6.0	1.117	0.895
6.5	1.128	0.887
7.0	1.138	0.879
7.5	1.148	0.870
8.0	1.160	0.862

Generally, sampling is conducted at approximately the same temperature and pressure as calibration in which case no correction for temperature and pressure is required and the sample volume reported to the laboratory is the volume actually measured. Where sampling is conducted at a substantially different temperature or pressure than calibration, an adjustment to the measured air volume may be required in order to obtain the actual air volume sampled. Tables 4.4 and 4.5 list correction factors for elevation and temperature differences respectively. Alternatively, if the elevation difference is greater than 1000 feet and a temperature differences is greater than 10, either Celsius or Fahrenheit, sampling site calibration is deemed necessary. The following equation can be used to correct for differences in pressure, volume, and temperature:

$$\frac{P_1 V_1}{T_1} = \frac{P_2 V_2}{T_2}$$
(4.8)

Where:

P = Pressure
V = Volume
T = Temperature, °K or °R

Table 4.5
Volume Correction Factors for Temperature

Temperature Difference	Sampling Location is Hotter	Sampling Location is Cooler
Temperature Difference in ° Celsius		
0	1.000	1.000
5	1.009	0.991
10	1.018	0.983
15	1.027	0.974
20	1.036	0.966
25	1.045	0.958
30	1.054	0.949
35	1.063	0.941
40	1.072	0.933
Temperature Difference in ° Fahrenheit		
0	1.000	1.000
10	1.010	0.991
20	1.020	0.981
30	1.030	0.971
40	1.040	0.962
50	1.050	0.953
60	1.060	0.944
70	1.070	0.935
80	1.080	0.926

E. Sample Handling

To prevent mix-ups, always label sample holders (tubes, cassettes, impingers) with a unique number before placing them in the sampling train. The last step is to prepare samples for shipment to the analytical laboratory. Sample seals are sometimes necessary. For example, OSHA compliance officers seal their samples prior to shipment. Seals, if properly constructed and applied, can provide proof that the samples were not tampered with during transit or storage. The seal should have a glue that will not allow removal of the seal without detection. Another purpose for seals is to keep the caps from coming off of sorbent tubes and the plugs from falling out of the inlet and outlet. Seals minimize the opportunity for contamination of samples. It is also important to seal the caps of any bottles or other containers being shipped to ensure that leakage does not take place. Samples should not be left in the work area, should not be easily accessible, and should not be kept at elevated temperatures such as in the trunk or glove compartment of an automobile.

When submitting samples to an analytical laboratory, there should always be an accompanying sampling data sheet. If the samples are incompatible, then a copy of the sampling data sheet should accompany each shipment. The sampling data sheet is often accompanied by or integrated into a chain-of-custody record.

V. AIR SAMPLING FOR GASES AND VAPORS

Integrated sampling refers to the methods of contaminated air collection onto sorbent media. It must be conducted over a certain period of time, usually 15 minutes to several hours, depending on the type and concentration of the contaminant that is present as well as on the sampling situation. Methods are either active or passive. The result is a sample representing the average concentration present in the air over the sampling period.

A. Integrated Sampling in Solid Sorbent Media

In sorbent tube sampling for gases and vapors, a pump is used to pull air through the solid sorbent. The commonly used solid sorbents are activated charcoal, silica gel, molecular sieves, and porous polymers. Typically, solid sorbents can be used for groups of chemical compounds but they are generally nonspecific. Sometimes, sorbents are manufactured with a special reagent coating to enhance collection of a specific compound. When a mixture of airborne contaminants is present, the amount of any individual air contaminant that can be collected is reduced. A reduction in sampling time or volume may be necessary because of the overall concentration of chemicals that may be present. Other factors that affect the sorption efficiency are the size and mass of sorbent granules.

The biggest concern in sorbent tube sampling for gases and vapors is whether breakthrough can occur. Breakthrough occurs when the front section of a tube is saturated and enough sampled contaminant accumulates in the backup section that it begins to exit the tube with the airstream. Breakthrough is defined as the presence of 20 to 25% of a contaminant in the rear portion of the sorbent tube. When results indicate breakthrough, the best interpretation is that actual concentrations are higher.

Another concern in sorbent tube sampling for gases and vapors is high humidity conditions during the sampling period. Water molecules present in the air may compete with the air contaminant to be sampled for sorbent sites in the sampling tube, with the result that concentrations will appear to be lower than they actually are.

The flow rate will also impact the ability to retain gases and vapor on a sorbent. An increase in the flow rate above the optimum flow rate specified by a method may result in a decrease in the ability of the sorbent to retain contaminants. Adsorption efficiency is also temperature dependent. At high temperatures, adsorption efficiency is reduced and the potential of breakthrough increases. Other factors that affect the ability of sorbent tubes to be efficient gas and vapor collectors include the size and mass of granules. A doubling of the sorbent mass doubles the ability of a tube to collect the contaminant being sampled, as well as increasing the concentration required for breakthrough.

1. *Activated charcoal tubes*

Activated charcoal is one of the most commonly used sorbents for a wide variety of organic compounds. Charcoal can be derived from a variety of carbon-containing materials, but most of the activated charcoal used for air sampling is coconut or petroleum based. Ordinary charcoal is "activated" by steam at 800 to 900°C, causing it to form a porous structure. The most common activated charcoal tube is manufactured in accordance with NIOSH recommended specifications which include fabrication of glass tubing 7 cm long with a 6-mm outside diameter and a 4-mm inside diameter (Figure 4.3). The charcoal is 20/40 mesh-sized. A 100-mg front section is separated from a 50-mg backup section by a piece of urethane foam. A second piece of foam sits at the outlet to prevent granules from being sucked out of the tube during sampling. A plug of glass wool is in the very front of the tube. An unused tube will have both ends flame-sealed. This structure and dimension is common to many tubes, although specialized tubes as well as scaled-up versions are also available. Analysis of activated charcoal tubes is generally done using carbon disulfide as the solvent to desorb the airborne contaminants collected on the activated charcoal sorbent.

Activated charcoal is not useful for sampling certain types of reactive organic compounds such as mercaptans and aldehydes due to its high surface reactivity with these compounds. Nonpolar organic compounds such as benzene, toluene, and 1,1,1-trichloroethane, which are preferentially sampled on charcoal, will displace more polar organic compounds such as ethyl cellosolve and dioxane in activated charcoal media. Competitive adsorption also occurs among polar organic compounds. Inorganic compounds such as chlorine, hydrogen sulfide, ozone, and sulfur dioxide react chemically with activated charcoal.

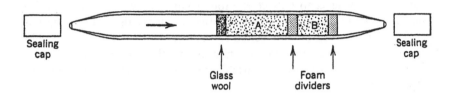

Sealing
cap

Sealing
cap

Glass
wool

Foam
dividers

Figure 4.3   Example of activated charcoal sampling tube.

Activated charcoal tubes have been shown to be affected by high humidity. Polar organic compounds are the most susceptible. Nonpolar organic compounds can be collected more reliably by limiting sample volumes where high humidity exists.

2. Silica gel tubes

Silica gel is considered a more selective solid sorbent than activated
charcoal. It is an amorphous form of silica derived from the interaction of
sodium silicate and sulfuric acid. Silica gel is the sorbent recommended for
collecting polar organic compounds such as aliphatic and aromatic amines.
Since the polarity of the adsorbed compound determines the binding
strength on silica gel, compounds of high polarity will displace compounds
of lesser polarity. Thus, in an attempt to collect relatively nonpolar compounds, the presence of coexisting polar compounds may interfere with
collection on silica gel. Examples of compounds that can be collected on
silica gel include aniline; chloroacetic acid; cresol; ethanolamine; hydrogen
bromide; hydrogen chloride; hydrogen fluoride; methanol; nitric acid; nitrobenzene; stibine; 1,1,2,2-tetrabromomethane; o-toluidine; and 2,5-xylidine.
The chief disadvantage of silica gel is its ability to adsorb water vapor
and displace collected components. It has been found that under high humidity conditions, the collected sample may be totally lost due to saturation
with water on the silica gel sorbent. Other factors that affect the dynamic
adsorption of materials onto silica gel include the size range of the gel particles, tube diameter and length, temperature during sampling, concentration
of contaminant being sampled, and duration of sampling.

3. Other solid sorbents

Molecular sieves are the carbon skeletal framework remains after pyrolysis of the synthetic polymeric or petroleum pitch precursors. The resultant material is a spherical, macroporous structure. The physical characteristics of the sieve, its pore structure, as well as particle size and shape, are determined by the precursor polymer or pitch. The diameters of the micropores and their number are responsible for the differences in sorption capacity, sorption coefficient, and retention volume. Molecular sieves are most commonly used to collect environmental samples of highly volatile nonpolar organic chemicals. Selected organic chemicals that can be collected by the use of carbon molecular sieves include acrylonitrile, allyl chloride, benzene, carbon tetrachloride, chloroform, ethylene dichloride, methyl chloroform, methylene chloride, trichloroethylene, vinyl chloride, and vinylidene chloride.

Porous polymers are another class of sorbent used for air sampling. Most porous polymers are copolymers in which one entity is styrene, ethylvinylbenzene, or divinyl benzene and the other monomer is a polar vinyl compound. Their wide variety offers a high degree of selectivity for specific applications

4. Coated sorbents

Many highly reactive compounds are unsuitable for sampling directly onto sorbents, either because they are unstable or cannot be recovered efficiently. In addition, some compounds may be analyzed more easily, or with greater sensitivity, by derivatizing them first, which can sometimes be achieved during the sampling stage. Methods have been developed that use coated sorbents, either sorbent tubes or coated filters. Selected organic chemicals that can be collected on coated sorbents include acetaldehyde, acrolein, arsenic trioxide, butylamine, formaldehyde, methylene dianiline, nitrogen dioxide, and toluene diisocyanate.

B. Integrated Sampling in Liquid Media

The absorption theory of gases and vapors from air by solution assumes that gases and vapors behave like perfect gases and dissolve to give a perfect solution. The concentration of the vapor in solution is increased during air sampling until an equilibrium is established with the concentration of vapor in the air. Absorption is never complete, however, because the vapor pressure of the material is not reduced to zero, but is only lowered by the solvent effect of the absorbing liquid. Continued sampling will not increase the concentration of vapor in solution once equilibrium is established. In general, the efficiency of vapor collection depends on (1) the volume of air sampled, (2) the volume of the absorbing liquid, and (3) the volatility of the contaminant being collected. Therefore, efficiency of collection can be increased by cooling the sampling solution, and increasing the sampling solution volume by adding two or more bubblers in series. Sampling rate and concentration of the vapor in air are not primary factors that determine collection efficiency.

Absorption of gases and vapors by chemical reaction depends on the size of the air bubbles produced in the bubbler, the interaction of contaminant with the sampling solution, the rapidity of the reaction, and a sufficient excess of sampling solution. If the reaction is rapid and a sufficient excess of reagent is maintained in the liquid, complete retention of the contaminant is achieved regardless of the volume of air sampled. If the reaction is slow and the sampling rate is not low enough, collection efficiency will decrease.

As discussed earlier, four basic types of bubbling or gas and vapor washing bottles are available: (1) simple bubblers, (2) helical absorbers, (3) fritted bubblers, and (4) glass bead columns (Figure 4.4). However, the midget impinger, a miniaturized simple bubbler, is probably the most widely used in this group. It is designed for impacting particles at a flow rate of 2.8 L/min. However, for industrial hygiene use as a bubbler, it is generally used with 10 to 20 ml of absorbing solution and a flow rate of 1.0 to 2.0 L/min. Air sampling is performed by connecting a personal pump to the outlet tube. The impinger is then attached to the worker's clothing. Care must be taken that the impinger does not tilt, which could result in a loss of absorbing solution into the sample pump. Spill-proof impingers have been designed to minimize this problem and are commercially available.

The helical absorbers may be used for collecting gases and vapors that are only moderately soluble or are slow reacting with reagents in the sampling solution. The spiral or helical structures provide for higher collection efficiency by allowing longer residence time of the contaminant within the tube. Slow-acting and less-soluble substances are given more time to react with the absorbing solution.

Gases and vapors that are sparingly soluble in the collecting medium may be sampled in fritted bubblers. They contain sintered or fritted glass. Air drawn into these devices is broken up into very small bubbles, and the heavy froth that develops increases the contact of gas and liquid. Frits come in various sizes and grades, usually designated as fine, medium, coarse, and extra coarse. A coarse frit is usually best for gases and vapors that are appreciably soluble or reactive. A medium porosity frit may be used for gases and vapors that are difficult to collect, but the sampling rate must be adjusted to maintain a flow of discrete bubbles. For highly volatile gaseous substances that are extremely difficult to collect, a frit of fine porosity may be required to break the air into extremely small bubbles and ensure adequate collection efficiency.

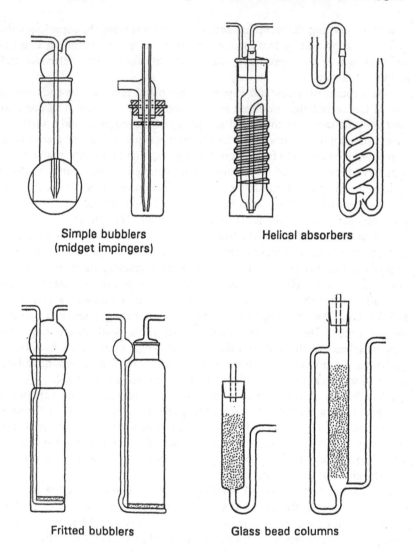

Simple bubblers
(midget impingers)

Helical absorbers

Fritted bubblers

Glass bead columns

Figure 4.4 Different types of absorbers.

Packed glass bead columns are used for special situations where concentrated solutions are needed. Glass beads are wetted with the absorbing solution and provide a large surface area for the collection of sample. It is especially useful when a viscous absorbing liquid is required. The rate of sampling is necessarily low, 0.25 to 0.5 L/min of air. Selected chemicals that can be collected in absorbing solutions include acetaldehyde, acetic anhydride, acetone, ammonia, benzaldehyde, benzene, bis(chloromethyl)-ether, butanol, carbon dioxide, carbon tetrachloride, cresol, diethyl mercury, ethyl benzene, formaldehyde, nitrogen dioxide, ozone, phenol, phosphine, styrene, sulfur dioxide, toluene diisocyanate, tricholorethylene, and vinyl acetate.

C. Passive Sampling for Gases and Vapors

Passive dosimeters are lightweight badge assemblies that rely on natural wind currents rather than pumps to move contaminated air to the collection surface. They accumulate an average dose and give an integrated measurement. Some dosimeters are specific for a given gas or vapor, such as acetone, ammonia, carbon monoxide, ethylene oxide, formaldehyde, nitrogen dioxide, nitrous oxide, ozone, and mercury; whereas others are nonspecific, such as organic vapors in general. Most of these passive dosimeters contain activated charcoal as the solid sorbent. Fig. 4.5 shows two different kinds of passive dosimeters. The obvious advantage is their ease and simplicity of use. They only need to be taken out of a package, the cover removed, and clipped onto an individual on the collar near the breathing zone, or a surface to be monitored. When used for personnel monitoring, they do not interfere with worker activity and are unlikely to affect the worker's normal behavior pattern, whereas wearing a pump with its tubing and sampling media might.

Passive devices rely on two basic collection principles: diffusion and permeation. With *diffusion*, the contaminant sample uptake is controlled by the length and diameter of the badge cavities as well as the physicochemical properties of the contaminant. The badge sampling rate is a function of the diffusion coefficient of the chemical being sampled and the total cross-sectional area of the badge cavity. With *permeation*, the contaminant sample uptake is controlled by the physicochemical characteristics of the membrane and the contaminant. The mass uptake is a direct function of the badge permeation sampling rate, the ambient concentration, and the sampling duration.

1. *Diffusion dosimeters*

On diffusion dosimeters, the gaseous or vapor contaminants pass through a barrier or draft shield that minimizes the effect of air currents to a stagnant air layer and then they are collected on the sorbent. A concentration gradient is created within the cavity of stagnant air and the amount of gas or vapor transferred is proportional to the ambient concentration. Each compound has a unique diffusion coefficient for each type of commercially available badge. The diffusion coefficient must be known in order to calculate the final concentration of the contaminant after analysis in the laboratory. This limits sampled materials to those for which this value is established. The sorbent surface area, path length from the badge's surface to the sorbent material, and badge sampling rate along with the desorption efficiency are also required in the calculation.

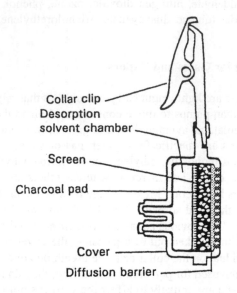

Collar clip
Desorption
solvent chamber
Screen
Charcoal pad
Cover
Diffusion barrier

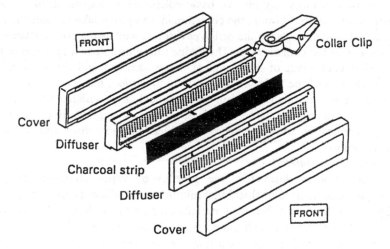

FRONT
Collar Clip
Cover
Diffuser
Charcoal strip
Diffuser
FRONT
Cover

Figure 4.5   Examples of different passive dosimeters.

2. *Permeation dosimeters*

On permeation dosimeters, the gaseous or vapor contaminants dissolve in a polymeric membrane and are then transferred to a collection medium. Permeation across the membrane is controlled by the solubility of the gas or vapor in the membrane material and by the rate of its diffusion across the membrane under a concentration gradient. The permeation constant, mass of compound collected, and exposure time determine the concentration collected. These constants vary, depending on the design of the monitor and the contaminants themselves. As with diffusion monitors, sampling rates must be determined for each compound and type of permeation monitor. Factors influencing permeation include thickness and uniformity of the membrane, affinity of the membrane for the contaminant, swelling or shrinking of the membrane, and possible etching by corrosive chemicals. The efficiency of these devices depends on finding a membrane that is easily permeated by the contaminant of interest and not by all others.

D. Sampling Bags for Gases and Vapors

Sampling bags offer alternatives in the collection of grab and integrated samples of gases and vapors, and they are particularly useful when representative samples are desired. They are also useful for the transport of calibration standards to sampling locations in the field. The bags come in different sizes, shapes, and materials. The materials include polyester, polyvinylidene chloride, high-density polyethylene, and polymers of fluorocarbons. In addition to the less reactive materials, bags that are black or opaque are now available for the collection of light-sensitive gases and vapors. In addition to the familiar rectangular shape, sampling bags are designed in other shapes to improve sampling techniques, minimize dead volume problems, and relieve stresses around fittings. Most bags have a valve to allow for filling using a pump and a septum for syringe needle. An array of valve types has been employed and is available from the different manufacturers.

All sampling bags should be leak tested, cleaned with pure compressed air, and conditioned before use. Three pump and cleaning cycles are recommended, unless condensable materials have deposited in the used sampling bag. The conditioning is first performed in the laboratory using test atmospheres. It is then repeated in the field before use by filling and emptying a bag several times at the sampling rate that will be used for taking the sample. Storage stability and decay curve for the intended gas or vapor must be determined.

E. Real-Time Instruments

The integrated air sampling instruments described so far are used to collect air samples for subsequent analysis. Instruments considered in this section are more complex. Sampling and analysis are carried out within the instrument, and the results of interest can be obtained in a matter of minutes, or even seconds. This type of air sampling is known as *real-time sampling*, which may be defined as *instantaneous sampling*. Another term often used for real-time instruments is *direct-reading instruments*, or DRIs.

Current direct reading instruments that are available with real-time measurement capabilities are largely the result of modern electronic components, such as laser illumination sources, high-sensitivity photometer or electrometer detectors, operational amplifiers, miniature power supplies, and microprocessors.

DRIs fall into several categories: **(1)** instruments designed to sample for a single chemical, **(2)** instruments that can be adjusted to sample for a group of chemicals and produce results specific for each chemical, and **(3)** instruments that are designed to sample for a large group of contaminants and produce results in total concentration. Obviously, there are more integrating sampling methods available for specific chemicals than there are real-time instruments for these chemicals. In a few cases, however, real-time instrumentation is the primary way of detecting certain chemicals such as carbon dioxide and carbon monoxide.

1. *Electrochemical instruments*

The operating principle of electrochemical instruments involves the measurement of electrical signals associated with chemical systems. These chemical systems are typically incorporated into electrochemical cells. Electrochemical techniques include instruments that operate on the principles of conductimetry, potentiometry, coulometry, and ionization.

Conductimetric sensors are electrochemical instruments that measure conductivity and rely on the fact that ions conduct electricity. Equally significant is the fact that at low concentrations, such as those concentrations typically found when these species are measured as workplace contaminants, conductivity is proportional to concentration. Conductimetric instruments are primarily used for detection of corrosive gases, e.g., ammonia, hydrogen sulfide, and sulfur dioxide. They are most effectively used in isothermal environments at or near ambient room temperature.

Potentiometric sensors are instruments that use a change in electrochemical potential as their principle of detection and are most commonly represented by the pH meter. Potentiometry is strictly defined as the measurement of the difference in potential between two electrodes in an electrochemical cell under the condition of zero current. Gases and vapors can react with reagents, resulting a redox reaction, the extent of which is proportional to the concentration of the reacting gas. Whereas potentiometry is basically a nonspecific technique, some degree of specificity may be obtained through the selection of the membrane through which the gaseous analyte must diffuse to enter the electrochemical cell, the selection of the reagent, and the type of electrodes used. Potentiometric instruments are used for the measurement of a variety of contaminants, including carbon monoxide, chlorine, formaldehyde, hydrogen sulfide, nitrogen oxides, oxygen, ozone, and sulfur oxides.

Coulometric sensors are coulometric analyzers that have as their principle of detection the determination of the quantity of electricity required to affect the complete electrolysis of the analyte of interest. The amount of electricity required is proportional to the amount of analyte present. This analyte may be the contaminant, or it may be a chemical with which the contaminant quantitatively reacts. The vast majority of coulometric analyzers are configured as oxygen or oxygen deficiency monitors, although coulometric analyzers are also available for carbon monoxide, chlorine, hydrogen cyanide, hydrogen sulfide, oxides of nitrogen, ozone, and sulfur dioxide.

Ionization sensors include a portable gas chromatograph (GC), which is a field instrument capable of separating and identifying many specific gases and vapors in the field. Most portable GCs come with a single detector. There are a number of detectors that can be used in GCs, but the most commonly used detectors are flame ionization detector (FID), photoionization detector (PID), and electron capture detector (ECD). The choice of detector will depend on the chemicals for which monitoring is needed, as well as on the nature of any other contaminants present and the sensitivity required. Both FIDs an PIDs are primarily nonspecific detectors used for the detection of organic compounds. However, the sensitivity of FIDs decreases with increasing substitutions of other elements such as oxygen, sulfur, and chlorine. PIDs can detect organic compounds such as aliphatic, aromatic, and halogenated hydrocarbons, as well as some inorganic compounds such as arsine, phosphine, and hydrogen sulfide, as well as nitric and sulfuric acids. ECD is a very sensitive technique, particularly for halogenated organic compounds, nitrates, conjugated carbonyls, and some organometallic compounds. It is also useful for sulfur hexafluoride and pesticide detection. All three detectors are available as stand-alone instruments.

2. Spectrochemical instruments

Instruments whose principle of detection is spectrochemical in nature include infrared (IR) spectrophotometers, ultraviolet (UV) and visible spectrophotometers, chemiluminescent detectors, and other photometric analyzers that use fluorescent and spectral intensity detectors. In general, spectrochemical analysis involves the use of a spectrum or some aspect of a spectrum to determine chemical species. A spectrum is a display of intensity of radiation that is emitted, absorbed, or scattered by a sample, versus wavelength. This radiation is related to photon energy via wavelength or frequency.

Infrared spectrophotometers are based on the principle that many organic compounds absorb light in the wavelength region of the IR spectrum. This selective absorption provides specificity by allowing for selection of an analytical wavelength at which potential interferences are not absorbed. The higher the concentration of molecules through which the IR beam passes, the greater the absorption. Table 4.6 lists specific IR absorption bands with respect to functional groups.

Table 4.6
Infrared Absorption Bands of Functional Groups

Functional Group		Absorption Band (u)
Alkanes	R—CH_2—CH_2—R'	3.35—3.65
Alkenes	R—CH=CH—R'	3.25—3.45
Alkynes	R—C≡C—R'	3.05—5.25
Aromatic hydrocarbons	(benzene ring)	3.25—3.35
Substituted aromatic hydrocarbons	(benzene ring with R substituent)	6.15—6.35
Chlorinated hydrocarbons	R—CH_2—Cl	12.80—15.50
Alcohols	R—CH_2—OH	2.80—3.10
Aldehydes	R—CHO	5.60—5.90
Ketones	R—CO—R'	5.60—5.90
Carboxylic acids	R—COOH	5.75—6.00
Esters	R—COO—R'	5.75—6.00

There are two categories of IR spectrophotometers: dispersive and nondispersive. Dispersive instruments use gratings or prisms to disperse the transmitted beam of radiation. The dispersed light is focused to a spectrum in the plane of the vertical exit slit. Rotation of the dispersing device causes the spectrum to move across the face of the exit slit and, ultimately, the detector. Dispersive instruments are used in the laboratory and frequently use dual beams of IR radiation for analysis. Most field instruments are nondispersive and do not incorporate prisms or gratings, so that the total radiation from the IR source passes through the sample.

Some IR instruments are designed as fixed wavelength monitors, whereas others are capable of scanning the entire IR spectrum. Some are designed as general detectors for organics and subgroups such as hydrocarbons; others are more specific monitors for compounds such as methane, ethylene, ethane, propane, butane, carbon monoxide, carbon dioxide, and several freons.

Ultraviolet detectors are used for compounds that absorb UV light, i.e., the portion of the electromagnetic spectrum having wavelengths from 10 to 350 nm. The actual spectral range for direct reading UV instruments is closer to 180 to 350 nm, which is termed "near-UV," in deference to its proximity to the visible spectrum. The visible spectrum has longer wavelengths than the UV (350 to 770 nm). Mercury absorbs strongly at 253.7 nm, thus allowing the UV principle to detect it. In addition to mercury, UV instruments have been designed to monitor ammonia, oxides of nitrogen, ozone, and sulfur dioxide. Some air contaminants form an intensely colored product with a reagent. The ability of this colored, liquid product to absorb light in the visible region of the spectrum is exploited.

Chemiluminescence, the emission of visible light, is the result of a specific case of chemical excitation wherein the excited-state reaction products are luminescent, thereby producing a reaction intermediate or product in an electronically excited state. The radioactive decay of the excited state is the source of luminescence. The amount of energy emitted at a specific wavelength during such a reaction is proportional to the number of molecules reacting. Chemiluminescent detectors have been developed to monitor oxides of nitrogen and ozone.

Fluorescence is the emission of photons from molecules in excited states when the excited states are the result of the absorption of energy from some source of radiation. If a molecule absorbs energy from a sufficiently powerful radiation source, such as a xenon arc lamp, the molecule will become excited, i.e., moving an electron to a higher energy state. When the electron returns to the lower energy state, it releases the absorbed energy in photons. A significant characteristic of fluorescence is that the emitted radiation is of a longer wavelength (lower energy) than the exciting radiation. Fluorescence instruments are available for carbon monoxide and sulfur dioxide.

The remaining real-time instruments in this category are simply referred to as *photometers*. These instruments have unique sampling characteristics and detection principles relative to the other instruments in this section, but they operate on spectrochemical principles nonetheless. The majority of these instruments allow for unattended sampling through the use of automated sampling media advance, i.e., tape samplers, rotating drum samplers, rotating disc samplers, and turntable samplers, or branched sequential sampling trains. These samplers typically involve a color change of the sampling medium and the analytic finish is the measurement of the light reflected from the sampling medium. These instruments are useful for area monitoring of such species as toluene diisocyanate, ammonia, phosgene, arsine, and hydrogen cyanide.

3. Thermochemical instruments

Gases and vapors have certain thermal properties that can be exploited in their analysis. Of the instruments available for industrial hygiene applications, one of two thermal properties, heat of combustion or thermal conductivity, is measured. *Heat of combustion* refers to the heat released by the complete combustion of a unit mass of combustible material. It is a measure of the maximum amount of heat that can be released by a certain mass of a combustible chemical. Table 4.7 lists the heats of combustion for some selected chemicals.

Table 4.7
Heats of Combustion of Selected Chemicals

Chemical	Heat of Combustion in kg-cal/mole at 25°C
Acetaldehyde	279
Acetone	428
Acetonitrile	302
Acetylene	311
Benzene	781
Cyclohexane	937
Diethyl ether	658
n-Hexane	995
Methane	213
Methanol	174
Methyl ethyl ketone	584
n-Pentane	839
Propane	531

Heat of combustion detectors, comprising the largest single class of real-time instruments for analyzing airborne gases and vapors, measure the heat released during combustion or reaction of the contaminant gas of interest. Virtually all heat of combustion instruments in use today are based on catalytic combustion. The released heat is a particular characteristic of combustible gases and may be used for quantitative detection. As the name implies, heat of combustion detectors are nonspecific, generic detectors for combustible gases. Some more specific heat of combustion detectors are available for carbon monoxide, ethylene oxide, hydrogen sulfide, methane, and oxygen deficiency. Most of these monitors read out in terms of percent of the lower explosive limit (LEL) or hundreds of ppm, and the limits of detection are a function of the analyte of interest. Table 4.8 lists the LELs of some organic compounds.

Table 4.8
Lower Explosive Limits of Selected Chemicals

Chemical	LEL (% in Air)
Acetaldehyde	4.0
Acetic acid 5.4	
Acetone	2.6
Acetonitrile	4.4
Acrolein	2.8
Allyl alcohol	2.5
Ammonia	12
Aniline	1.3
Benzene	1.3
Benzyl chloride	1.1
1,3-Butadiene	2.0
n-Butane	1.8
n-Butylamine	1.7
Carbon disulfide	1.3
Carbon monoxide	12.5
Chlorobenzene	1.3
Cyclohexane	1.3
Cyclohexene	1.0
Diborane	0.8
Dimethylamine	2.8
N,N-Dimethyl aniline	1.0
Ethane	3.0
Ethyl acetate	2.2
Ethyl benzene	1.0
Ethyl butyl ketone	1.4
Ethyl formate	2.8
Ethyl mercaptan	2.8
Ethylamine	3.5
Ethylene	3.1
n-Heptane	1.1
Hydrogen	4.1
Hydrogen sulfide	4.3
Isoamyl alcohol	1.2
Isobutane	1.8
Isopropyl alcohol	2.0
Methanol	6.7
Methyl acetate	3.1

Table 4.8 (cont'd)

Methyl ethyl ketone	2.0
Methyl formate	5.0
Methyl mercaptan	3.9
Nitrobenzene	1.8
m-Nitrotoluene	1.6
n-Octane	1.0
n-Pentane	1.5
Propane	2.2
Propylene	2.4
Propylene oxide	2.1
Styrene	1.1
Toluene	1.3
o-Toluidine 1.5	
Triethylamine	1.2
Vinyl chloride	3.6
m-Xylene	1.1

Thermal conductivity is another detection method used for explosive concentration that uses the specific heat of combustion of a gas or vapor as a measure of its concentration in air. It is used in instruments designed to measure very high concentrations, namely, percent of gas as opposed to percent of the LEL of a gas or vapor. When an instrument has dual scales for both percent gas and percent LEL, both a catalytic combustion and thermal conductivity sensor are incorporated into the instrument. Like heat of combustion detectors, thermal conductivity detectors are nonspecific detectors, and they are not sensitive to low levels. Because of the extremely high concentrations this instrument is capable of measuring, it must be used with extreme caution.

F. Colorimetric Indicator/Detector Tubes

Colorimetric indicator (detector) tubes are compact direct reading devices that are convenient to use for the detection and semi-quantitative estimation of gases and vapors in atmospheric environments. At present, there are tubes for several hundred atmospheric contaminants on the market. Indicator tubes have been widely advertised as being capable of use by unskilled personnel. While it is true that the operating procedures are simple, rapid, and convenient, many limitations and potential errors are inherent in this method. The results may be dangerously misleading unless the sampling procedure is supervised and the findings interpreted by a trained industrial hygienist.

Colorimetric indicator tubes are glass tubes filled with solid granular material, such as silica gel or aluminum oxide, which has been impregnated with an appropriate chemical reagent. The ends of the glass tubes are sealed during manufacture. The use of indicator tubes is extremely simple. After its two sealed ends are broken open, the glass tube is placed in the manufacturers holder, which is fitted with a calibrated squeeze bulb or piston pump. The recommended volume of air is then drawn through the tube by the operator. Adequate time must be allowed for each stroke. The observer then reads the concentration in the air by examining the exposed tube.

The earlier types of indicator tubes are provided with charts of color tints to be matched by the solid chemical in the indicating portion of the tube. Today, all indicator tubes are based upon producing a variable length of stain on the indicator gel. A scale is usually printed directly on the tube and the result of sampling can be obtained instantly. The range in the reading of results by different operators may be large, since in many cases the end of a stain front is not sharp.

A trained operator would take care to see that the pump valves and connectors are free of leaks. At periodic intervals, the flow rate of the apparatus should be checked and maintained within specifications. With most types of squeeze bulbs and piston pumps, the sample air flow rate is high initially and low toward the end when the bulb or pump is almost filled. This has been claimed to be an advantage because the initially high rate gives a long stain and the final low rate sharpens the stain front. The general certification requirement for the accuracy of indicator tubes is ± 25% of the true value when tested at one to five times the TLV. This accuracy requirement is modified to ± 35% at one half the TLV. At best, indicator tubes may be regarded as only range finding, and approximate in nature. Furthermore, many of the indicator tubes are far from specific. An accurate knowledge of the possible interfering gases that may also be present is very important.

VI. AIR SAMPLING FOR PARTICULATES

An advantage of integrated sampling for aerosols is that there are methods that can be used to identify, quantify, and differentiate between specific contaminants, whereas real-time methods are nonspecific and tend to measure particle counts or mass. As discussed earlier, an aerosol is a system consisting of airborne solid or liquid particles that are dispersed in a gas stream, usually the atmosphere. Dust, fumes, smoke, soot, mists, and fog are all terms used to describe certain types of aerosols. A unique aspect of aerosol sampling is the concern regarding particle size. The intent is to determine where the aerosol will interact with the respiratory system and the type of inhalation exposure, as mouth and nose breathing affect the composition of the aerosol inhaled as well as the amounts swallowed or absorbed into the lungs.

The most widely used particle-size definition for aerosols is *aerodynamic equivalent diameter (AED)*, based on the way a particle behaves when airborne in a force field such as air, rather than on its geometry under a microscope. When collection efficiency is discussed, aerodynamic diameter is commonly used rather than geometric diameter. For example, inertial impaction and gravitational forces are proportional to the particle mass and, in these cases, aerodynamic diameter is the particle size that governs particle motion. The cut diameter is the particle diameter for which collection efficiency and penetration equals 50%. This is useful for comparing efficiencies of various devices and efficiencies of the same device under various operating conditions even when penetration does not reach 50%.

A. Collection Techniques

For the most part, the general collection techniques discussed earlier for gases and vapors apply to aerosol collection as well. However, because of the wide range of airborne particulates confronting the industrial hygienist, there are several aspects of sampling that apply only for aerosols. The sampling train for aerosols consists of the following components: air inlet orifice, particle collection medium, flowmeter, flow rate control device, and pump. Of these, the most important by far is the particle collection medium used to separate the particles from the sampled air stream. Both the efficiency of the device and its reliability must be high. The pressure drop across the medium should be low, to keep the size of the pump to a minimum. The medium may consist of a single element, such as a cassette or impinger, or there may be two or more elements in series, to separate the particulate into different size ranges.

Typically, the cassette contains a membrane or fiber filter on which the aerosol is collected. Cassettes and filters come in many sizes, and some cassettes have an extension cowl such as is used in asbestos sampling to decrease electrostatic effects. In some cases, the cassette is very small, such as when it is used as a prefilter to screen unwanted material from gas and vapor samples. In other cases, the cassette is contained within another piece of sampling equipment, such as a cyclone, to provide size selection characteristics.

1. *Filters*

In the occupational environment, particulates are collected on filters, impactors, impingers, and elutriators. Filters in plastic cassettes are the most common media. Standard filters are 37 mm in diameter and are placed in closed-face cassettes, with a backup pad, to avoid contamination (Figure

4.6). The OSHA Asbestos Standard, however, is an exception. It requires a 25-mm filter and open-face (top removed) cassette and a 50-mm extension cowl on the cassette. The basic types of filters used for aerosol sampling are membrane filters and fiber filters.

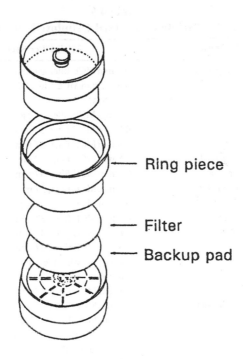

Figure 4.6   Example of "standard" filter cassette.

Membrane filters are made from mixed cellulose ester (MCE), polyvinyl chloride (PVC), or polytetrafluoroethylene (PTFE). Since most particle collection takes place at or near the surface, membrane filters are useful for applications where the collected particles will be examined under a microscope. MCE membrane filters dissolve easily with acid. Hence, they are used for collection of asbestos fibers, and also used for collection of metals and metal fumes for atomic absorption analysis. Pore sizes of 0.45 mm and 0.8 mm are the most commonly used. Table 4.9 lists selected compounds that are collected on MCE filters. PVC membrane filters are used for sampling carbon black, respirable silica, total and respirable dusts, and zinc oxide. Table 4.10 lists selected compounds that are collected on PVC filters. PTFE membrane filters come in a variety of pore sizes, 1 mm, 2 mm, and 5 mm, and are used for pesticides, alkaline dusts, and many other compounds. Selected compounds that are collected on PTFE filters include polycyclic aromatic hydrocarbons (e.g., anthracene, benzo(a)pyrene, chrysene, etc.); coal tar pitch volatiles; alkaline dusts; diborane; paraquat; methyl arsonic acid; pentachlorobenzene; o-toluidine; thiram; 1,2,4-trichlorobenzene; and warfarin.

Table 4.9
Chemicals Collected on Mixed Cellulose Ester Filters

Chemical	MCE Filter Specifications
Aluminum and compounds	0.8 mm MCE filter, 37 mm
Asbestos	0.8-1.2 mm MCE filter, 25 mm with cowl
Barium	0.8 mm MCE filter, 37 mm
Benzoyl peroxide	0.8 mm MCE filter, 37 mm
Beryllium and compounds	0.8 mm MCE filter, 37 mm
Cadmium and compounds	0.8 mm MCE filter, 37 mm
Calcium and compounds	0.8 mm MCE filter, 37 mm
Chlorinated diphenyl ether	0.8 mm MCE filter, 37 mm
Chromium and compounds	0.8 mm MCE filter, 37 mm
Cobalt and compounds	0.8 mm MCE filter, 37 mm
Copper dust and fume	0.8 mm MCE filter, 37 mm
Cyanides	0.8 mm MCE filter, 37 mm and potassium hydroxide solution in a bubbler
Dibutyl phthalate	0.8 mm MCE filter, 37 mm
Di-(2-ethylhexyl)-phthalate	0.8 mm MCE filter, 37 mm
Ethylene thiourea	0.8 mm MCE filter, 37 mm
Fluorides	0.8 mm MCE filter, 37 mm and sodium carbonate treated cellulose pad in series
Hydroquinone	0.8 mm MCE filter, 37 mm
Kepone	0.8 mm MCE filter, 37 mm and sodium hydroxide solution in an impinger
Lead	0.8 mm MCE filter, 37 mm
Mineral oil mist	0.8 mm MCE filter, 37 mm
Phosphorus	0.8 mm MCE filter, 37 mm and Tenax GC tube, 100/50
Sulfur dioxide	0.8 mm MCE filter, 37 mm and cellulose filter impregnated with potassium hydroxide, 37 mm, in series
Tungsten and compounds	0.8 mm MCE filter, 37 mm
Welding and brazing fume	0.8 mm MCE filter, 37 mm
Zinc and compounds	0.8 mm MCE filter, 37 mm

Table 4.10
Chemicals Collected on Polyvinyl Chloride Filters

Chemical	PVC Filter Specifications
Azelaic acid	5 mm PVC filter, 37 mm
Boron carbide	5 mm PVC filter, 37 mm
Carbon black	5 mm PVC filter, 37 mm
Chromium(V)	5 mm PVC filter, 37 mm
Dust (respirable)	10 mm nylon cyclone and 5 mm
	PVC filter, 37 mm
Dust (total)	5 mm PVC filter, 37 mm
Ethylene thiourea	5 mm PVC filter, 37 mm
Lead sulfide (respirable)	10 mm nylon cyclone and 5 mm
	PVC filter, 37 mm
Mineral oil mist	5 mm PVC filter, 37 mm
Silica, amorphous (respirable)	10 mm nylon cyclone and 5 mm
	PVC filter, 37 mm
Silica, crystalline (respirable)	10 mm nylon cyclone and 5 mm
	PVC filter, 37 mm
Vanadium oxide (respirable)	10 mm nylon cyclone and 5 mm
	PVC filter, 37 mm
Zinc oxide	0.8 mm PVC filter, 25 mm

The primary fiber filter is made of fiberglass. For occupational sampling, fiberglass filters are used to collect pesticides such as 2,4-dichlorophenoxyacetic acid (2,4-D). Table 4.11 lists selected compounds that are collected on glass fiber filters. Polycarbonate filters have superior strength and chemical and thermal stability. They are essentially transparent with a slight green tinge and are recommended for asbestos sampling for transmission electron microscopy.

Generally, the type of filter medium is specified by the analytical method; so in most cases, the industrial hygienist does not have to understand how to select a given filter medium. However, it is important to understand the limitations of each type of filter and why particular types are preferred for various uses.

2. Impactors

Impactors take advantage of a sudden change in direction in airflow and the momentum of the dust particles to cause the particles to impact against a flat surface. Usually, impactors are constructed in several stages, to separate dust by size fractions. Figure 4.7 shows the schematic diagram of a two-stage impactor. The particles adhere to the plate, which may be dry or coated with an adhesive, and the material on each plate is weighed or analyzed at the conclusion of sampling.

Table 4.11
Chemicals Collected on Glass Fiber Filters

Chemical	Glass Fiber Filter Specifications
Aldrin and	Glass fiber filter (type AE), 37 mm, isooctane in a bubbler
Benzidine and	Glass fiber filter (type AE), 13 mm, silica gel, 50
Chlorinated terphenyl	Glass fiber filter, 37 mm
Dibutyl tin *bis*(isooctyl-	Glass fiber filter, 37 mm, and XAD-2 tube, 80/40 mercaptoacetate
2,4-Dichlorophenoxyacetic acid 37	Binderless glass fiber filter (type AE), mm benzenethiophosphonate
Ethylene glycol and	Binderless glass fiber filter, 13 mm, silica gel tube, 520/260
Lindane and	Glass fiber filter (type AE), 37 mm, isooctane in a bubbler
Malathion	Glass fiber filter (type AE), 37 mm
Mercury	Glass fiber filter, 13 mm, and silvered Chromosorb P, 30
Mineral oil mist	Glass fiber filter, 37 mm
Naphthylamine	Glass fiber filter, 13 mm, and silica gel tube, 100/50
Parathion	Glass fiber filter, 37 mm
Polychlorinated biphenyl	Glass fiber filter, 13 mm, and fluorisil tube, 100/50
Pyrethrum	Glass fiber filter, 37 mm
Strychnine	Glass fiber filter, 37 mm
Tributyl tin	Glass fiber filter, 37 mm, and XAD-2 tube, 80/40
Tributyl tin chloride	Glass fiber filter, 37 mm, and XAD-2 tube, 80/40
2,4,5-Trichlorophenoxyacetic acid (type AE), 37	Binderless glass fiber filter mm
Tricyclohexyltin hydroxide	Glass fiber filter, 37 mm, and XAD-2 tube, 80/40

Cascade impactors have been widely used to separate an airstream of particulates into a particle size distribution that allows the investigator to characterize the primary size ranges according to their aerodynamic diameters. Particles enter the inlet jet and pass through a series of progressively smaller jets with which there is an associated collection plate. As the airstream of particulates moves through the plates, larger particles are deposited on the top stages and smaller ones are deposited near the bottom. If a series of plates are used, particles are collected in different size ranges. After the last impactor stage, remaining fine particles are collected on a filter or plate. Thus, the particles on each plate can be analyzed for total weight, particle count, and chemical composition. Airflow requirements are high with this type of instrument.

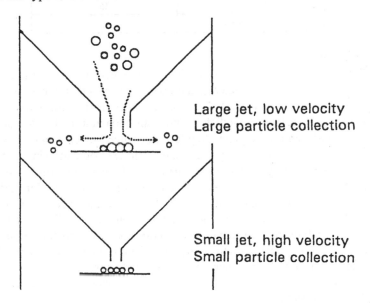

Figure 4.7 Two-stage cascade impactor showing large and small particle collection.

3. *Cyclone*

It is sometimes necessary to measure only the respirable fraction of dust. These are particles that are 10 mm or less in size and penetrate deep into the lungs. The cyclone was developed as a personal respirable dust sampling device. It is a centrifugal separator. Air enters the cyclone tangentially through an opening in the side of a cylindrical or inverted cone-shaped unit, and is drawn along a concentrically curved channel. Larger particles impact against the interior walls of the unit due to inertial forces and drop into a grit chamber in the base of the assembly. The lighter particles continue through and are drawn up through the center of the center of the cyclone, where they are collected on a membrane filter. Table 4.12 lists the collection characteristics of cyclone with respect to respirable dust.

Table 4.12
Collection Characteristics of Cylones

Particle Size in m	% Captured
2 or smaller	90
2.5	75
3.5	50
5.0	25
10 or larger	0

4. *Impingers*

Impingers operate much like an impactor, except that the sampled airstream jet is immersed in water at the bottom of the glass apparatus. The sampled airstream is accelerated in the impinger orifice and exits underneath the liquid surface immediately above an impaction plate. As the particles are impinged on the plate, they lose their high velocity and are wetted by the liquid. The particles trapped in the liquid are counted using a microscope. "Dust counting" is the determination of the particle number concentration, i.e., millions of particles per cubic foot of air (mppcf), for particles such as graphite, mica, and mineral wool fibers. In the past, the American Conference for Governmental Industrial Hygienists (ACGIH) used such dust concentration measurements to set more than a dozen threshold limit values (TLVs) for occupational exposure. These TLVs have since been converted from particle number concentrations to respirable mass concentrations, and impinger sampling for particles has been largely replaced by respirable mass sampling.

5. *Elutriators*

The elutriator is a separator or purifier. Elutriators are used in the front of a sampling train to remove coarse particulate matter, enabling a filter or other collection device to collect only the smaller sized particles. There are two types of elutriators: horizontal and vertical. Both use normal gravitational forces to separate particles. Vertical elutriators are typically used for sampling cotton dust.

B. Real-Time Instruments

Particulate sampling instruments described so far are used to collect particles for subsequent microscopic, gravimetric, or chemical analyses. Instruments considered in this section are more complex. Sampling and analysis are carried out within the instrument and the property of interest can be obtained immediately. Instruments of this type are called real-time, or direct reading, instruments, which are available to cover particles in the size range of 0.002 to 50 mm. These instruments have a fast time response and can follow rapid changes in both particle size and concentration. Good counting statistics can also be obtained because repeated measurements can be performed in a short time.

It must be pointed out that different aerosol properties are measured by different real-time instruments. Although many real-time instruments provide data in particle size, this size is derived from one of many possible particle properties, such as its gravimetric, optical, aerodynamic, mechanical, or force field mobility behavior. Thus, these particle sizes may not be directly compared without some correction of the data to account for these differences. Other aerosol properties determined by real-time instruments include aerosol number concentration, aerosol mass concentration, size distribution, opacity, and chemical composition.

The user of real-time instruments must also be aware of comparing properties of the same aerosol determined by several real-time instruments, particularly those using different principles, because this comparison is likely to give contradictory information. It is important to know what property is changed in the sensing zone and how this is assumed to be related to an aerosol property.

With a wide array of available commercial instruments, it is necessary for the investigator to understand the operating principles of the instruments and the aerosol system under study. Some of the criteria for selecting an appropriate instrument include the particle size range of interest, the system parameters to be studied, and the cost and compactness of the instrument.

For mass concentration measurement, a quick and easy method would be to indirectly measure the light-scattering intensity of the aerosol and to infer its mass concentration through calibration. The accuracy of the technique depends on the measured aerosols having nearly the same size distribution and differing only in concentration. For number concentration measurement, as required in many cleanroom applications, two classes of instruments may be used, depending on the size range of interest. For particle diameters less than 2 mm, several condensation particle counters may be used. For particle diameters larger than 0.1 mm, a large array of white light or laser optical particle counters may be used.

For size distribution measurement, the real-time instruments are divided into two major classes, depending on their measuring size ranges. For particles larger than 0.1 mm, the optical particle counters or the particle relaxation size analyzer may be used. The latter is capable of measuring the aerodynamic particle size, which is important for deposition studies in respiratory system. For submicron aerosols, the high-resolution, differential mobility particle sizes or the low-resolution diffusion battery and the electrical aerosol analyzer may be used.

VII. AIR SAMPLING FOR BIOAEROSOLS

Bioaerosols are airborne particles derived from microbial, viral, and related agents. They come in a wide variety of sizes, shapes and classifications. Human responses to bioaerosols range from innocuous effects to serious disease and depend on the specific agent and susceptibility factors within the person. Healthcare professionals increasingly recognize bioaerosols as a cause of preventable airborne infections and hypersensitivity disease, as well as related absences and lowered productivity in the workplace.

Bioaerosols can exist in both viable and nonviable states. Viable microorganisms such as bacteria, fungi, yeasts, and molds originate from sprays or splashes of media, from the agitations of dusts, and from sneezes and coughs of which only the small particles (<10 mm) remain in the air long enough to travel any distance. Spores, which can be formed by fungi and certain bacteria, can be both viable and nonviable and are capable of causing disease in both forms.

Bioaerosols can cause two basic conditions: infectious diseases such as Legionnaires' disease, tuberculosis, and histoplasmosis; and hypersensitivity reactions such as allergic asthma and rhinitis. Infections are generally the result of multiplication and growth of microbes inside humans, while hypersensitivity reactions are the result of exposures to antigens. Well-known diseases associated with occupational exposures include anthrax, brucellosis, and Q fever. Diseases for which concerns are increasing are those associated with health workers and include hepatitis and acquired immunodeficiency syndrome (AIDS). Antigens are capable of stimulating the production of antibodies that produce allergic reactions. Hypersensitivity reactions are the result of an antigen producing a response from the immune system. Sources of airborne antigens include bacteria, fungi, pollen, insect body parts, and dander, as well as saliva of animals. Sometimes, it is not the microbe itself that produces the harmful effect but the fact that it produces a toxin, e.g., endotoxins from Gram-negative bacteria and mycotoxins from fungi. Botulism is an example wherein the botulinum toxin is the responsible agent. When release of a toxin is involved, the organism can produce a disease without extensive multiplication or dissemination throughout the body.

The indoor environment usually contains a different variety of bioaerosols than the outdoor environment, e.g., there are greater numbers of airborne bacteria from humans indoors than outdoors, but more airborne fungi from environmental sources outdoors than indoors. Unlike the use of coliform bacteria to assess drinking water quality, there is no one group of "indicator organisms" to measure biological air quality. At present, gravimetric Threshold Limit Values exist for cotton dust and for some wood dust, both of which may be considered of biological origin. However, there are no TLVs for concentrations of culturable or countable bioaerosols. This is because culturable organisms or countable spores do not comprise a single entity and the measured concentrations of culturable and countable bioaerosols are dependent on the method of sample collection and analysis. It is not possible to collect and evaluate all of these bioaerosol components using a single sampling method.

A. Bioaerosol Sampling Considerations

The specialized characteristics of viable agents require specialized sampling instruments in order to preserve the organisms for laboratory culture, which is the primary means of identification. The fragility, temperature, moisture, and nutrient needs are the primary considerations when selecting a sampling device. While passive air sampling is simple and can be done by setting out plates containing culture media, it is not as effective as the use of active techniques involving the use of pumps. There are two basic methods for collecting these air samples: (1) air sampling trains incorporating a pump, rotameter, and media, such as used for integrated sampling for airborne chemicals, and (2) specialized bioaerosol sampling instruments. The specialized instruments can be used to house culture media and therefore, in most cases, are preferred to integrated sampling techniques. Area air samples are more commonly collected for bioaerosols than personal samples, regardless of the type of situation being monitored, due to the need to house culture media inside of instruments specialized for sampling viable bioaerosols.

At present, available instruments for sampling of bioaerosols tend to be somewhat less sophisticated than those for chemical particles. There are no real-time sampling instruments to detect airborne microorganisms or to identify them immediately after collection, except for particles that can be recognized by microscopic examination. Although automated identification systems for cultured bacteria are becoming more available and reliable, samples still require fairly time-consuming processing before these can be applied. Unique to sampling viable organisms is a requirement for aseptic equipment and specimen handling. Microorganisms are ubiquitous. Bioaerosol sampling results may be incorrect and misleading if the personnel collecting the samples do not appreciate the need to practice appropriate sterile techniques.

The investigator must consider possible bioaerosol sampling locations, number of samples, sample collection time, effects of seasonal and temporal variations in bioaerosol concentration, and assay system limitations. Incident airflow should be parallel and going in the same direction as the suction flow of the sampling equipment. All sample volumes should be adjusted to provide a sensitivity near 50 colony forming units (CFUs). Sampling times at each site should be varied from this maximum time and volume to prevent overloading, especially when culture plate samplers are used, and to allow for the logarithmic variability that these organisms display. Comparing to the collection of chemical samples, sampling periods for bioaerosols are much shorter, especially when viable organisms are being collected.

For indoor air studies, sampling should be done during work periods when occupants are present and also when the building is empty. A concern is that during off-periods, ventilation systems are often set to 100% recirculated air, whereas during the workday, a certain percent of outdoor air is utilized. In the indoor environment, samples should be collected in the supply and return air of rooms housing the affected occupants, as well as in other rooms where occupants have no complaints.

The investigator must also specify the desired level of identification and quantification of the recovered material. The last two factors are especially important because the biological, chemical, and physical condition of a bioaerosol is often critical for accurate identification and quantification. When studying bioaerosols, the only information that may be needed is the concentration of particles containing cells or spores without concern for whether the particles contain one or more individual units. Collection of the microorganisms directly on agar-based culture media, on which the microorganisms form discrete colonies, is the simplest method to determine the concentration of airborne particles carrying culturable microorganisms. In addition, some bioaerosols can be collected on slides and filters for microscopic examination, counting intact particles and estimating the number of individual cells or spores in each aggregate.

Prior to sampling, an appropriate laboratory should be selected for analyses. It should be pointed out that most clinical hospital laboratories have little experience in analyzing environmental air samples. It is imperative for the investigator to seek out competent laboratories for the analysis and identification of bioaerosol samples. Colony counts should be done in accordance with the manufacturer's specifications for the sampling instrument used. Such specifications should be provided along with the samples to the bioaerosol analytical laboratory. Table 4.13 lists sampling characteristics for various types of bioaerosol sampling instruments.

B. Integrated Sampling with Impactors

Currently, most integrated bioaerosol sampling methods rely on impaction. Impactors have short sampling times and allow for a wide variety of organisms to be sampled for over a short period of time, either because the sampled surface can be examined directly under a microscope, or a variety of sampling strips containing different culture media can be used sequentially. Collection by impaction onto an agar surface makes it possible to count viable particles and particle clumps as colony forming units.

Table 4.13
Air Samplers for Bioaerosol Sample Collection

Sampler	Operation	Sample Collection Rate, LPM	Sampling Duration, minutes
Slit-to agar	Impaction onto agar in a culture plate on a rotating turntable	28–50	1–60
Multiple hole impactors			
a. Single-stage	Impaction onto agar in a culture plate	10–28	1–30
b. Two-stage	Impaction onto agar in a culture plate	28	1–30
c. Six-stage	Impaction onto agar in a culture plate	28	1–30
d. Eight-stage	Impaction onto substrate or media in special trays	2	5–30
Liquid impingers			
a. Single-stage	Impingement to liquid	12.5	1–30
b. Three-stage	Impaction onto agar in a culture plate	10–50	1–30
Filters			
a. MCE filters	Filtration	1–5	1–30
b. Polycarbonate filters	Filtration	1–5	1–30
Settling plates			
a. Open culture plates	Gravity settling onto agar in a culture plate	Undefined	Day
b. Adhesive-coated surface	Gravity settling onto coated surface, e.g., microscopic slides	Undefined	Day

Slit-to-Agar Impactors. The slit-to-agar impactor is used for collecting screening samples of bioaerosols. An impaction orifice is located on top of the unit. Air is drawn through a slit and impacted directly onto an agar medium on a rotating turntable. Investigators should take caution not to overload stationary impaction slides and plates when sampling in heavily contaminated environments.

Multiple-Hole Impactors. The commonly used multiple-hole impactors come in different sizes: a single-stage, two-stage, six-stage, or even an eight-stage impactor. Depending on the design of the manufacturer, the number of holes can range from 100 to over 400. The single-stage instrument is useful in situations such as large office buildings, where it is important to sample simultaneously in many areas of a building in order to determine the relationship of the ventilation system and its variables with levels of viable microbes in various areas. In the multiple-stage impactor, the orifice size is constant for a given stage, but decreases for each following stage. Thus, the cascade impactor can be used to separate viable particles into respirable and nonrespirable particles. Directly below each stage is a glass petri dish containing a nutrient agar medium. Air is drawn through the holes and impacted directly onto the agar medium.

C. Integrated Sampling with Liquid Impingers

While collection by impaction onto an agar surface makes it possible to count particle clumps as colony forming units, the vigorous motion occurring in an impinger causes the clusters of cells to break apart and makes it possible to count total viable cells. Liquid impingers collect microbes directly into a liquid, providing some protection for the microorganisms, and allows immediate initiation of repair of any damage that may have been caused by the collection process itself. The sampled liquid can then be processed to detect a wide range of bioaerosol concentrations. Another advantage of having the sample in a liquid is that several types of media can be inoculated from the suspension or several plates of the same medium can be inoculated and incubated under different conditions.

Currently, there is an all-glass, single-stage liquid impinger available and an anodized aluminum alloy, or stainless steel three-stage liquid impinger. In both types of liquid impingers, air is drawn through a critical orifice and impacted against a wetted surface.

D. Integrated Sampling with Filters

Filtration is another method used for the collection of bioaerosols. Materials collected on filters can be washed into suspension for transfer to an assay system, or filters can be examined directly with a microscope. Mixed cellulose ester membrane filters are used for samples for light microscopy and for culturing, while polycarbonate filters are used for scanning electron microscopy or epifluorescence microscopy.

E. Passive Sampling for Bioaerosols

Passive sampling methods were probably the first attempt to collect airborne microorganisms and are still used today. In setting out culture plates, otherwise known as settling plates or gravity sampling, dishes of culture media or adhesive-coated glass slides are used that are placed in various areas and left open for air exposure for a number of hours. Then they are incubated and the number of colonies counted. The best application of passive sampling methods is as a screening technique when instruments for active air sampling are not available.

VIII. QUALITY CONTROL AND CALIBRATION TECHNIQUES

Proper interpretation of any environmental measurements depends on the quality of data, i.e., its accuracy and precision, and whether or not it is representative of the exposure conditions. Although considerations of the sampling location, sampling duration, and frequency of sampling are just as important in the evaluation of environmental hazards, they require knowledge of the process variables. It is important to document the nature and frequency of calibrations and calibration checks to meet legal, as well as scientific requirements. Measurements made to document the presence or absence of excessive exposures are only as reliable as the calibrations on which they are based. Formalized calibration audit procedures established by federal agencies provide a basis for quality assurance where they apply. They can also provide a systematic framework for developing appropriate calibration procedures for situations not governed by specific reporting requirements. The *Manual of Analytical Methods*, published by the National Institute for Occupational Safety and Health, recommends that sampling pumps be calibrated with each use, and that this calibration be performed with the sampler in-line. It also recommends that records of calibration be recorded with each unit.

A. Air Volume and Flow Rate Calibration

Accurate measurements of air volume and flow rate are an integral part of the calibration of most air sampling instruments. The various instruments and techniques in the calibration of air volume and flow rate can be divided into two general categories: primary and secondary standards. Primary standards use primary measurements which generally involve a direct measurement of volume on the basis of physical dimensions of an enclosed space. The primary standards have the least amount of error. Secondary standards are reference instruments or meters that trace their calibration to primary standards and which have been shown to be capable of maintaining their accuracy with reasonable handling and care in operation, though periodic recalibration with an appropriate primary standard is essential to ensure reliability and accuracy.

1. *Air volume calibration*

Many air sampling instruments utilize an integrating volume meter for measurement of sampled volume. Most of them measure displaced volumes that can be determined from linear measurements and geometric formulae. Such measurements usually can be made with a high degree of precision. Primary standards in air volume calibration include the spirometer (or gasometer) and the Mariotte bottle, as well as the soap bubble flowmeter. The spirometer and the Mariotte bottle measure displaced water instead of air.

When the valve at the bottom of the bottle is opened, water drains out of the bottle by gravity, and air is drawn into the bottle via a sample collector to replace it. The volume of air drawn in is equal to the change in water level multiplied by the cross-section at the water surface. At high flow rates, the accuracy of soap bubble meters declines because of potential gas permeation through the soap film. Secondary standards in air volume calibration include the dry gas meter and the wet test meter.

The electronic bubble meter is a recent innovation which involves incorporating a microprocessor and a printer to create an electronic bubble meter. These units are commercially available. A bubble generator produces a bubble film that is carried by a pump's airflow from the bottom to the top of a cylinder. As the bubble traverses past two infrared sensors, one mounted at the bottom and the other at the top of this flow tube, each sensor transmits a signal to a microprocessing unit that stores the time and performs the necessary calculation to determine the flow rate. Timers are usually capable of detecting a soap film at 80-msec intervals. Under steady flow conditions, this speed allows an accuracy of $\pm$ 0.5% of display readings. These units can provide almost instantaneous airflow readings and a cumulative averaging of multiple samples. The user can save calibration runs and average them. Electronic bubble meters are accepted by the Occupational Safety and Health Administration as a primary standard for volume calibration.

2. Flow rate calibration

Flow rate meters operate on the principle of conservation of energy. Each flow rate meter consists of a flow restriction within a closed conduit. The restriction causes an increase in the fluid velocity and therefore an increase in kinetic energy, which requires a corresponding decrease in potential energy. The flow rate can be calculated based on the pressure drops, the flow cross-section at the constriction, the density of the fluid, and the coefficient of discharge, which is the ratio of actual flow to theoretical flow and makes allowance for stream contraction and frictional effects. Commonly used standards in flow rate calibration include the rotameter, orifice meter, venturi meter, and the laminar flowmeter.

Electronic mass flow monitors can measure flows and generate a voltage output. These meters are hot wire devices in which measurement of the flow is based on thermal conduction. The temperature of the airstream depends on the mass rate of flow and the heat input. The readout is in actual flow units, so no extrapolation or calculations are required. If properly calibrated against a bubble buret, these units provide ease of use in the field, including a wide range of flow rates with a single instrument. Some electronic mass flowmeters are available with digital readouts.

B. Gas and Vapor Calibration

Proper interpretation of any industrial hygiene measurements depends on an appreciation of its accuracy, precision, and whether or not it is representative of the exposure condition. For the calibration of air sampling instruments, a known concentration of a specific contaminant is the best way to ensure accuracy. Methods of producing known concentrations are usually divided into two general classes: **(1)** static or batch mixture systems, and **(2)** dynamic or continuous flow systems. With static systems, a known amount of gas is mixed with a known amount of air to produce a known concentration, and samples of this mixture are used for calibration. Static systems are limited by two factors: loss of vapor by surface sorption and the loss of finite volume of the mixture. In dynamic systems, air and gas or vapor are continuously metered in proportions that will produce the final desired concentration. They provide an unlimited supply of the test atmosphere, and wall losses are negligible after equilibration has taken place.

1. Static calibration mixtures

Static calibration mixtures are prepared by introducing a known mass or volume of contaminant into a given volume of air. Generally the mixture is held in a container of fixed dimensions, but flexible containers may be used. Liquid contaminants are commonly introduced into the calibration system with a microsyringe or micropipet. Gaseous materials are handled with a gas-tight syringe. A pure gas sample may be obtained from a lecture bottle. The following equation is used to determine static calibration contaminant concentrations:

$$C = C_0 e^{(-V/V_0)} \tag{4.9}$$

or:

$$2.3 \log 10 \frac{C_0}{C} = \frac{V}{V_0} \tag{4.10}$$

Where:

C = Total contaminant, any time
V = Volume of sample withdrawn
C_0 = Original concentration
V_0 = Chamber volume

Generally, glass bottles and flasks make the best *rigid containers*. However, plastic and metal containers may be used if wall interactions with the contaminant gases are negligible. The volume of the container is obtained by direct measurement of the chamber boundaries, or volumetrically by filling with water. The usual size is on the order of 20 to 40 liters, which allows the removal of enough useful gas without causing excessive dilution by the replacement gas. A typical static calibration system is shown in Figure 4.8. The vessel is first purged with fresh air to remove any residual contamination; then the contaminant is injected directly into the vessel through the rubber septum. The gases and vapors are evaporated and mixed by an externally driven stirrer or magnetic stirrer. The gas mixture is then withdrawn after opening both valves. Placement of two or three of these vessels in series greatly reduces the dilution effect of the air as the sample gas mixture is removed.

Mixtures of air contaminants and air may also be prepared in plastic bags, generally made from Teflon, Mylar, and aluminized Mylar *non-rigid containers*. The bag is alternatively filled and deflated with air to remove residual contamination and final purge is accomplished by applying a vacuum to the system. The container is then connected to a flowmeter, a wet test meter, or a dry gas meter. After the bag is filled with a known quantity of air, the contaminant is introduced by a syringe.

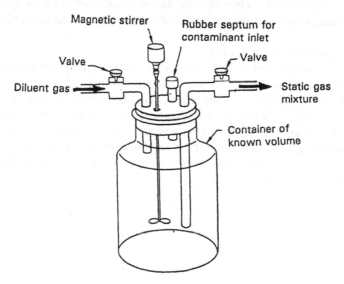

Figure 4.8   A "typical" static calibration system.

2. Dynamic calibration mixtures

In a gas and vapor calibration system, the term *dynamic* describes that the calibration mixture is continuously produced and fed into a calibration chamber following vaporization, dilution, and mixing. The principle of the use of permeation tubes as primary standards is based on diffusion of gas or vapor through a plastic membrane at very slow rates. A liquefied gas or volatile liquid sealed in a section of Teflon tubing placed in a metered airstream can be used as a dynamic calibration standard. The diffusion rate is a nonlinear function of temperature. Therefore, constant temperature conditions must be maintained for the permeation tube during gravimetric standardization and use as a calibration source. Also, the diffusion rate is influenced by the molecular weight of the carrier gas employed in the dilution. Dry nitrogen is commonly used as the carrier gas. The calculation employed to determine dilution calibration concentrations is:

$$C_t = C_o\, e^{-bt} \tag{4.11}$$

Where:

C_t = Concentration of agent in vessel at time t

C_o = Initial concentration at $t = 0$

e = Base of natural logarithms

b = Air changes in vessel per unit time

t = Time

Permeation tubes are available from several commercial sources and a wide range of permeant gas choices is available. The vendors of permeation tubes also offer custom-built permeation tubes, with the material designated by the customer. Perhaps the most important parameter in the standardization of permeation tubes is the time factor required for diffusion equilibrium to be reached. Two to five days should be allowed in a constant temperature oven under a flowing dry air or nitrogen stream for accurate standardization. The time is dependent upon diffusion rate and accuracy of weighing. In use, constant temperature should be maintained to $\pm 0.1°C$ for precision within $\pm 1\%$. Care must be taken to assure tight seals at the tube ends.

To a very good approximation, the output of a permeation tube is proportional to its active length, i.e., the length of tube in contact with the contained fluid. If the output of any given tube is known, then a rate per unit length can be calculated. To achieve any desired output, the length in cm of permeation tube required can be calculated from the following equation:

$$L = \frac{P}{R} \tag{4.12}$$

Where:

L = Required length of the permeation tube, cm

P = Required output, ng/min

R = Permeation rate, ng/min-cm

Table 4.14 lists the permeation rates of some chemicals. Thus, if an output of 5000 ng/min of H_2S is needed, using the tubes listed in Table 4.14, then the required length is $5000/250 = 20.0$ cm.

Table 4.14
Permeation Rate for Selected Chemicals

Chemical	Permeation Rate, ng/min-cm at 30°C
Ammonia	170
n-Butane	2
Chlorine	1500
Hydrogen sulfide	250
Methyl mercaptan	30
Nitrogen dioxide	1200
Propane	80
Sulfur dioxide	290

C. Real-Time Instrument Calibration

The primary concerns for quality control when using real-time instruments are understanding and minimizing sources of error through proper calibration, maintenance, and use of the instrument. Initial check and zero, and calibration procedures, should be performed daily and prior to use for monitoring. Of these common procedures, the most critical first step is proper calibration. Chemical calibration of real-time instruments ranges from factory calibration with several concentration ranges of different gases, to a specified concentration of a single gas done by the user of the instrument, to a field check with an unknown gas concentration. Small real-time instruments units are sometimes calibrated electronically rather than with a known gas or vapor. In electronic calibration, a voltmeter is generally used to generate a readout adjusted to match a number on the cell by turning a screw. Most instruments using vacuum pumps are designed to sample at a specified flow rate. The user is required to know that rate and what rate would represent a significant shift higher or lower than that flow. The manual of the instrument would most likely indicate what effect these changes could have on the instrument's response.

The preferable way to calibrate concentrations of gases and vapors is with the specific chemical that is going to be present during the survey. The precision of a real-time measurement is a function of the concentration of the contaminant present and the range of its linear measurements. The concentration of calibration gas or vapor should be similar to the concentration of contaminant to be monitored in the survey. Ideally, calibration before and after each use should be performed. It is also recommended that the calibration of most real-time survey instruments be carried out using two or more concentrations of the reference gas in order to verify the response over the full range of concentrations.

Prior to calibrating any instrument, it is a good practice to review the manufacturer's manual. Most manuals give step-by-step instructions for use with their specific instrument. In addition, any idiosyncrasies or problems encountered during calibration or in use, are often noted. Often, instrument manufacturers offer their own calibration kits. Always heed the manufacturer's instructions when selecting calibration methods for any given instrument.

Generally, most real-time instruments are a complete system and very few malfunctions can be easily corrected by the user. Thus, it is prudent for the user to check out an instrument in the laboratory prior to a survey in the field. A preventive maintenance program will also prevent or minimize problems in the field. Most reputable manufacturers have a toll-free hotline phone number for customers to use. Users are encouraged to call the manufacturer and use its services.

IX. CONCLUSION

Evaluating the impact of airborne chemical or bioaerosol contaminants necessitates the accurate determination of the amount of the contaminant present in a unit volume of air by a trained industrial hygienist. This can be accomplished either directly from the airstream or following collection upon a suitable medium. A successful operation involves four areas that are equally important in the accurate monitoring of airborne contaminants: (1) sampling, (2) calibration, (3) analysis, and (4) data interpretation. The accuracy, and conversely the error, in the determination of the concentration of a chemical or bioaerosol contaminant is a function of all four of these aspects of air sampling; and to consider one without the other three often results in the incorrect, or even improper determination of airborne concentrations.

X. CASE STUDY

An automotive manufacturer operates two engine assembly plants, Plant 1 and Plant 2, in Windsor, Ontario, Canada. Both plants are identical in layout and produce V-8 engines from precast iron or aluminum parts. The parts are ground and machined to the proper sizes on the production lines and then assembled into complete engines on the assembly line. Because of an oversupply of engines, Plant 2 was mostly shut down for 10 days, except for a small group of workers in two departments and they continued to work through the shutdown period. Two days after all workers returned to work and 8 o'clock in the morning, most of the workers in three departments in Plant 2 had become ill. One of those three departments was manned by the workers who worked through the shutdown period, while the workers in the other two departments had been off and just returned to work. The workers were complaining of headaches, severe body aches, high fever, and extreme fatigue. Plant 2 was immediately shut down, and an investigation began. The investigation found that 317 out of a total of 833 workers in Plant 2 became ill but none of the workers in Plant 1 were affected. Attack rates varied significantly by department in the plant. It decreased progressively from one end of the plant to the other end, suggesting a possible source, or cause, of the illness. The illness was short, with a median duration of 3 days, but severe enough to cause many of the ill workers to miss 2 days' work. There were no fatalities.

In Plant 2, the production departments and the assembly lines are serviced by several systems that produce aerosols or exhaust humidified air into the plant or on the roof: compressed air lines, parts washers, wet air cleaners, and coolant systems. The coolant systems lubricate, cool, and clean the grinding and machining surfaces. The individual coolant systems do not interconnect, and they range in capacity from 1000 to 30,000 gallons. The coolant is 88 to 99% water and 1 to 12% oil. Caustics are added as needed to keep the coolant's pH between 8.5 and 9.5 (except in systems servicing aluminum machining, for which the pH should not exceed 9), and biocides are added when the bacterial count exceeds 10^5 to 10^6 organisms per milliliter. The coolant, which also contains metal shavings, dirt, and other debris, is mechanically circulated through underground troughs from a main tank, to the machines, and back to the main tank. As the coolant is applied to the grinding or machining surface, large drops of coolant splash on and around the machine, and a very fine aerosol oil mist becomes suspended in the air.

Chemical investigation results indicated that all airborne concentrations of chemicals used in Plant 2 were well within the Canadian occupational exposure limits. Microbial investigation results of serologic tests were negative for the respiratory tract viruses of *Mycoplasma* and *Chlamydia*, as well as the known strains of *Legionella*. However, a Gram-negative, rod-shaped organism which did not grow on blood agar but required L-cysteine for growth, was isolated from the water-based coolant in Plant 2. The etiologic agent was subsequently determined to be a new species of *Legionella*, and it was named *Legionella feeleii* by the group of investigators who isolated them from Plant 2. The results of the investigation strongly suggested that the etiologic agent was airborne and spread through the plant from one end to the other. Aerosols from one particular coolant system were probably responsible for the outbreak. This system was not being circulated during the shutdown period because that particular department was not operating.

The outbreak in Plant 2 was very similar to the previously reported fever outbreaks in Pontiac, Michigan, and it was called Pontiac fever by the investigators because of its unknown etiology. Pontiac fever is a severe influenza-like illness associated with exposure to aerosols of contaminated water. The first recognized fever outbreak occurred in an Oakland County health department building in Pontiac, Michigan, and was traced to the airborne spread of *Legionella* from a contaminated air-conditioning system. The organism was isolated retrospectively from frozen samples of condenser water and from lung tissue of guinea pigs exposed to an aerosol of evaporative condenser water. Since this initial outbreak was recognized, several other common source outbreaks of Pontiac fever have been described. The attack rate for Pontiac fever is high, usually 95 to 100%, and the incubation period short, 5 to 66 hours. Cardinal symptoms include fever, chills, myalgias, headache, and malaise. Nonproductive cough, diarrhea, nausea, vomiting, chest pain, dizziness, and sore throat are also commonly reported in outbreaks of Pontiac fever. Physical examination is generally unremarkable except for fever and tachypnea. The illness is generally considered benign and self-limited, with spontaneous resolution after 1 to 5 days, and no fatalities have been reported.

Pontiac fever is clinically and epidemiologically distinct illness from Legionnaires' disease, which is caused by the bacterium *Legionnella pneumophila*. Legionnaires' disease, as it occurred in Philadelphia and elsewhere, is a multisystem illness involving most prominently the lungs. Other organs and systems such as the kidneys, the gastrointestinal tract, and the central nervous system are also involved. The incubation period averages 5 to 6 days and ranges from 2 to 10 days; 1 to 5% of exposed persons develop illness, and the case-fatality ratio averages 15%.

Pontiac fever should be suspected in circumstances where workers present with a flu-like illness following exposure to aerosolized water. The temperature of water in cooling towers, evaporative condensers, and whirlpools where outbreaks have been reported is usually 33 to 37°C, the temperature at which *Legionella* organisms grow well.

REVIEW QUESTIONS

1. The primary purpose for industrial hygiene air sampling is to:

    a.   Determine the status of compliance with OSHA regulations.
    b.   Measure the concentration of a hazardous substance in the air.
    c.   Respond to the complaint of workers.
    d.   Validate the calibration of air sampling equipment.

2. The aerodynamic equivalent diameter (AED) of airborne particulate matter is defined as the diameter of a hypothetical sphere of unit density having the same terminal settling velocity in air as the particle in question regardless of its:

    a.   Geometric size and shape.
    b.   Shape and true density.
    c.   Geometric size and true density.
    d.   Geometric size, shape, and true density.

3. In air sampling, the volume of sample collected is adjusted according to:

    a.   An estimate of the airborne concentration of the chemical.
    b.   The sensitivity of the analytical method.
    c.   The permissible exposure limit (PEL).
    d.   All of these.

4. The flame ionization detector (FID) and the photoionization detector (PID) are responsive to:

    a.   Polynuclear aromatic hydrocarbons.
    b.   A wide variety of organic compounds.
    c.   Aliphatic hydrocarbons only.
    d.   Aliphatic and aromatic amines only.

5. According to NIOSH recommendations, the suitable collection method for welding fumes is:

    a.   Midget impinger with a wetted surface.
    b.   Multiple-hole, multiple-stage impactor.
    c.   10-mm nylon cyclone.
    d.   Mixed cellulose ester membrane filter.

6. Activated charcoal is frequently the solid sorbent of choice in the collection of organic chemical vapors. The class of organic chemical that activated charcoal has the least affinity for is:

    a.   Polynuclear aromatic hydrocarbons.
    b.   Halogenated aliphatic hydrocarbons.
    c.   Substituted aromatic hydrocarbons.
    d.   Aliphatic and aromatic amines.

7. In the static mixture calibration of air sampling equipment by withdrawing a series of small volumes of samples from a rigid container:

    a. The concentration inside the container will become increasingly diluted.
    b. The change in the concentration of the static mixture can be calculated.
    c. Pure air replaces the volume drawn.
    d. All of these.

8. The spirometer is a device used to measure volume. When it is used as a calibration tool, it is considered a(n):

    a. Secondary standard.
    b. ASTM standard.
    c. NIOSH standard.
    d. Primary standard.

9. The dynamic system for producing known gas mixtures has many advantages over static methods. All of the following are advantages, except:

    a. The dynamic system is capable of producing reactive gas mixtures.
    b. Large volumes can be produced for extended time intervals.
    c. Loss of concentration due to wall and container adsorption is negligible.
    d. The cost is low and the method is simple.

10. The TLVs, as issued by the ACGIH, are recommendations and should be used as guidelines for good practices. Which of the following should not be longer than 15 minutes and should not occur more than four times a day?

    a. TLVTWA
    b. TLVSTEL.
    c. TLVC.
    d. TLVs with "skin" notation.

11. The principal particulate contaminant in the stack gases from a coal-fired electrical power generating plant is:

    a. Dust.
    b. Smog.
    c. Fumes.
    d. Fog.

12. An OSHA Compliance Officer is testing a workplace air for methane and discovers a concentration of 8%. Which statement is true?

    a. There is no permissible exposure limit (PEL) for methane.
    b. The atmospheric concentration of methane is potentially explosive.
    c. The concentration of methane is immediately dangerous to life and health (IDLH).
    d. The concentration of methane is asphyxiating.

13. Which of the following instruments is the best for the identification and quantification of airborne trichloroethylene?

   a. The electron capture detector (ECD).
   b. The photoionization detector (PID).
   c. The infrared (IR) spectrophotometer.
   d. The flame ionization detector (FID).

14. Which of the following will promote efficient collection of a vapor in an absorbing solution?

   a. Use coarse frit opening in a bubbler.
   b. Use two bubblers in series.
   c. Use a large absorber and increase the quantity of absorbing solution.
   d. Use a high sampling rate.

15. Which of the following may have an effect on the collection efficiency of silica gel tubes?

   a. Moisture in the air.
   b. Heat of vaporization of the chemical to be collected.
   c. Mixture of polar and nonpolar chemicals.
   d. All of these.

16. A two-stage impactor is used to collect an integrated sample of bioaerosol on two separate agar plates. The composition of the two fractions should be:

   a. Different.
   b. Dependent on the type of agar plate used.
   c. Dependent on the number of colony forming units (CFUs).
   d. Identical.

17. Beryllium and many other substances can be determined by fluorescence. Such analyses are based on the principle that fluorescing substances absorb energy at a certain wavelength and emit:

   a. Energy of a longer wavelength.
   b. Energy of the same wavelength.
   c. Energy of a shorter wavelength.
   d. Energy in the visible region of the spectrum.

18. In the pickling area of a rolling mill, a fine mist of sulfuric acid is being produced. The air is sampled by passing the air through a fritted glass bubbler containing a 0.01 N sodium hydroxide solution at the rate of 3 LPM for 3 hr. This is an example of:

   a. Integrated absorption sampling.
   b. Real-time absorption sampling.
   c. Integrated adsorption sampling.
   d. Real-time adsorption sampling.

19. Currently, several instrument manufacturers are marketing mercury vapor detectors. The method of mercury vapor detection in these portable instruments is:

    a.   Ultraviolet absorption.
    b.   Chemiluminescence.
    c.   Infrared absorption.
    d.   Different, depending on the manufacturer.

20. Which of the following statements about bioaerosols is true?

    a.   Bioaerosols exist in viable forms only.
    b.   The nonviable forms of bioaerosols are nonpathogenic.
    c.   The indicator organism to measure indoor biological air quality is *Legionnella*.
    d.   None of these.

21. Passive dosimeters rely on two basic collection principles. They are:

    a.   Adsorption and absorption.
    b.   Diffusion and permeation.
    c.   Permeation and absorption.
    d.   Diffusion and adsorption.

22. The preferred collection medium for a highly reactive chemical such as formaldehyde or acetaldehyde is:

    a.   Activated charcoal.
    b.   Silica gel.
    c.   A coated solid sorbent.
    d.   Molecular sieve.

23. Which of the following affects the collection efficiency of air samples on a solid sorbent material?

    a.   Size and mass of the sorbent granules.
    b.   Sampling temperature and humidity.
    c.   Airborne concentrations and mixture of contaminants.
    d.   All of these.

24. Which of the following statements is true of volatile organic chemicals (VOCs) at room temperature?

    a.   They have a vapor pressure >760 mmHg.
    b.   They have a vapor pressure of 10^{-7} to 1 mmHg.
    c.   They have a vapor pressure of 1 to 760 mmHg.
    d.   They have a vapor pressure of 10^{-7} to 760 mmHg.

25. Fumes are produced by certain chemical reactions or processes. Which of the following chemical reactions or processes can generate fumes?

    a.   Heat treatment of wet paint.
    b.   Running an automobile engine.
    c.   Welding.
    d.   All of these.

ANSWERS

1.	b.	14.	b.
2.	d.	15.	d.
3.	d.	16.	b.
4.	b.	17.	a.
5.	d.	18.	a.
6.	d.	19.	a.
7.	d.	20.	d.
8.	d.	21.	b.
9.	d.	22.	c.
10.	b.	23.	d.
11.	a.	24.	c.
12.	b.	25.	c.
13.	a.		

References

AIHA Analytical Chemistry Committee, *Quality Assurance Manual for Industrial Hygiene Chemistry*, American Industrial Hygiene Association Press, Fairfax, VA, 1988.

American Conference of Governmental Industrial Hygienists, *1992-1993 Threshold Limit Values for Chemical Substances and Physical Agents and Biological Exposure Indices*, ACGIH, Cincinnati, 1992.

Bisesi, M. S. and Kohn, J. P., *Industrial Hygiene Evaluation Methods*, CRC Press/Lewis Publishers, Boca Raton, FL, 1995.

Burge, H., Ed., *Bioaerosols*, CRC Press/ Lewis Publishers, Boca Raton, FL, 1995.

Caravanos, J., *Quantitative Industrial Hygiene: A Formula Workbook*, American Conference of Governmental Industrial Hygienists, Cincinnati, OH, 1991.

Clayton, G. D. and Clayton, F. E., Eds., *Patty's Industrial Hygiene and Toxicology, Volume I: General Principles*, 4th edition, Wiley Interscience, New York, 1991.

Cohen, B. S. and Hering, S. V., *Air Sampling Instruments*, 8th edition, American Conference of Governmental Industrial Hygienists, Cincinnati, OH, 1995.

Cox, C. S. and Wathes, C. M., Eds., *Bioaerosols Handbook*, CRC Press/Lewis Publishers, Boca Raton, FL, 1995.

Dinardi, Salvatore R., *Calculation Methods for Industrial Hygiene*, Van Nostrand Reinhold, New York, 1995.

Dinardi, Salvatore R., *The Occupational Environment, Its Evaluation and Control*, American Industrial Hygiene Association Press, Fairfax, VA, 1997.

Eller, P. M., Ed., *NIOSH Manual of Analytical Methods*, 3rd edition, National Institute for Occupational Safety and Health, Cincinnati, OH, 1984.

Hall, S. K., *Chemical Safety in the Laboratory*, CRC Press/Lewis Publishers, Boca Raton, FL, 1994.

Hall, S. K., Complying with the OSHA laboratory standard, *Pollution Engineering*, 24(10):57-60, 1992.

Hall, S. K., *Air Monitoring in the Chemical Laboratory*, in Improving Safety in the Chemical Laboratory, 2nd edition, edited by J. A. Young, John Wiley & Sons, New York, 1991.

Hall, S. K. and Lavite, S. A., Indoor air quality of commercial buildings, *Pollution Engineering*, 20(6):54-59 (1988).

Levin, L., *An Investigative Approach to Industrial Hygiene: Sleuth at Work*, Van Nostrand Reinhold, New York, 1996.

Lighthart, B. and Mohr, A. J., Eds., *Atmospheric Microbial Aerosols: Theory and Applications*, Chapman and Hall, New York, 1994.

Lippmann, M, Ch. F, Calibration of air sampling instruments, *Air Sampling Instruments for Evaluation of Atmospheric Contaminants, 7th ed.*, ACGIH, Cincinnati, 1989.

Lodge, J. P., Jr., Eds., *Methods for Air Sampling and Analysis*, 3rd edition, CRC Press/Lewis Publishers, Boca Raton, FL, 1989.

Maslansky, C. J. and Maslansky, S. P., *Air Monitoring Instrumentation*, Van Nostrand Reinhold, New York, 1993.

Morey, P., Feeley, J., and Otten, J., Eds., *Biological Contaminants in Indoor Environments*, American Society for Testing and Materials, Philadelphia, PA, 1990.

Ness, S. A., *Air Monitoring for Toxic Exposures*, Van Nostrand Reinhold, New York, 1991.

Occupational Safety and Health Administration, U.S. Department of Labor, *OSHA Technical Manual*, 4th ed., U.S. Government Printing Office, Washington, D.C., 1996.

Plog, B. A., Niland, J., and Quinlan, P., Eds., *Fundamentals of Industrial Hygiene*, 4th edition, National Safety Council, Itasca, IL, 1995.

Willeke, K. and Baron, B. A., Eds., *Aerosol Measurement: Principles, Techniques, and Applications*, Van Nostrand Reinhold, New York, NY, 1993.

Wright, G. D., *Fundamentals of Air Sampling*, CRC Press/Lewis Publishers, Boca Raton, FL, 1994.

Vincent, J. H., *Aerosol Science for Industrial Hygienists*, Elsevier Science, Tarrytown, NY, 1995.

Chapter 5

ANALYTICAL CHEMISTRY FOR INDUSTRIAL HYGIENISTS

Ilse Stoll, MS

I. OVERVIEW

The industrial hygiene laboratory is crucial to the industrial hygienist's ability to generate reliable compliance data. It is important for the industrial hygienist to understand the capabilities and the limitations of a laboratory and the analysis of an industrial hygiene sample. This chapter describes operations and decision-making in the industrial hygiene laboratory, and will give the industrial hygienist the necessary tools to evaluate and ensure that good data are being generated. This chapter also describes the calculations an analytical chemist must perform to generate industrial hygiene results.

The industrial hygienist has typically two questions for the analytical laboratory:

- What is it? This is called *qualitative* analysis.
- How much is there? This is called *quantitative* analysis.

A lot of information gathering is necessary to answer both of these questions. Both questions have in common the detection limits of a method. The detection limits in turn impact the industrial hygienist because the lower detection limit of today's analytical method may become tomorrow's exposure limit.

This chapter details the thought processes an industrial hygienist must go through to select the most appropriate sampling technique, and how this decision impacts the kind of analytical result a laboratory can generate from that sample. Many times the industrial hygienist must also make disposal and recycle decisions and therefore the screening of spent industrial chemicals is also briefly described. This chapter also describes typical and special analytical instrumentation used in an industrial hygiene laboratory. The instrumental analytical details will aid the industrial hygienist as well as the analytical chemist in the decision-making process to select the most prudent methods. Calculations necessary to generate meaningful data and other practical applications follow.

II. GENERAL CONSIDERATIONS

The industrial hygienist should think of a number of variables prior to sampling.

- What is the compound of interest? Is this known or must the laboratory make this determination, as in the case of unknown odors?
- How much of this compound may be present? Is this for Occupational Safety & Health Administration (OSHA) compliance verification, or is this an analysis for trace amounts of material, as in an indoor air quality type of situation?
- What method should be used for sampling, what type of sampler and what type of analysis, and where can the documentation of a method be found?
- How can the sampling method affect the final result?
- What does one need to know about an analytical laboratory before submitting a sample?
- What is the effect of sample stability on the final result?
- What other chemicals are present and will be sampled, and how may this affect the results?

Addressing these questions will help the industrial hygienist to collect representative information about chemicals in the workplace to assure a healthful environment for a worker.

A. Qualitative Analysis

What is the compound of interest? The first consideration of an industrial hygienist, when evaluating a potential exposure situation, is to determine to which chemical compounds a worker may be exposed. It is usually not as complex a problem when identifying compounds such as metals. Metallic compounds are typically analyzed for their metal content, and in most cases, it is not of importance to know what oxidation state the metal is in. An exception is chromium, because chromium (VI) is carcinogenic and the lower oxidation states of chromium are not as harmful to human health. Another exception is organometallic compounds. These compounds typically have lower exposure limits than the metal itself. But otherwise, it is the metallic component which is of toxicological importance and not the complete compound.

Organic compound names, on the other hand, can sometimes give reason for confusion. A *Material Safety Data Sheet* (*MSDS*) will usually give sufficient information about the compounds included in a product. For example, the MSDS for a spray paint may list solvent components which are part of the product at concentrations of 1% or greater. On the other hand, a solvent could be listed as a petroleum distillate or mineral spirit, which is a group of many, often more than 100, individual compounds.

Many times, additional information may need to be collected; for instance, information about byproducts generated during a chemical process. This is where the analytical laboratory can help to identify compounds, which means the laboratory will give qualitative information about a component. The identification gives the industrial hygienist information to develop potential control strategies.

The type of analytical instrumentation used to collect qualitative information for volatile organic compounds, generally a mass spectrometer (MS) in combination with a gas chromatograph (GC), commonly called a GC-MS, is different from that used commonly for the determination of quantitative information, typically a gas chromatograph with a flame ionization detector, commonly called a GC-FID. Often a different analytical method may be used to generate the quantitative information after a compound has been identified. Some instrumentation such as the GC-MS can be used as a qualitative and a quantitative tool. The lower detection limits of an instrument used for compound identification are usually several orders of magnitude higher in the total scan mode, which means it is less sensitive than instrumentation used for quantitative analysis. It is therefore of great importance to the industrial hygienist to gather any information about the potential compounds present before the sample is submitted for analysis and to share this information with the laboratory.

Resources for verification of a specific organic compound name or its synonym are the manuals of industrial hygiene analytical methods published by the National Institute for Occupational Safety & Health (NIOSH) and by OSHA. The NIOSH and OSHA methods typically list a compound or compounds and then reference a Chemical Abstract Service registry (CAS) number. This CAS registry number is very important when defining organic compounds, due to a number of common synonyms which may be used. An example is the compound with CAS # 107-98-2. This compound is commonly called: 1-methoxy-2-propanol, but also propyleneglycol monomethyl ether, or pgme. Another example is the compound with CAS # 111-76-2, which is called 2-butoxyethanol and also called butyl Cellosolve®. Acetone, with the CAS # 67-64-1, may also be called 2-propanone or dimethylketone. A specific compound has only *one* CAS registry number, no matter how many synonyms in any language may be in use. It is especially important to verify organic compounds by these CAS numbers to avoid any ambiguity prior to sampling and submission for analysis. The CAS number should be part of the request to the analytical laboratory by the industrial hygienist to assure that the appropriate analytical method and the correct chemical standard material is chosen for the analysis of organic compounds.

When there is an indoor air quality problem, no specific compound may need to be identified, based on the low concentration at which these compounds are typically present. This situation might involve acid or metal vapors, but usually it involves volatile organic compounds (VOCs). When determining the total amount of unidentified VOCs, this should be discussed between the industrial hygienist and the analytical laboratory to determine what type of unknown VOC may be in the sample. An example of one analytical approach is to add all of the volatiles found by gas chromatographic analysis and then calculate the quantity by using an average model compound such as xylene for the molecular weight and the analytical quantification. The reason why xylene may be chosen as a model compound is that xylene, with a molecular weight of 106, represents an average of the typical molecular weight range of volatile organic compounds which are detected in this analysis. For aliphatic and aromatic hydrocarbons, the upper molecular weight cutoff for this analysis is at about 350. Looking at average volatility, xylene, with a boiling point of 138 to 144°C, represents an average boiling point for this analysis, which covers compounds with a boiling range to about 350°C for volatiles and semivolatiles.

B. Quantitative Analysis

How much of a compound may be present? In most instances the industrial hygienist knows if the monitoring of a work situation is for trace amounts or if it is for larger amounts of a compound. If a small amount of a compound is expected and needs to be quantified, the limiting parameter becomes the lower detection limit of the analytical method used to quantify the compound. If a large amount of a compound is expected, the limiting parameter becomes the sampler capacity and breakthrough capacity.

If trace amounts of compounds are to be monitored, the amount of air sampled should be as large as practically possible. The industrial hygienist should be aware that the maximum air sample volume listed in a NIOSH or OSHA method is based on the loading of a sampler to about twice the PEL (personal exposure limit) level concentration. Therefore, this sampled air volume may have to be exceeded when sampling for trace amounts of a compound. In this case the sampling flow rate should not be varied, but the time of sampling should be increased to generate a larger total sample volume. The other variable of how much of a sample is needed is the lower detection limit of an analytical method. It is of value to obtain this lower detection limit from the analytical laboratory prior to sampling. The lower detection limit is the smallest quantity of a compound which can reliably be detected. This is described in more detail later in this chapter. The analytical method sensitivity is the slope of the response curve.

If a larger quantity of a compound is present, the sampling flow rate may need to be decreased or the sampling time may need to be shortened to avoid overloading the sampler. The sampler capacity is usually described in a NIOSH or OSHA method. The sampler capacity for metals on a filter may be limited by plugged filter pores causing a decrease in sampling flow rate during the sampling. The sampler capacity of organic solvents on adsorbents is limited by factors dependent on the type of compound, the amount of background compounds which may be co-sampled, as well as the type of adsorbent used in the sampler. Additionally, humidity usually has the effect of decreasing the sampler capacity because water molecules compete with the organic solvent molecules for the active sites of the adsorbent. The maximum loading of a sampler is usually described as two-thirds of the breakthrough capacity. When the capacity of a sampler is exceeded, the compound breaks through the adsorbent or the collection media of the sampler. To capture sample breakthrough most adsorbent samplers are constructed with front and back adsorbent sections. The front and back sections of a sampler are analyzed separately to determine potential breakthrough. Traditionally, when the concentration of the compound in the effluent of the front section of the sampler (determined from the back section of the sampler) exceeds 5% of the concentration found on the front section, then the capacity of the sampler has been exceeded. The breakthrough capacity has been defined as the mass of the analyte collected on a sampler when the effluent of the sampler is 5% of the total concentration of a compound present in the sampling environment.

Polar organic compounds, such as ketones and alcohols, as well as halogenated compounds show a lower adsorption affinity to charcoal than the lesser polar aliphatic or aromatic hydrocarbons. An example of breakthrough for ketones adsorbed on charcoal (100 mg/50 mg adsorbent tube size) is listed below in Table 5.1. The breakthrough is expressed in liters of sample volume, sampled at a high concentration of ketone, as listed.

Table 5.1
Breakthrough of Ketones Collected on Activated Charcoal

Compound	OSHA Exposure Limits (ppm)	Sampling Concentration (mg/m³)	5% Breakthrough at 0.2 L/min after sampling (liter)	mg per Sampler
Acetone	750	4500	4.3	19.4
Cyclohexanone	25	392	65	25.5
Diisobutyl ketone	25	582	44	25.6
2-Hexanone	5	790	>45	>35.6
MIBK	50	836	17	14.2
2-Pentanone	200	1570	19	29.8

The table shows that the charcoal tubes had an overall capacity of about 15 to 30 mg per ketone, which is due to the limited number of active sites on a fixed amount of charcoal adsorbent. If the breakthrough amount is not incorporated in the total amount detected, the analysis result will be reported as falsely low. If the breakthrough (the amount found in the back section of the sampler) is greater than 5% of the front section of the sampler, the analysis may be invalid because a significant amount of the sampled compound may have been lost during sampling due to incomplete adsorption. To avoid overloading of a sampler, a shorter sampling time or a larger sampler should be used, and the TWA or PEL should be considered when determining how much air to sample for a particular compound.

C. Resources

An industrial hygienist should have a method available prior to sampling to determine the adsorption media and sampling conditions. There are some excellent published methods available to the industrial hygienist, as well as the analytical chemist, specifically the *NIOSH Handbook of Analytical Methods* and the *OSHA Analytical Methods*. It should be noted that the newest editions of some publications may not include reissues of previously published methods. Therefore, back editions should also be considered when looking for the most appropriate method for sampling and analysis of less common compounds. Some of the newer editions of the NIOSH methods are available electronically, providing a convenient way to obtain useful information. An example of the summary page of a NIOSH method is shown in Figure 5.1 for analysis of the compound acrolein. Figure 5.2 provides a summary of the OSHA method for acrolein extracted from the OSHA website.

Other sources for analytical methods are the *American Industrial Hygiene Association (AIHA) Journal*, analytical chemists of the OSHA Laboratory for Methods Development in Salt Lake City, UT, and NIOSH Division of Physical Sciences and Engineering in Cincinnati, OH. The American Society for Testing and Materials (ASTM) *Methods of Air Sampling and Analysis* describes workplace air analysis analytical methods, and additionally EPA-TO-series methods, which are a reliable resource of analytical methods.

Additionally, there is a Methods Information Exchange Network administered through the American Industrial Hygiene Association (AIHA) Technical Affairs Department in Fairfax, VA. The caller is put in touch with a network of industrial hygiene analytical chemists, who have unpublished internal analytical methods for less common compounds.

D. Method Content

What is an analytical industrial hygiene method? It is a method which has been evaluated and documented for the analysis of industrial hygiene samples for a specific compound. As shown in Figure 5.1, the method typically includes the molecular weight (MW), exposure limits, properties of the compound, synonyms (mostly for an organic compound), sampling information such as sampler media, flow rate, minimum and maximum sample volume, and sample stability. The method may also offer statistical information such as the concentration range for which the method was evaluated

ACROLEIN   2501

$H_2C = CHCHO$ MW : 56.07   CAS: 107-02-8   RTECS: AS1050000

METHOD: 2501, Issue 2	EVALUATION: PARTIAL	Issue 1: 15 February 1984 Issue 2: 15 August 1994

	PROPERTIES:	
OSHA : 0.1 ppm NIOSH: 0.1 ppm; STEL 0.3 ppm; Group I Pesticide ACGIH: 0.1 ppm, STEL 0.3 ppm (1 ppm = 2.29 mg/m³ @ NTP)		liquid; d 0.8389 g/mL @ 20 °C; BP 52.5 °C; MP −88 °C; VP 28.5 kPa (214 mm Hg; 28% v/v) @ 20 °C; explosive range 2.8 to 31% v/v in air

SYNONYMS: 2-propenal; acrylicaldehyde; acrylaldehyde; allylaldehyde.

SAMPLING		MEASUREMENT	
SAMPLER:	SOLID SORBENT TUBE (2-(hydroxymethyl)piperidine on XAD-2, 120 mg/60 mg)	TECHNIQUE:	GAS CHROMATOGRAPHY, NITROGEN- SPECIFIC DETECTOR
FLOW RATE:	0.01 to 0.1 L/min	ANALYTE:	9-vinyl-1-aza-8-oxabicyclo[4.3.0]nonane
VOL-MIN:	13 L @ 0.1 ppm	DESORPTION:	2 mL toluene; ultrasonic bath for 30 min
-MAX:	48 L	INJECTION VOLUME: 1 μL	
SHIPMENT:	routine	TEMPERATURE-INJECTION:	230 °C
		-DETECTOR:	250 °C
SAMPLE		-COLUMN:	90 °C for 8 min;
STABILITY:	at least 4 weeks @ 25 °C [1]		20 °C/min; hold @ 200 °C for 11 min
BLANKS:	2 to 10 field blanks and 10 media blanks per set	CARRIER GAS:	Helium, 30 mL/min
		COLUMN:	2 m x 2-mm glass, 5% SP-2401-DB on Supelcoport (100-120 mesh)
ACCURACY			
RANGE STUDIED:	0.12 to 1.50 mg/m³ (24-L samples)	CALIBRATION:	acrolein spiked on coated sorbent
BIAS:	7.0%	RANGE:	3 to 36 μg per sample
OVERALL PRECISION ($\hat{S}_{rT}$):	0.111 [1]	ESTIMATED LOD: 2 μg per sample	
ACCURACY:	± 29%	PRECISION ($\hat{S}_r$): not determined	

APPLICABILITY: The working range is 0.05 to 0.2 ppm (0.13 to 0.5 mg/m³) for a 24-L air sample. For STEL measurements, the limit of the method is 0.9 ppm for a 15-min sample at 0.1 L/min.

INTERFERENCES: None known. Peaks for acetaldehyde and formaldehyde oxazolidines may be observed in the chromatogram, but they are resolved from the acrolein peak. Capacity of the sampler is reduced if sampling in air containing acids [1].

OTHER METHODS: This method was developed to give improved sample stability and to provide for ease of personal sampling compared to P&CAM 118 [2] and P&CAM 211 [2], which have not been revised.

NIOSH Manual of Analytical Methods (NMAM), Fourth Edition, 8/15/94

Figure 5.1 Summary of NIOSH Method 2501 for the analysis of acrolein.

and the precision, accuracy, and bias of the method. The heart of the method is a description of the analytical technique, including sample preparation, analytical instrumentation, and the lower detection limit. The method includes a section on the calculation to convert the raw analytical data into industrial hygiene results. Sometimes, interferences are listed in a method.

Acrolein

NM	:	Acrolein
REV	:	19920824
SYN	:	Acrylic aldehyde; Acrylaldehyde; Propenal; Allylaldehyde; Propenaldehyde; Diesel Exhaust Component
IMIS	:	0110
CAS	:	107-02-8
NIOSH	:	RTECS AS1050000; 4135
DOT	:	UN1092 Flammable Liquid
DESC	:	Clear colorless or yellowish liquid with a piercing, disagreeable odor, causes tears.
		MW: 56 BP: 52.7 C VP: 214 mm MP: -87 C
OSHA	:	0.1 ppm, 0.25 mg/m3
TLV	:	0.1 ppm, 0.23 mg/m3 TWA; 0.3 ppm, 0.69 mg/m3 STEL
REL	:	0.1 ppm TWA 0.3 ppm STEL
IARC	:	Group 3, Not classifiable as to its carcinogenicity to humans (Vol 63)
SYMPT	:	Eye, skin, mucous membrane irritation; abnormal pulmonary functioning; delayed pulmonary edema, chronic respiratory disease
HLTH	:	Irritation-Eye, Nose, Throat, Lungs, Skin---Marked (HE14)
		Mutagen (HE2)
		LD50 (oral, rabbit) 7 mg/kg; LD50
		(oral, rat) 46 mg/kg
ORG	:	Heart, eyes, skin, respiratory system
SLC1	:	MEDIA: Coated XAD-2 Tube (150/75 mg sections, 20/60 mesh)
	.	Coating is 10% (w/w) 2-(Hydroxymethyl)piperidine.
		ANL SOLVENT: Toluene
		MAX V: 48 Liters MAX F: 0.1 L/min (TWA)
		MAX V: 3 Liters MAX F: 0.2 L/min (STEL)
		ANL 1: Gas Chromatography; GC/NPD
	.	REF: 2 (OSHA 52) SAE: 0.12 CLASS: Fully Validated
SAM2	:	DEVICE: Detector Tube COMPANY: Kitagawa
		PART #: 136 RANGE: 0.005-1.8% CLASS: Mfg

Figure 5.2   OSHA Method for acrolein extracted from the OSHA homepage, Salt Lake City UT.

The lower detection limit of a method is important if low concentrations of a compound are to be measured. A calculation may be needed to assure that the sampling time used in combination with the sampling rate and the lower detection limit of a method will allow the industrial hygienist to look at a potential concentration of interest.

If there is more than one method available for monitoring a specific compound, the industrial hygienist needs to choose. The choice for a method may be based on the analytical sensitivity of a method, which is discussed below. The choice may be based on a sample's stability or the type of adsorbent recommended. The sample adsorbent should be readily available. Some methods may describe preparation of sampling media, particularly adsorbents. An example is described in the OSHA method for MDI (methylene biphenyl isocyanate) monitoring. For this method, glass fiber filters are treated with 1-(2-pyridyl)piperazine prior to sampling. However, the quality control involved with preparing sampling media for a specific sampling method may outweigh the ease and reliability of using commercially available sampling media which have been tested extensively by a reputable manufacturer of the industrial hygiene samplers.

E. Sampling Strategy

How may the choice of sampling method affect the analytical result? The analytical results for a sample are only as good as the method and technique used to obtain the sample. It is therefore very important to choose the sampling technique carefully. A variety of sampling methods are available.

Should a method be chosen which uses an impinger? An impinger or bubbler is sometimes the only reliable sampling method for a specific compound, but it is usually avoided by the industrial hygienist, wherever possible, because the impinger solutions may spill or a glass impinger may break during sampling or shipping.

The situation may warrant the use of diffusive samplers instead of sampling pumps. Diffusive samplers, also called passive dosimeters, passive monitors, or diffusion-controlled passive samplers are generally used for organic vapor sampling and have been tested by most of the diffusive sampler manufacturers. When diffusive samplers are used, it is important that the sampling rates are obtained from the manufacturer of the specific monitor, because these sampling rates are then utilized to determine the compound concentration in the sampled air. Diffusive samplers are very useful when sampling pumps are either not available, too expensive, or not reliable; when a large number of simultaneous samples have to be taken; when a charged sampling pump is not available or an electrical outlet can not be used; or when a very long-term sample is to be taken. It should be noted that the typical sampling rate for diffusive samplers is only about 15% or less of that of a sampling pump with charcoal adsorbent. A typical sampling method for an organic compound recommends sampling at 200 cm^3/minute, while a typical organic vapor monitor has an average sampling rate of only 30 cm^3/minute or less for an average VOC.

The diffusive sampler is therefore not practical when low concentrations of a compound are to be sampled for a short period of time. However, diffusive samplers have been used out of doors and over very long time periods, collecting very low concentration background samples. For example, they have even been used for a 1-month sampling period in order to determine trace amounts of organic compounds in clean rooms and in the makeup air from outside the building.

F. Laboratory Evaluation

What does the industrial hygienist need to know about an analytical laboratory? When comparing or evaluating an industrial hygiene laboratory, the industrial hygienist may want to know some facts. Is the laboratory accredited and by whom? What is the accreditation for? What instrumentation is available in the laboratory? If the request is for analysis of compounds which do not have a published method, is the laboratory experienced to develop an industrial hygiene analytical method?

1. *Instrumentation*

Before sampling it may be of value to confer with the analytical laboratory to determine if the analytical equipment described in a specific method is available. If qualitative analysis is necessary, the laboratory may not be equipped with a gas chromatograph coupled to a mass spectrometer, a GC-MS (typical instrument to identify unknown organic compounds), or inductively coupled plasma (typical instrument to obtain a metal screen) and therefore may not be readily able to do the work for compound identification. If less volatile compounds are to be quantified, such as MDI, a high-performance liquid chromatography (HPLC) system may not be available or the analytical expertise to perform HPLC analyses may be lacking. If oil mist in air is to be determined, the necessary infrared spectrometer may not be available.

2. *Certification/accreditation*

Determine if the laboratory is certified, under which program, and for what analyses is the laboratory certified. A laboratory may be certified for metals analyses but not for organic solvent analysis when you are interested in a volatile organic compound (VOC) analysis.

It is known that a chain is only as strong as the weakest link. For an industrial hygiene laboratory, this implies that the reliability of the results are only as good as the laboratory's quality control program. A good and consistent quality control program is usually practiced in an AIHA accredited laboratory. Accredited laboratories and the list of analytes for which the laboratories are certified are listed quarterly in the *AIHA Journal*. The accreditation program typically incorporates a known adherence to set standards consisting of:

- Participation in the Proficiency Analytical Testing (PAT) Program, administered by NIOSH, requiring the laboratory to analyze quarterly unknown quantities of known compounds and have the results evaluated against all participating laboratories results. A performance printout is issued and should be readily available to the industrial hygienist. Some laboratories may additionally participate in the National Institute of Standards and Technologies (NIST) program for National Voluntary Laboratory Accreditation (NVLAP) which indicates adherence to good analytical practices for that analysis.
- Inspection of the laboratory initially and then every 3 years to assure that the documented quality procedures are employed and to indicate areas of improvement.

- Generation of a quality assurance manual, self-audited annually, and which meets the AIHA standards and requirements for personnel training and experience. Additionally, documentation of instrument maintenance and repair, chain of custody procedures, record-keeping procedures, established analytical methods, and quality control data are required.
- The policies for meeting the AIHA accreditation are available through the AIHA. The accreditation may be used for a basis for court standing during litigation.

3. *Method validation*

Does the laboratory use validated methods? There are validated and nonvalidated analytical methods. A validated method is a method that has been verified through successful testing of various method elements that may include spiked samples, sample storage, sample analysis, and compound recovery. Some OSHA and NIOSH methods have been validated; however, the industrial hygiene laboratory may need to validate a non-validated method. An industrial hygienist may want to be aware when a nonvalidated method is used. Ideally, all industrial hygiene analyses should be performed using validated methods. However, this is not always possible. In the case of monitoring for novel chemical compounds, methods may not be fully established and a temporary or stopgap method may be sufficient to establish workplace concentrations of that compound during initial process monitoring.

If a laboratory does not show that a compound present can actually be recovered by a specific analytical method, this method is not valid. A quality control program using quality control samples and reference check samples reduces this concern. All AIHA accredited laboratories must follow a stringent quality control program. This is one of the main reasons why it is good practice for the industrial hygienist to choose an AIHA accredited laboratory.

G. Sample Stability

What is the sample stability prior to analysis? Some sampled compounds are very reactive and the sampling method may require stabilization of the compound during sampling. Whenever this is necessary, the detailed procedure for sampling is usually described in the respective NIOSH or OSHA method. It is of importance to determine the stability of the sample after sampling, to assure that the sample is still fully representative after shipment and storage at the time of analysis. Some methods may recommend monitoring a less stable compound directly and/or require cold shipment of a sampled compound. Some methods may require the addition of a stabilizer to the sample immediately after sampling, such as, for example, the addition of concentrated hydrochloric acid to silica after sampling for aminoethanol compounds. The industrial hygienist may not be in a situation where the handling of concentrated acid is practical or desired. However, if this step is omitted, there are chances for loss or degradation of the initial compound, thereby generating a falsely low final result. To avoid a false result, the industrial hygienist should employ methods which convert or derivatize the unstable compound to a stable compound. In the methods for aldehyde screening, for instance, unstable aldehydes are derivatized to stable oxazolidines. The oxazolidines can be shipped on the adsorbent at room temperature and have at least 1 week stability at room temperature prior to analysis.

H. Sample Matrix

Whenever an air sample is taken for a specific compound, it is impossible to isolate this compound from any surrounding material which may also get sampled on the same sampler. What are the other compounds which are present in the sampling area and will be sampled? This is a very important question that may not be immediately in the industrial hygienist's mind during sampling. Other compounds may interfere with the analysis for the compound of interest. Table 5.2 provided examples of interferences and their causes.

In the analysis of organic solvents, the gas chromatographic peak identification for a compound is achieved by matching the retention time of a sample peak to that of a standard peak. It may be possible to find two or more compounds that elute at the same or similar retention time, thereby giving a potential for misidentification. The analysis can be modified by the chemist if she or he is aware of the presence of other compounds. The modification of the analytical method for organic solvents may consist of an alternate temperature program, a different type of GC column, a different selectivity of detector, or switching to an HPLC method for the separation, detection, and quantification of a particular compound of interest.

Table 5.2
Example of Analytical Interferences

Analyte	Method	Interference	Cause
Cyanide	ISE (NIOSH 7904)	Chloride and sulfide: result will be lower	Formation of insoluble silver salts on the ISE
Alkaline dusts	Titration (NIOSH 7401)	Acid aerosols: result will be lower	Uncontrolled neutralization
Asbestos	Fiber count (NIOSH 7400)	Airborne non-asbestos fibers with length >5 μm and aspect ratio >3 to 1: result will be higher	The method is non-specific for fiber type

One particularly challenging analytical situation is the quantification of a specific organic solvent in the presence of petroleum-derived distillates such as mineral spirits or naphtha. These solvents are mixtures of compounds, often consisting of over 100 different volatile components. The analytical impact of petroleum distillate background is illustrated as an example when monitoring for trichoroethylene. The laboratory generates a chromatogram of the air sample. Figure 5.3 shows the chromatogram containing petroleum distillate in the absence and presence of trichloroethylene.

The asterisk (*) indicates the peak for trichloroethylene. All other peaks are part of the petroleum distillate. Quantification of a compound by gas chromatography is done by peak integration. The best condition for peak integration is a peak which is totally separated from any other peaks and has a defined start and a defined end of the peak for peak area integration. This sample contained 77 μg trichloroethylene and 617 μg petroleum distillate. Although these are not trace amounts, the quantification of a compound in this sample matrix is challenging. Even the unskilled observer can see that the quantification for the trichloroethylene is difficult due to the many other peaks surrounding it. Additionally, the result may become falsely high because other petroleum distillate compounds may co-integrate and be incorporated into the result.

This situation may not be avoidable for the industrial hygienist. The information of other potentially interfering compounds in the sample, however, will give the chemist information to reduce the interferences as much as possible.

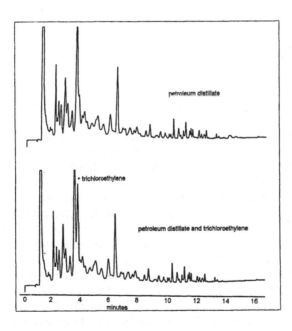

Figure 5.3 Gas chromatogram (GC-FID) of petroleum distillate in absence (a) and presence of trichloroethylene (b).

III. ANALYTICAL METHOD

As mentioned above, various publications of methods are available for performing the analysis of industrial hygiene samples. *The NIOSH Manual of Analytical Methods* and *OSHA Analytical Methods* are the most commonly used references. The method describes which analytical instrument and conditions to use, the sample preparation, the standard preparation to calibrate the instrument, quality control samples, determination of desorption efficiency where applicable, and calculations to obtain the final result. There are many variables to an analytical method which can impact the result.

Table 5.3
Analytical Method Variables

Variable	Impact
Linearity of a method	Accuracy of result
Sample preparation, desorption efficiency	Incomplete analyte recovery, result will be falsely low
Sensitivity of a method	Sampling strategy and lower detection limit
Precision of a method	Reproducibility of results
Reporting units	Comparison to published limits

A. Linearity

The *linearity*, which is also called the dynamic range of an analytical method, describes the range of concentration of an analyte that can be analyzed without dilution or preconcentration of the sample. This dynamic range is usually limited by the type of detector used. Some detectors have a wide range of linearity. For example, the flame ionization detector (FID) has a linear dynamic range covering 6 to 7 orders of magnitude, which is a remarkable range. This means that concentrations from about 0.001 mg/ml to above 100 mg/ml can be analyzed using the same calibration curve without dilution. An analytical method defines a concentration range within which the detector response is linear. It is this analytical concentration range that should be used when analyzing samples.

The linearity, or the linear portion of a calibration curve, is determined by plotting three or more different concentrations of an analyte and their respective detector responses. If a sample is analyzed outside the linear concentration range, the results will not be reliable unless a nonlinear mathematical model is applied. This approach is usually not recommended. The NIOSH and OSHA methods emphasize analysis within the linear range of a method.

If the sample concentration is too high and falls outside the linear range, the sample should be diluted and reanalyzed to assure quantification within the linear concentration range; otherwise, the result may be inaccurate.

B. Desorption Efficiency

Before a compound can be analyzed, it must be removed from the sampler. Metals are removed from a mixed cellulose ester filter (MCEF) by acid digestion of the filter and the collected particles. Acid vapor anions are extracted with an aqueous buffer solution. Both of these types of analytes usually show complete analyte recovery. Some organic compounds, when chemically desorbed with a solvent such as carbon disulfide during sample preparation, will not fully desorb. This effect, called *desorption efficiency* (DE), is due to the strong interaction of a compound with the adsorbent. The desorption efficiency of a compound with a 100% desorption efficiency, or a DE of 1, shows no desorption efficiency effect. The DE is defined by the following fraction:

$$DE = \frac{\textbf{Mass found}}{\textbf{Mass added}} \qquad (5.1)$$

The DE is unique for each compound and is dependent on the adsorbent as well as the desorbing solvent. When either of these parameters are changed, the DE must be determined for these new desorption conditions. The result of a sample is adjusted for the desorption efficiency, i.e., it is divided:

$$\textbf{mg in sample (DE corrected)} = \frac{\textbf{mg found}}{\textbf{DE}} \qquad (5.2)$$

This results in a mass larger than the uncorrected analytical value for compounds exhibiting a desorption efficiency effect. An example of desorption efficiencies for ketones is listed below in Table 5.4.

There have been studies performed to compare different techniques for the determination of the desorption efficiency. The recommended method by NIOSH is a liquid spike of a known quantity of compound onto the adsorbent (which is the sampling medium), followed by the chemical desorption after a specified period of time. The desorption efficiency is dependent on the amount of active adsorbent with respect to the adsorbing compounds, is therefore dependent on the concentration of a compound, and should therefore be determined over the analytical range of a method. If the desorption efficiency is not incorporated, a result may be reported lower than the actual concentration of a compound present at the sampling site.

Table 5.4
Desorption Efficiency of Ketones
Using Carbon Disulfide on Charcoal

Compound	Desorption Efficiency
Acetone	0.86
Cyclohexanone	0.82
Diisobutyl ketone	0.97
2-Hexanone	0.81
MIBK	0.91
2-Pentanone	0.90

C. Sensitivity

The sensitivity of a method is defined as the slope of the response curve. This is most easily visualized in colorimetry: a compound that is strongly colored can be more easily detected at different concentrations than a compound which has a faint color. Smaller differences in concentrations can be more reliably determined for the stronger colored compound than for the weaker colored compound. The method is therefore more sensitive to the stronger colored compound than it is to the weaker colored compound.

D. Lower Detection Limit

When a compound is not detected by an analytical method in the laboratory, the report typically shows the result as less than the *limit of detection* or *LOD*, also referred to as *method detection limit* or *MDL*. Some laboratories report less than the *lower limit of quantitation* or *LOQ*, also referred to as *method quantification limit* or *MQL*. Some laboratories may report the result as *none detected* (*ND*). It may be necessary to clarify with the laboratory which of these two important values are reported as the method limit. It may also be of value to the industrial hygienist to have the analyst clarify how the detection limit was determined. A report should not state that nothing was found without defining "nothing."

The LOD is defined as that amount of substance which causes a signal three times that of the noise signal. The noise signal is determined as the standard deviation of an instrumental analysis. Therefore, the LOD can also be defined as the blank value plus 3 standard deviations of a series of blank determinations. This is approximately equal to the upper value of the 99% confidence interval around the blank value. The LOD corresponds to a signal-to-noise ratio of approximately 3:1.

LOQ is defined as the blank value plus 10 standard deviations of a series of blank determinations. The LOQ corresponds to a signal-to-noise ratio of approximately 10:1. The noise level for instrumental methods can be determined based on instrument manufacturer recommendations. The attenuation may need to be greatly magnified, and when peak integration is part of the raw data gathering method, parameters such as peak width for integration have to be appropriately adjusted; otherwise, the standard deviation will be falsely minimized.

A practical method of obtaining the LOD for an analyte and its analytical method is to dilute a known sample to a concentration level which allows detection above the analytical background by the measuring system. A practical method of obtaining the LOQ for an analyte and its analytical method is to determine the lowest concentration of analyte recovered at >75%. Another approach to determine LOD and LOQ is to analyze a series of decreasing concentrations of standards. The lowest concentration which can still be detected by the instrument is called the LOD, and the lowest concentration still within the linear portion of a calibration curve of a standard is called the LOQ.

It should be noted that, using a specific analytical method for a specific instrument, the LOD for a compound can considered to be a constant in terms of detectable mass. (Note that the LOD can change with instrument performance and even sample matrix; however, these differences may be less than the differences obtained by using different instrumentation.) This LOD for a particular instrument and analytical method, however, changes when expressed as an air concentration, depending on the air volume sampled. This means that if a compound is to be sampled and quantified at trace levels, a large air volume should be sampled. If the exposure limit of a compound is very low, this becomes important. For trace level sampling the maximum volume stated in the analytical method may possibly be exceeded by up to 2 orders of magnitude, because the maximum volume in a method is mostly based on the limits of the adsorbent sampling capacity at the OSHA permissible concentration limits. However, before changing a published sampling method, the experts at OSHA in Salt Lake City or at NIOSH in Cincinnati should be consulted. The flow rate stated in a NIOSH or OSHA method for organic compounds, however, should *not* be exceeded since it is based on the actual rate of adsorption of a compound onto the adsorption medium. Whenever a special situation warrants a deviation from a method, the analytical laboratory and/or the OSHA Laboratory for Method Development should be involved in the discussion. The sensitivity of a method, in combination with the amount of air sampled, gives the value below which the compound of interest cannot be detected and, therefore, not be determined. It is always important to perform a preliminary calculation to determine the minimum sampling time required to detect the PEL concentration.

Sample Calculation: For example, when sampling for a compound such as chlorinated diphenyl oxide, which has a PEL of 0.5 mg/m³ (this is the same as 0.5 µg/L air), a maximum sampling flow rate (q_{max} of 1.5 L/minute) is recommended. The analytical quantification limit (LOQ) for this method is 0.2 µg per sample using gas chromatography with an electrolytic conductivity detector. The LOQ can be expressed as:

$$LOQ = PEL * q_{max} * T_{min} \tag{5.3}$$

Rearranging:

$$T_{min} = \frac{LOQ}{PEL * q_{max}} \tag{5.4}$$

Where:

LOQ = Limit of quantification, μg
PEL = Permissible exposure limit, $\mu g/L$
q_{max} = Maximum flow rate, L/min
T_{min} = Minimum sampling time, minutes (to detect the PEL concentration)

Solving:

$$T_{min} = \frac{0.2}{1.5 * 0.5}$$

$$T_{min} = 0.27 \text{ minutes}$$

This short minimum sampling time of less than one minute, actually about 15 seconds, should be very adequate to determine a short-term exposure. This is due to the relatively high flow rate and the highly sensitive electrolytic conductivity detector. It should be noted that such a short sampling time may not be very representative because it is very imprecise and it may take a longer amount of time to prime the sampling equipment than 15 seconds. If this analytical method, however, is not available and the compound is to be analyzed using a flame ionization detector with an estimated LOQ of 20 μg per sample (100 times more than 0.2 μg) for chlorinated diphenyl oxide, the minimum sampling time must be increased 2 orders of magnitude to 27 minutes.

Typically, concentrations of gases and vapors in air are established in terms of parts per million (ppm) of substance in air by volume. The ppm expressed concentration is not affected by barometric pressure and temperature; therefore, workplace concentrations of volatile compounds can be compared directly to the PELs or TLVs. PEL or TLV concentrations are established at 25°C temperature and 760 mm Hg barometric pressure in the U.S. However, some countries list the PELs as mg/m^3, and others may have the PEL corrected for altitude. When C_{ppm} is the desired unit of concentration, the molecular weight (MW) of a compound must be included in the calculation as well as the molar gas volume.

The molar volume of a gas at 0°C and 760 mm Hg at **Standard Temperature and Pressure** (STP) is 22.4 L per Mol. At **Normal Temperature and Pressure** (NTP), 25°C and 760 mm Hg, the molar volume is 24.45 L per Mol because the corrected molar volume (MV_{corr}) is:

$$MV_{corr} = \frac{V_{Mol} * T_{actual}}{T_{zero}} \qquad (5.5)$$

Where:

MV_{corr} = Molar volume, L/Mol
V_{Mol} = Molar volume at STP, 0°C and 760 mm Hg, 22.4 L/Mol
T_{actual} = Actual temperature, K
T_{zero} = Temperature at 0°C, 273K

At NTP:

$$MV_{corr} = \frac{22.4 * 298}{273} = 24.45 \text{ L/Mol} \qquad (5.6)$$

Sample Calculation. For example, formaldehyde has a molecular weight (MW) of 30 and as a suspected carcinogen has an action level (AL_{ppm}) concentration of 0.5 ppm in air by volume at NTP. This is first converted to µg/L as follows:

$$AL_{ppm} = \frac{AL_{µg/L} * 24.45}{MW} \qquad (5.7)$$

Rearranging:

$$AL_{µg/L} = \frac{AL_{ppm} * MW}{24.45} \qquad (5.8)$$

Where:

$AL_{µg/L}$ = Action level, µg/L
MW = Molecular weight, g/Mol
AL_{ppm} = Action level, parts per million

so:

$$AL_{µg/L} = \frac{0.5 * 30}{24.45}$$
$$AL_{µg/L} = 0.61 \text{ µg/L}$$

Sample Calculation. In the NIOSH method, the LOQ for formaldehyde is 1 µg per sample. The maximum flow rate for sampling is stated at 0.1 L/minute and the minimum sampling time (T_{min}) is derived as follows:

$$LOQ = AL_{µg/L} * q_{max} * T_{min} \qquad (5.9)$$

Re-arranging:

$$T_{min} = \frac{LOQ}{AL_{\mu g/L} * q_{max}} \qquad (5.10)$$

Where:

LOQ = Limit of quantification, μg

$\mathbf{AL}_{\mu g/L}$ = Action level, $\mu g/L$

$\mathbf{q_{max}}$ = Maximum flow rate for sampling, L/min

$\mathbf{T_{min}}$ = Minimum sampling time, minutes (to detect the AL concentration)

so:

$$T_{min} = \frac{1\,\mu g}{(0.61\ \mu g/L) * (0.1\ L/min)}$$
$$T_{min} = 16.3\ minutes$$

A minimum sampling time of about 16 minutes is necessary to allow the detection of formaldehyde at the action level.

In summary, the two equations used for the determination of minimum sampling time for the required detectable air concentration, based on the analytical lower detection limit, the exposure limits (EL), and the maximum flow rate of sampling are:

$$T_{min} = \frac{LOQ}{EL_{\mu g/L} * q_{max}} \qquad (5.11)$$

Where:

$\mathbf{T_{min}}$ = Minimum sampling time, minutes

LOQ = Limit of quantification, μg

$\mathbf{EL}_{\mu g/L}$ = Exposure limits, $\mu g/L$ (the concentration that must be detected)

$\mathbf{q_{max}}$ = Maximum flow rate, L/min

$\mathbf{mg/m^3}$ = $\mu g/L$

Rearranging using EL in ppm:

$$T_{min} = \frac{LOQ}{EL_{ppm} * q_{max}} * \frac{24.45}{MW} \qquad (5.12)$$

Where:

$\mathbf{T_{min}}$ = Minimum sampling time, minutes

LOQ = Limit of quantification, μg

$\mathbf{EL_{ppm}}$ = Exposure limits, ppm

$\mathbf{q_{max}}$ = Maximum flow rate for sampling, L/min

MW = Molecular weight, g/Mol

The exposure limit (EL) is stated in Equation 5.11 as μg/L or mg/m³, and in Equation 5.12 as ppm. Using Equation 5.11 and Equation 5.12 allows calculation of minimum sampling time necessary, knowing the lower detection limit of an analytical method and arriving at the exposure limit for a substance or compound. It should of course be noted that results should not be gathered at this "outer edge" of analytical performance. It is suggested to operate in concentration ranges of at least 1 order of magnitude higher, i.e., sampling for 10 times longer than calculated by these two equations. The sampling strategy should involve such a safety factor.

E. Precision

The analytical method may also list a precision for a result of a specific analyte. The precision of a method gives the industrial hygienist information about the reproducibility of an analytical result. Some laboratories may include with the result a measure of their method reproducibility, or precision. The precision is expressed as the result ± 1 to 3 standard deviations (SD). The SD is a measure of the confidence interval of a result:

1 SD represents a 66% confidence interval
2 SD represent a 96% confidence interval
3 SD represent a 99% confidence interval

The industrial hygienist can use this analytical precision and add it to the actual sampling precision. Precision is determined by collecting and analyzing multiple samples. With this information the industrial hygienist can then estimate the precision of a concentration value at a site. There are accuracy requirements for substances where specific standards have been established. Traditionally ± 25% is used at the PEL concentration. The OSHA method requirements are that the combined precision and accuracy of a sampling and analytical method is ± 35% of the actual concentration at the PEL concentration.

The precision of an analytical method is determined by repetitive analyses of the same sample. The precision may include:

- Day-to-day variability
- Instrument variability
- Operator variability

The precision of a method is usually concentration dependent and is usually poorer at concentrations near the LOQ. The precision is best in the midrange of a linear calibration curve.

F. Units of Results

The laboratory results are reported in units and then compared to the OSHA limits for air contaminants 29CFR1910.1000 or Threshold Limit Values (TLVs) published by the American Conference of Governmental Industrial Hygienists (ACGIH), listing the Permissible Exposure Limit (PEL), Time-Weighted Average (TWA), and Short-Term Exposure Limit (STEL) for chemical substances in air concentration units of mg/m^3 and also in air concentration units of ppm by volume, referenced to 25°C and 760 mmHg, Normal Temperature and Pressure, (NTP):

- For metal and inorganic acid analyses, the units are usually given in mg/sample or μg /sample. This is the analytical result. After the air sampled volume is incorporated into the result, the actual exposure concentration is given as mg/m^3 or μg/L, μg/m^3, or mg/1000 L. The atomic weight or the molecular weight of atoms, inorganic compounds, cations, or anions is not included in the calculation for concentration.

- For organic vapor analysis, the units of the analytical result are also usually obtained as mg/sample or μg/sample. However, after the sampling volume and the molecular weight of the compound are incorporated into the result, the actual exposure concentration is given as ppm at NTP but can also be given as mg/m^3.

$$C_{ppm} = C_{mg/m^3} * \frac{24.45}{MW} \tag{5.13}$$

Converting from ppm to mg/m^3:

$$C_{mg/m^3} = C_{ppm} * \frac{MW}{24.45} \tag{5.14}$$

Discussion with the analytical laboratory indicating the preferred unit designation can eliminate the need for recalculations by the industrial hygienist.

IV. ANALYTICAL INSTRUMENTAL TECHNIQUES

The analytical techniques used most commonly for industrial hygiene analyses are chromatography, spectroscopy, ion selective electrodes, titrations, gravimetry, and analytical microscopy.

A. Chromatography

Chromatography is a separation technique which is generally used for organic compounds. The separation allows the quantitation of several compounds in a single analysis. The separated compounds are passed through a detector sensing a difference in response from the carrier or background and converting this information to an electrical signal. The raw data output results in a chromatogram, which plots the detector response (y-axis) over the analysis time (x-axis). Quantitation of a compound is usually done by peak area integration of the detector signal from a chromatographic peak of a compound, and comparison to the peak area from known standards.

Sample preparation prior to analysis usually involves chemical desorption of the adsorbent. The charcoal air sampler is desorbed with carbon disulfide or another suitable solvent. The extract is then analyzed by chromatography.

There are three kinds of chromatographic techniques commonly performed for the analysis of industrial hygiene samples: gas chromatography (GC), high-performance liquid chromatography (HPLC), and ion chromatography (IC).

A summary of the chromatographic methods, applications, and LODs per IH sample is shown in Table 5.5.

Table 5.5
Comparison of Chromatographic Methods, Applications, and LODs

Method	Used for	Example	LOD* Estimated
GC	Volatile organics	Organic solvents	0.1–50 µg
HPLC	Volatile and nonvolatile organics	PAHs	0.05 µg
IC	Ionic species	Acid vapors	0.5–2 µg

* the LOD is based on the amount detectable on a sampler and is dependent on the GC detector and the compounds.

1. *Gas chromatography*

Gas chromatography is the technique of choice when analyzing for volatile organic compounds or solvents. A sample is injected into a hot injector port of a gas chromatograph, where it is volatilized and carried through a very long and narrow GC column with the aid of a carrier gas, usually helium. A mixture of compounds is separated by the GC column. During the separation, the GC column is heated in a GC oven under controlled conditions to volatilize the compounds and accelerate the elution.

The analytes may be liquids or even solids at room temperature. Compounds with a boiling point up to about 400°C can be separated and quantified by GC. The compounds eluting from the GC column are then passed through a detector and the detector response is monitored, resulting in a chromatogram. The time of elution, or retention time, defined as the time required for the compound to pass from the injector to the detector and register a response, is generally the identifier of a compound by comparison to the retention time of a known standard analyzed under identical conditions. The peaks in the chromatogram, representing the compounds detected in a sample, are then quantified by peak integration. A schematic of a gas chromatograph is shown in Figure 5.4.

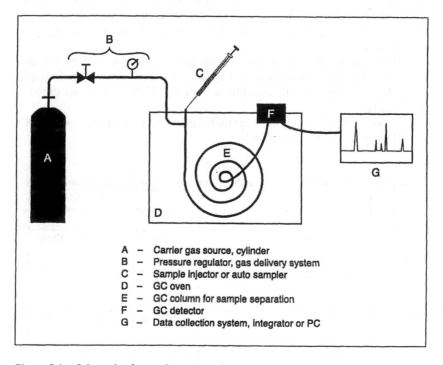

A – Carrier gas source, cylinder
B – Pressure regulator, gas delivery system
C – Sample injector or auto sampler
D – GC oven
E – GC column for sample separation
F – GC detector
G – Data collection system, integrator or PC

Figure 5.4   Schematic of a gas chromatograph.

- *GC column types*

There are different types of GC columns which differ in the sorbent, diameter, and length of the GC column. The GC column contains a chemical sorbent or coating that determines the separation characteristics of compounds. The chemical characteristics or the polarity of the column sorbent will determine the type of separation. For example, for the separation and analysis of several different alcohols, which are moderately polar compounds, a column with a coating having intermediate polarity is chosen.

For the separation and analysis of petroleum distillate components, which are nonpolar compounds, a nonpolar column is chosen. The GC column type is specified in the analytical method.

The sorbent may be a solid compound of large surface area (gas-solid chromatography–GSC); or it may be a nonvolatile liquid compound coated on an inert finely divided solid support (gas-liquid chromatography–GLC), or uniformly coated on the inner wall of long glass capillary columns as a thin film (open tubular chromatography). GSC and GLC use packed chromatography columns, currently being replaced by more efficient capillary and megabore columns that achieve a better separation of complex mixtures. Capillary or megabore columns (with an inner diameter of about 0.2 to 0.75 mm) are used for GC preferentially over glass columns (with an inner diameter of 2 mm or more), because they allow better peak separation in shorter analysis times, resulting in lower detection limits for a compound.

The length of a column affects the separation of compounds. A typical analytical capillary GC column is 30 to 50 m long. The column is bent in a circular coil, connected to an injector system and carrier gas, and resides in the GC oven during analysis. The greater the column length, the better the separation; however, the analysis time will also increase by the use of a longer column and a longer column is also more expensive. Also note that resolution only increases in proportion to the square root of the length, so longer columns give diminishing returns; it is better to change stationary phases and look for better selectivity, if possible.

Usually, a column is specified in an analytical industrial hygiene method. If methods for the analysis of volatile compounds or mixtures of compounds are not available in documented methods, a column of appropriate polarity is chosen and separation is demonstrated and documented as part of the method development. If a NIOSH or OSHA method recommends a GC column which is unavailable to a laboratory, the analysis must be demonstrated and documented on an alternate GC column before this column can be substituted for analysis of industrial hygiene samples. The OSHA analytical laboratory may be consulted prior to any substitutions.

- *GC temperature program*

The GC separation is carried out on the GC column in a GC oven and is performed under controlled conditions through a temperature program. Compounds with low boiling points need to be separated at low oven temperature and may even require subambient cooling to achieve good peak resolution from other interfering compounds. High-boiling compounds are analyzed at a higher GC temperature program. The temperature program is usually stated in a NIOSH or OSHA method, but may need to be modified to assure the best separation from potential interfering compounds in the sample and for best chromatographic performance.

- *GC detectors*

The detector senses the presence of components different from the carrier gas and converts this information into an electrical signal. The type of detector determines which compounds are detected and which are ignored. Universal detectors are nondiscriminant and detect almost all volatile organic compounds. Some detectors are very selective, responding only to certain kinds of compounds. Both types of detectors can be very useful for analytical problems. The GC detectors described here are:

Flame ionization detector (FID)
Thermal conductivity detector (TCD)
Electron capture detector (ECD)
Photoionization detector (PID)
Nitrogen–phosphorus detector (NPD)
Flame photometric detector (FPD)
Electrolytic conductivity detector (ELCD or ECN)
Mass spectrometer or mass selective detector (MS or MSD)
Infrared detector (IR)

A comparison of GC detectors with respective applications and LODs for an IH sample is shown in Table 5.6.

a. *GC-FID*

GC analysis with a flame ionization detector is frequently used when a mixture of organic solvents is to be analyzed. The FID is a nonspecific detector responding to most organic molecules with carbon-hydrogen bonds. These compounds ionize very well in an air/hydrogen flame. The FID will respond to aliphatic hydrocarbons (found, for instance, in mineral spirits), to aromatic hydrocarbons such as benzene, toluene, and xylenes, and to ethers and esters. A somewhat lower response is obtained from alcohols, and halogenated compounds such as methylene chloride, trichloroethylene, and tetrachloroethylene. A compound with a very low FID response, i.e., very low ionization in an air/hydrogen flame is carbon disulfide. This feature of the carbon disulfide FID response is actually used in many industrial hygiene analytical methods that use carbon disulfide as a desorbing solvent and thereby avoid a large solvent peak interference for the analysis. Other compounds with no, or extremely low, FID response are water, carbon tetrafluoride, and formic acid, which are used in industrial hygiene analyses as extracting solvents.

Table 5.6
GC Detectors, Applications and Their LODs

Detector	General Application	Example of IH Method	LOD * estimate (g)
FID (flame ioniza-tion detector)	Volatile organics	Organic solvents	1 - 10
TCD (thermal con-ductivity detector)	Volatile compounds	Universal gases and compounds	20
ECD (electron capture detector)	Chlorinated or oxygenated hydrocarbons	PCBs	0.01
PID (photo ioniza-tion detector)	UV absorbing compounds	Aromatic hydrocar-bons, tetraethyl lead	0.1
NPD (nitrogen phosphorus detec-tor)	Nitrogen or phosphorus containing hydrocarbons	Formaldehyde as oxazolidine	0.2
FPD (flame photo-metric detector)	Phosphorus and sulfur containing hydrocarbons	Pesticides	0.0004
ELCD (electrolytic conductivity de-tector)	Halogenated hydrocarbons	Chlorinated diphenyl oxide	0.1
MSD (mass selec-tive detector)	Any compound above a molecular weight of the carrier gas (MW >2)	Unknown byproduct analysis for identifi-cation, mixed alde-hydes	10
IR (infrared detec-tor)	Any compound except for diatomic gases	unknown compound analysis for identifi-cation, usually con-firmational to GC-MS	100

* LOD on a sampler.

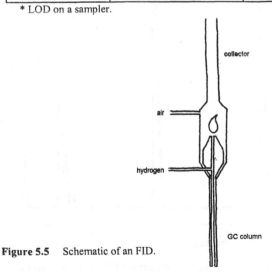

Figure 5.5   Schematic of an FID.

The detector uses purified air and hydrogen as a fuel to generate a flame. Organic compounds eluting from the GC column are ionized in the

flame and the ions generated during this process are attracted by a polariz-
ing voltage collector and produce a current. This current is directly proportional to the amount of sample component in the flame. Figure 5.5 shows
the schematic of an FID. The column effluent is mixed with hydrogen and
then enters the flame. The air is supplied separately. The shape of the
flame and its proximity to the detection zone has a strong effect on sensitivity.

The FID can typically detect 1 to 10 ng of an organic compound on a
sampler. This translates into a typical lower detection limit of about 1 to 10
µg in the desorption solvent for an aromatic or aliphatic hydrocarbon with a
1-µL injection, representing about 1 to 10 µg per air sample. Since the industrial hygiene methods usually involve desorption from the adsorbent
with 1 ml of solvent, the limit of 1 to 10 µg in an air sample relates to the
analytical detection limit of 1 to 10 µg/ml in the analyte if 1 µL is injected
into the chromatograph.

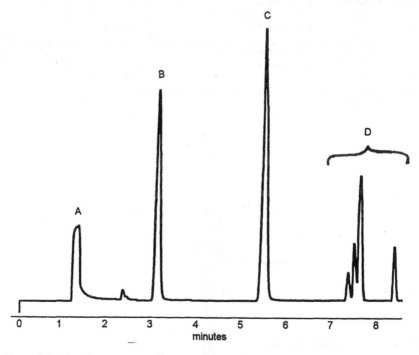

Figure 5.6   Gas chromatogram of benzene, toluene, and xylenes (BTX) in carbon disulfide.

A typical chromatogram for the analysis of organic solvent vapors collected on charcoal is shown in Figure 5.6. Peak A is carbon disulfide,
which is the desorbing solvent, Peak B is benzene, and Peak C is toluene.

The four peaks labeled D are isomers, from left to right, ethylbenzene, *p*-xylene, *m*-xylene, and *o*-xylene.

Figure 5.6 depicts the characteristics of a good chromatogram: the peaks are well separated. The *D Peaks* show a closer spacing, which is characteristic of xylene isomers. The peaks are almost symmetrical in shape, an indication of good column installation, appropriate column choice, injector port maintenance, and avoidance of detector overloading. The baseline of the chromatogram is almost flat, which is desirable. All of these factors contribute to good peak integration, which is very important in the quantitation of a compound analyzed by chromatography.

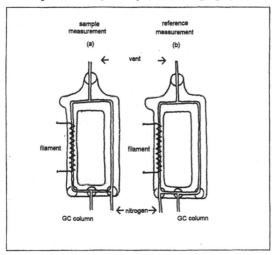

Figure 5.7   Schematic of a TCD.

b. *GC-TCD*

The thermal conductivity detector (TCD) actually predates chromatography, though it is still used because it is such a universal detector. The TCD responds to any compound whose thermal conductivity is different than that of the carrier gas. When helium, which has an exceptionally high thermal conductivity, is used as the carrier gas, most compounds give good sensitivity. When the analyte is present in the carrier gas, the thermal conductivity drops, and less heat is lost to the detector wall. Under constant applied voltage, a filament in the detector will heat up and its resistance will increase. This change is recorded and correlated to the concentration of the compound of interest. The TCD is a nondestructive detector and therefore can be used in series with more selective detectors. As shown in Figure 5.7, two filaments are incorporated in an electrical resistance bridge. It is balanced with pure carrier flowing through dual cells, with a reference filament in one cell. The detector filament's resistance, and its changes when compounds elute, is registered as a difference relative to the reference filament. In some TCD models, the filaments are replaced by transistors.

This detector can be used for the analysis of carbon dioxide. The TCD is about 2 orders of magnitude less sensitive than the FID, which results in a higher LOD for the TCD. However, the TCD is an excellent detector for gases that do not respond to other detectors.

c. *GC-ECD*

The electron capture detector (ECD) is used for analysis of halogenated compounds at very low concentrations. This detector is specifically used for analysis of polychlorinated biphenyls (PCBs) and chlorinated pesticides. The detector is also used for fluorinated compounds, nitroaromatics, and PAHs.

The detector works on the principle that electronegative species, species with electron affinities above a certain minimum value, capture thermal electrons to form negatively charged ions. The carrier gas is ionized by beta particles from a radioactive source. The resulting electron flow produces a small current which is measured. When the sample is introduced, electrons are captured by the sample, resulting in a decreased current, correlated to the quantity of analyte in the sample. Because the response of different compounds in the ECD varies widely, quantitation requires individual careful calibration. The detector is insensitive to many types of compounds, providing selectivity and the ability to detect traces of target species in complex matrices.

The detector is about 2 to 3 orders of magnitude more sensitive than an FID detector and most commonly uses ^{63}Ni as the radioactive source and nitrogen as the carrier gas. A schematic of an ECD is shown in Figure 5.8.

The detector responds to compounds with pairs of free electrons, such as those that contain halogens, oxygen, nitrogen, and sulfur. The ECD will also respond to double bonds in a molecule.

The lower detection limit for the ECD is typically about 0.1 µg/ml for PCBs (representing usually about 50 or more different compounds in a sample), which is 1 to 4 orders of magnitude more sensitive than the FID, depending on the molecular structure of the compound. The detector is also used for the industrial hygiene analysis of ethylene oxide. For ethylene oxide, the approximate LOD is about 1 µg per sample. The ECD should not be used when a sample has been desorbed with carbon disulfide because the strong ECD response to carbon disulfide will generate a very large peak, thereby overshadowing other peaks in that region of the chromatogram. The desorbing solvent of choice for this detector is a nonpolar solvent like an aliphatic hydrocarbon such as iso-octane or hexane.

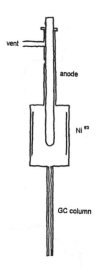

vent

anode

Ni ^{63}

GC column

Figure 5.8   Schematic of an ECD.

d. *GC-PID*

The photoionization detector is not commonly used for industrial hygiene analytical laboratory analysis. It is more commonly used in a portable GC system. The detector responds to any compound that can be ionized by the particular lamp installed in the PID. Different lamps are available, providing different UV wavelengths for irradiation. This gives some control over the classes of compounds that will respond. This provides some selectivity that can be an advantage. Very few NIOSH published methods suggest using GC-PID. One example of the use of GC-PID is for the analysis of tetraethyl lead. Figure 5.9 shows the schematic of a PID.

The detector operates on the principle that specific sample compounds are excited by the photons from a UV lamp and ionize. The PID consists of a UV lamp and an ionization chamber. The radiation from the lamp passes into an ionization chamber, where the sample absorbs the radiation and ionizes. Electrodes collect the ions and the current measurement is related to the sample concentration.

The sensitivity for this detector depends strongly on the UV absorbance of the compound of interest. It can detect about 0.1 µg of tetraethyl lead in a sample.

Desorbing solvents for this analysis should not have significant UV absorbance or ionization, which eliminates aromatic hydrocarbon solvents such as toluene or xylenes. A suitable desorbing solvent for PID analysis is methanol, hexane, or methylene chloride.

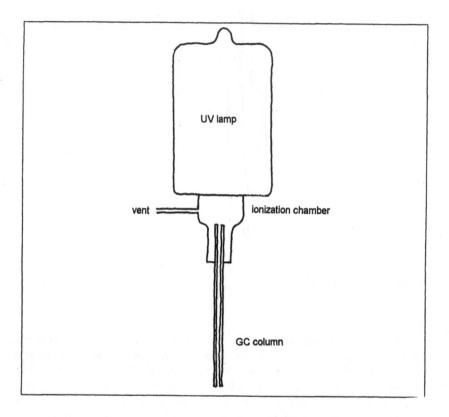

Figure 5.9   Schematic of a PID.

e. *GC-NPD*

The nitrogen/phosphorus detector (NPD) is a specific detector for the analysis of compounds containing nitrogen or phosphorus. It can be used for the analysis of amines. It can also be used in the analysis of acrolein and aldehydes. This analysis is a special case, where the unstable acrolein or aldehyde is derivatized or converted during sampling to form a stable oxazolidine using XAD-2 adsorbent coated with 2-(hydroxymethyl)-piperidine. The piperidine, and the oxazolidine, contain nitrogen and can therefore be selectively detected by the NPD.

The NPD is made selective for nitrogen- and phosphorus-containing compounds because a source of alkali salt is positioned above the jet. This introduces ions of alkali metal into the flame, minimizing hydrocarbon ionization and increasing the ionization of N- or P-containing compounds.

Many different solvents can be used for desorption. The detector does not detect inorganic nitrogen such as N_2 or ammonia, but it does detect elemental phosphorus, P_4.

The lower detection limit is about 1 order of magnitude more sensitive than the FID. About 0.1 to 0.5 µg of a nitrogen-containing compound can be detected using the NPD. The LOD depends, however, on the specific detector response of a compound. In the case of acrolein, the approximate LOD is about 2 µg per sample.

The NPD uses a jet and a collector similar to an FID. A schematic of an NPD is shown in Figure 5.10.

The other reason why an NPD is used for this type of analysis is that due to its specificity, it will not respond to the many potential other VOCs that may have been collected on the sampler. This specialized focus provides a clearer identification of the compound of interest in a   complex sample background matrix.

f. GC-FPD

The flame photometric detector (FPD) operates on the principle that when phosphorus- or sulfur-containing compounds are burned in an FID-type flame, they produce chemilumenescence. This emits light at wavelengths characteristic of the phosphorus or sulfur in the sample. An optical filter permits light of the desired wavelength to enter the sensor and produce a signal. A diagram of the operation of an FPD is shown in Figure 5.11.

This detector is specifically used in the analysis of certain sulfur- and or phosphorus-containing pesticides. The LOD of this detector is very much dependent on the compound. Phosphorus-containing compounds can be detected at lower levels than sulfur-containing compounds. Organophosphorus compounds show an approximate LOD of 0.4 ng per sample, which allows monitoring of atmospheres at concentrations as low as 0.0004 mg/m³. It should be noted that the operating conditions of the detector can influence the LOD strongly.

Pesticides may be sampled in the presence of petroleum distillates which could be used as a propellant or diluent. If the sample were analyzed by GC-FID, the petroleum distillate compounds would be detected, producing a complex chromatogram of unresolved peaks and the pesticide peak would be obscured. Using an FPD detector will result in a chromatogram of only the sulfur- and phosphorus-containing compounds. The FPD response is generally nonlinear.

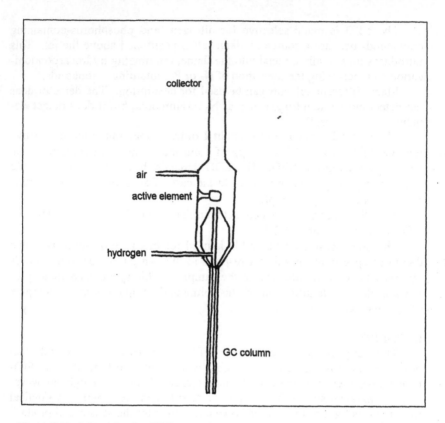

collector

air

active element

hydrogen

GC column

Figure 5.10   Schematic of an NPD.

g. *GC-ELCD*

The electrolytic conductivity detector (ELCD or ECN), or Hall™ detector is another selective detector used in gas chromatography. It is used in the analysis of halogenated hydrocarbons. Shown in Figure 5.12, the ELCD operates as follows: the GC column effluent is mixed with a reducing gas in a catalytic reaction tube. The reducing gas is usually hydrogen and the catalyst is nickel. The catalytic reaction causes breakage of the carbon-halogen bonds in the molecule by hydrogenation, thereby generating hydrocarbon and the respective halogenic acid. For example, chlorinated hydrocarbons produce hydrochloric acid. The acid is dissolved in *n*-propanol, mixed, and passed as a liquid through a conductivity detector. This conductivity is measured and correlated to the compound concentration. In industrial hygiene applications, the ELCD is used for the analysis of halogenated pesticides. The ELCD is more sensitive than the FID, showing an LOD of about 1 order of magnitude lower than that for the FID.

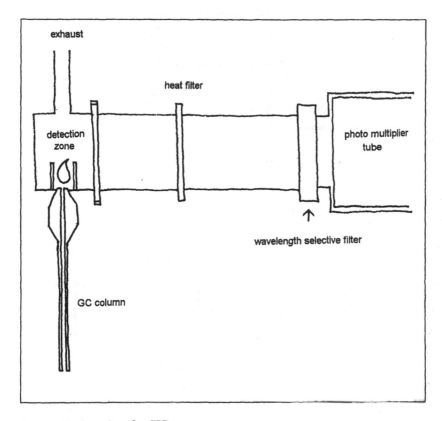

exhaust

heat filter

detection
zone

photo multiplier
tube

wavelength selective filter

GC column

Figure 5.11 Operation of an FPD.

h. *GC-MS*

The mass spectrometer (MS), sometimes also called a mass selective detector (MSD), is usually used to identify unknown organic compounds and rarely used for quantitation of industrial hygiene samples. Typically, an industrial hygiene sample is analyzed by GC-MS to identify its components and is then analyzed by GC-FID to quantify the identified compounds due to the better detector response of the FID. However, the MSD is more expensive, somewhat more complicated to operate, and has a smaller linear dynamic range. Not all industrial hygiene analytical laboratories have the capability to analyze by GC-MS, due to the cost of the instrument, as well as the special analytical skill needed for data interpretation. The schematics of a mass spectrometer are shown in Figure 5.13.

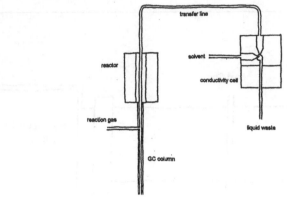

Figure 5.12 Schematic of an ELCD.

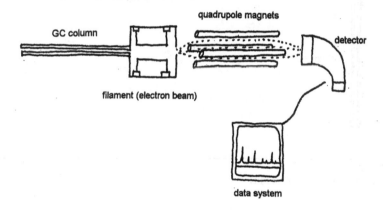

Figure 5.13 Schematics of a mass spectrometer.

The electron beam of the mass spectrometer bombards compounds eluting from the GC column. The electrons interact with the electrons of the compound and an energy transfer takes place. Sufficient energy is deposited in the molecules to exceed the ionization potential. The molecule expels an electron, becoming a positively charged radical. Usually, there is sufficient additional energy that, in the process of distributing that energy, bonds in the molecule and rearrangements of the structure of the initially formed ions occur. The fragments are then separated and detected on the basis of their mass-to-charge ratios. This signal response is recorded against the elution time, which generates a chromatogram. The ionized fragments of a particular compound are reproducible and predictable in their relative quantity, as well as their mass/charge ratio and are collectively called a mass spectrum. They are usually identified with the help of reference spectra.

The GC-MS is more frequently used in unknown odor or indoor air quality situations or for methods development to determine unknown by-products in new chemical processes. An example of a GC-MS chromatogram, along with the mass spectrum for one of the GC peaks, is shown in Figure 5.14.

The lower section of Figure 5.14 shows a portion of the total ion chromatogram, which is the sum of all ion signals as they are collected at any point in time during the analysis. The upper portion of Figure 5.14 shows the mass spectrum for the retention time of 4.929 minutes in the chromatogram. The spectrum corresponds to that of xylene.

The sensitivity of the MS in the total ion mode used to be about 2 orders of magnitude less than that of an FID. MS quadrupole instruments are the instruments of choice and magnetic sector instruments can also be used, both of which have better sensitivity. Another data collection mode of the mass spectrometer is the selective ion mode, which will decrease the detection limit by about 10–20x. The selective ion mode, however, can only be applied after the compounds of interest have been identified through the total ion mode. The strength of the mass spectrometer is its ability to differentiate compounds of a homologous series; for example, it can easily differentiate xylene (dimethylbenzene) from trimethylbenzene because the ion fragments of the higher molecular weight compound, trimethylbenzene, are also larger, resulting in different mass spectra. Most of the ions are the same; however, it is the detection of the high mass molecular ion and, in some cases, some existing higher mass fragment ions that allows differentiation. The MS, however, is not always a good tool to differentiate isomers such as o-xylene from m-xylene or p-xylene because the ion fragments of these isomers are similar, resulting in very similar mass spectra.

The NIOSH method for the screening of aldehydes recommends the use of a GC-MS and can be used as a guideline for GC-MS identification of other unknown volatile organic compounds. The lower detection limit for this method is estimated at 2 μg per sample.

i. GC-IR

The infrared (IR) detector, in combination with gas chromatographic separation, is even less frequently used for industrial hygiene analysis than the MS detector. It is, however, also an important detector for identification of unknown compounds and is an excellent complementary tool to the MS. The strength of the IR detector is its ability to differentiate isomers, which means it can differentiate between the three different xylenes (*ortho*, *meta*, and *para*) and ethylbenzene, but not very easily between homologous compounds such as pentane or hexane. The IR is also used to identify functional groups. For example, it can be used to determine and distinguish if a compound is an aliphatic alkane, an alcohol, a ketone, an aromatic compound, or has characteristics of several different functional groups.

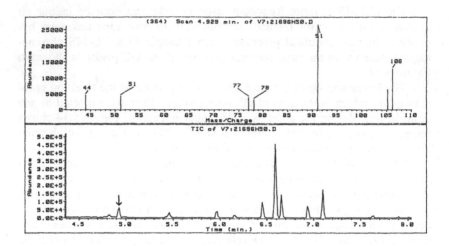

Figure 5.14 GC-MS spectrum of an indoor air quality sample.

The detector scans a compound as it elutes from the GC column over an energy range of about 4000 to 800 wavenumbers (infrared range) and generates an IR spectrum. The IR region of the electromagnetic spectrum lies between the visible and microwave regions. The principle of the detector is based on the fact that within molecules, the atoms vibrate and rotate with a few well-defined frequencies characteristic of the bonds within that molecule. These frequencies lie within the infrared region of the electromagnetic spectrum. When a sample is placed in a beam of infrared radiation, it absorbs energy at those frequencies. The resulting pattern is an infrared spectrum. A ketone will show a strong absorbance at about 1750 cm^{-1}, which is characteristic of the keto or C=O (carbonyl) group. This peak will also be present in esters and carboxylic acids, which both have the C=O functional group. Aliphatic hydrocarbon functional groups can be observed at about 2900 cm^{-1} wavenumber, usually as a doublet or triplet. The position of an aliphatic group on a disubstituted benzene ring will give a different spectrum for *ortho*, *meta*, and *para* substitution, thereby allowing differentiation between isomers such as xylenes.

The LOD for the IR is about 2 orders of magnitude higher than that of the MS, and the IR is generally not used to analyze personal industrial hygiene samples. It can be used to identify unknown bulk material such as process intermediates and complement the results of the GC-MS analysis. However, the IR is used without the GC as a quantitation tool for oil mists in industrial hygiene samples. See spectrometric methods below.

The IR is more commonly used by the industrial hygienist as a direct readout detector, in which case only one or a few key wavenumbers are read and not an entire spectral scan.

2. *Liquid chromatography*

As described above, compounds are separated in a gas chromatographic analysis after volatilization as they are carried through a GC column by a carrier gas. In liquid chromatography, in contrast, compounds are separated in their solubilized state as they are carried through a chromatographic column by a liquid eluent phase. The column is filled with very fine adsorbent particles which interact with the compounds coming through the column. The eluting compounds are sensed by a detector. Compounds do not need to be volatile, as is necessary for gas chromatography, but must be soluble in the eluent to be introduced into the system. Two types of liquid chromatography are commonly used for industrial hygiene analysis: high performance liquid chromatography and ion chromatography. A summary of the liquid chromatography detectors, applications, and some LODs is shown in Table 5.7.

Table 5.7
Liquid Chromatographic Detectors, Applications, and LODs

Detector	Used for (especially nonvolatiles)	Example	LOD Estimated
UV	Aromatic organic compounds	PAHs	0.2 µg
Fluorescence	Aromatic compounds	PAHs	0.05 µg
Visible	Colored nonvolatile soluble compounds		
RID (refractive index detector)	Generic for nonvolatile organic compounds		
Conductivity	Ionic species	Acid vapors	0.5–2 µg

a. *High-performance liquid chromatography*

High performance liquid chromatography (HPLC) is typically used for the analysis of nonvolatile organic compounds. These compounds cannot be analyzed by GC because of their high boiling point (>350°C) or their thermal lability. HPLC can be used for the analysis of any organic compound which can be dissolved. An HPLC system consists of six basic components. A schematic is shown in Figure 5.15.

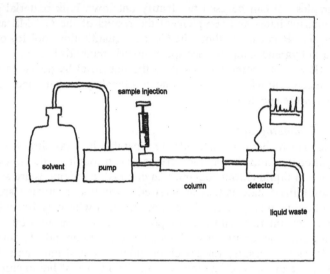

Figure 5.15   Schematic of an HPLC system.

For the analysis, a liquid solution of a sample is injected onto an HPLC column of very fine and uniform particles of a stationary solid adsorbent with the aid of a carrier solvent using a high-pressure pump operating at very constant pressure. The carrier solvent is also called the mobile phase. The compounds in the sample are separated by the HPLC column based on their differing affinity to the stationary phase and detected as they elute from the column. Very fine and tightly packed particles result in the high operating pressure of the system, usually about 2000–6000 psi, and are responsible for the resolution of the separation. The typical HPLC columns have particles from 3–10 µm in size. As the compounds are eluted they pass through a detector, sending the signal to an integrator, where the peak areas or peak heights are determined for quantitation.

The most common detector for HPLC is a fixed-wavelength UV detector typically set at a wavelength of 254 nm, specific to many aromatic compounds such as coal tar distillates or polynuclear aromatics (PNAs, also called PAHs for polynuclear aromatic hydrocarbons). Colored compounds can be detected in the visible light range. A variable-wavelength detector in the visible UV range operates from about 800–185 nm. This detector allows detection of a wide range of compounds. The UV/Vis detector is a selective detector. Based on the characteristics of a UV/Vis scan, an optimal absorbance wavelength is selected for the detector setting in the HPLC analysis.

An example is the HPLC analysis of MDI (4,4'-methylene-bisphenyl iso-cyanate) in the form of an acetylated MDI derivative. A UV/Vis scan of the MDI derivative is shown in Figure 5.16.

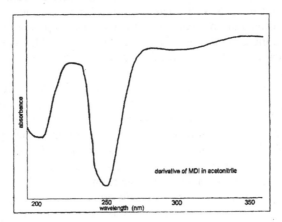

Figure 5.16 UV/Vis spectrum of an acetylated MDI derivative.

The strong absorbance between 250 and 260 nm is used in the analysis by setting the UV detector to 254 nm. This setting optimizes the sensitivity and contributes to the specificity of detection for the compound. If the UV detector was set at 225 nm, the compound would show almost no detector response based on the spectrum in the UV range shown in Figure 5.16.

An example of an HPLC chromatogram is shown in Figure 5.17.

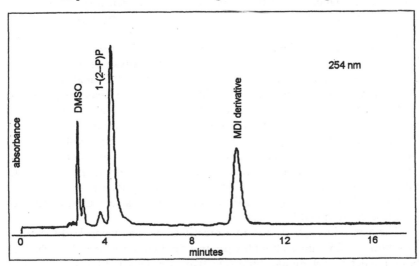

Figure 5.17 HPLC chromatogram of MDI analysis.

It can be seen that the peaks in Figure 5.17 are broader than those in the gas chromatogram shown in Figure 5.6. The peak width in Figure 5.6 is about 0.2 minutes and in Figure 5.17 is about 1 minute. This result is typical and can result in less efficient peak resolution by the HPLC method. Reduced peak resolution means that it is harder to separate peaks that may interfere with the quantitation of the peak of interest. This is of special importance when several compounds are present in a sample and need to be quantified.

Another detector used for HPLC analysis is the fluorescence detector. It shows good selectivity and sensitivity when analyzing aromatic compounds. The sensitivity is usually 1 order of magnitude better for this detector than the UV detector; however, this sensitivity strongly depends on the structural characteristics of a particular compound. The approximate LOD for the PNA benzo[a]pyrene is 0.2 µg per sample. A fluorescence detector operates at two wavelengths: the excitation wavelength, which is at a higher energy level (UV range), and the emission wavelength, which is at a lower energy level (visible range). It is the latter which is detected.

Another HPLC detector is the refractive index detector (RID), which is a very common HPLC detector. Each compound has a refractive index. The RID detects compounds based on their difference in refractive index from the mobile phase. There are no published industrial hygiene methods to date recommending the use of an RID with HPLC; however, special situations may suggest the use, i.e., if an organic compound is nonvolatile and has no UV; i.e., compounds without any double bonds or visible absorbance; compounds which are not colored, it can be separated by HPLC and quantified by an RID. The RID responds well to carbohydrate molecules such as sugars. The RID is not commonly the detector of choice because it is incompatible with gradient elution and is fairly insensitive relative to UV when a reasonably active chromophore is present. Derivatization is frequently used to make a compound UV-active or fluorescence-active so those detectors can be used. Of course, this is not always possible.

The most common separation column used for HPLC analysis is a reversed phase column, called a C18-column, containing octadecyl siloxane (ODS) as the column packing. This column, which is a nonpolar column, separates a wide range of medium-polar to nonpolar organic compounds. Where the gas chromatograph uses an oven to drive lesser volatile compounds through the GC column, here the HPLC uses the mobile phase, which consists of solvents of increasing polar strength to move soluble compounds through the column in order of increasing polarity.

Other HPLC columns are used for normal phase, size exclusion, ion exchange, and affinity HPLC analyses.

The approximate LOD for an aromatic compound using HPLC with a UV detector at 254 nm is about 0.2 µg on a field-collected sampler, which is about 1 order of magnitude more sensitive than the GC-FID detector. The HPLC is applicable for the analysis of polynuclear hydrocarbons, steroids, phenolics, dyes, drugs, herbicides, insecticides, petroleum residues, and other nonvolatile organic compounds.

b. *Ion chromatography*

Ion chromatography (IC) is the analytical method of choice when analyzing for acid vapors. It is a liquid chromatographic method as is HPLC; however, it is carried out at lower pressure, only about 1200 to 1900 psi for IC, versus 2000 to 5000 psi for HPLC. The components listed for the HPLC system in Figure 5.15 are the same ones used for IC. A sample is solubilized in the mobile phase and carried, after injection, through an ion exchange column, which is filled with fine particles (10 to 30 µm in size) of an exchange resin type of material. Interaction of the ions with the column material achieves separation, which is followed by background suppression, that can be carried out chemically or electrochemically. The eluting compounds are then detected with a conductivity detector. An ion chromatogram for the analysis of inorganic acids is shown in Figure 5.18. The inorganic acids are analyzed as anions.

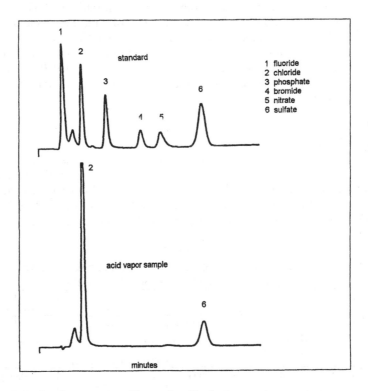

Figure 5.18 Ion chromatogram of inorganic acid anions.

The lower detection limit of the method is about 0.5 µg for fluoride or hydrofluoric acid and chloride or hydrochloric acid, about 0.8 µg for bromide or hydrobromic acid, nitrate or nitric acid, and sulfate or sulfuric acid, and about 2 µg for phosphate or phosphoric acid. Nitrate can also be detected using the UV detector, as discussed in the HPLC analysis section. This double detector method can sometimes be useful when other peaks in the sample may potentially interfere.

Ion chromatography can also be used for the quantitation of organic ions such as for the analysis of formic acid. Other anions and cations can be separated and quantified by this technique; however, the most commonly used industrial hygiene application is the analysis of the basic six inorganic anions of fluoride, chloride, phosphate, bromide, nitrate, and sulfate. This multispecies analysis of ions is an advantage over ion selective electrode (ISE) techniques. The IC technique is capable of analyzing at the parts per million level and, for some ions, even in the parts per billion concentration range.

It should be noted however that IC cannot identify ions directly. Ion identification is carried out, as in other chromatographic methods, by comparison of the retention time of a known standard with that of the unknown sample. Ions with similar exchange selectivity coefficients (eluting with similar retention times from the chromatographic column) interfere with each other. Samples with high organic content may swell the column packing and greatly reduce the column efficiency. A sample of interest must be soluble in aqueous media.

B. Spectroscopy
1. Metals analysis

Analysis for metals is usually carried out using spectrophotometric methods. Atomic adsorption (AA) and inductively coupled plasma analysis (ICP) are the techniques of choice. Many NIOSH and OSHA metal analysis methods are based on these two techniques. A general summary of atomic spectroscopy methods, applications, and LODs is shown in Table 5.8 below.

Table 5.8
Summary of Metals Analyses by Atomic Spectroscopy, Applications and LODs per IH Sample

Method	Used for	Example	LOD Estimated
AA-flame	Metals, one metal at a time	Solder fumes	0.01–10 µg
GF-AA	Trace metals	As Pb in drinking water	0.001–0.01 µg
Cold Vapor AA	Specifically for mercury	Hg	0.03 µg
Hydride AA	Specifically for As & Se	As & Se	0.2 µg As
ICP-AES	Simultaneous or sequential multi-metals analysis	Metal screening	0.001–5 µg

a. *Acid digestion*

For metals analysis by spectroscopy, the metal must first be solubilized. This is achieved by an acid digestion, carried out on either a hot plate or using a special microwave digestion system. The air sample of dust or fumes is usually collected on a mixed cellulose ester filter, removed from the sampling cassette, and then digested in concentrated acid. This digestion step is also carried out for wipe samples. A comparison of hot plate and microwave digestions actually shows better recovery results for the microwave digestion, although hot plate digestion is still more commonly used. The acid-digested solution is used to introduce the sample into the spectrometers.

b. *Atomic absorption*

Atomic absorption (AA) spectroscopy is a metal detection and quantification technique. The method is based on the principle that metallic elements in the ground state will absorb light of the same wavelength they emit when excited. Excitation is achieved by energizing the metal in a hollow cathode lamp (HCL) with a cathode made of the element or metal of interest. The HCL emission is used as an element-specific source of energy. Its radiation passes alternately through and around a flame into which the sample is aspirated. The solution is atomized and the metal atoms which are released absorb light at their characteristic wavelengths, proportional to their concentration in the flame. A photosensitive device measures the intensity of transmitted radiation from the two light paths. The atomization is carried out either by nebulization into a flame or introduction into a flameless device that is electronically heated. Either an acetylene/air or an acetylene/nitrous oxide flame is used as an energy source to dissociate the aspirated sample into the free atomic state.

The wavelength is specific for a metal, and the amount of light absorbed is a function of the concentration of the metal of interest.

The advantages of AA spectroscopy are sensitivity and selectivity. The sensitivity is high because only atoms in the ground state are detected. The detection limit is typically in the range of 0.05 to 1 µg. The high specificity is due to the very narrow resonance lines used. Disadvantages for the technique are that one lamp is needed for each element to be analyzed and that at least 2 ml of sample solution is needed for each determination. This limits the number of elements that can be determined on one collected filter to about 5 or 6. Keeping this in mind, the sample preparation for different metals may be different. For example, arsenic and selenium should not be determined on the same filter as copper and zinc.

Anions may interfere by combining with the metal of interest in solution and change the vaporization temperature. Sometimes, lanthanum nitrate is used as a releasing agent to overcome some of these limitations.

There are four different types of AA analyses: flame AA, graphite furnace AA, cold vapor AA, and hydride generator AA.

i. *Flame AA*

Flame AA is the most common mode of AA analysis, wherein the sample is aspirated into a flame. The limitation of this technique is the sensitivity for some elements, particularly lead, mercury, arsenic, and selenium. For higher sensitivity, graphite furnace AA, cold vapor AA, and AA with a hydride generator are used.

ii. *Graphite furnace AA*

Graphite furnace AA is also called electrothermal AA. The difference between flame AA and graphite furnace AA (GF-AA) is basically the temperature at which the metal is vaporized and ionized and the length of time during which the ionized metal is exposed to the light from the HCL. At the higher more controlled temperature of an electrically heated graphite furnace, more of the metal is ionized over a longer period of time than in the flame AA. The GF-AA dwell time is usually 0.1 to 5 seconds, while the dwell time of the flame-AA is about 1 millisecond. This is the reason why GF-AA is a more sensitive technique. GF-AA is used where very low concentrations of a metal are of importance. Lead in drinking water, which is regulated at ppb levels, is typically analyzed by GF-AA. The detection limits for flame AA and the graphite furnace AA are dependent on the metal to be analyzed. Lead, probably the most common metal analyzed for in the industrial hygiene laboratory, has a lower detection limit of only 2.6 µg per sample using the flame AA, but 0.02 µg per sample using the graphite furnace AA. Arsenic is also analyzed by GF-AA because it has very poor sensitivity by flame AA.

iii. *Cold vapor AA*

Cold vapor AA, which is also called flameless AA, is a special technique used for determination of mercury. It uses a chemical reduction. This method is extremely sensitive. The approximate LOD for mercury by this technique is 0.03 µg per sample. The cold vapor analysis is shown in a schematic in Figure 5.19.

This technique is the only recommended technique for low-level mercury analysis because the other analytical techniques for metal analysis, such as flame AA, ICP, and AA-GF, are not sensitive enough to detect mercury at industrial hygiene levels.

iv. *Hydride AA*

Hydride AA converts selenium and arsenic to their respective hydrides by use of sodium borohydride reagent. The hydrides are then aspirated into the AA flame and analyzed. This technique therefore has the advantage of being able to isolate these two elements from complex samples. Interferences are other metals which easily reduce, such as copper, silver, and mercury. The approximate LOD for arsenic by this technique is 0.2 µg per sample.

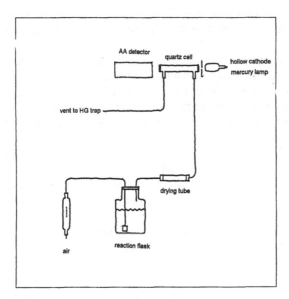

Figure 5.19   Cold vapor analysis system for mercury analysis.

c. *Emission spectroscopy*

Inductively coupled plasma–atomic emission spectroscopy (ICP-AES) is another analytical technique used for the determination of metals. This method is used when multi-element determinations are desired. The principle is the reverse of atomic absorption spectroscopy. Radiation is emitted at specific wavelengths by excited atoms corresponding to the energy released for an electronic transition from the excited state to ground state. With a focus on specific wavelengths, up to 30 elements can be measured in a single sample. ICP can be used in the sequential (one metal analyzed after another) or in the simultaneous (several metals analyzed simultaneously) detection mode. In this method, samples must also be solubilized. The instrument measures the characteristic emission spectra by optical spectroscopy. Samples are nebulized and the resulting aerosol is transported to the plasma torch produced by a radio frequency-induced coupled plasma with a temperature range of 5000 to 10,000°C.   The emission spectra produced are element specific. The spectra are dispersed by a grating spectrometer, and the intensities of the emission lines are monitored by photosensitive devices. The radiation is in the ultraviolet, visible, and near infrared regions of the electromagnetic spectrum. Most applications of ICP operate in the 190–800 nm region.

Sequented and simultaneous ICP are the two modes of operation. In the simultaneous mode, the radiation emitted by the excited atoms, as a result of the energy input of the source, is dispersed according to wavelength, detected and recorded using a spectrophotometer consisting of four parts:

- The entrance slit, permitting the light of mixed wavelengths emitted by the excited atoms to enter the instrument.
- The dispersion device, or grating, which "sorts out" the radiation and arranges it according to its component wavelengths.
- The detection system, which is a series of photomultiplier tubes which convert the light energy into an electrical signal, proportional to the amount of light detected and therefore to the concentration of the element that emitted that light.
- The recording system, which stores the electrical signals and converts the data from intensities to concentration and provides a printout.

Background radiation from other elements and from the plasma gases make it difficult at times to determine trace amounts of one metal in the presence of an abundance of another metal.

The lower detection limit for the ICP-AES for lead is about 1 µg per sample in a 25 ml digest. A broad analytical method for metals analysis in industrial hygiene samples is found in NIOSH Method 7300. It should also be noted that the sample must be in solution in order to introduce it into the high-temperature plasma region.

An additional ICP technique is ICP-MS. This technique couples ICP with mass spectrometry and is extremely sensitive and selective. This technique is generally not available in a commercial industrial hygiene laboratory because of its high cost of acquisition.

Table 5.9 shows the comparison of nominal detection limits for atomic absorption and emission techniques.

Table 5.9 shows the distinct difference in sensitivity for each analysis type for the listed elements. A method should be chosen with the analytical instrument's sensitivity in mind.

2. Other spectroscopic analyses

Spectroscopy is also used in infrared (IR) analysis, UV/Vis spectrophotometers, and X-ray diffraction analysis. A summary of spectroscopic methods and applications is shown in Table 5.10.

a. Infrared spectroscopy

The principle of an IR detector is described in the section for GC-IR. For industrial hygiene analyses, the IR detector is used in the determination of oil mist in air. The approximate LOD for this method is about 0.05 mg per sample. There are established OSHA limits for oil mist in air, which can arise from cutting oils, airborne mist of white mineral oil, cable oil, engine oil, hydraulic oils, machine oil, and transformer oils.

Table 5.9
Atomic Absorption and Emission Spectroscopy Detection Limits

Element	μg/sample (10 ml digest)		
	Flame AA	GF-AA	ICP-AES
Al	1	0.03	0.4
As		0.01	0.5
Be	0.05	0.002	0.001
Cd	0.02	0.001	0.04
Co	0.3	0.001	0.07
Cr	0.2	0.02	0.02
Cu	0.1	0.01	0.06
Mg	0.005	<0.005	3
Mn	0.1	0.002	0.02
Mo	1	0.01	0.08
Ni	0.2	0.01	0.15
Pb	0.5	0.01	0.4
Sb	0.7	0.03	0.3
Se		0.02	
Si	3	0.1	0.2
Sn	8	0.05	
Sr	0.3	0.01	0.005
V	2	0.025	0.08
Zn	0.05	0.015	0.02

Table 5.10
Summary of Spectroscopy Methods, Applications and LODs,
for other than Atomic Spectroscopic Methods.

Method	Used for	Example	LOD Estimated
MS	Identification of un-known organics	Indoor air, odor prob-lems, byproduct identi-fications	100 μg
IR	Identification of un-known organics, screen wastes for disposal	Complementary analy-sis to confirm MS re-sults, Oil mist in air	50–1000 μg
UV/Vis spectroscopy	Organic or inorganic compounds	Cr(VI), formaldehyde, ammonia	0.5 μg
XRD	Crystalline material	Silica	5 μg
XRF	Elemental analysis	Lead	0.1 mg/cm^2

In general, however, the IR is used more for qualitative analysis–in the identification of compound types. An example is the use of IR analysis for waste disposal screening to characterize waste material of unknown nature to allow proper disposal. Most waste-generating facilities separate chlorinated solvents from other hydrocarbon solvents. Aliphatic hydrocarbons have a significantly different IR spectrum from that of chlorinated solvents, allowing a quick screen prior to designation of disposal type. Another application of IR analysis is for the differentiation of used pump oils for environmentally responsible removal or recycling.

Industrial hygienists may need to address environmental issues such as disposal and recycling. IR analysis will help to clarify the decision-making for the industrial hygienist. Due to the significant cost difference of two types of pump oils shown in the IR spectra of Figure 5.20, the analysis can identify the different pump oils and aid the recycling of the more expensive material.

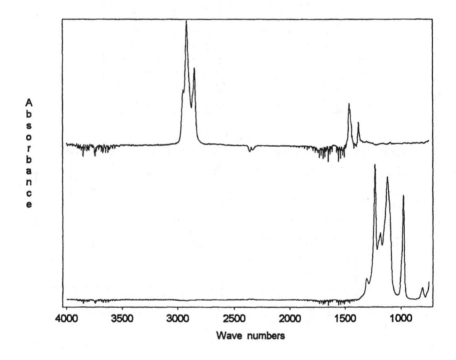

Figure 5.20   IR Spectrum of hydrocarbon pump oil and Krytox® pump oil.

The IR spectrum for hydrocarbon pump oil is shown in Figure 5.20 at the top, and Krytox® pump oil at the bottom, which is a perfluorinated high molecular weight hydrocarbon. The IR spectra are very different and easy to distinguish. The hydrocarbon pump oil shows absorbance peaks in the 3000–2800 wavenumber range. In this absorbance range, the Krytox oil does not absorb. The Krytox oil show peaks in the range below 1400 wavenumbers, where the hydrocarbon oil does not absorb. The pump oils were obtained from two high-vacuum process pumps.

b. Visible and ultraviolet spectroscopy

The UV/Vis spectrophotometer was discussed as an HPLC detector above. It is used in some industrial hygiene analyses. The electromagnetic region of energy from 185 to 380 nm is called the ultraviolet region and from 380 to 800 nm, it is called the visible region. When light is absorbed by a compound, there are resulting energy changes involving the valance electrons. The basic law of spectrophotometry is the Beer-Lambert law:

$$A = \varepsilon * c * b \qquad (5.15)$$

Where:

A = Absorbance

ε = Molar absorptivity, a constant dependent upon the wavelength of radiation, the molecular weight, and the nature of the absorbing material

c = Concentration, g/L

b = Interior cell thickness, mm (where the sample is placed in the lightpath, which can be varied to increase or decrease the analyte sensitivity)

This equation indicates that an instrument or a cell with twice the lightpath length has twice the sensitivity of analysis. Since the absorbance is a direct function of the concentration, the calibration curve of absorbance measured per concentration should be linear. For most UV/Vis spectrophotometers, the upper limit of linear absorbance is at about 2.0 absorbance units.

Transmittance measurements (T) are the negative exponent of the absorbance readout:

$$T = 10^{-\varepsilon * c * b} \qquad (5.16)$$

Where:

T = Transmittance

ε = Molar absorptivity, a constant dependent upon the wavelength of radiation, the molecular weight, and the nature of the absorbing material

c = Concentration, g/L

b = Interior cell thickness, mm (where the sample is placed is the lightpath and can be varied to increase or decrease the analyte sensitivity)

Transmittance measurements are exponential and therefore not linear and should not be used for calibration curves. Most spectrophotometers have absorbance and transmittance readouts.

A spectrophotometer is either of the prism or grating type. Inside the instrument, the light is broken down into its spectrum. A series of adjustable exit slits limits the wavelengths striking the sample.

Spectrofluorometers, which are special fixed-wavelength detectors, utilize the emission line spectrum from a single-element lamp. For a mercury lamp, for example, there are specific emissions due to the atomic spectroscopic emission characteristics of mercury: the mercury lamp emits light at 254, 313, 365, 405, 436, and 546 nm. The emission lines isolate these wavelengths with a simple filter generating monochromatic light. The strongest emission line of the mercury lamp is at 254 nm, coinciding with common absorbance wavelengths for aromatic compounds. There are numerous methods described for analyzing this type of compound at 254 nm.

The UV/Vis detectors use deuterium and tungsten lamps as a light source. In special cases it may be desirable to use a variable-wavelength detector to maximize the sensitivity for a given analyte. The schematic of a variable-wavelength detector is shown in Figure 5.21.

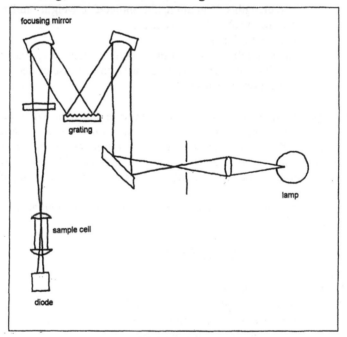

Figure 5.21 Schematic of a variable-wavelength detector.

An example of using the visible spectrophotometer is in the analysis of formaldehyde. In this case, a stable sulfite adduct is formed during the sampling of the unstable formaldehyde which is reacted with chromotropic acid and concentrated sulfuric acid in the laboratory to generate a purple color complex, which is then quantified in the spectrophotometer at 580 nm. The structure, and therefore the identity of the purple color complex is unknown. This method is specific for formaldehyde; other aldehydes do not interfere. The approximate LOD for this analysis is 0.04 ppm and the LOQ is 0.1 ppm based on a 4-hour sample, which is about 0.7 μg. The sensitivity of this method is limited by the low sampling rate of the diffusive sampler used in this method, which is 0.06 L/min. This method is used for exposures greater than 4 hours and not for STEL monitoring.

Other visible spectrophotometer analyses applications include the analysis for ammonia, which is complexed with EDTA and, after a color reaction, is measured at 630 nm. The approximate LOD for ammonia using this analysis is 0.5 μg per sample.

Advantages of UV/Vis spectrophotometric analysis are:

- Wide applicability to inorganic and organic species, and absorbing in the ultraviolet and visible ranges. Additionally, nonabsorbing compounds can be converted into absorbing species, as in the formaldehyde analysis.
- High sensitivity.
- Good accuracy, giving an average error in concentration measurement of 1 to 3%.
- Good selectivity, allowing the selection of a wavelength region where interferences are minimized.

Limitations are interferences stemming from changes in temperature and pH. Interference from stray light may cause deviation from Beer's law.

c. X-ray analysis
i. X-ray powder diffraction

X-ray powder diffraction (XRD) is a tool for the identification and quantification of inorganic crystalline materials such as silica and chrysotile, which is a specific form of the mineral asbestos. XRD is particularly advantageous because of its ability to analyze chemical compounds rather than elements. Any solid crystalline material diffracts an incidental beam of parallel, monochromatic X-rays whenever Bragg's law is satisfied for a particular set of planes in the crystal lattice.

$$\lambda = 2 * d * \sin \theta \tag{5.17}$$

Where:

- λ = Wavelength of the X-ray source
- **d** = Distance between the crystal lattice
- θ = Diffraction angle

By appropriate orientation of the sample relative to the incident X-ray beam, a diffraction pattern can be generated that will be uniquely characteristic of the structure of the crystalline phases present. Unlike other optical methods of analysis, XRD cannot determine crystal morphology and therefore cannot distinguish in asbestos analysis between fibrous and non-fibrous forms of serpentine and amphibole minerals. However, it can be used in conjunction with the microscopic methods for asbestos described below and can provide a reliable analytical method for the identification and characterization of asbestiform minerals in bulk materials.

In the XRD analysis of asbestos, the LOD for asbestos in talc or calcite is 0.2%; it is 0.4% in heavy X-ray absorbers such as iron (III) oxide by XRD analysis.

ii. *X-ray fluorescence spectroscopy*

X-ray fluorescence spectroscopy (XRF) is another X-ray technique, which is used to obtain elemental composition in a solid material. In XRF the sample is irradiated by an intense X-ray beam. The lines in the resulting secondary emission fluorescence spectrum are compared to those of a material with known elemental composition. The elements in the sample are identified by the energies (wavelengths) of their spectral lines, and their concentrations are determined by the intensities of these lines. Elemental data are obtained using X-ray fluorescence spectroscopy. This technique provides the capability for qualitative and quantitative elemental analyses for the range of sodium through uranium. The LOD is about 0.5 % and is matrix dependent. There are no NIOSH or OSHA published methods to date using this analytical method.

XRF is used in the *in situ* analysis of lead in paint with portable XRF analyzers. The method can distinguish lead in paint at concentrations from 0.1 to 90 mg/cm^2. Measurements below 1 mg/cm^2 are not very reliable. The Department of Housing and Urban Development (HUD) is defining lead-based paint at a lead content of equal or greater than 1.0 mg/cm^2 or 0.5 weight percent.

iii. *Proton induced X-ray emission*

Proton induced X-ray emission or PIXE analysis is another X-ray technique, which is used as a nondestructive analysis of elements from atomic number 11, which is sodium, to uranium, like XRF. Alpha particles and protons are generated by stripping the electrons from helium and hydrogen, respectively, in a plasma. The charged particles are then passed through a cyclotron, or an accelerator, such that they achieve extremely high kinetic energy. Using the electromagnetic fields, the charged particles are directed into a narrow beam which is directed into the sample. The sample is bombarded with protons from the proton accelerator and the resulting emissions are read in a spectrophotometer. The energy of each X-ray emission is characteristic of the element from which it is emitted. The number of X-rays emitted at a particular energy from a sample is a measure of the concentration of the emitting element. This technique is presently not used very much but is a very good tool when alternate methods of analyses are important, i.e., when there are interferences in the sample. The technique is simple and permits a large sample throughput. The cost, however, is high since it involves the use of an accelerator and vacuum equipment.

C. Other Analytical Techniques

Three other analytical techniques are commonly used in industrial hygiene analyses. They are: **(1)** ion selective electrode (ISE) analysis, **(2)** titration, and **(3)** gravimetric analysis, and are shown in Table 5.11.

Table 5.11
Other Techniques, Applications, and LODs

Method	Used for	Example	LOD Estimated
ISE	Specific anions or cations	Cyanide	2.5 µg
Titration	Acidic or alkaline species	Alkaline dust	30 µg
Gravimetry	Non-specific particles	Total dust	10 µg

1. *Ion selective electrodes*

Ion selective electrodes (ISE) are used in industrial hygiene analysis for the determination of cyanide and fluoride. ISE analysis can be used for other inorganic ions, for anions as well as cations. The ion concentration is determined potentiometrically in an aqueous solution by using a specific ion electrode (a cyanide electrode for cyanide analysis) in combination with a double-junction reference electrode and a pH meter having an expanded millivolt scale or a specific ion meter. The analysis technique is sometimes also referred to as potentiometry.

The cyanide ISE consists of a solid membrane containing a mixture of inorganic silver compounds bonded to the tip of an epoxy electrode body. When the electrode is in contact with a cyanide solution, silver ions dissolve from the membrane surface. Silver ions within the membrane move to the surface to replace the dissolved ions, setting up a potential difference that depends on the cyanide level in a solution. This solution is measured with a digital pH/mV or specific ion meter.

An important aspect of using ISE methods is to keep the ion strength and the pH the same for samples as well as for standards. This is achieved by adding a small amount of ion strength adjuster (ISA) to the sample as well as to the standards. The ISA for cyanide determination is a concentrated sodium hydroxide solution. The results are typically obtained as millivolts and are plotted on a calibration curve in a semi-log scale against the concentration. The plot should display a straight line in the mid-range concentration, and may show some curvature at very low and the very high concentration ranges.

The ISE must be used carefully and the results must be interpreted cautiously, because there can be some analytical interferences. Potential interferences for cyanide analysis by ISE may include:

- Transition metal cations, because they form very stable complexes with cyanide, which are not measured by the ISE.
- Sulfide, chloride, iodide, and bromide ions, because they form insoluble salts with the silver present in the ISE.
- Temperature changes, because they affect the electrode potential. For instance, an ISE calibrated at 22°C for 1.0 mg/L shows at 32°C only 0.64 mg/L.

The lower detection limit is about 2.5 µg cyanide per industrial hygiene sample.

2. Titrations

Titrations are used for some industrial hygiene analyses. The most common application is for the determination of alkaline dusts. Alkaline dusts occur in operations where alkaline material in liquid or dry spray is transferred and should be monitored. There is an OSHA limit for this air contaminant. The quantification of alkaline dusts consists of an acid/base titration using a pH electrode to determine the end point. The approximate LOD is about 0.03 mg per sample, representing 7×10^{-4} moles of alkalinity as NaOH. The sample, collected in an MCEF filter, is extracted in a known volume of 0.01 N HCl, then back-titrated to a neutral end point by adding 0.01 N NaOH. The difference between the amounts of HCl and NaOH indicate the amount of alkaline dust present, usually expressed as mg of sodium hydroxide.

3. Gravimetry

Some industrial hygiene analyses consist of nonspecific weight determination of dusts in air. For example, the analysis for carbon black, which is a major component in the toner of copy machines and printers, is a weight comparison. A matched pair of filters, commercially available, is used for the sampling. For the analysis, the pair of filters is separated and weighed on a high-precision analytical balance such as a five-place balance. Five place refers to five decimal places of the weight in grams, i.e., 0.00001 g or 0.01 mg or 10 µg can be determined. The difference in weight of the pair of filters is the analytical result of the amount of particles collected.

Gravimetry is also used in the quantitative analysis of asbestos in building materials or other bulk samples. In this method, the asbestos fibers are isolated and concentrated by selective dissolution and ashing of the other inorganic and organic fibrous matrix materials and then weighed. This method is usually followed by PLM, XRD, or TEM analysis, which are described in the microscopic analysis section.

Another application is total particulates in air.

D. Microscopic Techniques

Microscopic techniques are primarily used in industrial hygiene analysis of fibers, and most commonly used for the analysis of asbestos. These techniques require special skills and should only be performed by well-trained, experienced analysts of an asbestos analysis-certified laboratory. The analytical approach depends on the type of information desired from the analysis. A number of different microscopic techniques are available. There are optical and electron microscopic procedures used to determine fibers/fiber types in asbestos air or bulk samples.

1. *Optical microscopy*

Sufficient sample size is important for reliable analytical results. For bulk samples such a floor tiles, roofing felts, paper insulation etc., 3 to 4 cm^2 of layer material is an acceptable sample size. For materials such as ceiling tiles, loose-fill insulation, pipe insulation, etc., a sample size of approximately 1 cm^3 is sufficient. For samples of thin-coating materials such as paints, mastics, spray plasters, tapes, etc., a smaller sample size will suffice. As discussed throughout this chapter, the sample size affects the sensitivity of the analysis and the reliability in the quantitation steps.

If the asbestos type is to be determined, it is important that the appropriate laboratory is chosen. As described below, some asbestos methods can only determine quantity of fibers and not of fiber types. Standards are commercially available from NIST for the six regulated forms of asbestos: chrysotile, amosite, crocidolite, anthophyllite, actinolite, and tremolite.

For fibers in air analysis, 400 L represents a minimum air volume for 0.1 fiber/cm^3, and a maximum volume of optimum loading ranging from 100 to 1300 fibers per square millimeter.

a. *Stereo microscopy*

Bulk samples such as building materials are analyzed for asbestos fiber presence by a sequence of techniques. First, the bulk sample is examined with a low magnification stereo microscope (SM) to determine the presence of any fibers in the sample, and to determine the sample homogeneity. For hard samples, such as floor tiles, the broken edge is examined for exposed fibers. The morphology and physical characteristics of the fibrous components of the bulk materials are then identified. The microscopist reviews issues such as: Are the fibers straight and needle-like? Are they curly? Do the fibers occur in bundles? Do they exhibit electrostatic attraction? If fibers are found, the amount is estimated and the type of fiber is noted. This information must then be confirmed by polarized light microscopy.

b. *Polarized light microscopy*

Polarized light microscopy (PLM) is used for the identification of fibers. Bulk material may have to be ashed, acid-washed, or solvent-treated in order to separate the fibrous components from the sample matrix and its binders. Based on the fibers' physical characteristics which were examined under the stereo microscope, the analyst must then decide which refractive index (RI) liquid to immerse the fibrous material in. Although there are literally scores of commercially available refractive index liquids, the three most commonly used RI liquids for asbestos analysis have refractive indices of:

- 1.550 for chrysotile asbestos
- 1.605 for tremolite, anthophyllite, and actinolite asbestos
- 1.680 for amosite and crocidolite asbestos

Each of the RI liquids is specific for the identification of a different asbestos mineral type. The refractive index of a material is the ratio of the speed of light in a vacuum to the speed of light in that material. In the measurement of the optical characteristics of the fibers in the RI liquid, the analyst determines the fibers' bifringence, sign of elongation, extinction angles, refractive indices, and dispersion staining color. A variety of microscopic accessories are available to assist the microscopist in measuring the optical properties of the fibers present, such as red-1 compensator plate, double sets of polarizers, a rotating stage, or a central stop objective. Dispersion staining is a microscopic technique where color is imparted to a particle by evaluating the difference between the dispersion of the refractive index of a particle (fiber) and the liquid into which the particle is immersed. This is accomplished with a central stop objective and a reduced substage condenser aperture. Each asbestos variety will produce characteristic dispersion staining colors if immersed in the appropriate refractive index liquid.

Observation of specific colors indicates the presence of a specific asbestos type. An example of a microphotograph is shown in Figure 5.22. The colored strands are specific for specific asbestos mineral fibers. Crocidolite appears in red-magenta to blue-magenta parallel strands; amosite appears in bright yellow with red magenta to blue parallel cross fibers; and chrysotile has parallel strands of blue to blue-magenta color.

The quantification of asbestos in the bulk material is determined by comparison to known sample preparations. The estimated lower detection limit is below 1% of asbestos in a bulk sample. Figure 5.23 shows examples of amosite and of chrysotile in nonfibrous carbonate mineral matrix.

The optical properties of asbestos fibers are shown in Table 5.12.

c. Phase-contrast microscopy

Phase-contrast microscopy (PCM) is the technique of choice for quantitative analysis of total fibers in air. It is a fiber count method as described in excellent detail in NIOSH Method 7400. This method will only count fibers of 5 μm length or longer with a length to thickness ratio of greater than 3:1, which is called the aspect ratio. The method will distinguish those fibers from the other material which may have been collected in the sample.

This is the standard method for determining and quantifying fiber in air exposure. This method cannot differentiate asbestos fibers from other fibers. Fibers less than about 0.25 μm diameter will not be detected by this method.

The estimated lower detection limit is 7 fibers per square mm of filter area.

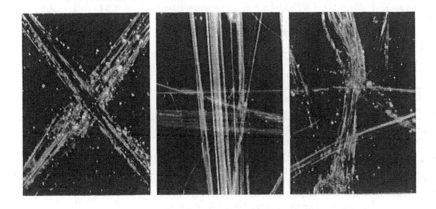

Figure 5.22   Microphotographs of different asbestos fibers.

The equipment needed for this analysis are a positive phase-contrast microscope at a 400 to 450X magnification, a Walton-Beckett graticule and a phase shift test slide. The graticule is a measuring device which is calibrated for a specific microscope and is used to determine the length and width measurements of fibers in a sample to determine the aspect ratio. Figure 5.24 shows the view through a PCM and a graticule.

The interpretation of a fiber count takes skill and experience, as is documented in NIOSH Method 7400. It is also referred to as the "A" rules. There are also "B" counting rules, which are applied for counting non-asbestos fibers. These rules must be used by an asbestos analysis-certified laboratory to assure consistent data interpretation during the analysis.

The NIOSH Method 7400 also contains detailed quality control steps which must be followed by the analyst to assure reliable data from an otherwise subjective method. Interlaboratory quality control is required of a laboratory as part of the NIOSH 7400 procedure.

Table 5.12
Optical Properties of Asbestos Fibers

Mineral	Morphology and Color	Refractive Index (Approximate Values) Perpendicular to Elongation	Parallel to Elongation
Chrysotile	Wavy fibers with kinks. Splayed ends on larger bundles. Colorless to light brown upon being heated. Non-pleochroic. Aspects ratio typically >10:1.	1.54	1.55
Amosite (Cummingtonite-Grunerite)	Straight fibers and fiber bundles. Bundle ends appear broom-like or splayed. Colorless to brown upon heating. May be weakly pleochroic. Aspects ratio typically >10:1.	1.67	1.70
Crocidolite (Rebeckite)	Straight fibers and fiber bundles. Long fibers show curvature. Splayed ends on bundles. Characteristic blue color. Aspects ratio typically >10:1.	1.71	1.70
Anthophyllite	Straight fibers and fiber bundles. Cleavage fragments may be present. Colorless to light brown. Nonpleochroic to weakly pleochroic. Aspects ratio typically <10:1.	1.61	1.63
Tremolite-Actinolite	Straight and curved fibers. Cleavage fragments common. Large fiber bundles show splayed ends. Tremolite is colorless. Actinolite is green and weakly to moderately pleochroic. Aspects ratio typically <10:1.	1.60–1.62 (tremolite) 1.62–1.67 (actinolite)	1.62–1.64 (tremolite) 1.64–1.68 (actinolite)

A summary of the techniques for fiber analysis is shown in Table 5.13.

Table 5.13
Summary of Analytical Techniques for Fiber Determination.

Method	Sample Type	Qualitative Information for Fiber (yes/no)	Fiber type (yes/no)	Quantitative
SM	Bulk	Yes	No	Est. percent
PLM	Bulk		Yes	Percent
PCM	Air	Yes	No	Count
TEM	Air or bulk		Yes	Count
XRD	Bulk		Yes*	Percent

* For chrysotile only

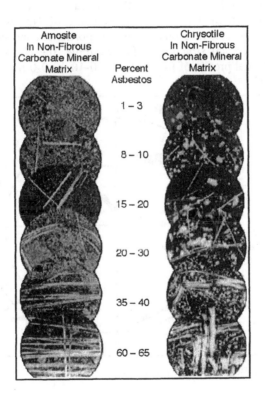

Figure 5.23 Examples of amosite and of chrysotile in nonfibrous carbonate mineral matrix.

2. *Electron microscopy*

Transmission electron microscopy (TEM), sometimes also called analytical electron microscopy (AEM), is a qualitative and quantitative method for the determination of asbestos fibers. This method is used to determine asbestos fibers in air samples and is a complementary method to phase contrast microscopy. AEM is also used as a reliable method for asbestos analysis in bulk materials that contain a large amount of interfering materials that can be removed by ashing and/or dissolution and contain asbestos fibers that are not resolved by PLM techniques.

In this technique the energy-dispersive X-ray (EDX) spectra of fibers with 0.25 and 0.5 μm diameter are generated. The portion of the fiber examined by EDX must be free of binder/matrix material. Therefore, sample preparation prior to analysis is critical. The elemental profile is obtained from the spectra for sodium, magnesium, silicon, calcium, and iron. Most minerals have a specific elemental profile and each of the asbestos minerals has a different elemental profile. The elements of Na-Mg-Si-Ca-Fe are common for all asbestos minerals, but the proportional concentrations differ due to different crystal structures. For example, the chrysotile structure shows an elemental profile of 0-5-10-0-0 to 0-10-10-0-0 for Na-Mg-Si-Ca-Fe, respectively.

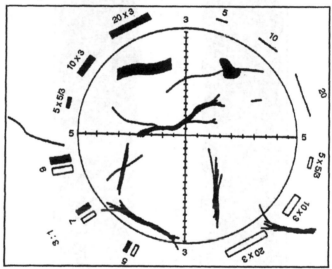

Figure 5.24 Walton-Beckett graticule with fibers.

The estimated LOD using TEM is 1 fiber above 95% of the expected mean blank value. Some other minerals of similar elemental composition may interfere by giving electron diffraction patterns or elemental profiles similar to that of the asbestos minerals. The LOD of asbestos in bulk material can be as low as 0.0001% depending on the sample preparation and concentration techniques.

E. Summary of the Applied Methods

A comparison of analyses for specific analytes by different industrial hygiene analytical methods using different instruments, methods, and detectors is shown in Table 5.14.

V. CHOOSING A LABORATORY

An analytical method has been chosen, the necessary instrumentation has been selected, and several analytical laboratories have been identified that could perform the analyses. How does the industrial hygienist make this final decision of where to have the samples analyzed?

Some guidelines could aid in the evaluation of laboratories:

- Assuring accreditation of the laboratory, as mentioned before, which provides confidence for consistent quality procedures. The industrial hygienist should be familiar with some key quality procedures.
- Performing a quality audit of the laboratory, which may include submitting a known blind sample and looking at quality control charts. The industrial hygienist should be able to conduct or oversee an audit.
- Determining the laboratory's general approach for analyses.
- Evaluating or deliberating an internal vs. external laboratory.
- Eliciting input from the industrial hygiene laboratory manager regarding the expected sampling conditions.

A. Quality Audit

An industrial hygienist may want to conduct a quality audit prior to choosing a laboratory for analysis of industrial hygiene samples. The following steps are possible steps to evaluate a laboratory:

- Does the laboratory have a quality manual?
- Can the laboratory recover a spiked amount in a "blind" sample?
- What do the quality control charts look like?

1. *Quality manual*

The audit should include obtaining the quality assurance manual of the laboratory. The manual should be specific to the laboratory where the samples will be analyzed. Some laboratories may use the AIHA accreditation application as a quality manual, which is fully acceptable, because it addresses the necessary quality elements for a good laboratory.

Table 5.14
Comparison of NIOSH and OSHA Methods for Various Detectors
and Their Detection Limits

Analyte	Analytical Method	LOD Estimated	NIOSH/OSHA Reference
Acetone	GC-FID	20 µg	N-1300
Acrolein	GC-NPD	2 µg	N-2501
Aldehydes	GC-MS screen	2 µg	N-2539
Alkaline dust	Titration	30 µg	N-7401
Ammonia	Visible spectrophotometry	0.5 µg	N-6015
Anthracene	HPLC-fluorescence/UV	0.05 µg	N-5506
Arsenic	GF-AA	0.06 µg	N-7901
Arsenic	Hydride AA	0.2 µg	N-5022
Arsenic	ICP	0.13 µg	N-7300
Benzene	GC-FID	1 µg	N-1501
Bromide	IC-CD	0.9 µg	N-7903
Carbon dioxide (in gas bag)	GC-TCD	1 ppm	N-6603
Chloride	IC-CD	0.6 µg	N-7903
Chlorinated diphenyl oxide	GC-ELCD	0.2 µg	N-5025
Cyanide	ISE	2.5 µg	N-7904
Ethylene oxide	GC-ECD	1 µg	N-1614
Fluoride	IC-CD	0.5 µg	N-7903
Formaldehyde	GC-FID of derivative	1 µg	N-2541
Formaldehyde	Spectrophotometry	0.7 µg	O-ID-205
Formic acid	IC-CD	2 µg	N-2011
Dust	Gravimetry	10 µg	N-0500
Lead	Flame AA	2.6 µg	N-7082
Lead	GF-AA	0.02 µg	N-7105
Lead	ICP	0.17 µg	N-7300
MDI	HPLC-UV	0.8 µg/m^3	O-47
Mercury	Cold vapor AA	0.03 µg	N-6009
Metals	ICP	1 µg	N-7300
Nitrate	IC-CD	0.7 µg	N-7903
Oil mist	IR	50 µg	N-5026
PCB	GC-ECD	0.1 µg	N-5503
Organophosphorus pesticides	GC-FPD	0.4 ng	N-5600
Phosphate	IC-CD	2.0 µg	N-7903
Silica	XRD	5 µg	N-7501
Sulfate	IC-CD	0.9 µg	N-7903
Tetraethyl lead	GC-PID	0.1 µg	N-2533

2. *Blind samples*

An additional tool for the evaluation of the performance of an industrial hygiene laboratory is the submission of a blind control sample for analysis.

Samples for metals on filters are commercially available from the National Institute of Standards and Technology (NIST) for a variety of metals. An NIST filter can be transferred into a blank methyl cellulose ester (MCEF) cartridge and submitted for quantitative analysis. The results for the laboratory are then compared to the certificate of analysis from NIST. A recovery within 10 to 15% of the label value should be expected from a good industrial hygiene laboratory.

Samples for organic compounds on charcoal are not as readily available for the field industrial hygienist. In this case the preparation of a QC sample is recommended as follows.

Using a microliter syringe, inject an exact amount ranging from one to 10 l of the compound of interest onto the charcoal in the adsorbent sampler tube or the diffusive sampler and reseal the sampler like a field sample. Calculate the amount added by multiplying the microliters added by the density of the compound. (Example: 1 l of toluene at a density of 0.867 g/ml or 0.867 mg/μl weighs 0.867 mg). A recovery within 10 to 20% (assuring proven injection techniques) is expected from the analysis of a good laboratory. In the case of some compounds, the desorption efficiency of the compound should have also been included by the laboratory in the final result.

An analytical laboratory may be willing to prepare quality control samples at various levels for the quality control evaluation by an industrial hygienist.

3. *Quality control charts*

Laboratory quality control can be assessed by evaluating the quality control data. Charts or database printouts should be current, dated, and show a relevant range of no more than +/-20% recovery of the expected value as 3 standard deviations (ceiling limit) for routine organic solvents and metals analysis. Some laboratories may keep QC data in tables, enabling sorting by analyst, analyte, concentration range, instrument and method. Either form is acceptable, as long as the chart shows when a control sample was analyzed, for what analyte, at which concentration, and what the recovery was. Charts should be dated and specific analytes identified.

B. Analytical Approach

How are samples analyzed by the laboratory? Different laboratories use a variety of approaches for analyzing industrial hygiene samples.

- Does the laboratory identify sample breakthrough from overloading the sampler with too much of a compound?

- Are the analyses carefully carried out to prevent carryover from previous analytes of contaminants in the analytical system?
- Will the laboratory share information about the sample as it may become available during the analysis, even if not requested by the industrial hygienist?

1. *Breakthrough*

Samples with a backup section, such as organic vapor adsorbents like charcoal, should be analyzed separately: sample preparation treats the front and the back sections as two individual samples and obtains analytical data for both. This separation is the only way an industrial hygienist can be assured that there was no sample breakthrough. More than 5% breakthrough can invalidate a result.

2. *Carryover*

Carryover can occur in all analytical systems if not carefully addressed. It may be in all cases due to overloading the system with a large amount of a compound, which may then linger in the detector, the transfer system, or the separation system and can falsify the results of the next sample.

In chromatography, carryover may also be associated with incomplete analysis. For example, should the analyst set the total analysis time too short, compounds may still elute from a previous sample after a new sample has already been injected. The peak identification by retention time is now invalidated. This carryover must be anticipated by the analyst. One approach is to perform duplicate analyses on each sample, thereby identifying if there is carryover. Another approach is to have a cleanup step follow each analysis, which could consist of an extended analysis time, during which all chromatographable compounds in the system will be eluted. For GC analysis this is achieved with a highend temperature bake-off and in HPLC this is achieved with a switch to a stronger mobile phase for clean-off.

In spectrometric analyses, as for metals analysis, the detector can get overloaded and thereby measure the subsequent analyses outside the linear range of the analytical method. In this case, the baseline absorbance increases and the analyst must clean the detector and rerun the samples.

3. *Unusual peaks*

How does a laboratory treat unexpected information about a sample? In gas chromatographic analysis, additional unexpected peaks may be observed by the analyst. These peaks may be of importance to the industrial hygienist. The laboratory should be asked how they approach unknown compounds of this type. If identification of a major unknown peak is required, the laboratory should have GC-MS capabilities, as discussed earlier.

C. External Versus Internal Laboratory

Many larger manufacturing facilities who have an industrial hygiene staff group may have an internal laboratory available to perform the industrial hygiene analysis. The internal laboratory should be expected to conform to the same standards as a commercial laboratory. For example, the laboratory should be certified to perform industrial hygiene analyses through the AIHA certification program, as remarked above, or other equivalent industrial hygiene laboratory accreditation/certification programs.

The advantages of an internal industrial laboratory may be:

1. *Proprietary and confidential information*

Although industrial hygiene data is of a fully disclosable nature to the employee and OSHA, the information about what chemicals are being used for proprietary processes will not have the public exposure when analyzed in an internal laboratory.

Analytical methods can be developed by the internal laboratory without potential disclosure of the use of proprietary chemicals.

2. *Customer service*

A dedicated internal laboratory can give the full attention and priority to an industrial hygiene problem because it serves only one customer. The followup identification of unknown compounds can be approached on a more informed level by an internal laboratory, based on knowledge of prior history, and of the chemical processes and raw materials typically used in that facility. Analytical methods can be developed in-house to address analyses of less common compounds for which there may not be published methods available. This may involve new compounds in new processes, for the purpose of workplace evaluations. A smaller commercial laboratory may not have the setup or staff for extensive methods development.

3. *Quality control*

The quality control exercised in an internal laboratory usually matches the overall corporate quality assurance policy of the company as a whole. An in-house laboratory is usually subjected to the same level of quality audits as other internal processes. This may not be a critical issue if AIHA certification is in place in the internal, as well as the external laboratory.

4. *Cost*

An external laboratory may be less expensive than an internal laboratory when a large number of routine samples need to be analyzed. Based on a larger volume of similar samples, a commercial laboratory may be able to perform routine analyses at a lower cost. But if only a limited number of analytes occur within an industrial setting, the cost may be comparable for both types of laboratories.

An accredited internal laboratory, on the other hand, may have the expertise to deal with challenges, such as methods development. For methods development and problem-solving analyses, an internal laboratory may be less expensive.

5. Conflict of interest

Use of an internal laboratory may create a potential conflict of interest when the laboratory is in the same organization as the areas to be monitored. To address this, the internal laboratory ideally should not be part of the manufacturing organization. However, if the laboratory follows the accreditation protocol, traceable analysis results will be generated, dependent on the quality control procedures used in the laboratory minimizing this concern.

VI. CALCULATIONS EMPLOYED DURING AIR SAMPLING AND ANALYSIS

A. Organic Vapors

The calculations performed for organic vapors in adjusting for sampling conditions, the temperature and pressure (T, P), in the workplace are as follows. The concentration C is defined (as usual) as the amount of mass (mg) present in any m^3-volume of air sampled at (T, P). If mass m is sampled in a volume V of air, then C is:

$$C = m / V \qquad (5.18)$$

C is then used directly, without any temperature or pressure correction, for comparison with the TLV or PEL.

Equivalently, the workplace concentration may be expressed in terms of "ppm at NTP" (at $T_o = 25°C$ and $P_o = 760$ mm Hg) for comparison (the PEL and TLV tables list both units). The unit, *ppm at NTP* is something of a misnomer, since it does not directly express the ratio of hazardous analyte to air molecules. Rather, *ppm at NTP* is equal to the volume which the mass m occupies at NTP, divided by the volume V (at T, P); *ppm at NTP* therefore differs from the usual concept of ppm as a ratio of molecule numbers at a single temperature and pressure. The volume occupied by the analyte molecules with mass m at NTP is equal to $(m / MW) * 24.24$ L, where MW is the analyte's molecular weight. Therefore, the concentration C_{ppm} in units of *ppm at NTP* is given by:

$$C_{ppm} = \frac{m}{MW} * 24.45 * \frac{10^6}{V} \qquad (5.19)$$

Where:

C_{ppm} = Concentration at NTP, ppm
m   = Mass, mg
MW = Molecular weight, g/Mol
24.45 = Molar gas volume at NTP, L/Mol
V   = Volume sampled at workplace, L

These calculations are somewhat further complicated since they differ when the organic solvent is collected with a sampling pump, rather than by a diffusive sampler. A pump, if calibrated correctly (e.g., at (T, P)) collects an air volume at the site pressure and temperature, and so the concentration C is directly obtained. However, the uptake rate of a diffusive sampler depends on the sampling site pressure and temperature. This is because diffusion is easier when the diffusing molecules are moving faster (higher temperature) or if they have farther to go prior to suffering a collision (lower pressure).

Specifically, in diffusion through air, the mass m sampled in time t is given in terms of the diffusion constant D at (T, P), the diffusion length L, and sampling area A by:

$$m = \frac{D * A * C}{L}$$

$$m = S_r * C \tag{5.20}$$

where S_r (cm^3/minute) is known as the *sampling rate*. The sampling rate is therefore equal to:

$$S_r = \frac{D * A}{L}$$

The diffusion coefficient D varies with the temperature (T)$^{3/2}$ and the pressure (P)$^{-1}$. Therefore, since A and L do not vary significantly with temperature and pressure, S_r has the same (T, P)-dependence as D. Therefore, if the sampling rate S_{ro} was specified at NTP, then the sampling rate S_r, is determined by a factor **corr** given by:

$$corr = \frac{T^{3/2}}{T_0} * \frac{P_0}{P} \tag{5.21}$$

in order to estimate C at (T, P) from m by inverting Equation 5.20.

Note that diffusion may not be entirely through air (e.g., if through a semipermeable membrane with an expansion coefficient different from that of air), and therefore the correction would in practice be given by the sampler manufacturer and may differ from Equation 5.21.

Note also that calculation indicates that if the concentration **C** is corrected to C_0, the concentration obtained by expanding or compressing, heating or cooling of the volume **V** to obtain NTP, then the workplace pressure **P** drops out of the calculation. Equivalently, the concentration expressed in terms of *ppm*, rather than *ppm at NTP* is dependent on **P**. This has been recognized by many researchers and has been a source of considerable confusion within the industrial hygiene community. It must be remembered, however, that at present it is the actual concentration **C**, expressed in units of mg/m³ or else *ppm at NTP*, which is compared to the PEL or TLV.

1. Organic vapors collected with sampling pumps

Sample Calculation: Calculate the acetone concentration in the workplace from the following information:

1.6 mg	= Collected acetone mass (uncorrected for desorption efficiency)
0.89	= Desorption efficiency of acetone for this analysis
1.5 L	= Air volume sampled with sampling pump at the workplace
32°C	= Workplace temperature
630 mmHg	= Workplace pressure
58 g/Mol	= Molecular weight of acetone

There are three steps involved to determine the acetone concentration in units of mg/m³ or ppm at NTP at the workplace.

Step 1:
    Correct the mass found in the analyte for the desorption efficiency. This step should always be carried out by the analytical laboratory.

$$m = \frac{mass}{DE} \tag{5.22}$$

Where:
    mass = mass analyzed, mg
    DE = desorption efficiency
    m = mass corrected for desorption efficiency, mg

So:
$$m = \frac{1.6 \text{ mg}}{0.89}$$

$$m = 1.80 \text{ mg}$$

The desorption efficiency-corrected mass is 1.80 mg.

Step 2:
Calculate the workplace air concentration C (mg/m³) from the sampled air volume and the DE-corrected laboratory analysis of the amount of acetone. This concentration is specific to the sampling conditions at the workplace and takes into consideration that the sampling pump was calibrated at the sampling site.

$$C = \frac{m * 1000 \text{ L/m}^3}{V} \qquad\qquad (5.23)$$

$$C = \frac{1.8 \text{ mg} * 1000 \text{ L/m}^3}{1.5 \text{ L}}$$

$$C = 1200 \text{ mg/m}^3$$

It should be noted that this calculation did not include any temperature or pressure calculations.

Step 3:
Now the workplace air concentration of acetone in units of ppm at NTP may be calculated using Equation 5.19:

$$C_{ppm} = \frac{m}{MW} * 24.45 * \frac{10^6}{V}$$

$$C_{ppm} = 1.8 \text{ mg} * \frac{1 \text{ Mol}}{58 \text{ g}} * \frac{0.001 \text{ g}}{1 \text{ mg}} * \frac{24.45 \text{ L}}{\text{Mol}} * \frac{10^6}{1.5 \text{ L}}$$

$$C_{ppm} = 506 \text{ ppm}$$

This is the acetone concentration which is then compared to the published PEL or TLV limits.

2. Organic vapors collected by diffusive sampler

With a diffusive sampler, the sampling rate at NTP (or other specified conditions) for each compound of interest must be supplied by the manufacturer of the sampling device. Due to the effect of pressure and temperature on the diffusion constant, both parameters need to be incorporated in the calculation of the concentration C if the sampling site conditions differ from NTP conditions.

Sample Calculation: Calculate the concentration of *n*-butylacetate in the workplace from the following information.

8 hours sampling time = 480 minutes
31.6 cm³/min = Sampling rate (S_{ro}) at NTP for *n*-butylacetate, supplied by the monitor manufacturer
2.3 mg = Measured amount of *n*-butylacetate
0.93 = Desorption efficiency of *n*-butylacetate by the analytical method
950 mm Hg = Pressure at sampling site
3°C temperature at sampling site = 276K
116 g /Mol = Molecular weight of *n*-butylacetate

Step 1:

Correct the mass found in the analyte for the desorption efficiency. This step should always be carried out by the analytical laboratory. Use Equation 5.22:

$$m = \frac{mass}{DE}$$

Where:

    mass = Mass analyzed, mg
    DE = Desorption efficiency using the diffusive sampler
    m = Mass corrected for desorption efficiency, mg

So:

$$m = \frac{2.3 \text{ mg}}{0.93}$$

$$m = 2.47 \text{ mg}$$

The desorption deficiency-corrected mass is 2.47 mg.

Step 2:

Determine the amount of **V** (effectively) sampled by the diffusive sampler, using **t** the sampling period (480 minutes) and the sampling rate Sr, as corrected to the sampling conditions **(T, P)**. To determine the amount of air sampled, the following equation is used:

$$V = t * S_{ro} * \frac{T^{3/2}}{T_0} * \frac{Po}{P} * \frac{1L}{1000 \text{ cm}^3} \tag{5.24}$$

So:

$$V = 480 \text{ min} * 31.6 \text{ cm}^3/\text{min} * (276/298)^{3/2} * (760/950) * (1 \text{ L}/1000 \text{ cm}^3)$$

$$V = 10.8 \text{ L}$$

Step 3:

Calculate the workplace air concentration C (mg/m³) at (T, P) from the corrected sampled mass m and air volume V effectively sampled. This is done by using equation 5.23.

$$C = \frac{m * 1000 \text{ L/m}^3}{V}$$

$$C = \frac{2.47 \text{ mg} * 1000 \text{ L/m}^3}{10.8 \text{ L}}$$

$$C = 229 \text{ mg/m}^3$$

A concentration for *n*-butylacetate of 229 mg/m³ was calculated for the concentration at the sampling site. Note that if the pressure and temperature corrections had not been considered, the result would have been 163 mg/m³. This is only 70% of the corrected concentration. However, if the sampling conditions had been at NTP, then the temperature and pressure corrections would not have been necessary and the effective air volume would have been 480 * 31.6/1000 or 15.2 L, instead of the 10.8 L found in step 2.

Step 4:

The concentration can be expressed as the workplace air concentration in ppm at NTP, exactly as for pumped sampling using Equation 5.19.

$$C_{ppm} = \frac{m}{MW} * 24.45 * \frac{10^6}{V}$$

$$C_{ppm} = 2.47 \text{ mg} * \frac{0.001 \text{ g}}{mg} * \frac{\text{Mol}}{116 \text{ g}} * \frac{24.45 \text{ L}}{\text{Mol}} * \frac{10^6}{10.8 \text{ L}}$$

$$C_{ppm} = 48.2$$

The overall equation for C_{ppm} in ppm at NTP may be expressed as:

$$C_{ppm} = \frac{10^6 * \text{mg mass} * 24.45 \text{ L/Mol} * 0.001 \text{ g/mg} * 1000 \text{ cm}^3/\text{L}}{MW \text{ g/Mol} * DE * t \text{ min} * S_{ro} \text{ cm}^3/\text{min} * (T/298K)^{3/2} * (760/P)} \quad (5.25)$$

B. Metal Analysis

The amount of metal found on a filter is expressed in mg or μg per filter. It is obtained analytically as a concentration of the acid digestate, which is the analyte, and is expressed as μg/ml (ppm). The result is multiplied by the digestion volume to obtain the μg metal present on a filter.

The equation to determine the μg of metal on a filter is:

$$\mu g \text{ metal/filter} = C_A * V_A \tag{5.26}$$

Where:

C_A = Concentration found in analyte, μg/ml
V_A = Digestion volume per filter, ml

To calculate the metal concentration as mg/m^3 in the sampled air, the μg per filter result is divided by the air volume, V in liters representing the sample.

$$C = \frac{\mu g \text{ metal/filter}}{V} \tag{5.27}$$

The overall equation for determining metal concentration in an air sample is:

$$C = C_A * \frac{V_A}{V} \tag{5.28}$$

Sample Calculation: For example, a filter was digested in 25 ml digestate and the digestate analyzed at 5.4 μg/ml lead. The total air volume sampled was 3.2 L; therefore, the concentration is:

$$C = 5.4 * \frac{25}{3.2} = 42.2 \text{ mg/m}^3 \text{ lead}$$

C. Acid Vapor Analysis

The results generated in the industrial hygiene laboratory for acid vapors are usually expressed in units of mg per sample. This value is obtained by the analysis of an acid vapor standard for each specific anion and comparing it to the analytical result of the sample. The analyst incorporates the sample extraction volume (which is typically 10 ml) and the concentration in the analyte (usually in ppm or μg/ml) as follows:

$$\text{mg/sample} = C_A * V_A * \frac{\text{mg}}{1000\mu g} \qquad (5.29)$$

Where:

C_A = Concentration found in analyte, $\mu g/ml$

V_A = Extraction volume per sample, ml

Incorporating the air volume representing the sample into this result, the actual concentration of anion in 1000 L or 1 m^3 air is determined as follows:

$$C_{mg/m^3} = \frac{\text{mg/sample}}{V} * \frac{1000L}{m^3} \qquad (5.30)$$

Where:

V = air volume sampled, L

Combining Equations 5.29 and 5.30 gives the overall equation for determining anion concentration in air as:

$$C_{mg/m^3} = C_A * V_A * \frac{\text{mg}}{1000\ \mu g} * \frac{1000\ L}{m^3} * \frac{1}{V} \qquad (5.31)$$

or simplifying:

$$C_{mg/m^3} = C_A * V_A * 1/V \qquad (5.32)$$

From this and the other industrial hygiene calculations, it can be seen that the sensitivity of a method is strongly dependent on the amount of air sampled.

Sample Calculation: For example, the lower detection limit for nitrate by this method is 0.3 mg/L in the analyte. If no nitrate was detected and a 3 L air sample was collected, then the report will state a lower detection limit of:

$$C_{mg/m^3} = \frac{0.3 * 10}{3} = 1 \text{ mg nitrate per } m^3 \text{ air}$$

Note: 10 ml is used as extraction volume.

If, however, 100 L of air were sampled, this value becomes:

$$C_{mg/m^3} = \frac{0.3 * 10}{100} = 0.03 \text{ mg nitrate per } m^3 \text{ air}$$

To convert the nitrate (MW = 62) to nitric acid (MW = 63), the result is multiplied by a factor of:

$$\frac{[HNO_3]}{[NO_3]} = \frac{63}{62} = 1.016$$

The concentration of 0.03 mg/m³ nitrate in air therefore becomes 0.0305 mg/m³ nitric acid. It would only introduce a 1.6% error into the result if the molecular weight correction from nitrate to nitric acid was not used.

D. Titrations

Titrations are most frequently used in the determination of alkaline dust in air. Standardization of the 0.01 N NaOH and the 0.01 N HCl is established as described in the NIOSH method. The sample concentration is established using the normality N and the volumes of NaOH (V_S) of the sample after the sample blank subtraction (V_B) incorporating the volume (V) of air sampled:

$$C_{mg/m^3} = [V_B - V_S] * \frac{N * MW}{V} * \frac{1000\ L}{m^3} * \frac{L}{1000\ ml} * \frac{1000\ mg}{g} \quad (5.33)$$

or

$$C_{mg/m^3} = [V_B - V_S] * \frac{N * MW}{V} * 1000 \quad (5.34)$$

Where:

V_B = Volume of NaOH sample blank, ml
V_S = Volume of NaOH sample, ml
N = Normality of sampled species, Mol/L
MW = Molecular weight, g/Mol
V = Air volume sampled, L

Sample Calculation: For example, if the MCEF blank used 4.93 ml of 0.01 N NaOH in the back-titration, the sample MCEF used 4.15 ml of 0.01 N NaOH, and a total of 578 L of air was sampled the following amount of NaOH was present in the air sampled:

$$C_{mg\ NaOH/m}{}^3 = (4.93 - 4.15) * \frac{0.01 * 40}{578} * 1000$$

$$C_{mg\ NaOH/m}{}^3 = 0.54\ mg/m^3$$

If the alkaline dust to be measured consisted of potassium hydroxide, the measurement and analysis is performed as described above. For the calculation, however, the molecular weight of 40 is replaced with that of KOH which is 56. This would change the analytical result from above to

$$C_{mg\ KOH/m}{}^3 = (4.93 - 4.15) * \frac{0.01 * 56}{578} * 1000$$

$$C_{mg\ KOH/m}^3 = 0.76\ mg/m^3$$

VII. ACKNOWLEDGMENTS

The author thanks the following people for their input, advice, help, and patience in putting this chapter together: Gerald Schultz (OSHA), David Bartley (NIOSH), Rolf Hahne (University of Washington), Martin Harper (SKC), Eric Johnson (3M), Richard D'Orazio (Lucent Technologies), Dave Parees, Sigi Koko, Christine Roysdon, Gary Fromert, Michael Hughes, Allen Zinnes, and Marci Koko.

REVIEW QUESTIONS

1. Toluene has a molecular weight of 92 and a TWA of 50 ppm. What is the TWA expressed as mg/m^3 at NTP?

   a. 13.3 mg/m^3 **b.** 11.2 mg/m^3 **c.** 188 mg/m^3 **d.** 50 mg/m^3

2. The lower limit of detection (LOD) of an analytical method is 1.5 µg per sample. How much is the approximate lower limit of quantification (LOQ)?

   a. 3.0 µg   **b.** 15 µg   **c.** 5 µg   **d.** 20 µg

3. What is the noise level of the method in question 2?

   a. 1 µg   **b.** 3 µg   **c.** 0.5 µg   **d.** 0.3 µg

4. A compound has a PEL of 0.1 ppm. It has a molecular weight of 100. The maximum flow rate (MFR) for sampling stated in the method is 0.1 L/minute. The analytical LOQ is 10 µg How long must the air sample be collected to determine 10 % of the PEL assuming NTP?

   a. 100 minutes   **b.** 2445 minutes   **c.** 409 minutes   **d.** 1000 minutes

5. What is the most sensitive technique for analysis of polychlorinated biphenyls (PCBs)?

   a. GC-MS   **b.** IR   **c.** GC-FID   **d.** GC-ECD

6. Which of these compounds will not exhibit any desorption efficiency effects in the analysis?

   a. Acetone   **b.** Methylisobutylketone   **c.** Naphthalene   **d.** Arsenic trioxide

7. Which compound cannot be analyzed by GC?

   a. Xylene   **b.** Naphthalene   **c.** Nitric acid   **d.** Tetrachloroethylene

8. The desorption efficiency of a compound is 72%. The analytical method determined 1.32 mg of the compound in the sample, uncorrected for desorption efficiency. The molecular weight of the compound is 60. The sample represents an air sample of 563 cm^3. What is the workplace concentration (in ppm) of the compound?

   a. 1327 ppm   **b.** 955 ppm   **c.** 2.344 ppm   **d.** 0.303 ppm

9. What is the best analytical technique to determine mercury in air?

   a. Flame AA   **b.** Graphite furnace AA   **c.** Hydride AA   **d.** Cold vapor AA

10. Which of these compound can be detected by GC-ELCD?

   a. Toluene   **b.** Chromium   **c.** Trichloroethylene   **d.** Acetone

11. Which GC detector is nonspecific for aliphatic, aromatic, halogenated, nitrogenated, phosphorilated, and sulfonated compounds and will detect them all?

   a. ECD   **b.** FID   **c.** FPD   **d.** NPD

12. A colorless liquid of unknown origin needs to be characterized for proper disposal as a chlorinated organic solvent mixture or an aliphatic hydrocarbon mixture. Which analytical technique is not appropriate to get this information?

    a. GC-FID   **b.** GC-MS   **c.** IR   **d.** GC-ECD

13. One of these methods does not operate in the electromagnetic range of UV light:

    a. GC-PID   **b.** ICP   **c.** GC-FID   **d.** HPLC-UV/Vis

14. A process may give workers a potential exposure to petroleum distillate and dichlorobenzene. Which analytical method would be a good choice to determine dichlorobenzene concentrations in the presence of petroleum distillate?

    a. GC-FID   **b.** ICP   **c.** ISE   **d.** GC-ELCD   **e.** GC-FPD

15. Cyanide is monitored and submitted to the laboratory. Which is a good analytical technique for quantification?

    a. XRF   **b.** ISE   **c.** GC-MS   **d.** IR

16. Arsenic is to be monitored at very low concentrations. Which is the method of choice?

    a. Flame-AA   **b.** GF-AA   **c.** Hydride-AA   **d.** ICP

17. A sample is collected with a pump sampling 50 L of workplace air at 600 mm Hg. The result shows that there are 0.05 mg of formaldehyde found on the sampler. What is the concentration at the workplace expressed in ppm? The molecular weight of formaldehyde is 30.

    a. 0.82 ppm   **b.** 1.03 ppm   **c.** 31 ppm   **d.** None of the above

18. A solder fume sample was digested in 10 ml of acid and analyzed. The acid digestate contained 2 µg/ml of lead. A total air volume of 10 L was collected. How much lead was in the air? The atomic weight of lead is 207.

    a. 407 mg/m^3 **b.** 40.7 mg/m^3 **c.** 20 mg/m^3 **d.** 2 mg/m^3

19. Alkaline dust is collected on a filter. Which analytical method is inappropriate for the quantification?

    a. ICP   **b.** HPLC   **c.** ISE   **d.** Titration

20. A new organic chemical process is generating a work atmosphere of unknown odor. What analytical technique is recommended to identify the unknown byproduct?

    a. GC-FID   **b.** GF-AA   **c.** GC-MS   **d.** GC-PID

ANSWERS

1.   c.   The formula for converting ppm to mg/m^3 is:
$$mg/m^3 = ppm * MW/24.45$$
$$mg/m^3 = 50 * 92/24.45 = 188$$

2.   c.   For noise = n

LOQ = 10 * n   and   LOD = 3 * n

If LOD = 1.5;

then   n = 1.5 / 3 = 0.5

LOQ = 10 * n = 5

3.   c.   n * 3 = LOD

n = LOD / 3 = 0.5

4.   b.   The formula for determining the minimum sampling time is:

sampling time = LOQ / MFR * 24.45 / MW / (PEL(ppm) * 0.1)

sampling time = 10 / 0.1 * 24.45 / 100 / 0.1/ 0.1

sampling time = 2445 minutes

5.   d.   (For PCBs, the sensitivity of detection is: ECD > FID > MS > IR)

6.   d.

7.   c.

8.   a.   The formula for concentration in ppm is derived from the sampling time equation:

sampling time = LOD / MFR * 24.45 / MW / PEL(ppm)

In this case the LOD is the actual analytical result and the

sampling time * MFR is the actual air volume; the PEL is the air concentration wanted:

ppm = [1.32 * 1000 (µg found)] / 0.563 (L) * 24.45 / 60 (MW)= 955 ppm

To correct for the desorption efficiency, divide the result by 0.72:

ppm = 955 / 0.72 = 1327 ppm

9.   d.

10.   c.

11.   b.

12.   a.

13.   c.

14.   d.

15.   b.

16.   b.

17.   b.   The molar gas volume is adjusted for the 600 mmHg sampling site pressure P

as:

V_{Mol} = 760 / 600 * 24.45 = 30.97 L/Mol

C_{ppm} = (mg found) * 1000 / (L sampled) / MW * V_{Mol}

= 0.05 * 1000 / 50 / 30 * 30.97

= 1.03 ppm

(Note: If the sample would have been collected on a diffusive sampler for formaldehyde, sampling the same air volume, and finding 0.05 mg, then no pressure correction would have been necessary and the subsequent result would be 0.815 ppm.)

18.   d.   The formula for determination of metal in air is:

mg/m³ = (µg/ml in analyte) * (ml digestion volume)/L of air volume

mg/m³ = 2 * 10/10 = 2 mg/m³

(The molecular weight is not included in metal concentration calculations.)

19.   b.

20.   c.

References

AIHA Journal 57, 208-215, 1996.

American Public Health Association, *Standard Methods for the Examination of Water and Wastewater*, APHA, 1995.

Beil, K., Crawford, W.W., Knolle, W.R., Stoll, I. AIHA Conference, New Orleans, LA, 1993.

Buffington, R. and Wilson, M.K., *Detectors for Gas Chromatography*, Hewlett Packard, 1987.

Cotton, F.A., et al., *Advanced Inorganic Chemistry, 2nd ed.*, Interscience Publishers, New York, 1966.

Eller, P.M., Chap. 3, Analytical methods, *Patty's Industrial Hygiene and Toxicology, 3rd ed., Vol 3, Part A*, John Wiley & Sons, New York, 1994.

Eller, P.M., Operational limits of air analysis methods, *Appl. Ind. Hyg.*, 1(2), 91, 1986.

EPA Method SW846-9213.

Kennedy, E.R., et al., *Guideline for Air Sampling and Analytical Method Development and Evaluation, NIOSH Publication No. 95-117*, May 1995

Lieckfield, R., Analytical methods, *Comprehensive Industrial Hygiene Review*, Clayton Environmental Consultants.

McCurry, JD, I. Stoll et al, Evaluation of desorption efficiency determination methods for acetone, *AIHA J.*, 50(10), 520-525, 1989.

National Institute for Occupational Safety & Health (NIOSH) Method 5025: Chlorinated diphenyl oxide by ELCD, *NIOSH Manual of Analytical Methods (MNAM), 4th ed.*, U.S. Department of Health, Education, and Welfare–NIOSH, Cincinnati, OH, August 15, 1994.

National Institute for Occupational Safety & Health (NIOSH) *NIOSH Manual of Analytical Methods (MNAM), 4th ed.*, U.S. Department of Health, Education, and Welfare–NIOSH, Cincinnati, OH, 1994.

National Institute for Occupational Safety & Health (NIOSH), Method 5503: Polychlorobiphenyls, *NIOSH Manual of Analytical Methods (MNAM), 4th ed.*, U.S. Department of Health, Education, and Welfare–NIOSH, Cincinnati, OH, August 15, 1994.

National Institute for Occupational Safety & Health (NIOSH), Method 1614: Ethylene oxide by GC-ECD, *NIOSH Manual of Analytical Methods (MNAM), 4th ed.*, U.S. Department of Health, Education, and Welfare–NIOSH, Cincinnati, OH, August 15, 1994.

National Institute for Occupational Safety & Health (NIOSH), Method 1300 for Ketones, *NIOSH Manual of Analytical Methods (MNAM), 4th ed.*, U.S. Department of Health, Education, and Welfare–NIOSH, Cincinnati, OH, August 15, 1994.

National Institute for Occupational Safety & Health (NIOSH), Method 2501: Acrolein by GC-NPD, *NIOSH Manual of Analytical Methods (MNAM), 4th ed.*, U.S. Department of Health, Education, and Welfare–NIOSH, Cincinnati, OH, August 15, 1994.

National Institute for Occupational Safety & Health (NIOSH), Method 5026: Mineral Oil Mist, *NIOSH Manual of Analytical Methods (MNAM), 4th ed.*, U.S. Department of Health, Education, and Welfare–NIOSH, Cincinnati, OH, August 15, 1994.

National Institute for Occupational Safety & Health (NIOSH), Method 5521: Monomeric isocyanates, *NIOSH Manual of Analytical Methods (MNAM), 4th ed.*, U.S. Department of Health, Education, and Welfare–NIOSH, Cincinnati, OH, August 15, 1994.

National Institute for Occupational Safety & Health (NIOSH), Method 2007: Aminoethanol Compounds, *NIOSH Manual of Analytical Methods (MNAM), 4th ed.*, U.S. Department of Health, Education, and Welfare–NIOSH, Cincinnati, OH, August 15, 1994.

National Institute for Occupational Safety & Health (NIOSH), Method 2539: Aldehyde screening by GC-FID and GC-MS as oxazolidines, *NIOSH Manual of Analytical Methods (MNAM), 4th ed.*, U.S. Department of Health, Education, and Welfare–NIOSH, Cincinnati, OH, August 15, 1994.

National Institute for Occupational Safety & Health (NIOSH), Method 2541: Formaldehyde by GC-FID with oxazolidine, *NIOSH Manual of Analytical Methods (MNAM), 4th ed.*, U.S. Department of Health, Education, and Welfare–NIOSH, Cincinnati, OH, August 15, 1994.

National Institute for Occupational Safety & Health (NIOSH), Method 1501: Aromatic Hydrocarbons, *NIOSH Manual of Analytical Methods (MNAM), 4th ed.*, U.S. Department of Health, Education, and Welfare–NIOSH, Cincinnati, OH, August 15, 1994.

National Institute for Occupational Safety & Health (NIOSH), Method 6603: Carbon Dioxide by GC-TCD, *NIOSH Manual of Analytical Methods (MNAM), 4th ed.*, U.S. Department of Health, Education, and Welfare–NIOSH, Cincinnati, OH, August 15, 1994.

National Institute for Occupational Safety & Health (NIOSH), Method 2533: Tetraethyl Lead by GC-ECD, *NIOSH Manual of Analytical Methods (MNAM), 4th ed.*, U.S. Department of Health, Education, and Welfare–NIOSH, Cincinnati, OH, August 15, 1994.

National Institute for Occupational Safety & Health (NIOSH), Method 5600: Organophosphorus pesticides by GC-FPD, *NIOSH Manual of Analytical Methods (MNAM), 4th ed.*, U.S. Department of Health, Education, and Welfare–NIOSH, Cincinnati, OH, August 15, 1994.

National Institute for Occupational Safety & Health (NIOSH), Method 5506: Polynuclear Aromatic Hydrocarbons by HPLC/FL/UV, *NIOSH Manual of Analytical Methods (MNAM), 4th ed.*, U.S. Department of Health, Education, and Welfare–NIOSH, Cincinnati, OH, August 15, 1994.

National Institute for Occupational Safety & Health (NIOSH), Method 7903: Inorganic Acids by IC, *NIOSH Manual of Analytical Methods (MNAM), 4th ed.*, U.S. Department of Health, Education, and Welfare–NIOSH, Cincinnati, OH, August 15, 1994.

National Institute for Occupational Safety & Health (NIOSH), Method 2011: Formic acid by IC, *NIOSH Manual of Analytical Methods (MNAM), 4th ed.*, U.S. Department of Health, Education, and Welfare–NIOSH, Cincinnati, OH, August 15, 1994.

National Institute for Occupational Safety & Health (NIOSH), Method 7082: Lead by Flame AA, *NIOSH Manual of Analytical Methods (MNAM), 4th ed.*, U.S. Department of Health, Education, and Welfare–NIOSH, Cincinnati, OH, August 15, 1994.

National Institute for Occupational Safety & Health (NIOSH), Method 7105: Lead by HGAAS, *NIOSH Manual of Analytical Methods (MNAM), 4th ed.*, U.S. Department of Health, Education, and Welfare–NIOSH, Cincinnati, OH, August 15, 1994.

National Institute for Occupational Safety & Health (NIOSH), Method 7901: Arsenic Trioxide by AA-GF, *NIOSH Manual of Analytical Methods (MNAM), 4th ed.*, U.S. Department of Health, Education, and Welfare–NIOSH, Cincinnati, OH, August 15, 1994.

National Institute for Occupational Safety & Health (NIOSH), Method 6009: Mercury by Cold Vapor AA, *NIOSH Manual of Analytical Methods (MNAM), 4th ed.*, U.S. Department of Health, Education, and Welfare–NIOSH, Cincinnati, OH, August 15, 1994.

National Institute for Occupational Safety & Health (NIOSH), Method 5022: Organic Arsenic by Hydride AA, *NIOSH Manual of Analytical Methods (MNAM), 4th ed.*, U.S. Department of Health, Education, and Welfare–NIOSH, Cincinnati, OH, August 15, 1994.

National Institute for Occupational Safety & Health (NIOSH), Method 7300: Elements by ICP, *NIOSH Manual of Analytical Methods (MNAM), 4th ed.*, U.S. Department of Health, Education, and Welfare–NIOSH, Cincinnati, OH, August 15, 1994.

National Institute for Occupational Safety & Health (NIOSH), Method 6015: Ammonia by Visible Spectrophotometry, *NIOSH Manual of Analytical Methods (MNAM), 4th ed.*, U.S. Department of Health, Education, and Welfare–NIOSH, Cincinnati, OH, August 15, 1994.

National Institute for Occupational Safety & Health (NIOSH), Method 7501: Silica by XRD, *NIOSH Manual of Analytical Methods (MNAM), 4th ed.*, U.S. Department of Health, Education, and Welfare–NIOSH, Cincinnati, OH, August 15, 1994.

National Institute for Occupational Safety & Health (NIOSH), Method 9000: Asbestos, Chrysotile by XRD, *NIOSH Manual of Analytical Methods (MNAM), 4th ed.*, U.S. Department of Health, Education, and Welfare–NIOSH, Cincinnati, OH, August 15, 1994.

National Institute for Occupational Safety & Health (NIOSH), Method 7904: Cyanides by ISE, *NIOSH Manual of Analytical Methods (MNAM), 4th ed.*, U.S. Department of Health, Education, and Welfare–NIOSH, Cincinnati, OH, August 15, 1994.

National Institute for Occupational Safety & Health (NIOSH), Method 7902: Fluoride by ISE, *NIOSH Manual of Analytical Methods (MNAM), 4th ed.*, U.S. Department of Health, Education, and Welfare–NIOSH, Cincinnati, OH, August 15, 1994.

National Institute for Occupational Safety & Health (NIOSH), Method for carbon black, *NIOSH Manual of Analytical Methods (MNAM), 4th ed.*, U.S. Department of Health, Education, and Welfare–NIOSH, Cincinnati, OH, August 15, 1994.

National Institute for Occupational Safety & Health (NIOSH), Method 7401: Alkaline Dusts, *NIOSH Manual of Analytical Methods (MNAM), 4th ed.*, U.S. Department of Health, Education, and Welfare–NIOSH, Cincinnati, OH, August 15, 1994.

National Institute for Occupational Safety & Health (NIOSH), Method 0500: Total dust in air, *NIOSH Manual of Analytical Methods (MNAM), 4th ed.*, U.S. Department of Health, Education, and Welfare–NIOSH, Cincinnati, OH, August 15, 1994.

National Institute for Occupational Safety & Health (NIOSH), Method 7400: Asbestos Analysis, *NIOSH Manual of Analytical Methods (MNAM), 4th ed.*, U.S. Department of Health, Education, and Welfare–NIOSH, Cincinnati, OH, August 15, 1994.

National Institute for Occupational Safety & Health (NIOSH), Method 9002: Asbestos (bulk) by PLM, *NIOSH Manual of Analytical Methods (MNAM), 4th ed.*, U.S. Department of Health, Education, and Welfare–NIOSH, Cincinnati, OH, August 15, 1994.

National Institute for Occupational Safety & Health (NIOSH), Method 7402: Asbestos by TEM, *NIOSH Manual of Analytical Methods (MNAM), 4th ed.*, U.S. Department of Health, Education, and Welfare–NIOSH, Cincinnati, OH, August 15, 1994.

NIOSH Measurements Services Section, Measurement Support Branch, Division of Physical Science and Engineering, *A Brief Description of Various Types of Analytical Instrument Techniques, Laboratory Analytical Instrumentation.*

Occupational Safety & Health Administration (OSHA), *OSHA Manual of Analytical Methods*, U.S. Department of Labor–OSHA, Salt Lake City, UT.

Occupational Safety & Health Administration (OSHA), OSHA Method 47 for Methylene Bisphenyl Isocyanate (MDI), *OSHA Manual of Analytical Methods*, U.S. Department of Labor-OSHA, Salt Lake City, UT.

Occupational Safety & Health Administration (OSHA), OSHA Method ID-205: Formaldehyde with Passive Badge Monitor, *OSHA Manual of Analytical Methods*, U.S. Department of Labor–OSHA, Salt Lake City, UT.

ORION (ATI), Cyanide Electrode Instruction Manual.

Perkins, RL and B.W. Harvey, Method for the Determination of Asbestos in Bulk Building Materials, EPA/600/R-93/116, 1993.

Personal communication with Gerald Schultz, OSHA Laboratory for Method Development, Salt Lake City, Utah.

Personal communication with David Bartley, NIOSH, Cincinnati, OH.

Personal communication with Rolf Hahne, University of Washington, Seattle, WA.

Personal communication with Martin Harper, SKC, Eighty Four, PA.

Personal communication with OSHA Laboratory for Method Development, Organic Methods Evaluation Branch, Salt Lake City, UT.

Sampling and Analytical Chemistry Committee of the AIHA, *Quality Assurance Manual for Industrial Hygiene Chemistry*, 1988.

Schultz, G.R., Comparing analytical results to OSHA permissible exposure limits, April 1985, USDOL–OSHA, Salt Lake City, Utah.

Schultz, G.R., How pressure and temperature affect sampling with diffusive samplers, March 1995, USDOL–OSHA, Salt Lake City, Utah.

Sheldon, L. et al., Biological Monitoring Techniques for Human Exposure to Industrial, PIXE Analysis Publication.

Silverstein, R.M., et al., Spectrometric Identification of Organic Compounds, *Patty's Industrial Hygiene and Toxicology, Vol. 3, Part A, 3rd ed.*, John Wiley & Sons, New York, 1994.

Smith, R.M., *Gas and Liquid Chromatography in Analytical Chemistry*, John Wiley & Sons, New York, 1988.

U.S. Code of Federal Regulations, *Title 29 of the CFR, Part 1910.1048*, December 1995.

U.S. Department of Labor, Occupational Safety and Health Administration, Salt Lake City, Utah, Internet homepage at http://www.osha-slc.gov/osha.html and the Chemical Sampling Information Internet site at http://www.osha-slc.gov/OCIS/toc_chemsamp.html.

Chapter 6

ENGINEERING CONTROLS: VENTILATION

D. Jeff Burton, BS, MS, PE, CIH, CSP

I. OVERVIEW

Industrial ventilation (IV) is the use of supply and exhaust ventilation to control chemical emissions and employee exposures in the industrial/manufacturing workplace. It can also be used to control hazardous temperature extremes. Heating, ventilating, and air conditioning (HVAC) systems are used to provide comfort, health, and well-being for employees in all occupancies, e.g., offices, hospitals.

A. Interaction of Air, Process Emissions, and Employees

When ventilation is being considered for emission or exposure control, the design team must know everything about the behavior of the emission source, the employee or occupant, the air and spacial characteristics, and their relationships to each other. Figure 6.1 depicts the relationship that must exist in any exposure problem, potential or real.

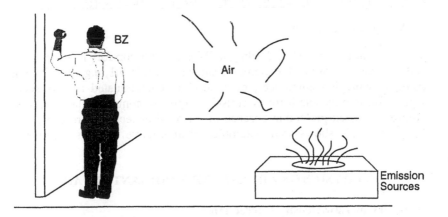

Figure 6.1   Emission, exposure, and spacial characteristics.

When ventilation is considered as a control, the information shown in the following table should be gathered.

Table 6.1
Ventilation Information Needs

Emission Source Behavior
Where are emission sources, or potential emission sources? [*Examples shown in brackets:* "floor sweeping," "process valve leak," "open access door."]
Which emission sources actually contribute to exposure? ["Floor sweeping because of the proximity of the emission source to the breathing zone."]
What is the relative contribution of each source to exposure? ["Sampling suggests floor sweeping contributes 50% of daily exposure dose."]
Characterize each contributor (e.g., chemical composition, temperature, rate of emission, direction of emission, initial emission velocity, continuous or intermittent, time intervals of emission). ["Silica dust, 70°F, every sweep, directly up into BZ, high velocity, intermittent."]
What is the relative toxicity of the emission source? ["Toxic at exposure level."]
Air Behavior
How does the air move (e.g., direction, velocity)? ["Very little movement; no open doors or windows; no exhaust ventilation during sweeping operation."]
Characterize the air (e.g., air temperature, mixing potential, supply and return flow conditions, air changes per hour, effects of wind speed and direction, effects of weather and season). ["70°F, 2 AC/H, no air speed or direction; summer time: door open."]
Employee Behavior
How do employees interact with emission sources? With the space? ["Employee within 4 feet of emission source during sweeping; at some distance from valve leak and open access door; valve leak to be corrected."]
Characterize employee involvement (e.g., location, workpractices, education and training, cooperation). ["Very near major source; poor workpractices; high school education and has no awareness of hazards of sweeping; could be cooperative."]

B. Alternative Controls

Maintaining employee health, well-being, and comfort involves controlling one or more of the elements shown in Table 6.1, e.g., eliminating sources, managing employees, or controlling contaminated air. As such, ventilation is only one form of emission/exposure control. Other forms of control are often preferable because of financial or technical reasons. The following table shows some examples of alternative controls commonly used.

II. PHYSICS OF AIR AND CONTAMINANTS IN AIR

A. Ideal Gas Law/Density Correction

Air has properties like any other physical material. The following paragraphs describe some of those properties.

1. *Density*

Air has a weight density of $\partial = 0.075$ lb/ft^3 or 1.20 kg/m^3 at STP.

Table 6.2
Types of Control

Type of Control	Example
Process change	Spray painting to paint dipping
Substitution	Hot soapy water for solvents (cleaning)
Isolation	Place employee in controlled environment Enclose process or equipment
Ventilation	Local exhaust (LE) ventilation General dilution ventilation
Administrative controls	Rotation of employees Training and education Management programs
Personal Protective Equipment	Respirators, gloves

2. *Standard conditions, STP*

Standard conditions, STP, for air are shown below for two standards-setting ventilation associations:

Table 6.3
Standard Conditions for ACGIH and ASHRAE

	U.S. units	SI units	Organization
Temperature	70° Fahrenheit 68° Fahrenheit	21° Celsius 20° Celsius	(ACGIH) (ASHRAE)
Barometric pressure	29.92 inches Hg	760 mm Hg	(ACGIH & ASHRAE)
Relative humidity	0% 50%	0% 50%	(ACGIH) (ASHRAE)
Weight density	0.075 lb/ft^3	1.20 kg/M^3	(ACGIH & ASHRAE)

3. *Molecular weight*

Air has a composite molecular weight of about MW = 29.0

4. *Ideal gas law*

All other conditions equal, changes in air density vary linearly with changes in absolute temperature. Similarly, air density varies linearly with the change in air pressure (e.g., if the pressure increases by 10%, the density will increase by 10%.) This linearity makes adjustments to ventilation calculations quite simple. These two relationships are combined in the *Ideal Gas Law*, which is used to calculate the *density correction factor d* (see Equation 6.1).

This factor can be used to correct measured air velocities to actual velocities, actual volume flow rates to standard volume flow rates, and so forth. At standard conditions, $d = 1.0$. At higher temperatures and altitudes, d is less than 1; at lower temperatures and altitudes below sea level, d increases above 1.

$$\text{Air density (actual)} = \text{Air density (standard)} \times d \qquad (6.1)$$

Where:

U.S. Units

$$d = \frac{530}{°F + 460} \times \frac{BP}{29.92} \qquad (6.2a)$$

SI units

$$d = \frac{294}{°C + 273} \times \frac{BP}{760} \qquad (6.2b)$$

Where:

d = Density correction factor, unitless
BP = Barometric pressure, U.S. = inches Hg; SI = mmHg
°F = Temperature, degrees Fahrenheit; °C is degrees Celsius

Refer to Table 6.4 for a summary of density correction factors

Table 6.4
Air Density Correction Factor, d

Elevation	Feet	0	1000	2000	3000	4000	5000	6000	7000
	Meters	0	305	610	915	1220	1525	1830	2135
Barometric pressure	*or* mmHg	760	733	707	681	656	632	608	587
	inches Hg	29.92	28.86	27.82	26.82	25.84	24.89	23.97	23.09
Air temperature	°C °F								
	-1 0	1.15	1.11	1.07	1.03	0.998	0.959	0.921	0.882
	0 32	1.08	1.04	1.01	0.969	0.933	0.897	0.861	0.825
	21 70	1.00	0.966	0.933	0.900	0.866	0.833	0.799	0.766
	38 100	0.946	0.915	0.883	0.851	0.820	0.788	0.756	0.725
	66 150	0.869	0.840	0.811	0.782	0.753	0.723	0.694	0.665
	93 200	0.803	0.776	0.749	0.722	0.696	0.669	0.642	0.625

Example: If the elevation and temperature at a plant are E = 1000 feet above sea level and T = 70°F, find d by using the appropriate column associated with E (in feet) and the row associated with T (in °F). When utilizing the correction factors for d presented in Table 6.4, d is determined to be 0.966.

B. Air Pressure

Pressure is defined as force per unit area. In the U.S. system, sea level air pressure is 14.7 pounds force per square inch absolute, or, *psia*. In more familiar terms, this means that a cross-sectional square inch of air reaching from the surface of the ocean to the top of the atmosphere weighs 14.7

pounds. This is the same weight as a similar 1-inch square column of water 407 inches in height, and a similar column of mercury 29.92 inches in height. In the SI system, sea level air pressure is 1.03 kilograms force per square centimeter. Again, this means that a cross-sectional square centimeter of air reaching from the surface of the ocean to the top of the atmosphere weighs 1.03 kilograms. This is the same weight as a similar 1-centimeter square column of mercury 760 mm in height, or 10.3 meters of water.

In ventilation, pressure is normally measured as:

U.S. units	SI units
inches of water	millimeters of water

Discussion of static pressure, velocity pressure, and total pressure is found in Section IV.

1. Aerosols in air

Emission sources often contain aerosols, or particulate matter. To make a ventilation system effective, you must know something about particles and their behavior. Let's review the basics. Particulate materials, or aerosols, come in two major physical forms: liquids and solids. The common types are shown in Table 6.5.

Table 6.5
Aerosols in Air

Aerosol Type	Natural State @ Room Temperature	Typical Source
Liquids		
Fogs	Liquid	Atomizing, spraying
Mists	Liquid	Spraying, acid baths
Smoke	Liquid, solid, gas	Combustion
Solids		
Dust	Solid	Mechanical action, vibration
Fume	Solid	Molten metal
Smoke	Liquid, solid	Combustion
Fibers	Solid	Natural, asbestos

C. Size and Shape

Aerosols vary in size and shape. Those of most importance are "respirable size," i.e., those most likely to be carried into the deepest regions of the lung. Typical sizes are shown in Table 6.6.

Table 6.6
Size of Typical Airborne Particles
-------------------------- Diameter, Micrometers ----------------------------

0.0003	0.001	0.003	0.01	0.03	0.1	0.3	1	3	10	20	50	100

respirable range

_ cond. nuc. _

_ _industrial fumes_ _

_ _ _____industrial particles

_ _ _ _____indoor *dust*

_____virus_____　　bacteria_____

gas molecules_____spores/pollen
　(odors)　　　　　　　_ _tobacco smoke_ _

invisible -----> - - <-----
visible

98% collection:　　　　　<----ULPA/HEPA---<----fabric filter----------<--cyclone----
<gravity

1. *Aerosol generators*

Aerosols are produced and introduced to the air by several basic methods, as described below:

Mechanical - Mechanical actions include wind blowing across a dusty surface, vibration, impaction of falling materials, grinding and drilling, mobile vehicle tires pounding against a dusty pavement, sweeping, and so forth. The mechanical action becomes the primary source of energy necessary to lift or throw particles into the air. Natural or secondary energy sources help keep the particles entrained and dispersed throughout the air space.

Chemical - Chemical actions include the formation of oxide fume, from evaporating liquid metal, welding and soldering fume, the chemical formation of smog particles, and so forth. Once an aerosol is in the air, it is influenced by two main forces: gravity and the mixing or drag forces of air movement. Gravity induces the particle to settle; air movement tends to delay settling and creates mixing of particles throughout the space.

It is important to know and remember that ambient air found in industrial settings is almost always in a constant state of movement. Turbulent and unpredictable mixing velocities exceeding 25-50 fpm are not uncommon in industrial environments. This turbulent mixing disperses small particles of health interest (submicrometer to 15 micrometers in diameter) throughout the space, and keeps particles from settling out quickly. Mixing turbulence can sometimes keep dusts in suspension in concentrations large enough to be explosive.

The settling velocity of particles, also known as the terminal velocity, is given by the following relationship (derived from Stokes law):

$$\mathbf{Vs = 0.0052 \ (S.G.) \ D^2} \qquad \mathbf{Vs = (26.4 \times 10^{-6})(S.G.) \ D^2} \qquad (6.3)$$

Where:
   Vs = Settling velocity, U.S. = feet per minute; SI = meters per second
   D = Particle diameter in microns (or the equivalent diameter for nonspherical particles)
   S.G. = Specific gravity, unitless (re: water = 1)
   [*Note*: the first equation uses U.S. units; the second, metric units.]

Table 6.7
Particulate Emission Source Terms

Particulate Emission Term	Definition
Terminal velocity	The final settling velocity of a particle in air.
Throw velocity	The initial velocity imparted to a particle by the emitting source, e.g., grinding wheel may throw particles off the wheel face at velocities exceeding 6000 fpm (or 30 mps.). Terminal velocity may be reached in several inches to several yards, depending on the particle mass.
Throw distance	The distance a particle travels after being emitted before reaching its terminal or settling velocity.
Null point	The position of the particle at the time it reaches settling velocity. The null point is often useful in determining the appropriate capture velocity.
Transport velocity	Also, scouring velocity. The minimum velocity necessary to create sufficient mixing turbulence to preclude permanent settling of particulates in the duct. Turbulent air mixing literally scours up dry particles that settle. The higher the velocity, the more carrying capacity.
Capture Velocity	The velocity of air in the vicinity of the emission source sufficient to capture the emission and carry it into the hood and ductwork.

Table 6.7 contains an overview of historical terms used to discuss particulate emission sources.

Refer to Figure 6.2 to view velocities associated with particle behaviors as discussed in Table 6.7.

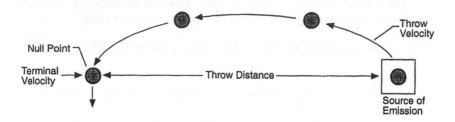

Figure 6.2   Velocities associated with particle behavior.

D. Vapors and Gases

Gases and vapors in air are single-molecule substances at room temperature. Gases are identified from the fact that they are normally gaseous at STP. Vapors are normally liquid at STP. However, the vapor pressure of most liquids allows for evaporation of the liquid and the buildup of a concentration of vapor-air mixture. For example, water at STP has a vapor pressure of vp = 18.5 mmHg. The vapor pressure creates evaporation of water vapor from the liquid. The higher the vapor pressure, the greater the evaporation. At sea level, water's vapor pressure at T = 212°F is vp = 760 mmHg, equal to atmospheric barometric pressure ("the water boils"). The highest concentration of a vapor-air mixture possible is determined *by partial pressure* rules, which are summarized below:

- At any temperature and pressure, air can hold or support a maximum amount of vapor.
- *Saturated air* is air holding the maximum concentration of vapor possible.
- *Warm air* holds more vapor than cooler air.
- If the partial pressure (pp) of a liquid is pp = 1.5 mmHg, then the maximum concentration by volume is C = pp/BP = 1.5/760 at STP, or 0.20% by volume, according to the following relationship:

$$C = pp/BP \times 100 \qquad\qquad (6.4)$$

Where:

pp = Partial pressure, mmHg
BP = Barometric pressure in consistent units
C   = Measured as %

1. *Gas and vapor generators*

In the work environment, vapors are generated from processes, equipment, fixtures and materials, people and human activities, remodeling and renovation activities, and outside sources. Vapor generation rates depend on the temperature, pressure, amount used, surface area exposed, the boiling point or vapor pressure, and other factors. Vapor generation formulae are provided in Section 6.3.

Solvent vapors and gases may be lighter or heavier than air. The traditional means of comparison is the specific gravity. The specific gravity of propane, for example, is 1.554, or 1.554 times as heavy as air (e.g., propane weighs 0.075 x 1.554 = 0.117 lb/cu.ft. at STP). Given the greater weight of propane, and in the absence of any mixing forces, propane and air should separate like water and oil, with air overlying the propane. But unlike oil and water, mixing does occur.

At concentrations in the parts per million (ppm) range, it is safe to assume that *complete* mixing occurs. At percent quantities of propane in air it is possible for separation to occur, depending on the level of air turbulence present. In very still, non-mixing air, highly concentrated vapors will have a tendency to settle. However, in industrial settings and at health-concern concentrations in air (i.e., ppm quantities), settling rarely occurs because of diffusion, and the fact that the settling rate is a function of the ratio of the density of air-vapor mixture to density of air. At percent quantities, this ratio becomes quite high, and it is possible to have separation of air and vapor.

III. DILUTION OR GENERAL VENTILATION SYSTEMS

Dilution ventilation is an ancient and effective method of exposure control. The approach allows chemical emissions to mix with the ambient air in a space. Contaminated air is diluted to acceptable concentrations. Dilution is the major basis of ventilation for control of indoor air quality (IAQ) problems and is widely used in industrial ventilation as well.

Sample Calculation: Two bypass-type lab fume hoods exhaust a total of 2400 cubic feet per minute. Makeup air is provided by the HVAC system. The hoods run all day long in a lab of dimensions 15 x 20 x 8 ft. What is the dilution volume flow rate? The number of air changes per hour in the lab?

The dilution ventilation volume flow rate must equal the exhaust rate. (Every cubic foot must be replaced.)

The number of air changes per hour (ACH) is given by:

$$N = \frac{Q \times 60}{Vol} \qquad (6.5)$$

Where:
N =   Number of air changes per hour, ACH
Q =   Volumetric flow rate of air, cfm
Vol =   Space volume, cubic feet (cubic meters)

so:

$$N = \frac{Q \times 60}{Vol}$$
$$N = \frac{2400 \times 60}{15' \times 20' \times 8'}$$
$$N = 60 \text{ ACH}$$

A. Selection Criteria and Applications

Certain conditions seem to suggest dilution ventilation. Table 6.8 lists the important considerations.

Table 6.8
Dilution Ventilation Selection Criteria

Favorable conditions	• Air contaminants are relatively nonhazardous or nontoxic.
	• Air contaminants are primarily vapors or gases, or respirable sized aerosols.
	• Emissions occur uniformly in time.
	• Emission sources are widely dispersed.
	• Emissions do not occur close to people.
	• Moderate climatic conditions prevail.
	• The dilution air is less contaminated than the inside air.
Unfavorable conditions	• Emission sources contain highly toxic materials.
	• Emitted materials are primarily large particulates.
	• Emissions vary widely over time.
	• Emission sources consist of large point sources.
	• People are in the immediate vicinity of emission sources.
	• The building is located in severe climates.
	• The dilution air is more contaminated than the inside air.
	• Resulting concentrations create irritation or complaints, even when exposures are below the TLV or PEL.

When the *no* conditions prevail, the volume of dilution ventilation will be high, thus raising operating and heating costs. Local exhaust ventilation or other forms of exposure control should be considered.

B. Decision Criteria

If preliminary indications point to dilution ventilation, follow these steps:

1. *Determine the acceptable exposure concentration, Ca.*
2. *Study emission sources.*
3. *Estimate the potential emission or evaporation rates.*

One useful approach uses the following formulae:

U.S. Units

$$q = \frac{387 \times lb\ evaporated}{MW \times Minutes \times d} \qquad (6.6a)$$

SI Units

$$q = \frac{0.0241 \times gram\ evaporated}{MW \times seconds \times d} \qquad (6.6b)$$

Where:
q   =   Volume of vapor flow rate (U.S. = acfm; SI = acms)
MW =   Molecular weight (molecular mass)
d   =   Density correction factor, unitless

4. *Characterize the space and air flow in the space.* Obtain all of the physical parameters of the space—width, height, length, barriers, obstructions, and so forth. Building plans are often helpful. Information on existing ventilation is also important (e.g., general ventilation, space heaters, open doors and windows, free standing "man-cooler" fans, local exhaust ventilation systems, and other dilution systems already in place). You should also study prevailing air movement in the space, including directions and velocities. Ask about other times of the day and other operations. ("What happens during graveyard shift?" "...when the furnace is turned on?")
5. *Estimate potential effects of climatic conditions.* Temperature and wind can have a big effect. Call the local weather service for information. Particularly useful are the *degree-day* estimates and the *wind-rose* averages. Degree-day information is used to estimate heating loads and costs. Wind-rose information is useful in estimating the effect of outside winds on the building. For example, if a south wind blows constantly against a building, dilution air should probably be designed to move with the prevailing winds.

Most industrial buildings are *open* to some degree. This may include open doors and windows, and almost always includes naturally occurring cracks and holes in windows and walls. Therefore, prevailing winds almost always have some effect on the airflow inside the building.

6. *Characterize worker exposure/work practices.* See Section I for approaches and information needs.

7. *Estimate dilution air volumes required.* The volume of air required to dilute a volume of vapor can be estimated as follows:

$$Q_d = \frac{q \times 10^6 \times K_{mixing}}{Ca\ (ppm)} \qquad (6.7)$$

Where:

Q_d = Volume flow rate of dilution air (U.S. = acfm; SI = acms)
q = Volume flow rate of vapor (U.S. = acfm; SI = acms)
(For q, please see Equation 6.6.)
Ca = Acceptable exposure concentration, ppm
K = Mixing factor to account for incomplete or poor mixing of the contaminant with the workroom air, unitless
K_{mixing} ranges from 1.5 to 4 in most cases

Sample Calculation: What is q, the volume flow rate of vapor formed, if 2.0 pounds of toluene are evaporated in a lab during an 8-hr period? What volume flow rate Q_d is required for dilution to Ca =10 ppm, if K_{mixing} = 1.5? (STP, MW = 92.1)

$$q = \frac{387 \times lb\ evaporated}{MW \times Minutes \times d}$$

$$q = \frac{387 \times 2\ lb\ evaporated}{92.1 \times 480\ minutes}$$

$$q = 0.0175\ scfm$$

$$Q_d = \frac{q \times K_{mixing} \times 10^6}{Ca\ (ppm)}$$

$$Q_d = \frac{0.0175 \times 10^6}{10\ ppm}$$

$$Q_d = 2600\ scfm$$

C. Good Practices

Dilution will be more effective, and lower values of the mixing factor K_{mixing} can be selected if the following conditions can be met:

- Route dilution air through the zone of emission. This may mean channeling or introducing makeup air by means of barriers, supply registers, and so forth.

- Distribute makeup air where it will be most effective; i.e., design and provide replacement air where you need it.
- Introduce dilution air so that employees are upwind of the dilution zone. As such, dilution will occur before emissions reach the employees' breathing zones.
- Position exhausts as close to emission sources as possible.
- Make use of auxiliary mixing fans to disperse emissions and to enhance dilution.

IV. LOCAL EXHAUST VENTILATION SYSTEMS

Local exhaust (LE) systems are usually designed by mechanical engineers under the supervision of management, e.g., the industrial hygienist. All exhaust ventilation systems should be designed, built, and operated to meet current codes and standards. The industrial hygienist can help assure this goal by becoming familiar with ventilation fundamentals, reviewing proposed plans and specifications, and conducting tests of installed systems.

A local exhaust system is used to control an air contaminant by collecting it at the source, as compared with a dilution ventilation system (which allows the contaminant to be diluted into the room air.)

Local exhaust is often preferred to dilution because it removes the contaminant from the work area, thus requires lower air volume flow rates. They are typically more energy efficient because they remove less tempered air from the building and use smaller fans and motors. An LE system, because of the greater concentration of contaminants, often has the ability to scrub or clean the contaminant from the exhaust air. As noted before, other emission controls may be more effective and should be explored before ventilation is installed (e.g., substitute less harmful materials, change the process, isolate the process from employees.)

A. Selection Criteria

A local exhaust system is usually appropriate for control when:

- Other more cost-effective controls are not available.
- Air sampling or employee complaints suggest that the air contaminant is a health hazard, a fire hazard, impairs productivity, or creates unacceptable comfort problems.
- Federal, state, local, building, or consensus standards require (or suggest) local exhaust systems. For example, OSHA's 29CFR1910.94 regulation calls for local exhaust for open surface tank operations, spray finishing, grinding, and abrasive blasting; the ANSI Z9.5 standard on lab ventilation requires chemical operations in a lab to be conducted in an exhausted lab fume hood. (See References at the end of the chapter for more information.)

- Improvements will be seen in production rates, housekeeping, employee comfort or morale, and the operation and maintenance of equipment.
- Emission sources tend to be large, few, fixed, and/or widely dispersed.
- Emission sources are close to the breathing zones of employees.
- Emission sources tend to vary with time.

B. Components

A local exhaust system consists of five important parts. (See Figure 6.3.)

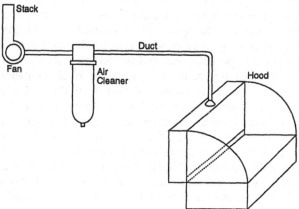

Figure 6.3   Components of a local exhaust ventilation system.

- *Hood:* Hoods capture, contain, and/or control the emission source.
- *Ductwork:* Ducts and piping carry the contaminant to the air cleaner and/or to the outside environment for dilution.
- *Air cleaner:* The air cleaner scrubs, separates, removes, or filters the air contaminant from the exhaust air, usually to meet permit requirements. It may also change the air contaminant to a less hazardous substance (e.g., using an afterburner to convert hydrocarbon vapors to carbon dioxide and water vapor).
- *Fan and motor:* The fan generates static pressure and moves air.
- *Stack:* The stack assists in dispersing remaining air contaminants.

C. Ductwork

Ductwork has the task of carrying air and any entrained contaminants away from the occupied space or exhausted enclosure for treatment and elimination from the building. Ductwork also provides a conduit for the static pressure to reach from the fan to the space or hood.

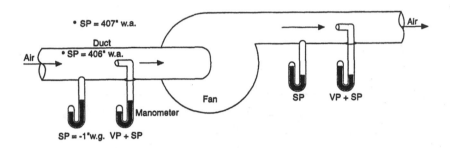

Figure 6.4   Pressure relationships of air flow in ductwork.

The basic terms and formulae associated with air flow in any duct system are:

- *Static pressure*: Static pressure is the potential energy of the ventilation system. It is converted into kinetic energy (velocity pressure) or other energy types (losses in the form of heat, vibration, and noise.) At sea level, the standard static pressure is 14.7 psia, 29.92 inches Hg, or 407 inches of water column. If the fan generates 1 inch of water of negative static pressure (SP = -1″   water gauge, *w.g.*), the absolute static pressure in the duct will be SP = 406″ w.g. See Figure 6.4.
- *Velocity pressure and velocity of air*: The velocity pressure is directly related to the velocity of air, as described by Bernoulli's equation:

$$V = 4005 \times (VP/d)^{0.5} \qquad\qquad \textbf{(6.8a)}$$

$$V = 4.043 \times (VP/d)^{0.5} \textbf{ (metric)} \qquad \textbf{(6.8b)}$$

Where:
    V = Velocity of air, feet per minute (fpm), or meters per second (mps)
    VP = Velocity pressure, inches or mm w.g.
    d = Density correction factor for non-standard conditions

Sample Calculation: The velocity pressure of an airstream in a lab fume hood duct is 0.33 inches w.g. What is the velocity if d =1?

$$V = 4005 \, (VP/d)^{0.5}$$
$$V = 4005 \, (0.33/1)^{0.5}$$
$$V = 2300 \text{ fpm}$$

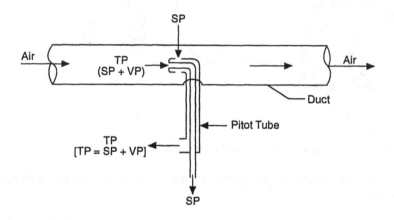

Figure 6.5   TP, SP, and VP in a system; measurements using a pitot tube.

- *Total pressure*: The *static pressure (SP)* and *velocity pressure (VP)* are related to *total pressure* through the following equation:

$$\text{TP} = \text{SP} + \text{VP} \tag{6.9}$$

The pitot tube is the most useful tool for measuring pressures in duct systems. Refer to Figure 6.5. Worksheet 6.1 is typical of those used to record data during the test.

Sample Calculation. Determine the velocity pressure, VP.

TP = -0.35" w.g   SP = -0.50" w.g.

TP = SP + VP

so

VP = TP - SP
VP = -0.35 - (-0.50)
VP = 0.15" w.g.

- *Volume flow rate:* Once a mass of air is identified, confined, and moving, it cannot give up or create new mass. Under steady flow conditions, the same mass flows past each point. If we assume incompressible flow (an important assumption made about flow of air in exhaust ventilation systems), the same volume flow rate passes each point. Air flow can be described mathematically by the equation shown below. The most common units are shown below Equation 6.10.

$$Q = V \times A \qquad\qquad (6.10)$$

Where:

Q = Volume flow rate, U.S. = cubic feet per minute (cfm); SI = cubic meters/second (cms)

V = Velocity, U.S. = feet per minute (fpm); SI = meters per second (mps)

A = Area, U.S. = square feet (sf); SI = sq. meters (m^2)

The units cfm and cms are usually written as scfm or scms (flow at standard conditions) and acfm or acms (flow at actual conditions). To enhance professionalism, write scfm or acfm (scms or acms).

Sample Calculation: The diameter of a duct is D = 12 inches. The average velocity of air flowing in the duct is V = 2250 ft. per minute. What is the flow rate, Q?

$$A = \frac{\pi \times D^2}{4}$$

$$A = \frac{\pi \times (1\ foot)^2}{4}$$

A = 0.7854 sq. ft.

Q = V x A

Q = 2250 fpm x 0.7854 sq. ft.

Q = 1770 scfm

Sample Calculation: The static pressure is measured in a 10-inch square duct at SP = -1.15 inch w.g. The average total pressure is TP = -0.85 inch w.g. Find the velocity and volume flow rate of the air flowing in the duct (STP; d = 1).

$$A = \frac{10''}{12''/ft} \times \frac{10''}{12''/ft}$$

A = 0.6944 sq. ft.

VP= TP - SP

VP= -0.85" - (-1.15")

$$VP = 0.30'' \text{ w.g.}$$

$$V = 4005 \, (VP/d)^{0.5}$$
$$V = 4005 \, (0.30/1)^{0.5}$$
$$V = 2194 \text{ fpm}$$

$$Q = VA$$
$$Q = (2194 \text{ fpm})(0.6944 \text{ sq. ft.}) = 1524 \text{ scfm}$$

D. Static Pressure Losses in Ductwork

Pressure differences are required to move air. In a local exhaust system, the fan creates a pressure lower than the atmospheric pressure on the upstream side of the fan. Since pressure is reduced in the hood, air is pushed into the hood by atmospheric pressure. Static pressure is the potential energy of the system. It is then converted to the kinetic energy of velocity, the heat energy of friction losses, and so forth.

$$\text{SP} \rightarrow \text{VP} + \text{heat, vibration, noise} \qquad (6.11)$$

Figure 6.6 shows static pressures in a local exhaust system. Notice that the static pressure increases in value as the air approaches the fan. Conversely, the static pressure approaches zero as we go toward the hood or stack.

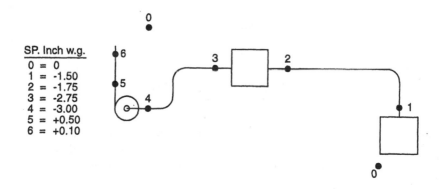

SP. Inch w.g.
0 = 0
1 = -1.50
2 = -1.75
3 = -2.75
4 = -3.00
5 = +0.50
6 = +0.10

Figure 6.6   Static pressures in a duct system.

Worksheet 6.1 Pitot Tube Data Sheet–Round Ducts

Basic Information: Name_____ Date_____ Sheet No. ___of___
Instruments: Manometer: Type_____ Brand_____ Pitot tube:____
General Location: Bldg: _____Room: _____AHU: _____Other: _____
Fan: Type _____ Size: _____ Brand: _____ RPM: _____
    SPin_____ SPout_____ Q, acfm _____ Fan Curves_____
Duct Location and Data: Circ._____ Dia _____ Material _____ SP _____
    Area, sq. ft._____ Settled material _____ Location of tap _____
Air Conditions: Temp _____ BP or Elev _____ R.H. _____ d _____
Sketch of duct location:

Traverse Data using Pitot Tube

Dis-tance, inches*	Horizon-tal Read-ing No.	VP, inches w.g.	Velocity, fpm	Vertical Read-ing No.	VP, inches w.g.	Velocity, fpm
	1			1		
	2			2		
	3			3		
	4			4		
	5			5		
	6			6		
	7			7		
	8			8		
	9			9		
	10			10		

Pitot tube marking:
*Distance from edge of duct wall measurement location inside round duct; mark pitot tube with these distances.

    Summary and Findings:
    Average velocity = _____, fpm (actual/STP)
    Duct area =   _____, sq. feet
    Q = VA =   _____, acfm/scfm
    Coefficient of flow = $(VP_{ave}/SP)^{0.5}$ = _____

    Duct wall thickness, T =_____

          Nearest 1/8th
  1. 0.019 x D + T = _____
  2. 0.077 x D + T = _____
  3. 0.153 x D + T = _____
  4. 0.217 x D + T = _____
  5. 0.361 x D + T = _____
  6. 0.639 x D + T = _____
  7. 0.783 x D + T = _____
  8. 0.847 x D + T = _____
  9. 0.923 x D + T = _____
10. 0.981 x D + T = _____

Source: D. Jeff Burton

There are several major loss categories. These are explained in more depth in the references, but include:

- Duct friction losses
- Duct losses in elbows, contractions, expansions, orifices
- Entry losses in branch entries, cleaner entries
- Hood entry losses due to turbulence and the vena contracta
- System effect losses at the fan
- Special fitting losses such as blast gates, valves, orifices, air cleaners, exhaust caps
- Others: special flow control equipment, stack losses, losses for noise control, and others

1. Loss = $f(V^2)$

Almost all losses are related to the velocity of the air stream. When air flows at low velocity down a duct, friction losses are low. As the quantity of air into the hood increases, or as the velocity of the air in the duct increases, the losses increase.

2. Loss = $f(VP)$

Static pressure losses have been found to be directly related to duct velocity pressure. (Fabric filters are one exception.) An equation can be written to reflect the relationship between static pressure loss and duct VP:

$$\text{Static Pressure Loss} = K_{loss} \times VP \times d \qquad (6.12)$$

Where:

SP_{loss} = Loss of static pressure, inches w.g. or mm w.g. at standard temperature and pressure (STP)

K_{loss} = Loss factor, determined experimentally, unitless

VP = Velocity pressure in the duct, inches w.g. or mm w.g.

d = Density correction factor, unitless

Some texts show:

"$SP_{loss} = K \times VP$" or "$SP_{loss} = F \times VP$"
(d is assumed = 1)   (where F = K)

Sample Calculation: Determine the friction loss in 100 ft. of 10-inch duct, where VP = 1.0" w.g. and K = 0.035 per foot (STP, d =1).

$$SP_{loss} = K \times VP \times d$$
$$SP_{loss} = 0.035/ft \times 100ft \times 1.0" \times 1$$
$$SP_{loss} = 3.5" \text{ w.g.}$$

Manipulating these equations, SP can be related to VP and V by employing the following relationships:

At STP: SP = TP - VP

$$SP = TP - (V_{average}/4005)^2 \qquad (6.13)$$

At STP, losses are routinely expressed in the following format:

At STP: $SP_{loss} = K_{loss} \times VP_{average}$

$$SP_{loss} = K_{loss} \times (V_{average}/4005)^2 \qquad (6.14)$$

3. Corrosion protection

Table 6.9 presents limited data on corrosion resistance in common duct materials (Hunter, 1992).

Table 6.9
Corrosion Resistant Properties of Duct Materials

Corrosive Chemical	Galvanized	Stainless Steel SS316	PVC	Vinyl Ester FRP	Phenolic FRP
Acetic acid	E	E	U	M	M
Acetone	E	E	U	E	E
Chlorobenzene	-	E	U	E	-
Hydrochloric acid	U	U	E	E	E
Hydrofluoric acid	U	U	M	E	-
MEK	M	E	E	E	E
Perchloric acid*	-	M	-	-	-
Phosphine	U	E	U	-	E
Phosphoric acid	M	M	E	U	E
Sodium hydroxide	E	M	E	E	-
Sulfuric acid/H_2O_2	M	E	-	-	-
Toluene	E	E	U	E	E
Xylene	E	E	U	-	E

E = Excellent;   M = Moderate;   U = Unsatisfactory;   - = N/A
* NFPA 45 requires stainless steel

4. Duct standards

The primary association for ductwork is SMACNA. (See References). SMACNA has established a number of important standards for round and square industrial exhaust ducts.

5. Flow in ducts

Air flows turbulently in all industrial ductwork. Because of the high air velocity (1000–4000 fpm is typically found in industrial ventilation systems), air is either totally turbulent or in the transition zone between laminar and turbulent flow. (In simple terms, laminar flow is flow in which no mixing occurs vertically between layers of flowing air. Air flowing in ductwork at less than 100 fpm may be laminar. Turbulent flow is characterized by vertical mixing of the air as it moves horizontally in the ductwork. Please see References for more information).

6. Transport or scrubbing velocity

There are recommended target velocities for a wide variety of different operations and dry particle sizes as detailed in charts contained in the References. For example, grinding wheel exhaust systems should be designed to provide a transport (scouring) velocity of 4500–5000 fpm. If aerosols in the exhaust air stream are sticky, mist, or wet, they will plate out in the duct-cleanouts or sloping ducts and drains should be provided. See Table 6.10 and Case Study 6.3.

Table 6.10
Transport or Scrubbing Velocities for Dry Particles in Round Ducts

Materials in Air in the Duct	Recommended Velocity	
	fpm	m/s (mps)
Very fine, very light dusts	2000	10
Fine dusts and powders	3000	15
Medium industrial dusts	3500	18
Coarse dusts	4000–4500	20–23
Heavy dusts	5000	25

Velocity profile: Please note the following about exhaust air in long, straight sections of duct:

- The theoretical velocity at the edge of duct is $V = 0$.
- The velocity at the centerline is the maximum encountered (usually, though there may be exceptions).
- The average velocity is somewhat less than the centerline velocity, V_{cl}. (The actual velocity at the centerline of the duct).

See Figure 6.7.

One way of estimating the average velocity in a long, straight section of duct is:

$$V_{ave} = 0.9 \times V_{cl} \qquad (6.15)$$

and if we substitute Bernoulli's equation for V, then:

$$P_{ave} = 0.81 \times VP_{cl} \qquad (6.16)$$

A long, straight section usually means at least five duct diameters downstream and two duct diameters upstream from any bend, obstruction, transition, etc.

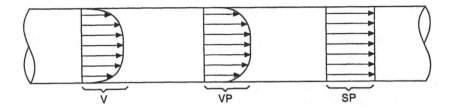

Figure 6.7   Velocity, velocity pressure, and static pressure profiles in ductwork.

Sample Calculation: The centerline velocity in a long, straight duct is V = 1800 fpm. What is the predicted average duct velocity?

$$V_{ave} = 0.9 \times V_{centerline}$$
$$V_{ave} = 0.9 \times 1800 \text{ fpm}$$
$$V_{ave} = 1620 \text{ fpm}$$

7. Duct shape

Use round ducts whenever possible for exhaust ventilation systems. The advantages of round ducts for exhaust systems are:

- It resists collapsing better than other shapes
- It provides better aerosol transport conditions
- It uses less metal than equivalent square/rectangular ducts

Square and rectangular ducts are often used in air-supply systems (e.g., heating and cooling systems).

Occasionally, square ductwork must be used in exhaust systems (e.g., where space is limited). For rectangular and square ducts, one formula for the circular equivalent is:

$$D_c = 1.3 \times [(ab)^5/(a+b)^2]^{0.125} \tag{6.17}$$

Where:

D_c = Circular equivalent, inches or cm
a = Length of one side, inches or cm
b = Length of the other side, inches or cm

[Source: ASHRAE]

E. Hoods and Exhausted Enclosures

The objective of the hood is to control emissions (and exposures). The hood controls an emission by:

- Enclosing or containing the emission
- Directing and capturing contaminated air
- Receiving the contaminant after it has been emitted

The three basic types of hoods are shown in Figure 6.8 and include:

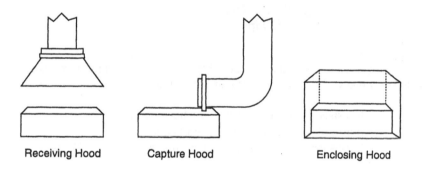

Receiving Hood   Capture Hood   Enclosing Hood

Figure 6.8   Three examples of hood types.

- *Receiving hood.* These hoods are designed to receive the emission source (which has some initial velocity imparted to it by the emitting source). A canopy hood is a receiving hood because it receives hot rising air and gases. A small hand-held tool hood may be a receiving hood. A push-pull hood is a receiving hood. Other names: passive hoods.
- *Capture hood.* These are hoods with one to three sides. A welding snorkel-type hood is typical. Others include sidedraft and downdraft. Other names: active or external hoods
- *Enclosing hood.* Examples are glove boxes, lab hoods, bench hoods, grinder hoods, and others with four or more sides.

1. *Hood static pressure*

Hood static pressure, SP_h, is measured in the duct serving the hood, often about 4-6 duct diameters downstream from the hood. VP, the average duct velocity pressure, can also be measured at the same location. The hood entry loss, H_e, is the sum total of all the losses from the hood face to the point of measurement in the duct:

$$H_e = K_{hood} \times VP \times d \qquad (6.18)$$

Substituting in the hood static pressure equation,

$$SP_h = VP + H_e$$
$$SP_h = VP + K_{hood} \times VP \times d$$
$$SP_h = (1 + K_{hood} \times d) \times VP \qquad (6.19)$$

Values of the loss factor K_{hood} have been determined for many hood shapes. References 1 and 2 provide loss factors for over 200 different hoods.

Sample Calculation: What is the hood static pressure SP_h when the duct velocity pressure is VP = 0.33 inches w.g., and the hood entry loss is H_e = 0.44 inches w.g.? What is the C_e?

$$| SP_h | = VP + He$$
$$| SP_h | = 0.33 + 0.44$$
$$| SP_h | = 0.77 \text{ inch w.g.}$$

so:

$$C_e = \frac{Q_{actual}}{Q_{ideal}}$$
$$C_e = [VP/| SP_h |]^{0.5}$$
$$C_e = [0.33/| 0.77 |]^{0.5}$$
$$C_e = 0.65$$

Sample Calculation. What is the entry loss H_e for a laboratory fume hood when the average velocity pressure in the duct is VP_d = 0.30" w.g.? (Assume K = 2.0, d = 1)

$$H_e = K \times VP_d \times d$$
$$H_e = 2.0 \times 0.30 \times 1$$
$$H_e = 0.60" \text{ w.g.}$$

The greatest loss normally occurs at the entrance to the duct, due to the *vena contracta* formed in the throat of the duct (a narrowing of the air stream as it flows through a duct fitting). Figure 6.9 illustrates a *vena contracta*. The center of the *vena contracta* is usually found about one-half duct diameter inside the duct.

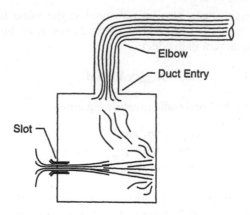

Figure 6.9   *Vena contracta.*

The efficiency of a hood can be described by the ratio of actual to ideal flow. This ratio is called the coefficient of entry, C_e.

$$C_e = \frac{Q_{actual}}{Q_{ideal}} \text{ unitless} \qquad (6.20)$$

Ideal flow could be achieved if all the hood static pressure were converted to velocity pressure, i.e., there was no hood entry loss. A similar but more useful formula for estimating C_e is shown below.

$$C_e = [VP/|\ SP_h\ |]^{0.5} \qquad (6.21)$$

Where:

VP = Average duct velocity pressure

SP_h = Positive value of the hood static pressure

Interestingly, C_e is also related to K_{hood}, the loss factor for the hood, by:

$$C_e = [1/1 + K_{hood}]^{0.5} \qquad (6.22)$$

The coefficient of entry, C_e, is a function of the shape of the hood. C_e never changes unless the hood shape (e.g., the sheet metal) changes. If C_e is measured at the time a hood is built, it can be used later with SP_h measurements to estimate flow rates and velocities. Equations of flow can be derived from the duct diameter, hood static pressure, and the coefficient of entry, as follows:

In U.S. Units:

$$Q = 4005 \, A \, C_e \, (SP_h/d)^{0.5}, \text{ in acfm} \qquad (6.23)$$

or:

$$Q = 4005 \, A \, [(SP_h/d)/(1 + K_{hood})]^{0.5}, \text{ in acfm} \qquad (6.24)$$

In the SI System:

$$Q = 4.043 \, A \, C_e \, (SP_h/d)^{0.5}, \text{ in acms} \quad (6.25)$$

or:

$$Q = 4.043 \, A \, [(SP_h/d)/(1 + K_{hood})]^{0.5}, \text{ in acms} \qquad (6.26)$$

The hood entry loss H_e and the coefficient of entry C_e have been determined for a number of standard hoods. See sources contained in References section.

Sample Calculation: Assume a bypass-type lab fume hood is exhausting 1000 cfm of air and the hood static pressure was measured at 0.75 inch w.g. Three months later, the hood static pressure is 0.50 inch w.g. Assuming continued standard conditions and no changes in the hood, air flow through the hood has been reduced by how much?

Initial Conditions	New Conditions
$SP_{h1} = -0.75$ inch w.g.	$SP_{h2} = -0.50$ inch w.g.
$Q_1 = 1000$ scfm	$Q_2 = ?$

Rearrange Equation 6.20, $Q = 4005 \, A \, C_e \, (SP_h/d)^{0.5}$, to solve for C_e and remember that C_e is the same in both cases:

$$C_e = \frac{Q_1}{[4005 \, A \, (SP_{h1}/d)^{0.5}]} = \frac{Q_2}{[4005 \, A \, (SP_{h2}/d)^{0.5}]}$$

so:

$$Q_2 = 1000 \text{ scfm } [-0.50 \text{ inch}/-0.75 \text{ inch}]^{0.5}$$
$$Q_2 = 817 \text{ scfm}$$

When selecting, designing, or troubleshooting a hood, we have these important parameters to determine:

- Its best shape
- The flow rate Q necessary to control the emission
- The loss factor K
- The coefficient of entry, C_e
- Velocities (face velocity, capture velocity, slot and plenum velocities, and duct transport velocity.)
- Slot dimensions

There are three basic ways to determine appropriate parameters: **(1)** using real models, **(2)** using available charts, and **(3)** using equations. Chart values can be found in sources listed in References 1 and 2.

Capture hood designers try to visualize how air will enter the hood. Basically, air approaches from all directions toward the source of negative pressure. Imagine a three-dimensional sphere around the end of a small, plain duct hood. See the next figure, a point-source hood. The velocity of air moving toward the opening is equal at all points on the surface of the sphere. The surface area of a sphere is given by Equation 6.27 in Figure 6.10.

$$A = 4 \pi X^2 \qquad\qquad (6.27)$$

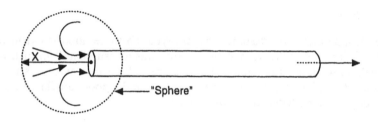

Figure 6.10 Air entering a small, plain duct hood.

When using this approach, approximate the three-dimensional shape, calculate the surface area at any distance of interest, and determine the required velocity. Then calculate the volume flow rate, Q = VA, required to move the air through the surface at that distance. The next figure shows some other two- and three-dimensional shapes for common hood types.

Sample Calculation: Air enters an ideal 4" plain duct hood. What is the required volume flow rate Q for capture 6" in front of the hood if we need V_c = 100 fpm?

$$Q = V_c \times A \qquad\qquad \text{where: } A = 4 \pi X^2$$
$$Q = 100 \times 4 \pi (0.5 \text{ ft})^2 \qquad \text{where: } 6 = 0.5 \text{ ft}$$
$$Q = 315 \text{ cfm}$$

The ACGIH IV Manual suggests the following simple empirical approximations for estimating Q:

Plain opening, $Q = V_c(10X^2 + A)$ $\qquad\qquad$ **(6.28)**

Flanged opening, $Q = 0.75V_c(10X^2+A)$ (6.29)

Slot opening, $Q = 3.7(LV_cX)$ (6.30)

Flanged slot, $Q = 2.6(LV_cX)$ (6.31)

Where:
Q = Volume flow rate: U.S. units, acfm; SI units, acms
X = Distance to source: U.S. units, feet; SI units, meters
V_c = Capture velocity at distance X: U.S. units, fpm; SI units, mps
L = Slot length: U.S. units, feet; SI units, meters
A = Hood open face area: U.S. units, sq. feet; SI units, sq. meters

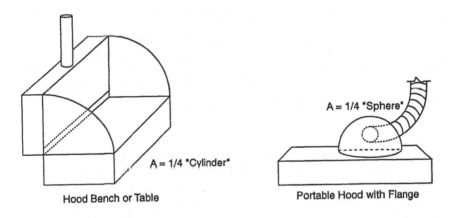

Figure 6.11   Shapes and areas for common hoods.

Side draft hoods almost always use a slot. A slotted hood is a compound hood; it requires air to "enter" twice: first through the slot into a plenum, where velocity is reduced, and then from the plenum into a duct. The slot velocity should be at least 2 to 4 times the plenum velocity. Slot hood entry loss factors are also compound, e.g.,

$$H_e = K_{slot}VP_{slot} + K_{duct}VP_{duct}$$ (6.32)

The loss associated with a slot is similar to the loss associated with a critical orifice. The value $K_{slot} = 1.78$ is almost always used. If the takeoff duct is directly behind the slot, the entry loss to the duct may be reduced, but this is rarely the case. The slotted hood distributes the plenum static pressure all along the slot. This helps maintain a constant velocity across the hood face.

The slot does not reach out to capture contaminants. The plenum-slot arrangement distributes plenum static pressure evenly across the back of the slot so that air flow into the hood is equalized across the face of the hood.

Finally, what is the predicted hood static pressure for a slotted hood? The traditional formula at STP is:

$$SP_h = H_e + VP_{duct} = KVP_{duct} + VP_{duct} \qquad (6.33)$$

For a slotted hood:

$$SP_h = \overline{K_{slot} \times VP_{slot}} + \overline{K_{duct} \times VP_{duct}} + VP_{slot} + VP_{duct} \qquad (6.34)$$

and since $K_{slot} = 1.78$:

$$SP_h = 2.78 VP_{slot} + (1+K)VP_{duct} \qquad (6.35)$$

This last equation works in most cases. Occasionally, the acceleration achieved in the slot can be carried on into the duct, but this occurs only rarely, e.g., when a carefully designed duct takeoff lies directly behind the slot. Most of the time, the acceleration of the slot is lost against the back wall of the plenum. Using the above equation errs on the side of safety.

V. FANS

Types of fans include centrifugal (forward curved, backward inclined, radial) and axial. See Figure 6.12. Forward curved fans are used most commonly in home furnace applications; backward inclined wheels are quiet and efficient and are widely used in commercial HVAC applications; radial fans are often referred to the "workhorse" of the dust control industry because these wheels are self-cleaning, rugged, and inexpensive.

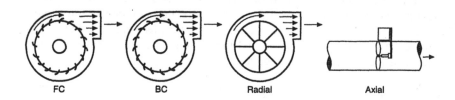

FC   BC   Radial   Axial

Figure 6.12 Types of fans.

Fan characteristic curves that plot volume flow rate Q against static pressure, horsepower, noise, and efficiency are depicted in Figure 6.13.

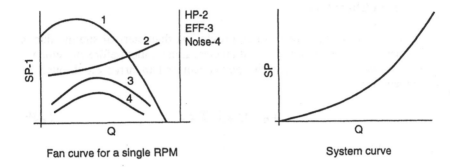

Figure 6.13   Fan curves.

A. Fan Specification

Fans are specified by *pressure* and flow rate. The pressure is normally measured across the fan: at the inlet and outlet of the fan in the ductwork.

B. Total Pressure

The *fan total pressure, FTP,* represents all energy requirements for moving air through the ventilation system. FTP is calculated by adding the absolute values of the average total pressures found at the fan. If the sign convention is followed, then a formula for FTP is:

$$\mathbf{FTP = TP_{outlet} - TP_{inlet}} \qquad (6.36)$$

Substituting for TP = SP + VP yields:

$$\mathbf{FTP = SP_{out} + VP_{out} - SP_{in} - VP_{in}} \qquad (6.37)$$

If VP (out) equals VP (in), i.e., if the average inlet and outlet velocities are equal, then the VP terms in the above equation cancel, leaving:

$$\mathbf{FTP = SP_{out} - SP_{in}} \qquad (6.38)$$

The FTP is often referred to as *the fan total static pressure drop*.

Sample Calculation: The inlet and outlet conditions at a fan are SP_{out} = 0.10", SP_{in} = -0.75". VP_{in} = VP_{out}. What is FTP?

$$FTP = SP_{out}\ SP_{in}$$
$$FTP = 0.10" \text{ wg.} - (-0.75")\text{ w.g.} = 0.85" \text{ w.g.}$$

C. System Effect Loss

To avoid system effect losses, provide six diameters of straight duct at the inlet of the fan and three duct diameters at the outlet before any elbows.

Air horsepower, ahp, is the power required to drive air through the ventilation system:

$$ahp = \frac{Q \times FTP \times d}{6356} \qquad (6.39)$$

Where:

Q = Volume flow rate through system
FTP = Fan total pressure
d = Density correction factor
   6356 is conversion factor

Brake horsepower, bhp, is the power required to move air through the system and the fan.

$$bhp = \frac{Q \times FTP \times d}{6356 \times ME} \qquad (6.40)$$

Where:
Q = Volume flow rate through system
FTP = Fan total pressure
ME = Fan mechanical efficiency (0.4–0.75)
d = Density correction factor
   6356 is conversion factor

Shaft horsepower is the power required at the shaft of the fan and includes drive losses.

$$shp = \frac{Q \times FTP \times d \times K_{dl}}{6356 \times ME} \qquad (6.41)$$

Where:
K_{dl} = Drive loss factor (1.05–1.30)
Q = Volume flow rate through system
FTP = Fan total pressure

ME = Fan mechanical efficiency (0.4–0.75)
d   = Density correction factor
      6356 is conversion factor

Rated horsepower, rhp, is horsepower available from the motor. For small systems (e.g., less than 25 horsepower), rhp is usually chosen to be about 1/3 greater than shp.

Sample Calculation: What is the required horsepower for the system, and what rated horsepower motor would you choose?

FTP = 2.0 inch w.g.   eff = 0.50
Q = 1300 scfm K_{dl} = 1.25

$$bhp = \frac{ahp}{eff}$$

$$bhp = \frac{FTP \times Q}{6356 \times eff}$$

$$bhp = \frac{2.0 \times 1300}{6356 \times 0.5}$$

$$bhp = 0.82 \ hp$$

so:

$$shp = bhp \times K_{dl}$$
$$shp = 0.82 \times 1.25$$
$$shp = 1.02 \ hp$$

rhp (rated horsepower) must be at least 1.02 hp. The next whole-sized motor is likely to be 1.5 rhp. Local exhaust systems often call for the rhp to be 33% more than shp.

D. Fan Laws

Fan laws are mathematical relationships that describe changes in Q, pressure, and hp when fan rpm (or diameter) change. The rpm fan laws are summarized below. The numbers 1 and 2 refer to conditions "one" and "two," or conditions before and after a change.

$$\frac{Q_2}{Q_1} = \frac{rpm_2}{rpm_1} \qquad (6.42)$$

$$\frac{SP_2}{SP_1} = (rpm_2/rpm_1)^2 \qquad (6.43)$$

$$\frac{HP_2}{HP_1} = (rpm_2/rpm_1)^3 \qquad (6.44)$$

Sample Calculation: What is the new flow rate, hood static pressure, and horsepower expected when the rpm is increased to 1125 rpm?

Initial Conditions	Conditions After Increase
RPM = 1000	RPM = 1125
Q = 1,000 scfm	Q = ?
SP_h = 0.85 inch w.g.	SP_h = ?
Shp = 0.50	shp = ?

$$rpm_2/rpm_1 = 1125/1000 = 1.125$$

Therefore:

$$Q_2 = Q_1 \times 1.125$$
$$Q_2 = 1,125 \text{ scfm}$$

$$SP_2 = SP_1 \times (1.125)^2$$
$$SP_2 = 1.08" \text{ w.g.}$$
$$shp_2 = shp_1 \times (1.125)^3$$
$$shp_2 = 0.71 \text{ hp}$$

VI. STACKS

A. HVAC Intakes and Exhausts

The placement of HVAC air intakes, exhausts, and ventilation stacks should minimize reentrainment of exhaust air. However, complete isolation is impossible. It has been estimated that even under ideal conditions, 1% of the exhausted air may be reentrained into the building. The reentry occurs at HVAC intakes, at makeup air units, or by infiltration along building walls. Under less than ideal conditions, much more may be reentrained.

1. *Infiltration/exfiltration*

Outside surfaces under positive pressure may experience infiltration through openings in the surface. Surfaces under negative pressure may experience exfiltration.

2. *HVAC intakes and exhausts*

D.J. Wilson (re: ASHRAE) has suggested the following rules of placement in order to avoid contamination of incoming air. Air intakes should be placed on the lower one-third of a building and exhausts on the upper two-thirds or on the roof. Intakes should not be placed near the ground, loading docks, or busy streets. Exhausts should preferably be placed on the predominant downwind side of the building, and intakes on the upwind side.

When exhausts are located on the roof, esthetic enclosures should be discouraged. If required, the enclosure should be of the open-louvered type, which allows horizontal winds to flush the enclosure. Intakes should not be located within the enclosure.

3. Exit velocity

Air exhaust velocities should be designed to exceed twice the wind speed. The average wind speed in the U.S. is 8 mph, so design for exit velocities = 15 mph (about 1300 fpm), or greater.

B. LEV Stacks

A ventilation exhaust stack should remove (for good) contaminants generated in the building. Unfortunately, a small amount will always be re-entrained, but good design attempts to minimize this problem.

1. Process exhaust

Stacks come in three varieties. Those which exhaust industrial process gases (e.g., flue gases from large metal-melting furnaces) are called *process exhaust stacks*. These are the *smokestacks* of yesteryear. Today these emissions are controlled by strict emission codes and usually involve extensive air cleaning and tall stacks.

2. LEV exhaust

The second type are called *fugitive emission stacks*, and are used to exhaust fugitive emissions from equipment and processes (e.g., copiers, IC engines). These are also called *local exhaust ventilation* (LEV) *stacks*.

3. Sanitary exhaust

The third type vents air from sewer lines, bathrooms, water heaters, boilers, and so forth. These *sanitary exhausts* may or may not have a fan attached. Many building codes call for these stacks and vents to be placed at least 10 feet from and 2 feet above (or below) an air intake.

The exhaust gas from local exhaust or sanitary systems usually consists almost entirely of air and typically small amounts of chemical contaminants (e.g., concentrations of gas or vapor in a typical lab fume hood rarely exceed 100 parts per million). The temperature is often close to room temperature. After the exhaust air leaves the stack, it is rapidly diluted in the ambient air.

The amount of dilution is proportional to the wind speed and the square of the distance away, and inversely proportional to the exhaust volume flow rate and the initial concentration in the exhaust air.

$$\text{dilution} = F\,[(V_{wind} \times L^2)/(Q_{exh} \times C)] \qquad (6.45)$$

At a distance of 50 feet, air exhausted from a stack will usually be diluted by at least a factor of 100X. In more favorable conditions (e.g., greater wind speed, longer distance, higher effective stack height), the dilution factor can be much higher. Unfortunately, it is very difficult to accurately predict the behavior of air after it leaves a stack, but a few expected trends are worth mentioning. The next figure shows a typical example of the distribution of an exhaust plume from a stack on a roof. In the figure, H is the building height, W is the building width parallel to air flow, h is the stack height, and h_{eff} is the effective stack height.

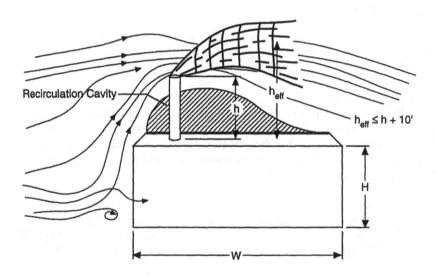

Figure 6.14 Air dispersion behavior at a stack.

The following good practices are generally accepted among designers of stacks and air handling equipment:

- Place the stack to avoid exhausting into the "recirculation cavity." (See Figure 6.14). Many designers feel 10 feet will provide enough height to breach the recirculation cavity on most roofs.

- Provide ample stack height. The Uniform Building Code suggests that sanitary stacks should be 10 feet away and 2 feet above an air intake. Check local building codes.

- A good rule of thumb says local exhaust stacks (e.g., lab fume hood stacks) within 50 feet of the roof line or air intake should be 10 feet tall.

VII. OVERVIEW OF HVAC SYSTEMS

A. Comfort

HVAC equipment is used for tempering, dehumidifying, and cleaning air for comfort, safety, and health. ASHRAE 55-1992 uses an 80/20/10 rule to establish satisfactory comfort performance. If 80% of the occupants are satisfied with all environmental conditions (e.g., temperature, humidity, odor, drafts), or if less than 20% are dissatisfied, then comfort requirements have been met. If more than 10% are dissatisfied with any one condition, then satisfactory performance is not achieved.

B. Zones

HVAC engineers talk in terms of *zones*. Each zone usually contains a thermostat to control temperature. The more zones there are, the better chance there is of providing satisfactory comfort conditions for more people.

In designing an air handling supply system, the user must choose combinations of volume flow rate, temperature, humidity, and air quality which will satisfy the needs of the space. Air handling systems generally consist of:

- Outside air plenums or ducts (OA)
- Supply fans (SF)
- Supply ducts
- Humidifying or dehumidifying equipment
- Distribution ducts, boxes, plenums, and registers
- Return air plenums and/or ducts (RA)
- Filters
- Heating and/or cooling coils
- Dampers
- Controls and instrumentation

C. Constant Air Volume Systems, CAV

Figure 6.15 is a schematic of a simple, constant air volume (CAV) commercial HVAC system, a single-zone constant volume system.

D. Filtration

Air filters are generally rated by two tests developed by ASHRAE. The first, called arrestance, measures a filter's ability to capture a coarse dust. The second, called dust spot efficiency, measures a filter's ability to control fine dusts. See References and ASHRAE Standard 52 for a more complete description of the tests, particle capture efficiencies, and a comparison between the methods (Note: as of the date of publication, ASHRAE was in the process of changing the standard. Contact ASHRAE to obtain latest information).

The *Economizer Cycle* means using outdoor air for cooling when the outside air temperature ranges from about 55–70°F.

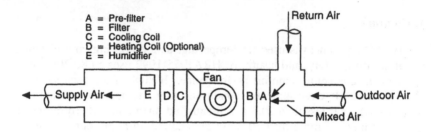

A = Pre-filter
B = Filter
C = Cooling Coil
D = Heating Coil (Optional)
E = Humidifier

Figure 6.15 CAV HVAC system.

E. Variable Air Volume Systems, VAV

The advantage to variable air flow is energy conservation and lower operating costs. Savings can be obtained if the VAV system actually reduces fan rpm. Air volume and fan rpm are linearly related. If the air flow can be cut in half, then the rpm is cut in half. But horsepower and rpm are related through a third-power relationship. (See Section 6.5.) If the rpm is reduced by half, the horsepower (and energy costs) are reduced by about 8 times.

Simply dampening down air flow with fan inlet dampers also results in lower motor operating costs, but not by the same factor as reducing fan rpm.

F. Troubleshooting

Common problems with HVAC systems include:

- Inadequate replacement air supply
- There is poor distribution of supply air in the space; certain areas receive little air exchange
- Stuffiness–not enough air delivery, or not delivered properly
- Improper pressure differences–doors hard to open because not enough air is supplied to the space
- Temperature extremes–too hot or too cold
- Humidity extremes–too dry or too humid
- Poor filtration–dirt, bugs, pollen in air delivery system
- Poor maintenance
- Energy conservation has become No. 1 priority

- Settled water in system
- Visual evidence of slime or mold
- Improper balance of distribution system
- Dampers at incorrect positions
- Supply terminal diffusers not at correct positions
- VAV systems capable of shutting down air delivery

G. Estimating Outdoor Air (OA) Delivery Rates

It is possible to estimate the percentage of outdoor air (OA) in the supply air (SA) by looking at the temperatures of the air (temperature changes are linear).

$$\%OA = \frac{T_{RA} - T_{MA}}{C_{RA} - C_{SA}} \times 100 \qquad (6.46a)$$

and:

$$\%OA = \frac{Q_{OA}}{Q_{SA}} \times 100 \qquad (6.46b)$$

Where:

T_{RA} = Temperature of return air (dry-bulb)
T_{MA} = Temperature of mixed return and outside air, the supply air (dry-bulb)
T_{OA} = Temperature of outdoor air (dry-bulb)
Q_{OA} = Volume of outdoor air, cubic feet per minute
Q_s = Volume of mixed return and outside air, the supply air, cubic feet per minute

Similarly:

$$\%OA = \frac{C_{RA} - C_{SA}}{C_{RA} - O_{OA}} \times 100 \qquad (6.47)$$

Where:

C_{RA} = Concentration of CO_2 in return air
C_{SA} = Concentration of CO_2 in supply air
C_{OA} = Concentration of CO_2 in outdoor air

Sample Calculation: Assume the following temperatures: $T_{RA} = 76°$, $T_{MA} = 66°$, $T_{OA} = 5°$. Estimate the percentage of OA.

$$\%OA = \frac{T_{RA} - T_{MA}}{C_{RA} - C_{SA}} \times 100$$

$$\%OA = \frac{76 - 66}{76 - 50} \times 100$$

$$\%OA = 38\%$$

1. *Tracer gases*

Carbon dioxide concentrations can be used to estimate the amount of OA reaching an area of the building. During the day, CO_2 builds up; after everyone leaves at 5 o'clock, the OA will dilute carbon dioxide. Knowing the initial and final concentrations, the time allows the use of purge formulas to predict the amount of OA delivered.

$$N = \frac{\ln C_i - \ln C_a}{\text{hours}} \qquad\qquad (6.48)$$

Where:

N   = Air exchange per hour, OA
C_i = Concentration of CO_2 at start of test, minus outdoor concentration, about 330 ppm
C_a = Concentration of CO_2 at end of test, minus outdoor concentration
hours = Time elapsed between start and end of test
ln   = Natural log

Sample Calculation: The carbon dioxide concentration is C_i = 1200 ppm at 5:30 p.m., when all of the people have departed a building. By 7:30 p.m., the concentration has been reduced to C_a = 400 ppm. The outside concentration is 330 ppm. How many air changes per hour of OA does this suggest? What is Q for a space volume of 50,000 cu. ft.?

$$N = \frac{\ln C_i - \ln C_a}{\text{hours}}$$

$$N = \frac{\ln (1200-330) - \ln (400-330)}{2 \text{ hours}}$$

$$N = 1.26 \text{ AC/hr}$$

so:

$$Q = \frac{N \times Vol}{60}$$

$$Q = \frac{1.26 \times 50,000}{60}$$

$$Q = 1050 \text{ cfm}$$

VIII. SYSTEM TESTING AND MONITORING

This section provides an overview of the basic measurements normally used to estimate containment of contaminants within the hood (which is directly related to employee protection). Objectives of testing normally include:

- Determination of the effectiveness of the ventilation system (emissions capture, particulate transport, employee and product protection, efficiency, air cleaning).
- Establishment of baseline or startup conditions.
- Monitoring of conditions throughout the life of system.
- Determining compliance with permitting agency requirements and management-determined performance criteria.
- Determining air distribution effectiveness.

A. Measurement Tools and Devices

1. *For containment*

Visible aerosols (e.g., smoke tubes and smoke generators for non-cleanroom areas; mechanical water vapor generators, liquid nitrogen, and dry ice vapor generators for cleanrooms.) Tracer gases (e.g., sulfur hexafluoride) and tracer gas detection equipment.

2. *For velocity*

Velometers, anemometers (e.g., swinging vane, rotating vane, thermal/hot wire anemometer, balometer®). Refer to Figures 6.16 and 6.17.

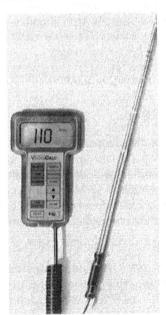 

Figure 6.16 Anemometer.   **Figure 6.17** Balometer.®
(Figs. 6.16. and 6.17, Courtesy: *TSI Incorporated,* St.Paul, MN).

3. *For pressure*

Pressure sensing devices (e.g., U-tube manometers, inclined manometers, electronic manometers, pitot tubes, anemometers (these instruments measure static pressure indirectly, bellows (e.g., Magnahelic gauges).

4. *For noise*

Noise and vibration monitoring equipment.

5. *For carbon dioxide*

Detector tubes and infrared-based instruments.

6. *For dimensions*

Measuring tapes and distance measuring devices.

7. *Other*

Rags, flashlight, drill, drill bits, mirror, tachometer, duct tape.

8. *Field kits*

A field kit might include the following items: a shoulder-strap carrying case, a smoke tube kit, a velometer, a pitot tube, flexible tubing, an electronic manometer, a portable drill, drill bits, a flashlight, a crescent wrench, a pair of pliers, a punch, a mirror, a roll of duct tape, a tape measure, a sound level meter, a pencil, a black marker pen, testing forms and charts, and a note pad.

B. General Approaches to Testing
1. *Physical measurements*

Typical measurements include the following:

- *Duct diameter, duct circumference.* These are measured for the purpose of calculating duct area. Inside duct diameter is the most important measurement, but an outside measurement is usually sufficient on sheet metal ducts. Duct measurements are often difficult. Measuring tapes seldom lay flat against the duct, ducts are often out of round, and they frequently vary from the nominal size by a small percent.

- *Duct length.* Lengths can be estimated from plans, drawings, and specifications. Measurements can be made with tapes or optical devices. If the duct is constructed of 2 1/2- or 4-foot sections, count the sections. Include elbows and tees in the length.

- *Radius of curvature.* Mostly eyeballing with measuring tape. The most common way of expressing radius of curvature is to divide the centerline radius by the duct width or diameter.

- *Sheet metal gauge and plastic thickness.* The "Mfrs Std. Gauges" for steel sheet metal is based on metal with a density of 490 lb/cu. ft. Galvanized metal is slightly thicker than nongalvanized. Industrial ventilation systems are often constructed of sheet metal, with typical construction in the 12-20 gauge range. Plastics (e.g., polypropylene) are typically 3/8" thick.

C. Capture and Face Velocities
1. Velocities
Velocities outside of the hood, or at the hood face or plane of penetration can be estimated with velometers.

2. Hood face
The following figure shows the most common method of estimating the average face velocity in a hood face or plane of penetration. Velometers and anemometers are suited to these measurements when the air is clean. All velometers have velocity correction (VC) requirements for nonstandard conditions. Check the instructions provided with the equipment. Often the correction factor is d, the density correction factor. However, this is not always the case. Knowing the average velocity, it is possible to estimate the volume flow rate from $Q = VA$.

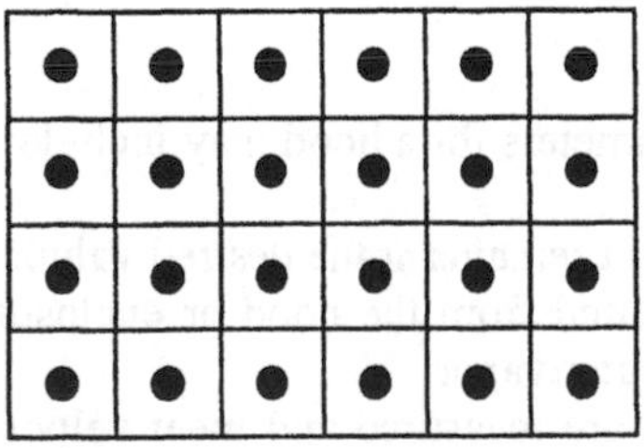

Hood face velocity measurement
- Mark off imaginary areas (the number of areas is a function of the accuracy desired; the more measurements, the greater the accuracy.
- Measure velocity at center of each area
- Average all measured velocities

Figure 6.18 Face velocity measurements.

D. Containment Measures (Quantitative and Qualitative)

The hood or exhausted enclosure should contain air contaminants within the hood to some specified value (e.g., *no visible emissions, 99.999% containment, PF = 8000, meets ASHRAE 110 at 8AM0.05, meets SEMI F15*, and so forth).

1. *Qualitative tests*

Smoke and visible vapor trails are useful because they are visible. Nothing convinces management or employees more quickly than to show smoke or vapor drifting away from the hood, moving the wrong way within the hood, or traveling up the operator's arm into the breathing zone.

For non-cleanroom locations, smoke tubes and smoke generators (typically based on titanium tetrachloride smoke) are quick, easy, and inexpensive. The smoke is irritating, so move employees out of the smoke path. For hoods installed and tested in cleanrooms, or for hoods which must be tested before shipment, ultrasonic water vapor generators create a steady, visible water vapor plume that can be used to demonstrate containment, air flow direction, and approximate air velocities. Dry ice and liquid nitrogen can also be used for containment testing.

2. *IH breathing zone (BZ) analysis*

Another approach is simply to measure chemical concentrations in the employee's breathing zone using traditional industrial hygiene testing methods. Exposures below some acceptable level (e.g.,1% of the PEL) are subjective evidence of satisfactory containment.

3. *Semiquantitative*

A more rigorous approach is to establish containment performance test parameters (CPP), compare them to acceptable standards, and assume containment if standards are met. For example, if CPP test results match the following accepted parameters, containment may be assumed (this is the method most often used for field testing after hoods are in use).

4. *CPP tests*

Typical containment performance parameters for a hood may include:

- The hood static pressure measurement remains at the desired value.
- No observed smoke or vapor is emitted from the hood or enclosure when tested with visible smoke or water vapor.
- Hood capture and/or face velocities are measured and meet velocity criteria (e.g., average face velocity of 80-120 fpm for lab fume hoods and open wet station hoods, capture velocity of 350 fpm for grinding operations, and so forth; management determines accept-

able face and capture velocities for each operation. Refer to Figure 6.19, which demonstrates how to collect velocity measurements at the face of a hood.

- No single measurement of face velocity exceeds some set value of the average face velocity (e.g., ±20% at wet sink station hoods).

Figure 6.19 Collection of face velocity measurements at the opening of a hood (Courtesy: *TSI Incorporated*, St. Paul, MN).

- Turbulent air mixing velocities at the face of the hood do not exceed some set value (e.g., 40 fpm at lab hoods with the hood exhaust turned off).
- Hood users understand and always follow correct work practices.
- The hood is configured as the manufacturer intended it to be.
- The operation has not changed from its original criteria (e.g., more toxic chemicals are not being used, additional equipment has not been added to the hood, and so forth.)

5. *Quantitative tests*

Tests similar to the ASHRAE 110 tracer gas test (1985 and revised draft, 1995) have been adopted by some as a required test method for approving some types of hoods. It is also the suggested test method for lab fume hoods (see ANSI Z9.5 and ASHRAE 110 in References).

6. *Approach*

In the ASHRAE 110 test, a tracer gas (e.g., SF6) is released at a known rate inside the hood from a single dispersion nozzle. Ten-minute samples of tracer gas are obtained in the breathing zone (BZ) of a mannequin standing at different positions in front of the hood. Results are reported as:

$$xx\ AU\ yyy \quad \text{or} \quad xx\ AI\ yyy \quad \text{or} \quad xx\ AM\ yyy$$

The xx refers to the release rate of tracer gas, typically 4–8 lpm. The yyy refers to the BZ concentration, typically 0.01ppm to 0.1 ppm. The AU, AI, and AM refer to "as used," "as installed" (before usage begins), and "as manufactured."

This method requires much time, effort, specialized equipment, calibration, and highly trained and motivated testing personnel. The test has not been adopted in any current codes or standards.

7. *PF*

Another useful approach is to estimate the ratio of the concentration of a contaminate in the exhaust air vs. the concentration in the breathing zone of a person actually working in front of the hood:

$$\text{Protection Factor (PF)} = \frac{C_{\text{exhaust air}}}{C_{\text{breathing zone}}} \qquad (6.49)$$

The PF approach is similar to the ASHRAE 110 and the methods can be compared when flow rates are known. Using the PF approach, it is easy to measure or predict concentrations in the field, and it provides a more real picture of containment during actual operations. Traditional IH sampling may be used to measure the concentration in the breathing zone of the hood user. The equations for dilution can be used to predict the concentration in the exhaust ductwork. Acceptable PFs should be determined by management for each enclosure, hood, and chemical contaminant, and usually range from 300 (lab hoods using routine chemicals) to 30,000 (open-front box hoods using toxic or radioactive chemicals).

Sample Calculation: Suppose a person is using isopropyl alcohol in a wet station hood. If the concentration in the exhaust air is $C_{\text{ea}} = 200$ ppm, and the concentration in the breathing zone is $C_{\text{bz}} = 0.2$ ppm, the Protection Factor is estimated as follows:

$$PF = \frac{200}{0.2}$$

$$PF = 1000$$

Sample Calculation: A wet station hood has passed the ASHRAE 110 test at the supplier test site with the following score: 8AM0.03 at an exhaust rate of 1100 acfm. What is the equivalent PF? (See Section 6.8.4 for details of test).

In this case 8 liters/min of tracer gas (SF6) is mixed with 1100 acfm of exhaust air. The concentration of tracer gas in the exhaust air is:

$$C = \frac{\text{Volume flow rate of gas}}{\text{Volume flow rate of air}}$$

$$C = \frac{8 \text{ lpm x } 0.0353 \text{ cfm/lpm}}{1,100 \text{ cfm}}$$

$$C = 256 \text{ ppm}$$

$$\text{Protection Factor (PF)} = \frac{C_{\text{exhaust air}}}{C_{\text{breathing zone}}}$$

$$PF = \frac{256}{0.03}$$

$$PF = 8600 \text{ (rounded)}$$

E. Hood Static Pressure, SP_h

SP_h is directly related to volume flow rate and can be used to monitor a hood's or enclosure's performance. Hood static pressure should be measured about 4 duct diameters downstream in a straight section of the hood takeoff duct. The measurement can be made with a pitot tube and a manometer, or from a static pressure tap in the duct sheet metal.

F. Duct Velocity and Volume Flow Rate Measurements

Air flow in exhaust ventilation ducts is almost always turbulent, with a small boundary layer at the surface of the duct.

1. *Velometers vs. pitot tubes*

In clean air streams, the velometer may be used to measure duct velocity. In air streams containing aerosols, dust, acid mist, and/or solvent vapors, it is best to use the pitot tube to measure velocity pressure. The pitot tube is the most suitable measurement instrument because it has no moving parts or electronics to be fouled by air contaminants.

Since velocity varies with distance from the edge of the duct, a single measurement is not usually sufficient. However, if the measurement is taken in a straight length of round duct, about 6 diameters downstream and 3 diameters upstream from obstructions or directional changes, then a rough approximation of the average velocity can be estimated at 9/10ths of the centerline velocity (the average velocity pressure is about 81% of center line velocity pressure).

A more accurate method is the traverse. In a traverse, six or ten velocity or velocity pressure measurements are made on each of two traverses across the duct, 90° opposed, or on each of three traverses, 60° opposed. Measurements are made in the center of concentric circles of equal area.

Convert velocity pressures to velocities as shown on Worksheet 6.1, which also provides a table for estimating distances.

IX. NONSTANDARD AIR CONDITIONS

Unfortunately, very few ventilation systems operate at standard conditions. Fortunately, the adjustments for nonstandard conditions are not difficult. The following paragraphs examine each variation in order.

A. Water Vapor in Air (Relative Humidity)

At near normal temperatures and pressures, one can ignore the effects of humidity (usually less than 2.5%). However, at high temperatures (air temperatures greater than 120°F) it may need to be taken into account. The discussion is beyond the scope of this book, however. Check the references for more detailed information.

B. Density Correction Factor, d

A simple method to adjust for nonstandard temperature and altitude conditions relies on the density correction factor, d.

C. Corrections for Air Flow

Density changes do not affect the formula $Q = VA$ for a single system where the temperature remains essentially unchanged, and where the static pressure changes in the system do not exceed 20 inches w.g. (500 mm w.g). Under these conditions we can assume incompressible flow without much error. However, we do need to know whether the flow occurs at standard or nonstandard conditions, and whether the velocity V is estimated from a measured value of VP (if V is a measured value, a density correction factor should be applied to the velocity measurement). If the temperature in the system changes (e.g., air passes through a furnace and rises 100°), then a correction to the flow must be made. See the next sample calculation.

Knowing the actual air flow is important because it helps correctly size ducts, fans, air cleaners, and so forth. If the air flow occurs at nonstandard conditions, the units of Q are *acfm* and *acms*: actual flow. For flow at standard conditions, write *scfm* and *scms*, or *acfm at STP* and *acms at STP*.

D. Corrections for Velocity Measurement

The formulas for estimating velocity from VP when the air is at non-standard conditions as detailed in Equations 6.8a and 6.8b, are given by:

$$V = 4005 \times (VP/d)^{0.5}$$

and:

$$V = 4.043 \times (VP/d)^{0.5}$$

Every commercial anemometer and velometer has a different density correction method. Check the individual instrument's operating instructions.

E. Transport and Capture Velocities

No correction is normally necessary. If very small particles (less than 1 micrometer in diameter) are carried in less dense air (e.g., d less than 0.75), increasing velocities 10–20% should overcome small slip losses.

F. Vapor Generation

When solvents evaporate under nonstandard conditions, the resulting vapor volume flow rate must be corrected. The formulas are:

$$q = \frac{387 \times lb \text{ evaporated}}{MW \times minutes \times d} \tag{6.50a}$$

and

$$q = \frac{0.0241 \times grams \text{ evaporated}}{MW \times minutes \times d} \tag{6.50b}$$

Where:

q   = Volume flow rate of vapor emission or leak, U.S. = acfm; SI = acms

MW = Molecular weight (or *molecular mass*)

d   = Density correction factor, unitless

G. Duct System Design

One simple procedure for handling nonstandard conditions during design is:

1. Calculate the density correction factor, d.

2. Size ducts, estimate pressure losses, and select fan for the desired Q as if the system was operating at standard conditions for the desired Q.

This means selecting the fan at actual flow (i.e., acfm or acms) and standard static pressure. See the next figure. It shows the system and fan curves at standard and one-half the standard conditions. Select at Point #1 but the fan-system combination will operate at Point #2. This means the measured values in the duct system will be less than estimated, by the density correction factor. Note that Q remains constant, but actual static pressure requirements are reduced.

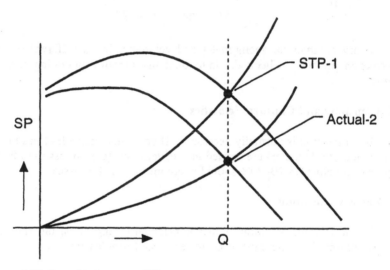

Figure 6.20 Fan selection at non-STP.

3. Reduce calculated air power (ahp and akW) requirements by the density correction factor, d. Actual pressures during operation will also be reduced by the density correction factor. For example:

$$\text{ahp (actual)} = \text{ahp (calculated)} \times d \qquad (6.51a)$$

and

$$\text{SP (actual)} = \text{SP (calculated)} \times d \qquad (6.51b)$$

4. At high altitudes (>5000 ft. or 1500 meters) increase motor rated horsepower by 10–15% to account for decreased motor cooling capacity of high-altitude air. Check with the motor supplier. For motors less than 25 hp (20 kW) in size, it is customary practice to increase rated horsepower by about 33% to allow for a 10% increase in fan rpm (and cfm) if the need arises after installation.

5. If an air cleaner is installed in the system, the rated static pressure drop may be reduced by d. For example, if a fabric filter were to be installed at Denver, at 5000 ft. (let d = 0.80):

Filter rated at STP: 1 to 4 inch w.g.
Rating at Denver: (1" to 4") x d = 0.80 to 3.2 inch w.g.

This means that if you want to have filters actually replaced at 4" w.g., then you may have to specify the maximum pressure drop during design at 4/d = 4/0.80 = 5.0" w.g.

If during design you use the manufacturer's value of 4" w.g. at STP, then the filter should be replaced at 3.2" w.g. in order to maintain the desired flow rate at d = 0.80.

H. System Design Where Air of Different Densities Are Mixed Together in the System

In this case (e.g., where branches of hot air mix with mains of cooler air), the actual density and density correction factors must be used as design proceeds. Every loss factor is adjusted using the density correction factor for the air at its actual density.

$$Loss\ factor = K \times d$$

At the end of the design, use the final density correction factor to estimate standard Fan Total Pressure. On standard fan curves and tables, fans are selected on the basis of actual flow rate and standard pressure.

This approach to design is more difficult and cumbersome, but it has the advantage of providing an estimate of the actual pressures to be found in the system after it is built. More information can be found in the literature.

X. CASE STUDIES

A. Case Study: Using Dilution Criteria

A San Jose, California-based computer components firm employed 20 people hand soldering defective terminal boards. Each solder station was about 10 feet apart in a large, closed building. Employee concern over potential lead exposure and occasional irritation by solder smoke had upset employees who were asking "for more air." Management was wondering what to do. Would dilution ventilation be appropriate?

Comparing conditions to the criteria defined in Table 6.8, lead is toxic, thus not a good candidate for dilution ventilation if emissions and exposures are high. Smoke particles were irritating even at low concentrations.

The final decision was also a function of the exposure (less than the OSHA Action Level of 30 micrograms per cubic meter), the distribution of sources (a lot, spread out), the time relationship of emissions (all day, varying with time), and the proximity to the breathing zone. Small dispersion and mixing fans (e.g., local exhaust) coupled with dilution ventilation solved the problem.

Nothing is easy in this world!

B. Case Study: Local Exhaust Ventilation

A local exhaust ventilation is being considered for an operation welding stainless steel in a confined space. Would this be an appropriate control?

The answer depends on the conditions described above:

Other controls:	*Not readily available or cost effective*
Hazardous exposures:	*Yes*
Codes:	*In the case of stainless steel, OSHA codes could be interpreted to require local exhaust*
Improvements will be seen:	*Yes*
Location of employee:	*Close to emission source*
Emission rate:	*Not constant; varies with time*
Emission source:	*Large compared to space*
Mobility:	*Could be mobile*

Based on the above data, local exhaust ventilation appears warranted.

C. Case Study: Transport Velocity

What transport velocity would be appropriate for welding operations in a shop facility? For solvent extraction systems in wafer fab facilities?

Welding: From literature sources we will find recommended $V_t = 3000$ fpm (15 mps). In addition to controlling welding fume, other operations conducted at the site: grinding, sanding, chipping, and so forth, that generate larger particles. Solvent vapors do not need transport velocity, but 2000 fpm is a cost-effective velocity, given initial and long-term operating costs. Many fabs are designed with oversized ductwork to accommodate future growth. That is acceptable where transport velocities are not required.

REVIEW QUESTIONS

1. The basic purpose of a local exhaust ventilation system is to:

   a. Contain, control, capture, and remove contaminants at their source
   b. Prevent re-entrainment of air contaminants at a stack
   c. Provide dilution ventilation of volatile substances
   d. Provide protective ventilation of microorganisms

2. At a point upstream from the fan in an exhaust system, the velocity pressure is VP = 1.00" w.g. and the static pressure is SP = -3.50" w.g.   What is the total pressure, TP?

   a. 2.50 inches of water
   b. -2.50 inches of water
   c. 4.50 inches of water
   d. -4.50 inches of water

3. Air flows out of a room through an open doorway at a rate of Q = 1000 scfm. The open area is 3' x 7'. What is the average velocity of the air through the doorway?

   a. 5000 fpm   b. 300 fpm   c. 50 fpm   d. 30 cfm

4. Room air at STP is exhausted at a rate of 100 cfm per enclosure for each of five enclosures. The air temperature rises to 600°F before leaving each enclosure. The exhaust duct is insulated to the fan housing. What volume of air must the fan be capable of handling?

   a. 4000 acfm at standard conditions
   b. 2000 acfm at 600°F
   c. 1000 scfm
   d   1000 acfm at 600°F

5. The hood static pressure downstream from a 24" wheel grinder hood is measured to be SP$_h$ = -1.69" w.g. The duct diameter is 6" and the hood shape is that of a standard grinder hood having a coefficient of entry of 0.78. What air flow at STP would this hood static pressure suggest? (Note: at STP, d = 1.)

   a. 400 cfm   b. 800 cfm   c. 1200 cfm   d. 1600 cfm

6. Round ducts on the exhaust side of a fan are considered better than square ducts because:

   a. They resist collapsing better
   b. The use less material for the same cross-sectional area
   c. They provide better conditions for consistent particle transport
   d. All of the above

7. Someone adjusts the slot widths on a side-draft hood from 2 inches to 1 inch (without any other modifications to the ventilation system). This action would have the following effect:

   a. Decrease capture velocities in front of the hood
   b. Reduce air flow through the hood
   c. Increase the hood entry losses
   d. All of the above

8. The sash of a conventional, non-bypass lab fume hood with 12" diameter ducting is lowered, blocking air flow. After lowering the sash, the absolute value of the hood static pressure will:

    a. Increase   **b.** Decrease   **c.** Stay the same   **d.** There is no way to know

9. A fan is rated in a manufacturer's catalog as delivering 8000 scfm when FTP = 3.0 inch w.g., and the fan speed is n = 400 rpms. The drive motor requires 6.2 shaft horsepower at these conditions. If the fan speed is increased to 500 rpm, determine the new Q, FTP, and shp.

    a.   Q = 14,250 scfm; FTP = 6.2 in. of water; shp = 13.6
    b.   Q = 10,000 scfm; FTP = 4.7 in. of water; shp = 12.1
    c.   Q = 12,253 scfm; FTP = 4.3 in. of water; shp = 11.6
    d.   Q = 11,158 scfm; FTP = 3.7 in. of water; shp = 10.4

10. Toluene is used in a glove box at the rate of 0.5 lb/hr. Assuming the density correction factor d = 0.88, how much air flow is needed to reduce the concentration to 10% of the LEL? (LEL = 1.27%, MW = 92, assume perfect mixing.)

    a. 31.4 acfm   **b.** 45.6 acfm   **c.** 58.1 acfm   **d.** 91.6 acfm

11. For all practical purposes, use of a pitot tube is limited to velocities greater than:

    a. 200–300 fpm   **b.** 400–500 fpm   **c.** 800–1000 fpm   **d.** Any velocity can be read easily

12. The velocity of air in a duct can be estimated by measuring the average velocity pressure. What instrument would be best for obtaining this measurement?

    a.   A vaned anemometer
    b.   A thermal anemometer
    c.   A pitot tube
    d.   A balometer

13. Which of the following statements concerning various types of air pressure is not true?

    a.   Velocity pressure may be a positive or a negative number according to the ACGIH convention for assigning signs
    b.   Static pressure may be a positive or a negative number according to the ACGIH convention for assigning signs
    c.   Static pressure is generally negative on the upstream side of the fan
    d.   TP = SP + VP

14. A hood has a coefficient of entry Ce = 0.82. If the duct area is 0.1964 square feet, and 630 scfm (d = 1) of air flows through the hood, the hood static pressure just downstream from the hood would be measured as:

    a.   -1.65 inches of water
    b.   -0.65 inches of water
    c.   -0.95 inches of water
    d.   -1.50 inches of water

ANSWERS

1. a.
2. b.   Total pressure = velocity pressure + static pressure
         TP = + 1.00 + (-3.50)
         TP = -2.50 inches
3. c.   V = Q/A
         V = 1000/21
         V = 50 fpm (rounded for significant figures.)
4. d. $Q_1 = 5 \times Q_{enc}$
         $Q_1 = 5 \times 100$ cfm
         $Q_1 = 500$ cfm entering at STP.
      Calculate d = 0.50 from ideal gas law:

$$d = \frac{530}{{}^\circ F + 460} \times \frac{BP}{29.92}$$

$$d = \frac{530}{1060} \times \frac{29.92}{29.92}$$

         d = 0.5
      Accordingly, the air out is half as dense and twice the volume:
         $Q_2 = Q_1/d$
         $Q_2 = 500/0.5$
         $Q_2 = 1000$ acfm
5. b. $Q = 4005 \times A \times C_e \times (SP_h/d)^{0.5}$
         $Q = 4005 \times 0.1963 \times 0.78 \times (1.69/1)^{0.5}$
         Q = 800 scfm (rounded)
6. d.
7. d.   VP in slots would increase which would increase the slot loss (where Loss =
   1.78VP); SP_h would be converted to loss instead of VP, resulting in less flow.
8. a.
9. b.   Flow varies directly with rpm. Pressure varies with rpm change squared. Power
         required varies with rpm change cubed.
         $rpm_2 = rpm_1 = 500/400 = 1.25$
         $Q_2 = Q_1 \times 1.25 = 10,000$ scfm
         $FTP_2 = FTP_1 \times (1.25)^2 = 4.7$" w.g.
         $shp_2 = shp_1 \times (1.25)^3 = 12.1$ hp
10. a.   q = 387 x lbs/(MW x min x d)
          q = 387 x 0.5/(92 x 60 x 0.88)
          q = 0.0398 acfm

      and:

          $Q = (q \times 106 \times K_{mix})/ppm$
          Q = (0.0398 x 106 x 1)/1270
          Q = 31.4 acfm
11. b.   VP (at 400 fpm) = 0.01" w.g., and is difficult to read on typical manometers until
          velocities exceed 400 to 500 fpm.
12. c.
13. a.   Velocity pressure is a measure of the kinetic energy in the system
          and hence must always be positive.
14. c.   V = Q/A
          V = 3210 fpm
          $VP = (V/4005)^{0.5}$
          VP = 0.64" w.g.
          $SP_h = VP/C_e^2 = -0.95$" w.g.
          (Negative on the upstream side of the fan.)

References

American Conference of Governmental Industrial Hygienists, *Industrial Ventilation Manual, 22nd ed.*, ACGIH, 1995.

American Glovebox Society (AGS), *Guideline for Gloveboxes (Draft)*, AGS, Denver, August, 1993.

American National Standards Institute, *ANSI Z9.5,1992, Laboratory Ventilation*, ANSI-AIHA, Fairfax, VA, 1992.

American National Standards Institute, *ANSI Z-9.7 (draft), Safety Code for Exhaust Systems-Recirculation*, ANSI-AIHA, Fairfax, VA, (Proposed Standard).

American Society of Heating Refrigerating & Air-Conditioning Engineers, Inc. (ASHRAE), *Handbook of Fundamentals*, ASHRAE, Atlanta, 1993.

American Society of Heating Refrigerating & Air-Conditioning Engineers, Inc. (ASHRAE), *ASHRAE Standard 62-1989-Ventilation for Acceptable Indoor Air Quality*, ASHRAE, Atlanta, 1989.

American Society of Heating Refrigerating & Air-Conditioning Engineers, Inc. (ASHRAE), *ASHRAE Standard 110, Method of Testing Performance of Laboratory Fume Hoods (under revision)*, ASHRAE, Atlanta, April, 1995.

American Society of Heating Refrigerating & Air-Conditioning Engineers, Inc. (ASHRAE), *ASHRAE Standard 52, Filter Testing Methods*, ASHRAE, Atlanta.

American Society of Heating Refrigerating & Air-Conditioning Engineers, Inc. (ASHRAE), *ASHRAE Guideline 1-1989, Guideline for the Commissioning of HVAC Systems*, ASHRAE, Atlanta, 1989.

Burton, D. J., *HVAC/IAQ Workbook*, IVE Inc., Bountiful, UT, 1995.

Burton, D. J., *Industrial Ventilation Workbook, 3rd ed.*, IVE Inc., Bountiful, UT, 1995.

Burton, D. J., *Lab Ventilation Workbook*, IVE Inc., Bountiful, UT, 1994.

Hunter, P., Effects of toxic, corrosive, pyrophoric, and flammable gases on ventilation system materials, *SSA Journal*, June 1992.

Institute of Environmental Sciences (IES), *Compendium of Standards, Practices, Methods and Similar Documents Relating to Contamination Control, (IES-CC-RP-009)*, IES, Mt. Prospect, IL.

National Fire Protection Association (NFPA), *NFPA 30, Flammable and Combustible Liquids Code*, NFPA, Quincy, MA.

National Fire Protection Association (NFPA), *NFPA 45, Fire Protection for Laboratories using Chemicals*, NFPA, Quincy, MA, 1986.

National Fire Protection Association (NFPA), *NFPA 90A, Standard for the Installation of Air Conditioning and Ventilating Systems* NFPA, Quincy, MA.

National Fire Protection Association (NFPA*), NFPA 91, Standard for the Installation of Blower and Exhaust Systems for Dust Stock and Vapor Removal or Conveying*, NFPA, Quincy, MA.

O'Brien, M., Evaluation of performance testing for fume hoods within the clean room environment, *SSA Journal*, June, 1993.

Rainer, D. and Quinn, W.E., Using flow-limiting restrictors, *AIHAJ*, August 1989.

SMACNA, *HVAC Systems–Testing, Adjusting, and Balancing*, SMACNA, Rockville, MD, 1983.

SMACNA, *Indoor Air Quality*, SMACNA, Rockville, MD, 1989.

Tubby, R. L., Tracer gas testing of secondary exhaust systems on hazardous gas enclosures, *SSA Journal*, June 1991.

Wilson, D. J., Contamination of air intakes from roof exhaust vents, *ASHRAE Transactions 82*, 1976.

Chapter 7

NOISE

Howard K. Pelton, PE

I. OVERVIEW

As extracted from OSHA: "Noise is one of the most pervasive occupational health problems. It is a byproduct of many industrial processes. Exposure to high levels of noise causes temporary and permanent hearing loss and may cause other harmful health effects as well. The extent of damage depends primarily on the intensity of the noise and the duration of the exposure."

The opening paragraph of the preamble to the OSHA Noise Standard 29 CFR Part 1910.95 Occupational Noise Exposure is very direct. It is well known that noise-induced hearing loss occurs in the workplace. In addition, many recreational activities such as, snowmobiles, motorcycles, as well as, noisy equipment found around the house, like power tools and small-engine driven equipment also contribute to noise-induced hearing loss. The question relating to noise, noise control, and hearing loss are very important to the Industrial Hygienist (IH). Noise can be managed as with other any other IH hazard. While hearing conservation is important, it should not be the first priority. This is stated in the OSHA 1910.95 par (b)(1) Noise standard. It states in part: "When employees are subjected to sound exceeding those listed in Table G-16, feasible administrative or engineering controls shall be utilized. If such controls fail to reduce sound levels within Table G-16, personal protective equipment shall be provided to reduce the sound levels within the levels of the table." Noise control must be the first line of defense. While hearing conservation can be used, it must not constitute the whole program. The purpose of this chapter will be to provide an overview and understanding of noise and its effects.

II. PHYSIOLOGICAL RESPONSE

A. Introduction

To understand the physiological response of the ear to noise and resulting hearing loss, it is necessary to be familiar with the anatomy of the ear and the function of its components. Study of the ear provides the insight and appreciation essential to the success of any noise control and hearing conservation program. First, such knowledge enhances the appreciation of the sensory function; second, those involved in a hearing conservation program will have a better knowledge of the hearing mechanism; third, those performing audiometric examinations can better understand the interrelationship of audiometric findings and the structural parts of the ear.

Cases of gunfire noise causing hearing loss aboard ships in the 1800s, as well as hearing loss caused by blacksmithing have been cited in previous works. The advent of the Industrial Revolution brought more machinery and equipment that made noise. An example is the steam engine used for primary power. Soon after World War I, studies done on animals and humans showed through histological samples, the deterioration of the *Organ of Corti*, since this could be directly related to noise. Not until World War II was there a serious effort made to conduct definitive studies to relate the cause and effect of noise and hearing loss. In addition, the construction of steam engine's boilers required riveting which indeed was quite loud; most people have heard the expression "boilermaker's ears." These workers had a great deal of noise-induced hearing loss.

One of the earliest *hearing conservation programs* (*HCPs*) was started by the Air Force in 1948. Other branches followed in 1950. During the 1950s and into the 1960s, it was recognized, documented, and defined by scientific studies that high levels of noise caused hearing loss. It was determined that HCPs were required to conserve hearing and reduce the impact of *noise-induced hearing loss* (*NIHL*).

The ear is a complex mechanism. There are many aspects of how the ear functions, particularly the working of the inner ear and pathways to the brain that are not fully understood. The ear enables one to detect sound waves from 20 to 20,000 Hz (Hertz or cycles per second). How well the ear detects these sounds depends on the hearing acuity. There has been an awareness of the connection between noise and hearing loss for a long time.

B. Anatomy of the Ear

The ear consists of three major divisions:

- *Outer ear*–This external part collects the sound energy and converts it into vibratory motion at the eardrum.
- *Middle ear*–This section mechanically couples the eardrum to the fluid-filled inner ear through three bones: the hammer, anvil, and stirrup.
- *Inner ear*–This portion converts the vibratory signal to electrical impulses, through the fluid-filled cavity and sensory hair cells, and transmits it to the brain for processing and interpretation.

1. Outer ear
The outer ear is divided into two parts. The external portion that is seen

and attached to the head is called the pinna. This is a delicately folded cartilaginous structure, with a few small muscles, covered by subcutaneous tissue and skin. The second part is the external auditory canal, also termed
meatus, a skin-lined pouch about 1.5" long. This is supported by cartilage
of the pinna in its outer one-third and by bone of the skull for the remainder
of the length. At the inner most end of the auditory canal is the *tympanic
membrane* or eardrum that separates the external from middle ear.

The hairs at the outer end of the canal keep out dust and dirt. Further
into the canal are wax-secreting glands. Normally, ear wax flows toward
the entrance of the ear canal carrying dirt and dust that accumulate in the
canal. Too much cleaning and prolonged use of ear plugs may cause the ear
canal to become occuled, creating *conductive hearing loss*. Any buildup of
wax should be removed very carefully and preferably by a doctor, or other
well-trained person, to prevent damage to the eardrum and middle ear
structure.

The eardrum is a very delicate and thin membrane that responds to minute changes in sound pressure level. It is seldom damaged by common
continuous high levels of noise. The eardrum can be damaged by trauma
such as explosions, firecrackers, or a rapid change in ambient pressure.
When the eardrum is ruptured, the attached middle ear bones may be dislo-
cated. The eardrum must be carefully examined after a trauma event to determine if surgical procedures may be required to repair the bones in the
middle ear. This does not usually result in the loss of hearing acuity, however.

2. Middle ear

This is an air-filled cavity between the outer and inner ear. The middle
ear contains three bones, the ossicles. These are the malleus (hammer), the
incus (anvil), and the stapes (stirrup). These are mechanically connected to
the eardrum in the outer ear canal and to the oval window at the entrance to
the inner ear.

The eardrum area has about 20 times the area of the oval window. This
produces a mechanical advantage of 20 to 1. The ossicle has a mechanical
advantage of about 3; therefore, the system has a mechanical advantage of
60 at the natural frequency of the system. This complex system also acts as
an ear protector through the small muscles in the middle ear that relax the
coupling efficiency of the bones. This is an involuntary relaxation, and the
reaction time is about 10 msec. This is why high impulse noise, like gunfire, can easily damage the ear. Since the rise time of gunfire noise is so
short, the muscles in the middle ear cannot react fast enough to protect it
from the rapid increase of sound pressure.

Infections are a common occurrence in the middle ear. This is a damp
dark cavity that become a breeding ground for bacteria. Prolonged middle
ear infections can result in hearing loss since the ossicles can be damaged.
If the *eustachian tube* is closed off by swelling due to infection or allergies,
the pressure in the middle ear cannot equalize. This can be painful when

traveling in an airplane or through the mountains.

Another middle ear problem is the abnormal growth of bones around the ossicles. This is called *otosclerosis*. There is a restriction of normal movement of the ossiles and the result is a conductive hearing loss. Correction is with hearing aids and surgery. One of the side benefits of a hearing conservation program is the early detection of these conditions during audiometric screening examinations.

3. Inner ear

This fluid-filled compartment is composed of two main parts: the *semicircular canals* containing a part of the body balance mechanism, and the *cochlea*, which contains the individual hair cells that are connected to the nerve endings and pathways to the brain. Interior to the cochlea is the *Organ of Corti* that contains the 20,000 to 30,000 minute hair cells. Another part of this is the *acoustic nerve* or the *eighth nerve*. This is a nerve bundle that contain 20,000–30,000 fibers, creating the pathway to the brain where sound is interpreted.

The vibrations transmitted from the middle ear to the oval window create pulsations in the fluid of the inner ear. The minute hair cells are set into motion and generate electrical impulses that travel to the brain. Different hair cells or groups of hair cells respond to different frequencies of sound. These electrical impulses are gathered in the eighth nerve and transmitted to the brain and we experience hearing.

C. Range of Hearing

The range of human hearing generally ranges from 16 to 16,000 cycles per second (Hertz or Hz). Children or young adults may be able to hear 20,000 Hz. Most adults are not able to hear above about 12,000 Hz. The general rule is that hearing acuity declines with age or with continuous noise insult hearing level decline in the 4000 to 8000 Hz range. If one can hear well from 500 to 2000 Hz, then most of everyday speech will be heard and understood. This is important for general communication. The ear is most sensitive from 1000 to 4000 Hz. This is why the high-pitched whine from airplane engines or sirens seems so annoying. The lower limit of hearing or audiometric zero is shown in Table 7.1.

Table 7.1
Lower Limit of Hearing at Given Frequencies

Freq Hz.	125	250	500	1000	2000	4000	8000
Level dB	52.4	38.8	24.5	16.5	17.1	14.5	28.3

The capacity of the ear to handle high-level sound is also limited. Table 7.1 illustrates the lowest levels that a healthy ear can hear, while the upper limits where sensation begins is about 120 dB. Above 120 dB to 130 and 140 dB, the sensation turns to pain. For some people, sound levels above 105 dB can be painful, though tolerance levels are individual-specific.

D. Normal Hearing

The audiometric zero, or normal hearing, is set 10 to 14 dB above acoustic zero. Instead of identifying the sound level an acute ear can hear, an average hearing threshold was established in an otological normal hearing group from 18 or 30 years of age. This is 10 to 14 dB above acoustic zero in the speech range 500 to 2000 Hz that the normal population can hear. This reference level was derived from a study performed by the U.S. Public Health Service in 1935. The hearing threshold varied by ±15 dB. Consequently, any hearing threshold within 15 dB of zero may be as good as the individual ever experienced.

E. Hearing Loss

Hearing loss is measured with an *audiometer* that measures the individual's hearing acuity above the standard audiometric zero. The actual extent of hearing loss can be accurately determined by comparing the subject's best hearing record with the current one. The subject is said to have normal hearing if the levels are within 15 dB of the standard audiometric zero used in the audiometer.

Hearing losses are of two types: *conductive* and *sensorineural*. Mixtures of these two are also encountered.

1. *Conductive loss*

This is a result of a blockage in the ear canal that interferes with normal transmission of vibration to the inner ear. These can be:

- Wax or foreign object in the ear canal
- Middle ear infections
- Malfunctions of the ossiles
- Trauma to the eardrum
- Blockage of the eustachian tube

This type of loss is seldom industrial in origin, is not caused by noise exposure, and may be identified in an otological examination. A typical graph of the conductive loss is shown on Figure 7.1. The graph is relatively flat across the spectrum.

2. *Sensorineural loss*

This type of loss is caused by damage to the nerves in the inner ear. This is a noise-induced loss caused by exposure to high levels of noise for extended periods. The hair cells are asked to respond to high vibration impulses in the inner ear fluid and they become fatigued. Continual insult on these minute hair cells causes them to respond less, they become atrophied, and even die. The first stage of this is a *temporary threshold shift (TTS)*. With rest, the hair cells recover and hearing is again normal. Continual insult on a daily basis over extended periods of time results in less and less recovery by the hair cells. This is known as a *noise-induced permanent threshold shift (NIPTS)*. The ability to hear sound with clarity is a distinct attribute of normal hearing. Damage to the outer and middle ear results in the perception of the sound intensity, but sensorineural damage reduces the clarity of sound due to the loss of hearing ability in the speech range. This type of hearing loss is shown on Figure 7.2.

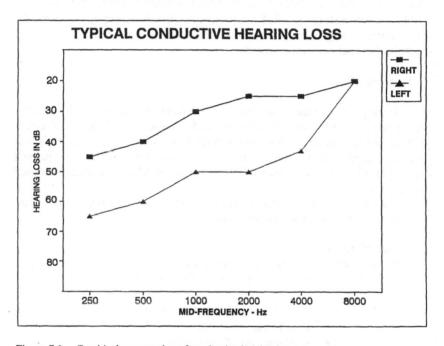

Figure 7.1   Graphical presentation of conductive hearing loss.

The typical 4000 Hz "notch," or drop, in hearing acuity at this frequency is indicative of NIPTS. People with this type of hearing loss lose the ability to hear consonants and have difficulty understanding speech, and music does not sound as "crisp." The hearing-impaired person is often frustrated by missing information that is vital for social and vocational functioning. In addition, background noise, such as radio, TV, or other people talking, is much more disruptive on the hearing-impaired individual than on the normal listener. This is because the normal-hearing person has the ability to separate background noise from the required information needed.

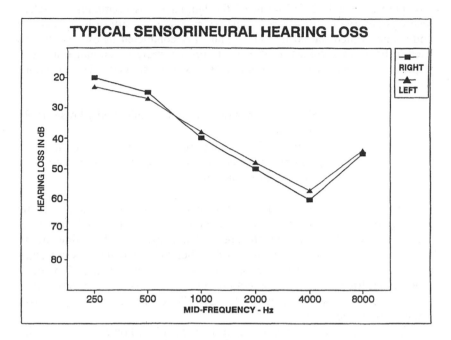

Figure 7.2   Graphical presentation of sensorineural hearing loss.

Social relationships become increasingly difficult, which can result in depression, isolation, suspicion, and withdrawal. Hearing loss can create other adverse effects. Warning shouts can be masked and create a safety hazard. As well, there have been reports of health-related problems such as stress-related illness, hypertension, and ulcers.

III. OSHA/ACGIH STANDARD/PERMISSIBLE NOISE LIMITS

A. Introduction

In 1969, the U.S. Government published the first noise regulation in the form of the Department of Labor's (DOL) Walsh-Healy Public Contracts Act. This applied to government contractors with contracts over $10,000. The noise limit was set at 90 dBA for an 8-hour exposure to noise. If the duration was halved, the level could be 95 dBA, and so on to a limit of 115 dBA. The ceiling peak value was 140 dB. If the levels were exceeded, then hearing protectors were provided or the noise was to be reduced through engineering means. This standard provided language providing for "a continuing effective hearing conservation program."

This became applicable to all industries with the enactment of the Occupational Health and Safety Act of 1970. A revision was made in 1983 to incorporate a detailed definition of the language: "A continuing effective hearing conservation program." This revision was a result of long hearings and studies which started in 1975. The EPA conducted its own study and tried to have the level set at 85 dBA to protect more of the noise-induced population. Technical and economic issues were studied. The questions were:

- Should the permissible noise limit (PNL) be set at 85 or 90 dBA?
- Should OHSA continue to require engineering and administrative control, or should hearing protection be allowed as a final solution?

The preliminary rules were promulgated in 1974 but did not define the language of "continuing effective hearing conservation program." The final rules were issued in 1983 which amended the regulation to define this in very specific terms after many years of deliberation. The result was a compromise for an *action level* set at 85 dBA as a time-weighted average (TWA) that triggers the development of a written hearing conservation program. The PNL was left intact at 90 dBA TWA with engineering controls implemented where required.

Engineering controls remained in the regulation; however, in the mid-1980s through a DOL administrative process, citations were not issued for failure to install engineering controls if there was a hearing conservation program in place. Hearing conservation programs were deemed satisfactory, must be in force, and shown to have not resulted in significant threshold shifts for the employees.

The hearings that OSHA conducted relative to this amendment in 1983 brought together the best thinking at the time on hearing conservation. Also presented were data on technical and economic feasibility for controlling noise from 85 to 90, and above 90 dBA. The results of these studies demonstrated that engineering control for some industries was not feasible, and for those where it was possible, it could be cost prohibitive. Finally, it was shown that if there were some amount of reduction possible, this can be helpful in allowing the hearing protective equipment to be more effective.

B. OSHA Noise Standards

The basic noise exposure standard is found in 29 CFR 1910.95 Occupational Noise Exposure Table G-16 and is shown below in Table 7.2.

Table 7.2
OSHA Table G-16, Permissible Noise Limits/Exposure Durations

Duration per Day, hours	Sound Level dBA Slow Response
8	90
6	92
4	95
2	100
1.5	105
1	105
0.5	110
0.25 or less	115

The following equation is extracted from the OSHA Noise Standard:

$$D = C_1/T_1 + C_2/T_2 + ... + C_n/T_n \tag{7.1}$$

Where:
   D = Dose
   C = Concentation
   T = Exposure duration

In the regulation there is a manual method for determining these values, but noise dosimeters that are worn by the employee are the most reliable and fastest method of determining the mixed exposure values. The percent dose is found by multiplying 100 times the fractional values in the above OSHA Noise equation (see Section VI, *Monitoring*).

The equation shown below can be used to calculate the TWA for an 8-hour day from the calculated Dose:

$$TWA = 16.61 \log_{10}(D/100) + 90 \text{ dBA} \tag{7.2}$$

Where:
   D = Dose

The regulation was revised in 1983 and issued in the Federal Register. This major change, more than 10 years in the regulatory process, was to provide a detailed interpretation of the original regulation, that said in effect, that if the noise levels exceed those values found in Table G-16, then administrative or engineering controls shall be utilized. It this fails then a continuing effective hearing conservation program shall be provided. This did not provide a uniform method for establishing a hearing conservation program. Paragraph (c) Hearing Conservation Program provides a detailed, proven, and uniform method to conserve hearing.

The hearing conservation section of the OSHA Noise Standard is quite long and, if followed, can provide a method to reduce the impact of noise upon hearing. The action level of 85 dBA TWA serves as a "trigger" to begin to review the noise in the facility in a serious way, and to begin establishing a mandatory hearing conservation program for all the exposed employees. All of these employees must be in the hearing conservation program from 85 to 90 dBA TWA. Above 90 dBA TWA, administrative controls or engineering controls must be utilized to reduce the noise exposure. The hearing conservation program will continue until the noise levels are reduced as far as feasible. If the noise exposure cannot be reduced below 90 dBA, then those employees will be in the HCP permanently. Many companies use an action level of 80 dBA TWA for an HCP, and as a trigger to seriously begin to review the noise levels in a facility. Also, a level of 85 dBA may be employed as the goal to reduce noise and to begin to implement engineering controls.

Paragraph (b) of the regulation has several subparagraphs that are important and must be considered. These include:

- (d) Monitoring
- (e) Employee notification
- (f) Observation of monitoring
- (g) Audiometric testing program
- (h) Audiometric test requirements
- (i) Hearing protectors
- (j) Hearing protector attenuation
- (k) Training program
- (l) Access to information and training materials
- (m) Recordkeeping
- (n) Appendices
- Appendix A: Noise Exposure Computation
- Appendix B: Methods of Estimation the Adequacy of Hearing Protector Attenuation
- Appendix C: Audiometric Measuring Instruments
- Appendix D: Audiometric Test Rooms
- Appendix E: Acoustic Calibration of Audiometers
- Appendix F: Calculation and Application of Age Corrections to Audiogram
- Appendix G: OSHA Onsite Consultation Project Directory
- Appendix H: Availability of Reference Documents
- Appendix I: Definitions

C. ACGIH Standards

The American Conference of Governmental Industrial Hygienists (ACGIH) is an organization devoted to the administration and technical aspects of occupational and environmental health. They also have a *Code of Ethics* for the practice of industrial hygiene.

In their handbook of *Threshold Limit Values (TLVs) and Biological Exposure Indices (BEIs)*, ACGIH states that the values are provided as a guideline for the practice of Industrial Hygiene and do not denote fine lines between safe and dangerous limits. They also state that any values provided should be used by trained personnel in the Industrial Hygiene discipline; and are the best available information at the time of printing. More detailed information can be found in their publication listed in the references.

The ACGIH recommended TLV for noise provides more protection than the OSHA standard. It is designed to protect the median population against noise-induced hearing loss of 2 dB, after 40 years of noise exposure at the 500, 1000, 2000, and 3000 Hz range. This assumes that a well-established and monitored hearing conservation program is implemented. Furthermore, any off-the-job noisy activities would also need to be accounted. The ACGIH TLVs are based on 3 dB doubling; i.e., 85 dBA for 8 hours and 88 dBA for 4 hours, as well as a lower value. The OSHA regulation has a 5 dB doubling; i.e., 85 dBA for 16 hours and 90 dBA for 8 hours. Many companies follow the ACGIH TLVs to provide that extra level of protection for noise-induced hearing loss. These limits will protect more of the noise exposed-population than the OHSA Regulation, and is considerably more conservative. Most of the later noise dosimeters allow the change in doubling rate and threshold limits; therefore, it is quite easy to monitor the noise exposure using these values.

Table 7.3 provides the ACGIH TLVs as compared to the OSHA values.

Table 7.3
ACGIH TLVs and the OSHA PNLs

Duration per Day	ACGIH Allowable dBA	OHSA Allowable dBA
8	85	90
4	88	95
2	91	100
1	94	105
1/2	97	110
1/4	100	115
1/8	103	Not allowed
1/16	106	Not allowed

The ACGIH method used to calculate noise exposure is the same as the OSHA Regulation's approach. Impulse noise is also addressed the same, and cannot be greater than 140 dB measured on the C-weighted scale. If an instrument is not available to measure C-weighting, then a measured level below 140 dB may be used to imply that the C-weighted levels are below 140 dB.

D. Noise Exposure Calculation Examples

Apart from the noise dosimeter used to monitor employee noise exposure automatically, the manual method of calculating noise exposure should be appreciated. Many times it is necessary for the industrial hygienist to check for noise exposure without the aid of a dosimeter. A quick check can be made by measuring the noise level at the employee workstation and estimating the time of exposure from interviews. This is not as definitive as the dosimeter but can be accurate enough when a job has changed, or some noise control device has been installed, or many other reasons. In Section VI there is a detailed discussion about noise dosimetry data evaluation. The following two examples illustrate this method.

Sample Calculation: The following noise levels (dBA) were measured at an employee's workstation with the time of the exposure determined from discussion with the employee:

dBA	Duration (hr)
95	2
90	2
92	1
85	3

From these values, is this employee being over-exposed to noise? What is the TWA of the exposure?

Using Equation 7.1 and the allowable time from the Table 7.2, the following Dose (D) can be calculated:

$$D = 2/4 + 2/8 + 1/6.2 + 3/16$$
$$D = 1.0988$$

Since the Dose exceeds 1, there is an over-exposure to this noise environment. The percent Dose can be simply calculated by multiplying 100 or 109.88%. Round this off to 110%.

Using Table 7.4, determine the TWA. This value is 90.7 dBA. This can also be easily calculated using the equation above; i.e.:

$$TWA = 16.61 \log_{10}(D/100) + 90 \text{ dBA}$$

This also results in a value of 90.7 dBA TWA. Since this is so close to 90 dBA, it is probably prudent to monitor with a dosimeter (See Section VI) for this employee. This method can provide valuable screening information, and if the noise sources can be associated with individual noise levels, then reducing the level of one or more sources can result in a reduced TWA to below 90 dBA or even below the Action Level of 85 dBA. Recheck the TWA by reducing the 95 dBA level to 90 dBA. The result is D = 1.04 or 90.3 dBA TWA. This is not enough. Reduce the 95 dBA level to 87 dBA. The result is D = 0.96 or 89.7 dBA TWA. This is enough to achieve the 90 dBA Engineering Control threshold. It is very important that a margin of safety be achieved. The result must be very dramatic. More than one noise source may require reduction. This method does provide a quick check of which noise source affects the noise exposure. A simple spreadsheet program can provide a good model for this method.

Sample Calculation: An employee in a utility plant located within a large manufacturing facility spends a portion of his day (6 hr) in a control room where the noise level has been measured at 65 dBA. His normal rounds and minor maintenance repairs take him into the utility plant for 2 hours each day where the noise level is 105 dBA. This, of course, can vary depending on the demands of the plant on any given day.

The employee is not exposed to noise while in the control room since the level is below 80 dBA. While in the plant, the employee can be exposed to 105 dBA for 1 hour. The dose calculation is:

$$D = 2/1 = 2 \times 100 = 200\%$$

Table 7.4
Abbreviated Table A-1 from OSHA Noise Standard
(Permissible Noise Limits)

Dose or Percent Noise Exposure	TWA
10	73.4
15	76.3
20	78.4
25	80.0
50	85.0
75	87.9
100	90.0
125	91.6
140	92.4
200	95.0
280	97.4
400	100.0
460	101.0
500	101.6
610	103.0
700	104.0
800	105.0
920	106.0
999	106.6

This is equal to a TWA of 95 dBA (refer to Table 7.4). Clearly, this requires a hearing conservation program. The control room is considered an engineering control since it protects the employee. For large utility plants, source or path noise control may not be feasible, but control rooms do provide an adequate method of reducing exposure. This example demonstrates that the TWA can be reduced by 10 dBA, which is considered significant. Due to the high noise level, it is prudent to conduct a detailed engineering noise survey to identify any potential sources that may be feasible to control. Employees in this situation require close observation through the HCP.

IV. DECIBEL (dB) SCALE

When the physical nature of sound is understood, the methods for control can be derived. Within the general field of physics, *acoustics* deals with the study of sound, its generation, propagation, and reception. Sound is all around us as a natural phenomenon and we usually perceive it through the physical changes in air pressure. It is a form of physical energy that can be as faint as the humming of a mosquito's *minute ripple* to the very high overpressure created by the firing of a large artillery piece's *tidal wave*. The amplitude of the vast majority of acoustic disturbances is very low, but the hearing mechanism for land beings are designed to detect these minute changes in sound pressure. Sound is omnipresent in the atmosphere through which the sound waves travel to reach our ear. This also is true in the ocean, but we do not experience this very often. Sound is transmitted through any solid, gas, or liquid at varying rates depending on the physical characteristics of the medium. For example, the speed of sound in normal air at 70° is 1130 fps, in water about 4800 fps, and in steel 20,000 fps. Sound is transmitted from one molecule to another; therefore, the closer the molecules are together, the faster the speed of sound will be transmitted through the medium. This is evident in the examples cited above by virtue of the material's density. This does not mean that the molecules in the medium travel at this rate; they do not. The individual molecules move an extremely small distance as they pass the acoustic energy on to the adjacent molecules. In this way, the wave of disturbance is transmitted through the medium.

The usual method of expressing the change in sound pressure above atmospheric pressure is the *decibel (dB)*. This is a nondimensional unit used to express the measured change of sound pressure above a reference–atmospheric pressure. A logarithmic scale is used to express this ratio. This is the *root-mean-square (RMS)* sound pressure related to the atmospheric pressure. The method of numeric notation is metric: *Pascal (Pa)*, *Newton per square meter*. This is a unit of pressure. Atmospheric pressure is *20 micro Pa (μPa)*. A microphone is used to measure this change in sound pressure level and convert it to dB read on the instrument. The following equation is used for this purpose:

$$dB = 20 \, Log_{10}(\text{pressure/reference pressure}) \qquad (7.3)$$

The range of sound pressures can be quite large: from atmospheric to the threshold of pain. For example, putting this in terms of pounds per square inch (psi), for convenience, the pressure of normal hearing is 3×10^{-9} psi to 3×10^{-2} psi, 0 to 130 dB, respectively. The sound pressure level is defined as the RMS pressure related to a reference pressure as:

$$L_p = 10 \, \log \, (P/P_{re})^2 \qquad (7.4)$$

Where:

   L_p = Sound pressure level, dB
   P = Measured pressure, Pa
   P_{re} = 20 µPa

This is was chosen as the reference pressure. It is lowest sound pressure average normal hearing young adult can perceive at 1000 Hz. Table 7.5 illustrates sound pressure levels for various items or activities:

The source of sound energy that causes the air molecules to vibrate is the acoustic power, or sound power. The sound power level describes the acoustical energy that radiates by a given source with respect to an international reference of 10^{-12} *watts*. The equation for sound power (L_w) is:

$$L_w = 10 \log (W/W_{re}) \tag{7.5}$$

Where:

   W = Sound power radiated by source, watts
   W_{re} = Reference power 10^{-12}, watts

Table 7.5
Typical Sound Levels - dB re 20 µPa

Decibel Intensity Level	Typical Source/Response
140	Threshold of pain
130	
120	Jet takeoff
110	Rock music in club
100	Jackhammer at 20 ft.
90	Heavy truck at 20 ft.
80	Freeway at 20 ft.
70	Vacuum cleaner at 10 ft.
60	Conversation at 3 ft.
50	Urban residence
40	Country area
30	Soft whisper at 6 ft.
20	North rim of Grand Canyon not wind
10	
0	Threshold of hearing at 1000 Hz

A source of 0.1 watts is 10 log(0.01/10^{-12}) = 10 log (10^{11}) = 110 dB. This is very small amount of acoustic power to generate this type of sound power level. Even though the sound power cannot be measured directly, it can be inferred by measuring the sound pressure and relating this to the environment where the measurements are made. The sound power causes the air particles to vibrate and transmit changes in pressure to the surrounding environment. Since sound pressure gives up energy through molecular vibration over distance and interaction with surfaces. Knowing the distance and something about the environment, inside or outside, the sound power can be determined. The relationship between the sound pressure level and sound power level is identified by the following equation:

$$Lp = Lw + Constant\ (K) \qquad (7.6)$$

Where:

   K = Constant value that is a function of the environment in which the source is located

The value of K is a function of the direct field of sound from a source plus the reverberant field that is reflected from the surfaces of the space. The equation is:

$$K = 10\ log(Q) - 10log(4\pi r^2) + 10log(4/R) + 10\ dB \qquad (7.7)$$

Where:

   Q = Directivity of source
   R = Room constant
   r = Distance to the source

When the source is located outside, or in the center of a large room, the 4/R term approaches zero.

An analogy that will put this concept in perspective is to picture a 100-watt light bulb shining in a closet painted black. The bulb seems dim, but still draws 100 watts of power. If the closet is painted white, the bulb seems to shine with more brilliance even though it is still 100 watts. The environment has changed and caused the resulting foot candles to increase due to the many reflections from the white walls. In the black closet, there are less reflections because the black closet does not reflect as much light. It is more absorptive. The same is true if a sound source of 0.1 watt is placed in a space such as a gymnasium. The resulting sound pressure level will be higher by 8 to 10 dB if all the surfaces are hard and reflective–white closet. If the walls and ceiling are covered with a sound absorbing material, the noise level will be much more normal, similar to that outside–black closet. This can be illustrated in Figure 7.3. This shows measurements made at one location under four conditions. The sound power is the same, but the sound pressure level changes with the effect of the environment.

A. Decibel Addition

The addition of decibels is done through logarithmic mathematics. The method uses the following equation and can be done on a scientific hand calculator:

$$Lp_{(total)} = 10\ log\ (10^{Lp1/10} + 10^{Lp2/10} + ... + 10^{Lpn/10}) \qquad (7.8)$$

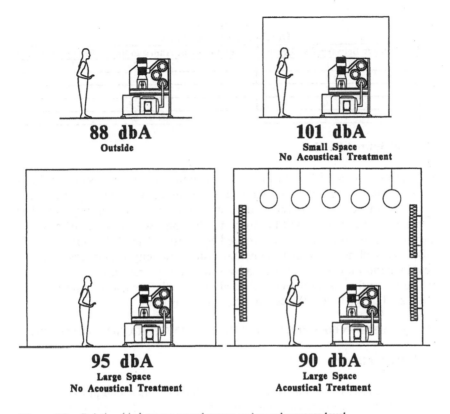

Figure 7.3 Relationship between sound power and sound pressure level.

A simpler method and as accurate as necessary for field conditions is to use the method shown in Table 7.6.

The method is as follows:

1. Determine the numeric difference between two levels to be added.
2. Find the number corresponding to the difference in the table.
3. Add this number to the higher of the two sound pressure levels to obtain the result.

Since the results of sound measurements can vary widely, in level it is always best to round off to the nearest whole number. Noise measurements are estimates at best, even though the new digital sound level meters provide the data in tenths of a dB. Field measurement repeatability of ± 2 dB is quite good. Listen to what is going on around when taking measurements. This will help make judgments about the accuracy of the data.

Table 7.6
Decibel Addition Table

Difference Between Two Levels	Amount To Be Added to the Higher of the Two
0 to 1	3
2 to 4	2
4 to 7	1
7 to 10	0.5
>10	0

B. A-Weighted Decibel

The metric used for most criteria involving hearing exposure, environmental criteria, and those levels where the subjective nature of the sound must be judged, is the A-weighted decibel, or dBA. This is a single number system that has been found to mimic the human response to a noise or annoyance, and how the ear is damaged by noise. The OSHA criteria is in dBA, as well as many environmental criteria for city ordinances, etc. In level notation it can be designated by L with a subscript a, or L_a. They can be added in the manner as described above. In the next section this will be discussed in relation to the center frequency spectrum.

V. DISCUSSION OF FREQUENCY/INTENSITY/VIBRATION/ABSORTION

A. Introduction

In Section IV, *Decibel Scale*, it was noted that the sound pressure above atmospheric pressure is called the *decibel (dB)*. The discussion in this section will focus on the other aspects of *Acoustics* that are necessary for a more complete understanding of the subject. All of the items in the title of this section combine and interact. While the dB scale denotes the level of sound, the frequency of that level describes the range of pitch over which the sound is heard. As sound spreads from a source, in an idealized spherical fashion, the intensity describes the energy over the area of that sphere spreading in terms of pressure and particle velocity at any given distance. As a surface vibrates, due to some energy input, it radiates and excites air particles next to the surface that transmit the energy on through the medium. The vibration can be controlled with isolation, damping, or changing the surface conditions.

Most materials are porous to some degree and allow the incident sound pressure to enter the surface. To a greater (or lesser) extent, the sound energy is either absorbed or reflected from the surface. The more dense the material, the more the energy is reflected, such as concrete. Materials like fiberglass are very porous and absorb most of the incident energy, depending on the frequency of the sound energy. Thus, to fully describe the sound or noise in a given environment, information about the level and frequency of the source is required.

B. Frequency

The *pitch* of a simple sound is defined as the number of vibrations the air particles make as they oscillate positively and negatively from their normal position. This is the number of vibrations made per unit of time, or *cycles per second*. The physical oscillation rate is called *frequency* and denoted by *f*. The unit of frequency is *Hertz* (*Hz*). The air particles are compressing and expanding, above and below atmospheric pressure. The expression for *f* can be assessed with the following equation:

$$f = 1/T \tag{7.9}$$

Where:

   T = Time period, in seconds, required for the air particles to move through one complete cycle

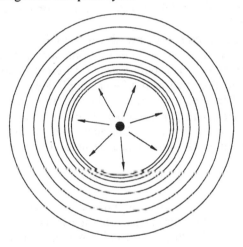

**SOUND WAVES SPREADING
FROM A SOURCE**

Figure 7.4   Simple sound wave.

The *amplitude* is the distance the air particles move above atmospheric pressure. The amplitude is determined by the intensity of the sound; i.e., how much energy has been imparted to the air particles to cause the movement. The distance between one complete cycle is the wavelength. The higher the frequency, the shorter the wavelength. Wavelength is given by:

$$\lambda = \frac{c}{f} \qquad\qquad (7.10)$$

Where:

λ = Wavelength

f = Frequency, Hz

c = Speed of sound, 1130 fps for air at 70°F

The simple sound wave is shown in Figure 7.4.

In general, the ear can distinguish between sounds that consist of periodic events that repeat on a regular basis and between sound that are aperiodic. The simplest periodic sound is a pure tone. This fluctuates sinusoidally at a fixed frequency. An example of this is a turbine compressor. This predominates based on the number of blades and the speed to produce a pure tone. For example, the fundamental frequency might be 2000 Hz, with harmonics at 4000 and 6000 Hz. Each harmonic is based on the fundamental frequency.

The noise of an engine or dishwasher is aperodic, i.e., random in nature and cannot be subdivided into a set of harmonic tones. It can, however, be described as an infinite set of pure tones of different frequencies an infinitesimal distance apart and all with different amplitudes.

The range of frequencies in the audible spectrum is from 20 to 20,000 Hz. The human ear can generally detect sound from 32 to 14,000 Hz. Some might be able to detect the full range. The audible spectrum is divided into nine contiguous parts in octave bands. These are:

31.5   63   125   250 500   1000   2000   4000   8000

Each of these can be divided by three or 1/3 octave bands. If more information is desired, the newer spectrum analyzers can divide the octave band into 12 parts or 1/12h octave bands. A more detailed discussion can be found in the Beranek reference.

C. Intensity

A sound source radiates power. Part of the power will flow through very small regions of the medium surrounding the source. A point source radiates in a spherical manner and is shown in Figure 7.5. As the spherical wave spreads, the spherical envelope increases in size and the small square area shown on the initial surface also increases in size, and the less power there is per unit area. This of course assumes no losses in the medium. The total power is the product of the intensity and the area: W = IS in watts. There is a distinct difference in the power per unit area in a plane wave; i.e., that flowing with a confined space of constant area and the spreading spherical wave from a point source. The difference is that the intensity, i.e., power per unit area remains constant as the wave travels; whereas, in a spherical wave, the area increases as a function of the square of the distance from the source. The surface area of the spherical wave is equal to 4 π times the square of the radius. For a constant power source, the power per unit area, decreases as the square of the distance from the source, because the product of the intensity and the area is equal to the constant power at the source. This is the basis of the *inverse square law* for intensity.

$$I = \frac{W}{4\pi r^2} \qquad (7.11)$$

Where:

I = Intensity, watts/m^2
W = Total acoustic power radiated, watts
r = Radius, meters

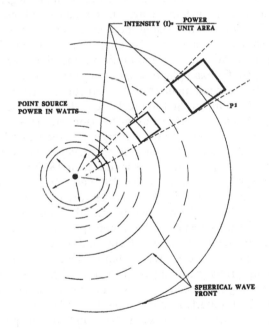

Figure 7.5   Point source radiating in a spherical manner.

D. Vibration

The discussions in this chapter on noise have omitted a common source of noise, and that is vibration. There are many sources of vibration in industry that create noise. Some examples are: blasting, forging hammers, engines, motors, punch presses, fluid turbulence in pipe and ducts, trucks, etc. The vibration of the equipment are troublesome to themselves and to anything to which they might be attached.

For example, a washing machine runs very smoothly in the spin cycle when it is properly balanced since the spinning tub is isolated from the frame; but if the load becomes unbalanced or the one leg is not solidly on the floor, then there is a great deal of vibration that is imparted into the structure and will create noise. Another example is a hydraulic pump bolted to an oil tank. The heavy oil in the tank provides inertia and resistance for the vibration of the pump; but if a light gauge control panel is bolted to the tank, then it will cause vibration and create a great amount of noise. Vibrations that generate noise require control because of one, or a combination of the following: complaints from building occupants that have real or perceived discomfort, lack of equipment maintenance, structural or mechanical failure.

The model for analysis of a single system is a single degree of freedom system where a mass vibrates in a constrained manner and only in the vertical direction. This means decreasing the transmission of the vibratory motion from one structure to another. Isolation then means placing a flexible material along the path to reduce and/or absorb the vibratory energy. This can be a rubber pad, for example, or a spring isolator can be used to support, as well as isolate the equipment from the structure. Flexible connections can also be used in a pipe or duct to break the path. The reduction of vibration energy depends on the efficiency of the isolator. This is a function of the driving frequency of the vibratory energy, and the natural frequency of the isolator when deflected. The driving frequency is based on the equipment speed. For example, a motor that rotates at 1800 rpm has a driving frequency of 30 cycles per second or Hz (1800/60). If the motor is driving a pump that has 7 vanes on the impeller, another driving frequency is the vane passing frequency. This is the number of times the vanes pass the cut-off in the pump case. This frequency is the pump rpm divided by 60 times the number of vanes on the impeller, or 210 Hz for this example.

The same approach is used with a fan or another piece of equipment when determining primary driving frequencies. These are known, in general, as blade passing frequencies (BPFs). For vibration isolation, the lowest speed is the most important, but if a pie from the pump is attached to a wall, the structure-borne vibration will radiate, and both the pump speed and BPF will become important. The details of vibration isolation can be found in various manufacturers' catalogs and texts.

VI. MONITORING TECHNIQUES

Noise monitoring is necessary for the following reasons:

- To identify employees who should be given audiometric tests and involved in a hearing conservation program.
- To identify those employees for whom hearing protection is mandatory under the provisions of the current OSHA regulations.
- To determine the amount of attenuation that hearing protectors need to provide.
- To instruct both employers and employees in the degree of noise hazard.

Monitoring refers to a program that includes calibration, measurement, and calculation of the noise dose, or noise exposure. This requires the employer to initially determine if any employee's noise exposure is over 85 dBA TWA for an 8-hour day, the action level that triggers a hearing conversation program. Other factors that should be used in determining if monitoring is required are: employee complaints about noise; difficulties in communication at a distance of 2 feet; any measurement with a sound level meter or dosimeter that indicates a level at, or above 85 dBA. If the employer makes a determination that employees are at 85 dBA TWA or more, then noise exposure monitoring must be carried out.

When a determination has been made that the levels are at 85 dBA TWA or more, the measurements must be repeated every 2 years. Measurements must also be made within 60 days of a process or equipment change or personnel assignment. Workers must be notified of the results within 21 days of the measurements.

A. Monitoring Procedures

The noise monitoring or measurements should be collected in several stages to define the nature of the problem. The first step is to conduct a general plant-wide survey. This can be done with a simple Type II sound level meter, placing the noise data on a plant layout at column lines. This will help define those areas of high noise, over 85 dBA, and those areas of hazardous noise, over 90 dBA. The grid data can be placed in a computer program to produce noise contours. This does allow the noise "to be seen." An example of this is shown in Figure 7.6. Once this has been completed, a more detailed survey should be undertaken in areas identified to have high and hazardous noise.

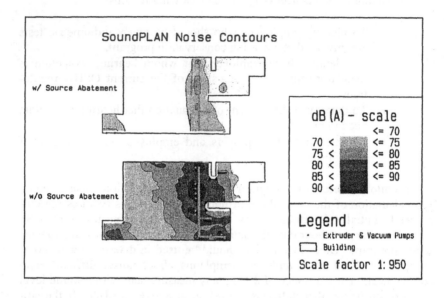

Figure 7.6 An example of computer-generated noise contours.

The area survey should be carried out in more detail within the areas identified and at employee workstations. The result of this will determine which employees need to be monitored with noise dosimeters. The dosimeters are placed on the employees (refer to Figure 7.7 for illustration of a typical dosimeter) to monitor the noise levels for the work shift. The microphone is placed near the ear, usually on the shirt collar. The monitor is worn on the belt for the entire shift. The results are transcribed at the end of the shift. If the noise levels vary, more than 1 day of monitoring should be completed to be sure that a valid number of samples are obtained. If a sound level meter is used, it should be placed near the head when taking samples. The third type of measurement is to place a monitor in the area at a fixed location. The first two methods can be used for assessing the employees noise exposure and is acceptable to OSHA. The area method is not acceptable for monitoring employee noise exposure, but can be used for planning noise control strategies. The following discussion describes the first method in detail.

B. Employee Noise Monitoring Discussion

In any physical measurement where repeated observations of a statistically controlled sample are made, fluctuations in the measured variable are usually found. In the case of industrial noise, fluctuations in the noise sources, as well as employee movements throughout the area, cause variation in noise exposure. Where these variables are random, statistical methods may be applied to the noise exposure in order to analyze the results.

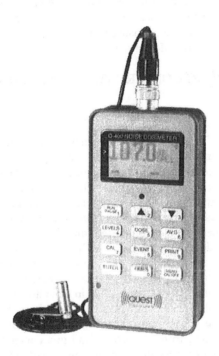

Figure 7.7   Dosimeter used for employee noise exposure monitoring (Courtesy: *Quest Technologies*, Oconomowoc, WI).

Noise exposure monitoring has been performed for quite some time. This has been based upon the Department of Labor Occupational Noise Exposure Standard 29 CFR 1910.95, Occupational Noise Exposure. Under this Standard, there was no requirement for noise exposure monitoring, but the use of the noise dosimeter has been the most expedient way to determine noise exposure. On January 16, 1981, an amendment to this Standard was published in the Federal Register, defining requirements for a hearing conservation program. As a part of this addition to the Standard, there is a requirement for employee noise monitoring to determine those individuals who should come under the hearing conservation program when their time-weighted average for an 8-hour exposure is 85 dBA or greater. The purpose of this discussion will be to outline a procedure for analyzing the results of the noise dosimeter measurements.

It has been OSHA's practice to monitor employees for one 8-hour period to determine their noise exposure. In some instances, this is entirely adequate; however, in other instances where the noise levels can be quite variable or the employees' work practice requires them to be exposed to a variety of noise levels throughout the work period, one noise dosimeter measurement may not adequately describe the employees' noise exposure accurately. The use of some basic statistical methods can be quite useful in analyzing a number of noise exposure measurements to determine if enough samples have been taken to describe the noise environment.

VII. HEARING CONSERVATION AND PROTECTION

A. Introduction

The purpose of implementing and maintaining an effective hearing conservation program (HCP) is to prevent permanent noise-induced hearing loss resulting from occupational noise exposure. The goal of most hearing conservation programs is compliance with the regulations. *But is this enough?* The best approach may be to pursue hearing loss prevention. This should be viewed as an overall business and performance issue. Hearing loss worker's compensation claims affect the "bottom line," and if made a part of loss prevention department, as well as safety and medical departments, there can be some control over these claims. Hearing health and hearing loss prevention are just as important a part of the business as are production and maintenance. If good hearing health prevents an accident, this will protect employees and save money. For example, the location of a sound in a plant is very important. If an employee cannot tell the direction a waning sound is coming from, this could lead to an accident. The ability to quickly and accurately assess the direction and level, and any message the warning is providing is critical for safety. Hearing health is an important part of this, as is the noise control and message system. The information in this section provides an overview for hearing conservation. When followed, this will assist in protecting employee health. To ensure compliance with the noise exposure regulations for a facility, the HCP should contain the following components:

1. Noise surveys
2. Education and sraining
3. Engineering and/or administrative controls
4. Hearing protection
5. Audiometric testing and evaluation

Table 7.4 contains the permissible noise limits (PNLs) that can be used to determine employee noise exposures. The PNLs are based on the best available information from industrial experience. The PNLs are sound pressure levels and exposure durations that represent conditions under which it is believed nearly all workers may be repeatedly exposed without adverse effect on their ability to hear and understand normal speech. The values should be used as a guide in the control of noise exposure and, due to individual susceptibility, should not be regarded as fine lines between safe and dangerous levels. These values apply to total duration of exposure per working day regardless of whether this is one continuous exposure or a number of short-term exposures. The equation and some example calculations of noise exposure are discussed in Section VI, *Monitoring*.

All employees exposed to noise at or above 50% of the PNL need to be included in the HCP. It is recommended that all personnel with borderline noise exposures, or those with infrequent exposures at or above 50% of the PNL, be included in the HCP (for example, maintenance personnel). For an 8-hour workday, a 50% noise exposure is equivalent to 85 dBA. This level is referred to as the time-weighted average (TWA) noise exposure. In those cases where work shifts differ from 8 hours, the TWA noise exposure for inclusion in the HCP shall be adjusted accordingly (e.g., 82 dBA is equivalent to 50% of the PNL for 12-hour workdays, etc.).

B. Noise Surveys

The first step is to assess the situation in order to define the problem. Three types of surveys can be used as the basis for the HCP: preliminary walk-through, area survey, and personnel monitoring.

1. *Walk-through survey*
The preliminary survey is a needs assessment. It provides an indication as to which areas or job activities need a more detailed survey. This survey involves a walk-through of the plant or installation using a sound level meter. Typically, if an area has any sound levels above 85 dBA (82 dBA where employees work 12-hour shifts), then it should be scheduled for an exposure survey. Major equipment changes or other work-site modifications may necessitate a new sound level meter survey. Otherwise, periodic spot tests to validate the previous survey(s) will be sufficient.

2. *Area sampling*
Area monitoring is the method of sampling with a sound level meter to determine the overall sound level at various locations throughout a facility. When indicated, personal monitoring should be conducted. However, if personal monitoring is impractical, the employees to be included in the HCP can be identified through area sampling. Area monitoring is the least accurate procedure from an exposure standpoint, yet it is the most conservative as far as including workers in the HCP. For purposes of identifying employees to be included in the HCP, a "worse-case" method is necessary. In

this situation, employees that work in areas with sound levels above 85 dBA (82 dBA for 12-hour shifts) would automatically be included in the program. The maximum area sound level is used to evaluate which hearing protectors are effective enough to attenuate the noise to 85 dBA or less. Methods for evaluating hearing protectors are described later in this section.

All monitoring and area sampling results are to be maintained for 30 years (per federal regulation). The medical facility or contractor service conducting the audiometric exams should be informed of the employee's most recent noise exposure assessment. All employees must be informed of the noise exposure determined for their job, along with an explanation of interpretation of the results.

3. Exposure survey

The results of the preliminary survey will indicate which employees should be included in the noise exposure survey. The purpose of this type of survey is to identify all employees to be included in the HCP and to provide data that will enable the safety and health professional to select the proper hearing protectors. A second objective is to determine actual or representative noise exposures for all employees whenever possible.

The sampling method to be used for the exposure survey should be flexible to allow the most effective, accurate, and efficient procedure to satisfy the objectives of the general survey. This survey can be completed with personal monitoring using a sound level meter and/or a noise dosimeter.

a. Personal monitoring with a sound level meter

The most accurate method of identifying employees exposed at or above 50% of the PNL is personal monitoring. This type of monitoring can be completed using either a sound level meter and a stopwatch, or a noise dosimeter. This type of survey requires the surveyor to obtain readings near the employee's ear (approximately 6–12 inches). Because sound level meters are only capable of taking spot readings at the instant measured, a sampling strategy needs to be developed. This strategy should include sufficient sound level readings taken at various times and locations for different noise levels. It is important that all varying noise levels encountered during the work shift be accounted for when taking measurements. The stopwatch is used to measure the time an employee being monitored actually spends at one location or is exposed to a specific sound level. After this information has been collected for the employee's workday, it is then used to calculate the noise exposure.

b. Noise dosimeters

The noise dosimeter is an electronic device worn by the employee. The dosimeter automatically averages varying sound levels during a given period (8-hour workday, etc.). This method is the most accurate procedure available for the assessment of noise exposure. Regardless of the personal monitoring method chosen, it is not necessary to sample every employee.

The use of representative monitoring is an efficient method which minimizes the burden of determining employee noise exposures. This procedure enables the surveyor to monitor a number of individuals from a group of employees, provided this group engages in a similar kind of work with exposure to similar noise sources. The result of the representative employees (noise exposure evaluations) is used to determine the noise exposure for that particular work group. To help ensure that the data is truly representative of exposure, it is recommended that monitoring be conducted at least twice before noise exposure assignments are made, depending on the statistical data evaluation as outlined in Section VI, *Monitoring*.

C. Hearing Protection Devices

Hearing protection devices are required to be worn by all workers whose noise exposure equals or exceeds a specific criterion level. There are situations when engineering and/or administrative control measures are neither feasible nor adequate, or they are being implemented. In those situations, hearing protectors are required until employee noise exposures can be reduced to a safe and acceptable level. Figures 7.8, 7.9, and 7.10 show three types of hearing protectors.

Employers must make hearing protectors available to all employees exposed at or above 50% of the PNL (action level) at no cost to the employees. Hearing protectors must be replaced free of charge when necessary at no direct cost to the employee. Employees must be presented with two or more different types of models of hearing protectors when making their selection. However, the employee is not permitted to select the size of protector they desire. The selection of the proper fit or size should be administered by an individual (e.g., nurse, physician, safety engineer, etc.) who is properly trained in sizing protectors. Employees must be shown how to use and care for their protectors, and must be supervised on the job to ensure that they continue to wear them correctly. This is also covered in training sessions.

The mandatory use of personal hearing protection devices is required for all affected employees whenever any of the following exist:

1. All workers that are exposed to or above 50% of the PNL (85 dBA for an 8-hour workday), and have experienced a standard threshold shift (STS).
2. All workers that are exposed to or above 100% of the PNL (90 dBA for an 8-hour workday).
3. "Hearing protection required" for all areas designated by the employer regardless of the amount of time spent in that area.
4. Any job that requires exposure to suspected elevated noise levels which has not been measured, and the job is not a recurring one (e.g., temporary construction or maintenance projects using air hammers, explosive guns, generator tests, etc.).

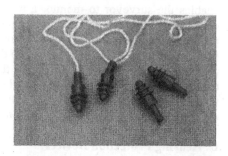

Figure 7.8   Ear plugs with
attached cord.

Figure 7.9   Ear plug inserts.

Figure 7.10   Ear muffs.

1. *Warning signs*

Management should ensure that all areas requiring the use of hearing protection devices are properly designated as such. When signs are used to designate an area, the following wording is recommended: "High Noise Area–Hearing Protection Required."

2. *Selection of hearing protection devices*

An important factor in establishing an effective HCP is the requirement by management that hearing protection devices (HPDs) be worn, and be worn properly by employees. Another critical element of a successful program is the proper selection of HPDs suitable for the specific noise environments in which they will be utilized.

The selection of HPDs should be such that the hearing protector attenuation will reduce an employee's noise exposure or TWA to 85 dBA or less. The most convenient method is to use the Noise Reduction Rating (NRR) developed by the Environmental Protection Agency. When using the NRR, the Hearing Conservation Amendment to the Noise Standard, 29 CFR 1910.95 (c) mandates the use of one of the following methods for assessing HPD adequacy:

- When using a dosimeter that is capable of C-weighted measurements: first obtain the employee's C-weighted dose for the entire work shift, and convert to a TWA; then subtract the NRR from the C-weighted TWA to obtain the estimated A-weighted TWA under the HPD.
- When using a dosimeter that is capable of A-weighted measurements: first convert the A-weighted dose to a TWA; subtract 7 dB from the NRR; then subtract the remainder from the A-weighted TWA to obtain the estimated A-weighted TWA under the HPD.
- When using a sound level meter set to the A-weighting network: obtain the employee's A-weighted TWA; subtract 7dB from the NRR; and then subtract the remainder from the A-weighted TWA to obtain the estimated TWA under the HPD.

It is important to remember that the calculated attenuation values used for the HPD to determine the NRR reflect realistic values only when the HPDs are properly fitted and worn.

Instead of using the NRR, the regulation allows employers to evaluate the adequacy of HPD attenuation by using one of the three methods developed by the National Institute for Occupational Safety and Health (NIOSH). These methods are described in the *List of Personal Hearing Protectors and Attenuation Data*, HEW Publication No, 76-120, 1975, pages 21-37.

D. Education and Training

A comprehensive employee education and training program is a critical component of an effective HCP. Employees will be better motivated to actively participate in the HCP and to cooperate by wearing their hearing protectors when the reasons for having a hearing conservation program are clearly explained and they understand the need to protect their hearing. All employees included in the HCP must be trained at least annually, and the training must include at least the following aspects:

1. The effects of noise on hearing.
2. The purpose, advantages, disadvantages, and attenuation of various types of hearing protectors.
3. The selection, fitting, and proper use and care of their protectors.
4. The purpose and procedures of the audiometric tests.

Training may be conducted in one session or as many separate sessions as necessary to cover all the topics. Each attendee to a training session should be required to acknowledge by signature that they received training, and this record should be maintained. In addition, the training materials (films, slide-tape program, etc.) used and a brief outline of the topics covered should be kept on file for documentation purposes.

E. Recordkeeping

The OSHA regulation requires that records of all noise data must be maintained for 30 years. All of the data must be made available to the employee and any outside contracting service used for audiometric testing. Any changes in noise levels must be monitored and made a part of the permanent records. All audiometric data, health history, off-the-job activities, etc. are also made a part of this record.

F. Audiometric Testing

This phase of the HCP is the true determination of the overall effectiveness of the program. Because the objective of the HCP is to prevent occupational hearing loss, the only mechanism available for measuring its success is through analysis of the audiometric test results. Analysis of the audiometric data provides several checks for the HCP effectiveness, such as:

1. Detecting significant threshold shifts in employees' hearing ability during the course of their employment.
2. Providing a record of an employee's hearing acuity.
3. Evaluating the effectiveness of engineering noise control measures by measuring the hearing thresholds of employees working near the treated equipment.
4. Identifying areas in the facility that require an engineering noise control study to help justify noise control expenditures.
5. Identifying weaknesses in the hearing protection program such as inadequate hearing protectors, lack of proper use, and/or ineffective education and training of employees.

They must have a baseline audiometric test, the baseline audiogram against which future hearing tests are compared for hearing conservation purposes. These employees must also have a periodic test, at least annually, after obtaining a baseline audiogram. It is best if these employees are away from workplace noise for at least 14 hours prior to their audiometric test. Off-the-job activities, such as shooting (refer to Figure 7.11) must also be noted; and if there is an indication that they have been exposed to high noise during the 14-hour quiet time, there might be a question about the validity of the test results. If the audiometric tests are to be conducted during the employees' work shift, then each employee to be tested should be instructed at the beginning of their work shift to wear hearing protection while on the job in "high-noise areas."

The medical professional responsible for the supervision of the audiometric testing program should determine the follow-up procedures necessary whenever an individual employee exhibits a standard threshold shift. Changes in hearing acuity that exceed an average of 10 dB or more at 2000, 3000, and 4000 Hz in either ear, relative to the baseline audiogram, are considered to be a standard threshold shift (STS).

When deemed practical, it is recommended that all employees included in the HCP have an audiometric examination prior to leaving, or retirement from the company, if their last audiometric test preceded the retirement date by more than 6 months.

Figure 7.11 Shooting is an off-the-job hobby that can contribute to hearing loss.

G. Audiometers

An audiometric examination (see Figure 7.12) is required to be performed using an audiometer that conforms to the requirements for wide-range, pure tone, discrete frequency audiometers prescribed by the *American National Standard Specifications for Audiometers*, ANSI S3.6-1969 (R-1973 or latest revision). If a pulsed tone audiometer is used, the on-time of the tone shall be at least 200 milliseconds. The instrument used must be either a manual audiometer or any other audiometer testing system of equal or greater accuracy and effectiveness.

Figure 7.12   Conducting an audiometric examination.

Calibration is required for an accurate test (see Figure 7.13). There are two types of calibration: *acoustical* and *biological*. Audiometer acoustic calibration is required to be checked at least annually to determine that the audiometer is within the tolerance permitted by ANSI S3.6-1969 (R-1973 or latest revision). This procedure should only be attempted by a properly trained and equipped individual; usually, the audiometer manufacturer can complete this calibration.

A biological calibration shall be made prior to each day's use of the audiometer. This procedure consists of:

1. At least one person must be tested that has a known stable audiometric curve that does not exceed a 10 dB hearing threshold level at any frequency and comparing the test results with the known curve.
2. This subject's response must be registered to distortions and/or unwanted sounds from the audiometer.

Figure 7.13 Calibration of an audiometer.

Whenever the results of the "daily use" biological calibration indicate hearing level differences greater than ±5 dB at any frequency, the signal is distorted; or if there are attenuator or tone switch transients (e.g., clicks, noises, hums, etc.), the audiometer shall be removed from service and subjected to an acoustical calibration prior to any further testing. Only after the problem with the audiometer is corrected to within permitted tolerances can it be put back into service.

H. Background Noise in Audiometric Test Areas

The area designated for audiometric testing shall be as free from noise and vibration as much possible. The sound pressure level in any octave band when measured in the audiometric booth, or room in its absence, where subjects are actually tested must not exceed the values given below:

Octave Bands Hz	500	1000	2000	4000	8000
dB Levels	21.5	29.5	34.5	42	45

Documentation is very important. Audiometer calibration records and background noise levels in test booths or rooms should be maintained with the audiometric test results. When contract services are used, these records must be provided by them and be kept on file.

Arithmetic mean is used as the measure of the central tendency of the measured distribution, and the standard deviation is the measure of dispersion about the mean. The calculated confidence limits measure the degree of certainty of the calculated mean. This will provide the range within which the noise exposure should fall 90% to 95% of the time, depending upon the degree of confidence that one wishes to use.

Arithmetic mean is simply the arithmetic average of a set of individual measurements, and is given by:

$$\overline{X} = \frac{\sum_{i=1}^{N}(x_i)}{N} \tag{7.12}$$

Where:

$\overline{X}$ = Arithmetic mean of levels
N = Total number of measurements in the set
X_i = Individual levels

The standard deviation, S, is the root mean square of the deviations from the mean and is given as:

$$S = [\sum_{i=1}^{N} (\overline{x} - x_i)^2]^{0.5}/N \tag{7.13}$$

The 90% confidence limit is the limit which the arithmetic mean may be expected to occur nine times in ten in a series of measurement samples. The confidence limit is equal to the standard deviation times a constant (*a*)s. The constant, *a*, is dependent upon the number of measurements made. Table 7.7 gives the values of *a*.

In order to simplify the analysis portion of the data, a sample table is included that can be utilized for summarizing and calculating the results. The table is set up so that job classifications can be entered and the equivalent noise level L_{OSHA} can be entered after computing this from the noise exposure.

When the work shift noise exposure is composed of two or more periods of noise at different levels, the total noise dose over the workday is given by:

$$D = 100 \, (C_1/T_1 + C_2/T_2 + \ldots + C_nT_n) \tag{7.14}$$

Where:

C_n = Total time of exposure at a specific noise level
T_n = Reference duration for that level as given by OSHA Table G-16a

Table 7.7
Factor "a" for Calculating
% Confidence Limits [X ± (*a*)s]

Sample Size N	90% Confidence Limit "a" Value	95% Confidence Limit "a" Value
2	4.50	9.00
3	1.70	2.50
4	1.15	1.59
5	.93	1.23
6	.81	1.04
7	.73	.93
8	.63	.83
9	.62	.76
10	.59	.72
11	.55	.69
12	.52	.63
13	.49	.60
14	.48	.58
15	.45	.55
16	.44	.53
17	.42	.51
18	.40	.49
19	.39	.48
20	.38	.47
21	.37	.45
22	.36	.44
23	.35	.42
24	.34	.42
25	.34	.41
26	.33	.40
27	.32	.39
28	.32	.38
29	.31	.37
30	.30	.36
40	.26	.31
60	.21	.25
120	.15	.18
N >120	$\dfrac{1.645}{\sqrt{N}}$	$\dfrac{1.96}{\sqrt{N}}$

The 8-hour time-weighted average sound level (TWA), in decibels, may be computed from the dose, in percent, by means of the following equation:

$$D = 16.61 \log_{10} [(D/100)/(480/T)] + 90 \qquad (7.15)$$

Where:
   D = Noise dose as measured by the dosimeter
   T = Time of noise measurement in minutes

The purpose of this last equation is to provide a means of computing the time-weighted average into equivalent dBA for summation in the statistical analysis of the data.

It should be noted from Table 7.7 that the more samples that are measured, the tighter the range will be, depending upon the standard deviation. The standard deviation for noise measurements should not exceed 2, and preferably should be less than 1; however, it is recognized that because the limitation of the dosimeter is ±2 dBA, that a standard deviation between 1 and 2 is satisfactory. In some instances, only four measurements may be required to have a standard deviation of 1 or less, while in other situations, it may require 8 measurements to achieve the standard deviation. The standard deviation will dictate total number of samples that will be required to reduce the confidence limits to an acceptable range. The simplest manner to compute these results using a statistical analysis, is with a hand calculator that has statistical analysis capabilities.

An example of the results of some noise exposure measurements are shown in Table 7.8, along with explanatory notes. A review of the data in this manner allows one to determine how much more data may be required to adequately describe the noise exposure of certain job classifications .

As explained in the amendment to the OSHA Noise Standard CFR1910.95 Noise Exposure, paragraph E, Monitoring 2, selected employees can be monitored without, in fact, measuring everyone in the area. This will help to reduce the need to monitor all employees and expedite the process.

The Noise Dosimeter Data Sheet will provide the required information for the individual employee that is being monitored. It is of importance to obtain this information for documentation and recordkeeping purposes.

On the lower portion of the sheet, sound level measurements can be recorded along with time and location. In particular, if an employee is mobile, a "walkaround" series of measurements should be taken to determine the items that make up the noise exposure. If an employee is at a fixed location, the data may show the variability of noise levels. This type of data will assist in evaluating noise exposure since most dosimeters provide only a single exposure result. It is recommended that at least eight measurements (more if possible) be taken throughout the day.

Table 7.8
Example Data Analysis

Area	Job Classi-fication	No. Samples	Mean dBA	Std. Dev. dBA	90% CI ±dBA	90% CI ±dBA	
A	1	5	80.7	4.9	4.6	85.3	76.3
	2	4	88.3	1.5	1.7	90.0	86.6
	3	6	88.6	3.5	2.8	88.6	85.7
B	1	5	88.4	3.1	2.9	91.3	85.5
C	2	3	90.1	1.3	1.2	92.3	87.9
	3	3	87.8	1.3	2.2	90.0	86.0
D	1	4	86.8	7.0	8.1	94.9	78.7
	2	4	94.2	0.7	0.7	95.0	93.5
	3	8	91.9	2.6	1.6	93.5	85.5
E	1	3	86.3	0.5	0.8	87.1	85.1
F	1	3	86.5	0.9	1.44	87.9	85.1
G	1	4	84.3	1.5	1.73	86.0	82.6
	2	4	86.8	0.5	0.6	87.4	86.2
	3	6	88.4	2.5	2.0	90.4	86.4
	4	6	86.0	4.1	3.4	89.4	82.6

NOTES:
1. All data taken with Type II Instrument GR 1954-9707 Noise Dosimeter for approximately 6 hours extrapolated to 8 hours.
2. All mean dBA levels are L_{OSHA} converted from % exposure dosimetry data.
3. Some data samples are not sufficient to provide a 90% confidence internal, while others are.
4. The results of this data set indicate that 9 out of 10 times the L_{OSHA} value should fall within the significant range shown. More data is required in some areas.
5. The exceptions to Note 4 are for Area D, Job Classification 1, the unit was shut down for maintenance turnaround before sufficient sample could be obtained, and Area A, Operator 1, the levels are 85 dBA or less.

Noise Dosimetry Monitoring Summary Data Collection Form

Dept.	Date	Job	Samples	% Exp.	Time (min.)	L_{OSHA}

Calculations	Results
Samples	
X	
Mean	
Std. Dev.	
"a"	
90 or 95% C.I.	
Significant range	

NOISE DOSIMETER DATA SHEET

Date:_____   Noise Dosimeter Model: _____

Name:_____   SS#:_____

Location of Person: _____

Area or Dept:_____   Job Title:_____

Monitor Make/Model:_____   Monitor S/N:_____

Calibration: *Before*_____   *After*_____

%Exposure:_____   LOSHA: _____ dBA

Time:_____   Min._____

Allowable Level Exceeds 115 dBA

TIME	LOCATION	dBA	TIME	LOCATION	dBA

NOTES:

Table 7.9
Typical "Loud" Industries and Noise Levels

Industry Type	Typical Equipment	dBA Range
Metal Working	Punch presses - 35 ton	95–105
	Punch presses - 400 ton	100
	Forging hammers - impact	115
	Swagging machines	110
	Cold headers	105
	Can and lid mfg equip	97–104
	Machining	97
Wood Processing	Head rigs	108–113
	Rip saws	97–105
	Planers	111–118
	Molding machines	97–101
Petro-Chemical Plants	Gas compressors trains	105
	High HP electric motors (1000)	103
	Small HP electric motors (100 to 500)	97–108
	Pipe flow - steam or gas	105-117
	Furnaces/heaters	95–100
	Gas turbine drives	100–110
	Steam turbine drives	98–105
Power Plants	Steam turbines/pipe flow	105–115
	Forced draft, induced draft, & primary air fans	95–120
	Air compressors	105–110
	Coal screening & sizing	100–110
Steel Production Plants	Furnaces	105
	Extrusion lines (mat'l hdlg)	95–100
	Cut-off saws	115
Foundries	Furnaces	95
	Sand casting	98–105
	Shake-out	115–120
	Cut-off saws	115
Cement Plants	Ball mills	95–100
	Compressor/blowers	100–105
Off-Shore Platforms	Engine rooms	110–120
	Drill floor	95–115
	Mud pumps	95
	Process areas	95–110
	Shipping pumps	95–105

Source: Author's Private File Data

VIII. "LOUD" INDUSTRIES/EQUIPMENT

"Loud" industries could arguably be considered with noise levels above the 95–100 dBA range or over 100 dBA continuously. The type of noise from these industries can be continuous, intermittent, or with high-impact noise. Table 7.9 illustrates these industries with typical equipment types, and the expected range of noise levels. This information is taken from the author's experience over the past 35 years in noise control engineering.

There are many more industries for which data could be provided. This tabulation is a general overview and range of levels. The approach for noise control should consider the frequency content of the noise source as well. Some industries, in certain areas, such as power plants, foundries, and offshore platforms, to name but three, may not have feasible engineering controls available, and hearing conservation programs must be used to prevent hearing loss.

IX. ADMINISTRATIVE CONTROLS

The OSHA 29CFR1910.95 regulation allows administrative means to control employee noise exposure. This provides a method of moving the employees from a noisy area to a quiet area to control the noise exposure levels. There are other interpretations of this as well. In addition to modification of an employee's schedule to reduce the TWA, administrative controls should also include establishing adequate equipment maintenance programs, administrative planning for noise exposure, defining noise limits for new equipment being purchased through a definitive purchase specification, and noise control planning for reducing employee noise exposure.

A. Controlling Employee Work Schedule

The daily work schedule can be changed to limit the noise exposure. This may be possible in some limited cases. One example is an employee working on a vinyl grinder for 4 hours with a noise level of 94 dBA, then moved to a job where the noise level was below 80 dBA. A level of 94 dBA allows a time of exposure of 4.6 hours. This provides a TWA of 89.6 dBA. This still requires the employee to be in a hearing conservation program and wear hearing protection. Other problems with this type, or any other type, of employee job change is the influence of unions, production schedules, cross training, etc. This type of administrative control is difficult and may not get to the heart of the matter: to protect the hearing of an individual employee if he or she is susceptible to hearing loss.

Another example that had an effect on the TWA values was to move an employee out of the high-noise zone of a concrete block machine operation to a quieter area below 90 dBA. The operator could still have control over the machine and see all aspects of the operation. This approach, again, is a special situation and may not work in all cases.

B. Maintaining Equipment as Administrative Controls

An *Acoustical Maintenance Program* can provide the elements that check the equipment noise levels, as well as point out the potential problems in overall equipment maintenance. Tight, well-maintained equipment operate quieter. There are well-established *Predicative Maintenance Programs* that monitor equipment bearing vibration levels and anticipate the time to change them to prevent unscheduled downtime. This always occurs at inopportune times and costs more that necessary.

An Acoustical Maintenance Program consists of the following elements:

- Baseline noise survey of equipment from which all future data can be compared.

- Periodically (bimonthly, if possible) conduct a maintenance noise survey of production equipment for comparison to the baseline. This is especially important in high-speed equipment that can become loose and more noisy.

- Conduct periodic visual inspections of the equipment for loose parts, etc. Maintenance personnel should be familiar with these procedures and know what to look for in areas of noise producing items. The operator who is most familiar with the equipment can tag the time to be repaired. A noise I.D. tag can be used to identify the items to be repaired. For example, air leaks are a major source of noise and cost money if left unattended. These too can be easily repaired as a part of the *Acoustical Maintenance Program.*

- Establish procedures for noise measurement of new equipment that should be followed in all cases for uniformity. This will provide consistent data that can be followed for all equipment in the plant.

- Predicative Maintenance for cost control due to unscheduled downtime.

C. Administrative Control Through Specification

The purchase specification that states "must comply with OSHA noise regulations" is not an acceptable method to direct a prospective supplier. This leaves too much to the imagination and there are too many unknowns. If a manufacturer is serious about the overall noise control program and intends to mange it properly, then a good and enforceable purchase specification is required. This can be established for new equipment, for remedial equipment purchases, and for new facilities. Each one of these has a different variation, but the result should be the same: *prevent hearing loss, meet the noise management goals, and comply with the OSHA noise regulations.* The specification approach must be integrated into the noise management process and be one of the cornerstones.

D. Administrative Control Through Planning

The noise control planning process can best be described as *Noise Control Management*. Any manufacturing facility conducts planning to assess future production and capacity requirements. A component of this can be noise control. Not only with existing equipment, but with expansion. This methodology provides a means of developing priorities using the noise survey data, exposure data, and cost of the noise control installation. The projects can help establish the annual noise control budget and be arranged to fit within this budget.

E. Summary

Administrative control can be a useful tool to control TWA noise exposure if it is thought of in broader terms, other than rearranging work schedules. Noise control planning and maintenance should be included in the Administrative Noise Control process.

X. ENGINEERING CONTROLS

The most direct method of reducing employee noise exposure is through implementation of engineering controls. Engineering controls should always be carefully considered whenever they are deemed feasible. Management should take into consideration the existing technology, economic factors, benefits, and practicality when evaluating the implementation of any controls. Another important factor will be the magnitude of the existing noise environment, which is determined by a comprehensive survey of work exposure and area sound levels.

Engineering controls involve the initial design or retrofit of existing equipment, treatment of the paths along which sound energy travels, and/or isolation of the worker(s). There are many books that provide more details about engineering controls.

A noise control engineering solution can be evaluated as one or combination of the following:

- Noise source
- Noise path
- Noise receiver

The following are examples of several different approaches to engineering controls.

A. Noise Source

The maintenance condition of a piece of equipment can result in a situation that is very noisy or much less noisy. The replacement or adjustment of worn and loose, or unbalanced parts of the machine may be required to reduce noise levels at the source. Also, lubrication of machine parts, the use of cutting oils, and sharp tools will keep noise sources at a lower level.

1. *Machine substitution*

Substitution of machines such as a rotating shears for square shears, hydraulic for mechanical presses, belt drives for gears, underwater pelletizers for standard pelletizers, etc. represent examples. Another example of source controls is to specify quiet electric motors. High-efficiency motors can be obtained with noise levels 3 ft. from the motor of 80 to 85 dBA. Typical motors found in process plants can range from 100 to 110 dBA.

2. *Process substitution*

Substitution of processes is another type of source control. These include examples such as compression for impact riveting; welding for riveting; hot for cold work; and pressing for rolling or forging. More examples of source control concern vibrating surfaces, such as reducing the driving force of vibrating surfaces. This may be minimized by reducing the forces; lowering rotational speed; or isolation of the surface from the source of vibration. The response of vibrating surfaces may be reduced by damping, further support of the piece, moving the vibrating part off the machine, increasing the stiffness of the material, or increasing the mass of vibrating members changing size to change resonant frequency. The sound radiation from the vibrating surfaces can be reduced by reducing the radiating area, reducing overall size of the piece, and perforating the surface.

3. *Fan blade substitution*

Fan blades can be designed to reduce turbulence. A good example of this is large fans for cooling towers. More efficient and aerodynamically designed fan blades can reduce noise levels by as much as 10 dBA. Another example is to substitute large low-speed fans for smaller high-speed fans. Reducing the velocity of fluid flow (air) when practical, increasing cross-section of streams, reducing the pressure, or reducing air turbulence are all methods of source control in fluid flow mediums.

4. *Other substitution control strategies*

Other source control not normally thought of are purchase specifications and reducing excessive plant noise levels through the initial planning stages for new facilities. The inclusion of noise specifications in purchase orders can be used successfully to obtain quiet equipment. Vendors supplying machinery and equipment should be advised that specified low noise levels will be considered in the selection process. Suppliers should be asked to provide information on the noise levels of currently available equipment.

B. Path Control

This method considers the path from the source to the receiver. Examples of these are barriers, silencers, vibration isolation, flexible connections, etc.

Barriers are used to great advantage, but are limited in their ability to reduce noise because of what they are: a shield. Typical reduction can be achieved from 10 to 15 dBA. If the source is low frequency in nature, then a barrier will be less effective at reducing noise intensity, than if the source is a high frequency noise. Refer to Figures 7.14 and 7.15. A typical equation for barriers will illustrate this point:

$$\text{Noise Reduction} \;=\; 10 \, \text{Log} \left(\frac{20 \text{H}^2}{\lambda(\text{R})} \right) \qquad\qquad (7.16)$$

Where:
   H = Height of barrier
   λ = Wavelength
   R = Distance to receiver from barrier

The barrier must be close to the source or receiver to be effective. The materials for barriers can be earth berm, concrete, wood, perforated metal panels, high-density plastic modular panels, or loaded vinyl curtains. Recycled materials such as automobile tires are also being used in barriers. Virtually any material can be used that has the mass and can serve as a shield.

Noise through solid connections can be reduced by using flexible mountings; flexible sections in pipe runs; flexible shaft couplings; fabric sections in ducts; and resilient flooring. These are also path controls that can be employed.

Vibration isolation can be employed if equipment is also a means of breaking the path. Isolators are neoprene pads or mounts, springs (See Figure 7.16) or pneumatic mounts. These work very well to reduce the energy into the structure, while reducing the overall radiated noise energy. The lower the equipment's rotational speed, the more deflection is required for good isolation efficiency. For example, an 1800 rpm motor might only need a ½" deflection from a neoprene pad, whereas a 500 rpm fan might need a 4" defection stable spring isolator with a large inertia mass to provide a stable base. Many manufacturers can provide the required information for this type of equipment.

Figure 7.14 Barrier surrounding noise source to reduce intensity.

Figure 7.15 Barrier surrounding noise source to reduce intensity.

Silencers can reduce sound produced by many types of gas flow such as air, steam, and other effluents. An air jet silencer is shown in Figure 7.17. Examples are inlet and discharge of gas turbines; pressure-reducing valves (see Figure 7.18), engine exhausts and inlets; and compressors of various types. An electric motor silencer is shown in Figure 7.19.

Complete enclosure of individual machines are, of course, a standard method of reducing noise for many types of equipment. Typical noise reduction can be achieved from 10 to 30 dBA. An example of this is shown in Figure 7.20. Use of absorptive materials should be used in machine enclosures. This reduces noise buildup in the space.

Another example of an enclosure is shown in Figure 7.21. This is a flexible acoustical blanket that serves as an enclosure for a vibratory cleaner. This can reduce noise 8 to 10 dBA. Other types of materials can be used to lag pipes, various types of vessels, or ductwork. The materials can be as simple as a preformed fiberglass insulation. If the noise source is high frequency in nature, this can provide 8 to 10 dBA of noise reduction. Adding a laminated aluminum cover over the fiberglass can increase the noise reduction 10 dBA more. If the outer cover is heavier and/or the fiberglass (rock wool can also be used) from 1 to 2, or even 4 inches, the noise reduction can be improved 10 to 20 dBA more. For higher level noise sources, a full enclosure might need to be installed for acoustical lagging.

The reflected noise path can be reduced by installing sound-absorbing materials on the walls and ceiling of a space. As a rule of thumb, the noise level can be reduced by 4 to 6 dBA if the ceiling height is not over 14 to 16 ft. and all the surfaces are hard. For this result, the material should cover 30 to 40% of the walls and surfaces, and be evenly distributed. It should be kept in mind that the absorption materials can only reduce the reflected path and *will not* reduce noise directly from any equipment. See Figure 7.22 for acoustical baffles, Figure 7.23 for functional panels, and Figure 7.24 for spray-on material.

C. Receiver Control

This can be accomplished by isolating the operator from the noise source by providing a work space designed to reduce noise to between 60 and 70 dBA. This will allow the operator or attendant to converse with fellow workers and on the telephone. A typical operator enclosure can reduce noise 25 to 30 dBA.

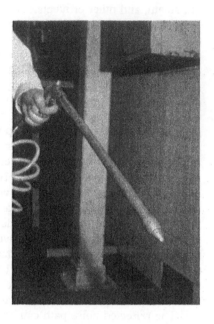

Figure 7.16 Vibration control using neoprene pads for springs.

Figure 7.17   Air jet silencer.

Figure 7.18 Pressure-reducing valves.   **Figure 7.19**   Electric motor silencer.

Figure 7.20 Complete equipment enclosure.

Figure 7.21 Acoustical blanket enclosing cleaner.

D. Summary

Most noise control problems can be solved with standard techniques if common sense is applied. The noise control problems must be defined and placed in a priority rating to budget the needed funds. Define the problem and determine the feasible and practical solutions that are compatible with the situation. Each noise problem can be divided into *noise source - noise path - noise receiver*. The problem(s) might seem daunting, and even overwhelming at first; but by dividing into parts that are manageable, and then looking at them from a *Noise Management* standpoint over a period of time, they can be handled in a straightforward manner much like any other engineering problem. The general steps for noise management program are: **1.** Identify the hazardous noise areas. **2.** Set reasonable goals in keep with policy and OSHA requirements. **3.** Conduct feasibility studies based on noise surveys. **4.** Select the compatible methods of noise control. **5.** Install the materials. **6.** Evaluate the installation. **7.** Make any changes required and evaluate for compliance. These are good engineering practices and they can easily be applied to noise issues as well as any other problem.

Figure 7.22 Acoustical baffles.

Figure 7.23 Functional panels.

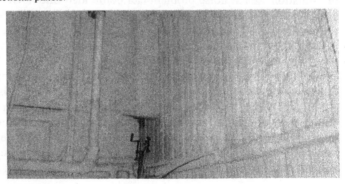

Figure 7.24 Spray-on material.

XII. CASE STUDY

A. Introduction

Noise sources on offshore drilling and production platforms are similar to those found on-shore, i.e., engines, turbines, generators, gearboxes, pumps, compressors, etc. Figure 7.25 is an aerial view of an offshore platform. The major difference is they are located on a steel structure that transmits noise and vibration energy very efficiently. For those that live aboard, for a crew shift of as much as 3 weeks, their exposure time may not be limited to the 12-hour daily work shift. Thus, areas like control rooms, laboratories, bunkhouses, dining and recreation rooms should have lower noise levels, be located away from noise sources, and/or isolated from structure-borne and airborne noise. This requires different criteria and design approach. The criteria for this type of noise control problem are related to speech interference and sleep interruption. Two case histories will be discussed that illustrate the scope of the problem, practical noise control solutions, and results. These include: (1) a semi-submersible combination drilling and production platform located in the Gulf of Mexico; (2) and smaller fixed leg flow station platforms off Indonesia. Before discussing the case histories, an overview of offshore platforms and noise issues will provide a basis to better understand them.

Figure 7.25 Aerial view of an off-shore drilling platform.

B. Typical Noise Control Sources and Levels

Offshore platforms can be divided into fixed and mobile marine structures. They usually operate as drilling or production facilities, or a combination of these two functions. In either event, these steel structures house all of the large equipment necessary to perform the desired function, plus house the crew for a week to a month.

Table 7.10
Typical Equipment Noise Levels

Equipment	dBA
Gas turbines	100–112
Reciprocating gas lift compressors	105–110
Diesel engine generator for drilling	100–120
Pressure-reducing and control valves	104–120
Drilling equipment	100–130
Water injection pumps	105–115
Circulating pumps	100–108
Oil transfer pumps	102–110
Separator areas	95–109
Production rooms	85–100
Ventilation fans	85–90
Air compressors	85–95

Sources: Judd, 1977; Pelton, 1973; Pelton, 1974; Pelton, 1987; Pelton, 1992; Ying, 1983.

Table 7.11
Typical Noise Levels in Support Areas

Support Areas	dBA Range
Production offices	62-72
Galleys and mess areas	62-76
Control and radio rooms	63-76
Recreation rooms	70-75
Bunkhouses	50-72

Sources: Judd, 1977; Pelton, 1973; Pelton, 1974; Pelton, 1987; Pelton, 1992; Ying, 1983.

Typical noise sources and levels found on these structures are shown in Table 7.10. These are indicated in a range since they have been compiled from the references at the end of this chapter.

The experience of being on an offshore platform is one of constant noise in the drilling and production areas. Since the crew, service personnel, and oil company personnel must work and live on board, some typical areas, control rooms, production offices, bunkhouses, galleys, mess areas, and recreation areas, should provide a quiet haven. These are shown in Table 7.11. In these cases, at least, there is little in the way of quiet areas to recuperate from the noise.

C. Case Histories

The following case histories illustrate some of the points outlined above. In one case, follow-up measurements were made; while in the other case, verbal field reports were all that was available. These will also provide an insight into the evaluation and design approach that was used.

1. *Evaluation of noise for the main control room for a semisubmersible drilling and production platform, Mississippi Canyon, Gulf of Mexico*

Figure 7.26 shows the helicopter approach to the offshore platform. This contains a variety of equipment and occupied spaces, including a control room and quarters. Figure 7.27 shows a plan layout of the control room with two large compressors in adjacent bays. Across the aisle from the control room is a production bay that is not shown in this figure. The objectives were to: (1) determine the source path of the noise coming into the control room; (2) evaluate the data; and (3) determine the solutions to reduce the levels for good speech communication.

A series of noise measurements was made in the control room and adjacent spaces. Shown in Figure 7.28 are three octave band measurements as measured in the control room. There is a great deal of low-frequency energy due to the adjacent compressors. In addition, what is not shown, but can be felt quite easily, is the amount of vibration that is in the structure. In order to begin an analysis, the octave band was A-weighted to remove the low-frequency components, which were determined to be from the compressors.

Upon inspection of the control room doors and construction in general, the room was "like a sieve" and noise poured in from all directions. In addition, the room, although it had an acoustical ceiling, was constructed of hard materials, and noise coming through the openings was being reflected and the level increased.

Figure 7.26 Helicopter approach to offshore drilling platform.

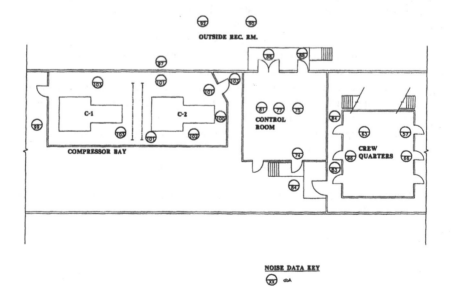

Figure 7.27 Layout of control room of offshore drilling platform.

In order to reduce the noise in the space, though knowing that the compressors could not be isolated or treated in any manner, it was necessary to improve the integrity of the surrounding walls and ceiling of the control room. It was determined that it was not feasible to treat the large compressors and other production equipment by modifying the source or path. The receiver (Control Room) could be treated, however.

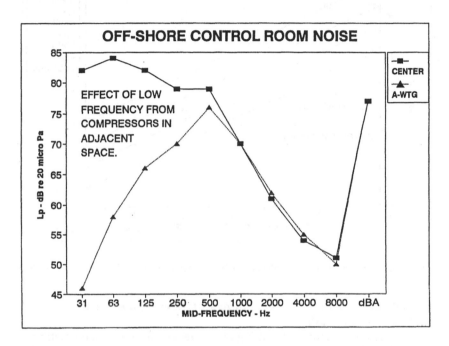

Figure 7.28 Octave band measurements from control room of offshore drilling platform.

This was done by providing acoustical doors, sealing up all the openings and cracks, and providing acoustical absorbing material in the control room to provide better sound conditioning. Figure 7.29 shows the before and after results in terms of speech interference reduction. Although the speech interference levels were reduced approximately 7 dB, there was a significant improvement in the subjective response. The vocal effort was reduced from a range of raised voice to very loud, to almost normal voice.

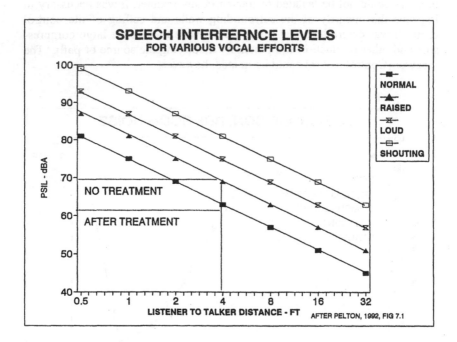

Figure 7.29   Speech interference reduction before and after sound conditioning was provided.

2. Evaluation of noise in the quarters of flow stations, Ardiuna Field, Java Sea, Indonesia

As a result of complaints of high noise levels on flow station platforms in the Java Sea, Indonesia, an investigation was made to evaluate noise sources on the platforms and in the crew quarters. These are much smaller than the open platforms. These platforms provide initial gas separation and transfer liquids to other locations. The crew quarters are typical mobile-type buildings stacked on the platform. In addition, the large separator vessels can be seen on the upper deck. The utilities are located on the cellar deck.

Typical noise levels for main and cellar decks are shown on Figures 7.30 and 7.31, respectively. The levels on the upper deck range from 92–106 dBA and on the lower deck around the generators and pumps from 103–111 dBA. In the quarters, the typical levels are in the range of 70–76 dBA. Intuitively, the problem seems obvious in that the quarters are not providing sufficient noise reduction to have a quiet haven to retreat from the noise. A quarter's building can only provide a limited amount of noise reduction. In this case, the reduction is about 30 dBA.

In order to evaluate the source and paths, a series of noise and vibration measurements were made. Figures 7.32 (sound levels) and 7.33 (vibration levels) show a 1/3rd octave band analysis of the noise levels in the dining area. Since there are a limited number of sources on the platform, they can be observed from the spectra made at various locations. Knowing the operating characteristics of the equipment, the peaks can be identified as seen in Figures 7.32 and 7.33.

The evaluation indicated that the low-frequency noise from the generators and pumps was from a structure-borne path, while the high-frequency noise from the separators was from an airborne path into the quarters. Since the quarters are inherently limited in noise reduction capabilities due to their lightweight construction, the basic recommendations were to provide a combination of methods to treat the equipment and the quarters. This is summarized as follows:

1. Acoustically lag the separators and piping and isolate piping from structure.
2. Install an additional silencer on the generators and replace the radiator cooling fans to slow them down, but with a different pitch to have the same cooling capacity. Isolate tailpipe from the structure and install an acoustical shroud on the AC generator.
3. Install motor silencers on the transfer pumps and isolate from the deck.
4. Seal all windows facing the production area in the quarters to improve the sound transmission loss of the whole wall. Relocate the galley exhaust fan to the "quiet" side of the quarters. Install a silencer on the fresh air fan used to pressure the quarters area.

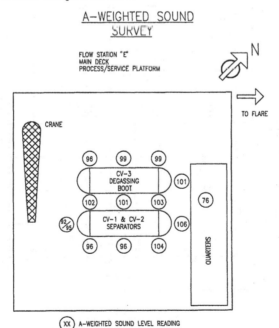

Figure 7.30 Typical noise levels for main deck.

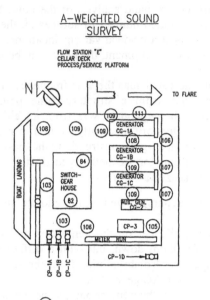

Figure 7.31 Typical noise levels for cellar deck.

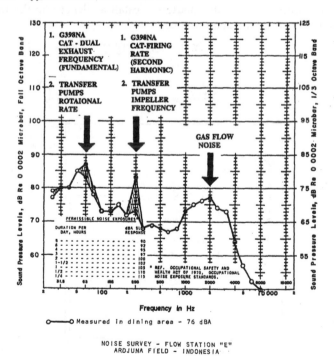

Figure 7.32 Sound levels in dining areas.

The basic complaint was a result of the wives of the production personnel. The elevated voice levels and other behavior problems, after coming home from their long shift of 7 to 14 days, caused tensions at home. Although there was not an opportunity to actually measure the results, the reports from the field were that the wives were much more satisfied after these noise control recommendations were installed. This was the group setting the "criteria" for the project.

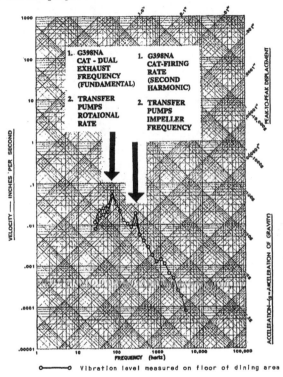

Figure 7.33 Vibration levels in dining area.

D. Summary

The case histories described have similar equipment found in on-shore process plants. The major difference with offshore platforms is that personnel are living on-board and in close proximity to the high noise producing equipment for extended periods of time. This offers some unique challenges for noise control management professionals to provide the proper environment for rest and work. Thus, different criteria and design goals, other than hearing damage risk, must be used to evaluate and solve these noise problems.

Disclaimer: The material in this chapter is provided only as a guide to the reader. The author accepts no responsibility in the application of any material found in this chapter in the furtherance of any noise control solution, hearing conservation program, or any other subject that is provided in this chapter.

REVIEW QUESTIONS

1. Which of the following has the highest velocity of sound?

    a. Air   **b.** Water   **c.** Wood   **d.** Steel

2. For sound measurements in air, the reference pressure is:

    a. 0.0002 n/m^2 **b.** 20 µPa **c.** 0.002 dynes/m^2 **d.** All of the above

3. The reference power for sound power is:

    a. 10^{-12} watts/m^2 **b.** 10^{-10} watts/m^2 **c.** 10^{-12} watts/cm^2 **d.** 10^{-12} watts

4. A sound power level of 120 dB, re 10^{-12} watts, would be equivalent to how much sound power in watts?

    a. 0   **b.** 1   **c.** 10   **d.** 1000

5. What part of the following equation is the directivity?
    $L_p = L_w + 10\log Q - 10\log(4r^2) + 10\log(4/R) + 10$ dB

    a. L$_w$ **b.** 10logQ   **c.** 10 dB   **d.** 10log(4/R)

6. The malleus, incus, and stapes are located in the:

    a. Outer ear   **b.** Oval window   **c.** Inner ear   **d.** Middle ear

7. How many hours of rest are required before an individual can be given an audiogram?

    a. 8   **b.** 14   **c.** 24   **d.** 36

8. The speech interference level is closely related to which weighting scale?

    a. A   **b.** B   **c.** D   **d.** Linear

9. Three motors each produce a noise level of 90 dBA at 3 ft. What is the total noise level for all three motors?

    a. 90 dBA   **b.** 270 dBA   **c.** 93 dBA   **d.** 95 dBA

10. Which of the following is not an acceptable engineering noise control measure?

    a. Spring isolator   **b.** Barrier   **c.** Ear plugs   **d.** Silencer

11. At what frequency will the noise-induced notch most likely occur in an audiogram?

    a. 2000 Hz   **b.** 500 Hz   **c.** 4000 Hz **d.** 6000 Hz

12. In order to comply with OSHA Noise Regulations, a sound level meter must be set at which of the following scales and meter responses?

    a.   A scale and fast response
    b.   Lin Scale and fast response
    c.   A scale and slow response
    d.   C scale and fast response

13. The OSHA regulations for noise and hearing conservation programs are found in the General Industry Standard as detailed in:

    a. CFR 1910.95
    b. CFR 1910.106
    c. CFR 1910.1000
    d. CFR 1910.1044

14. Which of the following can be used for vibration isolation?

    a. Spring mounts
    b. Neoprene pads
    c. Pneumatic mounts
    d. All of the above

15. When compressed air is used to wipe, clean, or blow a product, chips, or parts, the muffler should be placed:

    a. Downstream of the valve
    b. Upstream of the valve
    c. In series both upstream and downstream of the valve
    d. None of the above

ANSWERS

1. d.
2. d.
3. d.
4. b. $dB = 10\log(W/W_{ref})$
          $120 = 10\log(W/10^{-12})$
          Divide both sides by 10:
          $12 = \log(W/10^{-12})$
          Take antilog of both sides:
          $1^{12} = W/10^{-12}$
          $W = 1$ watt
5. b.
6. d.
7. b.
8. a.
9. d.   90 dBA + 90 dBA = 93 dBA
          90 dBA + 93 dBA = 95 dBA
10. c.
11. c.
12. c.
13. a.
14. d.
15. d.

References

American Conference of Governmental Hygienists, *1996 TLVs and BEIs, 2nd Printing*, ACGIH, Cincinnati, OH, 1996.

American Mutual Insurance Alliance, *Background for Loss of Hearing Claims,* Chicago, IL: American Mutual Insurance Alliance, 1964.

Beranek, L. L., *Noise and Vibration Control Revised Edition*, Washington. D.C. Institute of Noise Control Engineering. 1988, 586

Dobie, R., Ed., *Medical Legal Aspect of Hearing Conservation,* New York, Van Nostrand Rheinhold, 1992, 175.

Gasaway, D. C., *Hearing Conservation.* Englewood Cliffs, NJ: Prentice Hall, 1985.

Gasoway, D. C., *Hearing Conservation A Practical Manual and Guide*, Englewood Cliffs, NJ: Prentice-Hall, Inc., 1985, 73-74

Guild, R. S., War deafness and its prevention: report of the labyrinths of the animals used in testing of preventative measures, *J. Lab. Cli. Med*, 4:153-180, 1919.

Harris, C. M., Ed., *Handbook of Acoustical measurements and Noise Control, 3rd ed.,* New York: McGraw-Hill, Inc., 1991, 24.14, 24.15.

Irwin, J. D. and Graf, E. R., *Industrial Noise and Vibration Control,* Englewood Cliffs, NJ: Prentice Hall, 1979.

James Anderson Associates. Lansing MI. *The Noise Monitor,* Spring 1997. 4(2):1, 1977.

Judd, S. II., Noise exposure and control on fixed marine structures, *Sound and Vibration,* May,1977, 20-24

Lewis, A. B.. 1976. Some aspects of noise and vibration on board tankers, *Noise Control Engineering Journal*, 7(3)132-139, 1976.

Michael, P. L., Industrial noise and hearing conservation of hearing. *Patty's Industrial Hygiene and Toxicology, 4th ed., Vol 1.,* George D. Clayton and Florence E. Clayton, Eds., John Wiley & Sons, New York, 1991, 951-958.

Pelton, H. K., *Noise Control Report: King Salmon Platform Cook Inlet Alaska,* Private Report, 1973.

Pelton, H.K., *Noise Control Engineering Experience with Offshore Oil and Gas Platforms and Related Refinery Equipment.* ASME Paper 74-Pet-43, 1974.

Pelton, H. K., *Noise Control Report: Flow Stations "B", "E" & "F".* Private Report, 1974.

Pelton, H. K., *Noise Control Report: Evaluation and Reduction of Noise in Main Control Room for a Semi-submersible Drill and production Platform, Mississippi Canyon, Gulf of Mexico.* Private Report, 1987.

Pelton, H. K., *Noise control Design Using Criteria Other Than Hearing Damage Risk.* Presented to the American Industrial Hygiene Association, Session No. 46. Orlando, FL. May 1990.

Pelton, H. K., *Noise Control Management*, New York: Van Nostrand Reinhold. 1992, 154, 156-157.

Royster, J. D. and Royster, L.H., *Hearing Conservation Programs Practical Guidelines for Success*. Chelsea, MI: Lewis Publishers, Inc., 1990.

Sataloff, J. and Michael, P.L., *Hearing Conservation*, Springfield, IL. Charles C. Thomas Publisher, 1973.

Suter, A. H., Hearing Conservation. *In Noise and Hearing Conservation Manual*, Berger, E., Ward, W., Morrill, J., and Royster, L., Akron, OH: American Industrial Hygiene Association, 1986, 1-18.

U.S. Department of labor-Occupational Safety and Health Administration, *The William Steiger Occupational and Health Act of 1970*, 1970.

U.S. Department of Labor - Occupational Safety and Health Administration, Occupational Noise Exposure: Hearing Conservation Amendment. *Federal Register* 46(11; January 16, 1981):4078-4179, 1981.

Ying, S. P., Noise control for an offshore drilling facility. *Noise Control Engineering Journal*, 21(2): 66-73, 1983.

Chapter 8

IONIZING RADIATION

Wesley R. Van Pelt, PhD, CIH, CHP

I. OVERVIEW

For millions of years, all persons in the world have received a reasonably uniform radiation dose rate from natural sources.

Naturally radioactive elements in soil and rock such as uranium, thorium, radium, and potassium-40 contribute to the natural radiation exposure of humans. Gamma radiation from these sources irradiates humans with *external radiation*. Natural radioactivity is ingested and inhaled and results in *internal radiation* dose. A special case of internal radiation is the inhalation of airborne radon decay products which decay in the human lung giving an alpha radiation dose to the cells of the bronchial epithelium.

High-energy photons and charged particles from extraterrestrial (cosmic) sources bombard the earth's atmosphere and produce a continual dose rate on the earth's surface. This is the second source of external radiation exposure.

Artificial sources of radiation exposure add to the natural radiation dose. Persons exposed to artificially produced radiation include workers exposed occupationally, consumers, and medical patients.

Beginning with the first plutonium production reactors built during WWII, artificial radionuclides have been produced and used for everything from atomic bombs to cancer diagnosis and therapy. Ordinary matter is made radioactive in a nuclear reactor or a particle accelerator.

Commercial nuclear power reactors generate heat, and subsequently electricity, by bringing enough uranium together to sustain a chain reaction of nuclear fissions. *Fission* is an exothermic nuclear reaction which produces neutrons, heat, and radioactive fission fragments. The neutrons are absorbed by ordinary matter to produce radioactive *activation products*. The entire process covering uranium milling, fuel fabrication, reactor operation, spent fuel processing, and nuclear waste management is called the *nuclear fuel cycle*.

Accelerators produce radiation by accelerating charged particles in a vacuum through a potential drop of 10,000 to millions of volts or more. Accelerator is a generic term which includes linear accelerators for scientific research, ion implant devices used on the semiconductor electronics production, medical and dental X-ray machines, industrial radiography X-ray machines, cyclotrons to produce artificial radionuclides, cathode ray tubes used as TV screens and computer monitors, and electron microscopes.

Table 8.1
Annual Average U.S. Population Radiation Effective Dose Equivalent from Natural Sources

Natural Source Type	Effective Equivalent Dose, mrem/yr
Cosmic rays, at sea level, external	27
Terrestrial radioactivity, external	28
Terrestrial radioactivity, internal	35
Radon progeny inhalation dose to bronchial epithelium	240
Total Natural:	330

Table 8.2
Annual Average U.S. Population Radiation Effective Dose Equivalent from Man-Made and Artificially Enhanced Natural

Man-Made/Artificially Enhanced Source Type	Effective Dose Equivalent, mrem/yr
TV and video display terminals	<2
Domestic water supplies (radon)	1–6
Building materials	3.5
Other consumer devices (smoke detectors, radioluminous products, airport luggage inspection, agricultural products, coal and gas combustion)	<10
Nuclear fuel cycle including nuclear reactors	0.05
Diagnostic medical (X-rays and nuclear medicine)	55
Occupational radiation exposure	0.9
Total Man-Made and Artificially Enhanced Natural:	65

A *sealed source* is a quantity of encapsulated radioactive material designed to emit gamma or beta radiation for a useful purpose. Cs-137, Co-60, Am-241, Ir-192, and Kr-85 are typical of radionuclides used in sealed sources. Sealed sources are used for industrial radiography, gamma sterilization, level gages, density gages, and air ionization sources.

All of the above sources of ionizing radiation in our human environment potentially contribute to human radiation exposure. The U.S. average human annual radiation dose from natural and man-made radiation sources is summarized in Tables 8.1 and 8.2 (data from NCRP Report No. 93, Ionizing Radiation Exposure of the Population of the United States, 1987.)

The greatest average population radiation dose comes from naturally occurring sources (330 mrem/yr), while diagnostic medical procedures (55 mrem/yr) are the second greatest source. Although the average population occupational dose is less than 1% of the total dose, only a small fraction of the population is occupationally exposed, and the dose to individual radiation workers can be much higher than the U.S. population average.

II. RADIATION AND RADIOACTIVITY

A. Radiation

Ionizing radiation can be divided into four broad categories: electromagnetic radiation, energetic electrons (beta radiation), energetic heavy ions (e.g., alpha radiation, protons, fission fragments), and neutrons. To be classified as ionizing radiation, each photon or particle must have enough kinetic energy to ionize matter. About 12 eV is sufficient energy to remove an electron from an atom, creating an ion (ionization).

Electromagnetic radiation may be described as either waves or massless *quanta* called *photons*. Ionizing radiation begins in the high-energy portion of the ultraviolet light band, at about 12 eV (100 nm wavelength). Photons with greater energy (shorter wavelength) than ultraviolet light are called X-rays when they originate from the electron shell of an atom, and gamma rays when they originate from the atomic nucleus.

Beta radiation consists of high-energy electrons emitted from the nucleus during radioactive beta decay. Beta radiation can be negatively charged electrons or positively charged electrons (positrons).

Alpha radiation consists of high-energy particles ejected from the atomic nucleus during alpha decay. An alpha particle is made of two protons and two neutrons and is identical to the nucleus of a helium atom.

Energetic heavy ions such as protons and fission fragments are ionizing radiation, and are encountered in particle accelerators and nuclear reactors respectively.

Neutrons are subatomic particles with no electric charge and a mass of 1 atomic mass unit (amu). High-energy neutrons will ionize matter. A special characteristic of neutron radiation is that stable matter can become radioactive when neutrons are absorbed (*neutron activation*).

B. Radioactivity

All atoms can be divided into two categories: radioactive and nonradioactive (stable). Radioactive atoms are the same as nonradioactive matter except that they have an additional characteristic: they will spontaneously *decay* or *transmute*, releasing energy and emitting ionizing radiation.

To be clear, radioactivity is matter. Radiation is emitted when a radioactive atom decays. When a radioactive atom decays or transmutes, the resulting atom is (almost always) an atom of a different element.

When a species of atom is characterized by the numbers of neutrons and protons in its nucleus, it is called a *nuclide* and is described as:

$$_Z^A Y^N \qquad\qquad (8.1)$$

Where:
Z = Atomic number equals number of protons in the nucleus
N = Number of neutrons in the nucleus
A = Mass number equals $Z+N$
Y = Chemical symbol for the element with Z protons in the nucleus

Nuclides with the same Z, but with different N, are called *isotopes* of each other. Nuclides and isotopes that are radioactive are called, respectively, *radionuclides* and *radioisotopes*. It is common practice to leave off the Z and designate a nuclide as $^A Y$ or Y-A. For example, the isotope (which is also a radioisotope) of carbon with 6 protons and 8 neutrons in the nucleus is written as ^{14}C or C-14.

C. Units of Measure

The amount of radioactive material, called radioactivity or just *activity*, is measured in units of Curies (Ci) or Becquerels (Bq). A *Curie* is that amount of radioactive material that decays at a rate of 3.7×10^{10} nuclear transformations per second. One *Becquerel* of radioactive material decays at a rate of 1 nuclear transformation per second. Any amount of radioactive material can be converted between these units of measure using the relationship:

$$1 \text{ Ci} = 3.7 \times 10^{10} \text{ Bq}$$

The *Roentgen* is a measure of radiation exposure and applies only to gamma or X-ray radiation absorbed in air. Radiation ionizes air molecules and produces equal numbers of electrical charges of opposite sign. One Roentgen of exposure produces 2.58×10^{-4} coulomb of electric charge per kilogram of air.

The amount of energy deposited in matter by ionizing radiation is measured in *rads* and is called *absorbed dose*, or somewhat loosely, dose. One rad of ionizing radiation deposits 0.01 J per kg of absorbing medium. Note that the rad is a measure of physical energy absorbed in any type of matter, and does not take into account the biological effectiveness or damage the absorbed radiation can cause. Absorbed dose can also be converted to the newer International System of Units (SI) unit called the gray (Gy):

$$1 \text{ Gy} = 100 \text{ rad}$$

Ionizing radiation fields are described in rate units, e.g., R/hr, rad/m. It turns out that an exposure rate of 1 R/hr at a point in air will produce a dose rate of 0.97 rad/hr when an absorbing medium such as water or soft tissue is placed at that location. In radiation protection, Roentgens and rads are often used interchangeably, and the error in doing so is less than 5%.

Table 8.3 summarizes the definitions of activity, exposure, and absorbed dose.

Table 8.3
Basic Units of Activity, Exposure, and Absorbed Dose

Measure	Unit	Abbrevia-tion	Definition	Note
Amount of radioactivity (*activity*)	Curie	Ci	3.7×10^{10} trans-formation/sec	1 g of radium = 1 Ci
	Becquerel	Bq	1 transforma-tion/sec	New SI unit
Exposure	Roentgen	R	2.58×10^{-4} cou-lomb/kg of air	Only for gamma or X-rays in air
Absorbed dose (a physical unit)	rad	rad	100 ergs/gram of absorbing material	100 ergs/gram = 0.01 J/kg
	Gray	Gy	100 rads	New SI unit

In radiation protection, knowing the dose rate produced by a given amount of activity is often very important. Unfortunately, this relationship depends on many factors, including the type and energy of radiation(s) emitted, the distance from the source of radiation, the type of intervening material, the amount of scattered radiation from nearby objects, the nature of the absorbing medium, etc. However, an approximation ("rule of thumb") applicable only to gamma or X-ray radiation traveling through air from a point source of activity is:

$$D = 6CE \qquad (8.2)$$

Where:

D = Exposure rate at 1 foot distance from the point source of activity, mR/hr
C = Activity of point source, mCi
E = Average photon energy emitted per nuclear transformation, MeV

Sample Calculation: A small vial contains 10 mCi of Cs-137 which emits a 0.662 MeV gamma with an abundance of 0.85. The approximate gamma exposure rate at 1 foot away is:

$$D = 6CE$$
$$D = 6 \times (10 \text{ mCi}) \times (0.662 \text{ MeV}) \times 0.85$$
$$D = 34 \text{ mR/hr}$$

D. Inverse Square Law

A point source of radioactive material emits radiation in all directions with equal probability. As the distance, r, from the point source increases, the dose rate decreases. Neglecting attenuation in the intervening medium, the dose rate, D, can be written as:

$$D = \frac{k}{r^2} \qquad (8.3)$$

Where:
  D = Dose rate
  k = Constant
  r = Distance

The above relationship is called the *inverse square law* because the dose rate is inversely proportional to the square of the distance, r.

One often knows the dose rate, D_1, at a distance, r_1, and wants to calculate the dose rate, D_2, at another larger or smaller distance, r_2. Applying the inverse square law, the relationship becomes:

$$D_2 = D_1 \frac{r_1^2}{r_2^2} \qquad (8.4a)$$

or:

$$D_1 r_1 = D_2 r_2 \qquad (8.4b)$$

Where:
  D_1 = Dose rate of source at distance r_1
  D_2 = Dose rate of source at distance r_2
  r_1 = Distance of source at dose rate D_1
  r_2 = Distance of source at dose rate D_2

E. Half-life

Every radionuclide has a characteristic half-life. Table 8.4 lists half-lives for several more common radionuclides. Half-life, T_2, is the time required for half the radioactive atoms to decay, and is related to the decay constant, , by the relation:

$$T_{1/2} = \frac{\ln 2}{\lambda} = \frac{0.693}{\lambda} \qquad (8.5)$$

Where:

$T_{1/2}$ = Half-life of radionuclide
λ = Decay constant

The amount of radioactivity, A, at any time t, can be written by the exponential decay equation:

$$A = A_0 e^{-\lambda t} \qquad (8.6a)$$

or:

$$A = A_0 e^{(-0.693t/T_{1/2})} \qquad (8.6b)$$

Where:

A_0 is the initial activity at time zero. Figure 8.1 shows the form of the exponential decay equation.

F. Radioactive Decay Emissions: Alpha, Beta, and Gamma

When a radioactive atom spontaneously transmutes, or *decays*, it emits ionizing radiation in the form of an alpha particle, a beta particle, and/or gamma ray(s).

An alpha particle is identical to the nucleus of a helium-4 atom, consisting of two protons and two neutrons bound tightly together. The mass of an alpha particle is 4 amu and its electric charge is +2. Alpha emitters are always high atomic weight ("heavy") radionuclides such as plutonium, uranium, thorium, radium, radon, protactinium, bismuth, etc. An alpha particle has a single energy (monoenergetic) characteristic of the radionuclide emitting it. The Greek letter alpha, α, refers to an alpha particle.

A beta particle is identical to an electron and has a mass of 1/1837 amu and an electrical charge of -1 or +1. Some radionuclides emit a positively charged electron called a positron. A beta particle has a maximum energy characteristic of the radionuclide emitting it (called $E_{\beta,max}$), but each beta can have an energy between zero and $E_{\beta,max}$. The average beta energy per decay is equal to approximately 1/3 $E_{\beta,max}$. Negative beta particles are referred to as β^-, and positrons as β^+.

A gamma ray is a photon of electromagnetic energy with zero rest mass and zero charge. Gamma rays and X-rays are identical species, differing only in origin: gammas come from the atomic nucleus and X-rays originate outside the nucleus. When a gamma-emitting radionuclide decays, it emits one or more monoenergetic gamma rays characteristic of that radionuclide. Gamma rays are assigned the Greek letter gamma (γ).

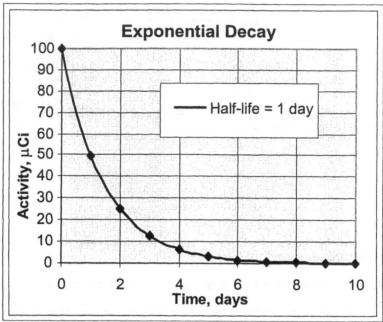

Figure 8.1   Exponential decay graphical presentation.

Table 8.4 presents examples of the emissions from several familiar radionuclides. As one can see, some radionuclides emit only an alpha or a beta particle with no gammas, while others also emit gammas. The abundance is defined as the probability that a particular particle or photon is emitted per decay. For example, the abundance of the 1.46 MeV gamma ray from the decay of a K-40 atom is 0.11 or 11%. Radionuclide decay schemes can be very complex, with dozens of different energies and abundances.

G. Fission and Fusion and Neutrons

Fission occurs when an atomic nucleus of high atomic weight absorbs a neutron and splits into two fission fragments with the release of energy. Fissionable nuclides, which fission when hit with fast neutrons, include Th-232 and U-238. Fissile nuclides, which fission after the absorption of a thermal neutron, include Th-233, U-235, and Pu-239.

Fission is vastly important because of two facts: **(1)** fissioning nuclei release very large quantities of energy (mostly in the form of kinetic energy of the fission fragments) per unit mass of nuclear fuel, and **(2)** fissioning nuclei release neutrons, which if absorbed by fissile or fissionable nuclides, will result in further fissions, hence forming a *chain reaction*. It is this large release of energy and chain reaction which forms the basis of nuclear power reactors and *atomic* bombs.

Table 8.4
Some Common Radionuclides and Their Radiological Properties

Radio nuclide	Half-life	Emission	Energy in MeV (Abundance)	ALI (μCi)	Comments
Hydrogen-3 (tritium)	12.6 yr	β⁻ γ	0.0186 (1.0) none	80,000	Used as tracer in biochemistry. Hydrogen bombs.
Carbon-14	5680 yr	β⁻ γ	0.156 (1.0) none	2000 (compounds)	Used as tracer in biochemistry. C-14 carbon age dating.
Phosphorus-32	14.3 d	β⁻ γ	1.71 (1.0) none	600 (ingestion)	Radiotracer in molecular biology.
Potassium-40	1.26x10⁹ yr	β⁻ β⁺ γ	1.314 (0.89) 0.483 (0.11) 1.46 (0.11)	400 (inhalation)	Naturally occurring.
Cobalt-60	5.26 yr	β⁻ γ₁ γ₂	0.31 (1.0) 1.17 (1.0) 1.33 (1.0)	30 (inhalation)	Neutron activation product in nuclear reactors. Sealed source of gamma radiation.
Iodine-131	8.05 d	β⁻₁ β⁻₂ β₃ β⁻₄ γ₁ γ₃ γ₅ γ₇ γ₈	0.82 (0.007) 0.608 (0.872) 0.33 (0.093) 0.25 (0.028) 0.08 (0.026) 0.284 (0.054) 0.364 (0.82) 0.637 (0.068) 0.723 (0.016)	50 (inhalation, thyroid gland)	Fission product in nuclear reactors. Thyroid seeker. Used in diagnostic and therapeutic nuclear medicine.
Cesium-137	30 yr	β⁻₁ β⁻₂ γ	1.176 (0.065) 0.514 (0.935) 0.662 (0.85)	200 (inhalation)	Fission product in nuclear reactors. Sealed source of gamma radiation.
Uranium-238	4.51x10⁹ yr	α₁ α₂ γ	4.2 (0.75) 4.15 (0.25) none	0.8 (inhalation, bone surface)	Naturally occurring. Principal component of "depleted uranium."

Fission fragments, the two or three pieces of a fissioned nucleus, are radioactive and are responsible for much of the high-level radioactive waste from nuclear power reactors and the "fallout" from atomic bomb explosions in the atmosphere.

When neutrons are absorbed by stable matter, they result in radioactive matter called *activation products*. Many commercially available radionuclides (e.g., C-14, P-32, I-125, Co-60) are activation products made by neutron activation of stable isotopes in a nuclear reactor.

Fusion takes place when very light nuclei (e.g., hydrogen or helium) fuse together. A very large amount of energy is released when two nuclei fuse. However, a high external energy (temperature) is necessary to overcome the repulsive electrical force between two nuclei before they can fuse. As a practical matter, fusion takes place only in stars, thermonuclear bombs (fission bombs which create a temperature high enough to trigger a hydrogen fusion reaction), and large experimental fusion devices such as Tokamaks which confine super-hot hydrogen plasmas with toroidal (doughnut-shaped) magnetic fields.

III. INTERACTION OF RADIATION WITH MATTER

A. Ionization and Excitation

When a particle of ionizing radiation passes through matter it interacts by producing both ionization and excitation. Ionization occurs when an electron is completely stripped from an atom, creating an ion pair. Excitation occurs when the kinetic energy delivered to a bound electron raises it to a higher orbit. On average, radiation produces one ion pair for every 35 eV of energy deposited. About half this energy produces ionization, the other half excitation. For example, a 0.7 MeV beta particle (the average energy of betas emitted by P-32) will create 700,000/35 = 20,000 ion pairs.

B. Alpha Particle Interaction with Matter

Alpha particles are massive charged particles which interact with atoms of absorbing material by colliding and transferring kinetic energy, and by knocking orbital electrons from atoms creating ions. Alpha particles are the least penetrating particles because of their positive charge (+2) and relatively high mass. Alpha particles have a maximum linear distance they will travel in matter called the range. The range of an alpha depends on its initial energy and the absorbing material, but typically is a few centimeters in air and a few micrometers (μm) in tissue. The specific ionization is also energy dependent, but is approximately 4000 ion pairs/mm in air for a typical energy alpha particle.

The short range of alpha particles in tissue has an important radiological health consequence. Since the thickness of the outer dead layer of the skin is about 0.007 cm (70 μm), external alpha irradiation of the human skin will not produce any dose to living cells. This means that alpha-emitting radioisotopes are hazardous only if taken into the body via inhalation, ingestion, or skin absorption.

C. Beta Particle Interaction with Matter

Beta particles interact with the electrons in matter by causing ionization and excitation. Since a beta particle is much less massive than an alpha, its specific ionization is less and its range is greater. Table 8.5 gives the range, in cm, for some common beta-emitting radionuclides in various materials.

Table 8.5
Beta Particle Range in Various Materials (Centimeters)

Material	Density (g/cm³)	H-3 $E_{\beta max} = 0.018$ MeV	C-14 $E_{\beta max} = 0.156$ MeV	Ca-45 $E_{\beta max} = 0.252$ MeV	P-32 $E_{\beta max} = 1.71$ MeV
Air	0.00129	0.5	25	51	600
Water	1	0.0006	0.04	0.06	0.8
Glass	2.6		0.02	0.025	0.35
Iron	7.9		0.004	0.008	0.07
Lead	11.4		0.002	0.005	0.04

Several practical radiological factors follow from these beta ranges. **(1)** Since the range of H-3 betas is very small, only 0.5 cm in air, H-3 surface contamination cannot be detected with any standard radiation survey meter. **(2)** For beta emitters with energies equal to that of C-14 and lower, external skin exposure will deposit most or all of the dose in the dead layer of skin. As such, these low-energy beta emitters are hazardous only when taken internally. **(3)** P-32, with the highest energy of the commonly used radionuclides, will be completely shielded by 0.8 cm of glass, water, and other common materials. Thicker shielding for P-32 beta radiation is superfluous. **(4)** P-32 deposits its entire beta energy within 0.8 cm in tissue, resulting in a very high local dose rate when the source is contamination on the skin or clothing.

If the beta particle is a positron, it undergoes all of the "normal" interactions with matter, but in addition it will combine with an electron (its antiparticle) and annihilate, forming two photons each with an energy of at least 0.51 MeV. While positron emitters are less common than beta-minus emitters, this additional annihilation radiation gives positron emitters more energy per decay.

D. Gamma Ray Interaction with Matter

Unlike particle radiation, gamma rays are photons of electromagnetic radiation and have no definite range in matter. A beam of parallel (or nearly parallel) photons will obey the following radiation attenuation formula:

$$I = I_0 e^{-\mu x} \qquad\qquad (8.7)$$

Where:

I_0 = Initial radiation intensity
I = Radiation intensity after passing through x distance of matter.
x = Distance over which attenuation occurs, cm
μ = Attenuation coefficient, cm^{-1}

The *attenuation coefficient* μ is a function of photon energy and the element(s) and density of the absorbing matter.

Sample Calculation: A 1-cm thick lead shield is placed in front of an I-131 source which emits a 0.364 MeV gamma. The attenuation coefficient for 0.364 MeV gammas in lead is 3.4 cm^{-1}. If the dose rate from this source was 1000 mrad/hr, the addition of 1 cm of lead shielding will reduce the gamma dose rate to what?

$$I = I_0 e^{-\mu x}$$
$$I = (1000 \text{ mrad/hr}) \, e^{-(3.4 cm^{-1})(1 cm)}$$
$$I = 33 \text{ mrad/hr}$$

Three principal interactions account for essentially all of the photon interactions with matter: photoelectric effect, compton scattering, and pair production.

The *Photoelectric Effect* occurs when a photon interacts with an inner shell electron in an atom, giving up all of its energy in the process. If the original photon's energy was greater than the binding energy of the electron, the electron is ejected with the remaining kinetic energy, leaving the atom as a positive ion. The probability that a photon interacts by the photoelectric effect goes as $1/E^3$, where E is the photon energy. The photoelectric effect is the predominant interaction mode at lower photon energies, below about 0.1 MeV.

When a photon hits a valence or loosely bound electron, *Compton Scattering* occurs. The original photon gives some of its kinetic energy to the electron and scatters in another direction and with a lower energy. The resulting Compton electron now behaves like a beta particle. Compton scattering is the predominant photon interaction between about 0.1 and a few MeV.

Pair Production occurs when a photon, with energy at least 1.02 MeV, passes through the electromagnetic field of a nucleus and spontaneously converts into a positron and an electron. In this reaction, 1.02 MeV of pure energy is converted into the rest mass of the positron and electron. Any additional energy of the original photon is shared as kinetic energy of the created pair. Pair production is the predominant interaction mode above a few MeV.

E. Neutron Scattering and Activation

Neutrons, particles with no electric charge, interact with the nuclei of atoms by elastic scatter or by neutron capture.

High-energy, or "fast" neutrons interact by elastic scatter, which means they "bounce" off nuclei giving some of their energy to the recoiling nucleus. After a series of elastic scatters, a fast neutron slows down to become a slow, and eventually, a thermal neutron. Since elastic scatter is most efficient with low Z material such as hydrogen, hydrogen-rich materials like water and paraffin are used as shielding materials for fast neutrons.

Neutron capture reactions tend to occur with slow or thermal neutrons. Neutron capture is the process where a neutron is absorbed into the nucleus and a second particle is emitted, such as a proton or alpha particle. Usually, the resultant nucleus is radioactive. Neutron capture, also called *neutron activation*, is the mechanism for the production of most of the commercially available radionuclides. Neuron activation also results in unwanted induced radioactivity in and around nuclear reactors. An example of neutron capture is the reaction which naturally creates C-14 in the earth's atmosphere from nitrogen and cosmic neutrons:

$$^{14}N(n,p)^{14}C \qquad (8.8)$$

IV. RADIATION "DOSE" IN RADIATION PROTECTION

The term "dose" is used loosely in radiation protection to mean some measure of ionizing radiation deposited in matter (often human beings). Dose is always proportional to the amount of ionizing radiation energy deposited per unit mass of absorbing material.

Radiation exposure (*Roentgen*) and the absorbed dose (rad or Gray) have already been defined. There are many other well-defined dose concepts which are important to the understanding of radiation protection.

The *Dose Equivalent*, measured in rems or Sieverts, is the product of the absorbed dose in tissue and the quality factor (QF). The quality factor is different for different types of radiation. (QF = 1 for X-rays, gamma and beta radiation. QF = 20 for alpha particles, multiply charged particles, fission fragments, and heavy particles of unknown charge. QF = 10 for neutrons of unknown energy and protons.)

The *Deep Dose Equivalent*, measured in rems or Sieverts, is the dose equivalent from external radiation measured at a tissue depth of 1 cm. It describes the dose from penetrating radiation as opposed to "weak," or nonpenetrating radiation.

The *Shallow Dose Equivalent*, measured in rems or Sieverts, loosely called *skin dose*, is the dose to the skin or an extremity from an external source and is defined as the dose equivalent as measured at a tissue depth of 0.007 cm (7 mg/cm^2) averaged over an area of 1 cm^2.

The *Effective Dose Equivalent (EDE)*, measured in rems or Sieverts, is a concept which attempts to average over all exposed tissues in the body using weighting factors for various tissues which are proportional to the radiosensitivity of that tissue for radiogenic cancer. If all tissues in the whole body are uniformly irradiated, the effective dose equivalent is equal to the dose equivalent. The EDE is the sum of products of the dose equivalent to organ or tissue and the weighting factors applicable to each body organ or tissue.

The *Committed Dose Equivalent (CDE)*, measured in rems or Sieverts, is the dose equivalent to organs or tissues that will be received from an intake of radioactive material by a person during the 50-year period following the intake. The CDE is defined only for internal radiation doses such as from inhaled or ingested radioactivity, and describes a person's total future dose after a radioactive material is taken into the body.

The *Committed Effective Dose Equivalent (CEDE)*, measured in rems or Sieverts, is the sum of the products of the weighting factors applicable to each of the body organs or tissues that are irradiated and the committed dose equivalent to these organs or tissues.

The *Total Effective Dose Equivalent (TEDE)*, measured in rems or Sieverts, is the sum of the deep dose equivalent (for external exposures) and the CEDE (for internal exposures).

A. Regulatory Limits on Radiation Dose

For radioactive material it regulates, the U.S. Nuclear Regulatory Commission (NRC) imposes maximum doses to persons occupationally exposed and members of the general public, as shown in Table 8.6.

Table 8.6
Regulatory Limits on Occupational and Public Annual Radiation Dose

Body Part	Occupational Dose Limit	General Public Limit
Whole body Total Effective Dose Equivalent (external plus internal dose)	5 rem/yr	0.1 rem/yr
Sum of Deep Dose Equivalent and Committed Dose Equivalent to any individual organ	50 rem/yr	na
Dose Equivalent to lens of the eye	15 rem/yr	na
Shallow Dose Equivalent to skin or extremities (hands, feet)	50 rem/yr	na

Occupationally exposed persons under age 18 are limited to doses of 10% of those for adult workers. The embryo/fetus of pregnant workers is limited to 0.5 rem TEDE during the entire pregnancy.

For assessing compliance with dose limits from internal exposure, two secondary limits have been derived from the fundamental dose limits: the DAC and the ALI. The *Derived Air Concentration (DAC)* is the air concentration, which if breathed for 1 year, will produce the maximum permissible dose. The DAC is expressed in units of microcuries of radionuclide per ml of air ($\mu Ci/ml$). For occupational exposure to inhaled radionuclides, 1 year's exposure is equal to 8 hr/d x 5 d/week x 50 weeks/yr = 2000 hours. Thus, if a worker were exposed to an average concentration equal to the DAC for a working year, his/her CEDE would equal 5 rem. The numerical value of the DAC is different for every radionuclide. Deriving the DAC is complex and depends on the type of radiation, its abundance and energy, physical and chemical form, fractional uptake by the body, distribution and metabolism in the body, radiological half-life, and biological elimination routes and rates.

The second derived unit is the *Annual Limit on Intake (ALI)*, which is the amount of radioactivity (μCi) which if taken internally would result in the maximum permissible annual dose.

For members of the general public, the NRC has established a maximum annual TEDE of 0.1 rem/yr from licensed radioactive material.

Secondary limits for environmental air and water concentrations have also been derived for assessing the dose limits to members of the public. These take the form of concentrations in air (or water) which, if breathed (or drunk) for 1 year would produce a CEDE of 0.1 rem. In deriving these environmental concentration limits, it is assumed that a member of the public breathes the contaminated air for 24 hr/d and 365 d/yr and derives all of his/her drinking water from the contaminated source.

Tables listing the DAC, ALI and environmental air and water concentration limits for more than 1000 radionuclides may be found in the U.S. NRC Regulation 10 CFR Part 20, Standards for Protection Against Radiation.

B. Biological Effects of Radiation

When ionizing radiation deposits energy in living tissue, it causes ionization of the atoms in the molecules composing that tissue. Direct radiation damage occurs when molecules are modified by the direct action of radiation. Indirect damage occurs when important molecules or structures are damaged by radiolysis products such as free radicals.

Scientists know that radiation causes harmful biological effects from animal studies, epidemiological studies of Japanese atomic bomb survivors, studies of medical patients treated with radiation, etc.

Acute or short-term effects take place within hours, days, or weeks after irradiation. Some examples of acute effects of radiation are skin reddening, hair loss, nausea, loss of appetite, and death. The dose to kill half of a population receiving that dose is defined as the LD_{50} (*Lethal Dose to 50%*). For humans and most mammals, the LD_{50} is about 400 rem delivered to the whole body over a short time. The lowest acute whole-body dose which results in any clinically measurable effect is about 10 to 30 rad. The clinical effects at 10 to 30 rad are chromosomal mutations in the circulating blood cells and decreased sperm count.

The chronic, or long-term effect of radiation is cancer. Radiogenic cancer occurs after a latent period of 5 to 30 years and is usually indistinguishable from "normal" cancers. For example, radiation is known to cause leukemia, bone, breast, GI track, lung, and thyroid cancers. The lifetime risk of cancer death from a single exposure of 10 rem is 1% of the naturally occurring cancer mortality. Thus, if the natural lifetime cancer mortality risk is 20%, a 10 rem radiation dose increases it to 20.2%.

Genetic mutation of the sperm and/or egg cells which then result in malformed offspring has not been confirmed in human populations (including the Japanese atom bomb survivors and their progeny) and is considered less important than it once was.

C. As Low As Reasonably Achievable (ALARA)

Cancer is the primary biological hazard from low-dose and/or chronic radiation. For radiation protection purposes, it has long been assumed that even a small radiation dose has a small probability of causing cancer. This concept is called the *linear non-threshold* hypothesis and it means that the adverse effect is linear with total radiation dose, and zero dose produces zero effect. That is, there is no positive threshold dose below which the effect is zero.

The practical consequence of the linear non-threshold hypothesis is the principle that all radiation doses should be kept *as low as reasonably achievable* (ALARA). The ALARA principle applies once all numerical dose limits have been complied with, and takes into account the economic and societal costs of further reducing radiation dose.

While the linear non-threshold hypothesis remains the conservative model for radiation protection purposes, it has recently been called into question because of almost no actual evidence of its existence at low doses. That is, many human populations exposed to sublethal radiation doses (e.g., about 50 rem) have cancer rates indistinguishable from normal, or even lower than normal expected rates.

V. DOSIMETRY

A. External Dosimetry

The entire profession of radiation safety depends on the ability to measure, estimate, and calculate a person's radiation dose. External dosimetry is a measure of the radiation dose to a person from external sources.

One can calculate, and/or using appropriate and calibrated survey meters, measure the dose rate, R, in mrem/hr at some location. If a person spends t hours there, his/her dose is R x t. This simple technique helps plan or reconstruct occupational radiation exposures.

Personal radiation dosimeters, loosely called *film badges*, measure the cumulative dose to a person during the wearing period. The radiation dosimeter is worn at the collar, lapel, or belt for accumulation periods of a week to 3 months. Personal dosimeters contain either a photographic emulsion, which gets developed like ordinary film (the darker the film the higher the dose), or a *thermoluminescent dosimeter* (*TLD*) material which is read out by heating, whereby it emits light proportional to the radiation dose it absorbed. Either type of dosimeter has a minimum sensitivity of about 10 mrem and will not detect weak beta radiation. The dosimeter case has radiation filters made of plastic, copper, and lead which allow the dosimetrist to estimate the type and energy of the radiation. Dosimeters can be obtained for wearing on the finger (ring dosimeter) or wrist for measuring extremity dose.

Electronic dosimeters containing miniature radiation detectors and digital memory are now available which will read accumulated dose at any time.

B. Internal Dosimetry

When radioactive material is inhaled, ingested, or otherwise taken into the body, internal radiation exposure occurs. Inhalation is the usual intake route in an occupational setting. A person's radiation dose from internal exposure depends on the type, energy, decay characteristics, and half-life of the radionuclide. It also depends on the distribution in the body, metabolism, uptake in various body organs and systems, elimination from those organs, and excretion rate and route.

Since alpha and beta particles do not travel more than a few microns or millimeters in tissue, they usually irradiate only the organ in which they reside. Gamma rays, on the other hand, can penetrate many tens of centimeters in tissue and can irradiate neighboring body organs and deposit their energy outside the body.

The concept relating the radiation dose to the intake of radioactivity is expressed by the Allowable Limit on Intake (ALI). The ALI is the derived limit for the amount of radioactive material taken into the body of an adult worker by inhalation or ingestion in a year. The ALI is the smaller value of intake of a given radionuclide in a year by the reference man that would result in either a CEDE of 5000 mrem or a CDE of 50,000 mrem to any individual organ or tissue. Table 8.4 lists the ALIs for a few selected radionuclides. Care must be taken in selecting and using ALIs because numerical values are different for different intake routes (inhalation vs. ingestion) and for different chemical forms.

The task of determining intake (and therefore dose) is done by whole-body counting and/or bioassay. Whole-body counting is the *in vivo* measurement of a person's internal radioactivity by placing a large radiation detector over the person and detecting gamma radiation emitted from internally deposited radioactivity. Shielding is used to reduce the background.

Whole-body counting systems are calibrated by constructing a "phantom," which simulates a whole human or sub-part such as a neck and thyroid, into which is loaded an accurately known amount of one or more radionuclides. The detector system is always a gamma spectrometer which can simultaneously measure multiple photopeaks and identify radionuclides.

Bioassay is the *in vitro* radiological analysis of a person's urine or feces to determine the intake of radioactivity. This technique is particularly useful for pure beta emitters which do not emit any penetrating radiation and cannot therefore be directly detected outside the body. Using standard biokinetic models, one or more bioassay samples can be related back to the intake at some prior time. Tables of excretion rate, in µCi/day, versus time after intake for many radionuclides can be found in NUREG/CR-4884, *Interpretation of Bioassay Measurements.*

VI. RADIATION PROTECTION ISSUES

A. Techniques for Radiation Protection

The three fundamental techniques for protection against radiation are:

- Time
- Distance
- Shielding

That is, decrease the time a person is exposed to radiation, increase the distance between the person and the radiation source, and increase the amount of shielding between the person and the radiation source, and the overall dose is thus reduced.

Other techniques used to protect against radiation include *ventilation* to reduce the concentration of airborne radioactivity; *source minimization* and *substitution* to decrease the amounts of radiation or radioactivity available for potential exposure; proper *facility design* to allow workers to operate safely and efficiently; *safety interlocks* and *warning devices,* such as automatic radiation level alarms; and *personal protective equipment* such as disposable clothing, respirators, gloves, and lead aprons.

B. Radioactive Waste Management and Disposal

In 1979 the governors of Washington state, Nevada, and South Carolina restricted the only three commercial radioactive waste disposal sites from taking radioactive waste from outside their respective states. This precipitated a radioactive waste disposal crisis in the U.S. stimulating Congress to pass the Low Level Radioactive Waste Disposal Act of 1980 and the Low-Level Radioactive Waste Policy Amendment Act of 1985 which put the responsibility for radioactive disposal on individual states or regional "compacts" of several states. Driven by public anxiety and political decisions, radioactive waste disposal has become unreliable and very expensive. Presently (1996), commercial disposal sites in Washington state, South Carolina, and Utah continue to accept radioactive waste for burial, but with certain restrictions on origin, waste form, concentration, radionuclide, etc.

In the past decade, generators of radioactive waste have been forced to manage their waste using a variety of techniques, including waste minimization, storage for decay, compaction, supercompaction, solidification of liquids, sewer disposal, incineration, and metal melting.

Mixed waste is waste which is both radioactive and also hazardous under EPA's Resource Conservation and Recovery Act (RCRA) regulations. Examples of mixed waste include radioactive compounds dissolved in organic solvents and radiologically contaminated lead. Storage and disposal of mixed waste must take into account all regulations covering both radioactive waste and RCRA hazardous waste. Because most commercial radioactive disposal sites are not permitted to accept RCRA hazardous waste, disposal options are limited. Many waste generators have programs to limit or prohibit the generation of mixed waste.

VII. RADIOLOGICAL SURVEYS

Radiation surveys are regularly scheduled or special measurements of radioactivity or radiation levels performed as part of a radiation protection program. Examples of surveys are radiation level surveys, surface contamination surveys, occupational air concentration surveys, environmental release air stack concentration measurements, sewer water concentration measurements, sealed source leak tests, and package surveys before and after shipping.

Every survey should have a written protocol which describes how the survey is to be conducted, by whom, where, when, how frequently, and how the results will be documented and interpreted. Every protocol should include action levels and the action to be performed at each level. For example: "any tools with removable contamination greater than 1000 dpm will be decontaminated by the user and resurveyed."

A. Instrumentation for Detecting and Measuring Radiation

Virtually all radiation detectors consist of a *sensitive volume*, which can be a gas, liquid, or solid, where energy from ionizing radiation is deposited. Some physical change or event in the sensitive volume is amplified and detected as an electrical voltage pulse or current which is then counted, or measured.

1. *Geiger probe*
One of the most common and versatile radiation detectors is the *Geiger probe*. It consists of a gas-filled tube with a central wire and outer conducting cylindrical shell. A high voltage of many hundred volts is applied across the tube. Ionizing radiation produces ions in the gas which initiate an ava-

lanche of ion pairs, an electric current, and a momentary drop in the voltage. For Geiger tubes, this voltage pulse is of the same size and shape for all detected radiation events, no matter how much radiation was deposited in the sensitive volume. A Geiger probe counts events but cannot tell the difference between a high- and low-energy photon. As a result, Geiger probes calibrated in dose rate (e.g., mrem/hr) at a specific photon energy will under-respond to higher-energy photons and over-respond to lower-energy photons. Geiger probes made of *metal* are used for *gamma* or high-energy *beta* detection. Geiger tubes made with *thin end* or *side windows* allow lower-energy *beta* and even *alpha* particles to enter the sensitive volume and be detected. A thin end-window Geiger probe with an approximately 1.5-inch diameter window and 0.5-inch cylinder height is often called a "pancake" probe because of its shape.

2. *Proportional counter*

Gas-filled detectors operated at lower voltage are called *proportional counters* because the height of the resulting voltage pulse is proportional to the energy deposited in the sensitive volume by the radiation event. Proportional detectors are often made with larger area windows than Geiger detectors.

3. *Scintillation detectors*

In *scintillation detectors*, the sensitive volume is usually a single crystal of solid sodium iodide doped with thallium. When ionizing radiation deposits energy in the crystal, it produces a flash of light which is amplified with a photomultiplier tube. The resulting pulse height is proportional to the amount of energy deposited in the detector. Scintillation detectors are more efficient in detecting *gamma* radiation than gas-filled detectors. This is because a gamma ray impinging on a solid crystal detector has a good probability of being entirely absorbed in the sensitive volume, whereas a gamma ray impinging on a gas volume has a high probability of passing through without interacting at all.

The radiation detector is connected to some device which reads out the voltage pulses produced by the detector. Typical output devices are ratemeters, scalers, and multichannel analyzers.

4. *Ratemeters*

Ratemeters simply read out the average rate at which voltage pulses are being generated by radiation events in the detector. Typical ratemeters read in units of counts per minute (cpm). Many detector-ratemeter combinations are calibrated in dose rate units such as mR/hr, rad/hr, or Sv/m. It is very important to know the type and energy of radiation used for the calibration, and to use the system for measuring only like radiation. For example, a thin-window Geiger detector and ratemeter calibrated in mR/hr with Cs-137 gamma radiation will also produce a meter reading with *beta radiation*, but will not read *beta dose* rate correctly.

5. *Scalers*

Scalers can be used instead of a ratemeter as the output device of a radiation detection system. A scaler is simply an electronic device which sums up the voltage pulses from a detector, usually over a preset time, and displays the result as an integer number of counts. Scalers are usually used in the lab rather than in the field and provide a more accurate estimate of the average count rate.

6. *Multichannel analyzers*

Multichannel analyzers (MCAs) are laboratory or field devices which sort the voltage pulses by pulse height into, say, 4000 channels, and display the resulting spectrum as a graph with channel number as the x-axis and counts per channel as the y-axis. Multichannel analyzers are used with detectors whose voltage pulse height is proportional to the energy deposited in the detector. When an MCA is coupled with such a detector, it is called a gamma ray spectrometer, and is capable of simultaneous qualitative and quantitative analysis of radionuclides.

7. *Ionization chambers*

Gas-filled *ionization chambers* consist of a gas, at atmospheric pressure or pressurized, with a voltage applied across the chamber. Radiation deposited in the gas produces ion pairs which induce an electric current across the chamber. This current is amplified and read out in units of dose rate. Ionization chambers must be calibrated in a known dose rate field. Ionization chamber detectors have the advantage of reading dose rate directly even if the radiation is of a different type or energy than that with which it was calibrated. Some ion chambers have a thin entrance window suitable for measuring low energy *alpha*, *beta*, or *gamma* radiation.

8. *Semiconductor detectors*

Semiconductor detectors are specialized reverse-biased semiconductor diodes with large depleted zones. Radiation deposited in the depleted zone produces an electrical current across the diode and is detected as a voltage pulse which is proportional to the energy deposited. *Germanium* detectors are used for *gamma ray* analysis and *silicon* for *alpha particle* analysis. Germanium detectors must be operated at very low temperatures and require a supply of liquid nitrogen to cool the detector. Semiconductor detectors are superior to scintillation detectors in that they have a better energy resolution. That is, with a semiconductor detector, monoenergetic gamma rays will give a higher and narrower peak in the energy spectrum.

A scintillation or semiconductor detector coupled with a multichannel analyzer is used for pulse height analysis of radiation. Evaluation of the resulting energy spectrum allows both qualitative and quantitative analysis of the radiation source. By examining the energy of gamma or alpha "peaks" in the spectrum, the radionuclide producing the peaks can be identified. Often, dozens of different radionuclides can be resolved and identified using semiconductor detectors and computer-based spectrum analysis.

9. *Liquid scintillation counters*

Liquid scintillation counters assay radioactivity in a sample vial where the sample and liquid scintillator are intimately mixed. The liquid scintillation solution consists of an organic scintillant dissolved in an organic solvent along with emulsifiers and other additives. The sample is added to the scintillation solution and mixed until it is dissolved or suspended. Liquid scintillation analysis is typically used for *beta* radiation but can also detect *alpha particles*. It is especially good for *low-energy beta emitters* such as H-3 and C-14, which cannot be easily assayed any other way. The vial containing the liquid is lowered into a dark shielded chamber. Radiation absorbed in the liquid produces a light flash which is detected by opposing photomultiplier tubes. The voltage pulse from the photomultipliers is amplified and sorted into three channels. Beta emitters are not monoenergetic but produce a broad range of beta energies with a maximum energy which is characteristic of the emitter. Therefore, beta spectra from different emitters overlap and cannot easily be resolved. However, computer-based analysis of liquid scintillation beta spectra can usually give quantitative results for several beta emitters whose maximum beta energies differ from each other sufficiently. For example, analysis of a single sample containing H-3, C-14, and P-32 will produce the activities of each radionuclide even though their spectra partially overlap. A practical disadvantage of liquid scintillation counting is the disposal of the spent vials containing waste organic chemicals mixed with radioactivity.

10. *Nuclear scalers*

When using a nuclear scaler, the fundamental measurement is a number of counts or events, n, detected over a counting time, t. Radioactive decay is random in that one can never predict when a single radioactive atom will decay. This leads to an inherent degree of uncertainty or variability when measuring radioactivity. The standard deviation, σ, of a number of counts, n, is given as:

$$\sigma = \sqrt{n} \tag{8.9}$$

and is a measure of the precision of a single measurement. In practice, a measurement consists of a background count, n_{bkg} (i.e., with no radioactive sample present), and a gross count, n_{gross} (i.e., total of sample counts plus background). The parameter of interest is $n_{net} = n_{gross} - n_{bkg}$, which represents the counts due only to the radioactive sample.

The standard deviation of n_{net} is σ_{net} which is the square root of the sum of the squares of the individual standard deviations:

$$\sigma_{net} = \sqrt{\sigma_{gross}^2 + \sigma_{bkg}^2} = \sqrt{n_{gross} + n_{bkg}} \qquad (8.10)$$

Sample Calculation: A sample is assayed with a proportional chamber detector and a scaler giving 27,304 gross counts in 10 min. A background count results in 16,891 counts in 10 min; net counts in the sample is:

$$n_{net} = n_{gross} - n_{bkg}$$
$$n_{net} = 27,304 - 16,891$$
$$n_{net} = 10,413 \text{ counts}$$
$$\text{counts per minute} = \frac{10,413 \text{counts}}{10 \text{ min}}$$
$$cpm = 1041.3 \text{ counts per minute (cpm)}$$

The standard deviation of the net counts is:

$$\sigma_{net} = (n_{gross} + n_{bkg})^{1/2}$$
$$\sigma_{net} = (27,304 + 16,891)^{1/2}$$
$$\sigma_{net} = (44195)^{1/2}$$
$$\sigma_{net} = 210 \text{ counts}$$
$$cpm = \frac{210 \text{ counts}}{10 \text{ min}}$$
$$cpm = 21.0 \text{ cpm}$$

(The result may be reported as 1041.3 ± 21.0 cpm)

VIII. REGULATORY AGENCIES

The Atomic Energy Act of 1954, as amended, gives the U.S. Federal government control of three categories of radioactive material: byproduct, source, and special nuclear material. *Byproduct material* is any radioactive material (except special nuclear material) made radioactive by exposure to the incident radiation, such as byproduct material associated with the process of producing or utilizing special nuclear material, and tailings or wastes from uranium, or thorium mining and ore processing. Byproduct materials include fission fragments and activation products formed in nuclear reactors. *Source material* is uranium or thorium in any physical and chemical form, or uranium or thorium equal to, or exceeding 0.05% of uranium plus thorium. *Special nuclear material* is plutonium, U-233, uranium enriched in U-233 or U-235, or any material artificially enriched in the foregoing.

Radioactive materials not included in the above categories are not regulated by the federal government and, by default, may be regulated by individual states. State-regulated radioactive materials are often referred to as *NARM* (*Naturally occurring and Accelerator produced Radioactive Material*). NARM must be received, licensed, used, and disposed according to the particular rules and regulations of the state in which the material resides.

The Atomic Energy Act includes provisions for allowing individual states to enter into agreements with the U.S. Nuclear Regulatory Commission (formerly part of the Atomic Energy Commission) to regulate byproduct material in that state. Currently (1996), 29 of the 50 states are so-called "Agreement States" which have adopted radiation control regulations which are essentially equivalent to those of the NRC.

Federal facilities involved in research, development, and production of nuclear weapons and related activities are regulated directly by the U.S. Department of Energy, which issues its own radiation protection rules in the form of DOE orders.

The EPA has some jurisdiction over the release of radioactive materials into the air environment under the National Emission Standards for Hazardous Air Pollutants (NESHAP) as authorized by the Clean Air Act. In 1995, the EPA exempted nuclear power reactors from compliance with NESHAP for radioactive emissions, but as of 1996, continues to regulate other NRC licensees.

Transportation of radioactive materials (and other hazardous materials) in interstate commerce is covered by the rules of the U.S. Department of Transportation. International shipping of hazardous materials is covered by the International Air Transport Association (IATA) and the International Civil Aviation Organization (ICAO). The basic principle is that the shipper must properly describe, package, mark, label, and manifest the radioactive material, and the carrier must properly transport, store, and deliver the material. U.S. DOT hazardous materials regulations are found in Title 49 of the Code of Federal Regulations.

Local and regional sewer authorities sometimes have rules governing the sewer disposal/discharge of radioactive materials. Nevertheless, their rules often refer to state and/or federal regulations on sewering of radioactive materials.

A. Licensing to Possess and Use Radioactive Materials

The NRC and state agencies issue licenses for the possession and use of radioactive materials. Small quantities and concentrations of certain radionuclides may be exempt from the requirement to obtain a license. Also, certain devices such as watch dials, marine compasses, and calibration sources may be exempt.

Most licenses are *specific licenses* where specific radionuclides in specific quantities are licensed for specific uses at specific locations. A *specific license of broad scope* is issued to large organizations like universities where broad discretion is given to the licensee over who may use what radioactivity where. Applicants for a specific license must describe their radiation protection program in detail.

A *general license* is permission to use radioactive materials based on a regulation granting a general license. No license application or approval is needed, but in some cases general licensees must register with the issuing agency. An example of a general license is the use of small quantities of H-3, C-14, I-125, I-131, Fe-59, and Se-75 for *in vitro* clinical or laboratory testing by physicians, veterinarians, clinical laboratories, or hospitals.

IX. MANAGING A RADIATION PROTECTION PROGRAM

A radiation protection program must always be customized to address the potential hazards associated with particular radionuclides, quantities, and uses. Obviously the program for safely using a single sealed source is much different from that for a nuclear power reactor or a major medical center. The following list of program elements can serve as a general checklist for anyone designing, managing or auditing a radiation protection program.

- Management's Responsibility
- License Administration
- Authorization Within an Organization
- Records Management
- Scaled Source Leak Test Survey
- Data Quality Assurance and Integrity
- Radioisotope Inventory
- Radioactive Waste Management
- Shipping of Radioactive Materials

- Incoming Packages of Radioactive Materials
- Radiation Survey Instruments: adequacy, use and calibration
- Radiological Surveys: environmental release sampling and monitoring; radiation level monitoring; contamination sampling and monitoring; occupational air sampling and monitoring

Personnel Dosimetry
- Training
- Critical Systems: Design, Operation, and Maintenance
- Communications
- Postings, Caution and Warning Signs, Labels
- Public Relations

- Security
- Emergency Planning and Response
- Special Classes of Personnel at Facility (contractors, minors, visitors)
- Declared Pregnant Females

- Annual Program Audit

- Bioassay

X. INDUSTRIES WITH RADIATION EXPOSURE HAZARDS

Many industries pose a potential for worker radiation exposure either because they produce or use radiation as part of their process, or radiation is a byproduct of the process.

The nuclear power industry has about 110 operating reactors in the U.S. which create radiation and radioactive waste materials as byproducts to energy (i.e., heat and electricity) production. Radiation protection is an integral part of the nuclear power industry.

Hospitals and medical clinics use radiation to diagnose and treat patients. Every hospital has medical X-ray machines and most have nuclear medicine departments and radiation cancer therapy (radiooncology) departments. Nuclear medicine involves injecting the patient with pharmaceuticals tagged with a radionuclide such as Tc-99m. The radiopharmaceutical accumulates in the target organ or tissue, which is then visualized with a specialized radiation detector (gamma camera). Radiation therapy beams can use radioisotopes such as Co-60 or medical accelerators which can provide intense electron or photon beams.

Radiation is used to kill all microorganisms (i.e., produce sterilization) in food, medical supplies, spices, etc. Commercial radiation sterilization facilities use about 1,000,000 curies of Co-60 which produce gamma radiation fields intense enough to kill a person in 10 seconds. Process material moves on conveyors into exposure vaults with shielding walls up to 6 feet thick.

Biomedical and pharmaceutical research institutions use radioisotopes as tracers for molecules. Typical tracer radionuclides include H-3, C-14, P-32, P-33, S-35, Ca-45, and I-125 and are found in both academia and industry.

Research in the physical sciences often uses machine-made radiation from linear accelerators, cyclotrons, X-ray machines, and other apparatus. Evaluation of the radiation hazard in these situations is often case-by-case because of the uniqueness of each apparatus.

Analytical X-ray machines are used in chemical analysis laboratories to measure elemental composition (X-ray fluorescence) and crystal structure (X-ray diffraction). These machines contain X-ray tubes capable of generating beams of up to 400,000 rem/min which can produce serious radiation burns in seconds. Normally, shielding and interlocked shutters contain the radiation within the instrument.

Industrial radiography is the process of taking an X-ray picture, or radiograph, with a radiation source and X-ray film. The radiographic pictures show the inner structure of such things as concrete structures, welds, metal castings, circuit boards, etc. A common type of industrial radiographic system uses an Ir-192 gamma source containing about 100 Ci of activity.

Radioactive sources are used in nuclear gages to measure the level, density, or thickness of process materials. A nuclear gage emits a beam of radiation through the process material and the attenuated beam is measured on the other side with appropriate radiation detectors. Nuclear gages are noncontact, nondestructive devices which are particularly useful for fast moving, viscous, or dangerous materials found in the paper, plastic, chemical, and food industries.

Atomic weapons production and maintenance potentially expose workers to plutonium, uranium, and critical nuclear reactions. Geological exploration and oil well management use well logging techniques where radiochemical tracers are injected into wells and radiation sources are lowered into wells for the purpose of analyzing underground structure and elemental composition.

Radioactive waste is generated in nuclear reactors and other industries when radioactive materials are no longer useful. Radioactive waste can be processed (compacted, solidified, melted, incinerated, etc.) before it is disposed in a licensed underground burial facility. Radioactive waste processing, transporting, and disposal workers have a potential for exposure to radiation.

While the above industrial uses of radiation all have potential for worker radiation exposure, a long history of very diligent radiation control has nearly eliminated serious overexposure in the U.S. workforce.

XI. THE RADIATION SAFETY PROFESSION

Radiation safety, radiological health, and health physics all describe the profession engaged in the protection of people and the environment from unnecessary exposure to radiation. The profession is concerned with understanding, evaluating, and controlling the health risks from radiation exposure. Professionals are called *health physicists*. Health physicists claim both ionizing and nonionizing radiation safety as part of their profession, but industrial hygienists have an equal claim on nonionizing radiation.

Founded in 1955, the Health Physics Society is open to all persons with an interest in radiation protection. It is the primary professional society in radiation protection. Plenary members must meet certain education and/or radiological protection experience qualifications. The HPS has more than 6000 members in more than 40 countries. It publishes the scientific journal *Health Physics* and a newsletter which all members receive. The administrative headquarters of the HPS are in McLean, VA. Local chapters of the HPS conduct regional meetings, hold seminars, and publish newsletters in support of their local members.

Health physicists who have met the education and experience requirements of the American Board of Health Physics (ABHP) and passed the certification exam are *Certified Health Physicists* and may use the initials *CHP* after their name. The Certified Health Physicist must remain active in the field, be knowledgeable of scientific and regulatory developments, be technically competent, and maintain professional ethics and integrity. CHPs must renew their certification every 4 years by attending courses, seminars, meetings, etc. In 1995, there were about 1100 CHPs. The process is entirely analogous to that run by the American Board of Industrial Hygiene, which certifies Industrial Hygienists.

XII. CASE STUDY: EMERGENCY RESPONSE TO MYSTERIOUS RADIATION SOURCE

Mike was the RSO for a large high tech research and development division of a multinational company. Since this facility dominated the small Illinois community in which it was situated, local officials often looked to the company for help with specialized issues. This was the case when Mike got a call from Police Sergeant O'Reilly who said that a citizen had found and turned in an item which might be radioactive. O'Reilly described it as a shiny metal cylinder about 3 inches long and 2 inches in diameter with female threads on one end and a three-bladed radiation caution symbol. It was now lying on O'Reilly's desk.

Thinking quickly, Mike said to keep everyone 10 feet away from the item and he would be right there. (Like most RSOs he had never been arrested so he had to ask for directions to the police station.)

Mike thought it might be a sealed radiographic source which could have a significant dose rate at close distances. Also, it could be leaking so he would have to be prepared for loose radioactivity. To be prepared to evaluate the situation and take immediate protective action, if necessary, Mike hurriedly gathered the following supplies and equipment:

- Thin-window Geiger probe ratemeter calibrated against Cs-137 gamma radiation.
- Ion chamber ratemeter with a thin window for measuring dose rate accurately.
- Tape measure.
- 12-inch remote grippers for holding and moving the item.
- Cotton swabs for leak testing the source and area for loose radioactivity.
- Radiation caution labels and signs.
- Paper towels, household spray cleaner (for cleaning surfaces), plastic bags.
- Impervious gloves, disposable shoe covers and lab coat.
- Safety glasses.
- His personal body dosimeter badge and ring dosimeter.
- Lead pig large enough to contain the 3-inch item.
- Notebook for recording data and information.
- Radiological Health Handbook.

Upon arriving at the police station, the sergeant and several police officers were gathered at one end of the room with the source on the now-abandoned desk. Mike, using the Geiger ratemeter to measure the dose rate as he approached the source, found 35 mR/hr at a distance of about 6 feet. Now knowing there was real radioactivity here, Mike stepped back and donned his shoe covers, lab coat, gloves, safety glasses, body dosimeter, and ring dosimeter. Moving forward again, he used the tape measure and ion chamber ratemeter to get accurate dose rate readings at accurate distances. Mike made notes of the measurements:

Distance, feet	Dose rate, mR/hr
1	125
2	65
3	42
6	20

At this point Mike decided to do a quick leak test. He took a cotton swab and wiped the desk around the source and then the source itself. Going into the next room (to get a low background), he placed the swab almost in contact with the thin-window Geiger probe. The reading was no different from background, so he figured the source was not leaking. Mike put the swab into a plastic bag for later analysis by liquid scintillation analysis.

Picking up the source with the 12-inch grippers, Mike could read *Ir-192* but not the activity. Mike then placed the source in a small plastic bag and put it in the lead pig, secured the cover on the pig, and placed a *Caution Radioactive Material* label on the pig. The dose rate at the surface of the pig was now 2.5 mR/hr. He asked the police temporarily to lock the pig in a small closet.

To be on the safe side, Mike then did a quick contamination survey of the immediate area, including O'Reilly's desk and hands. All clean.

Mike was unfamiliar with Ir-192 so he it looked up in the Radiological Health Handbook and found it to be a beta-gamma emitter with a 74-day half-life.

Mike then became concerned about the dose received by Sergeant O'Reilly. He interviewed O'Reilly and determined that he held the source in his hand for no longer than 5 minutes, and was 2.5 feet away for approximately 16 hours while at his desk. Since Mike had dose rate readings at 2 and 3 feet, he estimated the dose rate at 2.5 feet by interpolating between these readings and got about 54 mR/hr. However, to get the dose rate at contact with the source (0.25 inches from the center of the source), Mike used the inverse square law. He knew the dose rate was 125 mR/hr at 12 inches and wanted the dose rate at 0.25 inches:

$$D_2 = D_1 \, \frac{r_1^2}{r_2^2}$$

$$D_2 = 125 \times \frac{12^2}{0.25^2}$$

$$D_2 = 288{,}000 \text{ mR/h}$$

Mike knew that the inverse square law works for a point source, and at 0.25 inches the source appears as an extended volume rather than a single point. Therefore, Mike realized that 288,000 mR/hr is an overestimate of the dose rate to the sergeant's fingers, but decided to calculate the extremity dose based on this "conservative" dose rate. Mike recorded this dose evaluation in his data notebook:

Person: Police Sergeant O'Reilly		Date: 17 October 1996			
Body Part	Dose Rate, mR/hr	Exposure Time	Dose, mR	Occupational Dose Limit	Public Dose Limit
Whole body	54	16 hr	860	5000 mrem/yr	100 mrem/yr
Hands	288,000	5 min	24,000	50,000 mrem/yr	na
Lens of eye	54	16 hr	860	15,000 mrem/yr	na

Mike saw that O'Reilly was below the occupational dose limits, but since O'Reilly was considered an individual member of the public, he was above the 100 mrem/yr limit for Total Effective Dose Equivalent.

It seemed like days since Mike got the first call from the police, but it was only 1.5 hours. Mike told O'Reilly and the other police officers that they were not in any danger from radiation exposure but decided to leave any discussion of the numerical dose estimates to the proper authorities.

Ir-192 is a byproduct material, but since Illinois is an Agreement State, the state has jurisdiction over this type of radioactivity. Mike called the Illinois Department of Nuclear Safety, described the incident so far, and gave his findings. They promised to send an inspector to the police station within 4 hours.

Mike was going to offer to take the source back to his company and dispose of it with their other radioactive waste. But then Mike began worrying about regulatory and legal liability issues. After all, his company was not even licensed to possess Ir-192. He decided to leave the source with the police until the state regulatory people arrived.

Since the source was now shielded and secure, Mike decided to go back to his company. After all, he still had to survey nine incoming packages of radioactive materials and start up two environmental air sampling systems before 5:00 p.m.

REVIEW QUESTIONS

1. A Co-60 point source in air produces a gamma dose rate of 350 mrem/hr at a distance of 1 meter. What is the dose rate at 2 cm? Ignore the attenuation due to air.

   a. 875,000 mrem/hr   **b.** 0.14 mrem/hr   **c.** 350 mrem/hr   **d.** 21,000 mrem/hr

2. A container of P-32 arrives at your facility at 6:00 a.m. on Monday and is found to be exactly 1.25 mCi. What is the activity of P-32 in this container at 6:00 p.m. on Friday?

   a. 1.2 mCi   **b.** 1.0 mCi   **c.** 32 mCi   **d.** 0.20 mCi

3. Ionizing radiation interacts with the atoms in matter to produce what two fundamental effects?

   a. Neutrons and protons
   b. Fission and fusion
   c. Lewis and Clark acids
   d. Excitation and ionization

4. Which of the following beta emitters has a beta energy low enough that it has no external hazard and is only hazardous when taken internally?

   a. P-32   **b.** Ca-45   **c.** K-40   **d.** H-3

5. What are the two types of radioactive material created in a nuclear fission reactor?

   a. Fission fragments and activation products
   b. Atomic and nuclear residuals
   c. Direct and indirect thermoluminescence
   d. Cosmogenic and primordial radioactivity

6. The current NRC limit for an occupationally exposed non-pregnant person over the age of 18 is:

   a. 5 rem/yr Total Effective Dose Equivalent (TEDE)
   b. 0.1 rem/yr Total Effective Dose Equivalent (TEDE)
   c. 50 rem/yr Total Effective Dose Equivalent (TEDE)
   d. 15 rem/yr Total Effective Dose Equivalent (TEDE)

7. For a 2-cm thick iron radiation shield, which type of ionizing radiation will travel the farthest in the shield?

   a. Alpha   **b.** Beta   **c.** Gamma   **d.** Infrared

8. Which of the following basic units of measure apply to the *quantity* of radioactive material?

   a. rad and Gray (Gy)
   b. rem and Sievert (Sv)
   c. Becquerel (Bq) and Curie (Ci)
   d. All of the above

9. The lethal dose to 50% of humans (LD_{50}) exposed to acute whole-body gamma or X-ray irradiation is approximately:

    a. 350 mrem   **b.** 350 rem   **c.** 5 rem   **d.** 50 rem

10.   The acronym ALARA as used in radiation protection stands for:

    a.   Alarms Left Alarms Right Always
    b.   American Los Alamos RadioActivity
    c.   American Legal Atomic Radiation Act
    d.   As Low As Reasonably Achievable

11. For equal thicknesses of the following absorbers, which will produce the greatest attenuation of a beam of beta particles?

    a. Iron   **b.** Glass   **c.** Lead   **d.** Water

12. The radiation dose limits to the fetus and embryo of a declared pregnant worker are set at about what fraction of the adult occupational limit?

    a. 1%   **b.** 10%   **c.** 50%   **d.** Variable, depending on the age of the mother

ANSWERS

1.   a. $D_2 r_2^2 = D_1 r_1^2$
            $D_2 \times (2 \text{ cm})^2 = (350 \text{ mrem/hr}) \times (100 \text{ cm})^2$
            $D_2 \times (4 \text{ cm}^2) = (350 \text{ mrem/hr}) \times (10000 \text{ cm}^2)$
            $D_2 = \dfrac{3,500,000 \text{ mrem/hr-cm}^2}{4 \text{ cm}^2}$
            $D_2 = 875,000 \text{ mrem/hr}$

2.   b. $A = A_o e^{-\lambda t}$ or:
            $A = A_o e^{-0.693 t / T_{1/2}}$
            $A = 1.25 e^{-0.693 \times 108/343.2}$
            $A = 1.25 \times 0.804$
            $A = 1.0 \text{ mCi}$

3.   d.
4.   d.
5.   a.
6.   a.
7.   c.
8.   c.
9.   b.
10.   d.
11.   c.
12.   b.

References

Bureau of Radiological Health, U.S. Department of Health, Education and Welfare, Public Health Service, *Radiological Health Handbook, Revised ed.,* U.S. Government Printing Office, Washington, D.C., January 1970.

Cember, H., *Introduction to Health Physics*, 3rd ed., McGraw-Hill, 1996.

Eisenbud, M., *Environmental Radioactivity from Natural, Industrial, and Military Sources*, 3rd ed., Academic Press, New York. 1987.

Hendee, W. R., *Medical Radiation Physics*, Year Book Medical Publishers, Chicago, 1970.

Kathren, Ronald L., *Medical Physics Handbooks 16 - Radiation Protection*, Adam Hilger Ltd., Boston, 1985.

Lessard, E. T., et al., *Interpretation of Bioassay Measurements*, NUREG/CR-4884, Brookhaven National Laboratory, prepared for U.S. Nuclear Regulatory Commission, July 1987.

Miller, K. L., *Handbook of Management of Radiation Protection Programs*, 2nd ed., CRC Press, Boca Raton, FL, 1992.

National Council on Radiation Protection and Measurements, NCRP Report No. 93, *Ionizing Radiation Exposure of the Population of the United States*, Bethesda, MD, 1987.

Orn, M. K., *Handbook of Engineering Control Methods for Occupational Radiation Protection*, Prentice Hall, Englewood Cliffs, NJ, 1992.

Part 20 - Standards for Protection Against Radiation, *Title 10 U.S. Code of Federal Regulations*, U.S. Government Printing Office, updated continuously.

Shapiro, Jacob, *Radiation Protection, A Guide for Scientists and Physicians*, 3rd ed., Harvard University Press, Cambridge, MA, 1990.

Turner, James E., *Atoms, Radiation, and Radiation Protection*, Pergamon Press, 1968.

Chapter 9

Nonionizing Radiation

R. Timothy Hitchcock, CIH

I. OVERVIEW

Nonionizing radiation is composed of ultraviolet (UV), visible, infrared (IR), radio-frequency (RF) radiation, and extremely low-frequency (ELF) fields. Applications of nonionizing radiation have grown tremendously during the past 75 years. With this growth, there has been an increased understanding of the potential for overexposure to the various types of nonionizing radiation, and an awareness of the need to provide a comprehensive radiation safety program that includes nonionizing radiation.

II. THE ELECTROMAGNETIC SPECTRUM

All nonionizing radiation is electromagnetic radiation. *Electromagnetic radiation* is the propagation, or transfer, of energy through space and matter by time-varying electric and magnetic fields. As shown in Figure 9.1, these fields are generally depicted as being in time phase and space quadrature. *Space quadrature* means the electric and magnetic vector fields are mutually orthogonal. *Time phase* means that the waves have reached the same stage in their periodic oscillation with respect to time. *Electric fields* are produced by electric charges, while *magnetic fields* are produced by moving charges, or a current.

III. NONIONIZING RADIATION

Electromagnetic radiation is composed of a continuous spectrum of energies. The electromagnetic spectrum may be divided on the basis of photon energy into high-energy, ionizing radiation (gamma and X-rays), and lower energy, nonionizing radiation (UV, visible, IR, RF, and ELF). The generally accepted demarcation between ionizing and nonionizing radiation is a photon energy of 12 electronvolts (eV), with nonionizing radiation having energy less than this value.

Currently, laser radiation is restricted to ultraviolet, visible, and infrared radiation, although there is experimental work on X-ray lasers. Although the energy of the photons is less than 12 eV, interaction of some energetic (pulsed) laser beams may produce ionization, but this is due to multiphoton

1-56670-197-X/99/$0.00=$.50
© 1999 by CRC Press LLC

absorption during brief periods of exposure. Microwave radiation is a subset of radio-frequency radiation, while the extremely low-frequency band is categorized as sub-radio-frequency fields in the health professions.

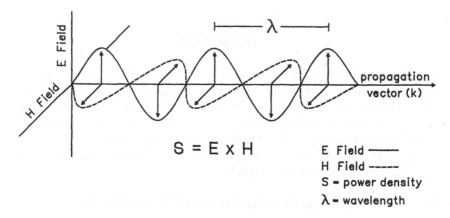

Figure 9.1   Electromagnetic radiation is characterized as two sinusoidal fields, electric and magnetic. Here, the electric field is vibrating up and down with the plane of the paper, while the magnetic field is vibrating in and out of the plane of the paper. This figure illustrates the concepts of time phase and space quadrature as discussed in the text.

Any location on the electromagnetic spectrum may be characterized by wavelength, frequency, and photon energy. By convention, photon energy is limited to ionizing radiation, while nonionizing radiation is described by either wavelength or frequency. This is done in the following manner. Wavelength is used for UV, IR, and laser radiation, while frequency is used for RF and ELF fields (see Table 9.1).

Wavelength, designated lambda (λ), is the distance between the ends of one complete cycle of a wave. *Frequency*, designated by the letter f or the Greek letter nu (ν), is the number of complete wave cycles that pass a point in space in 1 second (sec). Frequency and wavelength are related by the speed of light, c; as such:

$$f = c/\lambda \qquad\qquad (9.1)$$

Where:

f = Frequency, Hz
c = Speed of light, m/sec or cm/sec
λ = Wavelength, m or cm

Table 9.1
Fundamental Characteristics of Nonionizing Radiation

Region	Wavelength	Frequency
Ultraviolet	100–400 nm	–
UV-C[a]	100–280 nm	–
UV-B[b]	280–320 nm	–
UV-A	320–400 nm	–
Visible	400–770 nm	–
Infrared	770 nm–1 mm	–
IR-A	770–1400 nm	–
IR-B	1.4–3.0 mm	–
IR-C	3.0 mm–1 mm	–
Radio-frequency[c]		300 GHz-3 kHz
Extremely low frequecny[d]		3 kHz-3 Hz

Notes: **a.** Also defined as 100-290 nm; **b.** - Also defined as 290-320 nm; **c.** Definition used by the health professions. The RF band may also be divided into a number of order-of-magnitude bandwidths. See reference (Hitchcock and Patterson, 1995) for elaboration.; **d.** Also defined as 3 kHz to 30 Hz.

A number of quantities and units must be used to describe exposure to nonionizing radiation. These are shown in Table 9.2. In general, the dose is described in terms of some quantity of energy (joules, J), while the dose rate is in terms of power (watts, W) since 1 W = 1 J/sec.

Table 9.2
Critical Nonionizing Quantities and Units

Quantity	Unit Symbol	Unit Name	Abbreviation
Ultraviolet Radiation			
Irradiance	E	watt/meter square	W/m², mW/cm²
Radiant	H	exposure joule/meter square	J/m², mJ/cm²
Spectral irradiance	E_λ	watt/meter square per nanometer	W/m²/nm
Aphakic hazard[a]	–	watt/meter square and steradian	W/m²-sr
Laser Radiation			
Irradiance W/m²	E	watt/meter square	mW/cm²
Radiant exposure	H	joule/meter square	J/m², mJ/cm²
Visible Radiation			
Luminance[b]	L_v	candela/meter square	cd/m²
Blue-light radiance	L_{blue}	watt/centimeter square and steradian	W/cm²-sr

Table 9.2 (cont'd)

Retinal thermal hazard	–	watt/centimeter square and stera-dian	W/cm²-sr
Aphakic hazard	–	watt/meter square and steradian	W/m²-sr
Infrared Radiation			
Irradiance	E	watt/meter square	W/m², mW/cm²
Retinal thermal hazard	–	watt/centimeter square and stera-dian	W/cm²-sr
Radio-Frequency Radiation			
Contact current	I_c	milliampere	mA
Current density	J	ampere/meter square	A/m², mA/cm²
E-field strength	E	volt/meter	V/m, V²/m²
H-field strength	H	ampere/meter	A/m, A²/m²
Induced currents	I_i	milliamperes	mA
Power density	W,S	watt/meter square	W/m², mW/cm²
Extremely Low Frequency Fields			
Current density	J	ampere/meter square	A/m², mA/cm²
E-field strength	E	volt/meter	V/m, kV/m
Magnetic flux density	B	tesla	T (mT, mT)

Notes: **a.** Aphakic individuals have had the ocular lens removed (cataract surgery);
b. Photometric quantity based upon the response of the eye to visible radiation.

IV. REGULATORY ISSUES

Regulations addressing nonionizing radiation that may be of importance to health professionals are listed in Table 9.3. These have been promulgated by the Occupational Safety and Health Administration (OSHA), the Center for Devices and Radiological Health (CDRH) of the U.S. Food and Drug Administration, and the Federal Communications Commission (FCC). The CDRH standards address requirements for manufacturers of electronic products, and are emission-based standards. The FCC has established exposure limits for occupational and population exposures to radio-frequency radiation. OSHA promulgates occupational health and safety standards for general industry (29 CFR 1910), construction (29 CFR 1926), and shipbuilding (29 CFR 1915).

The OSHA standards for nonionizing radiation are not comprehensive. These standards address, in a limited fashion, microwave radiation safety, laser equipment in construction trades, and eye, face, and skin protection from radiant energies. However, there are no specific standards for UV, visible, IR, and ELF, and no modern standards on RF and laser radiation, and it is possible that some standards (e.g., 29 CFR 1910.97) may be revoked because they are obsolete.

Currently, OSHA inspectors may issue citations for recognized health hazards using consensus standards, such as ANSI Z136.1 for laser safety and IEEE C95.1 for RF safety. The basis for this was laid in court with the case of the International Union, United Automobile, Aerospace and Agricultural Implement Workers of America, and W.E. Brock vs. General Dynamics Land Systems Division. Here, the court ruled that if a specific standard is inadequate to protect employees from a known hazard, then the employer has a duty, under the general duty clause, to take measures to protect employees. The interpretation of the U.S. Department of Labor attorneys, based on this finding, was that OSHA inspectors may use state-of-the-art standards in issuing citations. To the knowledge of this writer, federal OSHA inspectors have issued citations dealing with RF and laser radiation, and recommending in the citation that companies gain compliance by using recognized consensus standards.

Table 9.3
Selected Regulations Involving Nonionizing Radiation

Agency	Citation	Name
OSHA	29 CFR 1910.97	Nonionizing Radiation
	1910.133	Eye and Face Protection
	1910.268	Telecommunications Standard
	1915.51	Ventilation and Protection in Welding, Cutting and Heating
	1926.54	Nonionizing Radiation
	1926.102	Eye and Face Protection
CDRH	21 CFR 1030.10	Microwave Ovens
	1040.10	Laser Products
	1040.20	Sunlamp Products and Ultraviolet Lamps Intended for Use in Sunlamp Products
	1040.30	High-Intensity Mercury Vapor Discharge Lamps
FCC	47 CFR 1.1310	Radiofrequency Radiation Exposure Limits

The CDRH has promulgated the federal laser product performance standard in 21 CFR subchapter J. This requires that, in the U.S., all manufacturers of lasers and products that contain, or are intended to contain, lasers must certify that their product(s) comply with federal requirements for such products. The first step in this process is to classify the laser or laser product. Based on the class of the laser, the regulations, contained in 21 CFR subchapter J, specify performance and labeling requirements for the product. Also, there are requirements for information that must be included in operations and service manuals, and in other documents that discuss the laser, such as marketing bulletins.

Universally, lasers and products containing lasers are classified according to their output power and their potential hazard. This is a numerical system with classes ranging from Class I to Class IV. The classification scheme of the CDRH is shown in Table 9.4. Class I is reserved for very low-power lasers that are not considered to pose a potential hazard under typical use conditions. Class IV, on the other hand, is restricted to powerful lasers that may be a potential hazard to the skin and eyes by exposure to direct or scattered radiation. However, it is possible that a Class I laser product contains a higher class laser embedded within its protective housing, such as an embedded Class IV laser. In this case, the laser would not be a hazard for its intended use, but may be a very grave hazard for individuals who service the laser and have access to Class IV levels of laser radiation.

Table 9.4
Laser Classification Used by the CDRH

Laser Class	Relative Output Power	Relative Hazard
I	Extremely low	None known
IIa	Very, very low	Very low
II	Very low	Low
IIIa	Low	Low-moderate
IIIb	Moderate	Moderate-high
IV	High	High

Although classification is a feature of most laser safety and laser product standards throughout the world, there are some differences among many of the standards. A discussion of these differences is beyond the scope of this chapter, but there are some useful sources as detailed in the references cited at the conclusion of the chapter.

V. TOXICOLOGICAL ISSUES

The biological and health effects associated with exposure to nonionizing radiation are compiled in Table 9.5. In general, the nonionizing radiation region of greatest concern is the UV region. This is because of the large number of documented health effects from sunburn to skin cancer. Laser radiation is a well-characterized ocular and dermal hazard, and there have been documented accidental overexposures. With RF radiation, there are some demonstrated hazardous effects with test animals, but reports of similar effects in humans are lacking. It has been shown that ELF electric fields can cause biological effects in test animals and humans, but the picture for magnetic fields is less clear. Currently, whether or not ELF fields can cause cancer is at the center of a public debate.

A. Ultraviolet Radiation

UV produces biological effects primarily by photochemical reactions. Photochemical effects follow photon absorption by a *chromophore* (absorbing structure of a molecule), which provides the activation energy for the chemical reaction. The photochemical reaction may form unwanted chemical products, and these products produce the damage. The most important chromophore is deoxyribonucleic acid (DNA).

Table 9.5
Biological and Health Effects

Spectral Region	Target Organ/System	Major Effects
Ultraviolet	Eyes	Photokeratitis and conjunctivitis, cataracts
	Skin	Erythema, dermal elastosis (aging), photosensitivity, cancer
	Immune	Suppression
Visible	Eyes	Retinal lesions
	Skin	Burns
Infrared	Eyes	Retinal lesions, cataracts
	Skin	Burns
Radio-frequency	Nervous	Behavioral disruption
	Reproductive	Congenital malformations and anomalies
	Eyes	Cataracts
	Skin	Burns/shock
ELF	Eyes	Phosphenes
	Nervous	Behavior
	Skin	Hair movement, burns/shock

Biological effects associated with exposure to UV have been established in both *in vitro*, and *in vivo* testing. *In vitro* testing has demonstrated that UV is cytotoxic and mutagenic. Various effects have been established in humans and laboratory animals. The target organs for UV are the skin, eyes, and immune system. The potential hazard to the skin and eyes varies with wavelength, as indicated by an action spectrum of biologically effective, wavelength-dependent doses. An action spectrum is composed of "reciprocal values of threshold doses for some effect."

1. Skin effects

The penetration depth of UV into tissues is dependent upon wavelength, tissue thickness, and pigmentation (melanin). To a large extent, wavelengths less than 300 nm are absorbed topically in the epidermis of the skin. Wavelengths longer than 297 nm penetrate more deeply into the dermis. UV transmission is greatest in Caucasian skin, intermediate for skin of the American Indian, and least for black skin. Studies with human epidermal tissues demonstrate an exponential decrease in transmission with increasing tissue thickness.

Acute effects that may be significant in the work environment include *erythema* (skin burns) and *photosensitization*. Chronic effects are skin aging, immune effects, skin cancer, and photosensitization.

Erythema is an inflammatory reaction of the superficial blood vessels involving dilation and increased permeability of the vessels, increased blood flow, and cellular exudation. A composite action spectrum shows that, for humans with moderate levels of pigmentation, the most effective wavelengths are between 250 and 300 nm. The dose necessary to produce erythema increases into the UV-A, where the erythemal doses are about a factor 1000 to 10,000 times greater. Erythema follows a dose-dependent latency period of 2 to 10 hours, although some physiological changes, e.g., vasodilation, may be detected very soon after exposure. In addition to redness associated with vascular changes, erythema produces cellular damage and may also produce edema and blistering. Following UV exposure, the epidermal layers of the skin thicken, *melanosomes* (pigment granules) migrate to the topmost layers of the skin, and the *melanocytes* produce more melanin. Both measures afford added UV protection to the skin.

Photosensitivity involves an abnormal skin reaction to UV in the presence of a chemical substance. Although a variety of chemical substances may be involved in photosensitive reactions, medications are prominent. Hence, photosensitivity may be more prevalent in elderly workers because they use medications more frequently. Two types of photosensitivity occur: *phototoxicity* and *photoallergy*. Phototoxicity, which is more common, is analogous to contact dermatitis. In general, it manifests as a sunburn-type response on sunlight-exposed parts of the body: face, arms, and hands. Photoallergy is an acquired response involving the immune system. The body's response may include unusual reactions such as urticarial or eczematous lesions. Photoallergens include some salicylanilides and antibiotics, hexachlorophene, cosmetics, and colognes containing musk ambrette.

Prolonged exposure to UV radiation is one causal factor in skin aging or dermal *elastosis*. This is characterized by damage to dermal connective tissues with a loss in elasticity. Signs include laxity of tissues, furrowing of the skin, and a leathery appearance. Skin changes in test animals have been caused by both UV-B and UV-A, although UV-B is more effective.

UV exposure has been linked to nonmelanoma skin cancer (NMSC) and melanoma. Regarding skin cancer, sunlight is viewed as both a tumor initiator and promotor. NMSCs, which are the most numerous skin cancers, include basal and squamous cell cancers. In general, the incidence of NMSC is increasing, while mortality rates are decreasing. NMSC appears to be induced primarily by exposure to solar UV-B, although UV-A (>340 nm) has produced NMSC in test animals. There is an inverse relationship between NMSC and latitude, and individuals living at higher elevations at a given latitude are at greater risk. The tumors are anatomically distributed to areas receiving the greatest exposure to solar UV: head, neck, arms, and hands. The incidence of NMSC increases with age, and is higher in males and individuals with fair skin. Factors that may predispose an individual to develop NMSC include: blonde or red hair, light eye color (blue), fair complexion, poor tanning ability, susceptibility to sunburn, and a history of repeated severe sunburn. The incidence rates of indoor and outdoor workers are similar until the age of 55, after which the outdoor rate increases sharply. Individuals with an indoor/outdoor occupation had an intermediate incidence rate.

Melanoma is a malignancy of pigment-producing cells of the epidermis, the melanocytes, where they are transformed into a neoplasm. An action spectrum for melanoma, determined in hybrids of platyfish and swordtails, indicated that this species of fish was most sensitive to UV-A and blue light. In man, melanoma accounts for the fewest skin cancers, but is the most virulent because the cancerous cells can metastasize. The disease is split about equally between the sexes. Melanoma occurs on sunlight-exposed parts of the body, and on parts of the body that receive little sunlight, such as the neck and trunk of men and the lower legs of women. It has a higher incidence in individuals with multiple cutaneous moles or dysplastic nevi, and individuals with birthmarks; blue, gray, or green eyes; with blonde, red, and brown hair; and with fair skin. There is an association with latitude and the intensity of the solar radiation, and there is a greater incidence in urban areas than rural areas. One theory links melanoma with intermittent exposure to high levels of UV from the sun.

2. Ocular effects
UV wavelengths less than about 300 nm are highly absorbed by the cornea. Wavelengths greater than 300 nm are increasingly transmitted through the cornea and absorbed by the lens. Two predominant features of absorption by the lens are a transmission window centered around 320 nm and an absorption band centered around 360 nm. Transmission through the cornea and lens may be deeper at other wavelengths in *aphakic* (lacking the eye's crystalline lens) or *pseudoaphakic* (receiving an implanted intraocular lens) individuals.

Photokeratitis and *conjunctivitis* (*photokeratoconjunctivitis*) are dose-dependent inflammations of the tissues of the cornea and conjunctiva, respectively. The most effective wavelengths are those in the 270–280 nm band. Typically, the onset of signs and symptoms follows overexposure by about 6 to 12 hr, although times may be as short as 30 min or as long as 24 hr. Signs and symptoms include pain, blepharospasm, tearing, congestion of the conjunctiva, photophobia, visual haze, and a scratchy sensation in the eyes, often described as feeling as if the eyes have sand in them. Erythema may develop on the eyelids and skin surrounding the eyes. The inflammation may be incapacitating for up to 2 days, but the effects are rarely permanent.

The most effective wavelengths for UV-induced cataracts are 290 nm to 310 nm, with a peak around 300 nm. Cataracts have been induced experimentally with exposures of high intensity/short duration and low intensity/long duration. UV-A may be effective in cataract formation, but high values of radiant exposure are necessary in long-duration exposures. Epidemiologic studies have revealed an association between environmental UV levels and the incidence of cataract. Figure 9.2 depicts a welding shield to protect workers from UV exposure during welding.

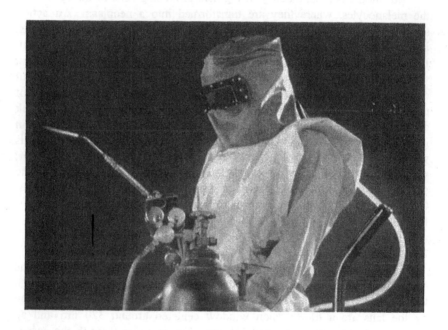

Figure 9.2   Worker wearing welding helmet during welding operations (Courtesy: *Mine Safety Appliances Company*, Pittsburgh, PA).

3. *Exposure guidelines*

Exposure guidelines have been published by the American Conference of Governmental Industrial Hygienists (ACGIH), the National Institute for Occupational Safety and Health (NIOSH), and the International Committee on Nonionizing Radiation Protection (ICNIRP, which used to be part of the International Radiation Protection Association). For the most part, these guidelines are identical. They apply to exposures of the skin and eye, with the eye considered the limiting organ because it lacks protective measures like the skin (thickening, tanning). The wavelength range is 180 to 400 nm, with the limiting value at 270 nm. This is based on the sensitivity of the eye to photokeratitis and conjunctivitis as discussed. The exposure limits may not be protective for photosensitive individuals or for aphakic individuals, who have had the ocular lens surgically removed.

B. Laser Radiation

The target organs for laser-induced biological effects are the skin and eyes. The scientific understanding of the potential hazards of laser radiation comes from studies of laboratory animals, although some human data from accidents is available.

Biological effects are primarily dependent on wavelength, irradiance, and exposure duration. The penetration depth is dependent on wavelength, while irradiance is the dose rate.

Exposure duration may depend on the temporal mode of the laser and behavior. Lasers may operate as *continuous wave* (*CW*) devices, or may be pulsed. In laser safety, CW lasers operate for times of 0.25 seconds and greater. When exposure is to the emission from a pulsed laser, the duration of a single pulse may be very brief, on the order of nanoseconds and less. If the laser is repetitively pulsed, dependent upon the pulse repetition frequency, a large number of pulses may be delivered in a short span of time. When the exposure duration is less than 0.25 seconds, the exposure is considered to be acute, while exposures greater than or equal to 0.25 seconds are chronic.

Exposures may be to the direct beam, or to reflections, which may be specular or diffuse. *Specular reflections* occur when the laser beam is incident on a mirror-like surface, which is a wavelength-dependent phenomenon. Diffuse reflections occur when the size of the surface irregularities are then greater than the incident wavelength. Often, a reflection of a laser beam may include both a specular and a diffuse component.

1. *Ocular effects*

In terms of adverse effects to the eye, the ocular structure at greatest risk of damage is determined by the wavelength of laser light, as shown in Table 9.6. In general, short-wavelength UV and long-wavelength IR are absorbed topically, which results in corneal damage. As the UV and IR wavelengths approach the visible part of the spectrum, penetration depth increases and the lens becomes a target. Near visible, UV-A wavelengths may be transmitted to the retina, as well as UV radiation around 325 nm, where the eye has a transmission window that allows UV to penetrate more deeply. Obviously, visible wavelengths are transmitted through the ocular media to the retina. However, so are IR-A wavelengths, although the latter do not produce the sensation of light upon absorption. Because of this, the wavelength region from 400 to 1400 nm, encompassing both the visible and IR-A, is called the retinal hazard region. Above 1400 nm, the retina is not at risk, while the anterior portions of the eye become potential target structures. Between about 1300 and 1400 nm there is a transition zone, where the primary ocular structure at risk changes from the retina to the anterior structures. In the mid- and far infrared, the structure at greatest risk is the cornea, although heat conduction may play a role in damaging adjacent structures.

Table 9.6
Ocular Structures at Greatest Risk of Laser-Induced Damage

Spectral Region	Target Structure	Relative Penetration Depth
UV-C	Cornea	Topical
UV-B	Cornea/Lens	Shallow
UV-A	Lens/Retina	Moderate
Visible	Retina	Maximum
IR-A	Retina/Lens	Maximum/Moderate
IR-B	Lens/Cornea	Shallow
IR-B	Cornea	Topical

The human aversion response to bright light is innate behavior which involves the blink reflex, but may also include turning the head or raising one's hand. The aversion response time is also 0.25 seconds. Studies with test animals have shown that, for visible beams, if an optical power of 1 mW is deposited on the retina for times less than the human aversion response time, there will be no damage. This information has been integrated into the definition of a Class II laser in the laser classification system. Class II laser beams have a maximum allowable power of 1 mW, and are considered safe to view for less than 0.25 seconds. However, if the beam from a Class II laser is viewed chronically (0.25 seconds), which requires some forethought since one must overcome the aversion response, there is an increased probability of retinal damage. If the power of a visible laser beam exceeds 1 mW, there is an increased probability of retinal effects for exposures that occur in less than 0.25 seconds.

2. Mechanisms

There are three mechanisms by which laser radiation may produce damaging effects: photomechanical, thermal, and photochemical mechanisms. *Photomechanical effects* are associated with brief pulses, typically times less than nanoseconds, that produce extremely high, local values of irradiance. Such brief pulses are generated by rapidly pulsed emissions from Q-switched and mode-locked lasers. Following absorption, a plasma is formed at the site of absorption. Rapid plasma expansion, changes of state, and thermal expansion can lead to shock waves and/or sonic transients that propagate rapidly (e.g., around 4 km/sec) from the site of absorption, disrupting tissue structure in its path. After a short distance, the pressure waves may terminate in cavitation, as the matter contained in the vapor bubble collapses. A number of accidents involving photomechanical effects have been documented, and these most often involve rapidly pulsed lasers. One pulsed laser that has been involved in a large number of the reported accidents is the neodymium:YAG (Nd:YAG) laser, which emits in the IR-A at 1064 nm.

Thermal effects occur in exposure times from microseconds to seconds. Here, the radiant energy is absorbed by a chromophore (absorbing structure) such as melanin, hemoglobin, or macular lutea. Absorption increases the random molecular motion and the total energy of the tissues. This is manifested as increased heat in the tissues, and produces damage by denaturation of protein and inactivation of enzymes. For threshold retinal thermal lesions, the damage is greatest at the center of the lesion, and diminishes toward the periphery. All types of laser radiation can produce thermal effects, although it may not be the predominate mechanism for lasers emitting in the UV region.

Photochemical effects occur primarily with exposure to UV laser light, but also with blue and green light, as well. Cellular studies with three types of excimer lasers that emit in the UV found that a wavelength of 248 nm (KrF excimer) was most effective in producing DNA damage resulting in mutagenic and cytotoxic effects. The relative effectiveness in producing DNA damage decreased at 308 nm (XeCl excimer), and again at 193 nm (ArF excimer). *In vivo* studies demonstrated photochemical effects from ocular exposure to blue and green light. These require relatively lengthy eye exposures, on the order of tens of seconds and longer. Threshold retinal photochemical lesions exhibit relatively uniform damage across the lesion, in contrast to retinal thermal lesions discussed above. Photochemical effects to the skin from exposure to blue and green laser radiation have not been reported.

3. Skin effects

The hazard to the skin is largely determined by wavelength, penetration depth, exposure duration, and, to some degree, the extent of skin pigmentation. In general, absorption is relatively high in the actinic UV and IR-B and -C. Penetration depth is relatively low, as the radiation is primarily absorbed in the epidermis. Penetration depth is greatest for visible and IR-A wavelengths. Reflectance is higher for fair skin, while absorption is higher for highly pigmented skin, at least for near-UV, visible, and near-IR wavelengths. Exposure to UV laser radiation may produce the same effects reviewed above: erythema, photosensitivity, skin aging, and skin cancer. Exposure to visible and infrared wavelengths produces thermal effects that may result in skin burns.

4. Non-beam hazards

In addition to hazards associated with beam access, there are a number of other potential hazards associated with lasers. Foremost among these non-beam hazards are electrical hazards, airborne contaminants, plasma radiation, fires, and explosions.

Although serious consequences result from overexposure to the beam, the primary hazard of laser systems is electrical hazards. In fact, the only deaths attributed to laser systems have been electrical fatalities.

Airborne contaminants may be generated when an energetic laser beam interacts with metals, concrete, plastics, glasses, wood, and tissues. Airborne contaminants may also be generated from the normal operation of gas or dye lasers. Some laser power supplies can produce X-rays, which are usually shielded by the laser enclosure. However, X-rays may be of concern during service activities when the housing is removed and the laser energized.

Plasma radiation occurs in the space above the site of interaction of an energetic laser beam with some metal. X-rays may be generated in the plasma of high-energy, rapidly pulsed lasers. A few studies of occupational exposures to plasma radiation have shown that the interaction of a CO_2 laser beam (IR-C radiation) with metals produces potentially hazardous levels of actinic UV and blue light near the site of interaction.

The optical power contained in a beam from a Class IV laser is a potential fire hazard if it interacts with combustible materials, such as clothing. This also may be an issue in the selection of shielding materials. Additionally, components of lasers and systems may be potential explosions hazards.

5. Exposure guidelines

Guidelines for laser radiation are recommended by the ACGIH, ANSI, and ICNIRP. Discussions of these guidelines and other international standards are available elsewhere. The ANSI standard (Z136.1) will be reviewed here. The exposure guidelines are called maximum permissible exposures (MPEs). These are referenced in terms of irradiance or radiant exposure, and are both wavelength and time dependent. Two sets of limits are specified, one for eyes and one for skin. The ocular MPEs for the retinal hazard region are specified: for viewing a small beam, but may be adjusted to exposure limits for an extended object by the application of a correction factor. Other wavelength-dependent correction factors include those for preretinal absorption, retinal absorption, blue-light, and for repetitively pulsed exposures.

In addition to the exposure limits, the ANSI standard provides detailed programmatic information covering areas like hazard evaluation, medical examinations, training, control measures, and non-beam hazards. For the most part, program requirements are determined as a function of the class of the laser.

The present laser exposure limits are based on damage thresholds, called effective doses. However, bright light, such as that emitted from lasers, may produce biological effects at levels that do not produce damage. One such effect is flash blindness. Although not a damaging effect, as noted, it has become a concern based on reports that airline pilots have been temporarily blinded when the beam from an entertainment laser has entered the cockpit during flight. Hence, laser safety standards, such as that recommended by the ANSI Z136 committee, are addressing limits for non-adverse effects.

C. Radio-Frequency Radiation

Human data are limited and present no clear trends. Therefore, scientists have relied on animals as models to establish biological effects. Animal studies have identified effects in most major systems including nervous, neuroendocrine, reproductive, immune, and sensory. Combined interactions of radio-frequency fields with neuroactive drugs and chemicals have been reported.

1. *Behavioral effects*

Presently, the exposure guidelines are based on a few well-established effects observed in studies with test animals. Reversible behavioral disruption in short-term studies is an effect often cited in the exposure guidelines. This is because this end point has been found to be a sensitive measure of RF exposure, and has been demonstrated in different laboratories, at various frequencies, and with more than one animal species. Generally, behavioral effects are a thermal effect due to significant increases in body temperature due to absorbed RF energy.

East European and Russian literature discusses certain nonspecific symptoms (headache, fatigue, insomnia, sweating, loss of sexual drive) in RF workers associated with the nervous system, with clinical signs extending to the cardiovascular system. Similar symptoms have been described in Western medical literature in two case reports of apparently high, acute overexposure to microwaves.

2. *Reproductive and developmental effects*

Epidemiological studies of reproductive end points have not demonstrated any significant trends, though reproductive and developmental effects have been demonstrated in laboratory research. Two studies reported effects on semen, but the small sample of people involved and lack of exposure data make interpretation difficult. In an hypothesis-generating study, researchers found a difference in the sex ratio of the offspring of female physical therapists who had higher indicators of RF exposure.

3. *Ocular effects*

RF-induced cataracts have been demonstrated in test animals, but not in man. A number of epidemiological investigations and clinical evaluations have been performed, but none has found an excess of cataracts in populations purported to have received RF and microwave exposure.

4. *Epidemiology*

Study populations include military personnel, microwave workers, physical therapists, and operators of VDTs. No differences were observed in cancer mortality in U.S. military personnel. In a study of Polish military personnel, Szmigielski reported statistically significant increases in the morbidity rates for all malignancies and cancers of the alimentary canal, nervous system, hematopoietic system and lymphatic organs. No differences were found between staff at the MIT Rad Lab who worked in radar development during World War II, and members of the U.S. population. A marginally significant difference in cancer of the gall bladder and bile ducts was reported when the Rad Lab workers were compared to a group of physician specialists. No other differences were reported. There has been no conclusive evidence that RF energies are carcinogenic to man. In general, there have been no trends and no meaningful reports related to exposure or dose.

5. *Exposure guidelines*

RF exposure guidelines are numerous, being recommended by municipalities, state governments, consensus committees, and various government agencies worldwide. In the U.S., the foremost guidelines are those published by the ACGIH and the Institute of Electrical and Electronics Engineers (IEEE). The Federal Communications Commission (FCC) promulgates regulations that incorporates elements of the guidelines recommended by the IEEE and the 1986 recommendations made by the National Council on Radiation Protection and Measurements (NCRPM).

The ACGIH guidelines are for occupational exposures between 300 GHz and 30 kHz, while the IEEE has a two-tier limit with a frequency range of 300 GHz to 3 kHz. The two-tier limits are based on whether or not people are in a controlled or uncontrolled environment. A key difference between the two environments is that people in the controlled environment are aware of the potential for RF exposure, while individuals in the uncontrolled have no knowledge or control of their exposure.

The fundamental quantity of these exposure criteria is the SAR, the specific absorption rate.

To evaluate free fields, there are three quantities that have been related to SAR: power density, electric-field strength, and magnetic-field strength. In general, power density is most meaningful at frequencies above 300 MHz (microwaves), while the guidelines require the evaluation of both E and H at frequencies less than 300 MHz.

The SAR does not vary with frequency, while the related quantities are frequency dependent. This is because the body's ability to absorb RF energy is frequency dependent, and each body has a resonance frequency that is a function of the body-to-wavelength ratio. This so-called geometrical resonance is illustrated in Figure 9.3 for different-sized bodies that are not in conductive contact with earth.

At frequencies less than 100 MHz, the guidelines require the evaluation induced and contact currents. Induced currents are RF-induced currents that flow within the body and out through the bottom of the foot as a short-circuit current. Contact currents are RF currents that are generated or stored in a conductor, with the potential for transmission to the grasping hand when contact is made.

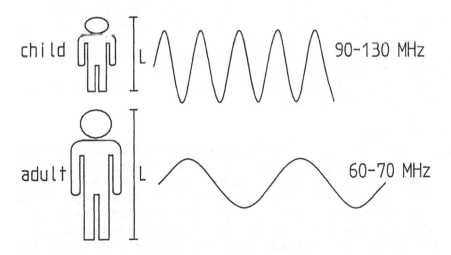

Figure 9.3   This illustrates the concept of geometrical resonance for adult- and child-sized bodies. The resonance frequency is inversely proportional to body height, so adults couple better to longer wavelengths. (The geometrical relationships shown here are not to scale, but are solely for purposes of illustration.).

D. ELF Fields

Biologic effects have been reported in animal studies, including effects on the eyes, behavior, reproduction and development, and the pineal gland. Some end points, such as cancer, have been evaluated in long-term studies. Studies with humans have primarily examined acute effects, cancer, and reproductive outcomes. A number of reviews are available.

1. *Phosphenes*

Phosphenes are visual sensations that have the appearance of flickering white light. They can be observed with the eyes open or closed, and appear to be generated through the stimulation of the retina. They are generated by stimuli other than light, including pressure, mechanical shock, sudden fright, chemical substances, and ELF electric and magnetic fields, which generate responses called electrophosphenes and magnetophosphenes, respectively. Phosphene sensitivity is reported to be greatest around 20 Hz. Magnetic flux density values of 10 to 12 millitesla (mT) have produced phosphenes. The current density necessary to generate electrophosphenes is reported to be 7 to 70 milliamperes per meter square (mA/m^2).

2. *Reproductive and developmental effects*

Human reproductive and developmental studies have evaluated substation workers, switchyard workers, VDT operators, male patients at an infertility clinic, and magnetic resonance imaging workers. The interpretation of the results from these studies is difficult because of limitations such as small study population, weak exposure assessment, differences in outcomes, lack of control for confounders, and no replication of results. Although there are studies, a general conclusion is that no pattern of deleterious effects has been found. Reviews are available.

3. *Melatonin*

Melatonin is produced by the pineal gland in a circadian rhythm, with levels higher at night and lower during the day. Exposure to light, to ELF fields, or to pulsed static magnetic fields can result in reduced levels of melatonin. Grota and colleagues observed a reduction in serum melatonin but not in pineal melatonin in rats exposed to ELF E fields, while others found no significant differences in melatonin levels. Some researchers have suggested that disruption of the melatonin circadian rhythm might be involved in female breast cancer and depression.

4. *Cancer*

A number of observations in *in vitro* studies may be significant in terms of ELF exposure and cancer. These include inhibition of human breast cancer cells with melatonin; reduction of the inhibitory effect of melatonin on human breast cancer cell proliferation; combined effects of 60-Hz magnetic fields and ionizing radiation in producing clastogenic changes in human lymphocytes; and an increase in the activity of the enzyme ornithine decarboxylase (ODC). An increase in ODC has also been reported in an *in vivo* study of rats exposed to a 50-Hz B field.

5. *Epidemiology*

Epidemiology of occupational cancer associated with ELF exposure can be separated into early studies and recent studies. Early studies (approximately 1982–1992) involved the use of job titles to assess exposure, a method which raises questions about the adequacy of exposure classification. In general, these studies demonstrated weak associations between presumed exposure in some electrical occupations and certain cancers, most specifically leukemia, brain cancer, melanoma of the skin and eye, and male breast cancer. For example, a meta-analysis found aggregate risk estimates (95% confidence interval) for total leukemias = 1.2 (1.1–1.3), acute leukemias = 1.4 (1.2–1.6), and acute myeloid leukemias = 1.5 (1.2–1.8). Theriault estimated the risk for total leukemias = 1.18 (1.09–1.29) and acute myeloid leukemia = 1.46 (1.27–1.65). A panel of U.S. scientists estimated a range of risk for total leukemias = 1.10–1.20. In a meta-analysis for brain cancer from studies published between 1983 and 1995, Kheifets and co-workers reported pooled average relative risks from different weighting methods that ranged from 1.08 to 1.30. Other reviews of the epidemiology are available.

More recent studies have attempted to address concerns with exposure assessment by integrating personal monitoring data for B-fields into analyses. Briefly, some point estimates of risk have been modestly elevated in these studies. Generally, the confidence intervals included unity, indicating that most estimates are not statistically significant and could be due to chance. The risk of all cancer types, brain cancer, and leukemia has been modestly elevated in a few studies, but the results were usually not statistically significant. Hence, no consistent pattern has surfaced for these diseases, as best indicated by three studies of the electric utility industry. Here, Sahl and colleagues found no associations for cancer, leukemia, and brain cancer, Theriault et al. reported significant associations for two leukemia subtypes but not for brain cancer, and Savitz and Loomis found an increase in risk of brain cancer, but not for leukemia or leukemia subtypes. In an evaluation of the part of the French-Canadian database in Theriault et al.'s research, a statistically significant increase was found in all leukemia and electric-field exposure at 345 V/m-years. Odds ratios for magnetic-field exposure and leukemia (including leukemia subtypes) were elevated, but not found to be statistically significant.

Stronger associations have been reported for leukemia subtype than for all leukemia, although there is no consistent observation in the various studies.

Theriault and colleagues found a significant increase in AML, but this was not observed by the other studies of power company workers. Floderus and co-workers and London et al. found increases in CLL. For CML, Theriault found no excess risk while London et al. found significant increases.

In conclusion, some positive, statistically significant associations have been reported. However, the findings of the best-designed modern studies have not provided consistent observations, and no clear trends have been demonstrated. Some point estimates of risk are elevated, with estimates weak to moderate in strength, but most are not statistically significant. Generally, exposure-response relationships have not been observed. When viewed in light of the equivocal findings in studies of cancer in animals and the lack of a proven mechanism of biological coupling with the magnetic field, it is fair to conclude that a cause-effect relationship between B-field exposure and cancer has not been established.

6. *Exposure guidelines*

Exposure guidelines for ELF electric and magnetic fields have been published by various organizations including the ACGIH and the International Commission on Nonionizing Radiation Protection. Both organizations utilize current density as the fundamental quantity of the exposure criteria, while electric-field strength and magnetic flux density are the measurable field quantities. The ACGIH recommended occupational limits are frequency dependent between about DC to 30 kHz, while the ICNIRP recommendations address both occupational and general public limits, but just for the power frequencies, 50 and 60 Hz. Both organizations allow higher magnetic-field exposure to the extremities, and have criteria for interference with cardiac pacemakers and medical devices.

VI. SOURCES

A. Ultraviolet Radiation

UV radiation is generated by electronic transitions in excited atomic systems. Some uses of UV include curing materials, photoluminescence, suntanning, chemical manufacturing, treating dermatological disorders, germicidal applications, nondestructive testing, UV spectroscopy, materials processing, and photomicrolithography. UV is often produced as a byproduct in welding, metal cutting, and glass manufacturing. Lists of selected occupations that may be associated with UV exposure are available.

The main source of UV is the sun. Global UV radiation has two components: the sun's beam and sky radiation. The latter component is diffuse and is caused by scattering in the atmosphere. Global UV-A is a stable component of global radiation. Short-term measurements indicate that UV-B is sensitive to the angle of the sun, being greatest when the sun is directly overhead. Solar UV-B wavelengths less than about 295 nm are attenuated by stratospheric ozone, although wavelengths as low as 286 nm have been detected.

Studies with mannequins have shown that the greatest sunlight exposures are received by the tops of the feet, shoulders and ears, back of the neck, and to the area around the nose, cheekbones, and lower lip. Although the solar irradiance is greatest around noon, ocular effects often do not develop because of shielding of the eye by the orbit, eyebrows, and eyelashes and by squinting.

A number of outdoor occupations and activities have been evaluated for UV exposure. These include fishermen, construction workers, farmers and gardeners, landscape workers, physical education teachers, outdoor sports activities, lifeguards, and airline pilots.

Welding and various lamps are the major man-made sources for exposure in the occupational sector. Some of the more hazardous welding processes include argon-shielded gas-metal arc welding (GMAW) with aluminum or ferrous alloys and helium-shielded gas tungsten arc welding (GTAW) with ferrous alloys. Parameters important in determining the UV irradiance include the arc current, shielding gas, electrode wire material, base metal, and joint geometry. Weld geometry and the degree to which the arc is buried in the weld also influence UV levels. Welding and some surface applications with carbon dioxide lasers may generate potentially hazardous actinic UV levels because of the interaction of the energetic far-infrared laser beam with metallic materials.

The following lamps may have emissions that contain relatively high levels of UV: metal halide, high-pressure xenon, mercury vapor, and other high-intensity discharge lamps. The spectral output of mercury vapor lamps varies with gas pressure. Low pressure lamps emit on the order of 95% of their UV radiation at a single wavelength, 253.7 nm, and are used primarily for germicidal applications. Overexposure and potential overexposures have been reported. Medium-to-high pressure lamps are used for area lighting, such as in gymnasiums and parking lots, and for curing. These lamps produce a bluish light and may be hazardous if viewed at close distances. Mercury lamps used for area lighting have two envelopes. The inner envelope is quartz, which is transparent to UV, while the outer envelope is made of materials that are largely opaque to UV. The output may produce hazardous exposures some distance from the lamp if the outer envelope is broken and the lamp continues to operate, which can result if the lamps are not self-extinguishing.

Lamps used for curing, polymerize UV-curable materials and monomeric photosensitizers to form a hard durable material. These lamps have a broadband output, with effective UV radiation in the 300 to 400-nm (predominantly 365 nm) range. Dentists may use UV and blue-light-curable materials. Generally, hand-held applicators that are properly designed, maintained, and used will minimize exposure to the dentist. However, improper use or excessive leaks may result in mouth burns to the patient.

Researchers have reported finding UV-A, -B, and -C from fluorescent lamps used in open fixtures. No UV-B or -C was found when an acrylic diffuser was used with the fixture. The highest UV-B and -C irradiances were associated with high output (HO) and super-high output (SHO) lamps. Fluorescent lighting has been investigated for an association between exposure and skin cancer. The International Radiation Protection Association (IRPA) has taken a position that the UV from fluorescent lights is not a causal factor in melanoma skin cancer.

Medical uses include phototherapy, photochemotherapy, tanning, and disinfection and sterilization. The spectral output from some lamps used in photomedicine has been reported, and it is possible that some exposures may be in excess of exposure limits. Phototherapy workers have a greater risk of dying from skin cancer than medical workers who do not use UV sources, but their risk is much less than medical workers who use ionizing radiation.

B. Laser Radiation

The radiation generated from lasers is coherent, nonionizing radiation where the electromagnetic waves are in phase with one another, both spatially and temporally. In general, laser radiation propagates as a highly directional, intense beam of optical radiation that is characterized by a small diameter, low divergence, and single wavelength.

Lasers have three major components: lasing medium, energy source, and optical cavity, as shown in Figure 9.4. The lasing medium is the medium that generates the laser light and may be gas, liquid, solid, or semiconductive. The medium is contained within the optical cavity which is terminated at either end by a mirror, one highly reflective and the other partially reflective. The latter mirror is called the output coupler mirror, and forms the aperture where the laser radiation is emitted from the optical cavity. The energy source may be electrical, electromagnetic, or chemical in nature.

Laser radiation is generated when the lasing medium absorbs energy, producing an excited state in the atomic structure of the medium. The excited species emit photons as they deexcite to the ground state. Some of the photons propagate down the axis of the optical cavity, and interact with other excited atoms. This "stimulates" these atoms to emit photons, a process described by Einstein in 1917, called *stimulated emission*. Stimulated emission can cause a large amplification in the number of photons propagating between the mirrors, a photon cascade, and these photons are emitted as a beam of radiation. Hence, the process by which laser radiation is generated is "light amplification by the stimulated emission of radiation:" Laser.

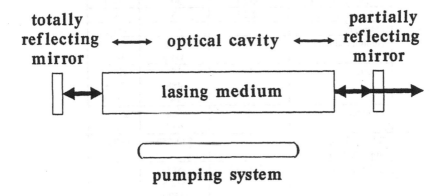

Figure 9.4 The basic components of a laser.

The name of most lasers is derived from the lasing medium. Some common lasers are compiled in Table 9.7. A family of lasers with UV output that is gaining popularity is the excimer. The uses of semiconductor lasers are also growing, due to their small size and an increase in useful wavelengths.

A useful feature of some lasers is wavelength tunability. The first type of tunable laser was the dye laser, where a powder is dissolved in an organic solvent, then pumped into the optical cavity. Vibronic lasers are tunable solid-state lasers, such as alexandrite and Ti:sapphire lasers. Color-center lasers are another type of tunable, solid-state laser.

Laser applications touch on many areas of modern life, including medicine, dentistry, science, industry, communications, agriculture, construction, commerce, education, entertainment, consumerism, and military. One of the major applications of laser radiation is material processing, where the interaction of a powerful laser beam is used to weld, cut, treat the surface, or remove material from a variety of substances such as metals, glasses, ceramics, semiconductors, wood, plastics, cloth, and tissues. The major types of lasers used for materials processing includes Nd:YAG, CO_2, and excimer. HeNe and diode lasers are used to scan information contained in universal product codes in industry and commerce, and for alignment and positioning in industry, construction, and agriculture. Krypton and argon lasers are used in the entertainment industry, while some semiconductor lasers are significant in optical fiber communication systems.

Table 9.7
Some Common Lasers and Their Wavelengths

Lasing Medium	Abbreviation	Wavelength (nm)	Spectral Region
Excimer			
Xenon chloride	ArF	193	UV-C
Krypton fluoride	KrF	248	UV-C
Xenon chloride	XeCl	308	UV-B
Xenon fluoride	XeFl	351	UV-A
Nitrogen	N	337	UV-A
Helium-cadmium	HeCd	325	UV-A
		441.6	Blue
Argon	Ar	351	UV-A
		488	Blue
		514.5	Green
Krypton	Kr	568	Yellow
		647	Red
Rhodamine 6G (tunable dye)	–	570-650	Yellow-red
Helium-neon	HeNe	543	Green
		632.8	Red
		1152.6	IR-A
		3392	IR-C
Ruby	–	694.3	Red
Titanium:sapphire (tunable solid state)	Ti:sapphire	660–1180	Red - IR-A
Alexandrite (tunable solid state)	–	700–800	Red - IR-A
Gallium arsenide	GaAs	850	IR-A
Neodymium:glass	Nd:glass	1060	IR-A
Neodymium:YAG doubled	Nd:YAG	1064	IR-A
		532	Green
Holium:YAG	Ho:YAG	2100	IR-B
Carbon dioxide	CO_2	10,600	IR-C

C. Radio-Frequency Radiation

RF radiation is generated by the acceleration of charge in oscillatory circuits. Generators include power grid tubes (triodes, tetrodes, pentodes), microwave vacuum tubes (magnetrons, klystrons, traveling-wave tubes), and solid-state devices (semiconductor diodes and triodes).

Naturally occurring background sources of RF include terrestrial, extra-terrestrial, atmospheric electrical discharges (lightning), and even the human body. The major sources of exposure to members of the general population include emissions and/or leakage from microwave communications (including cellular phones and other wireless devices), VHF and UHF broadcasts, microwave ovens, CB radio, and televisions and video display terminals

(VDTs). The major occupational sources of exposure include dielectric heaters, induction heaters, plasma processing units, broadcasting, communications, radar, VDTs, diathermy units, hyperthermia units, and electrosurgical devices.

Dielectric heaters may be used to weld, seal, emboss, mold, dry, cure, bake, and heat dielectric materials such as plastics, wood products, glues, and resins. Heaters operate between 10 and 70 MHz, with 27.12 MHz encountered most often. There are some CW dielectric heaters, but most units operate for short periods of time, generally less than 10 sec, and have a duty cycle. A number of workplace evaluations have demonstrated the potential for overexposure of individuals who work with dielectric heaters.

Induction heating is used to heat conducting materials that are placed within a coil that is connected to the RF generator. These units may operate at frequencies as low as 50–60 Hz up to 27 MHz, though most RF units operate between 250 kHz and 488 kHz. Typically, units operate with an on-off cycle, and measurements will require duty-cycle correction. High values of magnetic-field strength, in excess of exposure guidelines, may be found near these units. However, the intensity of the fields diminishes rapidly with distance from the source.

Uses of plasma processing may include chemical milling, nitriding of steel, synthesis of polymers, modifying polymeric surfaces, deposition (sputtering) and hardening of coatings and films, and etching, cleaning, or stripping photoresist. RF operational frequencies may span 100 kHz up to 27 MHz, with many units operating at 13.56 MHz, and some operating at more than one frequency. Typically, evaluations have demonstrated that leakage from well-designed, installed, and maintained units is low. However, viewing ports may be a potential problem area if they are not shielded.

Broadcasting includes standard amplitude-modulation (AM) audio broadcasting, frequency-modulation (FM) broadcasting, and educational and commercial television (TV). The frequencies are allocated by the Federal Communications Commission. An area of concern is accessible areas on tower structures near energized FM and TV antennas, especially if there are multiple antennas. If maintenance personnel must service an antenna occupying a high position on the tower, it is possible that they could be exposed to intense fields associated with energized antennas located lower on the tower. The hands and feet of climbing personnel may receive high exposures, especially if the transmission line is located near the ladder. In addition to receiving exposures to high field strengths, workers may also be in areas susceptible to spark discharge and sustained contact currents.

Communication systems may be either mobile or fixed, or a combination of these two. Fixed systems are used in the telecommunications industry and include high-frequency (HF) radio, tropospheric scatter, satellite communications (SATCOM), and microwave radio (point-to-point radio relay) systems. Mobile systems include small portable transceivers like radios, walkie talkies, and vehicular units. Combination systems may be used for paging or cellular radio. Evaluations of HF radio, tropospheric scatter, long-haul telephony, microwave radio, and SATCOM systems have not demonstrated potential overexposures of ancillary and operational personnel. Maintenance personnel who work on high-power systems, such as SATCOM systems, have the greatest potential for overexposure if they do not follow proper lock-out, tag-out procedures. In certain cases, it may be possible for personnel who maintain cell-site antennae to be exposed to relatively high levels, for example while working very near an energized antenna on a tower or building. It is possible that the power density near hand-held, portable transceivers may exceed the values of power density recommended in exposure guidelines. However, the power deposition will not exceed the criteria for the whole-body-average SAR. Typically, power deposition will be less than the spatial-peak SAR for partial-body exposure, but some researchers have reported maximum values of local SAR that exceed the IEEE recommendations for the uncontrolled environment.

Most radar units operate in the microwave spectral region, specifically in the EHF, SHF, and UHF bands. Lower frequencies, 5 to 300 MHz, and submillimeter and millimeter microwaves may also be used. Radar units usually operate in a pulse-modulated mode, at high peak transmitter powers. Therefore, the duty cycle is usually an important factor. Evaluations of commercial radar (airport surveillance, airport approach traffic control, etc.) have not revealed potential overexposures during normal operation. It is possible that maintenance activities may produce relatively high power densities. Evaluations of aircraft radar have indicated that there is the potential for overexposure of maintenance personnel if the beam is accessed near the aircraft. The EPA found that the typical beam location was 1.8 meters above the ground. Exposures near marine radars have not been reported to be problematic. Evaluation of potential exposures to police traffic-control radar have demonstrated that levels are very low. Overexposure to RF energy from military radar units has been reported.

Diathermy units utilize ultrasonic, microwave (915 or 2450 MHz), or shortwave (13.56 or 27.12 MHz) frequencies. Diathermies may be continuous wave or pulsed. The leakage field around the applicator depends upon the type of applicator used. Relatively high field strengths may be found in the vicinity of the cables. If the therapist adjusts the equipment during operation, the greatest exposure will be to the hands. Service personnel may be exposed during maintenance of an energized system. Fields near the back of an energized unit that had the access cover removed for servicing were approximately 1000 V/m and 3 A/m.

Electrosurgical units (ESUs) are used for cauterizing or coagulating tissues. Frequencies of operation typically span 500 kHz to 2.4 MHz, with some units operating up to 100 MHz. ESUs may be CW or amplitude modulated. Evaluations of spark-gap and solid-state units demonstrated that field strengths increased with increasing output power, and levels were higher for solid-state units. Levels near the probe and unshielded leads may exceed exposure criteria. Relatively high levels have been reported near the eye/forehead region, about 20 cm from the active lead. E-field strength was higher in the coagulation mode than in the cutting mode. Between 0.5 and 2 MHz measurements of induced body currents were 20 to 40 mA.

D. ELF Fields

ELF fields are generated by the periodic oscillation of electric charge. When the charge distribution is widely separated, it produces an electric field. When the charge is moving, it is called an electric current and it produces a magnetic field.

Naturally occurring fields are associated with global and local weather processes, and geological, solar, and stellar processes. Some uses of ELF fields include degaussing, induction heating, induction processes involving coils, metal detection, communications, and medical uses such as magnetic resonance imaging and bone healing. In some environments, the ELF electric and magnetic fields are a byproduct, such as in power generation, transmission, distribution, and use.

The major utility sources are power plants, power lines, and substations. The power frequency in North America is 60 Hz, and 50 Hz in the rest of the world. Power is generated, then transmitted on high-voltage lines to a step-down transformer located at a substation. Primary distribution lines lead from the substation to a (local) distribution step-down transformer. Secondary lines from this transformer connect to the commercial and residential end user. Peak field levels at power plants may be relatively high, but average values are much lower. The E and B fields are highest near power lines, decreasing with distance from the lines. E fields are typically highest at midspan, because the conductors are closest to the earth. Ground-level B fields near pole-mounted, step-down transformers are local in nature, extending only a few meters from the source.

Electric welding, arc welding, resistance welding, and electroslag refining use 50- or 60-Hz current to heat the components to be welded. Operating currents may range from the tens to thousands of amperes, although measurement data has not demonstrated a correlation between current and field strength for arc welding. One report found a range of 0.1 to 10 mT (1 to 100 G) in electric welding processes, with the highest values near spot-welding machines. B fields may be locally high near conductive cables, but decrease rapidly with distance. Relatively high exposures were found to the hand, but the highest average exposures for arc welders were at the waist and gonads. Average E-field values were greatest at the head and gonads.

Electric furnaces, such as ladle, arc, induction, and channel furnaces, are used for hardening, smelting, and heat treating conductive materials. Typical

B-field levels in the electro-steel industry in Sweden were less than exposure criteria, with the highest levels near induction and ladle furnaces. Near two-phase electric-arc melting furnaces and in control booths 60-Hz B-field values were within exposure guidelines.

VII. MONITORING AND EVALUATION

A. Ultraviolet Radiation

Radiometers are used to measure radiant power and have calibrated output in irradiance or radiant exposure. They may be portable, battery-operated units for field use, or fixed-position laboratory units. The optical sampling train may include input optics, band filters/monochromators, detectors (photodiodes, thermopiles, photomultiplier tubes), and the electronics package. Diffusers may front the detector so that it is uniformly illuminated, and to make the detector alignment with the source less critical.

A common radiometric error is susceptibility to stray source and/or ambient light. Ambient lighting and broadband source emissions may exceed the UV intensity one is trying to measure. Unless the instrument can reject this light, there may be considerable error associated with the measurement data. Stray light may be controlled by filtration or with monochromators.

Interference filters, neutral-density filters, and special filters may be used. A bandpass filter used with some portable radiometers provides spectral weighting so that transmission closely matches the UV-B and -C exposure criteria. A similar filter is available for UV-A. Monochromators may be used with scanning spectroradiometers. These transmit a narrow spectral region to the detector, without the use of filters.

Prior to measurement, the instrument should be zeroed and any contributions from sources of visible emissions adjusted for, or subtracted by filtration. Data should be collected under conditions of normal use and maintenance, and at locations that represent worker exposure. Data collected at other locations may be useful in identifying and controlling leaks, but should not be interpreted as potential human exposure.

The effective irradiance either is read directly or is calculated. For broadband measurement data in the UV-B and -C, the allowable time of exposure may be calculated as the quotient of the allowable radiant exposure at 270 nm (limiting energy dose) and the measured value of effective irradiance, as:

$$T_{(s)} = \frac{30 \text{ J/m}^2}{E_{eff}} \qquad (9.2)$$

Where:

$T_{(s)}$ = Allowable time, seconds

E_{eff} = Effective irradiance, W/m^2

At times, it may be necessary to estimate the irradiance at other locations by use of the inverse square law:

$$E_1 = E_2 r_2^2 / r_1^2 \qquad (9.3)$$

Where:

E_1 = Irradiance of source at location, distance r_1, W/m^2
r_1 = Distance from source, feet
E_2 = Irradiance of source at location, distance r_2, W/m^2
r_2 = Distance from source, feet

Sample Calculation: For example, if the irradiance, E_2 is 10 W/m^2 at 2 feet, r_2, from the source, what is E_1 at 1 ft, r_1?

$$E_1 = E_2 r_2^2 / r_1^2$$
$$E_1 = [(10 \text{ W/m}^2) (2 \text{ ft})^2]/(1 \text{ ft})^2$$
$$E_1 = [(10 \text{ W/m}^2) (4 \text{ ft}^2)]/(1 \text{ ft}^2)$$
$$E_1 = \text{is } 40 \text{ W/m}^2$$

B. Laser Radiation

Laser hazard evaluation is addressed in ANSI Z136.1 and in a number of useful reviews. In general, industrial hygienists (as laser safety officers, LSOs) are discouraged from making measurements of laser radiation because of the complexity of making meaningful measurements, and the potential for accidental exposure when attempting to align a small diameter beam with a small entrance aperture (e.g., 7 mm) that would cover a laser power detector.

Instead of measuring, hazard evaluation techniques may be utilized by employing numerical methods to analyze potential laser hazards. A major method is to calculate the nominal hazard zone (NHZ) of the laser application. The NHZ is the zone around the laser where the beam intensity exceeds the MPE. This is demonstrated in Figure 9.5, where the laser beam divergence reduces the irradiance to the MPE at some distance from the laser.

The ANSI Z136.1 standard recommends four numerical models for determining the NHZ: intrabeam, lens-on-laser, fiber optics, and diffuse reflection. The ANSI Z136.2 standard includes similar models for optical fiber communications systems, which is different than the fiber optics models in the Z136.1 standard.

The largest NHZ will be determined for the case of intrabeam viewing. This is because the use of a lens or fiber optics, or the reflection of laser light from a diffusing surface, reduces the NHZ in comparison to the intrabeam condition. The equation used to calculate the intrabeam NHZ is:

$$r = \frac{1}{\phi}[(\frac{4P}{\pi \, MPE}) - a^2]^{1/2} \qquad (9.4)$$

Where:

r	= Intrabeam nominal hazard zone, millimeters
P	= Optical power, watts
ϕ	= Beam divergence, milliradians, (radians)
a	= Exit beam diameter, millimeters
MPE	= Maximum permissible exposure, watts/cm^2

Here, P is the optical power of the laser beam in watts, ϕ is the beam divergence, and a is the exit beam diameter, usually in millimeters. All units must be consistent, which can be determined by dimensional analysis. Note that laser beam divergence is usually in terms of milliradians, but all calculations must be in radians. Also, since MPEs are largely time dependent, it is possible to calculate more than one hazard zone (distance) for a given laser.

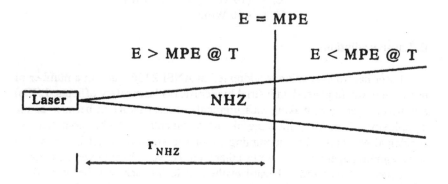

r = distance
E = irradiance (exposure)
T = exposure duration

Figure 9.5   The concept of the nominal hazard zone for directly viewing the beam is shown here. Note that the irradiance is greater than the MPE within the NHZ, but less than the MPE outside the boundary of the NHZ. Also, because the MPEs are time dependent, the NHZ is also time dependent.

C. Radio-Frequency Radiation

Most instruments used for RF measurements are broadband receivers. These instruments have a frequency-independent response over a large part of the RF spectrum. These instruments may measure E-field strength and/or H-field strength, or power density. Most designs allow the user to select from multiple probes so that the instruments can be used for different ranges of intensities and/or frequencies, or so that they can be used for E or H fields.

Elements of broadband measurement systems include antennas, detectors (thermocouples or diodes), connecting cable or optical fiber, electronics package, and case. E-field antennas are usually monopoles or dipoles, while H-field antennas are usually loop. Antennas may be encased in RF-transparent materials that are opaque to infrared and visible radiation, insulate against thermal shock of the sensing elements, and serve as a spacer. Arrays with three mutually orthogonal antennas are most common in broadband instruments. Such an array provides spatial coverage that produces a response that is independent of direction and polarization of the field (i.e., isotropic). Figure 9.6 illustrates a worker measuring broadband RF waves.

Instruments to measure induced currents may be stand-on or clamp-on designs. Stand-on devices measure foot current (short-circuit current), while clamp-on devices usually measure current in an appendage such as the ankle or wrist. Stand-on instruments are a type of parallel-plate capacitor, having two parallel, external metal sheets. A person stands on the top plate, while the bottom plate contacts the floor or ground. A review of the performance of two manufacturers' makes of stand-on meters is available. Clamp-on devices encircle the appendage with a current transformer. In this configuration, the body is the primary circuit while the measurement device includes the secondary circuit and a detector. The readout module may be on the transformer or may be remote. Reviews of clamp-on devices are available.

Presently, only one model of contact current monitor is commercially available. Monitoring is performed with a hand-held probe that is connected to a ground plate with a connective cable. The instrument has the impedance equivalent of the human body when making grasping contact with the hand, and output is in mA. The performance of this instrument has also been reviewed.

The measurements required by IEEE C95.1-1991 are shown in Table 9.8. Prior to collecting data, make sure that the operational frequency of the source is within the calibrated frequency range of the instrument. If it is not, meaningless measurement data will result from the evaluation. For broadband measurement of free fields, examine the interaction of the operator with the source, then survey the source to determine emissions.

Figure 9.6   Worker measuring broadbrand radio frequency nonionizing radiation (Courtesy: *Holaday Industries, Inc.* Eden Prairie, MN).

Table 9.8
RF Measurements Required by C95.1-1991

Frequency	Measurement
300 GHz to 300 MHz	E or H or W: spatial average
300 MHz to 100 MHz	E & H: spatial average
100 MHz to 3 kHz	E & H: spatial average [a] Induced currents Contact currents [b]

Notes: **a.** Consideration should be given to remote monitoring of electric fields, since the human body may significantly perturb these fields; **b.** Only if there are conductive objects in the workplace that can contain/store RF energy as an electrical current.

To determine spatial average exposures as specified in IEEE C95.1-1991, use a 200-cm dielectric "stickman" made from tubular PVC or CPVC tubing, using no conductive fasteners. Set this at the location of the operator, and measure at a minimum of 10 points on the stickman located 20 cm apart. Calculate an arithmetic average using power density, or the square of the E-field strengths, or the square of the H-field strengths.

It is possible to numerically estimate the intensity for some sources with source characteristics. For example, to calculate the maximum power density on the beam axis, W, in the near field of an aperture antenna, use:

$$W = 16P\eta 1/\pi D^2 \qquad (9.5a)$$

and:

$$W = 4P\eta 2/A \qquad (9.5b)$$

Where:

W = Power density on beam axis in the near field of an aperature antenna
P = Antenna power, watts
η = Antenna efficiency
D = Diameter of antenna
A = Cross-sectional area of antenna

To calculate the power density in the far field, use:

$$W = GP/4\pi r^2 \qquad (9.6a)$$

or:

$$W = GP/\pi r^2 \qquad (9.6b)$$

Where:

W = Power density on beam axis in the near field of an aperature antenna
G = Far field antenna gain
P = Antenna power, watts
r = Distance from the circular antenna

Equation 9.6b assumes 100% ground reflection, and will produce slightly higher estimates. Gain, G, may be determined from gain tables or calculated from:

$$G = 4\pi A/\lambda^2 \qquad (9.7)$$

Where:

G = Antenna gain
A = Effective area of antenna
λ = Wavelength

Since the near-field equations are independent of distance, assume that the calculated power density exists between the source and the boundary between the near and far fields, R, which is estimated by:

$$R = \frac{2 D^2}{\lambda} \qquad (9.8)$$

Where:

R = Power density
D = Antenna diameter
λ = Wavelength

To estimate the far-field hazard distance of an aperture antenna, substitute the appropriate exposure limit (EL), then rearrange to solve for distance (r). For FM and TV antennas, hazard distance formulas without and with ground reflections are, respectively:

$$r = (GP/4\pi EL)^{1/2} \qquad (9.9a)$$

or:

$$r = (GP/\pi EL)^{1/2} \qquad (9.9b)$$

Where:

r = Distance, meters
G = Abosolute gain of the antenna
P = Power, mW
EL= Exposure limit, V/m

To measure induced currents, place stand-on instruments at the operator position. Have the operator stand on the instrument and perform their normal function while you monitor the output. If the source is pulsed, be sure and monitor a sufficient number of work cycles to determine operator exposure. For clamp-on devices, clamp the current transformers around the appendage. Locate the data logger on the belt and secure the optical fiber so that it is out of the way. Switch the units on, and determine exposure.

D. ELF Fields

Instruments are available to measure electric and/or magnetic fields. One type of E-field instrument involves a displacement current sensor, while most B-field instruments utilize coil detectors or Hall-effect sensors.   Some instruments are broadband, while others have a narrow bandpass that is centered around 50/60 Hz. Figure 9.7 is an EMF instrument that detects and evaluates both E and H field strength.

Instruments with three-axis probes are available, while others have a single-axis probe. The tri-axis designs determine the root-mean-square (rms) field. Single-axis probes are direction sensitive.   However, to account for spatial and temporal changes in the B field, the resultant field may be determined by orientation of the sensor in three orthogonal planes in space (x,y,z). This is used to calculate the resultant from vector addition as:

$$M_r = (M_x^2 + M_y^2 + M_z^2)^{1/2} \qquad (9.10)$$

Where:

The applicable E- or B-field value may be substituted for M.

Figure 9.7   EMF field survey instrument (Courtesy: Holaday Industries, Inc., Eden Prairie, MN).

Some E-field instruments come with dielectric handles or nonconductive tripods. This allows the instrument to be separated from the human body. This is important because the body is highly conductive at these low frequencies and perturbs the E field, which can be a source of measurement error.

Personal exposure monitors are available for both E and B fields. However, due to the conductive nature of the body as mentioned above, these are more useful in measuring B fields in close proximity to the body because the body is relatively transparent to low-frequency B fields, i.e., it does not perturb the B field as it does the E field.

VII. CONTROL MEASURES

Measures that are amenable to the control of nonionizing radiation include engineering, administrative, and personal protective equipment (PPE). In general, engineering controls include isolation or shielding, and associated safety subsystems, such as interlocks and alarms. Administrative controls include pre-purchase reviews, medical approval, and information and training. PPE is most useful for UV and laser radiation, with more limited utility for RF radiation.

A. Ultraviolet Radiation

Some materials useful in the manufacture of barriers or enclosures include polyester films, cellulose acetate, methyl methacrylate-based polymers, polycarbonate, and standard glass (actinic UV). Commercially available welding curtains may be made of materials that are either opaque or transparent to visible wavelengths. Doors or access panels to barriers or enclosures that contain UV radiation in excess of the exposure limit should be interlocked. Visual and/or audible alarms should be used where workers may access a noninterlocked enclosure that contains hazardous levels of UV.

Sunscreens and sun blocks may be used to protect the skin from sunburn. Blocks usually contain zinc oxide or titanium dioxide, and reflect the UV radiation. Sunscreens contain photoactive agents that absorb and react with UV radiation. It is important that a broad spectrum active agent or agents are used, so protection is afforded not only to actinic UV but to UV-A. The ability of these materials to protect the skin is characterized by the sun protection factor (SPF). Generally, an SPF = 15 is a minimum recommendation. It is important to note that the SPF is dependent on the amount of material used. So, it is possible to use an insufficient amount of sunscreen and not achieve the rated SPF.

Factors affecting UV protection of fabrics include weave, color, stretch, weight, quality, and water content. One evaluation showed that UV leaks through openings in the mesh structure of woven fabrics. Fabrics made of continuous films exhibited good attenuation. Generally, laundering of fabrics increased the attenuation.

Optical density (OD) is used to specify the ability of protective eyewear to attenuate optical radiation, as detailed in the following equation:

$$OD = \log_{10}(ML/EL) \qquad (9.11)$$

Where:
    OD = Optical density
    ML = Measured level
    EL = Exposure limit
    Note: For UV radiation, the ML may be irradiance or radiant exposure, and EL must be in the same units.

Construction materials commonly used in protective eyewear include glass, polycarbonate, and CR-39. Glass and polycarbonate absorb actinic UV well, but all three materials may not provide sufficient attenuation in the near-UV at wavelengths above about 350 nm, unless UV-absorbing additives have been used. Even ordinary sunglasses provide some UV protection, although there may be substantial transmission windows in the UV. Results are also available on studies of UV-protective eyewear, prescription eyewear, children's sunglasses, and contact lenses.

B. Laser Radiation

In general, it is a good practice to enclose high-power (Classes IIIb and IV) beams. If it is necessary to view the beam in industrial applications, this may be done remotely with closed-circuit TV or video cameras. For benchtop work, beams should not be delivered at eye level, nor should they be directed at doorways or windows. Unauthorized personnel should be excluded from the area of the beam through the use of appropriate barriers, such as walls, curtains, or even rope barriers.

If open-beam systems are necessary, one should consult the ANSI Z136.1 standard. Control measures, as a function of the laser class, are addressed in this standard. Foremost among these is the establishment of a laser safety officer (LSO), an individual who has the responsibility and authority for laser safety.

The ANSI standard requires the establishment of laser controlled areas where there is access to Class IIIb and IV levels of laser radiation (open-beam systems). Class IV laser controlled areas require some form of entryway controls. The standard allows the LSO to select from three options that include non-defeatable interlocks on the entryway, defeatable entryway controls, and procedural entryway controls. Non-defeatable entryway controls are the simplest method, but this is often unworkable since individuals who are approved to work with the laser must have access to the area while the laser is working. The other two methods allow access to an area where there is an open beam, which requires authorization, training, additional engineering controls (e.g., light baffles at the entryway), and the proper use of PPE at the entrance. There is also a requirement for a temporary laser controlled area. This usually occurs where service activities require access to the beam from a powerful laser that is embedded within a Class 1 application. Often, barrier curtains are used to exclude unauthorized individuals from a temporary laser controlled area.

An area of particular concern in laser safety is beam alignment. In general, this should be done with the beam power as low as practical, or with an alignment laser which is usually a low-power laser with a visible beam that is coaxial with the application beam. Only diffuse reflections from beams should be viewed, and this should be performed while wearing the appropriate type of laser protective eyewear. Invisible beams may be viewed while using viewing cards. IR beams may be observed with infrared viewers that produce a visible image that may be seen through IR-protective eyewear.

There are many different types of protective eyewear available because of the large number of laser types (wavelengths) and output powers. Hence, it is important to make sure that the proper eyewear is prescribed and used. Eyewear may be either spectacles or goggles, while faceshields may be used to protect the face and eyes from scattered UV radiation or possibly from reflected CO_2 laser radiation. Eyewear may reflect or absorb radiation, while holographic filters have been evaluated. Useful guides to the selection of laser protective eyewear have been published.

Laser protective eyewear is characterized by optical density (OD) and wavelength. In general, an OD of 6, which represents a reduction of the incident radiation by a factor of 1 million, should be considered a practical ceiling on useful values of OD. This is because it is possible for the eyewear to fail (crack, melt, bleach) when exposed to high-intensity laser beams.

Eyewear must be marked with OD and wavelength so users may select the proper eyewear. Obviously, this becomes an important issue when more than one type of eyewear is used in an area with more than one type of laser. Eyewear should be used as directed, and stored properly. It should be inspected routinely to determine if there is pitting, crazing, or solarization of the lenses, and if goggle straps or spectacle sideshields are in good condition.

The ANSI standard also requires medical approval and training for users of Class IIIb and IV lasers. Standard operating procedures are required for Class IV lasers, and recommended for Class IIIb lasers. These should address alignment procedures, since a large proportion of the reported laser accidents have occurred during alignments.

C. Radio-Frequency Radiation

Shielding effectiveness (SE) is used to describe the effectiveness of a material as an RF shield. SE, in decibels, results from the contributions of reflection (R), absorption (A), and internal reflection (B).

E-field shielding materials include silver, copper, gold, aluminum, brass, bronze, tin, lead, and conductive polymers. These materials may be combined or machined to produce plated plastics, composite plastics, laminates and film coatings, clad metals, conductive paints, and arc-sprayed metals. Meshes, other woven textiles, and perforated materials may be used. H-field materials are iron, some stainless steels, mile steel, and nickel-iron and cobalt-iron alloys.

A control measure that is useful when openings are required in an enclosure for conveyance of product or for ventilation, is the *waveguide-beyond-cutoff*. A *waveguide* is a metal tube (circular, rectangular, or square) that is used to confine and guide electromagnetic waves. Although most waveguides are sized to allow optimum propagation of RF energy, a waveguide-beyond-cutoff is sized to minimize propagation of RF waves that are below a specific (cutoff) wavelength.

For frequencies near the human, whole-body, grounded-resonance frequencies, about 10 to 40 MHz, it has been shown that the SAR may be reduced by separating the body from the ground plane by a small distance. This breaks the connection with earth, shifting the body into the "free-space" absorption rate mode, which effectively reduces the absorption rate at (now) subresonant frequencies for free-space absorption. This effect has been demonstrated by simulating an air gap between the test subject and the ground with expanded polystyrene and hydrocarbon resin foam. The same effect has been observed with everyday footwear and socks, where the SAR has been modified when volunteers were exposed on a ground plane at frequencies between 10 and 40 MHz. The use of shoes and socks reduced the SAR in comparison to barefoot volunteers by the following: 15 to 45% for nylon socks and thin leather-soled shoes, and 35 to 75% for wool socks and rubber soles.

RF protective suits have been a controversial topic in RF safety programs. In general, suit materials and entire suits should be tested at the frequencies and power levels to which they will be exposed. In addition to determining their SE, attention must be paid to the potential for leakage through suit openings and zippers, arcing, and flammability. Information and some evaluations of various suits are available.

The ANSI C95.2 subcommittee recommends the design and color scheme of a warning symbol to be used for RF radiation between 300 kHz and 100 GHz. The symbol is a point-source antenna with emanating wavefronts. Recommendations for inclusion of this symbol onto a sign are made by ANSI.

D. ELF Fields

There are a number of control options for ELF fields, including shielding, field cancellation, and distance. E fields are easily shielded by many materials that are commonly used in construction such as soil, metal framing, and wood products. However, most common materials are transparent to low-frequency B fields, with the exclusion of magnetizable materials such as ferrous materials and special alloys containing nickel and cobalt. Hence, a good approach to control B field, is to provide for field cancellation in the design of the power line, appliance, motor, or transformer.

Because the intensity of these low-frequency fields decreases rapidly with distance, movement of workers a short distance will usually reduce exposure. For example, a mushroom pedestal controller can be moved away from an application, thereby reducing worker exposure.

VIII. CASE STUDY

Some sources may generate more than one type of nonionizing radiation. For example, a microlithography unit was used in an exposure application in the electronics' industry. This device generated both RF (microwave) and UV radiation. Microwave radiation was produced at 2.45 GHz by twin magnetrons. The magnetrons were suspended above mercury-vapor lamps. The microwave energy emitted by the magnetrons was used to energize the lamps, which then emitted UV radiation. The magnetrons and lamps were located within a partially shielded (metallic) enclosure. The box-like enclosure completely shielded both microwave and UV radiation, except at its base.

Here, there was a slot-like opening that was used to provide ventilation.

The purpose of the tool was to expose photosensitive materials to the UV radiation. The product was loaded into an aluminum "boat" by a laboratory technician who sat at the work station. The boats were then automatically introduced into a cavity located at the base of the box-like enclosure, where they were exposed to UV radiation. Employees in the area wore clean-room smocks and hoods, gloves, and safety glasses.

During the initial inspection, the industrial hygienist recognized that there was the potential for exposure to both RF and UV. Therefore, measurements were performed using broadband instruments for UV (irradiance) and microwave radiation (electric-field strength). The results showed that microwave radiation leakage was minimal, being much less than the recommended value for 2.45 GHz in the exposure guidelines. However, the much shorter UV wavelengths were emitted out of the slot, and posed a potential for overexposure to the hands of the operator. Also, there was the possibility of relatively high levels of UV radiation being reflected from the aluminum boats, and this could reach the face and eyes.

The capability of the safety glasses and the clean-room gloves to attenuate UV radiation was determined by measurement, at the time of the initial evaluation. This involved the placement of the material under test (one lens or one layer of glove material) over the detector. The detector was spatially oriented at typical use distances from the source, and at locations that would be occupied by either the eye or the hand. This was compared to a measurement at the same location, but without the test material covering the detector. The safety glasses, which had polycarbonate lenses, reduced exposure below the limits of detection of the instrument. However, the gloves reduced the UV irradiance by only a factor of about two, which was judged to be insufficient.

Therefore, it was necessary to provide additional control measures for UV exposure. The priority was to do so through the use of engineering controls. After brief study, it was determined that it was possible to modify the base of the enclosure with a light baffle. This was made from a rectangular metal piece that was added to the exterior of the existing enclosure. The metal was angled, at 90°, on two sides so that it not only shielded the area directly in front of the operators chair, but also shielded either side of the existing enclosure, locations where the boats entered and exited the process. This could be done without affecting the process, and the baffle could be painted the same color as the rest of the tool, allowing the client to achieve the desired appearance.

After construction, the baffle was mounted in a position where it was slightly offset from the existing enclosure. This allowed for a minimum impact on airflow. Post-measurement data showed that UV leakage was greatly reduced, which was attributed to scattering and containment within the baffled enclosure.

REVIEW QUESTIONS

1. By convention, nonionizing radiation is characterized by:

   a. Frequency or photon energy
   b. Photon energy or field strength
   c. Wavelength and frequency
   d. Permittivity and field strength

2. In nonionizing radiation protection, the dose is in some terms of:

   a. Energy in joules
   b. Power in watts
   c. Penetration depth in centimeters
   d. Wavelength in nanometers or micrometers

3. The target organs/systems for UV-induced health effects are:

   a. Skin, eyes, and blood
   b. Eyes, immune system, and skin
   c. Endocrine system, immune system, and eyes
   d. Eyes, skin, and endocrine system

4. Two quantities important in the evaluation of UV exposure are the:

   a. Radiometric and photometric systems
   b. Radiance and irradiance
   c. Power density and dose rate
   d. Irradiance and radiant exposure

5. The acceptable duration of exposure to a broadband source of UV, where the effective irradiance is 5 microwatts per centimeter square, is:

   a. 10 min   b. 60 sec   c. 10 sec   d. 8 hr

6. The retinal hazard region is composed of wavelengths from:

   a. 400 to 1400 mm
   b. Visible through IR-B
   c. 320 to 1400 mm
   d. UV through visible

7. The risk of laser-induced damage to the eyes and skin is greatest for lasers in:

   a. Classes IIIa and IIIb   b. Classes IIIb and IV   c. Class II   d. Class IIIb

8. Lasers most often used in materials processing applications (e.g., welding, cutting, drilling, etc.) include the:

    a. Excimer, carbon dioxide, and neodymium:YAG
    b. Carbon dioxide, neodymium:YAG, helium-neon
    c. Argon, helium-neon, and carbon dioxide
    d. Neodymium:YAG, carbon dioxide, and diode

9. The nominal hazard zone for a 0.25-sec exposure to a 20-watt argon laser (beam divergence = 1 milliradian; exit diameter = 1 millimeter; exposure limit = 0.0025 watts per centimeter square) is:

    a. 10,009 cm   b. 33,400 ft.   c. 3330 ft.   d. 3000 m

10. The effect most often sited as the biological basis for the RF exposure guidelines is:

    a. Reproductive effects in physical therapists
    b. Cataracts in test animals
    c. Reversible behavior in disruption in test animals
    d. Testicular effects in microwave workers

11. The fundamental quantity of the RF exposure criteria is the:

    a. Specific absorption rate
    b. Power density
    c. Field strength
    d. Specific adsorption ratio

12. A source of RF radiation that has been often demonstrated as a potential source of overexposure is the:

    a. Induction heater   b. Dielectric heater   c. Video display terminal   d. SATCOM system

13. The on-axis power density in the near field of a microwave dish antenna (power = 50 watts; antenna diameter = 2 meters; frequency = 3 GHz) is:

    a. 6.4 milliwatts per centimeter square
    b. 640 milliwatts per meter square
    c. 100 milliwatts per meter square
    d. 2.1 milliwatts per centimeter square

14. In the evaluation of exposure to low-frequency RF fields, it may be necessary to evaluate the following quantities:

    a. Power density and specific absorption
    b. E and H field strength, induced currents, and contact currents
    c. Power density and induced currents
    d. Current density and power density

15. The ELF exposure guidelines are based on initiating the value of endogenous:

    a. Specific absorption rate
    b. Magnetic-field strength
    c. Magnetic flux
    d. Current density

16. The ELF magnetic-flux density in three concentric, mutually orthogonal planes is 15 mG, 500 mG, and 33 mG. What is the resultant field strength?

    a. 501 mG   **b.** 50 mG   **c.** 548 mG   **d.** 32 mG

ANSWERS

1. c.
2. a.
3. b.
4. d.
5. a. $t = \dfrac{0.003 \text{ J/cm}^2}{5 \times 10^{-6} \text{ W/cm}^2} \times \dfrac{1 \text{ min}}{60 \text{ sec}}$

    $t = 10 \text{ min}$

6. a.
7. b.
8. a.
9. c. $r_{NHZ} = [(1/0.001) \times ((4 \times 20\text{W}/\pi \times 0.0025 \text{ W/cm}^2 3) - 0.01 \text{ cm}^2)^{0.5}] \times 1 \text{ ft}/30.28$
cm

    $r_{NHZ} = 3333 \text{ ft.}$

10. c.
11. a.
12. b.
13. a. $W = \dfrac{4 \times 50,000 \text{ mW}}{(\pi/4) \times (200 \text{ cm})^2}$

    $W = 6.4 \text{ mW/cm}^2$

14. b.
15. d.
16. a.

References

Adams, A. J. and Skipper, J. E., Safe use of infrared viewing cards, *Health Phys.*, 59, 225, 1990.

Adrian, D. J., Auditory and visual sensations stimulated by low-frequency electric currents, *Radio Sci.*, 12(6S), 243, 1977.

Amato, J. A., Use protective clothing for safety in RF fields, *Mobile Radio Technol.*, 12(4), 40, 1994.

American Conference of Governmental Industrial Hygienists, *1996 TLVs and BEIs*, ACGIH, Cincinnati, OH, 1996, 121.

American Conference of Governmental Industrial Hygienists, *1996 TLVs and BEIs*, ACGIH, Cincinnati, 1996, 97.

American Conference of Governmental Industrial Hygienists, *1996 TLVs and BEIs*, ACGIH, Cincinnati, OH, 1996, 113.

American Conference of Governmental Industrial Hygienists, *1996 TLVs and BEIs*, ACGIH, Cincinnati, OH, 1996, 118.

American National Standards Institute, *Radio Frequency Radiation Hazard Warning Symbol* (ANSI C95.2-1982), Institute of Electrical and Electronics Engineers, New York, 1982.

American National Standards Institute, *For the Safe Use of Lasers* (ANSI Z136.1-1993), Laser Institute of America, Orlando, FL, 1993.

Anderson, V. and Joyner, K. H., Specific absorption rate levels measured in a phantom head exposed to radio frequency transmissions from analog hand-held mobile phones, *Bioelectromagnetics*, 15, 60, 1995.

Anderson, W. J. and Gebel, R. K. H., Ultraviolet windows in commercial sunglasses, *Appl. Opt.*, 16, 515, 1977.

Appleton, B., Hirsch, S. E., and Brown, P. V. K., Investigation of single-exposure microwave ocular effects at 3000 MHz, *Ann. N.Y. Acad. Sci.*, 247, 125, 1975.

Armanini, D., Conti, R., Mantini, A., and Nicolini, P., Measurements of power-frequency electric and magnetic fields around different industrial and household sources, paper 36-107, in *CIGRE, Proceedings of the 33rd Session of the International Conference on Large High Voltage Electric Systems*, Paris, August, 1990.

Armstrong, B. K. and Kricker, A., Epidemiology of sun exposure and skin cancer. *Cancer Surveys*, 26, 133, 1996.

Austin, L. and Brackman, U., Dye lasers and laser dyes, in *The Photonics Design and Applications Handbook* (Book 3), Laurin Publishing Co., Pittsfield, MA, 1991, H-204.

Balzano, Q., Garay, O., and Steel, F. R., A comparison between the energy deposition in portable radio operators at 900 MHz and 450 MHz, in *Twenty-Eighth IEEE Vehicular Technology Conference* (IEEE Cat. No. 78CH1297-1VT), Institute of Electrical and Electronics Engineers, Piscataway, NJ, 1978, 46.

Balzano, Q., Garay, O., and Steel, F. R., Heating of biological tissue in the induction field of VHF portable radio transmitter, *IEEE Trans. Veh. Technol.*, VT-27, 51, 1978.

Baroncelli, P., Battisti, S., Checcucci, A., Comba, P., Grandolfo, M., Serio, S., and Vecchia, P., A health examination of railway high-voltage substation workers exposed to ELF electromagnetic fields, *Am. J. Ind. Med.*, 10, 45, 1986.

Bart, R. S. and Schnall, S., Eye color in darkly pigmented basal-cell carcinomas and malignant melanomas, *Arch. Dermatol.*, 107, 206, 1973.

Beers, G. J., Biological effects of weak electromagnetic fields from 0 Hz to 200 MHz: a survey of the literature with special emphasis on possible magnetic resonance effects, *Mag. Res. Imag.*, 7, 309, 1989.

Beitner, H., Norell, S. E., Ringborg, U., Wennersten, G., and Mattson, R., Malignant melanoma: aetiological importance of individual pigmentation and sun exposure, *Brit. J. Dermatol.*, 122, 43, 1990.

Belforte, D. A., High-power Nd:YAG lasers shine in industrial processing, *Laser Focus World*, 28(9), 69, 1992.

Bellossi, A., Desplaces, A., and Morin, R., Effect of pulsed magnetic field on tumoral C3H/Bi female mice, *Cancer Biochem. Biophys.*, 10, 59, 1988.

Beniashvili, D. Sh., Bilanishvili, V. G., and Menabde, M. Z., Low-frequency electromagnetic radiation enhances the induction of rat mammary tumors by nitrosomethyl urea, *Cancer Lett.*, 61, 75, 1991.

Benoit, H., Clark, J., and Keon, W. J., Installation of a commercial excimer laser in the operating room, *J. Laser Appl.*, 1(3), 45, 1989.

Beral, V., Shaw, H., Evans, S., and Milton, G., Malignant melanoma and exposure to fluorescent lighting at work, *Lancet*, 2 (Aug. 7), 290, 1982.

Berman, E., Carter, H., and House, D., Observations of Syrian hamster fetuses after exposure to 2450-MHz microwaves, *J. Microwave Power*, 17, 107, 1982.

Berman, E., Carter, H., and House, D., Reduced weight in mice offspring after *in utero* exposure to 2450-MHz (cw) microwaves, *Bioelectromagnetics*, 3, 285, 1982.

Berman, E., Chacon, L., House, D., Koch, B. A., Koch, W. E., Leal, J., Lovtrup, S., Mantiply, E., Martin, A. H., Mild, K. H., Monahan, J. C., Sandstrom, M., Shamsaifer, K., Tell, R., Trillo, M. A., Ubeda, A., and Wagner, P., Development of chicken embryos in a pulsed magnetic field, *Bioelectromagnetics*, 11, 169, 1991.

Berne, B. and Fischer, T., Protective effects of various types of clothes against UV radiation, *Acta. Derm. Venereol.*, 60, 459, 1980.

Bersin, R. L., A survey of plasma-etching processes, *Solid State Technol.*, 19, 31, 1976.

Betz, A., Retzbach, M., Alber, G., and Prange, W., Automated laser systems for high volume production, *ICALEO 1990*, 627, 1993.

Birngruber, R., Hillenkamp, F., and Gabel, V. P., Theoretical investigations of laser thermal retinal injury, *Health Phys.*, 48, 781, 1985.

Bissett, D. L., Hannon, D. P., and Orr, T. V., An animal model of solar-aged skin: histological, physical, and visible changes in UV-irradiated hairless mouse skin, *Photochem. Photobiol.*, 46, 367, 1987.

Bitran, M. E., Charron, D. E., and Nishio, J. M., *Microwave Emissions and Operator Exposures from Traffic Radars used in Ontario* Ontario Ministry of Labour, Weston, Ontario, 1992.

Boettner, E. A. and Wolter, J. R., Transmission of the ocular media, *Invest. Ophthalmol.*, 1, 776, 1962.

Boettrich, E. P., "Hazards Associated with Ultraviolet Radiation in Academic and Clinical Laboratories," Masters thesis, Department of Radiation Biology and Biophysics, University of Rochester, Rochester, NY, 1985.

Boettrich, E. P., Hazards Associated with Ultraviolet Radiation in Academic and Clinical Laboratories, presented at the American Industrial Hygiene Conference, Orlando, FL, 1990.

Boldrey, E. E., Little, H. L., Flocks, M., and Vassiliadis, A., Retinal injury to industrial laser burns, *Ophthalmol.*, 88, 101, 1981.

Bonneville Power Authority, *Electrical and Biological Effects of Transmission Lines* (DOE/BP-945), BPA, Portland, OR, 1993.

Bonomi, L. and Bellucci, R., Considerations of the ocular pathology in 30,000 personnel of the Italian telephone company (SIP) using vdts, *Bollettion Di Oculistica*, 68 (S7), 85, 1989.

Borkman, R. F., Cataracts and photochemical damage in the lens, in *Human Cataract Formation*, Pitman, London, 1984, 88.

Bos, A. J. J. and De Haas, M. J., On the safe use of a high power ultraviolet laser, in *Human Exposure to Ultraviolet Radiation: Risks and Regulations*, Passchier W. F. and Bosnjakovic, B. F. M., Eds., Elsevier Science Publishers, New York, 1987, 377.

Bowker, K. W., Hazards associated with sources of ultra-violet radiation used in a research environment, in *Human Exposure to Ultraviolet Radiation: Risks and Regulations*, Passchier, W. F. and Bosnjakovic, B. F. M., Eds., Elsevier Science Publishers, New York, 1987, 371.

Bracken, T. D., Exposure assessment for power frequency electric and magnetic fields, *Am. Ind. Hyg. Assoc. J.*, 54, 165, 1993.

Bracken, T. D., Occupational exposure assessment for electric and magnetic fields in the 10-1000 Hz frequency range, in *Proceedings of the Scientific Workshop on the Health Effects of Electric and Magnetic Fields on Workers* (DHHS (NIOSH) Publication No. 91-111), Bierbaum, P. J. and Peters, J. M., Eds., National Institute of Occupational Safety and Health, Cincinnati, OH, 1991.

Bracken, T. D., Rankin, R. F., Senior, R. S., Allredge, J. R., and Sussmn, S.S., Magnetic field exposure among utility workers, *Bioelectromagnetics*, 16, 216, 1995.

Brandao, F. M., de Castro, J. C., and Pecegueiro, M.., Photoallergy to musk ambrette, *Contact Dermatitis*, 9, 332, 1983.

Brandt, L. and Nielsen, C. V., Congenital malformations among children of women working with video display terminals, *Scand. J. Work Environ. Health*, 16, 329, 1990.

Breysse, P. N., Matanoski, G. M., Elliott, E. A., Francis, M., Kaune, W., and Thomas, K., 60 Hertz magnetic field exposure assessment for an investigation of leukemia in telephone lineworkers, *Am. J. Ind. Med.*, 26, 681, 1994.

Breysse, P., Lees, P. S. J., McDiarmid, M. A., and Curbow, B., ELF magnetic field exposures in an office environment, *Am. J. Ind. Med.*, 25, 177, 1994.

Bridges, T. J., Strand, A. R., Wood, O. R., Patel, C. K. N., and Karlin, D. B., Interaction of carbon dioxide laser radiation with ocular tissue, *IEEE J. Quant. Electron.*, QE-20, 1449, 1984.

Brill, M. L., Taking advantage of laser properties to enhance demonstrations and student laboratories, *J. Laser Appl.*, 7(1), 58, 1995.

Brookner, E., Phased-srray radars, *Sci. Am.*, 252, 94, 1985.

Brownell, A. S., Parr, W. H., and Hysell, D. K., Skin and carbon dioxide laser radiation, *Arch.*

Environ. Health, 18, 437, 1969.

Bruls, W. A. G., Slaper, H., van der Leun, J. C., and Berrens L, Transmission of human epidermis and stratum corneum as a function of thickness in the ultraviolet and visible wavelengths, *Photochem. Photobiol*, 40, 485, 1984.

Byus, C. V., Peiper, S. E., and Adey, W. R., The effects of low-energy 60-Hz environmental electromagnetic fields upon the growth-related enzyme ornithine decarboxylase, *Carcinogenesis*, 8, 1385, 1987.

Carey, Jr., W. P., Thompson, C. J., Synnestvedt, M., Guerry, D., Halpern, A., Schultz, D., and Elder, D. E., Dysplastic nevi as a melanoma risk factor in patients with familial melanoma, *Cancer*, 74, 3118, 1994.

Carpenter, R. and Van Ummersen, C., The action of microwave radiation on the eye, *J. Microwave Power*, 3, 3, 1968.

Carstensen, E. L., Buettner, A., Genberg, V. L., and Miller, M. W., Sensitivity of the human eye to power frequency electric fields, *IEEE Trans. Biomed. Eng.*, BME-32, 561, 1985.

Carstensen, E. L., Magnetic fields and cancer, *IEEE Eng. Med. Biol. Mag.*, 14(4), 362, 1995.

Carts, Y. A., Scientific dye lasers employ a variety of technologies, *Laser Focus World*, 26(2), 57, 1990.

Cartwright, C. E., Breysse, P. N., and Booher, L., Magnetic field exposures in a petroleum refinery, *Appl. Occup. Environ. Hyg.*, 8, 587, 1993.

Challoner, A. V. J., Corless, D., Davis, A., Deane, G. H. W., Diffey, B. L., Gupta, S. P., and Magnus, I. A., Personal monitoring of exposure to ultraviolet radiation, *Clin. Exp. Dermatol.*, 1, 175, 1976.

Chatterjee, I., Gu, Y.-G., and Gandhi, O. P., Quantification of electromagnetic absorption in humans from body-mounted communication transceivers, *IEEE Trans. Veh. Technol.*, VT-34, 55, 1985.

Chou, C.-K., Guy, A. W., and McDougall, J. A., Shielding effectiveness of improved microwave-protective suits, *IEEE Trans. Microwave Theory Tech.*, MTT-35, 995, 1987.

Chou, C.-K., Guy, A. W., Kunz, L. L., Johnson, R. B., Crowley, J. J., and Krupp, J. H., Long-term, low-level microwave irradiation of rats, *Bioelectromagnetics*, 13, 469, 1992.

Cleary, S. F. and Pasternack, B. S., Lenticular changes in microwave workers, *Arch. Environ. Health*, 12, 23, 1966.

Cleary, S. F., Pasternack, B. S., and Beebe, G. W., Cataract incidence in radar workers, *Arch. Environ. Health*, 11, 79, 1965.

Cleveland, R. F. and Athey, T. W., Specific absorption rate (sar) in models of the human head exposed to hand-held UHF portable radios, *Bioelectromagnetics*, 10, 173, 1989.

Coelho, A. M., Easley, S. P., and Rogers, W. R., Effects of exposure to 30-kV/m, 60 Hz electric fields on the social behavior of baboons, *Bioelectromagnetics*, 12, 117, 1991.

Cole, C., Forbes, P. D., Davies, R. E., and Urbach, F., Effect of indoor lighting on normal skin, *Ann. N.Y. Acad. Sci.*, 453, 305, 1985.

Conover, D. L., Moss, C. E., Murray, W. E., Edwards, R. M., Cox, C., Grajewski, B., Werren, D. M., and Smith, J. M., Foot currents and ankle sars induced by dielectric heaters, *Bioelectromagnetics*, 13, 103, 1992.

Conover, D. L., Murray, W. E., Foley, E. D., Lary, J. M., and Parr, W. H., Measurement of electric- and magnetic-field strengths from industrial radio-frequency (6-38 MHz) plastic sealers, *Proc. IEEE 68*, 17, 1980.

Conover, D. L., Murray, W. E., Lary, J. M., and Johnson, P. H., Magnetic field measurements near RF induction heaters, *Bioelectromagnetics*, 7, 83, 1986.

Cox, C., Murray, W. E., and Foley, E. P., Occupational exposures to radiofrequency radiation (18-31 MHz) from RF dielectric heat sealers, *Am. Ind. Hyg. Assoc. J.*, 43, 149, 1982.

Cristofolini, M., Franceschi, S., Tasin, L., Zumiani, G., Piscioli, F., Talamini, R., and La Vecchia, C., Risk ractors for cutaneous malignant melanoma in a northern Italian population, *Int. J. Cancer*, 39, 150, 1987.

Culkin, T. J. and Kugler, T. R., Industrial laser processing, in *The Photonics Design and Applications Handbook* (Book 3), Laurin Publishing Co., Pittsfield, MA, 1991, H-216.

Curtis, R. A., Occupational exposures to radiofrequency radiation from fm radio and tv antennas, in *Nonionizing Radiation--Proceedings from a Topical Symposium*, American Conference of Governmental Industrial Hygienists, Cincinnati, OH, 1990, 211.

D'Andrea, J. A., Microwave radiation absorption: behavioral effects, *Health Phys.*, 61, 29, 1991.

D'Andrea, J., Gandhi, O., and Lords, J. L., Behavioral and thermal effects of microwave radiation at resonant and nonresonant wavelengths, *Radio Sci.*, 12(6S), 251, 1977.

Davey, J. B., Diffey, B. L., and Miller, J. A., Eye protection in psoralen photochemotherapy, *Brit. J. Dermatol.*, 104, 295, 1981.

Davis, H. P., Mizumori, S. J. Y., Allen, H., Rosenzweig, M. R., Bennett, E. L., and Tenforde, T. S., Behavioral studies with mice exposed to dc and 60-Hz magnetic fields, *Bioelectromagnetics*, 5, 147, 1984.

Davis, R. L. and Mostofi, F. K., Cluster of testicular cancer in police officers exposed to hand-held radar, *Am. J. Ind. Med.*, 24, 231, 1993.

de Lorge, J. O. and Grissett, J. D., Behavioral effects in monkeys exposed to extremely low frequency electromagnetic fields, *Int. J. Biometeor.*, 21, 357, 1977.

de Lorge, J. O., Disruption of behavior in mammals of three different sizes exposed to microwaves: extrapolation to larger mammals, in *Proceedings of the 1978 Symposium on Electromagnetic Fields in Biological Systems*, International Microwave Power Institute, Edmonton, Canada, 1978, 215.

de Lorge, J. O., The effects of microwave radiation on behavior and temperature in Rhesus monkeys, in *Biological Effects of Electromagnetic Waves* (HEW Publication (FDA) 77-8010), Johnson, C. C. and Shore, M. L., Eds., Bureau of Radiological Health, Rockville, MD, 1976, 158.

de Lorge, J. O., The thermal basis for disruption of operant behavior by microwaves in three animal species, in *Microwaves and Thermoregulation*, Adair, E. R., Ed., Academic Press, New York, 1983, 379.

Delgado, J. M. R., Leal, J., Monteagudo, J. L., and Gracia, M. G., Embryological changes induced by weak, extremely low frequency electromagnetic fields, *J. Anat.*, 134, 533, 1982.

Dennis, J. A., Muirhead C. R., and Ennis, J. R., *Human Health and Exposure to Electromagnetic Radiation* (NRPB-R241), National Radiological Protection Board, Chilton, Oxon, UK, 1992.

Dennis, J. E., Amending the CDRH performance standard for laser products, *J. Laser Appl.*, 6, 49, 1994.

Dietz, A. and Bradford, E., Safe handling of excimer gases, in *The Photonics Design and Applications Handbook* (Book 3), Laurin Publishing Co., Pittsfield, MA, 1991, H-240.

Diffey, B. L. and McKinlay, A. F., The UVB content of 'UVA fluorescent lamps'and its erythemal effectiveness in human skin, *Phys. Med. Biol.*, 28, 351, 1983.

Diffey, B. L. and Oakley, A. M., The onset of ultraviolet erythema. *Brit. J. Dermatol.*, 116, 183, 1987.

Diffey, B. L. and Roscoe, A. H., Exposure to solar ultraviolet radiation in flight, *Aviat. Space, Emviron. Med.*, 61, 1032, 1990.

Diffey, B. L., Kerwin, M., and Davis, A., The anatomical distribution of sunlight, *Brit. J. Dermatol.*, 97, 407, 1977.

Diffey, B. L., Larko, O., and Swanbeck, G., UV-B doses received during different outdoor activities and UV-B treatment of psoriasis, *Brit. J. Dermatol.*, 196, 33, 1982.

Diffey, B. L., Tate, T. J., and Davis, A., Solar dosimetry of the face: the relationship of natural ultraviolet radiation exposure to basal cell carcinoma localisation, *Phys. Med. Biol.*, 24, 931, 1979.

Dimbylow, P. J. and Mann, S. M., SAR calculations in an anatomically realistic model of the head for mobile communication transceivers at 900 MHz and 1.8 GHz, *Phys. Med. Biol.*, 39, 1537, 1994.

Doyle, D. J. and Kokosa, J. M., Chemical by-products of laser cutting of Kevlar, *Polymer Preprints*, 27, 206, 1986.

Easley, S. P. Coelho, A. M.., and Rogers, W. R., Effects of a 30-kV/m, 60 Hz electric field on the social behavior of baboons: a crossover experiment, *Bioelectromagnetics*, 13, 395, 1992.

Easley, S. P. Coelho, A. M.., and Rogers, W. R., Effects of exposure to 60-kV/m, 60 Hz electric fields on the social behavior of baboons, *Bioelectromagnetics*, 12, 361, 1991

ECRI, Laser safety eyewear, *Health Devices*, 22(4), 159, 1993.

Elder, J. A. and Cahill, D. F., Eds., *Biological Effects of Radiofrequency Radiation* (Report No. EPA-600/8-83-026F), National Technical Information Service, Springfield, VA, 1984.

Ellingson, O. L., The characterization of a black light device: a hazard evaluation process, *Am. Ind. Hyg. Assoc. J.*, 47, 488, 1986.

Engel, D., Laser generated metal dust explosive potential, in *Proceedings of the 1992 International Laser Safety Conference*, Laser Institute of America, Orlando, FL, 1993, 4I-21.

English, D. R., Rouse, I. L., Zhong, X., Watt, J. D., D'Arcy, J. H., Heenan, P. J., and Armstrong, B. K., Cutaneous malignant melanoma and fluorescent lighting, *J. Natl. Cancer Soc.*, 74, 1191, 1985.

Envall, K. R. and Lamanna, A., Engineering Control Measures for Blocking Ultraviolet Radiation, poster session presented at the American Industrial Hygiene Conference, Detroit, MI, 1984.

Envall, K. R. and Lamanna, A., Engineering Control Measures for Blocking Ultraviolet Radiation, poster session presented at the American Industrial Hygiene Conference, Detroit, MI, 1984.

Environmental Protection Agency, *Ultraviolet Radiation and Melanoma With a Special Focus on Assessing the Risks of Stratospheric Ozone Depletion* (EPA 400/1-87/001D), EPA, Washington,

D.C., 1987.

Evans, J. A., Savitz, D. A., Kanal, E., and Gillen, J., Infertility and pregnancy outcome among magnetic resonance imaging workers, *J. Occupat. Med.*, 12, 1191, 1993.

Faber, M., Ultraviolet radiation, in *Nonionizing Radiation Protection* (Second Edition, World Health Organization Regional Publications, European Series No. 25), Seuss, M. J. and Benwell-Morison, D. A., Eds., World Health Organization, Copenhagen, 1989, 13.

Fam, W. Z., Long-term biological effects of very intense 60 Hz electric field on mice, *IEEE Trans. Biomed. Eng.*, BME-27, 376, 1980.

Fears, T. R., Scotto, J., and Schneiderman, M. A., Mathematical models of age and ultraviolet effects on the incidence of skin cancer among whites in the United States, *Am. J. Publ. Health*, 66, 461, 1976.

Federal Communications Commission, Guidelines for evaluating effects of radiofrequency radiation (47 CFR Parts 1, 2, 15, 24, and 97), *Federal Register*, 61(153), 41006, 1996.

Feero, W. E., Electric and magnetic field management, *Am. Ind. Hyg. Assoc. J.*, 54, 205, 1993.

Fischer, T., Alsins, J., and Berne, B., Ultraviolet-action spectrum and evaluation of ultraviolet lamps for psoriasis healing, *Int. J. Dermatol.*, 23, 633, 1984.

Fisher, P. D., *Microwave Exposure Levels Encountered by Police Traffic Radar Operators* (MSU-ENGR-91-007), Michigan State University Press, East Lansing, MI, 1991.

Fisher, P. D., Microwave exposure levels encountered by police traffic radar operators, *IEEE Trans. Electromag. Compat.*, 35, 36, 1993.

Fleeger, A. and Moss, C. E., Airborne emissions produced by the interaction of a carbon dioxide laser with glass, metals, and plastics, in *Proceedings of the International Laser Safety Conference*, Laser institute of America, Orlando, FL, 1991, 3-23.

Floderus, B., Persson, T., Stenlund, C., Wennberg, A. Ost, A., and Knave, B., Occupational exposure to electromagnetic fields in relation to leukemia and brain tumors. a case-control study, *Cancer Causes Control*, 4, 465, 1993.

Folkes, J. A., Developments in laser surface modification and coating, *Surface Coatings Technol.*, 63, 65, 1994.

Forman, S. A., Holmes, C. K., McManamon T. V., and Wedding, W. R., Psychological symptoms and intermittent hypertension following acute microwave exposure, *J. Occup. Med.*, 24, 932, 1982.

Freeman, R. L., *Telecommunication Transmission Handbook*, 2nd ed., John Wiley & Sons, New York, 1981.

Frey, A. H. and Wesler, L. S., Interaction of psychoactive drugs with exposure to electromagnetic fields, *J. Bioelect.*, 9, 187, 1990.

Gandhi, O. P., Ed., *Biological Effects and Medical Applications of Electromagnetic Energy*, Prentice Hall, Englewood Cliffs, NJ, 1991.

Gandhi, O. P., Some numerical methods for dosimetry: extremely low frequencies to microwave frequencies, *Radio Sci.*, 30, 161, 1995.

Gandhi, O., Chen, J.-Y., and Riazi, A., Currents induced in human beings for plane-wave exposure conditions 0-50 MHz and for RF sealers, *IEEE Trans. Bio-Med. Eng.*, BME-33, 757, 1986.

Garland F. C., Shaw, E., Gorham, E. D., Garland, C. F., White, M. R., and Sinsheimer, P. J., Incidence of leukemia in occupations with potential electromagnetic field exposure in United States Navy personnel, *Am. J. Epidemiol.*, 132, 293, 1990.

Garrison, L. M., Murray, L. E., and S. Green, A. E. S., Ultraviolet limit of solar radiation at the earth's surface with a photon counting monochromator, *Appl. Opt.*, 17, 683, 1978.

Geissinger, L. G., Waller, P., Chartier, V. L., and Olsen, R. G., Electric and magnetic field reduction and research: a report to the Washington state legislature, *Proceedings of the American Power Conference 1993* (Vol. 55 - II), Institute of Technology, Chicago, IL, 1993, 1674.

Gellin, G. A., Kopf, A. W., and Garfinkel, L., Basal cell epithelioma, *Arch. Dermatol.*, 91, 38, 1965.

Gies, H. P., Roy, C. R., Elliott, G., and Zongli, W., Ultraviolet radiation protection factors for clothing, *Health Phys.*, 67, 131, 1994.

Gies, H. P., Roy, C. R., Toomey, S., MacLennan, R., and Watson, M., Solar UVR exposures of three groups of outdoor workers on the sunshine coast, Queensland, *Photochem. Photobiol.*, 62, 1015, 1995.

Gorham, B. J., Safety aspects in the use of outdoor and surveying lasers, *Opt. Laser Technol.*, 27, 19, 1995.

Grossweiner, L. I., Photodynamic therapy, *J. Laser Appl.*, 7(1), 51, 1995.

Grota, L. J., Reiter, R. J., Keng, P., and Michaelson, S., Electric field exposure alters serum melatonin but not pineal melatonin synthesis in male rats, *Bioelectromagnetics*, 15, 427, 1994.

Gulvady, N. U., UV keratoconjunctivitis vs. established dose effect relationships, *J. Occup. Med.*, 18, 573, 1976.

Guy, A. W., Chou, C.-K., Kunz, L. L., Crowley, J., and Krupp, J., *Effects of Long-Term Low-Level Radiofrequency Radiation Exposure on Rats Volume 9. Summary* (Report USAFSAM-TR-85-64), United States Air Force, School of Aerospace Medicine, Brooks Air Force Base, TX, 1985, 16.

Guy, A. W., Lin, J. C., Kramar, P. O., and Emery, A. F., Effect of 2450-MHz radiation on the rabbit eye, *IEEE Trans. Microwave Theory Tech.*, MTT-23, 492, 1975.

Hack, H. and Neuroth, N., Resistance of optical and colored glasses to 3-nsec laser pulses, *Appl. Optics*, 21, 3239, 1982.

Haferkamp, H., Goede, M., Engel, K., and Witebecker, J.-S., Hazardous emissions: characterization of CO_2 laser material processing, *J. Laser Appl.*, 7, 83, 1995.

Haifeng, L., Guanghuang, G., Dechang, W., Guidao, X., Liangshun, S., Jiemin, X., and Haibiao, W., Ocular injuries from accidental laser exposure, *Health Phys.*, 56, 711, 1989.

Ham, Jr., W. T. and Mueller, H. A., Ocular effects of laser infrared radiation, *J. Laser Appl.*, 3(3), 19, 1991.

Ham, Jr., W. T. and Mueller, H. A., Retinal sensitivity to damage from short wavelength light, *Nature*, 260, 153, 1976.

Ham, Jr., W. T. and Mueller, H. A., The photopathology and nature of the blue light and near-UV retinal lesions produced by lasers and other optical sources, in *Laser Applications in Medicine and Biology* (Volume 4), Wolbarsht, M. L., Ed., Plenum Press, New York, 1989, 191.

Ham, Jr., W. T., Mueller, H. A., and Ruffolo, Jr., J. J., Retinal effects of blue light exposure,

SPIE, 229, 46, 1980.

Ham, Jr., W. T., Mueller, H. A., Ruffolo, Jr., J. J., and Clarke, A. M., Sensitivity of the retina to radiation damage as a function of wavelength, *Photochem. Photobiol.*, 29, 735, 1979.

Ham, Jr., W. T., Ruffolo, Jr., J. J., Mueller, H. A., and Guerry, D., The nature of retinal radiation damage: dependence on wavelength, power level and exposure time, *Vision Res.*, 20, 1105, 1980.

Handcock, M. S. and Kolassa, J. E., Statistical review of the henhouse experiments: the effects of a pulsed magnetic field on chick embryos, *Bioelectromagnetics*, 13, 429, 1992.

Hankin, N. N., *An Evaluation of Selected Satellite Communication Systems as Sources of Environmental Microwave Radiation* (EPA-520/2-74-008), Washington, D.C., Environmental Protection Agency, 1974.

Hankin, N. N., *The Radiofrequency Radiation Environment: Environmental Exposure Levels and RF Radiation Emitting Sources* (EPA 520/1-85-014), Washington, D.C., U.S. Government Printing Office, 1986.

Hansan, T. and Khan, A. U., Photoxicity of tetracyclines: photosensitized emission of singlet delta dioxygen, *Proc. Natl. Acad. Sci.*, 83, 4604, 1986.

Hansson, H.-A., Purkinje nerve cell changes caused by electric fields–ultrastructural studies on long-term effects on rabbits, *Med. Biol.* 59, 103, 1981.

Hansson, H. A., Rozell, B., and Stemme, S., Effects of experimental exposure to power–frequency fields on the nervous system, in *Interaction of Biological Systems with Static and ELF Electric and Magnetic Fields*, Anderson, L. E., Kelman, B. J., and Weigel, R. J., Eds., Battelle Pacific Northwest Laboratory, Richland, WA, 1987, 297.

Harris, M. G., Dang, M. Garrod, S., and Wong, W., Ultraviolet transmittance of contact lenses, *Opt. Vis. Sci.*, 71, 1, 1994.

Harvey, S. M. and Sherar, M. D., The effect of distribution transformer location on street level magnetic fields and residential ground currents (Report No. 90-272-K), Ontario Hydro Research Division, Ontario, Canada, 1990.

Hatch, M., The epidemiology of electric and magnetic field exposures in the power frequency range and reproductive outcomes, *Ped. Perinatal Epidemiol.*, 6, 198, 1992.

Hawk, J. L. M., Photosensitivity in the elderly, *Brit. J. Dermatol.* 122 (Suppl. 35), 29, 1990.

Heath, B. A. and Kammerdiner, L., Plasma processing for vlsi, in *VLSI Handbook*, Einspruch, N. G., Ed., Academic Press, New York, 1985, 487.

Heath, C. W., Electromagnetic field exposure and cancer: a review of epidemiologic evidence, *Cancer J. Clin.*, 65, 29, 1996.

Heinrichs, W. L., Fong, P., Flannery, M., Heinrichs, S. C., Crooks, L. E., Spindle, A., and Pedersen, R. A., Midgestational exposure of pregnant BALB/c mice to magnetic resonance imaging conditions, *Mag. Res. Imaging*, 6, 305, 1988.

Henkes, H. E. and Zuidema, H., Accidental laser coagulation of the central fovea, *Ophthalmologica*, 171, 15, 1975.

Herzfeld, P. M., Fitzgerald, E. F., Hwang, S.-A., and Stark, A., A case-control study of malignant melanoma of the trunk among white males in upstate New York, *Cancer Detect. Prev.*, 17, 601, 1993.

Heynick, L., *Critique of the Literature on Bioeffects of Radiofrequency Radiation: A*

Comprehensive Review Pertinent to Air Force Operations (Report No. USAFSAM-TR-87-3), USAF School of Aerospace Medicine, Brooks Air Force Base, TX, 1987.

Hietanen, M. and Von Nandelstadh, P., Scattered and plasma-related optical radiations associated with industrial laser processes, in *Proceedings of the International Laser Safety Conference*, Laser Institute of Ameica, Orlando, FL, 1990, 3-105.

Hietanen, M., Honkasalo, A., Laitinen, H., Lindross, L., Welling, I., and von Nandelstadh, P., Evaluation of hazards in CO_2 laser welding and related processes, *Ann. Occup. Hyg.*, 36, 183, 1992.

Hill, D. A., Effect of separation from ground on whole-body absorption rates, *IEEE Trans. Microwave Theory Tech.*, MTT-32, 772, 1984.

Hill, D. G., "A Longitudinal Study of a Cohort with Past Exposure to Radar: The MIT Radiation Laboratory Follow-up Study," (Ph.D. dissertation), Johns Hopkins University, University Microfilms International, Ann Arbor, MI, 1988.

Hill, S. M. and Blask, D. E., Effects of the pineal hormone melatonin on the proliferation and morphological characteristics of human breast cancer cells (MCF-7) in culture, *Cancer Res.*, 48, 6121, 1988.

Hintenlang, D. E., Syngeristic effects of ionizing radiation and 60 Hz magnetic fields, *Bioelectromagnetics*, 14, 545, 1993.

Hitchcock, R. T. and Patterson, R. M., *Radio-Frequency and ELF Electromagnetic Energies: A Handbook for Health Professionals*, Van Nostrand Reinhold, New York, 1995.

Hitchcock, R. T., McMahan, S., and Miller, G. C., *Extremely Low Frequency (ELF) Electric and Magnetic Fields*, American Industrial Hygiene Association, Fairfax, VA, 1995.

Hitchcock, R. T., *Ultraviolet Radiation*, American Industrial Hygiene Association, Fairfax, VA, 1991.

Hjeresen, D. L., Kaune, W. T., Decker, J. R., and Phillips, R. D., Effects of 60-Hz electric fields on avoidance behavior and activity of rats, *Bioelectromagnetics*, 1, 299, 1980.

Hjeresen, D. L., Miller, M. C., Kaune, W. T., and Phillips, R. D., Behavioral responses of swine to a 60-Hz electric field, *Bioelectromagnetics*, 3, 443, 1982.

Hoke, J. A., Burkes, E. J., Gomes, E. D., and Wolbarsht, M. L., Erbium:YAG laser effects on dental tissue, *J. Laser Appl.*, 2(3-4), 61, 1990.

Hollows, F. and Moran, D., Cataract–the ultraviolet risk factor, *Lancet*, II-1981, 1249, 1981.

Holmberg B. and Rannug, A., Magnetic fields and cancer development in animal models, *Radio Sci.*, 30, 223, 1995.

Huuskonen, H., Juutilainen, J., and Komulainen, H., Effects of low-frequency magnetic fields on fetal development in rats, *Bioelectromagnetics*, 14, 205, 1993.

IEEE Magnetic Fields Task Force, A protocol for spot measurements of residential power frequency magnetic fields, *IEEE Trans. Power Del.*, 8, 1386, 1993.

IEEE Magnetic Fields Task Force, Measurements of power frequency magnetic fields away from power lines, *IEEE Trans. Power Del.*, 6, 901, 1991.

Igawa, S., Kibamoto, H., Takahaski, H., and Arai, S., A study on exposure to ultraviolet rays during outdoor sports activity, *J. Therm. Biol.*, 18, 583, 1993.

Institute of Electrical and Electronics Engineers, *IEEE Standard for Safety Levels with Respect to*

Human Exposure to Radio Frequency Electromagnetic Fields, 3 kHz to 300 GHz (IEEE Std. C95.1-1991), IEEE, New York, 1992.

Institute of Electrical and Electronics Engineers, *IEEE Standard Procedures for Measurement of Power Frequency Electric and Magnetic Fields from AC Power Lines* (ANSI/IEEE Std. 644-1987), IEEE, New York, 1987.

International Commission on Nonionizing Radiation Proection, Recommendations for minor updates to the IRPA 1985 guidelines on limits of exposure to laser radiation, *Health Phys.*, 54, 573, 1988.

International Commission on Nonionizing Radiation Protection, Guidelines on limits of exposure to laser radiation of wavelengths between 180 nm and 1 mm, *Health Phys.*, 49, 341, 1985.

International Radiation Protection Association, Fluorescent lighting and malignant melanoma, *Health Phys.*, 58, 111, 1990.

International Radiation Protection Association, Guidelines on limits of exposure to ultraviolet radiation of wavelengths between 180 nm and 400 nm (incoherent optical radiation), *Health Phys.*, 49, 331, 1985.

International Radiation Protection Association, Interim guidelines on limits of exposure to 50/60 Hz electric and magnetic fields, *Health Phys.*, 58, 113, 1990.

International Union, United Automobile, Aerospace and Agricultural Implement Workers of America, W.E. Brock v. General Dynamics Land Systems Division 815 F 2d 1570 (District of Columbia Cir. 1987).

Jaeger, J. L. and Carlson, R. T., Laser communication for covert links, *SPIE*, 1866, 95, 1993.

Joyner, K. H. and Bangay, M. J., Exposure survey of civilian airport radar workers in Australia, *J. Microwave Power*, 21, 209, 1986.

Joyner, K. H., Copeland, P. R., and MacFarlane, I. P., An evaluation of a radiofrequency protective suit and electrically conductive fabrics, *IEEE Trans. Electromag. Compat.*, 31, 129, 1989.

Juutilainen, J. and Saali, K., Development of chick embryos in 1 Hz to 100 kHz magnetic fields, *Radiat. Environ. Biophys.*, 25, 135, 1986.

Kallen, B., Malmquist, G., and Moritz, U., Delivery outcome among physiotherapists in Sweden: is non-ionizing radiation a fetal hazard? *Arch. Environ. Health*, 37, 81, 1982.

Kalliomaki, P. L., Hietanen, M., Kalliomaki, K., Koistinen, O., and Valtonen, E., Measurements of electric and magnetic stray fields produced by various electrodes of 27-MHz diathermy equipment, *Radio Sci.*, 17(5S), 29S, 1982.

Kato, M., Honma, K.-I., Shigemitsu, T., and Shiga, Y., Effects of exposure to a circularly polazrized 50-Hz magnetic field on plasma and pineal melatonin levels in rats. *Bioelectromagnetics*, 14, 97, 1993.

Kaune, W. T. and Zaffanella, L. E., Assessing historical exposures of children to power-frequency magnetic fields, *J. Expos. Anal. Environ. Epidemiol.*, 4, 149, 1994.

Kaune, W. T., Assessing human exposure to power-frequency electric and magnetic fields, *Environ. Health Perspect.*, 101(Suppl. 4), 121, 1993.

Kavet, R., Silva, J. M., and Thornton, D., Magnetic field exposure assessment for adult residents of Maine who live near and far away from overhead transmission lines, *Bioelectromagnetics*, 13, 35, 1992.

Kestenbaum, A., Coyle, R. J., and Sloan, P. P., Safe laser system design for production, *J. Laser Appl.*, 7, 31, 1995.

Kheifets, L. I., Afifi, A. A., Buffler, P. A., and Zhang, Z. W., Occupational electric and magnetic field exposure and brain cancer: a meta-analysis, *J. Occup. Environm. Med.*, 37, 1327, 1995.

Klein, R. M., Cut-off filters for the near ultraviolet, *Photochem. Photobiol.*, 29, 1053, 1979.

Kligman, L, H., Akin, F. J., and Kligman, A. M., The   contributions of UVA and UVB to connective tissue damage in hairless mice, *J. Investig. Dermatol.*, 84, 272, 1985.

Knave, B., Electric and magnetic fields and health outcomes–an overview, *Scand. J. Work Environ. Health*, 20, 78, 1994.

Knave, B., Gamberale, F., Bergstrom, S., Birke, E., Iregren, A., Kolmodin-Hedman, B., and Wennberg, A., Long-term exposure to electric fields, *Scand. J. Work Environ. Health*, 5, 115, 1979.

Knickerbocker, G. G., Kouwenhoven, W. B., and Barnes, H. C., Exposure of mice to a strong ac electric field--an experimental study, *IEEE Trans. Power Apparat. Sys.*, PAS-86, 498, 1967.

Kochevar, I. E., Biological effects of excimer laser radiation, *Proc. IEEE*, 80, 833, 1992.

Kokosa, J. M. and Eugene, J., Chemical composition of laser-tissue interaction smoke plume, *J. Laser Appl.*, 1(3), 59, 1989.

Kokosa, J. M., Hazardous chemicals produced by laser materials processing, *J. Laser Appl.*, 6, 195, 1994.

Kolmodin-Hedman, B., Mild, K. H., Hagberg, M., Jonsson, E., Anderson, M. C., and Eriksson, A., Health problems among operators of plastic welding machines and exposure to radiofrequency electromagnetic fields, *Int. Arch. Occup. Environ. Health*, 60, 243, 1988.

Koontz, M. D. and Dietrich, F. M., Variability and predicitability of children's exposure to magnetic fields, *J. Expos. Anal. Environ. Epidemiol.*, 4, 287, 1994.

Kowalczuk, C. I., Robbins, L., Thomas, J. M., Butland, B. K., and Saunders, R. D., Effects of prenatal exposure to 50 Hz magnetic fields on development in mice. implantation rate and fetal development, *Bioelectromagnetics*, 15, 349, 1994.

Kromhout, H., Loomis, D. P., Mihlan, G. J., Peipins, L. A., Kleckner, R. C., and Savitz, D. A., Assessment and grouping of occupational magnetic field exposure in five electric utility companies, *Scand. J. Work Environ. Health*, 21, 43, 1995.

Kues, H. and Lutty, G., Dyes can be deadly, *Laser Focus*, 11(4), 59, 1975.

Kurzel, R. B., Wolbarsht, M. L., and Yamanashi, B. S., Ultraviolet radiation effects on the human eye, in *Photochemical and Photobiological Reviews*, Smith, K. C., Ed., Plenum Publishing, New York, 1977, 133.

Kuster, N. and Balzano, Q., Energy absorption mechanism by biological bodies in the near field of dipole antennas above 300 MHz, *IEEE Trans. Veh. Technol.*, 412, 17, 1992.

Lai, H., Horita, A., Chou, C. K., and Guy, A. W., A Review of Microwave Irradiation and Actions of Psychoactive Drugs. *IEEE Engineering in Medicine and Biology Magazine*, 6(1), 31–36 1987.

Lancranjan, I., Maicanescu, M., Rafaila, E., Klepsch, I., and Popescu, H. I., Gonadic function in workmen with long-term exposure to microwaves, *Health Phys.*, 29, 381, 1975.

Landry, R. J. and Andersen, F. A., Optical radiation measurements: instrumentation and sources of error, *J. Natl. Cancer Inst.*, 69, 155, 1982.

Lapiere, C. M., The ageing dermis: the main cause for the appearance of 'old' skin, *Brit. J. Dermatol.*, 22 (Suppl 35), 5, 1990.

Larko, O. and Diffey, B. L., Occupational exposure to ultraviolet radiation in dermatology departments, *Brit. J. Dermatol.*, 114, 479, 1986.

Larko, O., and Diffey, B. L., Natural UV-B radiation received by people with outdoor, indoor, and mixed occupations and UV-B treatment of psoriasis, *Clin. Exp. Dermatol.*, 8, 279, 1983.

Larsen, A. I., Olsen, J., and Svane, O., Gender-specific reproductive outcome and exposure to high-frequency electromagnetic radiation among physiotherapists, *Scand. J. Work Environ. Health*, 17, 324, 1991.

Lary, J. M., Conover, D. L., Foley, E. D., and Hanser, P. L., Teratogenic effects of 27.12 MHz radiofrequency radiation in rats, *Teratology*, 26, 299, 1982.

Lary, J. M., Conover, D. L., Johnson, P. H., and Burg, J. R., Teratogenicity of 27.12-MHz radiation in rats is related to duration of hyperthermic exposure, *Bioelectromagnetics* 4, 249, 1983.

Lary, J. M., Conover, D. L., Johnson, P. H., and Hornung, R. W., Dose-response relationship and birth defects in radiofrequency-irradiated rats, *Bioelectromagnetics*, 7, 141, 1986.

Lee J. M., Stromshak, F., Thompson, J. M., Hess, D. L., and Foster, D. L., Melatonin and puberty in female lambs exposed to EMF: a replicate study, *Bioelectromagnetics*, 16, 119, 1995.

Lee J. M., Stromshak, F., Thompson, J. M., Thinesen, P., Painter, L. J., Olenchek, E. G., Hess, D. L., Forbes, R., and Foster, D. L., Melatonin secretion and puberty in female lambs exposed to environmental electric and magnetic Fields, *Biol. Reprod.*, 49, 857, 1993.

Leffell, D. J. and Brash, D. E., Sunlight and skin cancer, *Sci. Am.*, 275(1), 52, 1996.

Lerman, S., Human ultraviolet radiation cataracts, *Ophthalmic. Res.*, 12, 303, 1980.

Levine R. L., Dooley, J. K., and Bluni, T. D., Magnetic field effects on spatial discrimination and melatonin levels in mice, *Physiol. Behav.*, 58, 535, 1995.

Levine, R. L., Dooley, J. K., and Bluni, T. D., Magnetic field effects on spatial discrimination and melatonin levels in mice, *Physiol. Behav.*, 58, 535, 1995.

Liburdy, R. P., Sloma, T. R., Sokolic, R., and Yaswen, P., ELF magnetic fields, breast cancer, and melatonin: 60 Hz fields block melatonin's oncostatic action on breast cancer cell proliferation, *J. Pineal Res.*, 14, 89, 1993.

Lindbohm, M.. L., Hietanen, M., Kyyronen, P., Sallmen, M., von Nadelstadh, P., Taskinen, H. Pekkarinen, M., Ylikoski, M., and Hemminki, K., Magnetic fields of video display terminals and spontaneous abortion, *Am. J. Epidemiol.*, 136, 1041, 1992.

Lindbohm, M.. L., Kyyronen, P., Hietanen, M., and Sallmen, M., Magnetic fields of video display terminals and spontaneous abortion (letters to the editor), *Am. J. Epidemiol.*, 138, 902, 1993.

Lobraico, R. V., Schifano, M. J., and Brader, K. R., A retrospective study on the hazards of the carbon dioxide laser plume, *J. Laser Appl.* 1(1), 6, 1988.

London, S. J., Bowman, J. D., Sobel, E., Thomas, D. C., Garabrant, D. H., Pearce, N., Bernstein, L., and Peters, J., Exposure to magnetic fields among electrical workers in relation to leukemia risk in Los Angeles County, *Am. J. Ind. Med.*, 26, 47, 1994.

Lorenz, A. K., Gas handling safety for laser makers and users, *J. Lasers Appl.*, 6(3), 69, 1987.

Loscher W. and Mevissen, M., Animal studies on the role of 50/60-hertz magnetic fields in carcinogenesis, *Life Sci.*, 54, 1531, 1994.

Loscher W., Mevissen, M., Lehmacher, W., and Stamm, A., Tumor promotion in a breast cancer model by exposure to a weak alternating magnetic field, *Cancer Lett.*, 71, 75, 1993.

Lovsund, P., Oberg, P. A., and Nilsson, S. E. G., ELF magnetic fields in electrosteel and welding industries, *Radio Sci.*, 17(5S), 35S, 1982.

Lovsund, P., Oberg, P. A., and Nilsson, S. E. G., Influence on vision of extremely low frequency electromagnetic fields, *Acta Ophthalmol.*, 57, 812, 1979.

Lovsund, P., Oberg, P. A., Nilsson, S. E. G., and Reuter, T., Magnetophosphenes: a quantitative analysis of thresholds, *Med. Biol. Eng. Comput.*, 18, 326, 1980.

Lubinas, V. and Joyner, K. H., Measurement of induced current flows in the ankles of humans exposed to radiofrequency fields (Report 8000), Telecom Research Laboratories, Clayton, Victoria, Australia, 1991.

Lund, D. J., Edsall, P. R., Fuller, D. R., and Hoxie, S. W., Ocular hazards of tunable continuous-wave near infrared laser sources, *SPIE*, 2674, 53, 1996.

Lundsberg, L. S., Bracken, M. B., and Belanger, K., Occupationally related magnetic field exposure and male subfertility, *Fertil. Steril.*, 63, 384, 1995.

Lyon, T. L. and Marshall, W. J., Nonlinear properties of optical filters–implications for laser safety, *Health Phys.*, 51, 95, 1986.

Lyon, T. L., Hazard analysis technique for multiple wavelength lasers, *Health Phys.*, 49, 221, 1985.

Lyon, T. L., Laser measurement techniques guide for hazard evaluation (tutorial guide - part 1), *J. Laser Appl.*, 5(1), 53, 1993.

Lyon, T. L., Laser measurement techniques guide for hazard evaluation (tutorial guide - part 2), *J. Laser Appl.*, 5(2&3), 37, 1993.

Lytle, C. D., Hitchins, V. M., and Beer, J. Z., Estimation of Carcinogenic Risk from Lamps Which Emit Ultraviolet Radiation, presented at the Seminar on Human Exposure to Ultraviolet Radiation: Risks and Regulations, Amsterdam, The Netherlands, 1987.

Maffeo, S., Brayman, A. A., Miller, M. M., Carstensen, E. L., Ciaravino, V., and Cox, C., Weak low frequency electromagnetic fields and chick embryogenesis, failure to reproduce positive findings, *J. Anat.*, 157, 101, 1988.

Maffeo, S., Miller, M. M., and Carstensen, E. L., Lack of effect of weak low frequency electromagnetic fields on chick embryogenesis, *J. Anat.*, 139, 613, 1984.

Magnante, D. B. O. and Miller, D., Ultraviolet absorption of commonly used clip-on sunglasses, *Ann. Ophthalmol.*, 17, 614, 1985.

Magnus, I. A., Biologic action spectra, introduction and general review, in *The Biologic Effects of Ultraviolet Radiation*, Urbach, F., Ed., Pergamon Press, New York, 1969. 175.

Magnus, K., Habits of sun exposure and risk of malignant melanoma: an analysis of incidence rates in Norway 1955-1977 by cohort, sex, age, and primary tumor site, *Cancer*, 48, 2329, 1981.

Magnus, K., Incidence of malignant melanoma of the skin in the five Nordic countries: significance of solar radiation, *Int. J. Cancer*, 20, 477, 1977.

Mainster, M. A., Spectral transmittance of intraocular lenses and retinal damage from intense light sources, *Am. J. Ophthalmol.*, 85, 167, 1978.

Mainster, M. A., White, T. J., and Allen, R. G., Spectral dependence of retinal damage produced by intense light sources, *J. Optical Soc. Am.*, 60, 848, 1970.

Majewska, K., Investigations on the effect of microwaves on the eye, *Pol. Med. J.*, 38, 989, 1968.

Marino, A. A., Becker, R. O., and Ullrich, B., The effect of continuous exposure to low frequency electric fields on three generations of mice: a pilot study, *Experientia*, 32, 565, 1976.

Marks, R., Jolley, D., Dorevitch, A. P., and Selwood, T. S., The incidence of non-melanocytic skin cancers in an Australian population: results of a five-year prospective study, *Med. J. Australia*, 150, 475, 1989.

Marshall, J., Eye hazards associated with lasers, *Ann. Occup. Hyg.*, 21, 69, 1978.

Marshall, W. J. and Conner, P. W., Field laser hazard hazard calculations, *Health Phys.*, 52, 27, 1987.

Marshall, W. J. and Van DeMerwe, W. P., Hazardous ranges of laser beams and their reflections from targets, *Appl. Optics*, 25, 605, 1986.

Marshall, W. J., Comparative hazard evaluation of near-infrared diode lasers, *Health Phys.*, 66, 532, 1994.

Marshall, W. J., Determining hazard distances from non-Gaussian lasers, *Appl. Optics*, 30, 696, 1991.

Marshall, W. J., Laser hazard evaluation method for middle infrared laser systems, *J. Laser Appl.*, 8, 211, 1996.

Marshall, W. J., Laser reflections from relatively flat specular surfaces, *Health Phys.*, 56, 753, 1989.

Marshall, W. J., Understanding laser hazard evaluation, *J. Laser Appl.*, 7, 99, 1995.

Martin, A. H., Development of chicken embryos following exposure to 60-Hz magnetic fields with differing waveforms, *Bioelectromagnetics*, 13, 223, 1992.

Maxwell, K. J. and Elwood, J. M., UV radiation from fluorescent lights, *Lancet*, 2(Sept. 3), 579, 1983.

Mazzaferro, J., Laser-based communications systems emit radiation, an eye hazard, *Occupat. Health Safety*, Feb. 1995, 59.

McGiven, R. F., Sokol, R. S., and Adey, W. R., Prenatal exposure to a low-frequency electromagnetic field demasculinizes adult scent marking behavior and increases accessory sex organ weights in rats, *Teratology*, 41, 1, 1990.

McKinlay, A. F. and Diffey, B. L., A reference action spectrum for ultra-violet induced erythema in human skin, in *Human Exposure to Ultraviolet Radiation: Risks and Regulations*, Passchier, W. F., and Bosnjakovic, Eds., Excerpta Medica, New York, 1987, 83.

McKinlay, A. F., Artificial sources of UVA radiation: uses and emission characteristics, in *Biological Responses to Ultraviolet A Radiation*, Urbach, F., Ed., Overland Valdenmar Publishing Company, Overland Park, Kansas, 1992, 19.

McLean, J. R. N., Stuchly, M. A., Mitchel, R. E. J., Wilkinson, D., Yang, H., Goddard, M., Lecuyer, D. W., Schunk, M., Callary, E., and Morrison, D., Cancer promotion in a mouse skin model by a 60–Hz magnetic field. II. tumor development and immune response,

Bioelectromagnetics, 12, 273, 1991.

Meinel, H. and Rembold, B., Commercial and scientific applications of millimetric and submillimetric waves, *Radio Electronics Eng.*, 49, 351, 1979.

Menzel, E. R., Laser applications in criminalistics, *J. Laser Appl.*, 3(2), 39, 1991.

Mevissen M., Kietzmann, M., and Loscher, W., *In vivo* exposure of rats to a weak alternating magnetic field increases ornithine decarboxylase activity in the mammary gland by a similar extent as the carcinogen DMBA, *Cancer Lett.*, 90, 207, 1995.

Mevissen, M., Buntenkotter, S., and Loscher, W., Effects of static and time-varying (50-Hz) magnetic fields on reproduction and fetal develppment in rats, *Teratology*, 50, 229, 1994.

Mevissen, M., Stamm, A., Buntenkotter, S., Zwingelberg, R., Wahnschaffe, U., and Loscher, W., Effects of magnetic fields on mammary tumor development induced by 7,12-dimethylbenz(a)anthracene in Rats. *Bioelectromagnetics*, 14, 131, 1993.

Michaelson, S. M., Howland, J. W., and Deichmann, W. B., Response of the dog to 24,000 and 1285 MHz microwave exposure, *Ind. Med.*, 40, 18, 1971.

Mild, K. H. and Lovstrand, K. G., Environmental and professionally encountered electromagnetic Fields, in *Biological Effects and Medical Applications of Electromagnetic Energy*, Gandhi, O. P., Ed. Prentice Hall, Englewood Cliffs, NJ, 1990, 48

Mild, K. H., Occupational exposure to radio-frequency electromagnetic fields, *Proc. IEEE*, 68, 12, 1980.

Milham, S., Mortality from leukemia in workers exposed to electrical and magnetic fields, *New Eng. J. Med.*, 307, 249, 1982.

Miller, A. B., To, T., Agnew, D. A., Wall, C., and Green, L. M., Leukemia following occupational exposure to 60-Hz electric and magnetic fields among Ontario electric utility workers, *Am. J. Epidemiol.*, 144, 150, 1996.

Miller, G., Industrial hygiene concerns of laser dyes, in *Proceedings of the International Laser Safety Conference*, Laser Institute of America, Orlando, FL, 1991, 3-97.

Minder, Ch. E. and Pfluger, D. F., Extremely low frequency electromagnetic field measurements (ELF-EMF) in Swiss railway engines, *Radiat. Prot. Dosimetry*, 48, 351, 1993.

Miserendino, L., Recommendations for safe use in dentistry, *J. Laser Appl.*, 4(3), 16, 1992.

Mora, R. G. and Burris, R., Cancer of the skin in blacks: a review of 128 patients with basal-cell carcinoma, *Cancer*, 47, 1436, 1981.

Morgan, M. G., *Measuring Power-Frequency Fields*, Carnegie Mellon University, Pittsburgh, PA, 1992.

Moseley, H., Ultraviolet and visible radiation transmission properties of some types of protective eyewear, *Phys. Med. Biol.*, 30, 177, 1985.

Moss, C. E. and Gawenda, M. C., *Optical Radiation Transmission Levels Through Transparent Welding Curtains* (DHEW/NIOSH Pub. No. 78-176), National Institute for Occupational Safety and Health, Cincinnati, OH, 1978.

Moss, C. E., *Hazard Evaluation and Technical Assistance Report HETA 89-284-L2029 Technical Assistance to the Federal Employees Occupational Health Seattle, Washington* (NTIS # PB91-107920), National Technical Information Service, Springfield, VA, 1990.

Moss, C. E., personal communication, 1993.

Moss, E., Murray, W., Parr, W., and Conover, D., Radiation, in *Occupational Diseases A Guide to their Recognition*, Key, M. M., Henschel, A. F., Butler, J., Ligo, R., and Tabershaw, I. R., Eds., U. S. Government Printing Office, Washington, D.C., 1977, 475.

Murray, W. E., Hitchcock, R. T., Patterson, R. M., and Michaelson, S. M., Nonionizing electromagnetic energies, in *Patty's Industrial Hygiene & Toxicology*, Vol.3, Part B, 3rd ed., Cralley, L., Cralley, L., and Bus, J., Eds., John Wiley & Sons, New York, 1996.

Murray, W. E., Ultraviolet radiation exposures in a mycobacteriology laboratory, *Health Phys.*, 58, 507, 1990.

N. W. Couch, personal communication, 1985.

National Council on Radiation Protection and Measurements, *Biological Effects and Exposure Criteria for Radiofrequency Electromagnetic Fields* (NCRP Report No. 86), National Council on Radiation Protection and Measurements, Bethesda, MD, 1986.

National EMF Measurement Protocol Group, Power Frequency Magnetic Fields, *A Protocol for Conducting Spot Measurements Part One: Background and Instructions*, American Public Power Association, Washington, D.C., 1992.

National EMF Measurement Protocol Group, Power Frequency Magnetic Fields, *A Protocol for Conducting Spot Measurements Part Two: Protocol Data Forms*, American Public Power Association, Washington, D.C., 1992.

National Institute for Occupational Safety and Health, *Occupational Exposure to Ultraviolet Radiation* (HSM 73-11009), NIOSH, Cincinnati, OH, 1972.

Nelemans, P. J., Groenendal, H., Kiemeney, L. A. L. M., Rampen, F. H. J., Ruiter, D. J., and Verbeek, A. L. M., Effect of intermittent exposure to sunlight on melanoma risk among indoor workers and sun-sensitive individuals, *Environ. Health Perspect.*, 101, 252, 1993.

Nielsen, C. V. and Brandt, L., Spontaneous abortion among women using video display terminals, *Scand. J. Work Environ. Health*, 16, 323, 1990.

Nilsson, S. E. G., Lovsund, P., Oberg, P. A., and Flordahl, L.-E., The transmittance and absorption properties of contact lenses, *Scand. J. Work Environ. Health*, 5, 262, 1979.

Nishikawa, U., Biological effects of pulsing electromagnetic fields (pemfs) on ICR mice, *J. Jpn. Orthop. Assoc.*, 61, 1413, 1987.

Nishikawa, U., Hirotani, H., and Tanaka, O., Study on postnatal development in mice exposed to electromagnetic fields (pemfs) during their prenatal period, *Congenit. Anomal.*, 26, 219, 1986.

Nordstrom, S., Birke, E., and Gustavsson, L., Reproductive hazards among workers at high voltage substations, *Bioelectromagnetics*, 4, 91, 1983.

O'Conner, L., Making light work of fabric cutting, *Mechanical Eng.*, 115(11), 64, 1993.

Okuno, T., Thermal effect of visible light and infra-red radiation (i.r.-a, i.r.-b and i.r.-c) on the eye: a study of infra-red cataract based on a model, *Ann. Occup. Hyg.*, 38, 351, 1994.

Olsen, R. G., James, D. C., and Chartier, V. L., The performance of reduced magnetic field power lines theory and measurements of an operating line, *IEEE Trans. Power Del.*, 8, 1430, 1993.

O'Neill, F., Turcu, C. E., Xenakis, D., and Hutchinson, M. H. R., X-ray emission from plasmas generated by an XeCl laser picosecond pulse train, *Appl. Phys. Lett.*, 55, 2603, 1989.

ORAU: *Health Effects of Low Frequency Electric and Magnetic Fields* (NTIS # ORAU/92 F-9, Oak Ridge Associated Universities Panel for The Committee on Interagency Radiation Research and Policy Coordination). Springfield, VA, 1992, National Technical Information Service.

Patel, R. S. and Baisch, G., Single pass laser cutting of polymers, *J. Laser Appl.*, 5(1), 13, 1993.

Pathak, M. A., Molecular aspects of drug photosensitivity with special emphasis on psoralen photosensitization reaction, *J. Natl. Cancer Inst.*, 69, 163, 1982.

Patonay, G., Antoine, M. D., and Boyer, A. E., Semiconductor lasers in analytical chemistry, *SPIE*, 1435, 52, 1991.

Paul, M., Hammond, K., and Abdollahzadeh, S., Power frequency magnetic field exposures among nurses in a neonatal intensive care unit and a normal newborn nursery, *Bioelectromagnetics*, 15, 519, 1994.

Pavlik, R. E., Measurements for Sarasota County Sheriff's Office, Division of Safety, Bureau of Consultation and Enforcement, Tampa, FL, 1991.

Pavlik, R. E., Measurements for St. Petersburg Police Department, Division of Safety, Bureau of Consultation and Enforcement, Tampa, FL, 1991.

Paz, J. D., Potential ocular damage from microwave exposure during electrosurgery: dosimetric survey, *J. Occup. Med.*, 29, 580, 1987.

Peak, D. W., Conover, D. L., Herman, W. A., and Shuping, R. E., *Measurement of Power Density from Marine Radar* (DHEW Publication (FDA) 76-8004), U.S. Government Printing Office, Washington, D.C., 1975.

Petersen, R. C. and Testagrossa, P. A., Radio-frequency electromagnetic fields associated with cellular-radio cell-site antennas, *Bioelectromagnetics*, 13, 527, 1992.

Petersen, R. C., Electromagnetic radiation from selected telecommunications systems, *Proc. IEEE*, 69, 21, 1980.

Petersen, R. C., Levels of electromagnetic energy in the immediate vicinity of representative microwave radio relay tower, in *ICC '79 Conference Record,* Vol. 2, IEEE International Conference on Communications, Boston, MA, 1979.

Pettit, G. H., Lasers take up residence in the surgical suite, *Circuits Devices*, 8(3), 18, 1992.

Pfau, R. G., Hood, A. F., and Morison, W. L., Photoageing: the role of UVB, solar-simulated UVB, visible and psoralen UVA radiation, *Brit. J. Dermatol.*, 114, 319, 1986.

Philen, D. L., Petersen, R. C., Wakefoose, A., and Edwards, J., Safety and liability issues in optical fiber communications, *Right of Way*, Oct. 1986, 23.

Phillips, K. L., "Characterization of Occupational Exposures to Extremely Low Frequency Magnetic Fields in a Health-Care Setting," Master's thesis, School of Public Health, University of Texas, Houston, TX, 1993.

Phillips, K., Morandi, M., Oehme, D., and Clouthier, P., Characterizartion of occupational exposure to extremely low frequency magnetic fields in a hospital, presented at the American Industrial Hygiene Conference & Exposition '93, New Orleans, LA, 1993.

Piltingsrud, H. V. and Stencil, J. A., A portable spectroradiometer for use at visible and ultraviolet wavelengths, *Am. Ind. Hyg. Assoc. J.*, 37, 90, 1976.

Pitts, D. G. and Tredici, T. J., The effects of ultraviolet on the eye, *Am. Ind. Hyg. Assoc. J.*, 32, 235, 1971.

Pleven, C., A description of fourteen accidents caused by lasers in a research environment, in *Lasers et Normes de Protection (First International Symposium on Laser Biological Effects and*

Exposure Limits), Court, L. A., Duchene, A., and Courant, D., Eds., Centre de Recherches du Service de Sante des Armees, Paris, 1986, 406.

Polk, C. and Postow, E., Eds., *Handbook of Biological Effects of Electromagnetic Fields*, CRC Press, Boca Raton, FL, 1986.

Portet, R. and Cabanes, J., Development of young rats and rabbits exposed to a strong electric field, *Bioelectromagnetics, 9*, 95, 1988.

Poulsen, J. T., Staberg, B., Wulf, H. C., and Brodthagen H., Dermal elastosis in hairless mice after UV-B and UV-A applied simultaneously, separately, or sequentially, *Brit. J. Dermatol.*, 110, 531, 1984.

Puliafito, C. A. and Steinert, R. F., Short-pulsed Nd:YAG laser microsurgery of the eye: biophysical considerations, *IEEE J. Quantum Electron.*, QE-20, 1442, 1984.

Rannug A., Holmberg, B., Ekstrom, T., and Mild, K. H., Rat liver foci study on coexposure with 50 Hz magnetic fields and known carcinogens, *Bioelectromagnetics*, 14, 17, 1993.

Rannug, A., Ekstrom, T., Mild, K. H., Holmberg, B., Gimenez-Conti, I., and Slaga, T. J., A study on skin tumour formation in mice with 50 Hz magnetic field exposure, *Carcinogenesis*, 14, 573, 1993.

Rannug, A., Holmberg, B., Ekstrom, T., Mild, K. H., Gimenez-Conti, I., and Slaga, T. J., Intermittent 50 Hz magnetic field and skin tumour promotion in SENECAR mice, *Carcinogenesis*, 15, 153, 1994.

Raugi, G. J., Storrs, F.J., and Larsen, W.G., Photoallergic contact dermatitis to men's perfumes, *Contact Dermatitis*, 5, 251, 1979.

Ready, J. F., *Industrial Applications of Lasers*, Academic Press, New York, 1978.

Ren, Q., Keates, R. H., Hill, R. A., and Berns, M. W., Laser refractive surgery: a review and current status, *Opt. Eng.*, 34, 642, 1995.

Repacholi, M. H., Sources and applications of radiofrequency (rf) and microwave energy, in *Biological Effects and Dosimetry of Nonionizing Radiation Radiofrequency and Microwave Energies*, Grandolfo, M., Michaelson, S., and Rindi, A., Eds., Plenum Press, New York, 1981, 19.

Repacholi, M. H., Grandolfo, M., Ahlbom, A., Bergqvist, U., Bernhardt, J. H., Cesarini, J. H., Court, L. A., McKinlay, A. F., Sliney, D. H., Stolwijk, J. A. J., Swicord, M. L., and Szabo, L. D., Health issues related to the use of hand-held radiotelephones and base transmitters, *Health Phys.*, 70, 587, 1996.

Richardson, B. A., Yaga, K., Reiter, R. J., and Morton, D. J., Pulsed static magnetic field effects on *in vitro* pineal indoleamine metabolism, *Biochimica et Biophysica Acta*, 1137, 59, 1992.

Richter, P. I., Air pollution monitoring with lidar, *Trends Anal. Chem.*, 13, 263, 1994.

Rivas, L., Oroza, M. A., and Delgado, J. M. R., Influence of electromagnetic fields on body weight and serum chemistry in second generation mice, *Med. Sci. Res.*, 15, 1041, 1987.

Robinette, D., Silverman, C., and Jablon, S., Effects upon health of occupational exposure to microwave radiation (radar), *Am. J. Epidemiol.*, 112, 39, 1980.

Rockwell, Jr., R. J. and Moss, C. E., Optical radiation hazards of laser welding processes part II: CO_2 laser, *Am. Ind. Hyg. Assoc. J.*, 50, 419, 1989.

Rockwell, Jr., R. J., Laser accidents: are they all reported and what can be learned from them?, *J. Laser Appl.*, 1(4), 53, 1989.

Rockwell, Jr., R. J., Laser accidents: reviewing thirty years of incidents: what are the concerns–old and new?, *J. Laser Appl.*, 6, 203, 1994.

Rockwell, Jr., R. J., Learning from case studies: how to avoid laser accidents, in *Expert Strategies for Practical and Profitable Management*, Breedlove, B. and Schwartz, D., Eds., American Health Consultants, Inc., Atlanta, 1985, 57.

Rockwell, Jr., R. J., Smith, J. F., and Ertle, W. J., Playing it safe with industrial lasers, *Photonics Spectra*, 29(4), 118, 1995.

Rockwell, Jr., R. J., Utilization of the nominal hazard zone in control measure selection, in *Proceedings of the International Laser Safety Conference*, Laser Institute of Ameica, Orlando, FL, 1990, 7-25.

Rommereim, D. N., Kaune, W. T., Anderson, L. E., and Sikov, M. R., Rats reproduce and rear litters during chronic exposure to 150 kV/m, 60-Hz electric fields, *Bioelectromagnetics*, 10, 385, 1989.

Rommereim, D. N., Kaune, W. T., Buschbom, R. L., Phillips, R. D., and Sikov, M. R., Reproduction and development in rats chronologically exposed to 60-Hz electric fields, *Bioelectromagnetics*, 8, 243, 1987.

Rommereim, D. N., Rommereim, R. L., Sikov, M. R., Buschbom, R. L., and Anderson, L. E., Reproduction, growth, and development of rats during chronic exposure to multiple field strengths of 60-Hz electric fields, *Fund. Appl. Toxicol.* 14, 608, 1990.

Rosenthal, F. S. and Abdollahzadeh, S., Assessment of extremely low frequency (ELF) electric and magnetic fields in microelectronics fabrication rooms, *Appl. Occup. Environ. Hyg.*, 6, 777, 1991.

Rosenthal, F. S., Bakalian, A. E., and Taylor, H. R., The effect of prescription eyewear on ocular exposure to ultraviolet radiation, *Am. J. Pub. Health*, 76, 1216, 1986.

Rosenthal, F. S., Bakalian, A. E., Lou, C., and Taylor, H. R., The effect of sunglasses on ocular exposure to ultraviolet radiation, *Am. J. Pub. Health*, 78, 72, 1988.

Rosenthal, F. S., Phoon, C., Bakalian, A. E., and Taylor, H. R., The ocular dose of ultraviolet radiation to outdoor workers, *Investig. Ophthalmol. Vis. Sci.*, 29, 649, 1988.

Rothman, K. J., Chou, C.-K., Morgan, R., Balzano, Q., Guy, A. W., Funch, D. P., Preston-Martin, S., Mandel, J., Steffens, R., and Carol, G., Assessment of cellular telephone and other radio frequency exposure for epidemiologic research, *Epidemiology*, 7, 291, 1996.

Ruggera, P. S., and Schaubert, D. H., *Concepts and Approaches for Minimizing Excessive Exposure to Electromagnetic Radiation from RF Sealers* (HHS Publication (FDA) 82-8192), U.S. Government Printing Office, Washington, D.C., 1982.

Ruggera, P. S., *Measurements of Emission Levels During Microwave and Shortwave Diathermy Treatments* (HHS Publication (FDA) 80-8119), U.S. Government Printing Office, Washington, D.C., 1980.

Ruggera, P. S., Near-field measurements of RF fields, in *Symposium on Biological Effects and Measurement of Radio Frequency/Microwaves* (HEW Publication (FDA) 77-8026), Hazzard, D. G., U.S. Government Printing Office, Washington, D.C., 1977, 104,

Sahl, J. D., Kelsh, M. A., and Greenland, S., Cohort and nested case-control studies of hematopoietic cancers and brain cancer among electrical utility workers, *Epidemiology*, 4(2), 104, 1993.

Sahl, J. D., Kelsh, M. A., Smith, R. W., and Aseltine, D. A., Exposure to 60 Hz magnetic fields in the electric utility work environment, *Bioelectromagnetics*, 15, 21, 1994.

Santini, R., Hosni, M., Deschaux, P., and Pacheco, H., B16 melanoma development in black mice exposed to low-level microwave radiation, *Bioelectromagnetics*, 9, 105, 1988.

Saunders, R. D., Sienkiewicz, Z. J., and Kowalczuk, C. I., *Biological Effects of Exposure to Non-Ionising Electromagnetic Fields and Radiation* (NRPB-R240). National Radiological Protection Board, Chilton, Oxon, UK, 1991.

Saunders, R. D., Sienkiewicz, Z. J., and Kowalczuk, C. I., Biological effects of electromagnetic fields and radiation, *J. Radiol Prot.*, 11, 27, 1991.

Savant, G., Broadband near IR laser hazard filters (AD-B160619), National Technical Information Service, Springfield, VA, 1990.

Savitz, D. A. and Calle, E. E., Leukemia and occupational exposure to electromagnetic fields: review of epidemiological surveys, *J. Occup. Med.*, 29, 47, 1987.

Savitz, D. and Loomis, D., Magnetic fields in relation to leukemia and brain cancer among U.S. utility workers, *Am. J. Epidemiol.*, 141, 123, 1995.

Schiller, R. L., *An Evaluation of an Ultraviolet Radiation Survey Meter* (Air Force Report No. CI 79-66T), Air Force Institute of Technology, Wright-Patterson Air Force Base, 1977.

Schmidt, R. E. and Zuclich, J. A., Retinal lesions due to ultraviolet laser exposure, *Invest. Ophthalmol. Vis. Sci.*, 19, 1166, 1980.

Schnorr, T. M., Grajewski, B. A., Hornung, R. W., Thun, M. J., Egeland, G. M., Murray, W. E., Conover D. L., and Halperin, W. E., Video display terminals and the risk of spontaneous abortion, *New Engl. J. Med.*, 324, 727, 1991.

Scotto, J. and Nam, J., Skin melanoma and seasonal patterns, *Am. J. Epidemiology.* 111, 309, 1980.

Scotto, J., Fears, T. R., and Fori, G. B., *Measurements of Ultraviolet Radiation in the United States and Comparisons with Skin Cancer Data* (DHEW No.[NIH] 76-1029). National Institutes of Health, Bethesda, MD, 1975.

Scotto, J., Fears, T. R., and Fraumeni, J. F., *Incidence of Nonmelanoma Skin Cancer in the United States* (NIH Publication No. 83-2433), Bethesda, MD, National Institutes of Health, Bethesda, MD, 1983.

Scotto, J., Kopf, A. W., and Urbach, F., Non-melanoma skin cancer among caucasians in four areas of the United States, *Cancer*, 34, 1333, 1974.

Sebo, S. A., Caldecott, R., and Kasten, D. G., Magnetic field reduction options and techniques for ac substations, in *Ninth International Symposium on High Voltage Engineering,* Vol. 8, Institute of High Voltage, Garz, Austria, 1995, 8389-1.

Selmaoui, B. and Touitou, Y., Sinusoidal 50-Hz magnetic fields depress rat pineal NAT activity and serum melatonin. Role of duration and intensity of exposure, *Life Sci.*, 57, 1351, 1995.

Setlow, R. B., Grist, E., Thompson, K., and Woodhead, A. D., Wavelengths effective in induction of malignant melanoma, *Proc. Natl. Acad. Sci.*, 90, 6666, 1993.

Seto, Y. J., Majeau-Chargois, D., Lymangrover, J. R., Dunlap, W. P., Walker, C. F., and Hsieh, S. T., Investigation of fertility and *in utero* effects in rats chronically exposed to a high-intensity 60-Hz electric field, *IEEE Trans. Biomed. Eng.*, BME-31, 693, 1984.

Shaw, G. M. and Croen, L. A., Human adverse reproductive outcomes and electromagnetic field exposures: review of epidemiologic studies, *Environ. Health Perspect.*, 101 Suppl. 4, 107, 1993.

Shaw, H. M., McCarthy, W. H., and Milton, G. W., Changing trends in mortality from malignant melanoma, *Med. J. Austral.*, 2, 77, 1977.

Sheikh K., Exposure to electromagnetic fields and the risk of leukemia, *Arch. Env. Health*, 41, 56, 1986.

Shmaenok, L. A., Simanovskii, D. M., Gladskikh, A. N., and Bobashev, S. V., Soft X-rays emitted by a laser plasma created by two consecutive laser pulses, *Tech. Phys. Lett.*, 21, 920, 1995.

Siekierzynski, M., Czerski, P., Gidynski, A., Zydecki, S., Czarnecki, C., Dziuk E., and Jedrzejczak, W., Health surveillance of personnel occupationally exposed to microwaves. III. Lens translucency, *Aerospace Med.*, 45, 1146, 1974.

Sienkiewicz, Z. L., Robbins, L., Haylock, R. G. E., and Saunders, R. D., Effects of prenatal exposure to 50 Hz magnetic fields on development in mice. II. Postnatal development and behavior, *Bioelectromagnetics*, 15, 363, 1994.

Sikov, M. R., Rommereim, D. N., Beamer, L. J., Buschbom, R. L., Kaune, W. T., and Philips, R. D., Developmental studies of Hanford miniature swine exposed to 60-Hz electric fields, *Bioelectromagnetics*, 8, 229, 1987.

Silverman, C., Epidemiology of microwave radiation effects in humans, in *Epidemiology and Quantitation of Environmental Risk in Humans from Radiation and Other Agents*, A. Castellani, Ed., Plenum Press, New York, 1985, 433.

Silverman, C., Nervous and behavioral effects of microwave radiation in humans, *Am. J. Epidemiol.*, 97, 219, 1973.

Silverstone, H. and Searle, J. H. A., The epidemiology of skin cancer in Queensland: the influence of phenotype and environment, *Brit. J. Cancer*, 24, 235, 1970.

Sisken, B. F., Fowler, I., Mayaud, C., Ryaby, J. P., Ryaby, J., and Pilla, A. A., Pulsed electromagnetic fields and normal chick development, *J. Bioelectricity*, 5, 25, 1986.

Skotte, J. H., Exposure to power-frequency electromagnetic fields in Denmark, *Scand. J. Work Environ. Health*, 20, 132, 1994.

Sliney, D. and Wolbarsht, M., *Safety with Lasers and Other Optical Sources*, Plenum Press, New York, 1980.

Sliney, D. H. and Clapham, T. N., Safety with medical excimer lasers with an emphasis on compressed gases, *J. Laser Appl.*, 3(3), 59, 1991.

Sliney, D. H., Benton, R. E., Cole, H. M., Epstein, S. G., and Morin, C. J., Transmission of potentially hazardous actinic ultraviolet radiation through fabrics, *Appl. Ind. Hyg.*, 2, 36, 1987.

Sliney, D. H., Eye protective techniques for bright light, *Ophthal.*, 90, 937, 1983.

Sliney, D. H., *Guide for the Selection of Laser Eye Protection* (3rd ed.), Laser Institute of America, Orlando, FL, 1993.

Sliney, D. H., Infrared laser effects on the eye: implications for safety and medical applications, *SPIE*, 2097, 36, 1993.

Sliney, D. H., Laser effects on vision and ocular exposure limits, *Appl. Occup. Environ. Hyg.*, 11, 313, 1996.

Sliney, D. H., Laser eye protectors, *J. Laser Appl.*, 2(2), 9, 1990.

Sliney, D. H., Laser safety, *Lasers Surg. Med.*, 16, 215, 1995.

Sliney, D. H., Moss, C. E. Miller, C. G., and Stephens, J. B., Semitransparent curtains for control of optical radiatiion hazards, *Appl. Opt.*, 20, 2352, 1981.

Sliney, D. H., Ocular injuries from laser accidents, *SPIE*, 2674, 25, 1996.

Sliney, D. H., The IRPA/INIRC guidelines on limits of exposure to laser radiation, in *Lights, Lasers, and Synchrotron Radiation*, Grandolfo, M., Ed., Plenum Press, New York, 1990, 341.

Sliney, D. H., The merits of an envelope action spectrum for ultraviolet radiation exposure criteria, *Am. Ind. Hyg. Assoc. J.*, 33, 644, 1972.

Smith R. F., Clarke, R. L., and Justesen, D. R., Behavioral sensitivity of rats to extremely-low-frequency magnetic fields, *Bioelectromagnetics*, 15, 411, 1994.

Smith, J. F. and Dennis, J. E., Toward harmonized laser product standards, *J. Laser Appl*, 8, 9, 1996.

Sparks, R. A., Missle guidance electromagnetic sensors, *Microwave J.*, 26, 24, 1983.

Spicer, M. S. and Goldberg, D. J., Lasers in dermatology, *J. Am. Acad. Dermatol.*, 34, 1, 1996.

Sterenborg, H. J. C. M. and van der Leun, J. C., Tumorigenesis by a long wavelength UV-A source, *Photochem. Photobiol.*, 51, 325, 1990.

Sterenborg, H. J. C. M., de Gruijl, F. R., Kelfkens, G., and van der Leun, J. C., Evaluation of skin cancer risk resulting from long term occupational exposure to radiation from ultraviolet lasers in the range from 190 to 400 nm, *Photochem. Photobiol.*, 54, 775, 1991.

Stern, J. L., Broadcast transmission practice, in *Electronics Engineers' Handbook* 3rd ed., Fink D. G., and Christiansen, D. Eds., McGraw-Hill, New York, 1989, 21-2.

Stevens, R. G. and Davis, S., The melatonin hypothesis: electric power and breast cancer, *Environ. Health Perspect.*, 104 Suppl. 1, 135, 1996.

Stevens, R. G., Electric power use and breast cancer: a hypothesis, *Am. J. Epidemiol.*, 125, 556, 1987.

Stiglitz, M. R. and Blanchard, C., Over-the-horizon backscatter radar, *Microwave J.*, 33(5), 32, 1990.

Stuchly, M. A. and Lecuyer, D. W., Exposure to electromagnetic fields in arc welding, *Health Phys.*, 56, 297, 1989.

Stuchly, M. A. and Lecuyer, D. W., Induction heating and operator exposure to electromagnetic fields, *Health Phys.*, 49, 693, 1985.

Stuchly, M. A., McLean, J. R. N., Burnett, R., Goddard, M., Lecuyer, D. W., and Mitchel, R. E. J., Modification of tumor promotion in the mouse skin by exposure to an alternating magnetic field, *Cancer Lett.*, 65, 1, 1992.

Stuchly, M. A., Repacholi, M. H., Lecuyer, D. W., and Mann, R. D., Exposure to the operator and patient during short wave diathermy treatments, *Health Phys.*, 42, 341, 1981.

Stuchly, M. A., Repacholi, M. H., Lecuyer, D., and Mann, R., Radiation survey of dielectric (rf) heaters in Canada, *J. Microwave Power*, 15, 113, 1980.

Svedenstal B. M. and Holmberg, B., Lymphoma development among mice exposed to x-rays and pulsed magnetic fields, *Int. J. Radiat. Biol.*, 64, 119, 1993.

Swearengen, P. M., Vance, W. F., and Counts, D. L., A study of burn-through times for laser protective eyewear, *Am. Ind. Hyg. Assoc. J.*, 49, 608, 1988.

Sweet, J., Effective laser applications in high school physics, *J. Laser Appl.*, 6, 248, 1994.

Swope, C. H., The eye--protection, *Arch. Environ. Health*, 18, 428, 1969.

Szmigielski, S. A., Szudzinski, A., Pietraszek, A., Bielec, M., Janiak, M., and Wrembel, J., Accelerated development of spontaneous and benzopyrene-induced skin cancer in mice exposed to 2450 MHz microwave radiation, *Bioelectromagnetics*, 3, 171, 1982.

Szmigielski, S., Cancer morbidity in subjects occupationally exposed to high frequency (radiofrequency and microwave) electromagnetic radiation, *Sci. Total Environ.*, 180, 9, 1996.

Szydzinski, A., Pietraszek, A., Janiak, M., Wrembel, J., Kalczak, M., and Szmigielski, S., Acceleration of the development of benzopyrene-induced skin cancer in mice by microwave radiation, *Dermatol. Res.*, 274, 303, 1982.

Tarroni, G., Melandri, C., De Zaiacomo, T., Lombardi, C. C., and Formignani, M., Characterization of aerosols produced in cutting steel components and concrete structures by means of a laser beam, *J. Aerosol Sci.*, 17, 587, 1986.

Taskinen, H., Kyyronen, P., and Hemminki, K., Effects of ultrasound, shortwaves, and physical exertion on pregnancy outcome in physiotherapists, *J. Epidemiol. Comm. Health*, 44, 196, 1990.

Tell, R. A. and Nelson, J. C., Microwave hazard measurements near various aircraft radars, *Rad. Data Rep.*, 15, 161, 1974.

Tell, R. A., *Engineering Services for Measurement and Analysis of Radiofrequency (RF) Fields* (FCC Report No. OET/RTA 95-01), U.S. Government Printing Office, Washington, D.C., 1995.

Tell, R. A., Hankin, N. N., and Janes, D. E., Aircraft radar measurements in the near field, in *Operational Health Physics, Proceedings of the Ninth Midyear Topical Symposium of the Health Physics Society*, Health Physics Society, McLean, VA, 1976.

Tell, R. A., *Induced Body Currents and Hot AM Tower Climbing Assessing Human Exposure in Relation to the ANSI Radiofrequency Protection Guide* (PB92-125186), National Technical Information Service, Springfield, VA, 1991.

Teppo, L., Pukkala, E., Hakama, M., Hakulinen, T., Herva, A., and Saxen, E., Way of life and cancer incidence in Finland, *Scand. J. Soc. Med.*, S19, 50, 1980.

Theriault, G. P., Health effects of electromagnetic radiation on workers: epidemiologic studies, in *Proceedings of the Scientific Workshop on the Health Effects of Electric and Magnetic Fields on Workers*, Bierbaum, P. J. and Peters, J. M., Eds., National Institute for Occupational Safety and Health, Cincinnati, 1991, 93.

Theriault, G., Goldberg, M., Miller, A. B., Armstrong, B., Guenel, P., Deadman, J., Imbernon, E., Chevalier, A., Cyr, D., and Wall, C., Cancer risks associated with occupational exposure to magnetic fields among electric utility workers in Ontario and Quebec, Canada, and France: 1970-1989, *Am. J. Epidemiol.*, 139, 550, 1994.

Thomas, D. K., Often overlooked electrical hazards common in many lasers, in *Proceedings of the 1992 International Laser Safety Conference*, Laser Institute of America, Orlando, FL, 1993, 4I-41.

Thomas, J. R., Schrot, J., and Liboff, A. R., Low-intensity magnetic fields alter operant behavior in rats, *Bioelectromagnetics*, 7, 349, 1986.

Thorington, L., Spectral, irradiance and temporal aspects of natural and artificial light, *Ann. NY Acad. Sci.*, 453, 28, 1985.

Titus-Ernstoff L., Ernstoff, M. S., Duray, P., H., Barnhill, R. L., Holubkov, R., and Kirkwood, J. M., A relation between childhood sun exposure and dysplastic nevus syndrome among patients with nonfamilial melanoma, *Epidemiology*, 2, 210, 1991.

Tolliver, D. L., Plasma processing in microelectronics–past, present, and future, *Solid State Technol.*, 23, 99, 1980.

Truong, H. and Yellon, S. M., Continuous or intermittent 60 Hz magnetic-field exposure fails to affect the nighttime rise in melatonin in the adult Djungarian hamster (abstract), *Biol. Reprod.*, 51, 72, 1995.

Tubbs, R. L., Moss, C. E., and Fleeger, A., *Health Hazard Evaluation Report HETA 89-364-2202 ARMCO Advanced Materials Corporation Butler, Pennsylvania*, (NTIS # PB92-221027), National Technical Information Service, Springfield, VA, 1992.

Tyndall, D. A. and Sulik, K. K., Effects of magnetic resonance imaging on eye development in the C57BL/6J mouse, *Teratology*, 43, 263, 1991.

Ubeda, A., Leal, J., Trillo, M. A., Jimenez, M. A., and Delgado, J. M. R., Pulse shape of magnetic fields influences chick embryogenesis, *J. Anat.*, 137, 513, 1983.

Urbach, F. and Wolbarsht, M. L., Occupational skin hazards from ultraviolet (UV) exposures, *Proc. SPIE*, 229, 21, 1980.

Urbach, F., Geographic pathology of skin cancer, in *The Biological Effects of Ultraviolet Radiation*, Urbach, F., Ed., Pergamon Press, New York, 1969, 635.

Urbach, F., Ultraviolet radiation and skin cancer, in *Topics in Photomedicine*, Smith, K. C., Ed., Plenum Press, New York, 1984, 39.

Vann, C. S., Bliss, E. S., and Murray, J. E., Target alignment in the national ignition facility, *Fusion Technol.*, 26, 833, 1994.

Varanelli, A. G., Electrical hazards associated with lasers, *J. Laser Appl.*, 7, 62, 1995.

Ventzas, D. E. and Angelopoulos, A., Laser particle size analysis, *Model. Meas. Control*, 54(4), 17, 1994.

Viola, M. V. and Houghton, A., Melanoma in Connecticut, *Conn. Med.*, 42, 268, 1978.

Vitaliano, P. P. and Urbach, F., The relative importance of risk factors in nonmelanoma carcinoma, *Arch. Dermatol.*, 116, 454, 1980.

Vogt, D. R. and Reynders, J. P., An experimental investigation of magnetic fields in deep level gold mines, in *Seventh International Symposium on High Voltage Engineering 1991*, Dresden University of Technology, Dresden, Germany, 1991, 63.

Vos, J. J., Certainities and uncertainities about safe laser exposure limits, *SPIE*, 2052, 467, 1993.

Weiner, R., Expect laser safety standards revisions to impact users, *J. Laser Appl.*, 7, 66, 1995.

Weinstock, M. A., Nonmelanoma skin cancer mortality in the United States, 1969 through 1988, *Arch. Dermatol.*, 129, 1285, 1993.

Werner, J. S., Children's sunglasses: caeat emptor, *Opt. Vis. Sci.*, 68, 318, 1991.

Wetterberg, L., Lighting nonvisual effects, *Scand. J. Work Environ. Health*, 16 Suppl. 1, 26, 1990.

Weyandt, T. B., *Evaluation of Biological and Male Reproductive Function Responses to Potential Lead Exposures in 155 MM Howitzer Crewmen*, National Technical Information Service (AD-A247 384), Springfield, VA, 1992.

Wilkening, G. M., Nonionizing radiation, in *Patty's Industrial Hygiene and Toxicology General Principles*, 4th ed., Clayton, G. D. and Clayton, F. E., Eds., John Wiley & Sons, New York, 1991, 657.

Wilkinson, T. K., Health aspects of laser use: air contamination and lasers, *Arch. Environ. Health*, 18, 443, 1969.

Williams, P. and Mild, K. H., *Guidelines for the Measurement of RF Welders* (Undersokningsrapport 1991:8; Rapportkod: ISRN AI/UND-91-8-SE), National Institute of Occupational Health, Umea, Sweden, 1991.

Williams, R. A. and Webb, T. S., Exposure to radio-frequency radiation from an aircraft radar unit, *Aviat. Space Environ. Med.*, 51, 1243, 1980.

Wilson, B. W., Chess, E. K., and Anderson, L. E., 60-Hz electric-field effects on pineal melatonin rhythms: time course for onset and recovery, *Bioelectromagnetics*, 7, 239, 1986.

Wilson, B. W., Chronic exposure to ELF fields may induce depression, *Bioelectromagnetics*, 9, 195, 1988.

Wilson, D. F., Barrier curtains for laser hazard control, in *Proceedings of the International Laser Safety Conference*, Laser Institute of America, Orlando, FL, 1990, 5-71.

Wolbarsht, M. L., Low-level laser therapy (lllt) and safety considerations, *J. Laser Appl.*, 6(3), 170, 1994.

Wolfe, J. A., Laser retinal injury, *Military Med.*, 150, 177, 1985.

Wood, A. W., Possible health effects 50/60 Hz electric and magnetic fields: review of proposed mechanisms, *Australasian Phys. Eng.*, 16, 1, 1993.

World Health Organization, *Ultraviolet Radiation* (Environmental Health Criteria 14), World Health Organization, Geneva, 1979, 43,80.

Wu R. Y., Chiang, H., Shao, B. J., Li, N. G., and Fu, Y. D., Effects of 2.45-GHz microwave radiation and phorbol ester 12-O-tetradecanoylphorbol-13-acetate on dimethylhydrazine colon cancer in mice, *Bioelectromagnetics*, 15, 531, 1994.

Wuebbles, B. J. Y. and Felton, J. S., Evaluation of laser dye mutagenicity using the Ames/Salmonella microsome test, *Environ. Mutagen.*, 7, 511, 1985.

Yaga, K., Reiter, R. J., Manchester, L. C., Nieves, H., Sun, H.-H., and Chen, L. D., Pineal sensitivity to pulsed static magnetic fields changes during the photoperiod, *Brain Res. Bull.*, 30, 153, 1993.

Yost, M. G., Lee, G. M., Duane, D., Fisch, J., and Neutra, R. R., California protocol for measuring 60 Hz magnetic fields in residences, *Appl. Occ. Environ. Hyg.*, 7, 772, 1992.

Zhao-Zhang, L., Jia-Nu, W., Bao-Kang, G., and Yan, Z., Damage thresholds of skin irradiated by ultraviolet lasers, *Health Phys.*, 56, 683, 1989.

Zhao-Zhang, L., Jia-Nu, W., Bao-Kang, G., and Yan, Z., Ultraviolet erythema of laser radiation, *Lasers Life Sci.*, 2, 91, 1988.

Ziegler, A., Jonason, A. S., Leffell, D. J., Simon, J. A., Sharma, H. W., Kimmelman, J., Remington, L., Jacks, T., and Brash, D. E., Sunburn and p53 in the onset of skin cancer, *Nature*,

372, 773, 1994.

Ziegler, A., Jonason, A., Simon, J., Leffell, D., and Brash, D. E., Tumor suppressor gene mutations and photocarcinogenesis, *Photochem. Photobiol.*, 63, 432, 1996.

Zigman, S., Datiles, M., and Torczynski, E., Sunlight and human cataracts, *Invest. Ophthalmol. Visual Sci.*, 18, 462, 1979.

Zigman, S., Spectral transmittance of intraocular lenses, *Am. J. Ophthalmol.*, 85, 878, 1978.

Zuclich, J. A., Hazards to the eye from UV, in *Nonionizing Radiation–Proceedings of an ACGIH Topical Symposium*, American Conference of Governmental Industrial Hygienists, Cincinnati, OH, 1979, 129.

Zuclich, J. A., Schuschereba, S., Zwick, H., Cheney, F., and Stuck, B. E., Comparing laser-induced retinal damage from IR wavelengths to that from visible wavelengths, *SPIE*, 2674, 66, 1996.

Zwick, H. and Beatrice, E. S., Long-term changes in spectral sensitivity after low-level laser (514 nm) exposure, *Mod. Probl. Ophthal.*, 19, 319, 1978.

Chapter 10

HOT AND COLD ENVIRONMENTS: TEMPERATURE EXTREMES

Peter Bellin, PhD, CIH

I. OVERVIEW

This chapter will discuss the adverse effects, evaluation, and control of extreme temperature environments. A range of adverse effects may be expected from working under elevated and low temperatures. These effects may be completely prevented by proper work practices discussed in this chapter.

The human body has a variety of mechanisms to maintain a stable body temperature, at approximately 37°C. This temperature varies slightly during the day, and to some extent, between individuals. Temperature-related illness can develop if the core body temperature departs significantly from this normal level.

II. HIGH TEMPERATURE HAZARDS

Hot working conditions may be found in many locations in modern industry. Industrial operations may be classified as hot and dry, or hot and wet, depending on the environmental conditions present. Hot, dry environments are those locations where radiant sources or high temperature operations are present, with little water vapor added to the environment; examples include: steel mills, forge shops and glass manufacturing operations. In these conditions, the body can be effectively cooled by evaporation of perspiration.

Hot, wet environments, in contrast, have significant amounts of water vapor present with significant heat sources. Examples include: laundries, paper manufacture, and mining. In these environments, the body's cooling ability is compromised somewhat due to high humidity; sweat is not as readily evaporated from the skin.

Outdoor work is affected primarily by the weather conditions; thus, it can be either humid or dry during heat waves. Extreme temperatures result from environmental conditions more than from work operations. Examples here include construction activity, field work, traffic control, or athletics.

Another source of heat stress conditions, particularly those that might otherwise not be a problem, is the expanding field of hazardous waste control, including lead and asbestos abatement. Although environmental conditions may be moderate, the required use of whole-body personal protective equipment can result in significant heat strain primarily from the reduced availability of cooling by evaporation of perspiration. It is particularly important to recognize that chemical cleanup work may occur in hot conditions, where the resulting heat strain is accentuated by the chemical protective clothing that must be worn.

A. Heat Illnesses

A variety of heat illnesses are associated with work under hot conditions. If core body temperature exceeds 38°C, the body is considered at risk of developing heat illnesses. Due to inter-individual variances in the body's reaction to heat, core body temperature may not always be elevated in a heat illness situation.

1. *Heat stroke*

The classic symptoms of *heat stroke* include: elevated body temperature, above 41°C; unconsciousness or convulsions; and a cessation of sweating (a hot dry skin). Although cessation of sweating is a classic symptom of heat stroke, it may be present in exertional heat stroke. Often, the heat stroke victim will display altered mental status, ranging from irrational behavior, to poor judgment and confusion. Nausea and vomiting may be present.

Heat stroke is a dangerous condition. It is a medical emergency, as the central nervous system, the liver and kidneys, and other body systems are affected. Without prompt cooling of the body core, and medical treatment, this condition may be fatal. In some cases, delayed response to an episode of heat stroke can occur in the short term. Thus, any known or suspected heat stroke episode must be referred for medical attention. During severe heat waves, it is common for a significant number of heat stroke fatalities to occur among elderly or persons in poor health. In an occupational setting, heat stroke occurs as a result of physical work under hot conditions.

2. *Heat exhaustion*

Heat exhaustion is a condition where, in response to exertion under hot conditions, an individual experiences headache, nausea, dizziness, weakness, and extreme fatigue. It is associated with a slightly elevated body temperature, up to 39°C. Heat exhaustion may be considered an early form of heat stroke; workers may progress from heat exhaustion to heat stroke if they continue to work while experiencing the above symptoms. In fact, it is wise to evaluate the worker carefully to exclude heat stroke if these symptoms are present.

Unacclimatized workers are more likely to succumb to heat exhaustion than acclimatized workers. Heat exhaustion may be generated from salt or water depletion. Heat exhaustion represents a reversible syndrome if the body is allowed to cool, and is hydrated. Victims of heat exhaustion should also be referred for medical attention.

3. *Heat cramps*

Heat cramps commonly occur in the same muscles used heavily while working in the heat. They are severe muscular spasms, and their incidence is associated with heavy work in acclimatized workers. Heat cramps may occur due to salt imbalance if heavy amounts of water are ingested without replacement of the electrolytes lost in perspiration. It is more common for heat cramps to occur in acclimatized workers.

Heat cramps are best treated with rest and electrolyte replacement. The salts should be in solution, rather than in the form of salt tablets. Dietary salt could also be a source of electrolytes. Prevention can be practiced by ensuring that workers have adequate water and salt intake.

4. *Heat rashes*

Heat rash, *miliaria*, occurs in several forms, resulting from the interaction of sweat on the skin, and is generally transitory in nature.

Prickly heat, *miliaria rubra*, is characterized by red papules, in areas where clothing binds to the skin. It is experienced as a prickly sensation during work (and sweating). It may be caused as excess sweat is absorbed by the keratinous layers of the skin.

Miliaria crystallina occurs in damaged skin. Small clear vesicles form in the surface layers of the skin. It rapidly resolves upon cessation of work (and of sweating), and has no long-term effects.

Miliaria profunda occurs when water collects deep in the skin. The appearance is of small white to grey bumps (like goosebumps), again occurring in areas of damaged skin.

These heat-related skin conditions do not have long-term implications. However, if these conditions are occurring in a worker or a group of workers, they should be taken as an indication that the hot work is not under control. The application of safe work practices, discussed below, should help prevent these heat disorders. Table 10.1 contains an overview of the types of heat illness that may adversely impact a worker, key signs and symptoms, applicable First Aid treatment techniques, and prevention mechanisms.

B. Physiology of Thermal Regulation

Normal body functioning is maintained, in part, by the body's ability to maintain a relatively constant core temperature. The body is able, through a variety of mechanisms, to balance heat gain (generated both externally and internally) and heat loss to hold body temperature to $37 \pm 1°C$. Heat is generated internally by metabolic activity, ranging from 1.5 kcal/min (for a 70 kg man) at rest, to as much as 9 kcal/min under a heavy work load. The central nervous system, utilizing the hypothalamus, is responsible for thermoregulation in humans. Through neural transmitters, the hypothalamus is able to sense temperature changes in skin, muscle, stomach, and other tissue in the body. In turn, the hypothalamus can signal either heat-conserving mechanisms (shivering, cutaneous vasoconstriction) or heat-dissipation mechanisms (increased perspiration, cutaneous vasodilation).

Table 10.1
Heat Illnesses: Signs and First Aid Treatment Needed

Heat Illness	Signs and Symptoms	First Aid Treatment	Prevention
Heat stroke	Hot, dry skin Rectal temperature above 40.5 °C Confusion, loss of consciousness, convulsions.	Immediate and rapid cooling by immersion in chilled water with massage or by wrapping in wet sheet with vigorous fanning. Medical attention needed.	Medical screening of workers, acclimatization for 5-7 days, monitoring workers during high heat situations.
Heat exhaustion	Fatigue, nausea, headache, giddiness. Skin clammy and pale.	Remove to cooler environment, rest in a reclined position, administer fluids. Rest until rehydrated.	Acclimatize workers 5 - 7 days. Ensure adequate hydration by making sure workers have several cups of water per day. Salt supplements with caution.
Heat cramps	Painful spasms of muscles used during work. Onset during or after work hours.	Salted fluids by mouth, or IV infusion.	Adequate salt intake with meals. Unacclimatized workers supplement salt intake at meals.
Heat rashes	Miliaria rubra: red papules, prickling sensation. Miliaria profunda: gooseflesh appearance on damaged skin.	Miliaria rubra: drying lotions, skin cleanliness. Miliaria profunda: no direct treatment; return to cooler environment.	Miliaria rubra: allow skin to dry between heat exposures. Miliaria profunda: treat heat rash, sunburn (avoid damaging skin). Periodic relief from sustained heat.

Acclimatization to heat is an important factor in human reaction to elevated temperatures. On initial exposure to hot conditions, workers readily show signs of distress. Core body temperature may increase in response to work in hot conditions, and heat exhaustion or fainting may occur. On repeated exposures, workers will develop a reduced blood flow and increased sweat rate in response to the heat. The conditions that initially may cause distress will not be so uncomfortable. Acclimatization occurs with as little as 1 to 2 hours daily exposure in 7 to 10 days. The response is effective for either dry or wet heat exposure. Heat exposure guidelines limiting work intensity and establishing rest periods generally assume acclimatization has occurred, and the physiologic response should be integrated to programs protecting workers from heat stress illnesses.

Heat is gained from or lost to the external environment by mechanisms of *convection*, transfer of heat between moving air and the skin; *conduction*, transfer of heat between solid surfaces and the body when the two are in physical contact; *radiation*, transfer of heat between solid surfaces and the body when the two are not in physical contact; and *evaporation of sweat*, which results in cooling to the body. Factors that affect these inputs include air temperature, wind velocity, relative humidity, radiant sources, and the temperature of surfaces in contact with the body. Usually, heat loss or gain from body contact with surfaces and heat loss through breathing are minor components of heat balance in the work environment.

Convection, radiation, metabolism, and evaporative cooling are considered together in a heat balance equation:

$$\text{Heat gain/loss} = (M-W) \pm C \pm R - E \qquad (10.1)$$

Where:

(M-W)	= Total metabolism - external work, kcal/hour
C	= Convective heat exchange, kcal/hour
R	= Radiative heat exchange, kcal/hour
E	= Evaporative heat loss, kcal/hour

If the components–metabolism, convection, radiation, and evaporation sum to a positive number, heat load on the body will increase, and heat stress is possible.

1. *Evaluation of metabolic load*

The human body is constantly generating internal heat. Consumption of oxygen occurs constantly to maintain body temperature, body function, and food digestion. Excess heat energy is released to the external environment. This basic level of oxygen consumption is referred to as the resting metabolic rate. It is measured as minute volume of oxygen, or the liters of oxygen consumed per minute (VO_2). It does not vary much between individuals. However, as physical work is conducted, oxygen consumption increases, and the heat released in the body also rises. Factors such as age, physical conditioning, and job design affect the oxygen required to perform a task, and is variable between people.

Each liter of oxygen consumed releases about 4.8 kcal of heat energy inside the body. Oxygen consumption is considered an effective index for determining the level of work being conducted. At rest, the "average man" consumes 0.3 liters of oxygen per minute, equivalent to 1.4 kcal/min heat energy in the body.

Work intensity is usually graded into several categories, from light to heavy, based on caloric demands of job activities. For example, light machine work may have an energy cost of 3.3 kcal/min, while a task like slag removal may have an energy cost of over 11 kcal/min. Reference tables have been generated that classify jobs into the broad categories of light, moderate, or heavy work. These are useful guidelines, are not labor intensive, but are not precise and may be inaccurate, particularly in comparison with task analysis or physiologic measurements. Table 10.2 lists average energy expenditures for a number of jobs published by Durnin and Passmore.

In order to deliver oxygen to the working muscles, the heart rate will increase in direct association with oxygen consumption. Heart rate will reflect oxygen consumption up to a point; it is a less effective measure at high work rates. The heart rate-VO_2 relationship should be calibrated for each individual for the best accuracy in estimating VO_2 from observing heart rate. Evaluation of metabolic load may utilize measurement of oxygen consumption, heart rate (as a surrogate for oxygen consumption), or task analysis.

Table 10.2
Caloric Demands of Some Jobs

Job	Average (Range), kcal/min
Office Work	
Typing	1.5 (1.5–1.6)
Desk work	1.6 (1.3–1.8)
Moving around	1.9 (1.5–3.3)
Light Industry	
Machine sewing	2.9 (2.7–3.0)
Pressing clothes	4.0 (3.6–4.5)
Auto repair	4.1 (3.6–4.6)
Cabinet maker	5.6 (5.5–5.7)
Carpenter - assembler	3.9 (2.9–5.0)
Punch press operator (electrical industry)	4.2 (3.3–5.0)
Punch press operator (machine tool industry)	5.7 (5.6–5.8)
Stock room operator	4.2 (3.8–4.5)
Lathe operator	3.4 (3.2–3.6)
Spray painting	3.4 (3.2–3.5)
Welding	3.4 (2.7–4.1)

2. Oxygen consumption

Theoretically, oxygen consumption is an ideal way to classify work intensity. The amount of oxygen consumed is directly related to the intensity of muscular work, and thus metabolic heat generation.

Indirect calorimetry evaluates metabolic heat by measuring oxygen consumption. The worker breathes workroom air, but exhaled air is collected in a flexible container. The volume of air exhaled, as well as its oxygen content can be measured. Thus, oxygen consumption can be measured in the field. This can be conducted for specific tasks, or for specified time periods during a work day. However, this process is highly impractical in field situations.

3. Heart rate monitoring

Since oxygen consumption is not easily measured in the field, heart rate may be used as a surrogate for oxygen consumption. Heart rate increases with oxygen consumption, but again, this relationship is affected by age and conditioning. Heart rate changes will be somewhat specific for each individual. A more physically fit individual will achieve a lower heart rate for a specific task than one who is less physically fit. Maximal heart rate and maximal minute volume for oxygen consumption can be measured for individual workers. Maximal heart rate (MR) can be estimated by the formula:

$$MR = 220 - Age \text{ (years)} \qquad (10.2)$$

Workers can maintain the maximal heart rate for only short periods of time. This information can be used to evaluate the intensity of job tasks being performed.

Workers can readily check their own pulse rate. One method proposed by Bernard and Kenney measures body temperature and average heart rate to estimate heat strain. Their method involves the use of dermal temperature sensors and self-assessment of heart rate to evaluate heat strain. The method is intended for use as part of an overall heat stress protective program.

Fuller and Smith recommended a method involving periodic measurement of oral temperature and heart rate. For example, if heart rate is measured at 1 minute and 3 minutes after a task, these values can be evaluated (along with oral temperature) to rate a job for heat strain. The criterion for heat strain is an oral temperature less than 37.5°C, with pulse rate at 3 minutes less than 90 beats per minute. However, if the oral temperature exceeds 37.5°C, or the change in pulse rate from 1 to 3 minutes of rest is less than 10 beats per minute, heat strain is too high for an individual.

These types of analyses may be conducted simply, but are only feasible when the rest and evaluation periods do not interfere with the work day. Heart rate measurement may be cumbersome to perform in the field if continuous measurement is desired. The best accuracy can only be achieved when the heart rate–VO_2 relationship is calibrated for an individual worker. The procedure will most likely be useful in manufacturing establishments with a stable work force and relatively constant working conditions.

4. Task analysis

Task analysis is an indirect guideline to classify jobs into the categories of light, moderate, or intense work. Task analysis involves no instrumentation other than a stopwatch, and a reference table. The American Conference of Governmental Industrial Hygienists (ACGIH) published a Threshold Limit Value (TLV) that provides a simple method that may be used. Table 10.3 lists energy costs of some activities that can be used in task analysis. An observer may document the body position and movement conducted by a worker, and the type of work performed (hand work, work with one or two arms, work with body), assigning estimates of work load to each activity, in kcal/min. With some simple calculations, an estimate of metabolic load can be calculated, which in turn can be used to classify the work intensity as light (<200 kcal/hr), moderate (<350 kcal/hr), or heavy (up to 500 kcal/hr). Although error rates may be as high as ±30%, the accuracy can be as good as ±10% with a well-trained observer. However, this method is widely and easily applied to a variety of working environments, and can provide adequate protection if its limitations are recognized and accepted.

Table 10.3
Assessment of Work Load

Body Position and Movement	kcal/min
Sitting	0.3
Standing	0.6
Walking	2.0–3.0
Walking up hill	add 0.8 per meter rise

Type of Work	Average kcal/min (Range)
Hand work - light	0.4 (0.2–1.2)
Hand work - heavy	0.9
One arm - light	1.0 (0.7–2.5)
One arm - heavy	1.7
Both arms - light	1.5 (1.0–3.5)
Both Arms - heavy	2.5
Whole body - light	3.5 (2.5–15.0)
Whole body - moderate	5.0
Whole body - heavy	7.0
Whole body - very heavy	9.0

5. Evaluation of convective load

One potential source of heat load to the body is heat exchange between the air and the skin of a worker. It is a function of air temperature, mean skin temperature, and wind speed. Stated algebraically for a 'standard worker' wearing customary work clothes:

$$C = 7.0 \, V_{air}^{0.6} \, (t_{air} - t_{skin})$$ (10.3)

Where:

C = Convective heat exchange, kcal/hour
V_{air} = Air velocity, meters per second
t_{air} = Air temperature, °C
t_{skin} = Mean weighted skin temperature (often assumed to be 35°C)

From this relationship, it can be seen that if air temperature is above 35°C, convection will add heat to the body; and if air temperature is below 35°C, convection will remove heat from the body. The convective load becomes most significant in heat wave situations and in hot-wet work environments. When evaluating convective load as a potential contributor to heat stress, the benefit from cooling (see below, evaluation of evaporative cooling) should be measured as well, since the benefit from cooling may outweigh heat gain from convection if the relative humidity is less than 50%. Dryer air will allow more evaporation of sweat. Measurement of wind speed and air temperature should be straightforward, unless wind speed is highly variable.

Sample Calculation. Assume wind speed is 4.5 m/sec and air temperature is 38°C; calculate the convective heat exchange, C.

$$C = 7.0 \, V_{air}^{0.6} \, (t_{air} - t_{skin})$$
$$C = 7.0 \, (4.5 \text{ m/sec})^{0.6} \, (38°C - 35°C)$$
$$C = 51.9 \text{ kcal/hr}$$

6. Evaluation of radiative load

Heat radiation is a function of the difference between the mean radiative temperature of an environment and the average skin temperature. This radiant heat can elevate surface temperature without directly heating the air separating them; thus, it must be evaluated as well:

$$R = 6.6 \, (t_{radiant} - t_{skin}) \tag{10.4}$$

Where:

R = Radiant heat exchange, kcal/hour
$t_{radiant}$ = Mean radiant temperature of solid surfaces, °C
t_{skin} = Mean weighted skin temperature, usually assumed to be 35°C

Mean radiant temperature can be evaluated by measuring globe temperature. Globe temperature is measured using a black globe thermometer: a 15-cm (6 in.) hollow copper sphere painted a matte black, with a temperature sensor placed in the center of the globe. This device is often referred to as the Vernon globe thermometer, and the resultant reading the Vernon globe temperature (VGT). The globe thermometer must be given sufficient time to stabilize in a particular environment or measurements may be in error. Usually, a minimum of 15 minutes is required for this probe to reach stability, although the ACGIH recommends a 25-minute stabilization period.

Globe temperature is used to calculate the Mean Radiant Temperature ($t_{radiant}$, above); as such:

$$t_{radiant} = VGT + (1.8 \, V_{air}^{0.5})(VGT - t_{air}) \tag{10.5}$$

Where:

VGT = Vernon Globe temperature, °C
V_{air} = Air velocity, meters/second
t_{air} = Air temperature, °C

Sample Calculation: To evaluate radiant load, let us assume the following conditions: $t_{air} = 38°C$, VGT $= 39°C$, $V_{air} = 4.5$ m/sec.

$$t_{radiant} = VGT + (1.8 \ V_{air}^{0.5})(VGT - t_{air})$$
$$t_{radiant} = 39°C + 1.8(4.5 \text{ m/sec})^{0.5}(39°C - 35°C)$$
$$t_{radiant} = 54.3°C$$

so:

$$R = 6.6(t_{radiant} - t_{skin})$$
$$R = 6.6(54.3°C - 35°C)$$
$$R = 127.4 \text{ kcal/hr}$$

7. Evaluation of evaporative cooling

The evaporation of sweat from the body is the primary mechanism of cooling the body. The maximum potential for evaporation is a function of air velocity and the vapor pressure in the surrounding air. One external indicator that the potential for evaporation has reached its maximum is if sweat is dripping from the body. Evaporative loss can be estimated by the following formula:

$$E = 14 \ V_{air}^{0.6}(p_{skin} - p_{air}) \tag{10.6}$$

Where:

E = Evaporative heat loss, kcal/hr
V_{air} = Air speed, meters/second
p_{air} = Water vapor pressure of ambient air, mmHg
p_{skin} = Vapor pressure of water on skin, assumed to be 42 mmHg at a 35°C skin temperature.

The water vapor pressure in air can be measured using a sling psychrometer. This is a device containing two thermometers. One is covered with a wick. The instrument is slung in a circle for approximately 1 minute, and the dry and wet bulb temperatures are read. If the instrument is slung a second time and no change in the wet bulb temperature is noted, the observations can be used to obtain relative humidity and vapor pressure of water in the air, using a psychrometric chart. A psychrometric chart (Figure 10.1) is used to determine various measures of water content in air, based on simple measurements of wet bulb and dry bulb temperatures.

Sample Calculation: Assume that $p_{air} = 37$ mmHg, $V_{air} = 4.5$ m/sec; calculate the evaporative heat loss.

$$E = 14 \ V_{air}^{0.6}(p_{skin} - p_{air})$$
$$E = 14(4.5 \text{ m/sec})^{0.6}(42 \text{ mmHg} - 37 \text{ mmHg})$$
$$E = 142.6 \text{ kcal/min}$$

C. Evaluation of Heat Stress

1. *Heat stress index*

In the 1950s, Belding and Hatch proposed a method of evaluating heat stress involving estimation of metabolic, radiant, and convective load, and the maximal evaporational cooling anticipated. These values can be estimated using Equations 10.3 through 10.6. The maximal evaporation is compared to the required evaporation, which translates to the algebraic sum of metabolic, convective, and radiative loads. The Heat Stress Index can be calculated by:

$$HSI = (E_{req}/E_{max}) \times 100 \qquad (10.7)$$

The heat stress index may be evaluated using the guidelines in Table 10.4.

Table 10.4
Evaluation of Heat Stress Index

Heat Stress Index	Physiologic and Hygienic Implications of 8-hour Exposures to Heat Stress Index
-20	*Mild cold strain.* This condition frequently exists in areas where recovery from exposure to heat occurs.
0	*No thermal strain.*
+10	*Mild to moderate heat strain.* For a job that involves higher intellectual function, dexterity, or alertness, subtle to substantial decrements in performance are expected. In the performance of heavy physical work, little decrement is expected unless the ability of individuals to perform such work under marginal stress would be expected to be detrimental.
+40	*Severe heat strain,* involving a threat to health unless workers are physically fit. Break periods are required for workers not previously acclimatized. Some decrement in performance of physical work is to be expected. Medical selection of personnel desirable because these conditions are unsuitable for those with cardiovascular or respiratory impairments or with chronic dermatitis. These working conditions are also unsuitable for activities requiring sustained mental effort.
+70	*Very severe heat strain.* Only a small percentage of the population may be expected to qualify for this work. Personnel should be selected (a) by medical examination and (b) by trial on the job (after acclimatization). Special measures are needed to assure adequate water and salt intake. Amelioration of working conditions by any feasible means is highly desirable and may be expected to decrease the health hazard, while increasing efficiency on the job. Slight 'indisposition' that in most jobs would be insufficient to affect performance may render workers unfit for this exposure.
+100	*Maximum strain* tolerated by fit, acclimatized young workers.

Source: Mutcher, J.B., Chap. 20, Its effects, measurement, and control, in *Patty's Industrial Hygiene and Toxicology*, Vol. 1, G.D. Clayton and F.E. Clayton, Eds., 1978.

2. *Wet bulb globe temperature*

The evaluation of *Heat Stress Index*, although it is useful because the measurement process provides guidance in designing engineering control, is somewhat complicated to evaluate because the method requires a careful evaluation of metabolic load and the environmental factors that affect required and maximal evaporative cooling.

The Wet Bulb Globe Temperature (WBGT) was developed in 1957 to protect soldiers training in hot conditions. This index has become widely accepted and provides the basis for guidelines proposed by ACGIH, the National Institute for Occupational Safety and Health (NIOSH), and the International Standards Organization (ISO). The measurement involves a dry bulb reading, a natural wet bulb reading, and a globe temperature, reading and allows for a level of protection similar to that obtained by measuring corrected effective temperature. The wet bulb temperature is obtained by placing a wet wick around the bulb of a bare thermometer. The globe temperature is the Vernon Globe Thermometer described above. WBGT can be obtained using standard thermometers and laboratory stands. This setup is inexpensive, although cumbersome due to the size of the equipment.

Alternatively, a variety of vendors can supply electronic, portable devices that can measure WBGT, with the added convenience of providing a time-weighted average reading and computer interface. These devices are simple to operate and can provide a time history of heat conditions. Figure 10.1 shows a typical heat stress monitor, and Figure 10.2 shows a personal heat stress monitor.

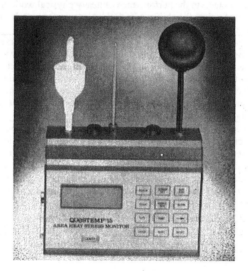

Figure 10.1   Area heat stress monitor (Courtesy: *QUEST Technologies*, Oconomowoc, WI).

Figure 10.2 Personal heat stress monitor (Courtesy: *QUEST Technologies*, Oconomowoc, WI).

WBGT can be readily calculated using one of the following formulas:

<u>Outdoors with Solar Load:</u>
$$WBGT = 0.7\ NWB + 0.2\ GT + 0.1\ DB \qquad (10.8a)$$

or:

<u>Indoors or Outdoors with No Solar Load</u>
$$WBGT = 0.7\ NWB + 0.3\ GT \qquad (10.8b)$$

Where:
    NWB = Natural wet bulb
    DB   = Dry bulb
    GT   = Globe temperature

WBGT is used to estimate work-rest regimens for work intensity levels. The index is intended for use in a comprehensive heat stress management plan, as is discussed in the section, Control of Heat Stress.

3. *Physiologic monitoring*
An alternative to the evaluation of environmental conditions and esti-mating work load is to conduct direct measurement of body temperature and heart rate; heat illnesses can be prevented if core body temperature is main-tained below 38°C. Rectal temperature is considered the most accurate way to evaluate body temperature, but tympanic and oral temperatures can be

used, with reduced precision. Monitoring core body temperature is particularly important when chemical protective clothing is worn. Studies of work in laboratory conditions show that tympanic temperature will typically be lower than rectal temperature. Oral temperature is also typically lower and more variable than rectal temperature. At the current state of the art, tympanic, or aural canal, temperature can provide a good, but not optimal, method of monitoring individual heat strain. External factors, such as ambient temperature and wind velocity, can affect both oral and tympanic temperatures; both must be adjusted to adequately predict rectal temperature. With further development and better definition and control of environmental factors such as wind speed and air temperature that may confound the results, monitoring body temperature can be a useful tool in protecting against heat stress.

D. Control of Heat Stress

Heat stress illnesses can be prevented by careful application of work guidelines, worker training, and preparation for heat emergencies (such as during heat waves). The ACGIH, NIOSH, and ISO have developed guidelines that should help prevent heat stress in most situations.

The ACGIH TLV is designed to prevent heat illnesses in healthy, acclimatized workers. The TLV is not intended to be applied to unacclimatized workers; the ACGIH recommends reducing the TLV by 2.5°C for unacclimatized workers conducting moderate work. The ACGIH recommends medical evaluation of workers before they work in hot conditions. Table 10.5 (ACGIH TLV) shows WBGT limits for work at light, moderate, and heavy work loads. As WBGT increases, increasing rest (decreasing work) is indicated. The TLV is based on the assumption that environmental conditions at rest are similar to the environmental conditions at work. The TLV includes specific instructions for WBGT measurement and metabolic load evaluation, as well as adjustment for clothing type. The ACGIH TLV is expressed as a 1-hour time-weighted average.

Table 10.5
ACGIH Heat Stress Threshold Limit Values –
WBGT Threshold Limit Values, °C

Work-Rest Regimen	Light Work 200 kcal/hr	Moderate Work 350 kcal/hr	Heavy Work 500 kcal/hr
Continuous	30.0	26.7	25.0
75% Work 25% Rest	30.6	28.0	25.9
50% Work 50% Rest	31.4	29.4	27.9
25% Work 75% Rest	32.2	31.1	30.0

Note: For unacclimatized workers, performing moderate work, the TLV should be reduced by approximately 2.5°C. Source: American Conference of Governmental Industrial Hygienists (ACGIH), *Threshold Limit Values and Biological Exposures Indices for 1991-1992*, 1991.

NIOSH also recommends the use of WBGT, with separate criteria for acclimatized and unacclimatized workers. This allows protection of workers as they adjust physiologically to the hot conditions. The NIOSH Recommended Exposure Limit (REL) includes specific medical evaluation and worker training components for worker protection. Table 10.6 shows the NIOSH REL. NIOSH does not recommend the application of WBGT for workers wearing vapor- and air-impermeable chemical protective clothing (CPC). Instead, it recommends physiologic measurements every 2 hours when the adjusted air temperature exceeds 20°C, to as often as every 15 minutes if the adjusted air temperature exceeds 32°C.

Table 10.6
NIOSH Recommended Exposure Limit -
Acclimatized (Unacclimatized), (WBGT in °C)

Work-Rest Regimen	Light Work 200 kcal/hr	Moderate Work 350 kcal/hr	Heavy Work 500 kcal/hr
Continuous	30 (27.5)	27 (25)	25 (21)
75% work 25% rest	31 (29)	28 (26)	26 (23)
50% work 50% rest	32 (30)	29 (28)	27.5 (26)
25% work 75% rest	33 (31)	31 (29.5)	30 (29)

Source: National Institute for Occupational Safety & Health (NIOSH), *Criteria for a Recommended Standard: Occupational Exposure to Hot Environments*, 1986.

Table 10.7
International Standards Organization Guidelines

Metabolic Rate (W/m²)	WBGT Reference Value			
	Person Acclimatized to Heat, °C		Person Not Acclimatized to Heat, °C	
M < 65	33		32	
65 < M < 130	30		29	
28			26	
	No Sensible Air Movement	Sensible Air Movement	No Sensible Air Movement	Sensible Air Movement
200 < M < 260	25	26	22	23
M > 260	3	25	18	20

Source: Parsons, K.C., International heat stress standards, *Ergonomics*, 1995.

The International Standards Organization (ISO) standards recommend the use of the WBGT as an initial assessment of potential heat strain. This standard involves initial evaluation using WBGT over time and height (relative to the body). Wind speed is also incorporated in the evaluation of WBGT. If the WBGT exceeds the reference values listed in the ISO standard (see Table 10.7), then the heat stress should be reduced, or a more detailed assessment of heat stress should be made. This more detailed assessment includes evaluation of metabolic, convective, and radiant loads, as well as evaporative cooling. ISO publications describe methodology for evaluating WBGT and evaluation of required sweat rate.

1. Engineering controls

The application of the above indices demonstrate situations where potential exists for heat illnesses. Engineering controls can be used to alter environmental or metabolic factors that contribute to heat stress. Evaluation of the convective, radiative, and evaporative loads, as well as metabolic demand, can help indicate what control methods will be most effective in preventing heat stress illnesses.

Convective heat control becomes increasingly important as air temperature increases above skin temperature. Thus, convective heat load can be moderated by reducing air temperature to below skin temperature, or to a point where evaporative cooling exceeds convective loading. If environmental air velocity is low, less than 2.5 m/sec, increasing air velocity can effectively increase evaporative cooling. Above this velocity, little benefit from increasing air velocity is expected. For example, with a wind speed of 2.5 m/sec and an air temperature of 42°C, evaporative cooling will equal convective heat loading at a relative humidity of about 60%. In some situations, spot cooling can reduce convective load to an individual worker. Spot cooling can be accomplished by directing centrally cooled air to a shop location, or with the use of portable air conditioners. This is particularly important when the entire workplace cannot be controlled. This method must be used with caution, as spot cooling can interfere with local ventilation for control of toxic substances.

Radiant heat control requires knowledge of which hot surfaces can radiate heat to the workers. Shielding is simple, and often is effectively used to control radiant heat load. Reflective shielding, in the form of barriers or protective clothing, is typically employed. In some cases, insulating material may be placed on hot surfaces. Remote operation, relocation of heat sources in the work area, or reducing the operating temperatures may also be considered. Some materials, such as aluminum or aluminized paint, have reduced *emissivity* and generate less radiant heat than other materials at similar internal temperatures.

In some situations, it may be possible to increase cooling from evaporation. This can be done by increasing wind speed and reducing humidity. This could be accomplished by dehumidification as well as by controlling incidental sources within a facility (from leaks, condensation, steam vents, etc.).

2. Personal protection

Personal cooling systems are another way to provide increased cooling from evaporation and/or heat transfer. These systems involve the use of ice packs, gels, circulating cooled air or liquids, and other mechanisms that serve to transfer heat from the body to the cooling system. Water-cooled garments circulate cooled water around the head, torso, arms, or legs. The water is cooled in an ice pack. Air-cooled garments utilize a source of compressed air supplied via an air-line to the worker. Both of these garment types limit the workers' mobility somewhat.

Ice packs or cold gels can be used, with the cold material placed in pockets around the torso. These devices can provide cooling for several hours at a time. Wetted overgarments are available, in the form of scarves, caps, or bandannas. These rely on evaporative cooling of water held in the fabric or in gels contained in the garments. These types of personal protection can provide significant cooling to individual workers, and may alleviate some of the strain imposed by encapsulating garments.

In the case of strong radiant load, reflective garments are effective. These are used when particularly strong sources of radiant heat are present. They may be a necessary part of worker protection in certain heat stress situations.

a. *Special concerns: hazardous waste workers and the effects of clothing*

The ACGIH TLV and NIOSH REL are valid and effective for work with regular cotton clothing. These guidelines are not intended to protect workers who must wear chemical protective clothing (CPC). This clothing is intended to prevent dermal contact with liquids, sometimes aqueous liquids. Thus, it is obvious that the cooling from evaporation of sweat is likely to be compromised. Several studies have demonstrated the increased heat load placed on individuals wearing chemical protective clothing. Hazardous waste work and asbestos abatement activities have shown similar heat stress from clothing intended to protect against contact with dust and fibers. A recent literature review supported a 10°C reduction in the ACGIH TLV for workers wearing vapor-impermeable CPC. See Table 10.8 for WBGT corrections for different clothing ensembles.

NIOSH recommends correcting the REL for heavy clothing that is more than one layer, or CPC materials. NIOSH suggests a correction of 2°C for two-layer clothing and 4°C when partially air- and vapor-impermeable clothing is worn. NIOSH recommends against WBGT for CPC (encapsulating ensembles) situations. Instead, NIOSH states the use of adjusted air temperature (dry bulb temperature adjusted for solar radiation) supplemented by physiologic monitoring (oral temperature and/or pulse rate). The need to protect workers at hazardous waste sites while wearing CPC ensembles is the driving force behind the development of personal heat monitors that measure aural or tympanic temperature and/or heart rate.

Personal cooling devices may be quite effective in protecting workers wearing CPC ensembles. Several studies have shown that increased performance and reduced heat strain are expected among workers wearing cooling vests under CPC material.

Table 10.8
ACGIH Corrections of WBGT for Clothing

Clothing Type	Clo Value	WBGT Correction, °C
Summer work uniform	0.6	0
Cotton coveralls	1.0	-2
Winter work uniform	1.4	-4
Water barrier, permeable	1.2	-6

Note: Clo insulation value of clothing. One Clo unit = 5.55 kcal/m²/hr of heat exchange by radiation and convection for each °C of temperature difference between the skin and adjusted dry bulb temperature.

Whether workers wearing CPC, particularly encapsulating CPC, are protected by adjusting the WBGT index, or by supplementary personal cooling devices, particular attention should be paid to the potential for heat strain. It is wise to emphasize the warning signs of heat strain in training programs, and some level of stress monitoring should be implemented. This could consist of monitoring aural temperature (potential measurement error should be accounted for) or pulse rate at appropriate periods.

E. Summary

To summarize, protection against heat stress illnesses requires a comprehensive heat stress management program. These programs should be in place whenever environmental conditions are severe enough to create the potential for heat illnesses. Factors such as work load, protective clothing, acclimatization, and heat wave conditions should be considered when deciding if a heat stress management program is, or may be, required.

III. LOW TEMPERATURE HAZARDS

Cold stress may be experienced in a variety of occupational situations, from outdoor work in cold regions to work in cold storage areas. Thus, construction workers, farmers, fishermen, utility workers, public safety workers, oil platform workers, cold storage room workers, and a number of other occupations may experience exposure to conditions that could lead to cold stress or cold-weather injuries.

A. Cold Temperature Injuries

If exposed to cold environmental conditions that are severe enough for a sufficient time, hypothermia occurs. *Hypothermia* is excessive lowering of the core body temperature. It can develop when heat is lost from the body at a rate that is greater than metabolic warming. It can result from exposure to cold air (in the outdoors or in cold storage areas), or immersion in cold water (18°C for a sufficient period can cause hypothermia). Hypothermia has been defined as a core body temperature at or below 35°C; severe hypothermia has been defined as a core body temperature below 33°C.

The onset of hypothermia can be insidious; shivering is a common response to cold conditions, and maximizes at a body temperature of 35°C. As body temperature decreases, confusion, impaired coordination, slurred speech, drowsiness, weakness, and lethargy may develop. Unconsciousness may occur between 30 and 32°C. Coma occurs below 28°C. Hypothermia can be fatal if not treated, and is the most serious cold-weather injury.

Frostbite is localized freezing of the skin. Skin freezes at about -2.2°C. If the frostbite is more than superficial, there is tissue loss. Gangrene may develop from tissue loss and compromise blood flow. Frostbite can be a serious condition, depending on the extent of tissue damage that occurs. If the freezing is superficial, the condition may be referred to as frostnip.

Trench foot (immersion foot) results from prolonged immersion in cold water. The name derives from military experience in World War I. The foot becomes swollen, numb, with a white or cyanotic color. Then pain, tingling, swelling, and blistering may develop. After several weeks, the condition improves.

Frostbite and trench foot are vascular injuries and localized in nature. Unless secondary infection occurs, these are transient conditions. Nevertheless, they should be prevented, and require medical treatment if such injury develops.

B. Evaluation of Cold Conditions

The primary determinants of cold injury are air temperature and wind speed. A common method of evaluating the potential for skin freezing is the *Equivalent Temperature* (*Wind Chill Index*). This is a combination of air temperature and wind speed, and can readily be looked up in Table 10.9. If the equivalent temperature is below -1°C, frostbite may develop.

The evaluation of cold stress should also include consideration of motor tasks being performed, the duration of exposure to the cold (for example in cold storage work), and the health of the worker. Medical supervision should be considered part of the cold stress prevention program. The potential for wetting of clothing should be of particular concern, since wet clothing (from sweat or splashing) will provide poor insulation against the cold.

C. Control of Cold Stress

The ACGIH TLV (Tables 10.10a and 10.10b) provides a range of recommendations for protection from cold-related injuries. Another recommendation for protection of workers from cold conditions has been published by Holmer. The upcoming discussion follows the ACGIH TLV's recommended approaches. At air temperatures below 2°C, workers immersed in water should be provided immediate change of clothing. Wet clothing not only provides poor insulation, evaporation of water will increase cooling of the body. At temperatures below 16°C, consideration should be given to warming the hands if they are exposed. This is intended to promote manual dexterity for fine motor work.

Table 10.9
Cooling Power of Wind on Exposed Flesh Expressed as Equivalent Temperature

Estimated Wind Speed, mph	Actual Temperature Reading, °F											
	50	40	30	20	10	0	-10	-20	-30	-40	-50	-60
	Equivalent Chill Temperature, °F											
Calm	50	40	30	20	10	0	-10	-20	-30	-40	-50	-60
5	48	37	27	16	6	-5	-15	-26	-36	-47	-57	-68
10	40	28	16	4	-9	-24	-33	-46	-58	-70	-83	-95
15	36	22	9	-5	-18	-32	-45	-58	-72	-85	-99	-112
20	32	18	4	-10	-25	-39	-53	-67	-82	-96	-110	-121
25	30	16	0	-15	-29	-44	-59	-74	-88	-104	-118	-133
30	28	13	-2	-18	-33	-48	-63	-79	-94	-109	-125	-140
35	27	11	-4	-20	-35	-51	-67	-82	-98	-113	-129	-145
40	26	10	-6	-21	-37	-53	-69	-85	-100	-116	-132	-148

Wind speeds greater than 40 mph have little additional effect	*LITTLE DANGER*, in < hr with dry skin. Maximum danger of false sense of security	*INCREASING DANGER*, Danger from freezing of exposed flesh within 1 minute	*GREAT DANGER*, Flesh may freeze within 30 seconds
	Trenchfoot and immersion foot may occur at any point on this chart		

Source: U.S. Army Research Institute of Environmental Medicine.

Consideration should be given to direct contact with cold surfaces, by thermal insulating material or gloves. This is important at air temperatures below -1°C.

Work under cold conditions should be conducted under a set of guidelines or work rules designed to prevent hypothermia. At temperatures at or below -12°C, the following work rules should be observed:

- Have a buddy system in effect. Have two workers on each job or have close supervision. This system allows a second person to observe the symptoms of excessive shivering, fatigue, drowsiness, irritability, or other indicators of hypothermia. These could be missed by self-monitoring.

- The work rate should not be so high as to produce heavy sweating that could result in wet clothing. If the clothing becomes wet for any reason, it should be changed immediately.
- For new workers, a few days should be allowed for acclimatization to the cold conditions, and to enable workers to get accustomed to the protective clothing.
- The weight and bulkiness of the protective clothing should be included in estimating work capacity.
- Work should be scheduled so that workers are in motion; standing still for long periods will reduce metabolic heating. Workers should not be in contact with cold metal surfaces, or sit in metal chairs.
- The workers should be trained in safety and health procedures for cold work:

> Proper re-warming procedures
> Proper clothing
> Proper eating and drinking
> Warning signs of frostbite and hypothermia
> Safe work practices

Table 10.10a
ACGIH Threshold Limit Values Work/Warm-up Schedule for 4-Hour Shift

Air Temperature: Sunny Sky		No Noticeable Trend		5 mph Wind		10 mph Wind	
°C (approx.)	°F (approx.)	Max. Work Period	No. of Breaks	Max. Work Period	No. of Breaks	Max. Work Period	No. of Breaks
-26 to -28	-15 to -19	Norm. Breaks	1	Norm. Breaks	1	75 min	2
-29 to -31	-20 to -24	Norm. Breaks	1	75 min	2	55 min	3
-32 to -34	-25 to -29	75 min	2	55 min	3	40 min	4
-35 to -37	-30 to -34	55 min	3	40 min	4	30 min	5
-38 to -39	-35 to -39	40 min	4	30 min	5	Non-emergency work should cease	
-40 to -42	-40 to -44	30 min	5	Non-emergency work should cease			
-43 and below	-45 and below	Non-emergency work should cease					

Table 10.10b
ACGIH Threshold Limit Values Work/Warm-up Schedule for
4-Hour Shift (Cont'd)

Air Temperature: Sunny Sky		15 mph Wind		20 mph Wind	
°C (approx.)	°F (approx.)	Max. Work Period	No. of Breaks	Max. Work Period	No. of Breaks
-26 to -28	-15 to -19	55 min	3	40 min	4
-29 to -31	-20 to -24	40 min	4	30 min	5
-32 to -34	-25 to -29	30 min	5	Non-emergency work should cease	
-35 to -37	-30 to -34	Non-emergency work should cease			
-38 to -39	-35 to -39				
-40 to -42	-40 to -44				
-43 and below	-45 and below				

In most cold work environments, regular breaks should be scheduled to allow warming, and maximum work periods established. Tables 10.10a and 10.10b indicate suggested work/rest regimen from the ACGIH.

Protective clothing should be made available when temperatures are below 4°C. Consideration should be given to the comfort and functionality of the clothing. There are a wide range of products available for protection from the cold. In some cases, such as in cold storage work, workers may resist wearing protective clothing as they will have to constantly enter and leave the units; proper clothing selection will help address this problem.

Some companies supply personal warming devices packets that react with air to generate heat. These can be used to warm the hands without breaking to a rest area. These should only be used if workers will not forego required warming breaks, or in situations where the principal concern is keeping the hands warm.

Engineering control in the external environment is obviously quite difficult. However, in many situations, temporary wind shields or spot heaters can provide protection, if workers are near a fixed structure or work in a fixed location.

IV. CASE STUDY

The owner of a small foundry has asked you to evaluate heat stress concerns in one section of her plant. Two workers are included in the survey, a general duty worker and a welder. Both workers experience the same environmental conditions. Both work indoors all day. You collect the following information:

Time	0800	1000	1200	1400	1600
DB °C	21	28	32	33	22
NWB °C	20	26	28	30	22
GT °C	21	29	34	36	24
V (m/sec)	0.7	1.0	1.3	1.3	0.5

You are able to observe the general duty worker during the workday, and record his activities:

Walking around area - 20% of time
Standing, sorting small parts, two arms - 20% of time
Heavy work, whole body - 25% of time
Light work, whole body - 35% of time

You do not make observations of the welder's activity.

Questions:
1. Calculate the WBGT inside and outside. Discuss the results and their implications for heat stress potential.
2. Calculate the heat load on each worker, using the formulas to predict convective load, radiant load, metabolic load, and evaporative loss.
3. What recommendations, if any, would you make to the business owner to protect her employees from heat stress?

Case Study Answers.

1. Calculation of WBGT indoors and outdoors:

Time	0800	1000	1200	1400	1600
$WBGT_{in}$	20.3	26.9	29.8	31.8	22.6
$WBGT_{out}$	20.3	26.8	29.6	31.5	22.4

Evaluation of work load:

General worker:
Refer to Table 10.1: Energy requirements of some activities:

1. Walking around area, 3.0 kcal/min
2. Standing, sorting small parts, two arms, 0.6 + 2.0 kcal/min
3. Heavy work, whole body, 7 kcal/min
4. Light work, whole body, 3.5 kcal/min

Calculate overall work intensity by multiplying work intensity (kcal/min) by fraction of work time at that intensity:

(3.0 x 0.2) + (2.6 x 0.2) + (7 x 0.25 + (3.5 x 0.35) = 4.795

Add 1 kcal/min for basal metabolism, for total intensity = 5.795 kcal/min
Hourly rate = 347.7, round to 350 kcal/hr

Classify general worker as heavy work.

Welder, no work observations; use approximation from Table 10.2: 3.4
kcal/min, 204 kcal/hr; classify welder as moderate work load.

Then compare WBGT$_{in}$ (work is conducted indoors), and results indicate
between the hours of 1000 and 1400, both workers are at risk for heat stress
illnesses. The general laborer is severely overexposed to heat.

2. Calculation of heat strain, based on calculation of convection, radiation,
   and metabolic loads and evaporative cooling.

Time	0800	1000	1200	1400	1600
DB	21.0	28.0	32.0	33.0	22.0
NWB	20.0	26.0	28.0	30.0	22.0
GT	21.0	29.0	34.0	36.0	24.0
MRT	21.0	30.8	38.1	42.2	26.5
VP$_{water}$	18.9	24.1	26.5	30.2	18.8
V	0.7	1.0	1.3	1.3	0.5
WBGT$_{in}$	20.3	26.9	29.8	31.8	22.6
WBGT$_{out}$	20.3	26.8	29.6	31.5	22.4
C	-79.1	-49.0	-24.6	-16.4	-60.0
R	-92.4	-27.7	20.5	47.2	-55.8
E	261.1	250.6	254.0	193.4	214.3
M (laborer)	350.0	350.0	350.0	350.0	350.0
M (welder)	204.0	204.0	204.0	204.0	204.0
C+R+M-E					
Load Laborer	-82.6	22.7	91.9	187.5	19.9
Load Welder	-228.6	-123.3	-54.1	41.5	-126.1

This analysis shows the laborer to be at extreme risk of heat stress injury; potential for injury can be better estimated if the time that different
work tasks occur are known. Also, radiant load contributes about 10% of
the heat load. Heat stress can be reduced by lowering work load, increasing
wind velocity, and adjusting work schedule.

REVIEW QUESTIONS

1. The most severe heat-related illness is:

   a. Heat stroke   **b.** Heat syncope   **c.** Heat exhaustion   **d.** Miliaria profunda

2. The risk of developing serious heat illnesses increases when body temperature exceeds:

   a. 37°C   **b.** 38°C   **c.** 39°C   **d.** 40°C

3. Heat cramps occur:

   a. Most commonly in unacclimatized workers.
   b. Acclimatized workers with excessive electrolyte intake.
   c. Acclimatized workers with inadequate electrolyte intake.
   d. With equal likelihood in acclimatized and unacclimatized workers.

4. Which of the following is the best measure of body temperature?

   a. Rectal temperature   **b.** Oral temperature   **c.** Aural temperature   **d.** Dermal temperature

5. Cooling of the body occurs primarily through:

   a. Convection   **b.** Radiation   **c.** Evaporation   **d.** All of the above

6. Each liter of oxygen consumed releases how much heat energy in the body?

   a. 0.3 kcal   **b.** 1.4 kcal   **c.** 4.8 kcal   **d.** 8.0 kcal

7. A work load of 300 kcal/min should be classified as a _____ work load.

   a. Resting   **b.** Low   **c.** Moderate   **d.** Heavy

8. Why is monitoring heart rate useful in preventing heat stress?

   a. Workers must not be allowed to work near their maximal heart rate for any period of time.
   b. Heart rate can be used as a surrogate for oxygen consumption, as an index of work intensity.
   c. It requires workers to take rest breaks they might not otherwise take.
   d. Heart rate can be used as a surrogate for measuring body temperature.

9. What temperature measurements are included in WBGT?

   a. Natural wet bulb, dry bulb, and globe temperatures
   b. Natural wet bulb, radiant temperature, and dry bulb temperature
   c. Wet bulb, dry bulb, and effective temperature
   d. Natural wet bulb, dry bulb, and corrected effective temperature

10. How many days are required for a worker to become acclimatized to hot conditions?

    a.   1 or 2 days
    b.   7 to 10 days
    c.   21 to 24 days
    d.   Acclimatization is rarely effective in heat stress prevention.

11. What correction to the ACGIH TLV is recommended for workers wearing encapsulating CPC garments?

    a. 2°C   **b.** 5°C   **c.** 7°C   **d.** 10°C

12. At what body temperature does unconsciousness from hypothermia occur?

    a. 18°C   **b.** 35°C   **c.** 31°C   **d.** 28°C

13. Why is wet clothing of particular concern in cold conditions?

    a. Frost bite is more likely.
    b. Wet clothing provides poor insulation.
    c. Wet clothing interferes with job performance.
    d. Wet clothing is heavier than dry clothing, increasing metabolic demand.

14. Manual dexterity in cold conditions can be compromised when air temperature falls below:

    a. 25°C   **b.** 18°C   **c.** 10°C   **d.** 0°C

15. The 'buddy system' and other protective measures should be instituted when temperatures fall below:

    a. 0°C   **b.** -5°C   **c.** -12°C   **d.** -20°C

ANSWERS

1.	a.	**9.**	a.
2.	b.	**10.**	b.
3.	c.	**11.**	d.
4.	a.	**12.**	c.
5.	c.	**13.**	b.
6.	c.	**14.**	a.
7.	c.	**15.**	c.
8.	b.		

References

American Conference of Governmental Industrial Hygienists (ACGIH), *Threshold Limit Values and Biological Exposure Indices for 1991-1992*, ACGIH, Cincinnati, 1991.

Astrand, P.O. and Rodahl, K., *Textbook of Work Physiology*, McGraw-Hill Book Co., New York, 1970.

Beaird, J.S., Bauman, T.R., and Leeper, J.D., Oral and tympanic temperatures as heat strain indicators for workers wearing chemical protective clothing, *Am. Ind. Hyg. Assoc. J.*, 57:344-347, 1996.

Bernard, T.E. and Kenney, W.L., Rationale for a personal monitor for heat strain, *Am. Ind. Hyg. Assoc. J.*, 55:505-614, 1994.

Durnin, J.V.G.A. and Passmore, R., *Energy, Work and Leisure*, Heinemann Educational Books, Ltd., London, 1967.

Evenson, E., Chap. 7, Cold environments, in *Physical and Biological Hazards of the Workplace*, Wald, P.H. and Stave, G.M., Eds., Van Nostrand Reinhold, New York, 1994.

Fuller, F.H. and Smith, P.E., Evaluation of heat stress in a hot workshop by physiological measurements, *Am. Ind. Hyg. Assoc. J.*, 42:32-37, 1981.

Gullickson, G.M., Chap. 6, Hot environments in *Physical and Biological Hazards of the Workplace*, Wald, P.H. and Stave, G.M., Eds., Van Nostrand Reinhold, New York, 1994.

Helander, M., *A Guide to the Ergonomics of Manufacturing*, Taylor and Francis, London, 1995.

Holmer, I., Cold stress part I. Guidelines for the practitioner, *Int. J. of Ind. Ergo.*, 14:139-149, 1994.

Holmer, I., Cold stress Part II. The scientific basis (knowledge base) for the guide, *Int. J. of Ind. Ergonomics* 14.151-159, 1994.

International Standards Organization, *Hot Environments - Analytical Determinations and Interpretation of Thermal Stress Using Calculation of Required Sweat Rate.* ISO Publication 7933, 1989.

International Standards Organization, *Hot Environments - Estimation of the Heat Stress on Working Man, Based on the WBGT-Index (Wet Bulb Globe Temperature).* ISO Publication 7243, 1989.

Malchaire, J., Methodology of investigation of hot working conditions in the field, *Ergonomics*, 38, 183-192, 1995.

Montoye, HJ, HCG Kemper, WHM Saris, RA Washburn, *Measuring Physical Activity and Energy Expenditure*, Human Kinetics, Champaign, IL, 1996.

Mutcher, J.B., Chap. 20, Heat stress, its effects, measurement and control in *Patty's Industrial Hygiene and Toxicology*, Vol. 1, Clayton, G.D. and Clayton, F.E., Eds., John Wiley & Sons, New York, 1978.

National Institute for Occupational Safety and Health (NIOSH), *Criteria for A Recommended Standard: Occupational Exposure to Hot Environments. Revised Criteria,* [DHHS (NIOSH) pub no.86-113] Cincinnati, OH, U.S. Department of Health and Human Services, Public Health Service, Centers for Disease Control, NIOSH, 1986.

Parsons, K.C., International heat stress standards, *Ergonomics,* 38:6-22, 1995.

Plog, B.A., Niland, J., and Quinlan, P.J., Eds., *Fundamentals of Industrial Hygiene,* 4th ed., National Safety Council, Chicago, 1996.

Reneau, P.D. and Bishop, P.A., A review of the suggested WBGT adjustment for encapsulation protective clothing, *Am. Ind. Hyg. Assoc. J.,* 57:158-161, 1996.

Reneau, P.D. and Bishop, P.A., Validation of a personal heat stress monitor, *Am. Ind. Hyg. Assoc. J.,* 57:650-657, 1996.

Chapter 11

PERSONAL PROTECTIVE EQUIPMENT (PPE)

S. Zack Mansdorf, PhD, CIH, CSP, PE

I. OVERVIEW

When occupational hazards cannot be eliminated, personal protective equipment (PPE) can be used as an effective control measure. Nevertheless, it should not be relied upon unless other control measures are not feasible. This is because the hazard will still exist and effective PPE control relies on proper fit, use, and maintenance by the wearer. It is important to remember that the failure of personal protective equipment through improper selection, use, or maintenance is likely to result in an injury or illness.

PPE is that equipment which is worn by a worker and affords protection by altering the external environment, such as with respirators, or provides a barrier through which the hazard cannot pass, such as with welding goggles. PPE can range from simple and relatively inexpensive plastic "safety glasses" that prevent eye injuries, to relatively expensive fully encapsulating body suits with integrated respiratory protection, such as those worn by the rocket fuelers at NASA.

The remainder of this chapter will address protocols detailing how to conduct hazard assessments for the selection of PPE, respiratory protection, protective clothing, and other protective equipment. This is followed by case studies on the selection and use of PPE and questions. The section on hazard assessments for selection of PPE follows.

II. HAZARD ASSESSMENTS FOR THE SELECTION OF PERSONAL PROTECTIVE EQUIPMENT

PPE is considered the "last line of defense," as discussed in the introduction. Therefore, a decision to rely on personal protective equipment should be based on a thorough understanding of the hazards involved and the effectiveness of proposed control measures. Put another way, the risks to the worker need to be fully investigated before the final choice of control measures is made. This evaluation is called a *hazard assessment* by OSHA.

A. OSHA Requirements

Overall requirements for PPE are found under 1910.132 of the General Industry Standards. Specific requirements are contained under other sections of the OSHA General Industry Standards or within the sector standards such as the Construction Standards and Maritime Standards.

The general requirements include:

- Maintenance of the equipment can be accomplished in a reliable and sanitary manner.
- Equipment is of a safe design and construction for the work to be performed.
- A hazard assessment is completed before selection of the equipment.
- Defective or damaged equipment is not used.
- The employee has received training and understands when the equipment is needed, the proper use and maintenance of the equipment, how to properly put on, wear, and take off the equipment, its useful life, and how to properly dispose of the equipment.

Hazard assessments are fundamental to the practice of industrial hygiene. These assessments encompass the four tenets of the practice (anticipation, recognition, evaluation, and control). Therefore, we will just briefly review the process and the essence of the OSHA requirements. OSHA, in a nonmandatory appendix, suggests that the hazard assessment start with a walk-through survey to identify sources of hazards to workers. These hazards might include impact, penetration, compression, chemicals, heat, harmful dusts, electricity, and other hazards. They suggest that the data obtained from the walk-through be followed by analysis of the information collected. Analysis would include type of hazard, level of risk, and seriousness of potential injury (or illness). This would be followed by evaluation of selection guidelines and references from expert sources and equipment suppliers. From this information, the selection would be made and the devices fitted to the user. Another element not specifically mentioned but necessary is an assessment of the effectiveness of the device after selection (i.e., during use). OSHA does suggest that hazards be reassessed for changes as necessary. OSHA also does not specifically state that the hazard assessment be done in writing; however, they do require the employer to certify that the assessment has been done. A written assessment (recommended professional practice) would meet this requirement.

Additional professional practice guidelines as well as OSHA requirements are specific to the type of personal protective equipment used and the work environment. These requirements or suggested guidelines are covered in the sections on protective clothing, respiratory protection, and other protective devices.

III. PROTECTIVE CLOTHING

In the United States (U.S.), dermatological disorders are one of the National Institute for Occupational Safety and Health's (NIOSH) top ten leading occupational health problems. These disorders are primarily a result of unprotected exposures to harmful biological, physical, and chemical agents. Most of these injuries and disease risks can be prevented or reduced through the appropriate selection and use of protective clothing or other control methods.

A. An Overview of Hazards Affecting the Skin

There are three major categories of hazards affecting the skin. They are biological, physical, and chemical hazards as shown in Table 11.1. Other work-related skin hazards can exist that are either rare, or closely related to those in the table (such as animal and insect bites). Only typical or characteristic hazards by category will be included in our discussion.

Table 11.1
Examples of Hazard Categories Affecting the Skin

Hazard	Examples
Biological	Human pathogens Animal pathogens Environmental pathogens
Physical	Trauma producing Thermal hazards (hot/cold) Fire Vibration Radiation
Chemical	Irritants Allergens Corrosives Dermal toxins Systemic toxins Cancer-causing agents

Source: Adapted from *The Occupational Environmental-Its Evaluation and Control*, published by AIHA.

1. *Biological hazards*

Biological hazards include infection from agents and disease common to humans, those common to animals, and those common to the work environment. Biological hazards common to humans include pathogenic microbes such as those causing AIDS, hepatitis, TB, Legionnaires disease and others. Work in the healing arts can involve exposure to blood or body fluids containing human pathogens. This work usually requires some type of respiratory protection, liquid-resistant garments, and gloves. In the U.S., OSHA requires that appropriate protective clothing be used for work situations where there is likely exposure to body fluids under their Bloodborne Pathogens regulation (29 CFR 1910.1030). Diseases can also be transmitted from animals (including fish) through handling. For example, North Sea fisherman can develop a skin condition called Bogger Bank Itch, while there are common large animal zoonoses which can be passed to humans; examples include: Brucellosis, Anthrax, Tularemia, and others. Most of these zoonoses have a long history of recognition and require protective measures similar to those used for handling body fluids and wastes from humans. Work environments that can present a hazard from biological agents include sewage treatment plants, composting facilities, clinical and microbiological laboratories, genetic research facilities, as well as other special work environments.

2. Physical hazards

As noted in Table 11.1, physical hazards can be categorized to include those that are trauma producing (produce injuries such as cuts), from thermal effects, vibration, and radiation. Trauma to the skin from physical hazards (cuts, abrasions, etc.) is common for many jobs and industrial sectors such as construction and meat cutting. Thermal hazards to the skin include the adverse effects of extreme cold and heat such as those from molten metals and handling cryogenic liquids. The protective attributes of clothing for these hazards is related to the insulation provided (generally increases with thickness), whereas protective clothing for flash fire and electric flash over requires flame resistance properties. Protection from some forms of both ionizing and nonionizing radiation can be achieved using protective clothing. In general, protective clothing for ionizing radiation is based on the principle of shielding (e.g., lead lined aprons and gloves) or preventing particulate radionuclides and liquids from direct contact with the skin. Clothing for electromagnetic radiation, such as microwave, is based on grounding, whereas protection from light (UV, Visible, and IR) is dependent upon the wavelength of the radiation. For example, IR protection is usually afforded by reflective clothing such as aluminumized coverings. Excessive vibration can have several adverse effects on body parts, primarily the hands. Occupations such as mining (hand-held drills) and road repair (pneumatic hammers or chisels) are two examples where excessive vibration can lead to loss of circulation and bone degeneration in the hands (Raynaud's phenomenon). Specialized protective clothing (e.g., water-proof gloves with urethane, gels, or foam linings) can help to damp the vibration received by the body and to keep the hands warm and dry.

3. Chemical hazards

Many chemicals may present more than one type of dermal risk (for example, a substance such as benzene is both toxic and flammable). There are four major factors to be considered when assessing the dermal risk that chemical hazards pose. These are:

- Routes of exposure, including inhalation, ingestion, dermal, and injection
- Potential adverse effects of unprotected exposure
- The exposure potential (likely dose)
- Physical hazards (e.g., fire)

Some exposure situations simply present a cleanliness issue (e.g., oil and grease), while other situations (e.g., skin contact with hydrofluoric acid) could present a situation which is immediately dangerous to life or health.

As shown in Table 11.1, adverse effects of skin contact with chemicals can be generally categorized as causing irritation, an allergic response, corrosion (chemical burns), skin toxicity, systemic toxicity (permeation through the skin), and promotion of cancer of the skin or other body cancers. As an example of a chemical which normally presents the greatest risk by the dermal route, the pesticide Parathion has significant toxicity due to its skin permeability. This characteristic resulted in the deaths of some agricultural workers when an over-spray from the aerial application of Parathion in an orchard occurred while workers were collecting fruit. This is only one of many instances where the dermal route of entry presents a significant risk. Chemicals which have the potential to significantly contribute to a worker's overall dose by the dermal route are identified in the U.S. by OSHA in their permissible exposure limits (PEL) and the American Conference of Governmental Industrial Hygienists (ACGIH) in their threshold limit values (TLV) by a "skin" notation. However, many other substances which do not normally present inhalation hazards can have significant adverse effects on unprotected skin. The inorganic acids are an example of substances with low vapor pressures (hence, normally a minimal inhalation hazard) which are hazardous to the skin due to their corrosive nature. As a worst-case example, a single unprotected skin exposure to anhydrous hydrofluoric acid (above 70% concentration) can be fatal. In this case, as little as a 5% burn can result in death from the both the corrosion and the effects of the fluoride ion. On the opposite end of the spectrum, an example of a material which has high human toxicity but little skin toxicity is inorganic lead. In this case, the concern is contamination which could later lead to ingestion or inhalation since inorganic lead dusts will not permeate intact skin.

4. Chemical permeation of barriers

Research showing the diffusion of solvent through "liquid-proof" protective clothing barriers has been extensively published. Acetone, for example, has been shown to go through neoprene rubber (of normal glove thickness) usually within 30 minutes of direct liquid contact on the normal outside surface of the barrier. This movement of a chemical through a protective clothing barrier is called permeation. The permeation process consists of the diffusion of chemicals on a molecular level through protective clothing. The permeation process occurs in three steps:

- Absorption of the chemical at the barrier surface
- Diffusion through the barrier
- Desorption of the chemical on the normal inside surface of the barrier

The permeation of a barrier, since it is essentially diffusion, is directly related to temperature, while inversely related to barrier thickness. This relationship can be expressed in terms of the time it takes a chemical to fully permeate a protective barrier. Movement of the chemical from the outside surface of the barrier to the inside surface is known as *breakthrough*. Hence, an increase in temperature will shorten the time to breakthrough, while an increase in thickness of the barrier will increase breakthrough time.   It should also be noted that for permeation to occur, continuous contact with the challenge material is not required. The time elapsed from the initial contact of the chemical on the outside surface until detection on the inside surface is called the breakthrough time.

The *permeation rate* is the steady-state rate of movement (mass flux) of the chemical through the barrier after equilibrium is reached. The permeation rate is normally reported in mass per unit area per unit time (e.g., $\mu g/cm^2/min$) and may be normalized for thickness. Most current laboratory testing done for permeation resistance is for periods of up to 8 hours to reflect normal work shifts

5. Permeation testing

A standard test method for determining permeation of protective clothing is F739-91, *Test Method for Resistance of Protective Clothing Materials to Permeation by Liquids or Gases Under Conditions of Continuous Contact* published by the American Society for Testing and Materials. In brief, the test consists of placing the barrier material between a reservoir of the challenge chemical (liquid or gas) and a collection cell connected to an analytical detector. It should also be noted that the actual reported breakthrough time is also related to the sensitivity of both the analytical method and system (i.e., collection system). Variations of this method can also be performed for determining permeation associated with intermittent contact. A novel approach measures evaporation through the barrier by weight loss for chemicals that have a vapor pressure of at least 10 mmHg.

Aside from the permeation process just described, there are two other chemical resistance properties of concern to the safety and health professional. These are *degradation* and *penetration*. Degradation is a deleterious change in one or more physical properties of a protective material caused by contact with a chemical. For example, the polymer polyvinyl alcohol (PVA) is a very good barrier to most organic solvents but swells and is degraded by water. Latex rubber, which is widely used for medical gloves, is readily soluble in toluene and hexane as another example. Many times, degradation can be assumed if the barrier swells or has a change in physical appearance (wrinkles, burns, changes color, etc.). Degradation may not always be visible, such as when solvents dissolve the plasticizer in some polymers, causing them to become brittle without visible evidence of the change.

Penetration is the flow of a chemical through zippers, weak seams, pinholes, cuts, or imperfections in the protective clothing on a nonmolecular level. Even the best protective barriers will be rendered ineffective if punctured or torn. Penetration protection is important when the exposure is unlikely or infrequent and the toxicity or hazards are minimal. Penetration usually is a concern for splash protection garments.

Given the lack of published skin permeability and dermal toxicity data, the approach taken by most safety and health professionals is to select a barrier with no breakthrough for the duration of the job or task, usually 8 hours, which is essentially a no-dose concept. This is an appropriate conservative approach; however, it is important to note that there is no protective barrier currently available which provides permeation resistance to all chemicals. For situations where the breakthrough times are short, the safety and health professional should select the barrier(s) with the best performance (i.e., longest breakthrough time and/or lowest permeation rate) as well as considering other control measures (such as a clothing change).

There are several guides that have been published listing chemical resistance data (many are also available in an electronic format). In addition to these guides, most manufacturers also publish current chemical and physical resistance data for their products.

IV. TYPES OF PROTECTIVE CLOTHING

Protective clothing includes all elements of a protective ensemble (e.g., garments, gloves, boots, etc.). Protective clothing items can range in complexity from a finger cot providing protection against paper cuts to a fully encapsulating suit requiring a self-contained breathing apparatus like those used for emergency responses to hazardous chemical spills.

Protective clothing can be made of natural materials (e.g., cotton, wool, leather), man-made fibers (e.g., nylon), or various polymers (rubbers and plastics such as neoprene rubber, nitrile rubber, butyl rubber, polyvinyl chloride, chlorinated polyethylene, etc.). Materials which are woven, stitched, or are otherwise porous (not resistant to liquid penetration or permeation) should not be used in situations where protection against a liquid or gas is required. Specially treated or inherently nonflammable porous fabrics and materials are commonly used for flash fire and electric arc (flashover) protection (e.g., petrochemical industry and electric industries), but usually do not provide protection from any lasting heat component. It should be noted here that fire fighting requires specialized clothing that provides flame (burning) resistance, a water barrier, and thermal insulation (protection from heat), while entry into fuel fires (strong IR component) also requires reflective clothing (e.g., aluminumized overcover) for protection against the infrared component. Table 11.2 summarizes typical physical, chemical, and biologic performance requirements and common protective materials used by hazard.

Table 11.2
Common Protective Clothing Performance Requirements

Hazard	Performance Characteristic Required	Common Protective Clothing Materials
Chemical	Permeation, penetration, degradation resistance	Polymeric materials
Biological	Protection against microbes	Polymeric materials
Radiological	Provide shielding or not allow liquid penetration	Polymers and lead-lined protective clothing
Thermal	Insulation	Thick natural or man-made fabrics or with insulating layers
Fire	Insulation and fire resistance	Aluminized gloves; flame-resistant treated gloves; aramid fiber and other special fabrics
Mechanical abrasion	Abrasion resistance; tensile strength	Heavy fabrics; leather with metal studding
Cuts and punctures	Cut resistance	Metal mesh; aromatic polyamide fiber and other special fabrics
Vibration	Damping	Natural or polymeric gloves with elastomeric linings

Source: Adapted from *The Occupational Environment-Its Evaluation and Control,* published by AIHA.

Protective clothing configurations vary greatly, depending on the intended use. However, normal components are analogous to personal clothing (i.e., pants, jacket, hood, boots, gloves) for most physical hazards. Special-use items for applications such as flame resistance in the molten metals industries can include chaps, armlets, and aprons constructed of treated natural and synthetic fibers and materials. Clothing for cut protection can range from garments of aramid fibers to chain mail gloves of metal construction (including titanium), or special fiber batting for protection against chain saws. Protection from cold extremes usually includes multiple components of high insulating values which allow for the wicking and/or evaporation of perspiration. Chemical protective clothing can be even more exotic in terms of construction.

A. Gloves

Chemically protective gloves are usually available in a wide variety of single polymers and combinations such as polymer-coated natural fiber (manufactured using a dipping process). Some of the new foil and laminate gloves are only two dimensional (flat) and hence do not fit the hand as well as gloves made from human forms in a dipping process, but are highly chemical resistant. These gloves typically work best when a form-fitting outer polymer glove is worn over the top of the inner flat glove (called double gloving) to conform to the shape of the hands. Polymer gloves are available in a wide variety of thicknesses, ranging from very lightweight (<2 mm) to heavyweight (>5mm), and with and without inner liners or substrates (called scrims). Gloves are also commonly available in a variety of lengths, ranging from approximately 30 cm for hand protection to gauntlets of approximately 80 cm extending from the worker's shoulder to the tip of the hand.

It should also be noted in this section that latex rubber gloves can cause irritant contact dermatitis, chemical sensitivity dermatitis, or severe allergic reactions in some workers. Natural latex rubber gloves have poor chemical resistance properties but are extensively used for biological hazards, especially in the health care fields. Options available for these situations include use of gloves made of other rubbers or plastics (such as thin nitrile), use of hypoallergenic latex gloves (not effective against allergic reactions to latex), and use of non-powdered latex gloves.

B. Boots

Chemically resistant boots are available in only a limited number of polymers, since the boot heel and sole require a high degree of abrasion resistance. Common polymers and rubbers used in chemically resistant boot construction include PVC, butyl rubber, nitrile, and neoprene rubber. Specially constructed laminate boots using other polymers can also be obtained but are quite expensive and limited in polymer choices at the present time.

C. Garments

Chemical protective garment types range from one-piece fully encapsulating (gas-tight) suits with attached gloves and boots, to simple single components, such as jackets, pants, and hoods. Some protective barriers will have multiple layers or laminants. Layered materials are generally required for polymers that do not have good inherent physical integrity and abrasion-resistant properties (e.g., Teflon). Common support fabrics are nylon, polyester, aramide fibers, and fiberglass. These substrates are coated or laminated by polymers such as polyvinyl chloride (PVC), Teflon, polyurethane, polyethylene, and other proprietary materials (see Figure 11.1). Some suits use layering of different polymers to improve the overall chemical resistance of the article. For example, one protective suit could use an outside layer of neoprene, nylon in a middle layer for support and strength, and butyl rubber on the inside layer.

Figure 11.1 Example of a saran-coated suit (Courtesy: *Kappler Safety Group*, Guntersville, AL).

Over the last decade there has been enormous growth in the use of relatively inexpensive nonwoven polyethylene and other materials for disposable suit construction. The spun bonded suits, sometimes incorrectly called *paper suits*, are made using a special process where the fibers are bonded or pressed, and glued together rather than woven (refer to Figure 11.2). Uncoated microporous (called *breathable* because they allow some water vapor transmission and hence are less heat stressful) and spun bonded garments have good applications for protection against particulate and fiber hazards, but are not normally chemical nor liquid resistant. Spun bonded garments are also available with various coatings such as polyethylene, PVC, and Saran, which provide good chemical resistance.

The fully encapsulating gas-tight suit of one-piece construction provides the highest level of protection provided by chemical protective clothing. In the majority of these configurations, the respiratory protection device (air-line or SCBA) is also protected by the suit, since it is worn within the suit. Protection factors for suits of this type are typically higher than those for the respiratory protection (protection factor in excess of 10,000), provided the suit has appropriate chemical resistance to the challenge.

Figure 11.2 Example of a spun-bonded suit (Courtesy: *Kappler Safety Group*, Guntersville, AL).

V. HUMAN FACTORS IN SELECTION OF PROTECTIVE CLOTHING

In most cases, use of protective clothing and equipment will *decrease* productivity and *increase* worker discomfort. Exceptions to this general rule would include cold environments and work involving abrasion resistance. The use of protective clothing may also lead to decreased quality, since error rates increase with the use of protective clothing. For chemical protective clothing, and some fire-resistant clothing, there are some general guidelines that need to be considered concerning the inherent conflicts between worker comfort, efficiency, and protection. Thicker is better because of the decrease in breakthrough of chemicals, and for certain applications, the improved insulation values. Nevertheless, the thicker the barrier, the more it will decrease ease of movement and user comfort. Thicker barriers also increase the potential for heat stress. Barriers which have excellent chemical resistance will also increase the level of worker discomfort and heat stress since they also act as barriers to the evaporation of sweat. One of the primary ergonomic constraints for chemically resistant suits and ensembles is the issue of heat stress. Once the worker dons the suit, the inside environment quickly approaches 100% humidity and, at least, body temperature. The amount of heat retained within the suit will depend on the metabolic work load. Nevertheless, almost all suits that are chemically resistant, and cover all or most of the body, present heat stress challenges.

In general, the higher the overall protection of the clothing, the more time the job will take to accomplish, the higher the level of potential heat stress, and the more likely there will be errors. It should also be noted that in a few situations, the use of protective clothing could actually increase risk to the user (e.g., around moving machinery, where risk of heat stress is greater than the chemical hazard, etc.). The work situation must always be considered in the selection of the protective clothing for the job. The optimum solution is to select the minimum level of protective clothing and equipment that is necessary to safely do the job.

A. Selection of Protective Clothing

As stated earlier in this chapter, OSHA regulations (29 CFR 1910.132) require a hazard assessment for personal protective equipment be conducted before assignment.

The overall approach to the selection of protective clothing for most situations can be illustrated using a four-step process which incorporates the required OSHA hazard assessment. These steps are as follows:

1. Completing the hazard assessment
2. Determining performance characteristics needed for protection
3. Determining the need for decontamination (as applicable)
4. Assessing all factors and making the selection

B. The Hazard Assessment

The process should begin with a determination of the chemical hazards of the process or work task, likely exposures and routes of exposure, and extent of exposure. The best method to determine the likely worker exposures and potential routes of exposure is to actually inventory the chemicals used and observe the task or work assignment. This is commonly done by what OSHA and most others call a walk-through survey. A second but less informative approach is to have the work process described or to evaluate it from a written description. Coupled with knowledge of the chemicals used in the process such as their physical characteristics (e.g., vapor pressure, physical state, etc.) from the material safety data sheet (MSDS), an assessment of the most likely routes of exposure can be made. Likewise, an assessment of the dermal hazards presented by the chemicals used can be obtained from the MSDS and other reference sources.

Once the evaluation of the likely exposures, route(s) of entry, and toxicity of the materials has been completed, an assessment of the extent of potential exposure needs to be determined. That is, what is the nature and extent of worker contact? For liquid chemicals as an example, is the nature of potential contact simply from an inadvertent splash or do workers have direct contact with the chemical? For those scenarios where the material is highly hazardous (e.g., liquid sodium cyanide), although the likelihood of contact is remote, the worker must obviously be provided with the highest level of protection available. For situations where the exposure represents a very minimal risk (e.g., a nurse applying rubbing alcohol to a patient), the level of protection does not need to be absolute. This selection logic is essentially based on an estimate of the adverse effects of the material combined with an estimate of the likelihood of exposure.

Once the hazards and worker risks are known, the first consideration should always be whether or not the job or task can be safely done without the use of protective clothing. It has long been the philosophy of NIOSH that personal protective clothing should not be used before consideration of other control options. These include substitution with a less hazardous material, use of mechanical means for accomplishing the task, and the use of engineering controls (e.g., ventilation) or non-engineering controls (e.g., administrative controls such as working in the mornings or evenings for hot work, etc.).

C. Determining Protective Clothing Performance Requirements

The physical and chemical hazards of the job will define the performance characteristics of the protective clothing. Many times, these performance characteristics will require compromise, since no single selection may meet all of the performance criteria. For example, there is no commercially available glove that provides good chemical and thermal/fire resistance. The performance characteristics which are needed must be rank-ordered so that the greatest risks are resolved before the lesser risks are addressed.

D. Determining the Need for Decontamination

For protective clothing that is used with hazardous chemicals, decontamination must be considered even if the clothing is disposable. This is because cross-contamination can occur with doffing of the protective clothing. Not all situations will require decontamination, however; there are many case histories of workers needlessly contaminated because decontamination was not considered. This is especially important for strong acids and bases, pesticides, heavy metals, and other materials where cross-contamination might be a concern. It is also an important consideration if the work clothing is stored with street clothing, reused, or taken home.

Methods of detection of contamination and decontamination will vary depending on the level and type of contaminant. For water-soluble chemicals, a simple water wash may suffice, while most organic contaminants require elaborate methods of detection and removal.

E. Assessing the Factors and Making the Selection

After consideration of all of the factors that are important, including ergonomic and heat stress limitation and costs, a final selection can be made. These factors will include all of those previously mentioned based on input extracted from the use of available reference guides, manufacturer literature, and recommendations from chemical manufacturers and others. In some cases, it may be necessary to redesign the work tasks to better fit the ergonomic requirements of performance in protective clothing. For some tasks, it may even be concluded that the work cannot be efficiently or safely performed using protective clothing. In these cases, a different approach to accomplish the task must be developed.

F. Worker Education and Training for Protective Clothing Use

Adequate education and training for users of protective clothing is essential. Training and education should include:

- The nature and extent of the hazard(s)
- When protective clothing should be worn
- What protective clothing is necessary
- Use and limitations of the protective clothing to be assigned
- How to properly inspect, don, doff, adjust, and wear the protective clothing
- Decontamination procedures, if necessary
- Signs and symptoms of overexposure or clothing failure
- First aid and emergency procedures
- The proper storage, useful life, care, and disposal of protective clothing

This training should incorporate all of the elements listed above and others of pertinence that have not already been provided to the worker through other programs. For those training subjects and topics already provided to the worker from other training programs, an assessment of the need for refresher training should be undertaken. Finally, the workers should have an opportunity to try out the protective clothing before a final selection decision is made and significant quantities ordered.

G. Developing and Managing a Protective Clothing Program

A good professional practice would be to develop a written program. This will reduce the chance for error, increase worker protection, and establish a consistent approach for the selection and use of protective clothing. A model program should contain the following elements:

- An organization scheme and administrative plan
- A risk assessment methodology
- An evaluation of methods other than protective clothing to protect the worker
- Performance criteria for the protective clothing
- Selection criteria and procedures to determine the optimum choice
- Purchasing specifications for the protective clothing
- A validation plan for the selection with medical surveillance, as appropriate
- Decontamination and reuse criteria, as applicable
- A user training program
- An auditing plan to assure procedures are consistently followed

There are several examples of misuse of protective clothing that can be commonly seen in industry. Misuse is usually the result of a lack of either a management and/or worker misunderstanding the limitations of the protective clothing assigned. An example is the use of nonflame-resistant protective clothing for workers handling flammable solvents or in situations where open flames, burning coals, or molten metals are present. The protective clothing made of polymeric materials such as polyethylene will support combustion and actually melt, causing a more severe burn. A second common example is the reuse of protective clothing (including gloves), where the chemical has contaminated the inside of the protective clothing and resulted in the worker increasing his exposure on each subsequent use. It is very common to see another variation of this problem where workers use natural fiber gloves (e.g., leather or cotton) or personal (leather) shoes to work with liquid chemicals. If the chemicals are spilled on, or contact the natural fibers, they will be retained for long periods of time and migrate to areas of direct skin contact. These are only a few of the more prominent examples of the misuse of protective clothing. Many of these problems can be overcome by simply making sure the wearer understands the proper use and limitations of the protective clothing.

VI. RESPIRATORY PROTECTION

Respirators are widely used as protective devices. NIOSH has estimated that as many as 7 million workers uses respirators at some time each year. Respiratory protection is extensively regulated by OSHA. The major requirements under OSHA are described in the next section.

A. OSHA Requirements

OSHA regulates respiratory protection in the General Industry Standards under 29 CFR 1910.134; under 29 CFR 1926.103 of the Construction Industry Standards; as well as coverage detailed in several sections of the Maritime Standards. Specific requirements for respirators are also found in the substance-specific standards (such as asbestos and lead) and some of the requirements from the General Industry Standards overlap, or are incorporated in, the Construction Industry Standards.

In summary, the respiratory protection section of the General Industry Standards requires that employers:

- Use control methods other than respirators unless not feasible
- Provide employees with suitable respirators for the purpose intended
- Implement a respiratory protection program
- Require that employees use the respirators properly
- Use respirators that have NIOSH approval
- Choose respirators in accordance with ANSI standard Z88.2
- Provide Grade D breathing air for supplied air respirators
- Prepare standard procedures for respiratory protection selection, use, care, and maintenance
- Provide for fit testing of respirators
- Provide worker training in respiratory protection and use

The above list is only a short summary of the major OSHA requirements. The main regulatory focus and good industrial hygiene practice is contained in the respiratory protection program requirements. A summary of the OSHA requirements for a written respiratory protection program follows:

- Establish written operating procedures for respirator selection and use
- Base respirator selection on the hazards to which the worker is exposed
- Instruct and train employees on using the respirator properly and recognizing its limitations
- Clean and disinfect respirators regularly and clean those used by more than one worker after each use
- Store respirators in a convenient, clean, sanitary place

- Inspect and repair, as necessary, routinely used respirators during cleaning
- Inspect emergency respirators at least once per month and after each use
- Maintain surveillance of work area conditions and degree of employee exposure or stress
- Regularly inspect and evaluate the continuing effectiveness of the program
- Have a physician determine the physical ability of the user to wear respiratory protection with periodic medical review
- Use only NIOSH approved respirators (Bureau of Mines approvals have been combined with NIOSH approvals, except for mine rescue)

Additional requirements and good practices are contained in the remainder of this section on respiratory protection, including more detail on the elements of a respiratory protection program.

B. Overview of Respiratory Hazards

The most common route of entry for hazardous materials in the industrial environment is by the respiratory route. There are many processes which can generate airborne particulates, vapors, and gases of concern. These include the general categories of manufacturing, maintenance, and construction, as well as chemical spills and fire-fighting. There are also work situations and processes which can result in potentially oxygen-deficient atmospheres.

Airborne particulate contamination can include dusts, fibers, fumes, and mists. Dusts are generated by the mechanical breakup of solids through such processes as grinding and cutting. Air contamination can also be created during the application of particles, such as spray painting and blasting to clean parts, or through the use of powders in processes such as mixing of compounds, regrinding, etc. Airborne fibers can be generated through industrial processes such as the manufacture of fiber products (e.g., carbon and glass fibers) and through construction work and demolition. Fumes are common contaminants of welding and cutting of metals through the condensation of heated solids as they cool. Mists are common to heated processes such as plating, metals descaling, and other processes where vapors are condensed, or through the mechanical action of applicators where small suspended liquid droplets are produced. Vapors are produced when liquids (or solids with low vapor pressures) evaporate and will occur whenever there is an open reservoir of any liquid with a significant vapor pressure. There are many processes which can result in vapor production, including almost any process that uses volatile solvents, such as paint manufacture, chemical batch operations, paint stripping, degreasing, and others. Gaseous contaminants are generated when gases are used or are a product or byproduct of the industrial process or natural decomposition and not completely captured. Examples include the production of gases, use of gases such as in the semiconductor industries or as sterilants (e.g., ethylene oxide), brewing (e.g., carbon dioxide), and refining (e.g., hydrogen sulfide).

Oxygen-deficient atmospheres can be created naturally through oxygen displacement by decomposition, the depletion of oxygen through oxidation, the addition of other gases, or through other means. This situation is most common in confined spaces which do not allow normal diffusion and gas mixing, or in high-altitude situations.

C. Types of Respirators

Respirators can be classified into two general categories:

1. Air purifying
2. Air supplied

The air purifying respirators filter or otherwise "purify" the ambient air to remove contaminants; however, they do not provide breathing air. Air supplied systems actually provide breathing air from either a remote natural source or from compressed sources. A special class of respirators actually produce breathing air (oxygen) from a chemical reaction. These respirators are used in special applications such as escape, and some military applications (e.g., submarine applications).

Respiratory protection configurations range from small devices that are inserted into the mouth, to the self-contained breathing apparatus (SCBA) in which the breathing apparatus is self-contained and carried by the wearer.

The simplest form of respirators are those which are air purifying. These fall into three categories: (1) particulate respirators, (2) chemical cartridge respirators, and (3) respirators that incorporate a combination of the particulate and chemical cartridges. Particulate respirators may be powered or unpowered. Powered air purifying respirators (PAPRs) utilize a blower to pump air to the user, whereas nonpowered air purifying respirators rely on the negative and positive atmospheric pressures generated by the user's respiration to move air across the filters or chemical cartridges and exhalation valves. Respirators that use large canisters instead of the smaller cartridges are commonly called gas masks. Respirators may also be designed for disposable use or for reuse (cartridges or canisters still require changing).

Air supplied respirators consist of those that provide breathing air to a loose-fitting face piece, such as a sandblasting hood, those providing air to a tight-fitting face piece, and the SCBA. SCBAs use either compressed air sources (tanks) or are rebreathers (provide oxygen and scrub carbon dioxide) such as the astronauts or some military divers wear.

Respirator face pieces (also called masks) may partially, or fully cover the face. These are commonly referred to by the amount of the face they actually cover, ranging from one-quarter to a full face piece.

1. *Air purifying respirators*

NIOSH has just recently changed its certification process for particulate respirators based, in part, on the efficiency of the filters. This is largely a result of the enormous growth of disposable particulate respirator use, from the earlier use of quarter face piece or half face piece rubber respirators that used cartridges or filter pads (see Figure 11.3). NIOSH has estimated that employers annually purchase over 110 million disposable respirators. These new-generation disposable respirators are generally easy to wear, relatively inexpensive, and lightweight. They are typically manufactured as one piece of filter media molded to the face, although these units are available in other configurations. The certification process for particulate respirators was also changed because of a recently developed, inexpensive particulate respirator that employs nonwoven hydrophobic filter media to fit cartridge-style respirators.

Figure 11.3 Full face piece air purifying respirator w/chemical cartridges (Courtesy: *Mine Safety Appliances Company*, Pittburgh, PA).

Particulate respirators are now classified by three levels of efficiency based on testing against 0.3-micron sodium chloride or dioctylphthalate (DOP) aerosols. The new classifications have three-letter designations and three levels of filter efficiency within each letter designation. The letter designations, called a series by NIOSH, are N, R, and P. The letter designation N represents *not resistant to oil*. The letter designation R represents *resistant to oil*, while P represents *oil proof*. The efficiency designations are 95%, 99%, and 99.97% although the respirator labels will use 100 for 99.97% (e.g., filter labels for the R series would be R95; R99; R100).

N series filters are tested against sodium chloride and are effective for solid and water-based particles, but not oil-based aerosols which could degrade the filter media. The R and P series are tested against DOP and thus can be used against any particle or fiber contamination as appropriate for the level of efficiency selected (i.e., 95, 99, or 99.97). The P100 filter meets the HEPA (high efficiency particulate air) criteria and will be the only filter of those described that will have the magenta color coding used for absolute (HEPA) filters.

Table 11.3
Color Codes for Chemical Cartridges and Canisters by Contaminant

Color	Contaminant
White	Acid gases
White with a ½-inch green stripe	Hydrocyanic acid gas
White with a ½-inch yellow stripe	Chlorine gas
Black	Organic vapors
Green	Ammonia gas
Blue	Carbon monoxide gas
Yellow	Acid gases and organic vapors
Brown	Acid gases, ammonia, and organic vapors
Red	Acid gases, ammonia, carbon monoxide, and organic vapors
Magenta (purple)	Radioactive materials (except tritium and noble gases) and asbestos

Air purifying respirators that use chemical cartridges and canisters (gas masks) are available as half face piece or full face piece disposable units, reusable units, and vary in configurations from single to twin cartridges that are attached to the face piece or can be attached. Gas masks that utilize larger chemical sorbent beds are full face piece respirators with a hose attached to a large chemical canister. Gas masks are most commonly used for escape purposes. NIOSH has established a color coding requirement for chemical cartridges and canisters. This color coding is very helpful in determining that the right cartridge or canister is being used for the chemical contaminant present. Table 11.3 lists the color coding by contaminant.

It should be noted that the magenta color indicates a HEPA filter for use with highly hazardous particulates, such as asbestos, heavy metals (lead, cadmium, etc.), microbial agents. The HEPA filter *is not* a chemical sorbent.

Powered air purifying respirators work on the same principle as those that are not powered. However, because they use a blower to supply air (at least 4 cubic feet per minute) to the user through a hose to a half or full face piece, loose-fitting hood, or loose-fitting helmet, they offer higher levels of protection to the user. This is because of the positive pressure supplied, hence the lower likelihood of inward leakage of the contaminant to the breathing zone.

2. Air supplied respirators

Air supplied respirators encompass all of those that provide breathing air from either a remote source or from compressed gas. This includes respirators that have a combination of air purifying cartridges and supplied air. Air supplied respirators are required to deliver air that is free from contaminants and oil (Grade D per the requirements of the Compressed Gas Association). Air-line respirators represent the simplest of the air supplied respirators. Air-line respirators have a hose that supplies air from an air pump, compressor, or bottled (compressed air) source through a regulator to a hood or face piece on the wearer. These systems are of three types. They are either demand, pressure demand, or continuous flow. Demand air supplied systems provide air with the negative pressure developed by inhalation from the user. Pressure demand systems work in a similar fashion, except the pressure in the face piece is kept positive relative to the ambient atmosphere outside of the face piece, and is increased on inhalation. Continuous flow systems provide a continuous flow of air to the user under positive pressure of at least 4 but not exceeding 15 cubic feet per minute. All of these systems may include an air purifying cartridge for use for escape should the supply of air falter. For those situations where the effectiveness of the air purifying respirator element could be exceeded, or where there could be an oxygen deficiency, a supplemental air bottle worn by the user must be provided for escape purposes. OSHA and NIOSH have a number of requirements for these systems, including limits to pressures, hose lengths, air quality, and type of compressor that may be used, as well as other requirements. The OSHA regulations on respiratory protection should be consulted for these details.

The self-contained breathing apparatus (SCBA) with tight-fitting full face piece represents the highest level of protection for the wearer, but at a cost (refer to Figure 11.4). These units are expensive and require the air source to be worn by the user, adding complexity, weight, and limited time of use based on the extent of the air supply. These are of two types: (1) open circuit or (2) closed circuit. Open-circuit devices exhaust the exhaled air to the atmosphere, while the closed-circuit units "scrub" exhaled carbon dioxide and add oxygen to replace oxygen that has been used. The open-circuit units usually consist of a tank of compressed air (Grade D or better) of 30 to 120 minutes use duration that is fed through a pressure regulator to the respirator face piece. Closed-circuit devices have smaller containers of highly compressed or liquid oxygen with a scrubbing system (usually a chemical such as sodium hydroxide), allowing up to 4 hours of use. These units may be used in atmospheres that are immediately dangerous to life and health (IDLH). Some of the open-circuit units also permit the use of air-line connections to increase the useable time and allow for escape with the SCBA.

Figure 11.4 Self-contained breathing apparatus, *SCBA* (Courtesy: *Mine Safety Appliances Company*, Pittburgh, PA).

D. Selection and Fit-Testing of Respirators

Respirators should be selected based upon the hazard assessment discussed previously. Once the conditions of exposure are known, the required protection factor needed can be determined. Protection factors have been determined experimentally by measuring the ambient concentration of contaminant outside the respirator and the concentration inside the respirator (Refer to Figure 11.5 for electronic fit-test instrument to determine protection factors). Dividing the outside concentration by the inside concentration yields the protection factor as:

$$\textbf{PF = outside concentration/inside concentration} \qquad \textbf{(11.1)}$$

Minimum expected PFs have been assigned for each class of respirator even though higher values may be found with quantitative fit-testing. These are experimental values that will vary based on the actual fit of the respirator to the user and the efficiency of the respirator actually selected. Nevertheless, it is good practice and a requirement that only the minimum values be used when determining the correct respirator to assign.

Sample Calculation: If the air concentration of a dust were 20 mg/m³ and a half mask particulate respirator with a PF of 10 were used, what would the theoretical exposure to the worker be?

PF   = Outside concentration/inside concentration
10   = (20 mg/m³)/(inside concentration)
Inside concentration = 2 mg/m³

Using the data from the previous sample calculation, could a half mask particulate respirator be used for protection against a dust with a permissible exposure limit of 1 mg/m³ if the air concentration were 20 mg/m³?

PF   = Outside concentration/inside concentration
10   = (20 mg/m³)/(inside concentration)
Inside concentration = 2 mg/m³

Figure 11.5 Electronic fit-test instrument to determine protection factors (Courtesy: *TSI Incorporated*, St. Paul, MN).

Since the inside concentration, or exposure, exceeds the contaminant's PEL, a full face piece respirator would need to be used, which has a PF of 100, hence the worker would have a theoretical exposure level of 0.2 mg/m³ which is below the PEL.

Table 11.4 shows typical protection factors for common classes of respirators. The protection factors are used to calculate the maximum expected exposure to the worker.

For air purifying respirators, there are also other limitations on their selection. Since the chemical cartridges have a finite capacity to absorb or otherwise remove the contaminant and breakthrough is possible, an odor threshold or other warning property, such as irritation, must exist that is below the exposure limit. For some substances, such as the isocyanates, this eliminates their use since there is no odor threshold at the PEL. Even when there are adequate odor warning properties, a schedule of changing the chemical cartridges or canisters should be established based on the capacity of the units being used. This can be calculated using the capacity information provided by the manufacturer and the expected airborne concentration. It should also be noted that most chemical cartridges contain either activated charcoal or other sorbents which will passively absorb the contaminant if exposed to the ambient air. Once the seal to the cartridge or canister is removed, the life of the cartridge or canister could be reduced. Therefore, they should remain sealed until ready for use. Particulate filters usually do not have this limitation since the dust loading will create additional breathing resistance, signaling the user to the need to change the respirator and generally increase the efficiency of the filter. However, the increased breathing resistance can result in a greater likelihood of leakage around the face piece. A schedule to change the filters or replace the respirator (disposable respirators) should also be established.

Table 11.4
Protection Factors for Common Respirator Classes

Respirator Type	Protection Factor
Air purifying half face piece respirators	10
Powered air purifying half face piece respirators	50
Air supplied half face piece respirators	50
Air supplied half face piece respirators	50
Full face piece air purifying respirators	100
Full face piece powered air purifying respirators (tight-fitting face piece or hood)	1000
Air supplied full face piece, respirators positive pressure	1000
Self-contained breathing apparatus, full face piece-pressure demand/rebreathers	10000

Fit-testing of respirators is required by the OSHA regulations for tight-fitting respirators and is a good practice. This is because of the differences in the geometry and sizes of faces among users, fit patterns, and respirator sizes among manufacturers. Needless to say, anything that would prevent the sealing of the respirator to the face would prevent a proper fit. This would include the use of normal eyeglasses for full face piece respirators and beards or other facial hair at the sealing areas. There are three types of fit-testing. These are the negative and positive pressure fit-test, qualitative and quantitative fit-testing. The negative and positive pressure test should routinely be used by the wearer before use of the respirator each day and after each change of cartridges or other change to the respirator. The negative and positive pressure fit-test is only applicable to those respirators that have inhalation and exhalation valves (not the disposable particulate respirators). It consists of covering the filter cartridge or canister opening with the hands and inhaling in a normal fashion. If the face piece does not leak, this should result in a negative pressure, causing the face piece to conform to the face. If there is leakage, a new face piece or a different respirator should be tried. The positive pressure phase of the test is done by covering the exhalation valve (may require removal of protective covers) and exhaling normally. The respirator should move from the face evenly since one-way valves in the cartridge ports prevent exhaling through the cartridges. These tests can also be performed on full and half face piece air supplied respirators by simply covering the hose connection from the face piece to the air supply. This is a cursory user test that should be performed each time the respirator is worn.

Qualitative fit-testing involves the use of a stimulant to actually test the fit of the respirator under use conditions. There are three commonly used stimulants. These are irritant cold smoke, isoamyl acetate (banana oil), and saccharin mist. Irritant smoke can be obtained from ventilation smoke tubes (caution should be exercised if an acid mist is used), banana oil can be obtained from chemical supply sources, and saccharin mist generators are available through respirator manufacturers and safety supply sources. In this test procedure, the subject is first tested for sensitivity to the stimulant. If sensitivity is demonstrated, the test is performed. After the respirator has been properly fitted with the applicable cartridge (particulate for smoke and saccharin, and organic vapor for banana oil) and the negative and positive pressure tests performed, the subject is placed in an environment of the stimulant with the respirator on (usually using a hood or similar device to hold the stimulant). The subject then is asked to perform a number of exercises to simulate a work environment. These usually include the following:

1. Normal breathing
2. Breathing deeply
3. Breathing normally while turning the head side to side and up and down
4. Running in place for a few moments .
5. Reading or reciting a passage such as the OSHA *Rainbow Passage*
6. Breathing normally

Once the test is completed, the respirator user is usually asked to remove the respirator to determine if he/she is still sensitive to the stimulant. At any point in the test procedure during which the user detects the stimulant, the test should be stopped and another respirator (different size or manufacturer) tried.

Quantitative fit-testing is the most accurate measure of the level of protection provided the user since it actually provides a protection factor by measuring the ambient and inside-the-respirator concentration of contaminant or a stimulant. It can be performed using several different techniques. These include the use of sodium chloride, DOP (considered potentially carcinogenic and no longer commonly used), corn oil (replaced DOP), ambient aerosols, and negative pressure measuring devices. These tests require the use of relatively expensive equipment but follow similar protocols as those described in the qualitative fit-test. From these tests, the worker is usually given a card identifying the specific respirator used, protection factor found, and period for which the fit-test is valid (usually 1 year). Typically, protection factors exceeding those established by NIOSH, OSHA, and ANSI will be achieved. Nevertheless, OSHA will only allow protection factors that have been developed from laboratory tests, as noted in Table 11.4.

E. Respiratory Protection Programs

ANSI also has a standard for respiratory protection programs (American National Standard for Respiratory Protection, Z88.2). This voluntary standard is considered a more comprehensive approach to respirator selection and use to include the design of a respiratory protection program. Both the OSHA standard and the ANSI recommendations should be consulted when developing a respiratory protection program.

The OSHA requirements for a respiratory protection program were highlighted in the introduction to this section. A more detailed description for each element of the OSHA requirements follows.

Establishment of standard operating procedures (SOPs) for respirator selection and use. OSHA requires these procedures be written. Procedures can be written to describe the entire program. However, a better approach would be to develop a document that details the overall program to include an SOP for each job task that requires use of a respirator. The SOP for each job or task requiring a respirator might include the following information:

- A description of the operation or task requiring use of a respirator
- Nature and type of hazards requiring respirator use (including expected air concentrations)
- Warning properties or signs and symptoms should the respirator fail
- Types of respirators that should be used
- Maintenance of the respirators, including cleaning and storage
- Emergency procedures

A written selection logic or procedure should be established that bases respirator selection on the hazards to which the worker is exposed. This would include a hazard assessment (exposure assessment), as discussed earlier in the chapter. For respiratory hazards, this would require a quantitative evaluation of the level of airborne hazards. The level must be determined to permit selection of a respirator with the required protection factor. This usually requires air sampling for time-weighted average concentrations, short-term peak exposures, and a determination of oxygen content if there is a potential for oxygen deficiency.

Instruction and training for employees using respirators must be accomplished. The training of respirator users should include:

- An explanation of the nature and extent of the hazard requiring the use of the respirators. This should include the information required under the Hazard Communication Standard (e.g., acute and chronic effects of exposure, warning signs, emergency procedures, etc.). An explanation of other control methods being considered or other efforts to reduce employee exposures should also be presented at this time.
- An explanation of the respirator selection process and OSHA requirements.
- Demonstrations on the proper fitting, donning, wearing, and removing of the respirator, including an opportunity for hands-on experience with the negative pressure test and other fit-testing procedures.
- An explanation of the limitations, capabilities, and operation of the respirator.
- An explanation and demonstration of the proper maintenance and storage of the respirators.
- An explanation and demonstration on how to inspect the respirator for defects.
- Information on how to recognize emergencies and what to do in these situations.

This information and training should emphasize practical hands-on experience for the respirator user. It is also advisable to conduct a written and/or performance evaluation of the worker's knowledge. These records should be maintained with the employee's training records.

OSHA requires that the employer establish a program to clean and disinfect respirators regularly, and clean those used by more than one worker after each use. Disposable respirators will not require this element, provided they are not worn more than once. Improper cleaning of respirators is a common deficiency in many respirator programs. It is common for employees who have assigned respirators to keep them in their work lockers or tool boxes. This can easily lead to contamination on the inside of the face piece. Respirator programs that include the collection, cleaning, and disinfecting of the respirators after each use resolve this concern. Manufacturers of respirators provide instructions for cleaning, disinfecting, and storage, which should be followed.

Store respirators in a convenient, clean, sanitary place. Some rubber face pieces and straps are susceptible to excessive heat and cold, ozone, ultraviolet radiation, and other factors which might degrade the respirator. The manufacturer's instructions should be followed for storage. Some organizations utilize sealed plastic bags for storage after cleaning. This maintains the sanitary conditions of the respirator and gives the users a level of confidence that the respirator has been properly cleaned and maintained.

Inspect and repair, as necessary, routinely used respirators during cleaning. Inspections of the face piece, straps, lens, and other parts of the respirator should be performed by the user before and after use. Inspection of the face piece should include the valves (one-way inhalation and exhalation valves) and other features (e.g., hoses for gas masks and supplied air respirators, etc.), plus supporting equipment (e.g., SCBA tanks and regulators), and should be performed by trained personnel on a regular basis.

OSHA requires the inspection of emergency respirators at least once per month, and after each use. This would include devices that are used for escape and those used to respond to emergencies. Generally, these respirators are maintained in marked cabinets mounted on the wall or are housed in a similar fashion.

Maintain surveillance of work area conditions and evaluate the degree of employee exposure or stress. The effectiveness of protection provided by respirators is dependent on the type and level of contaminant and the work environment that the respirator is designed to control. If the contaminant were to change or the ambient concentration were to increase, the respirator might not be effective. Hence, a program of routine evaluation of the work area is required. Typically, this means a quarterly air monitoring program to assure that the respirator is still effective. Medical surveillance can also be used when there is a biomarker or means of determining the dose the worker is actually receiving (e.g., blood lead levels). However, most contaminants do not have reliable biological indicators. Secondly, finding overexposures during medical surveillance means that the employee has already suffered the effects of overexposure. Therefore, medical surveillance should not be relied upon as the sole means of assuring that respiratory protection is effective.

Regularly inspect and evaluate the continuing effectiveness of the program. It is good practice to routinely determine the effectiveness of all safety and health programs. This is accomplished in most major companies through both internal and external auditing of programs similar to the long-held practice of financial auditing. This is essentially an organized means to determine if OSHA requirements, SOPs, internal company standards, and other written policies are being followed. The program administrator for the respiratory protection program would normally provide routine surveillance of the program while someone outside of the assigned responsibility for the program would perform an annual audit. Typically, the annual audit might be performed by someone from a corporate office or by an outside consultant. This provides a level of confidence to all concerned that the program is effective.

Have a physician determine the physical ability of the user to wear respiratory protection, with periodic medical review. While this requirement has been interpreted in different ways relative to the type of medical examination and the medical staff performing it, a good practice is to have a Board Certified Occupational Health Physician examine the proposed respirator user for his/her ability to use the respirator. All respirators place an additional level of stress on the worker. Workers with respiratory difficulties (e.g., asthma, lung disease or dysfunction, etc.) or other physical limitations (e.g., heart disease, etc.) are at risk. Obviously, this medical evaluation requires the industrial hygienist to provide information to the physician to include the type of respirator to be assigned, the work conditions and level of contaminants, and other relevant information prior to the examination. It is also recommended that the medical evaluation be written to include any limitations for respirator use and these records be maintained with the worker's other medical records. Periodic reviews of the ability of the worker to wear the respirator assigned should also be conducted. The frequency and type of re-examination should be determined by the physician and is normally related to physical condition and age. Many companies perform these examinations and fit-testing annually.

Use only NIOSH approved respirators (Bureau of Mines approvals have been combined with NIOSH approvals, except for mine rescue). It is a requirement that only NIOSH approved respirators be used. Nevertheless, it is a common practice to provide nuisance dust respirators to workers doing dusty jobs that do not exceed recommended or required exposure limits (e.g., sweeping and cleaning jobs). These disposable respirators are less expensive and generally more comfortable than the NIOSH approved respirators. However, this is not considered a good practice. NIOSH approval of respirators may be determined from the required label on the respirator or the NIOSH published listing of approved respirators and systems.

VII. OTHER PERSONAL PROTECTIVE EQUIPMENT

Common personal protective equipment other than protective clothing and respirators includes head protection, eye and face protection, and hearing and foot protection. Hearing protection is discussed separately in the chapter on noise.

A. Head Protection

Head protection may be needed to protect workers from falling or flying objects, bumping into objects, electrical shock, and getting long hair caught in moving machinery. Head protection is required in certain occupations such as construction work, shipbuilding, mining, electric power line maintenance, food service, and others. The type of head protection necessary will depend on the nature of the inherent hazards and the tasks to be accomplished. Common forms of head protection include helmets, bump caps, hair nets, and caps. OSHA regulates head protection under their General Industry Standards (1910.135), Construction Industry Standards (1926.100), and Maritime Standards (refer to Figure 11.6 for examples of head protection).

1. Helmets

Helmets, or as they are commonly called *hard hats* or *safety hats*, are required when there is a risk of injury from falling or moving objects, or overhead hazards that may come in contact with the workers head. A special class of head protection is that used to insulate the head from electric shock. Hard hats can also serve to prevent injury from moving into low-hanging objects (same function as the bump cap). Hard hats will not prevent injuries to the neck or lower body parts.

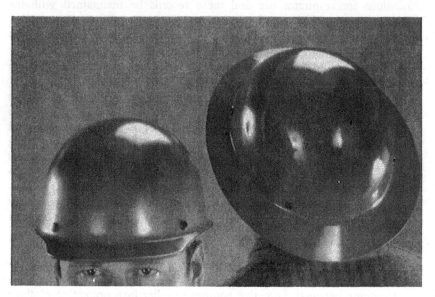

Figure 11.6 Examples of head protection styles (Courtesy: *Mine Safety Appliances Company*, Pittsburgh, PA).

Hard hats are available in a variety of shapes, sizes, and colors. Differences in color are commonly used for designating the role of the worker (e.g., white hard hats for supervisors). Hard hats can be constructed of molded thermoplastic, fiberglass, or metal. Most hard hats today are made of impact-resistant plastic. OSHA regulates the general characteristics and performance of hard hats by requiring that they meet the specifications of the American National Standards Institute (ANSI) standard Z89.1. This ANSI standard specifies three classes of hard hats. These are:

- Class A: Helmets intended to protect the head from the force of impact of falling objects and from electric shock during contact with exposed low-voltage conductors.
- Class B: Helmets intended to protect the head from the force of impact of falling objects and from shock during contact with exposed high-voltage conductors.
- Class C: Helmets intended to protect the head from the force of impact of falling objects.

Helmets come with a multiple-point suspension system to keep the helmet from direct contact with the head. Since the purpose of the suspension within the helmet is to distribute the force of any blow to the head, the headbands and suspension webs inside the helmet should have at least 1 inch of clearance to the helmet shell. Configurations that are available vary but can include those with face shields and hearing protection integrated into the helmet. Additionally, ANSI and OSHA require the helmets to be labeled as to the helmet class and manufacturer name.

All helmets should be routinely inspected for cracks, excessive wear, broken suspensions, chemical damage, and other signs of potential defects. Damaged helmets should be replaced.

2. Bump caps

Bump caps are not helmets since they typically fail to meet the ANSI standards for protection. They are principally used to prevent injury from bumping into stationary objects rather than falling objects, or are used for promotional purposes to identify an affiliation such as the employer or a team. They are usually lighter, may not have suspension systems, and may have brims.

3. Hair nets and caps

Hair nets and caps are used to prevent long hair from getting caught in moving machinery or to prevent contamination of the product (food or other products, such as electronics) from human hair. Most food processing operations require the use of hair nets or caps for sanitation. Caps and other similar devices do not provide protection from falling objects or chemicals. In situations where the hazard includes hot sparks, open flames, or other fire hazards, the protective device should be fire or flame resistant.

There have been a number of tragic incidents where workers have become entrapped or lost their scalp to moving machinery due to their hair being caught. While hair nets and caps are a simple protective measure, they can prevent serious injury. As an example of the need for this simple protective measure, a maintenance worker at a large metropolitan hospital with long hair had his head close to a moving shaft to listen to the gears and was killed when his long hair became entangled. In this case, a simple hat could have saved the worker's life. Obviously, the hair must be completely covered by the protective apparel.

VIII. EYE AND FACE PROTECTION

OSHA regulates eye and face protection under the General Industry Standards (1910.133), Construction Industry Standards (1926.102), or Maritime Standards. Eye and face protection is required if there is a risk of injury to the worker by flying objects or particles, chemical splashes, chemical vapors, ultraviolet radiation, or other hazards to the eye. The nature of the hazard will determine the type of protection required; however, the general classes of eye protection include:

- Eyeglasses
- Shaded eyeglasses
- Goggles
- Face shields
- Welding helmets

As noted earlier, full face piece respirators can also provide eye protection but are not included in this section.

A. Eyeglasses

Eyeglasses or spectacles (commonly called "safety glasses") are used to protect the wearer against the frontal impact of airborne objects that can be generated during operations that include machining, grinding, hammering, sawing, and other similar activities that might result in flying particles. The lenses may be constructed of either tempered glass or shatter-resistant plastic. Again, OSHA uses an ANSI standard to establish the specifications and performance requirements of this protective device. ANSI Standard Z87.1 establishes the impact resistance requirements for the lenses and the frames. Safety glasses should be used with side shields to prevent entry of flying objects from the side of the eyeglass frame. Side shields may be made of plastic, wire mesh, or other materials. They should completely cover the eye, from approximately the temple of the head to the eyeglass lenses. For employees that require prescription lenses, the glasses must incorporate that prescription unless they can be worn over the regular prescription lenses.

B. Shaded Eyeglasses

Shaded lenses are used in special applications, such as protection from lasers and other forms of intense radiation. Shaded lenses can also be used to protect against glare from the sun and snow for outdoor jobs. To be effective, the configuration of the glasses and attachments should protect the eye from stray radiation at all angles. The selection of shading and color of the lenses is determined by the spectrum of the radiation which presents the hazard.

C. Goggles

Goggles provide all of the protection of eyeglasses plus protection against dusty environments and liquid splashes. Goggles should be used for work environments where liquid splashes of materials that are hazardous to the eyes are a possibility since glasses could allow entry. Goggles come in a variety of styles and configurations. These range from tight-fitting eyecups to chemical goggles. Ventilation for the goggles is usually provided to prevent fogging. However, goggles used for protection against chemical splash should have vents that do not allow the penetration of liquids.

D. Face Shields

Neither glasses nor goggles provide protection for the face. When the entire face must be protected, a face shield is the appropriate choice. For example, face shields should be used for protection against potential chemical splashes of corrosive materials and other hazards that could adversely affect the skin, as well as the eyes. Face shields usually do not meet the ANSI Z87 requirements for impact resistance; consequently, safety glasses may also need to be worn under the face shield (as detailed in Figure 11.7). Face shields may be constructed of clear thermoplastic or wire mesh that is curved to protect the face from approximately the temples forward. Wire mesh face shields are commonly used in the molten metals industries.

E. Welding Helmets

Welding requires the protection of the entire face from both hot particles as well as radiation hazards (see Figure 11.8). Welding torches and arc welders can emit both heat (infrared) and strong ultraviolet radiation (especially electric arc welding). The degree of protection needed from the shaded glass lens in the welding helmet is determined by the type of welding being performed. It should also be noted that welding may also require respiratory protection and hearing protection, depending on the nature of the work being performed.

Figure 11.7 Face shield covering worker's face protected with safety glasses (Courtesy: *Mine Safety Appliances Company*, Pittsburgh, PA).

Figure 11.8 Welding helmets used for welding operations (Courtesy: *Mine Safety Appliances Company*, Pittsburgh, PA).

Figure 11.9 Metatarsal guards used to protect the top of the foot operations (Courtesy: *Mine Safety Appliances Company*, Pittsburgh, PA).

IX. FOOT PROTECTION

Foot protection is necessary for those tasks or jobs where there is a potential for injury to the feet from falling or rolling objects, objects piercing the sole of the foot, electrical hazards, and other hazards to the feet. OSHA regulates foot protection under their General Industry Standards (1910.136), Construction Industry Standards (1926.96), and Maritime Standards. The OSHA regulations incorporate the requirements of the ANSI Z41.1 standard for protective footwear. Protective footwear includes:

- Shoes
- Boots (of varying heights)
- Chemical or waterproof boots

Shoes with protected toe areas (usually by including a metal cover over the toe area inside the shoe) are commonly called *safety shoes*. Shoes can also include a steel or metal plate incorporated in the sole to prevent penetration injuries, as well as slip-resistant and chemically resistant soles. They may also be obtained with nonconductive soles for certain industries where static charges are a problem.

In general, boots may be obtained with all the features mentioned above. In some industries, such as the steel industry, metatarsal guards may also be required (refer to Figure 11.9). Metatarsal guards are worn over the boots to protect the upper part of the foot from crushing injuries. They are somewhat similar to the guards worn over skates by hockey players. Water-resistant or "water-proof" boots may be necessary for wet work, but chemically resistant boots should be obtained based on specific resistance to the chemicals being used (as discussed in the section on chemical protective clothing).

X. CASE STUDIES

The following case studies illustrate problems that are common to the use of personal protective equipment in a work environment.

A. Case Study One: Proper Use of PPE for Hydrofluoric Acid Work

An Eastern refinery used anhydrous hydrofluoric acid in a process for treating the product stream for gasoline. Anhydrous hydrofluoric (95% concentration) is an extremely reactive and corrosive acid. Anhydrous hydrofluoric acid is extremely dangerous to humans for two reasons: (1) its corrosive nature; (2) the fluoride ion dissociates after contact with human skin and can be fatal due to its effects on the Krebs cycle. A 5% body burn has been shown to be fatal. Because of this hazard to workers, the refinery had extensive safety checklists, and the requirement for use of a suit that had demonstrated resistance to the effects of the acid. During a fairly routine maintenance operation, two workers were assigned to block and bleed a pump line that fed hydrofluoric acid. The task involved placing blinds (solid metal plates) on both sides of the pump to allow the pump to be removed for maintenance. The process lines had been checked for pressure and all safety checklists were completed. The workers were outfitted with full face piece air-line respirators, a one-piece suit consisting of a hydrofluoric acid-resistant polymer with a hood per standard company procedures. The workers approached the lines which were approximately four feet in elevation, with a diameter of approximately 24 inches. A standby worker was assigned to the team at about 25 feet away with a fire hose, should there be an unexpected release (to knock down any vapor cloud formed from the hydrofluoric acid reacting with water vapor in the air). The workers started the unbolting of the flanges to install the blinds. After removal of the fifth bolt on one flange, a pressurized spray of hydrofluoric acid struck the workers. Immediately, the spray formed a vapor cloud which enveloped the workers. Because the air-line hoses to their respirators were fed under mid-chest level hoods, channels formed from the path of the hoses to the inside of the hoods. This allowed hydrofluoric acid vapors to move inside the hoods where the respirators were worn. Because the respirator face piece covers the face but not the neck, back of the head, nor ears, the workers were being burned and could not see because of the vapor cloud formed. They immediately ran, amid shouts from their standby worker and other co-workers. The hose lengths did not allow for much travel away from the site of the incident. One worker ran in the direction of help; however, the other did not. Both eventually ran out of hose and tore off the respirators. Both workers survived but both suffered extensive neck, head, and ear damage.

In this actual incident, what was done right and what was done improperly? What would have prevented the serious injuries to the workers?

B. Case Study Two: Fiberglass Boat Manufacture: Acetone Usage

A fiberglass boat-building company used acetone to clean the fiberglass resin from tools and other contaminated parts. The process involved spraying fiberglass into a mold and then covering it with the sticky resin. Layers were built up using this process until the boat was essentially formed. Because the resin was sticky, workers were given spun-bonded polyethylene suits to keep their clothes clean. The lightweight suits consisted of pants with an elastic band and a zippered jacket. The workers also wore gloves when applying the resin. The spray guns would need to be cleaned in acetone after use. Acetone applied to rags was also used to clean up any excess resin overspray. The workers kept the acetone in "safety cans" (self closing metals cans) because of the fire hazard, as acetone is extremely flammable. Workers also would use acetone to clean their work clothes worn under the suits should the resin penetrate the suits. Acetone cleaning of worker clothes under the suits was not a company-sponsored practice, but was commonly done.

A young worker needed to fill his acetone can. He went outside to the filling area, which was very similar to a gasoline pump commonly used to put gas in a car. He filled his can and also put some acetone on a rag to clean a spot of resin that had gotten on his blue jeans under the suit near his knee. He pulled the suit pants down, rubbed the spot clean, and pulled the polyethylene suit pants back up. While walking back to his work station, he snapped the elastic band against his belt buckle (a large Western-style belt buckle). This caused a spark from static charge in the suit and he ignited. The worker was engulfed in flames from the explosive atmosphere created by the evaporating acetone off of his blue jeans under the suit. The suit burned and the melted plastic stuck to exposed skin. He suffered third degree burns but survived.

From this actual incident, what practices could have prevented this incident? Was the protective clothing assigned appropriate for the work?

C. Case Study Three: Plastics Manufacture

A small plastics regrinding and manufacturing company employed an itinerant worker who spoke little English, nor was capable of understanding most written instructions. He worked the evening shift and was responsible for the operation of a banbury. A banbury is a device that is used for mixing (usually rubber). It typically has a small input door that is self-closing where the feedstock is fed to a rotating shaft with mixing blades. The completed mix comes out of a door on the bottom of the device. Banburys can range from a few feet in length to over 30 feet in length. This one was about 10 feet in length (vertical orientation). The company would receive spent plastic, regrind it, and add various pigments based on the color needed for the new product. The pigments were all lead chromate based. His job was to put in the plastic mix, then add bagged pigment by slitting and then dumping the 50-pound bags into the banbury door. The worker was assigned a full face piece air purifying cartridge respirator. The cartridges were HEPA filters (99.97% effective against a 0.3-micron aerosol). He did not receive training other than being shown how to wear it. During a routine examination, he was found to have lead poisoning and was chelated. He told his foreman that he routinely wore the respirator. He was off work for a short time and a followup examination 4 months later showed blood lead levels well above the OSHA standards (almost twice). It was found that he would clean the banbury after his shift using compressed air. He would do the cleaning by placing a high-pressure air hose inside the banbury door to remove all unused pigment. He would do this while visually inspecting for pigment on the walls of the banbury. Since he could not see well, nor fit his head into the banbury with the respirator on, he took it off to do the compressed air cleaning.

What went wrong in this actual situation (why was he overexposed to lead)? What should have been done?

REVIEW QUESTIONS

1. Before selection of personal protective equipment, OSHA requires:

   a. A hazard assessment
   b. Analysis of data
   c. Cost evaluations
   d. Proper fitting of the device

2. Bogger Bank Itch is an example of:

   a. A skin disorder transmitted from ticks
   b. A skin order transmitted from fish
   c. A skin disorder transmitted from cows
   d. A disease of the lungs

3. The protection afforded by clothing for thermal hazards is related to:

   a. Heat reflectiveness
   b. Resistance to scorch
   c. Thickness
   d. Shielding

4. Protective clothing for IR (infrared) would normally be:

   a. Black in color
   b. Aluminized
   c. Thick
   d. Yellow in color

5. Raynaud's phenomenon is caused by:

   a. High noise levels
   b. Radiation
   c. Exposure to UV
   d. Excessive vibration

6. Parathion is an example of a chemical which presents a significant:

   a. Dermal hazard
   b. Respiratory hazard
   c. Physical hazard
   d. Fire hazard

7. OSHA and ACGIH identify chemicals with a potential for significant exposure by the dermal route with the notation:

   a. Caution
   b. PEL
   c. TLV
   d. Skin

8. Acetone has been shown to permeate neoprene rubber of normal glove thickness in approximately:

    a. 8 hours
    b. 4 hours
    c. 30 minutes
    d. 3 minutes

9. Permeation is the term used to describe the process of a chemical crossing a barrier by:

    a. Osmosis
    b. Movement on a molecular basis
    c. Reverse osmosis
    d. Conduction

10. Changes in the physical appearance of a protective barrier are commonly an indication of:

    a. Permeation
    b. Penetration
    c. Degradation
    d. Diffusion

11. Which is best, in terms of chemical resistance?

    a. A long break-through time
    b. A short break-through time
    c. A short break-through time and high permeation rate
    d. A short break-through time and low permeation rate

12. Natural latex rubber gloves are effective against:

    a. Organic solvents
    b. Organic acids
    c. Biological hazards
    d. Physical hazards

13. Breathable protective garments are those that allow:

    a. Air exchanges
    b. Water vapor transmission
    c. Enough room for the user to breathe
    d. For using respiratory protection

14. Use of polyethylene garments in open flame environments:

    a. Could increase the danger of burns to the worker
    b. Could decrease the danger of burns to the worker
    c. Will have no effect
    d. None of the above

15. The minimum frequency for inspection of emergency respirators is:

    a. Daily
    b. Weekly
    c. Monthly
    d. Annually

16. Which particulate filter should not be used with oil mists?

    a.   N99
    b.   R99
    c.   P99
    d.   P95

17. The black color code on a respiratory cartridge means the respirator is for protection against:

    a.   Ammonia
    b.   Hydrogen cyanide
    c.   Radioactive materials
    d.   Organic vapors

18. What respiratory protection system may be used in an IDLH atmosphere?

    a.   Air purifying
    b.   Powered air purifying
    c.   SCBA
    d.   None of the above

19. The protection factor is:

    a.   Calculated by dividing the outside concentration by that within the respirator
    b.   Calculated by dividing the concentration inside the respirator by that outside
    c.   Calculated by estimating the efficiency of the filters used
    d.   Calculated by dividing the size of the cartridge by the area of the face piece

20. If a respirator has a PF of 10 and the ambient air concentration is 100 ppm, the workers theoretical exposure would be:

    a.   1000 ppm
    b.   100 ppm
    c.   10 ppm
    d.   1 ppm

21. Helmets meeting ANSI Z89.1 requirements are categorized into three classes:

    a.   A, B, C
    b.   B1, B2, B3
    c.   A1, A2, A3
    d.   None of the above

22. From the list in Question 21, which class of helmet does not provide protection from electrical shock?

    a.   C
    b.   B1
    c.   A3
    d.   B

23. To protect against a potential splash of 30% sulfuric acid onto a worker:

a. Safety glasses would be required
b. Safety glasses with side shields would be required
c. Chemical goggles would be required
d. A face shield would be required

ANSWERS

1.	a.	**13.**	b.
2.	b.	**14.**	a.
3.	c.	**15.**	c.
4.	b.	**16.**	a.
5.	d.	**17.**	d.
6.	a.	**18.**	c.
7.	d.	**19.**	a.
8.	c.	**20.**	c.
9.	b.	**21.**	a.
10.	c.	**22.**	a.
11.	a.	**23.**	d.
12.	c.		

References

ASTM, Test Method for Resistance of Protective Clothing Materials to Permeation by Liquids or Gases Under Conditions of Intermittent Contact (Method F1383), American Society for Testing and Materials, West Conshohocken, 1992.

ACGIH, Dermal absorption, in *Documentation of Threshold Limit Values and Biological Exposure Indices*, American Conference of Governmental Industrial Hygienists, Cincinnati, 1992.

ASTM, Test Method for Resistance of Protective Clothing Materials to Permeation by Liquids or Gases Under Conditions of Continuous Contact (Method F739), American Society for Testing and Materials, West Conshohocken, 1991.

ASTM, Test Method for Resistance of Protective Clothing Materials to Liquid Permeation—Permeation Cup Method (Method F1407), American Society for Testing and Materials, West Conshohocken, 1995.

Brown, P. L, Protective clothing for health care workers: liquidproofness versus microbiological resistance, in *Performance of Protective Clothing*. STP 1133, American Society for Testing and Materials, West Conshohocken, 1992, 65-82.

Colton, C., Birkner, L., and Brosseau, L., Eds., *Respiratory Protection, A Manual and Guideline*, 2nd ed., American Industrial Hygiene Association, Fairfax, VA, 1991.

Coletta, G. C., Mansdorf, S. Z., and Berardinelli, S. P., Chemical protective clothing test method development. Part II. Degradation test method, *Am. Ind. Hyg. Assoc. J.*, 41, 1980, 26-33.

Davies, J., Conductive clothing and materials, in *Performance of Protective Clothing*, ASTM STP 989, American Society for Testing and Materials, West Conshohocken, 1988, 813-831.

Day, M. A., Comparative evaluation of test methods and materials for thermal protective performance, in *Performance of Protective Clothing*, ASTM STP 989, American Society for Testing and Materials, West Conshohocken, 1988, 108-120.

Forsberg, K. and Keith, L., *Chemical Protective Clothing Performance Index Book*, 2nd ed., John Wiley & Sons, New York, 1997.

Forsberg, K. and Mansdorf, S. Z., *Quick Selection Guide to Chemical Protective Clothing*, 3rd ed., Van Nostrand Reinhold, New York, 1997.

Grandjean, P., *Skin Penetration: Hazardous Chemicals at Work*, Taylor and Francis, New York, 1990.

Henry, N. and Schlatter, N., Development of a standard method for evaluating chemical protective clothing to permeation of hazardous liquids, *Am. Ind. Hyg. Assoc. J.* 42, 1981, 202-207.

Johnson, J. and Anderson, K., Eds., *Chemical Protective Clothing*, Vol. II, American Industrial Hygiene Association, Fairfax, VA, 1990.

Johnson, J., Schwope, A., Goydan, R., and Herman, D., *Guidelines for the Selection of Chemical Protective Clothing*, 1991 Update, (U.S. Dept. of Energy, Office of Environment, Safety and Health), National Technical Information Service, Springfield, VA, 1992.

Johnson, J. S., and Stull, J., Measuring the integrity of totally encapsulating chemical protective suits, in *Performance of Protective Clothing*, ASTM STP 989, American Society for Testing and Materials, West Conshohocken, 1988, 525-534.

Linch, L. L., Protective clothing, in *CRC Handbook of Laboratory Safety*, CRC Press, Boca Raton, FL, 1971, 124-137.

Mansdorf, S. Z., Anhydrous hydrofluoric acid, *Am. Ind. Hyg. Assoc. J.* 48, 7, 1986.

Mansdorf, S. Z., Risk assessment of chemical exposure hazards in the use of protective clothing–an overview, in *Performance of Protective Clothing*, ASTM STP 900, American Society for Testing and Materials, Philadelphia, 1986, 207-213.

Mansdorf, S. Z., Chap. 12, Personal protective equipment, in *Complete Manual of Industrial Safety*, Prentice Hall, Englewood Cliffs, NJ, 1993,

Mansdorf, S. Z., Industrial hygiene assessment for the use of protective gloves, in *Protective Gloves for Occupational Use*, CRC Press, Boca Raton, FL, 1994.

Mickelsen, R. L. and Hall, R. A breakthrough time comparison of nitrile and neoprene glove materials produced by different manufactures, *Am. Ind. Hyg. Assoc. J.* 48, 941-947, 1985.

Mickelsen, R. L., Roder, M., and Berardinelli, S. P., Permeation of chemical protective clothing by three binary solvent mixtures, *Am. Ind. Hyg. Assoc. J.* 47, 189-194, 1986.

NIOSH, *Preventing Allergic Reactions to Natural Rubber Latex in the Workplace*, Publication 97-135, U.S. Department of Health and Human Services, Public Health Service, Centers for Disease Control, National Institute of Occupational Safety and Health, Atlanta, 1997.

NIOSH, *Proposed National Strategies for the Prevention of Leading Work-Related Diseases and Injuries*, U.S. Department of Health and Human Services, Public Health Service, Centers for Disease Control, National Institute of Occupational Safety and Health, Atlanta, 1988.

NIOSH: Report to Congress on Workers' Home Contamination Conducted Under the Workers Family Protection Act, DHHS, NIOSH, Cincinnati, 1995.

OSHA: General Industry Standards, 29 CFR 1910.132. Occupational Health and Safety Act of 1970 (84 Stat, 1593). Revised 1996.

Perkins, J. L., Chemical protective clothing (Vol. I): Selection and use, *J. Applied Ind. Hyg.* 2, 222-230, 1987.

Perkins, J. L., Solvent-polymer interactions, *Chemical Protective Clothing*, Vol. I. American Industrial Hygiene Association, Fairfax, VA, 1990, Chap. 4.

Plog, B., Niland, J., and Quinlan, P., Eds., Respiratory protection, *Fundamentals of Industrial Hygiene*, 4th ed., National Safety Council, Itasca, IL, 1996, Chap. 22.

Sansone, E. B. and Tewori, Y. B., The permeability of laboratory gloves to selected solvents. *Am. Ind. Hyg. Assoc. J.* 39, 169-174, 1978.

Slater, K., Comfort or protection: the clothing dilemma, *Performance of Protective Clothing*, 5th Vol., STP 1237, American Society for Testing and Materials, West Conshohocken, 1996, 486-497.

Stull, J., Connor, M., and Heath, C., Development of a combination thermal and chemical protective ensemble for U.S. Navy fire fighting applications, *Performance of Protective*

Clothing, 5th Vol., STP 1237, American Society for Testing and Materials, West Conshohocken, PA, 1996, 408-427.

Chapter 12

ERGONOMICS

James Kohn, PhD, CIH, CSP, and Celeste Winterberger, PhD, CIT

I. OVERVIEW

According to the Bureau of Labor Statistic's (BLS) report titled *Annual Occupational Injury/Illness Survey: Workplace Injuries and Illnesses in 1994*, the work-related injury/illness frequency rate has declined steadily during the period between 1992 and 1994. The Bureau of Labor reported that in 1992 the injury/illness rate was 8.9 cases per 100 full-time workers. In 1993 the rate had declined to 8.5 cases and again in 1994 the injury/illness rate further decreased to 8.4 cases per 100 full-time workers. While the overall occupational health and safety statistics appear to be decelerating, ergonomic-related incidences were found to be actually accelerating. Ergonomic incidents in the form of repeated trauma increased by over 15% between 1992 and 1994 (Bureau of Labor Statistics, 1995). Considering Bureau of Labor data, it is obvious why ergonomics has been called the occupational injury/illness epidemic of the 1990s.

This ergonomic epidemic phenomenon, however, is not limited to the United States. Reports from Australia, Canada, Germany, New Zealand, Sweden, and the United Kingdom indicate that this is a global problem. According to a recent German medical journal study, carpal tunnel syndrome was reported to be the most common compression syndrome in Germany accounting for almost 20% of all nerve lesions. New Zealand, for example, reported that in the year ending March 31, 1989, over $16.5 million in compensation had been paid to 6200 recipients filing ergonomic-related claims. The Ontario Workers' Compensation Board reports that soft-tissue musculoskeletal disorders represented one-third of the disabilities among Ontario construction workers and accounted for two-thirds of their compensation costs.

Health and safety professionals are aware of the growing magnitude of the ergonomic problem. Review any occupational health and safety publication and the odds of finding ergonomics-related articles are good. Suggestions for improving workplace layouts; modifications of mercantile delivery vehicles; examination of sitting, standing, and combination workstations and their effects upon employee health; management responsibilities associated with ergonomics; surveys of ergonomic management activity; ergonomic issues in the construction industry; and exercise and cumulative trauma disorders are just a few of the hundred topics recently published that examine issues associated with ergonomics. The magnitude of the ergo-

1-56670-197-X/99/$0.00=$.50
© 1999 by CRC Press LLC

nomics problem justifies the time that professionals are spending on these issues.

In response to the ergonomic epidemic, some state and federal agencies have proposed the enactment of ergonomic legislation. For example, California has proposed legislation to address repetitive motion injuries in the occupational environment. *California Title 8; General Industry Safety Orders, Article 106, Section 5110,* was proposed as a result of the California Division of Labor Statistics and Research study of repetitive motion injuries. This study indicated that over 28,000 employers reported disorders associated with repetitive trauma. The federal government has not ignored the ergonomic problem. The proposed ergonomic standard has been in draft form for several years, but the moratorium on all new federal legislation has detained its passage. This standard was intended to address the repetitive motion injury problem as well as lifting and vibration issues.

Regulators are aware of the ergonomics problem. Health and safety professionals are also aware of the ergonomic epidemic. Even the public has been made aware of these issues through newspaper and television articles on the topic. Professionals and lay-persons all agree that ergonomics is a problem that must be addressed.

This chapter will focus on *occupational ergonomics*. It will examine the need for studying ergonomics and go into greater detail examining the costs and benefits associated with addressing the ergonomic problem. The physiological basis of ergonomics will then be presented. This will be followed by an in-depth study of repetitive injuries in various workplace environments. In addition, options for the elimination and/or control of ergonomic hazards will be presented. This chapter concludes with the review of several case studies that demonstrate options used to eliminate occupational ergonomic hazards.

A. Terminology

Ergonomics is the discipline that examines the capabilities and limitations of people. The term ergonomics is based upon two Greek words: *ergos* meaning "work," and *nomos* meaning "the study of" or "the principles of." In other words, ergonomics refers to "the laws of work." The goal of ergonomics is to design the workplace to conform with the physiological, psychological, and behavioral capabilities of workers.

Most ergonomics problems arise out of pre-existing operations. It is then important for specially trained professionals to anticipate, recognize, and identify ergonomic hazards. Evaluation and control measures would be some of the activities that would have to be performed and implemented to eliminate ergonomic hazards and minimize ergonomic risk factors.

According to the OSHA meatpacking guidelines, *ergonomic hazards* refer to workplace conditions that pose a biomechanical stress to the worker. Such hazardous workplace conditions include, but are not limited to, faulty workstation layout, improper tools, excessive tool vibration, and job design problems. They are also referred to as (ergonomic) *stressors*. The meat-packing guidelines defines *ergonomic risk factors* as "conditions of job, process, or operation that contribute to the risk of developing CTDs." Examples include repetitiveness of activity, force required, and awkwardness of posture." In addition, an *ergonomist* or *ergonomics professional* is defined in that same publication as "a person who possesses a recognized degree or professional credentials in ergonomics or a closely allied field (such as human factors engineering) and who has demonstrated, through knowledge and experience, the ability to identify and recommend effective means of correction for ergonomic hazards in the workplace."

1. *Need for study*

As mentioned in the overview, ergonomics has become an occupational problem of major proportions. National and international data point to ergonomic injuries and illnesses as the primary health issue of the 1990s. The medical and legal costs associated with ergonomically related dysfunctions are spiraling with no apparent end in sight. Ergonomic costs to employers are rising and are impacting their ability to compete in the global marketplace. An examination of ergonomic incidents and related costs seems appropriate.

B. Incident Prevalence

The Bureau of Labor Statistics (BLS) published 1994 workplace injury and illness statistics on December 15, 1995. This annual survey provided estimates of the frequency and associated incident rates of workplace inju ries and illnesses based on OSHA 200 logs maintained by employers and submitted to BLS. The BLS survey indicated that 65% of all illness in 1994 were disorders associated with repeated trauma. Approximately 332,100 new repetitive motion cases were reported that year, resulting in an incidence rate of 0.41 cases per 100 full-time workers. BLS indicated that this was an increase of almost 10% over 1993 statistics and more than a 15% increase over 1992 survey results. According to this survey, "the private industries with the highest incidence rates of disorders associated with repeated trauma in 1994 are meatpacking plants (12.6 cases per 100 full-time workers), knit underwear mills (10.53 cases per 100 full-time workers), motor vehicles and car body plants (9.6 cases per 100 full-time workers), and poultry slaughtering and processing (8.3 cases per 100 full-time workers). Refer to Table 12.1 for the top 15 private industries with the highest incidence rates of disorders associated with repeated trauma in 1994.

Table 12.1

The 15 Private Industries with the Highest Repeated

Trauma Incident Rates with SIC Codes for 1994

Industry	SIC Code	Incident Rate (per 100 full-time workers)
Meatpacking plants	2011	12.6
Knit underwear mills	2254	10.6
Motor vehicles and car bodies	3711	9.6
Meat products	201	8.8
Poultry slaughtering and processing	2015	8.3
House slippers	3142	7.3
Motor vehicles and equipment	371	5.6
Motorcycles, bicycles, and parts	375	5.3
Men's and boy's underwear and nightwear	2322	5.0
Engine electrical equipment	3894	4.8
Potato chips and similar snacks	2096	4.6
Automotive stampings	3465	3.8
Household refrigerators and freezers	3632	3.8
Men's and boy's work clothing	2326	3.6
Vehicular lighting equipment	3647	3.6

Source: Bureau of Labor Statistics, 1995.

Between 1990 and 1993, repetitive motion illnesses increased by 63.1% in the United States. During that same period, compensation cases increased by 27.6% in Texas, 11.6% in Washington state, 16.2% in British Columbia, Canada, and 25.4% in Ontario, Canada. Official Swedish statistics revealed that the hands are the body part most affected by occupational injury (30% of all cases reported). Almost 71% of the occupational diseases reported were musculoskeletal in nature. Occupational diseases to the hand and wrist resulted in an average of approximately 60 workdays lost, except for older workers who reported more workdays lost on average.

Specific industries are especially affected by ergonomic problems. For example, sprains and strains caused 37.6% of all lost-workday injuries in construction. One-quarter of those injuries affected the back. In general, 31.8% of all workers compensation costs in the United States are attributed to back injuries.

II. COSTS ASSOCIATED WITH ERGONOMIC HAZARDS

A. Medical Costs

In a recent publication, it was reported that more than one-third of all workers' compensation costs, over $10 billion annually, goes to cumulative trauma disorder cases. It is estimated that between 20 and 50 billion dollars are spent annually for back injuries. The American Academy of Orthopedic Surgeons has also estimated that cumulative trauma injuries have totaled $27 billion yearly in medical bills and lost workdays. Dr. Steven J. Barrer estimated that the average company with high repetitive motion risks would spend approximately $25,000 per carpal tunnel syndrome case. NIOSH estimates that the average carpal tunnel case costs about $3000 in benefits and $40,000 in medical costs. The typical cost of carpal tunnel surgery is approximately $18,000. Back injuries have been reported to cost an average of $9000 in worker's compensation and medical expenses. Medical expenses and Worker's Compensation are expected to rise during the next decade. Ergonomic injuries and illnesses are expected to result in a greater proportion of a company's health expenses.

B. Legal Costs

Companies may incur ergonomically related legal costs in a variety of ways. If an organization experiences a high frequency of repetitive motion or lifting-related injuries, it could be cited by OSHA under the general duty clause. Citations could then result in substantial fines, depending upon a variety of factors, including the size of the company, the number of violations cited, an employer's willful disregard of employee safety, or the good faith effort an employer demonstrates in attempting to abate ergonomic hazards. General Motors, for example, recently agreed to pay $420,000 in penalties to resolve ergonomic-related citations associated with repeated motion and overexertion problems. Lehman Brothers Inc., a much smaller company, was fined $4650 for alleged word processing-related wrist, back, and neck injuries in its New York office. Other companies cited for alleged ergonomic related violations include IBP Inc. ($3.1 million), Pepperidge Farm Inc. ($1 million), and a United Parcel Post facility ($140,000).

The more common legal costs associated with ergonomic injuries is associated with employee's lawsuits. CTDNews reported that a Long Island Railroad ticket clerk was awarded $55,000 for an alleged repetitive motion injury. The injury was claimed to have developed while using a manual ticket vending machine (Ryan *vs.* Long Island et al., No. 92 Civ. 3029 (EDNY). The jury initially awarded the plaintiff $100,000, but reduced the award, claiming that she was partially responsible for her injury. In another ergonomically-related case, a former medical director presented arguments to the U.S. Court of Appeals for the Seventh Circuit (Chicago), claiming his former employer failed to accommodate his disability, carpal tunnel syndrome. Filing under the Americans with Disabilities Act, the plaintiff claimed that the company fired him because he could not perform the essential functions of his job. No decision has been made to date.

C. Costs to Organizations

In addition to the various ergonomic-related costs already mentioned, organizations may experience losses associated with poorer productivity, indirect costs associated with all injuries and illnesses such as staff investigation time and increased absenteeism. Ergonomic programs have repeatedly been pointed out to make good economic sense. Incentives pointed out that justifies the implementation of an ergonomic program include: a reduction in the number of errors caused by poor working conditions, lower absenteeism and employee turnover, and improved overall productivity.

An example of lost productivity associated with ergonomic problems is the loss of a computer programmer as a result of repetitive motion trauma. The programmer could be on disability for 6 months or more, resulting in project delays. In addition, the company may have to hire less-competent temporary workers to replace the company-trained programmer, resulting in decreased productivity.

Another example of stealth costs to companies is manual materials handling. It is estimated that more than 60% of an employee's time is spent in material handling. If product flow and handling could be improved, more time could be spent in production-related activities. An example of improved productivity associated with reduced material handling is the assembly of medical components. Through automation of the assembly and conveyor systems, one company realized a savings of approximately $100,000 and a 2.3-month payback period for the initial financial investment. Automation had eliminated tasks that had resulted in over $95,000 in workers' compensation medical expenses and increased productivity at the same time.

1. Benefits

According to a recent survey, companies are observing several benefits when implementing an *ergonomics program*. Benefits most frequently cited included improved worker morale, improved productivity, better ergonomic hazard control, fewer complaints, fewer injuries reported, and improved health awareness. Brown has reported an 85% increase in productivity with a cost-benefit ratio of 1 to 10 with the implementation of ergonomic programs. Brown pointed out that the combination of increased production, along with the reduction of workers' compensation expenses, reaped significant rewards when compared to the minimal costs incurred during the redesign of the workplace equipment and facility. Webb has also reported substantial gains in productivity as a result of modest investments in ergonomic programs and equipment. Numerous benefits have been demonstrated with the implementation of ergonomic programs. These benefits can positively impact both employees and employers.

III. ERGONOMIC DOMAINS

When examining the human element associated with ergonomics, there are four broad areas of concern: physiological factors, psychological factors, behavioral factors, and psychosocial factors. Physiological factors include anthropometric and biomechanical variables that influence an individual's ability to perform work-related tasks. Anthropometrics is the study of human physical dimensions such as height, forward arm reach, or eye height in the sitting position. Biomechanics can be defined as the study of the mechanical operation of the human body. It is the science of motion and force in living organisms.

Through the application of anthropometric and biomechanical principles, it is possible to reduce the physiological stresses placed on the worker's body via the redesign of tools, equipment, and facilities. This, in turn, reduces the likelihood of strain and sprain injuries prevalent in poorly designed occupational environments. The physiological factors of the ergonomic model will be discussed in greater detail later in this chapter. Psychological, behavioral, and psychosocial factors are the worker variables that contribute to various nonphysiological reactions that adversely affect worker performance. These will be discussed at this time.

A. Psychological

Psychology is the science that studies human behavior. Some of the psychological factors that have been found to contribute to ergonomic hazards in the occupational environment include: attention, memory, fear, boredom, fatigue, job satisfaction, and occupational stress. An example of a psychological factor that can adversely influence worker performance and health would be fear. Individuals have been involved in traumatic accidents that have resulted in debilitating reactions not from the initial injuries sustained, but from the fear generated when those individuals returned to the accident scene. A master electrician with over 15 years of experience received second and third degree burns across his hands, arms, and feet when he made contact with an energized pad mount transformer. He returned to work approximately 6 months later only to find that every time he entered the substation where the accident took place, he became very nervous. His hands would shake and he could not perform the work assignment. This example of a psychological ergonomic reaction points to health-related dysfunctions that go beyond the frequently cited sprains and strains.

B. Behavioral

Behavioral factors refer to changes in worker activity that are observable and measurable. Reaction time, response accuracy and appropriateness, adaptation, and endurance are just a few examples represented by this category. An example of a behavioral factor might be response accuracy, reaction time and endurance of an emergency response clean-up worker during the hot temperatures of summer. While the psychological factor of fatigue and the physiological factor of exhaustion are readily apparent, behavioral factors under these conditions must also be considered. How quickly will an individual react to an unplanned change in the environment while using Level A (whole body, eye, and respiratory) personal protective equipment in temperatures approaching 100°F? The environment and equipment used can adversely affect the worker's ability to react to an emergency or perform a complex sequence of tasks.

C. Psychosocial

Psychosocial factors are worker behaviors that are influenced by co-workers, supervisors, or the organization. It refers to worker behavior in a group environment. Concepts such as leadership style, employee motivation, organization reward systems, and attitude formation and change are just a few of the elements studied under this worker ergonomic category. An example of a psychosocial factor could be the influence of co-workers versus management upon the use of personal protective equipment. A new employee wearing hearing protection could be in conflict with co-workers if the majority of the employees do not perceive a need for the protection. This could result in isolation and ridicule. There is less likelihood of the new employee following organizational guidelines for the use of personal protective equipment (PPE) under these conditions. If, on the other hand, new employees were trained in the hazards associated with their job, the company communicated its PPE policy, and co-workers urged compliance, there would be an increased use (behavioral factor of appropriate performance) in the required equipment.

IV. METHODS OF ANALYSIS

There are a variety of tools and techniques used in the analysis of ergonomic hazards in the occupational environment. Anthropometric and biomechanical measurements are common starting points for the ergonomist. The health and safety professional would be concerned with the dimensions of workers as well, as the dimensions of the workstations where tasks are performed. Observation of biomechanical body motions would then be conducted to determine if unnatural motions are required to perform required work activities.

Time and motion studies as well as behavioral sampling strategies are also of great importance in determining frequency, duration, force, and pace of the various motions. Quite often, videotaping is required to closely study these variables. Job activity analysis, work task analyses, and work cycle analysis are just a few of the numerous methods of analysis that may be conducted in a thorough ergonomic risk assessment.

In addition, specific activities may be analyzed, such as lifting, using the *NIOSH Lifting Guidelines*. Surveys of employee opinions, health, and organizational activities are often a part of the analysis of ergonomic risk factors. The design of facilities and tools would also be analyzed to determine if they contribute to the ergonomic hazards identified in the workplace. Besides the processes, tools, equipment, facility, organizational variables, and other possible factors, environmental stressors must also be studied.

Environmental monitoring of stressors such as temperature, lighting, noise, humidity, and air contaminants is another area of measurement required to obtain the ergonomic "big picture."

An in-depth review of all ergonomic analytical methods is not possible in this chapter. Readers are urged to review the references cited for more details concerning ergonomic methods of analysis.

V. PHYSIOLOGICAL RESPONSE

It is very important for industrial hygienists to understand the physiological responses of the human body stressors in order to more effectively implement ergonomic solutions. These responses can be visible; however, most are invisible. Some of the more visible responses include: sweating and/or bulging muscles. These outward manifestations, while important, do not represent an accurate picture of the stress placed on the human body. It is the invisible responses which give a much better picture of the overall functioning of the human body. This section will give the practitioner a basic overview of those physiological systems typically affected by ergonomic stressors.

Following this discussion of basic physiology, musculoskeletal disorders and their effects on the physiological systems of the human body, as well as some common physiological disorders, will be discussed (refer to Table 12.2 for organ systems typically implicated in ergonomic injuries).

Table 12.2
The Basic Physiological Systems Typically
Affected by Ergonomic Stressors

Cell
Cardiovascular system
Nervous system
Musculoskeletal System

A. Micro Physiological Response

Although the *cell* represents one of the smallest units of the body, it is at the root of all physiological responses (refer to Table 12.3). Each human being starts with one cell, which then divides countless numbers of times to produce a human being. All cells are composed of essentially the same parts. However, it is important to note that most cells perform highly specialized functions within our separate organ systems. The cell itself is composed of a cell wall. This cell wall performs two functions: it forms a protective layer between other cells and it allows for the transport of materials in and out of the cell. Proteins and protolipids are the most important elements found in the cell wall. If the cell wall is exposed to certain chemicals, it may break down and allow the further transport of the hazardous material. Inside the cell are a number of structures. Cytoplasm is the gel-like portion of the cell where the different structures exist. The nucleus contains the genetic material (called DNA) which is reproduced during the process of mitosis. Chemicals classified as mutagens or teratogens can alter the DNA of either the parent or the fetus.

Table 12.3
Several Structures Found in the Cell

Endoplastic Reticulum	Contains RNA and ribosomes and plays an important part in cellular contractile systems, including muscle contraction
Golgi Apparatus	Helps the cell in secretion and retention of protein materials
Lysosomes	Makes enzymes which break down excess material in the cytoplasm
Mitochondria	Produces the energy needed by the cell

The *cardiovascular system* is used to deliver blood, which contains nutrients and oxygen, to the cells. It also removes carbon dioxide and other waste materials from the cells. Carbon dioxide is expelled out of the body by the lungs, while other waste materials are taken to the kidney for excretion from the body. This system consists primarily of the heart, lungs, veins, and arteries. The heart is the engine of the body, pumping blood through the arteries. First, the blood is taken to the lungs via the pulmonary artery. Oxygen is transferred into the red blood cells at the alveoli while carbon dioxide is passed off into the lungs for exhalation. Blood at this stage is bright red in color. Once the blood is taken to the cellular level, it transfers the oxygen and other nutrients to the cells and removes carbon dioxide and other waste products (at this point the blood is dark red in color). This blood is then returned to the heart and the process begins again. The most important thing for hygienists to remember about the cardiovascular system is that it reaches every cell of the body, allowing toxic materials full access to all the body's organ systems.

There are two basic components of the *nervous system*, the central nervous system (CNS) and the peripheral nervous system. The CNS is composed of the brain and the spinal cord. In the brain, electrical impulses are sent to the various lobes (refer to Table 12.4).

Table 12.4
The Lobes of the Brain Where Electrical Impulses Are Sent

Cerebral Cortex	Where voluntary nervous impulses, sensory perception, and sophisticated mental events occur
Thalamus	Which is a relay station for synaptic input and has some role in motor control
Hypothalamus	Which control the basic functions of the body
Cerebrum	Which controls muscle tone and muscle coordination for the more skilled activities
Brain Stem	Which controls most of the autonomic nervous system

From the brain, nervous impulses are carried through the spinal cord for distribution to the rest of the body. The spinal cord is approximately 45 cm (18 inches) long and 2 cm (0.79 inches) in diameter. There are five different sections of spinal nerves. Eight cervical nerves can be found in the neck region. Twelve thoracic nerves can be found in the trunk area. Five lumbar nerves compose the lower portion of the spine. Finally, five sacral nerves make up the lowest portion of the spine. A coccygeal nerve can be found at the end of the spine, which is often called the "tail bone." To protect the highly fragile spinal cord, there is a bony structure composed of vertebrae. Each vertebra is separated from each other by cartilaginous structures called disks. The peripheral nervous system takes nervous impulses from the CNS and distributes them throughout the body. From there, the peripheral nervous system can be further divided into the afferent and efferent divisions. The afferent division carries information to the CNS on the status of body systems. Efferent division nerves carry nervous system commands to the different body systems. There are two types of nervous system response. The first is a voluntary response. A voluntary response is caused when an afferent nerve fires in response to some external stimuli. For example, look at a simple wave. The eye will detect that someone is moving away, which sends an impulse to the brain which will initiate a response by efferent nerves to the muscles of the arm to begin the wave function. In voluntary responses, the individual has control of the action. There is a second type of nervous response which is controlled by what is known as the autonomic nervous system (ANS). For example, when an eye blinks, the individual does not have to think of moving the muscles which control that function. It just happens. The heart beats with a rhythm which is controlled by the nerves located in that organ. When a person exerts him- or herself, the heart begins to beat faster, but not because of a conscious effort made by that person. Then, there are some responses known as reflex actions. Reflex actions are a learned response to a certain type of stimulus. The jerking back of the finger when one touches a very hot object is one example of a simple reflex. When humans are born, these connections are not immediately made, but just let a child touch a hot object a few times and they will learn to reflexively move when they touch a hot object.

The *musculoskeletal system* consists of bones and muscles. Bones form the skeleton which protects organs and organ systems within the human body. The vertebrae protect the spinal column, the ribs protect the heart and lungs, and the pelvis protects the reproductive and abdominal organs. Bones also are used to support the weight of the human body. The bones of the legs and feet must be able to withstand the force of gravity and the constant pounding of walking, running, and lifting. In order for humans to achieve movement, these bones must be connected by contractual tissue. This tissue is known as muscle. Muscles consists of a number of fibers bundled together to form one mass. There are three types of muscles: voluntary, involuntary, and heart (cardiac muscle). Voluntary muscles are called skeletal or striated muscle and connect the bones of the body. Involuntary muscles are found in the internal organs and are known as smooth muscle. It is important to note that muscles can only pull. Therefore, any muscle must have a restretch mechanism. In cardiac muscle, the restretch mechanism is accomplished by new blood entering the chambers, smooth muscles are restretched when material enters the organ and skeletal muscles work in pairs so as one muscle contracts the other is stretched. This arrangement of skeletal muscle pairs is known as antagonistic (for example, biceps and triceps). Skeletal muscle bundles are connected to the bone by tendons. The end of the muscle attached to the more stationary joint is called its origin, while the attachment at the moving part is called its insertion. However, since humans are flexible, there must be one additional mechanism available which allows for movement. These articulations are called joints. The bone and joint form a lever system with the bone being the lever and the joint being the fulcrum. There are six types of body movements which can occur around joints (refer to Table 12.5).

Fluid-filled sacs called bursa (disks in the back) cushion the bones between the joints so the bones do not rub together causing pain. Ligaments stabilize the bones around a joint.

Table 12.5
The Six Types of Body Movements Which Can Occur Around Joints

Flexion	Decreasing the angle of a joint such as bending the elbow
Extension	Increasing the angle of the joint such as straightening the leg at the knee
Abduction	The moving of a body segment away from the midplane of the body such as raising the arm up and away from the side of the body
Adduction	Moving a body segment toward the midplane of the body such as lowering the arm from a horizontal position to a vertical position next to the body
Rotation	Moving in a body segment in a circular motion around a joint such as "windmilling" the arm around the shoulder joint

Source: Sanders and McCormick, 1993.

B. Macro Physiological

The effects of poor ergonomic design is typically not limited to one organ or organ system. Many times, a number of systems will be affected, with the failure of one system frequently initiating the failure of another. Macro physiological disorders are a result of one or more physiological systems being impacted by a stressor(s). These will be covered by two broad categories: musculoskeletal disorders and physiological stress disorders.

C. Musculoskeletal Disorders

Musculoskeletal disorders focus primarily on the skeletal muscles and their attachments to the bones. Since nerves play a major role in muscle contraction and feeling, they also have an important role in these types of disorders. There are three common forms of musculoskeletal disorders an industrial hygienist should be concerned with: cumulative trauma disorders, back injuries, and segmental and whole-body vibration injuries.

The most common of musculoskeletal disorders involve the back. Nearly, 50% of all back injuries are caused by material handling. Of those material handling injuries, 50% are caused by lifting objects, 9% occur while pushing and pulling objects, and 6% occur while holding, wielding, throwing, or carrying objects. Table 12.6 shows data compiled on high-risk industries for the most common incident types. Please note that injuries caused by overexertion (defined as incidents caused by maneuvering objects) have the highest rate of occurrence.

Table 12.6
Industries Having the Highest Incident Rates for Common Injuries per 10,000 Workers in 1994

Event or Exposure	Industries	Rates
Overexertion	Nursing homes/Air transport	318/307
Struck by object	Logging/Wood containers	241/227
Fall on same level	Roofing/Water supply	121/118
Transportation incident	Taxicabs/School buses	114/102
Repetitive motion	Hats, millinery/Men's suits	104/89
Assault by person	Residential care/Nursing homes	40/37

Source: Bureau of Labor Statistics (BLS), *Characteristics of Injuries and Illnesses Resulting in Absences from Work, 1994*, 1996.

Often, during the lifting process, the back is used as a lever with the hips joints being the fulcrum. This is known as the stoop lifting method. Even using a proper lifting technique can create an enormous stress on the fifth lumbar (L5) and first sacral (S1) disk in the back. One factor known to increase the stress on the L5/S1 disk is the horizontal position of the load. It has been found that a simple inverse relationship exists between maximum weight lifted and the horizontal position of the load. This can be calculated using the following formula:

$$W = K(1/H) \tag{12.1}$$

Where:
 W = The maximum weight lifted
 K = A constant which depends on gender and vertical location of the load
 H = Horizontal location of the load

According to Sanders and McCormick (1993) and NIOSH (1981), an object which normally weighs 44 pounds and is held 8 inches from the body exerts a force of 400 pounds on the L1/S5 disk. Increase the distance to 30 inches and the force becomes 750 pounds. Muscles in the back and abdomen are also affected. Some workers who have been performing lifting tasks for a number of years may even experience a herniated or "slipped" disk. This injury will require surgical intervention, resulting in an employee who is often partially disabled and under medical care for an extended period of time. Pre-employment screening will lower the initial risk to the employer. However, this does not absolve the employer from training employees in proper lifting techniques (straight back/bent knee lifts) and designing workstations that limit the risk of back injury. There are a number of sources of anthropometric data for correct lifting heights. In addition, the revised NIOSH lifting equation (1991) can be used to determine the recommended weight limit (RWL) for manual material handling tasks. To calculate the RWL for a task, one would use the following equation:

$$RWL = (LC)(HM)(VM)(DM)(AM)(FM)(CM) \qquad (12.2)$$

Where:

RWL = Recommended weight limit

LC = Load constant = 51 lb.

HM = Horizontal multiplier calculated as 10/H, where H is the horizontal of the hands from the floor; measured from the midpoint between the ankles

VM = Vertical multiplier calculated as:
$[1 - (0.0075 | V-30 |)]$, where V is the vertical location of the hands from the floor measured at the origin and destination of the lift

DM = Distance multiplier calculated as: $[0.82 + (1.8/D)]$, where D is the vertical travel distance between the origin and the destination of the lift

AM = Asymmetric multiplier calculated as $[1 - (0.0032A)]$, where A is the angle of symmetry, the angular displacement of the load from the sagittal plane; measure at the origin and destination of lift

FM = Frequency multiplier as determined from NIOSH table, where F is the average frequency rate of lifting measured in lifts/min; duration is defined to be: ≤ 1 hour; ≤ 2 hours; or ≤ 8 hours, assuming appropriate recovery allowances

CM = Coupling multiplier as determined from NIOSH table

Using this formula and the following data, one can determine the RWL for a lifting task. Given the following data:

$$H = 6 \text{ inches}$$
$$V = 0 \text{ inches}$$
$$D = 0 \text{ inches}$$

A = 30°
F = 0.5 lifts/min
coupling multiplier = 1 (assume good coupling)
LC = 51 lb.
HM = 10/H = 10/6 = 1.67
VM = [1 - (0.0075|V-30|)] = [1 - (0.0075|0-30|)] =
[1 - (0.0075(30)] = [1 - 0.225] = 0.775
DM = [0.82 + (1.8/D)] = [0.82 + (1.8/70)] =
[0.82 + 0.0257)] = 0.846
AM = [1 - (0.0032A)] = [1 - (0.0032(30))] =
[1 - 0.096] = 0.904
FM = 0.5 lifts per minute for 8 hours with a V of 70
from the table = 0.81
CM = 1.0

RWL = (LC)(HM)(VM)(DM)(AM)(FM)(CM)
RWL = (51)(1.67)(0.775)(0.846)(0.904)(0.81)(1)
RWL = 40.9 lbs

Finally, research on the use of back belts has not found any benefit to their use. In autumn 1992, the National Institute for Occupational Safety (NIOSH) formed a working group to review the literature regarding back belts and their use.

The group's mission was to find any data which supported the premise that back belts reduce injuries. Because of the lack of scientific data regarding the efficacy of using back belts, the NIOSH working group concluded that back belts do not prevent lifting injuries and do not reduce the hazards to employees caused by repeated lifting, pushing, pulling, twisting, or bending.

Cumulative trauma disorders (CTDs) or *repetitive stress disorders (RSDs)* are a growing problem in the workplace. According to Bureau of Labor Statistics, CTDs accounted for 6.8% of the of nonfatal injuries and illnesses in U.S. business during 1994. This was up from 4.5% in 1993. Most of these injuries occur in the manufacturing sector. Cumulative trauma can be defined as injuries that are the result of repeated mechanical stresses. Table 12.7 summarizes typical symptoms associated with cumulative trauma cases.

TABLE 12.7
Typical Symptoms of Cumulative Trauma

Pain
Restriction of joint movement
Soft tissue swelling
Loss of feeling
Reduction in manual dexterity

There are three categories of CTDs: tendon disorders, nerve disorders, and neurovascular disorders.

The most famous type of nerve disorder is carpal tunnel syndrome. This occurs when the tunnel containing the tendons, nerves, and blood supply to the hand is collapsed by repeated pressure to the underside of the wrist. After repeated compression of the carpal tunnel, the median nerve becomes compressed, resulting in pain, numbness, and tingling in the hand. Eventually, if the individual continues to perform the same repetitive task, permanent loss of hand function may occur.

The most common neurovascular disorder is thoracic outlet syndrome. This is caused by the compression of nerve and blood vessels between the neck and the shoulder. It has symptoms similar to that of carpal tunnel syndrome, including numbness in the arm and finger.

TABLE 12.8
Tendon Disorders

Disorder	Cause
Tendinitis	This occurs when the muscle and tendon is repeatedly tensed and the tendon becomes inflamed. Pianists and punch press operators tend to get this affliction.
Tenosynovitis	This occurs when repeated movements cause an excessive production of synovial fluid (a tendon lubricant); when the fluid accumulates, the synovial sheath will swell, causing pain. Buffing and grinding operations tend to cause this affliction to occur.
DeQuervain's disease	A form of stenosing tenosynovitis which affects the tendons on the side of the wrist and at the base of the thumb, pulling the thumb back and away from the hand. Meatpacking operations tend to cause this affliction to occur.
"Trigger finger"	This occurs when the tendon sheath surrounding the finger is so swollen that the tendon becomes locked, causing jerking and snapping movements of the finger; typically this occurs in employees using tools with sharp handles such as drills, radial arm saws, and rivet guns.
Ganglion cysts	This is caused by excess synovial fluid building up to form a bump under the skin.
Unsheathed tendons	These are usually found in the elbow and shoulder when the joint is used for impact or throwing motions; the irritated tendons will then cause a shooting pain from the joint and down the arm. Activities such as tennis, pitching, or bowling can cause this affliction to occur.

Vibration syndrome, Raynaud's syndrome or "white finger", is caused by the use of vibrating hand tools in cold environments. In its early stages, vibration syndrome may cause numbness and tingling in the fingers, but if it is allowed to continue there will be a loss of all sensation and control in the hands and fingers.

About 8 million individuals are subjected to some sort of vibration on a regular basis. The whole-body vibration frequencies of most importance to the industrial hygienist are from 0.1 Hz to 20 Hz with accelerations of from 0.2 to 4 g. These are typically caused by motors, compressors, and the impact of uneven elements of road/track. Segmental vibration frequencies of importance range from 8 to 500 Hz, with accelerations of 1.5 to 80 g. Typically, segmental vibration is caused by held-held power instruments such as chain saws, jackhammers, drills, and torque. Additionally, when a industrial hygienist deals with vibration, he/she must understand the concepts of resonance and damping. Resonance occurs at a frequency where an object will vibrate at its maximum amplitude which is greater than the amplitude of the original vibration. In essence, the object will act as a magnifier at resonance frequencies. Each substance has its own unique resonance frequency. "Since the body members and organs have different resonant frequencies, and since they are not attached rigidly to the body structure, they tend to vibrate at different frequencies." Generally, the larger the mass of an object, the lower its resonance frequency. Damping is a natural or artificial method used to absorb vibration. A good example of a natural damping is when an individual uses the bending or straightening of the legs in response to movements within 1 Hz to 6 Hz range. One artificial method to used control vibration exposure in the workplace is to have the operator stand on a foam pad.

The Eastman Kodak Company (1993) states that there are three forms of vibration which the industrial hygienist should be concerned with:

Whole-body vibration	These are caused by transportation vehicles which are traveling over rough roads which create vibrations that are primarily in the vertical plane.
Whole-body vibration	These are caused by production machinery. The vibration characteristics of equipment is based on the individual machine and should be measured by accelerometers.
Segmental vibration	This is associated with the use of hand tools. It occurs primarily in the arms and hands of an individual and is based on the type of power tool being used.

Whole-body vibration is typically concerned with frequencies below 100 Hz, while segmental vibration occurs in the range of 40 Hz to 1000 Hz. Although there is some overlap of frequencies between the two forms, the major difference between the two is that segmental vibration causes problems with specific body parts such as fingers, wrists, elbows, shoulders, or backs, while whole-body vibration deals with the effects on the entire human system. Over 1 million individuals are exposed to segmental vibration each year. Some of the injuries which have been associated with segmental vibration include:

- Small areas of decalcification seen in the X-rays of the small bones of the hand
- Injuries to the soft tissues of the hand
- Osteoarthritis of the joints of the arms

The effects of whole-body vibration vary, depending on several parameters. Specific effects are discussed relative to a range of frequency, although there appear to be specific peak frequencies that can lead to resonance effects. The type of physiological effects of vibration on a given individual are primarily determined by the intensity, frequency and/or duration of exposure. In the 2–20 Hz range at 1 g acceleration, examples of the physiological effects of low-frequency vibration include:

- Loss of equilibrium
- Chest pain
- Abdominal pain
- Shortness of breath
- Nausea
- Muscle contractions

D. Physiological Stress Disorders

While most of the research available seems to key on repetitive stress disorders such as carpal tunnel syndrome, there are other long-term physiological disorders which are associated with on-going emotional stress. For example, unrealistic work schedules may cause stress for employees who are unable to keep up or employees could be stressed when they are asked to perform delicate assembly tasks in low light conditions.

It is well known that stress causes problems with the gastrointestinal system. Irritable bowel syndrome or chronic diarrhea can be caused by stress. If these symptoms continue, it could have serious consequences, including the possibility of cancer. Another portion of the gastrointestinal system, the stomach, can also be affected by stress. When individuals are subjected to stress, the stomach produces excess stomach acid which, over time, causes a deterioration of the stomach lining. This often results in a sore called an ulcer. If the ulcer breaks completely through the stomach lining, it is called a perforated ulcer which requires immediate surgical intervention.

The deleterious effects of stress can also be found in the cardiovascular system. Heart attacks are the major cause of death for many Americans. One of the contributing factors in heart attacks is stress. When an individual is subjected to stress, adrenaline, a powerful stimulant, is released by the adrenal glands. Adrenaline causes both respiration and heart rates to increase. The muscle will tense. For short periods of time, this can cause an individual to be more productive. However, as the heart rate increases, so does the blood pressure; the pressure at which blood is pumped through the body. Chronic high blood pressure, if unchecked, can lead to heart attacks.

These are only a few of the occupational stress disorders which can affect people in the workplace. As medical research continues in this area, it will surely uncover other physiological disorders that are caused by on-the-job stress. Ergonomically designing workstations for the individual can reduce, if not eliminate, some of the causes of work-related stress.

VI. INDUSTRIES WITH REPETITIVE INJURIES/EQUIPMENT

Disorders such as Carpal Tunnel Syndrome, Tendinitis, Tenosynovitis, DeQuervain's Syndrome, and Thoracic Outlet Syndrome are just a few of the many dysfunctions that rarely result from single incidents. These problems tend to occur as a result of what has been termed "microtrauma." Microtraumas are the result of frequent repetitions of motions and forces associated with workplace manipulative tasks. Many terms have been used to describe the syndromes that result from microtraumas. Terms such as repetitive motion injuries, repetitive strain injuries, cumulative trauma disorders, and even occupational overuse syndrome (the term currently used in New Zealand) attempt to emphasize the chronic nature of these "illnesses."

Physicians admit that there are numerous questions associated with the development of cumulative trauma disorders. A factor that compounds the difficulty of determining cause is that there are just as many non-occupational causes of repetitive motion problems as there are occupational related causes. Repetitive motion problems may be caused by non-occupational disease factors, including:

> **System diseases:** Rheumatoid arthritis, acromegaly, gout, diabetes, myxoedema, ganglion formation, and certain forms of cancer.
> **Congenital defects:** Bony protrusions into the carpal tunnel, anomalous muscles extending into or originating in the carpal tunnel, and shape of the median nerve.
> **Tunnel size:** The smaller the diameter of an individual's carpal tunnel, the greater the chance the individual will experience carpal tunnel syndrome if working a repetitive task involving use of the wrist, hands, and/or arm.

Gynecological state: Pregnancy, use of oral contraceptives, menopause, and gynecological surgery.

Occupationally related causes result when workers are required to perform tasks that involve forceful and repetitive motions. Synergistic factors such as demanding awkward postures and unnatural positions of the upper extremities compound the risks. In addition, the lack of muscle strength recovery time may also contribute to the microtrauma. Workers involved in these types of tasks will often experience symptoms such as tightness, stiffness, or pain which can start in the fingers, hands, wrists, forearms, and elbows. These symptoms can progress to sensations of tingling and numbness in the hands. The loss of strength and coordination along with radiating pain associated with inflammation of the median nerve can follow. These problems can be "productivity killers" since carpal tunnel syndrome has been found to result in 30 median days away from work per case for workers in general industry, according to 1993 BLS statistics. Tendinitis is another repetitive motion syndrome that has been found to result in an average of 10 days away from work per case for workers in general industry, according to those same 1993 BLS statistics.

Certain occupational environments have increased repetitive motion risks. Every workplace has unique tools or equipment that contribute to the ergonomic problem. A review of these conditions follows.

A. Meatpacking

As cited earlier in the chapter, the BLS reported that in 1994 meatpacking plants experienced 12.6 cases of carpal tunnel syndrome (CTS) per 100 full-time workers *Plants* (refer to Table 12.9 for tasks impacting the meatpacking industry). This significant incident rate has remained stable over the past several years, and is responsible for the Occupational Safety and Health Administration publishing the document titled *Ergonomic Program Management Guidelines for Meatpacking*.

TABLE 12.9
Ergonomic Repetitive Motion Problems Identified
in *The Ergonomic Program Management Guidelines*
for Meatpacking Plants

Lifting and twisting activities associated with material and product handling
Use of cutting tools such as knives
Power tools and their triggers
Vibration associated with power tools

The repetitive motion problem with the use of knives was associated with the handle design. Most meatpacking employees used a straight-handled knife. As the carcasses went by on an overhead conveyor, employees made vertical slices on the carcass. The motion required by these slices caused the wrist to bend, increasing the likelihood of pinching the median nerve. In addition, employees were required to grasp the knives tightly for the duration of the shift. To reduce the ergonomic risk found in this task, a pistol grip-handled knife was introduced into the workplace. This allowed the employees to keep their wrists straight while performing the cutting task. In addition, a band was added to handle, permitting the employee to hold onto the knife without having to grasp tightly. This band allowed the employees to release their grip and relax the muscle and tendons of the hand when not performing the cutting task.

Postures, positions, and work methods were also analyzed. Static loading of the arm and shoulder muscles (contraction of muscles to support the position of body components) were ergonomic problems identified. One recommendation suggested to eliminate this problem was automation. Articulated arms and counter balances were also introduced so tools could be suspended by mechanical means. This modification meant that the employee used less muscular strength and force to hold the tools. Power tools were also introduced into the workplace in place of manually operated tools to reduce the repetitive motions required to perform many of the cutting tasks.

B. Office Environment

New office equipment has contributed to the rising numbers of repetitive motion problems. Microcomputers are the most common contributor to the cause of carpal tunnel syndrome. Other equipment contributing to the office ergonomic epidemic are adding machines or cash registers that require repetitive finger extension for data entry. These are motions that are similar to the keystrokes required when using a computer. The examination of office ergonomics will focus upon the computer.

With the increasing number of computers being introduced into the office environment, industrial hygienists have witnessed a steady increase in the prevalence of repetitive motion-related cases. By replacing the typewriter with the computer or wordprocessor, office workers have witnessed the elimination of several tasks that served as "breaks" for the keystroke motions required when typing. Inserting and removing typing paper, using the carriage return at the end of each line, using corrective fluids, and pausing to look up the spelling of words were activities that interrupted the keystroke typing activity. Using computers in the modern office environment can require workers to perform over 23,000 keystrokes in a single work period without hand motion breaks.

Does the frequency of keystroke motions increase the risk of repetitive motion syndromes? If the answer to this question is yes, then one would expect that occupations requiring high frequencies of keystroke motions would have higher incidences of repetitive motion problems. A 1990 study conducted in California and funded by NIOSH found that cases of carpal tunnel syndrome were reported in approximately 50 different occupations.

In this same study, 23% of the cases were reported in administrative support occupations and 13% were from cashiers.

Examination of the interface between people and workstations can provide the clues necessary to identify potential ergonomic problems. Correct typing posture and position, as well as reduced frequency, pace, and force, have been repeatedly reported to be the ergonomic solutions to computer-elicited repetitive motion problems. Job observations should be conducted to ensure that hips and thighs as well as forearms are parallel to the floor. Wrists should also be kept in a straight and level neutral position. While actually typing, the worker's wrists should be straight and not supported. Furniture such as desks, computer monitor stands, and chairs should be adjusted to fit the anthropometric measurements of the worker. Refer to checklists for the evaluation of workstations and worker posture to determine if deviations are contributing to the ergonomic problem. Figure 12.1 demonstrates a properly designed computer workstation.

Figure 12.1 Ergonomically designed workstation (Courtesy: *Proformix*, Branchburg, NJ).

C. Labor Intensive Industries

There are a wide variety of work environments that require workers to perform repetitive motions to accomplish work-related tasks. Repetitive motions typically include lifting motions, as well as hand, wrist, and arm motions. Repetitive motions and related injuries in several different industries will be examined.

1. *Construction*

Unlike other industries that have experienced upper extremity repetitive motion problems, the majority of the ergonomic hazards in the construction industry are related to manual materials handling. The BLS reported that only 1.5% of all lost workday illnesses in the construction industry were due to repetitive motion. However, construction was second only to transportation as the industry with the highest incidence rates for sprains and strains resulting in lost-workday injuries. Ergonomic hazards abound in the construction environment.

Some of the ergonomic hazards that exist in the construction industry include: static positions, repetitive motions, material handling, awkward postures, overhead work, and exposure to vibrating tools or equipment. These are commonly associated with a wide variety of tasks, from site excavation to internal structure finishing such as painting, carpet laying, and trimming.

For example, during site excavation activities, the use of heavy earthmoving equipment poses several ergonomic hazards. Whole-body vibration while operating earthmoving equipment, uncomfortable postures resulting from poorly designed seats, repetitive motion associated with body twisting, and turning while operating vehicles in reverse are just a few hazards common in this environment.

During bricklaying activities, material handling is the primary ergonomic hazard of concern. A worker may lift up to 1000 bricks during a typical day. This translates to between 6600 to 8800 pounds (about 3000 to 4000 kilograms) daily. In addition, the bricklayer may perform 1000 trunk-twist flexions during the course of a normal work shift. These activities have a substantial impact upon the cardiovascular, nervous, and musculoskeletal systems. The cardiovascular system is under stress during this activity because oxygen and nutrients are being distributed by the heart and circulatory system to the muscle groups performing the work. The cellular metabolic process in these muscles converts the oxygen and nutrients into mechanical energy used during brick handling. As muscles work harder, the cardiovascular system must work harder to provide the much-needed energy. If the worker is in poor health, cardiovascular problems can occur.

The nervous and musculoskeletal systems are also under substantial stress during material handling activities. During bricklaying, trunk-twist flexions may result in pinched nerves in the L5/S1 area of the spinal column as well as overexertion muscle injuries such as strains of the arm (triceps and brachioradialis), shoulder (deltoids and trapezius), and back muscles (latissimus dorsi, infraspinatus, teres minor and major).

Other studies have found that concrete-reinforcement workers experience high frequencies of back pain. This ergonomic problem results from spending significant portions of a shift bent over to tie reinforcement rods, commonly referred to as rebar. Electricians were also found to spend a considerable amount of time in awkward postures and positions while installing cable, wiring, or conduit. However, upper extremity hazards associated with the use of pliers and screwdrivers posed potential repetitive motion threats for these workers as well.

Low-cost solutions have been recommended for many of the problems identified in the construction industry. First, evaluation of tasks must be performed to reduce muscular effort, as well as the frequency of lifting and climbing. In addition, repetitive motions observed should be reduced by way of work-site modifications and the use of alternative tools. For example, the use of height-adjustable work platforms was found to reduce the frequency of bending for bricklayers. Concrete mixers required to open and pour bags of mortar into a mixer could use an inexpensive stand to raise the height of the bags and reduce the frequency of bending. Smaller bags of product that weigh less is another low-cost solution to ergonomic material handling problems in the construction industry.

2. Manufacturing/assembly

On April 26, 1994, the BLS released a flyer titled *Work Injuries and Illnesses by Selected Characteristics, 1992*. In this flyer, repetitive motion characteristics were profiled. The report indicated that women accounted for two-thirds of the nearly 90,000 repetitive motion injuries and illnesses cases. This report indicated that 18% of the repetitive motion cases reported resulted from repetitive use of tools. 31% of the repetitive motion cases reported resulted from placing, grasping, or moving objects other than tools. Approximately 56% of these cases occurred in manufacturing. Occupations most frequently cited were operators, fabricators, and laborers (51%), machine operators (24%), and assemblers (8%).

As revealed in the above-mentioned statistics, manufacturing operations have experienced significant losses associated with ergonomic hazards in the workplace. Manual material handling and repetitive motion are the two leading ergonomic causes in this environment. The material handling statistics are just as staggering as the repetitive motion statistics. In 1993, approximately 28% of the total estimated days away from work in this country, or 21,000,000 days, resulted from lifting injuries. When viewed together, ergonomic-related losses from material handling and repetitive motion cost companies billions of dollars every year.

Most jobs in the manufacturing environment involve manual material handling. Workers handle raw materials and/or finished products. They may also be required to carry tools and/or containers. In addition, packing operations go on continuously at most facilities.

To avoid material handling injuries, many companies have employed the use of scissor lifts, carts, dollies, and powered hand trucks. Palletizing materials and using forklifts are additional solutions to material handling problems.

Repetitive motion problems pose a greater challenge for the health and safety professional. The National Safety Council reports that 2,925,000 days are lost each year as a result of repetitive motion injuries. The frequency of repetition of motions that can be tolerated without experiencing injury varies by age, gender, health, and a variety of other factors. No specific safe repetitive motion threshold limits have been established to date. However, the proposed OSHA ergonomic standard recommended modification of work processes if one of the following attributes was observed:

- Tasks requiring over 2000 manipulations per hour (such as keystrokes during word-processing)
- Manual task work cycles that are 30 seconds or less in duration (such as pinch grip repetitive motions during small component assembly)
- Repetitive tasks which exceed half of the worker's shift (a data entry clerk who performs calculation keystrokes during 90% of the shift or inspectors who manipulate parts during inspection 85% of their shift)

The use of hand and power tools contributes to repetitive motion problems. Using a power grip to hold a file or wrist deviations associated with soldering are just some of the ergonomic hazards that have caused repetitive motion injuries. Applying labels to equipment manually or packaging finished products are additional causes of ergonomic upper extremity hazards.

Repetitive motion problems are also associated with the loading and unloading of power presses and riveting machines. For example, employees are often required to use a pinch grip (grasping a component between the thumb and forefinger or thumb and remaining fingers) when removing small components from a storage bin and placing the part into the press die. In addition to the repetitive pinch grip, dorsiflexion and palmar flexion may result during the component handling as well as during power press manual stock feeding and removal activities. Wherever possible, modify the task or the process to eliminate the frequent repetitions, stressful postures, or muscle exertion requirements. For example, one company required that finished product be lifted over the side and placed into a cardboard box for sealing and labeling. By reducing the height of the work table and placing the box on its side, workers could roll the product into the box. This eliminated the lifting motion as well as the bending of the workers' wrists. At another location, product was removed from a conveyor at the end of an assembly line and placed on pallets that were on the shop floor. The company introduced a scissor lift that supported the pallets. This eliminated the need to bend when loading the finished product onto the pallet. The scissor lift is a mechanism that raises and lowers as stock is loaded or unloaded. In this way, the pallet height is maintained near knuckle height for the standing employee, reducing the need to bend or extend overhead during the palletizing process.

VII. CONTROL

There are a number of ways that the ergonomic hazards discussed above can be mitigated. This section will discuss four different methods for controlling the ergonomic environment of a workplace. In addition, different industrial scenarios for each method will be given.

A. Human Control Methods

The best human control method is proper training. For example, employees performing manual lifting tasks should be trained in proper lifting methods. Assuring that employees know how to lift properly can reduce the number of back injuries far better than using PPE such as back belts. Training on the proper use of special ergonomic equipment such as adjustable chairs can help ensure that employees actually use those features which makes their job easier.

Another human control method is exercising prior to beginning the work shift. Most individuals who work out know that stretching the muscles before strenuous exercise can lower the risk of injury. Many companies, including Wal Mart, have started using stretching exercises to prevent injuries to their employees during the workday. Some research has been done regarding the successfulness of these exercise programs in reducing workplace injuries. In a review of eight research studies, McGorry and Courtney (1995) found that not enough data exists to support the contention that on-site exercise programs alone will reduce the number of CTDs in the workplace. They did conclude that studies of exercise programs coupled with engineering controls and/or administrative controls showed positive results. However, they suggest that more controlled studies be performed to determine whether exercise alone will be effective in reducing CTDs.

Wellness programs have also become very popular in the past few years. Many companies feel that promoting the physical fitness of their employees may reduce the number of injuries on the job. GlaxoWellcome, at their Zebulon, NC, manufacturing facility has a fitness center on-site where employees can work out 24 hours a day. Health Maintenance Organizations (HMOs) such as Healthsource North Carolina, Inc. have offered special incentives to their subscribers who exercise on a regular basis. Some of these incentives include lower cost memberships at local health clubs, as well as rebates for the purchase of home exercise equipment.

B. Human Environmental Control Methods

Whether it is temperature, noise, lighting, and/or vibration, most of the environmental stressors at the workplace place significant demands on the physiological systems of the body. Methods for reducing the effects of each of these stressors will be discussed in this section.

Extremes in temperature cause problems with the regulation of the core temperature. Since the normal body core temperature is from approximately 96.8° to 98.6°F, any temperature stressor that has the potential to raise or lower these values can be dangerous. One of the best ways to control the effects of temperature is to allow the employee's body to acclimate to the environment. For warm climates, allow the worker 2 weeks to adjust. This may be achieved by having the employee work in the warmer environment for increasing periods of time during the 2-week period. There is some doubt that humans can ever acclimate to cold climates; however, the same practice of increasing work exposures over a defined period of time may allow workers to somewhat acclimate to the colder temperatures. Clothing can also be used to control temperature extremes. Insulated clothing, including gloves and hats, can be used in cold environments. An important concept to remember in cold environments is to use clothing with natural fibers, especially close to the skin. If the clothing near the body becomes saturated with sweat because the fibers do not allow the moisture to move away from the body, the individual may become cold even if they wear multiple layers of clothing. For warm environments, loose, light-colored clothing works best. However, some operations, such as arc welding, require the worker to wear special protective clothing. In these cases, cooling vests may be used to keep the body core temperature within normal limits.

Noise can be dangerous, as well as annoying. High noise levels can cause permanent hearing damage. Noise may also be a barrier to effective communication. The best way to control environmental noise is to isolate the employee from the noise source. This may call for the use of sound-attenuating barriers, personal protective equipment such as ear plugs or muffs, in addition to controls which minimize exposure duration.

Inadequate or improper lighting can place an employee under stress. This includes either too little or too much light. Proper monitoring methods should be used to determine the level of illumination in a given area. If the illuminance is not at the correct level for the task, efforts should be made to adjust the illuminance levels. One way of increasing illumination is the use of task lighting. Other methods may involve the use of different light sources. This may include the use of halogen lights, sodium vapor lights, or fluorescent lights. Glare can also create stress for the employee. Efforts should be made to reduce the effects of glare, especially on glass-covered controls and computer workstations. Glare can be reduced by using diffuse overhead lighting sources and task lighting.

Segmental vibration can be controlled by isolating the worker from the source. This can be done by using hand tools that are ergonomically designed or by providing the employee with gloves. Whole-body vibration can be reduced by having the employee stand and/or sit on a cushioning material.

C. Equipment/Facility Control Methods

Most ergonomics texts will indicate that engineering controls (equipment, workstation, and facility design) are the best method of reducing or eliminating ergonomic disorders. There is an abundance of anthropometric data available for industrial hygienists on the design of equipment, workstations, and facilities. Van Cott and Kincade's text (1972) contains a wealth of anthropometric data regarding the design of equipment and controls. All military contractors are required to comply with the anthropometric data found in MIL-STD 1472D when they design equipment, workstations, and facilities for use by U.S. military forces. NASA is also an excellent source for design data, most of which can be easily accessed using the World Wide Web. In addition, Web sites, such as ErgoWeb (http:\\ergoweb.mech.utah.edu\) or CTD News *Online* (http:\\CTD News. com\) are other sources of information which can be used in design applications.

With all these sources of data and the continuing research on equipment, workstation, and facility design, it would seem that these control methods could be easily implemented. Unfortunately, this is not the case. With most industries trying to remain competitive by downsizing, economics begins to mean more than ergonomics.

Many tool, equipment, and furniture manufacturers now offer ergonomically designed products. One cannot open a safety or engineering related magazine or journal which does not have a least one advertisement for companies selling ergonomic products. Sears even carries some hand tools which have been designed for ergonomic use. Many Web sites also exist for companies marketing ergonomic equipment. In addition, many consultants are offering their services to industry for the design and implementation of ergonomic solutions for equipment, workstations, and facilities.

D. Organizational Control Methods

If design efforts fail to reduce ergonomic disorders, organizational control methods (sometimes referred to as administrative controls) is the second-best control method available. Most of the organization controls involve limiting the amount of time that a worker is exposed to environmental stressors, which may include:

- Providing rest breaks
- Using job rotation

- Reducing the work rate
- Requiring the use of personal protective equipment

Organizations can also create a corporate environment where ergonomics is considered important to the overall success of the company. Workers can be encouraged to submit suggestions for solutions to ergonomic problems found either in their workstation or department. Successful implementation of these ideas would result in the employee sharing part of the first year's savings. The use of safety committees composed of employees from all levels is another way that organizations can encourage the use of ergonomics.

VIII. CASE STUDIES

The two case studies discussed in this section represent examples of ergonomic task and work layout hazards. Readers should relate the hazards identified in these case studies to personal experiences. Consider alternative options that could be implemented to eliminate the ergonomic risk factors identified.

A. Case Study: Repetitive Motion

Jane was a 31-year-old female construction worker who began experiencing numbness in her arm and shoulder after a night of sleep. Concerned that these symptoms were not the result of sleeping improperly on her arm and shoulder, Jane made an appointment with her physician. After doing a complete medical work-up, the doctor told Jane she was suffering from carpal tunnel syndrome in both hands.

In 1987, few doctors had much experience with this disorder. In fact, this doctor was so uninformed that he did not tell Jane that she was eligible for workers' compensation. So, because she needed the money, Jane went back to work.

The construction site was at a highway overpass. Jane's job was to use a 30-pound jackhammer to drill holes for bolts in hardened concrete. Because she often worked while tethered to scaffolding, Jane typically balanced herself by holding on with one hand while running the drill with the other. The job also involved holding this 30-pound piece of equipment at shoulder height or higher. In addition, Jane was responsible for manually inserting, tightening, and torquing the bolts using a large ratchet wrench. This second process also had Jane working at heights above shoulder level.

Once the construction season was over, Jane began working for a printing company. Although her job required some keyboarding, she mostly did paste-up work. However, Jane was still experiencing numbness and she sometimes lost her grip on tools. At this point, she consulted a hand surgeon who suggested that Jane have orthroscopic surgery to correct the carpal tunnel in her right hand.

Immediately after the surgery, Jane was not experiencing any pain in her hands so she went back to work for another printing company. For this job, Jane was required to do a significant amount of keyboarding. She again started to experience pain and numbness, only this time the pain was located in her neck and left shoulder. Concerned that she might have a different medical problem, Jane consulted with another doctor. The doctor indicated to Jane that the neck and shoulder pain she was experiencing was consistent with carpal tunnel syndrome. This time it was in her left hand. She contacted a second hand surgeon who indicated that a more invasive form of surgery would be necessary to correct the problem. So 1½ years after the initial diagnosis, Jane underwent surgery on her left hand.

Jane stated that to her knowledge the construction job she was performing was never remediated. She also stated that at no time did the original doctor suggest she just stop doing that job for awhile. Jane spent nearly 3 years on workers' compensation and finally received a settlement from the company a year ago. She now works for a large wood manufacturing company as a supervisor. Using a computer or vibrating tools still causes her to experience a loss of feeling in her finger tips, as well as the sudden loss of grip strength. Because of the sudden loss of strength, Jane has been cautioned never to climb ladders and to limit her use of vibratory tools. Whenever she uses a computer, Jane must make sure that she has the chair at the proper height and uses a wrist rest for keyboard and mouse.

The most important lesson that can be learned from Jane's experience is that surgery is not a cure for carpal tunnel syndrome. Perhaps the best solution to the problem of repetitive motion disorders is to redesign jobs so they do not require the worker to use tools at awkward angles and heights, keeping flexion of the wrist at a minimum. If the job cannot be redesigned, the worker should be moved to another job.

B. Case Study: Computer Workstation

A certified professional ergonomist (CPE) was hired by an agricultural cooperative to assess ergonomic hazards at its corporate headquarters. An ergonomic audit instrument was used to collect computer workstation data. (Note: this data collection instrument was similar to the office ergonomic audit instrument found in Bisesi and Kohn (1995)).

The audit revealed numerous ergonomic hazards. Two contrasting conditions existed that contributed to these hazards. In one instance, some of the locations had computers placed on traditional office furniture and desks not designed to accommodate computers. In other workstation environments, new chairs and tables were provided. However, the "ergonomically designed" furniture had not been adjusted to fit the operators using the workstations.

The majority of offices audited had computers that were placed on traditional office furniture. In addition, older style chairs were in use at those locations, resulting in an undesirable condition. These chairs failed to provide adequate back support and the height adjustment for effective worker-keyboard interface was nonexistent. Common ergonomic hazards that were found in these locations included:

1. Stands, tables, and supports for the keyboard that were not adjustable. With the use of standard office desks, body posture during computer operation was unsafe. This resulted in dorsiflexion (the movement of the hand that decreases the angle between the back of the hand and the arm) of the operators' wrists increasing the likelihood of repetitive motion injuries among operators required to perform data entry or word processing activities for prolonged periods of time.
2. Computer monitors were set at improper angles. In some instances, computer monitors were placed upon system units resulting in an operator head posture which facilitated neck, shoulder, and back fatigue.
3. Arm, head, and feet supports were not available. In a majority of the observed workstations, arm, hand, and feet supports were not used. This resulted in improper posture increasing the potential for physiological fatigue in the short term and possible injury in the long term.
4. Monitors were located in front of windows or immediately below and forward of overhead lighting fixtures. As a result of improper equipment placement, glare-related conditions existed for equipment operators. This condition could be expected to result in eye fatigue and strain.
5. Work area surfaces were inadequate and document holders unavailable or poorly placed when available. In some instances, working surfaces or document holders that were used resulted in operators bending, twisting, and turning their necks. This could be expected to increase neck, shoulder, and back fatigue.
6. Chairs that failed to provide adequate ergonomic support for computer operators. In many instances, chair back and shoulder support was not available and chairs could not be easily adjusted to provide necessary support.
7. Operators could not properly adjust tables and chairs. In many instances, operators set up make-shift computer equipment layouts, including placing keyboards in open desk draws or on laps. These conditions would increase the probability of fatigue and physiological strains.

Ergonomic problems existed in office locations that had adjustable furniture as part of the computer workstation. Several offices audited had computer workstations with "state-of-the-art" office furniture. Despite the availability of good equipment, undesirable conditions existed from an ergonomic perspective. Common ergonomic hazards in these locations included:

1. Chairs and keyboard stands incorrectly adjusted or improperly used for the workstation operator.
2. Computer workstations placed in locations that promoted screen glare.
3. Operators did not correctly adjust blinds to reduce glare and screen contrast problems.
4. Operators unaware of proper computer workstation operating positions and how to correctly adjust chairs, keyboard supports, stands, and tables.
5. Ideal placement of documents to reduce neck and upper torso movements.
6. Strategies that can be used to reduce repetitive motion-related problems, including but not limited to exercise, change in activities, and periodic breaks.

The consultant recommended the following strategy to reduce the risk of ergonomic related hazards:

1. Purchase foot and arm/wrist rests to promote proper posture among operators requiring additional physiological support.
2. Prioritize computer workstation location by percent time spent on the units by operators. Replace chairs and furniture with sound ergonomically designed equipment in those locations where the most significant risks exist.
3. Consider job enrichment activities for computer operators who spend more than 50% of their time on keyboard-related tasks. Employ alternative tasks to minimize time spent in repetitive motion activities.
4. Ensure adequate breaks to allow computer operator stretching and exercise activities for the reduction of fatigue and stress.
5. Evaluate all computer workstations and modify workplace environments to minimize illumination-related problems.
6. Evaluate all computer workstations to ensure that layouts comply with acceptable ergonomic guidelines.
7. Ensure the availability and use of document holders to minimize computer operator twisting and bending motions.
8. Perform periodic audits, looking for proper computer operator posture, and recommend modifications necessary to minimize back, neck, and arm strain.
9. Conduct ergonomic training for all computer operators.

This organization accepted the recommendations presented by the consultant, including the ergonomic training classes. Forty-five minute ergonomic training classes were conducted for all computer users. The goal of the training program was to ensure participant knowledge of ergonomics, proper computer workstation layout and design, and repetitive motion injury prevention methods. The training program consisted of the following objectives:

1. Definition of ergonomic terms
2. Application to workplace and home equipment and furniture design
3. Symptoms associated with poorly designed environments
4. Discussion of specific concerns expressed by course participants
5. Identification of ergonomic hazard countermeasures
6. Commitment by the course instructor to observe all interested participants and evaluate their workstations.

After all of the operators were trained, approximately 20 department representatives were involved in a train-the-trainer program. This was conducted to ensure that new employees would be trained in the principles presented by the consultant.

When each training session was completed, the consultant accompanied by a department representative, visited each workstation where training program participants worked. The intent was to help each operator adjust the furniture and reposition the workstation to eliminate ergonomic hazards. In addition, if changes could not be made, a list of required furniture or equipment (like wrist and foot rests) was compiled. This list was then submitted to management.

Of interest to the consultant was that although participants knew the correct way to set up computer workstations following the training program, they had no idea how to personally adjust their own furniture and equipment. The consultant and department representative then worked with each employee to correctly adjust his/her workstation.

Implementation of most of the recommendations and the ergonomic training program reduced the number of employee complaints by over 50%. Employee morale was reported by department supervisors to have improved and no cases of repetitive motion injuries were reported for an 18-month period following implementation of this program.

Table 12.10
The Benefits of Implementing Ergonomics Programs

• Decreased errors and product defects	• Decreased loss of customers
• Decreased time required to perform tasks (as a result of increased visibility and easier handling of stock, finished product, and tools)	• Improved company efficiency resulting from smaller workforce (resulting from reduced absenteeism and lost time injuries/ illnesses)

Table 12.10 (cont'd)

• Reduced training and associated costs (as a result of a more stable workforce)	• Reduced management and supervision costs (fewer incident investigations and time spent solving related issues)
• Reduced hidden costs (such as disability salaries and insurance premiums)	• Increased labor pool (older, less fit, and disabled workers can be employed in ergonomically designed workplaces)
• Improved morale	• Decreased ergonomically related litigation
• Reduced worker discomfort	• Reduced disruption of work teams
• Reduced fatigue-related costs	• Reduced productivity fluctuations resulting from late-shift or late-week operations
• Improved hazard identification and control	• Increased occupational health and safety awareness
• Improved quality	• Improved general health awareness
• Improved organizational performance	• Financial savings

REVIEW QUESTIONS

1. According to the Bureau of Labor Statistic's 1994 survey, 65% of all illnesses were disorders associated with:

   a. Overexertion   b. Dermatitis   c. Repeated trauma   d. Respiratory Illness

2. The industry with the highest incidence rate of disorders associated with repeated trauma in 1994 was:

   a. Meatpacking   b. Automobile   c. Textile   d. Pharmaceutical

3. Examples of occupational legal costs associated with repetitive motion or lifting-related injuries include all of the following except:

   a. OSHA citations
   b. Employee lawsuits
   c. Medical malpractice suits
   d. Litigation associated with violations of the Americans with Disabilities Act

4. All of the following are examples of benefits associated with the implementation of ergonomics programs except:

   a. Increased productivity
   b. Decreased Worker's Compensation fraud
   c. Increased morale
   d. Improved ergonomic hazard control

5. The study of human physical dimensions is a definition for which discipline?

   a. Biomechanics   b. Ergonomics   c. Physiology   d. Anthropometry

6. Reaction time and endurance are examples of topics associated with the _____ ergonomic domain.

   a. Physiological   b. Psychological   c. Behavioral   d. Psychosocial

7. Time and motion studies and behavioral sampling are of great importance in ergonomic assessment for determining:

   a. Frequency of motions
   b. Duration of tasks
   c. Force
   d. Pace
   e. All of the above

8. The brain and spinal column make up the:

   a. Central nervous system
   b. Autonomic nervous system
   c. Peripheral nervous system
   d. Both a. and c.

9. Disk herniation associated with manual material handling and poor lifting methods occurs in the _____ vertebrae area of the spinal column.

    a.   Fifth lumbar (L5) and first sacral (S1)
    b.   Fifth sacral (S5) and first lumbar (L1)
    c.   Twelfth thoracic (T12) and first lumbar (L1)
    d.   Twelfth lumbar(L12) and first thoracic (T1)

10. Back belts are viewed by ergonomists as:

    a.   The back injury prevention method of choice
    b.   Not beneficial
    c.   A possible tool to be used in conjunction with engineering and administrative controls
    d.   Both b and c

11. Carpal tunnel syndrome is a repeated trauma dysfunction associated with the compression of the:

    a. Median nerve   b. Ulnar nerve   c. Radial nerve   d. Tendon sheaths

12. Numbness, blanching of fingers, and loss of muscular strength and finger control resulting from vibrating equipment and aggravated by cold environments is a definition for which repeated trauma syndrome?

    a.   Carpal tunnel syndrome
    b.   Tenosynovitis
    c.   DeQuervain's disease
    d.   Raynaud's syndrome

13. The forms of vibration which are of ergonomic concern in the occupational environment include:

    a.   Whole-body vibration caused by transportation vehicles
    b.   Whole-body vibration caused by production machinery
    c.   Segmental vibration which is associated with the use of hand tools
    d.   All of the above

14. Examples of non-occupational causes of repetitive motion problems may include:

    a.   Rheumatoid arthritis
    b.   Bony protrusions into the carpal tunnel
    c.   Use of oral contraceptives
    d.   All of the above

15. Organization controls used to minimize exposure to ergonomic stressors may include:

    a.   Providing rest breaks
    b.   Using job rotation
    c.   Reducing the work rate
    d.   All of the above

ANSWERS

1.	c.	9.	a.
2.	a.	10.	b.
3.	c.	11.	a.
4.	b.	12.	d.
5.	d.	13.	d.
6.	c.	14.	d.
7.	e.	15.	d.
8.	a.		

References

Banham, R., The new risk in ergonomics solutions, *Risk Management*, May 1994, 22-30.

Barrer, S., Gaining the upper hand on carpal tunnel syndrome, *Occupational Health & Safety*, January, 1991, 38-43.

Berne, R. and Levy, M., Cardiovascular system in *Principles of Physiology*, Berne, R. and Levy, M. Eds., Mosby, St. Louis, MO, 1990, 188-311.

Bisesi, M., and Kohn, J., *Industrial hygiene evaluation methods*, CRC/Lewis, Boca Raton, FL, 1995.

Bridger, R., *Introduction to Ergonomics*. McGraw-Hill, New York,1995.

Brown, R., Todd, G., and McMahan, P., Ergonomics in the U.S. railroad industry, *Human Factors Bulletin*, 34, 1991.

Bureau of Labor Statistics, *Characteristics of injuries and illnesses resulting in absences from work, 1994* [On-line]. Available: ftp://stats.bls.gov/pub/news.release/osh2.txt, May 8, 1996.

Bureau of Labor Statistics, *Bureau of Labor Statistics annual occupational injury/illness survey: Workplace injuries and illnesses in 1994*, [On-line]. Available: http:// ergoweb. mech. utah.edu: 80/Pub/info/bls.html, December 15, 1995.

Bureau of Labor Statistics, *Workplace injuries and illnesses by selected characteristics, 1992. Repetitive motion profile*, [On-line]. Available: http://dragon.acadiau.ca/~rob/rsi/usflier.html, April 26, 1994.

Bureau of Labor Statistics releases 1994 repeated trauma numbers, *CTDNewsOnline*. [On-line]. Available:http:\\ctdnews.com\bls94.hmtl, 1995.

Carson, R., Key egonomic tips for improving your work area design, *Occupational Hazards*, 43-46, August, 1994.

Davies, J.R., Automation and other strategies for compliance with OSHA ergonomics, *Industrial Engineering*, February, 1995, 48-51.

Dessoff, A., Seek simple solutions for ergonomics problems in construction, *Safety and Health*, January, 1996, 62-65.

Eastman Kodak Company, *Ergonomic design for people at work,* Vol.1, Van Nostrand Reinhold, New York, 1983.

Figura, S., Dissecting the CTS debate, *Occupational Hazards*, November, 1995, 28-32.

Fine, D., A break now saves money later, *Infoworld*, 54, May, 1995.

Freeman, A., Cellular function and fundamentals of physiology. *Basic Physiology for the Health Sciences*, 2nd ed., Selkurt, E., Ed., 3-29, Little, Brown, and Company, Boston, 1982.

Grandjean, E., *Fitting The Task To The Man*, Philadelphia:Taylor & Francis, 1988.

Kohn, J., Evaluating ergonomic progress, *Ergonomic News*, April, 1996.

Kohn, J., Friend, M., and Winterberger, C., *Fundamentals of Occupational Safety and Health*, Government Institute Press, Rockville, MD, 1996.

Kroemer, K., *Ergonomics of VDT Workplaces*. American Industrial Hygiene Association: Akron, OH, 1983.

Kroemer, K., Kroemer, H., and Kroemer-Elbert, K., *Engineering physiology: Bases of human factors/ergonomics*, 2nd ed., Van Nostrand Reinhold, New York 1990.

Kutchai, H., Cellular physiology. *Physiology 2nd ed.,* Berne, R. and Levy, M, Eds., 5-65, Mosby, St. Louis, MO, 1988.

MacKinnon, L., Construction: Building a safer industry. *OH&S Canada*, July/August, 1995.

McGorry, R. W. and Courtney, T.K., Exercise and cumulative trauma disorders: The jury is still out. *Professional Safety,* 40, June, 1995, 22-25.

Meiss, R., Muscle: Striated, smooth, and cardiac. *Basic Physiology for the Health Sciences* 2nd ed., E. Selkurt, E., Ed., pp. 3-29, Little, Brown, and Company, Boston, 1982.

National Council on Compensation Insurers, *Workers compensation back claim study,* NCCI, Boca Raton, FL, 1992.

National Safety Council (NSC), *Accident facts 1994 ed.*, NSC Itasca, IL, 1994.

National Institute for Occupational Safety and Health, *Carpal Tunnel Syndrome: Selected References,* Washington: U.S. Department of Health and Human Services, March, 1989.

Occupational Safety and Health Administration, *Ergonomics program management guidelines for meatpacking plants,* U.S. Department of Labor, Washington, D.C., 1991.

Official Statistics of Sweden, National Board of Occupational Safety and Health, *Occupational Diseases and Occupational Accidents*, 1991 ed., Stockholm: National Board of Occupational Safety and Health, 1993.

Owensby, G., Carpal tunnel syndrome, *Ohio Monitor*, March/April, 1993, 8-11.

Parker, K., and Imbus, H., *Cumulative Trauma Disorders: Current Issues and Ergonomic Solutions-A Systems Approach*. Lewis Publishers, Boca Raton, FL, 1992.

Petersen, D., Streamline your workflow. *American Printer,* June, 1993, 18-21.

Putz-Anderson, V., *Cumulative Trauma Disorders: A Manual for Musculoskeletal Diseases of the Upper Limbs*. Taylor & Francis, Philadelphia, 1988.

Ramsey, R., What supervisors should know about ergonomics, *Supervision*, August, 1995, 10-12.

Rowan, M. P., and Wright, P.C., Ergonomics is Good for Business, *Work Study*, Vol. 43, No. 8, 1994, 7-12.

Rogers, S., *Ergonomic Design for People at Work Vol. 2*, Van Nostrand Reinhold, New York, 1986.

Sand, R.H, Firestone wins an ergonomics battle, *Employee Relations Labor Journal*. Vol. 21, No. 1, 139-144, Summer, 1995.

Sanders, M.S. and McCormick, E. J. *Human Factors in Engineering and Design,* 7th ed., McGraw-Hill, New York, 1993.

Sanders, R.E., Bad vibrations. *Workplace Ergonomics*, May/June, 1995, 1, 22-25, 28-29.

Sherwood, L., *Human physiology: From cells to systems* 2nd ed., West Publishing, Minneapolis/St. Paul, MN, 1993.

Sweeney, M., Gardener, L., Parker, J., Walters, T., Flesch, J., Huduck, S., and Smith, S., *Workplace use of back belts,* [On-line], Available:http:\\ergoweb.mech.utah.edu\Pub\Info\Std\ backbelt.hmtl.

Thornburg, L., Workplace Ergonomics makes economic sense, *HR Magazine*, October, 1994, 58-59.

Tsimberov, D., Guidelines warn when the rattling can cause harm. *Workplace Ergonomics*, 1, October, 1994, 26-32.

Van Cott, H., and Kincade, R., eds., *Human Engineering Guide to Equipment Design.* U.S. Government Printing Office, Washington, D.C., 1972.

Webb, R., A feeding frenzy, *OH&S Canada*, 5, 1989, 91-92.

Willis, W., Jr., Nervous system, *Principles of Physiology*, Berne, R. and Levy, M., eds., 56-151, Mosby, St. Louis, MO, 1990.

Chapter 13

AIR POLLUTION

Jack Daugherty, CIH

I. OVERVIEW

A Native American saying goes, "This land we live on is not inherited from our parents, it is on loan to us from our grandchildren." That sentiment surely includes the air we breathe as well. The measurement and control of air pollution has long been the concern of industrial hygienists, and the methods approved by the U.S. Environmental Protection Agency (EPA) for measurement of air pollution were derived from industrial hygiene measurement techniques that have been developed over time since modern science began.

Air pollution is neither recent, nor entirely *anthropogenic* (human-made). Volcanoes pump many more tons of particulate matter (PM) and acid rain precursors into the atmosphere during a 24-hour period than anthropogenic sources can. Violent land bound storms, such as tornados or even exceptional thunderstorms, stir up a great deal of PM compared to industrial air pollution. However, natural causes of air pollution are episodic, having peaks of generation that trail off to zero pollution allowing the atmosphere to recover from the insult.

Anthropogenic air pollution, while pale in comparison to natural episodes, is pumped out constantly, ever increasing in direct proportion to population growth. The human insult to the atmosphere never stops, never lets up. In the fifteenth century, the cities of Spain suffered from air pollution. One queen of England refused to live in the palace because the London air was foul with pollution. In 1930, steel mill emissions in Belgium's Meuse River Valley accounted for about 60 deaths during a 6-day period. Donora, Pennsylvania, experienced 20 deaths and 6000 illnesses attributed to steel mill and industrial emissions trapped by an inversion layer on Halloween in 1948. Hydrogen sulfide emissions in Paza Rica, Mexico, caused twenty deaths and about 320 illnesses in 1950. Three thousand deaths are attributed to a London smog episode during a 2-week period in 1952.

The purpose of this chapter is to provide the novice industrial hygienist with an overview of the practice with respect to air pollution and to provide computational guidance as appropriate. The status of U.S. laws and regulations concerning air pollution will be briefly explained and computational methods will be presented where necessary. Many of the calculations an industrial hygienist will use in air pollution control are the fundamental methods learned in chemistry and physics. The methods presented in this chapter are those that are not obvious to one trained in chemistry and physics.

II. HISTORY OF THE CLEAN AIR ACT

In the U.S., the prime law for the control of air pollution is the Clean Air Act (CAA) and its 1990 Amendments (CAAA), the purpose of which is to protect and enhance air resources so as to promote public health and welfare and the productive capacity of our population. Federal intervention was necessary because most of the nation's population is located in urban areas crossing state boundaries, where air pollution increasingly threatens public health and welfare. Table 13.1 shows six major headings (or Titles) of the CAA as amended throughout the years. Table 13.2 summarizes the development of the CAA.

Table 13.1
Critical Clean Air Act Titles (As Amended)

CAAA Title	Description
I.	Air Pollution Prevention and Control
II.	Emission Standards for Moving Sources
III.	General
IV.	Acid Deposition Control
V.	Permits
VI.	Stratospheric Ozone Protection

Table 13.2
History of the Clean Air Act

Clean Air Act of 1955
The Motor Vehicle Air Pollution Control Act of 1963 (Title II)
Air Quality Act of 1967
Clean Air Act Amendments of 1970
Technical Amendments to CAA 1973, 1974
Clean Air Act Amendments of 1977
Technical Amendments to CAA of 1977, 1978, 1980, 1981, 1982, 1983
PL 101-549 of 1990 (CAAA)

The original law was ineffective because it required each state to determine and enforce air quality standards within its borders; yet in too many cases the sources of pollution were across state lines, out of jurisdiction. In 1963, Congress added Title II, formally called *The Motor Vehicle Air Pol-*

lution Control Act, to deal with the most prevalent source of air pollution, the family car. The Ralph Nader Commission on Air Pollution Control reported to Congress, in 1968, that other changes were needed to make the law more effective and, in 1970, a totally restructured federal/state scheme of pollution control was passed. The establishment of the EPA in 1971 ensured that the amended CAA had some teeth. Table 13.3 highlights the strengthening factors of the 1977 amendments.

Table 13.3
Strengthening Provisions of the 1977 Amendments

1. Prevention of significant deterioration. 2. Stringent standards imposed on areas failing to attain National Ambient Air Quality Standards. 3. Restricted use of dispersion techniques. 4. Strengthened enforcement.

Hence, this law evolved in hodgepodge fashion over a 40-year period. The major programs of the CAA before and after 1990 are shown in Table 13.4 and discussed below.

A. CAAA of 1990

The Clean Air Act Amendments of 1990 (CAAA) corrected some of the problems of the original CAA. The CAAA adds programs for Acid Deposition (called *acid rain*) and Stratospheric Ozone Protection, as shown in Table 13.4, and orders the EPA to change the way it issues regulations to control hazardous air pollutants (HAPs). From 1971 to 1990, the EPA issued only eight NESHAPs based on risk assessment (see Table 13.5), so Congress expressed its impatience in the CAAA and required the EPA to provide technology-based standards for a minimum of 189 chemicals that were listed in the law. These include heavy metal species, toxic organics, acids, oxidizers, asbestos, phosgene, radionuclides, fine mineral fibers, and other compounds Congress considered a threat to the American public health.

Table 13.4
Major Programs of the Clean Air Act

Before 1990	1990 and After
National Ambient Air Quality Standards (NAAQS)	Adds acid deposition
New Source Performance Standards (NSPS)	Adds stratospheric ozone protection
National Emissions Standards for Hazardous Air Pollutants (NESHAP)	Changes hazardous air pollutants (HAP)
Monitoring and Recordkeeping	Establishes NAAQS compliance schedules

Table 13.5
NESHAPS Issued Pre-CAAA (40 CFR 61)

Inorganic arsenic
Asbestos
Benzene from fugitive sources
Beryllium
Radionuclides
Radon
Mercury
Vinyl chloride

B. National Ambient Air Quality Standards

Under Title I, Congress establishes ambient air quality standards for certain pollutants called *criteria* pollutants. Each state is required to achieve these *NAAQS* or National Ambient Air Quality Standards, presumably by placing controls on industrial emissions and vehicular emissions. Hence, air permits are easier to get in some localities than in others. Table 13.6 lists the criteria pollutants.

Table 13.6
Criteria Pollutants

Carbon monoxide (CO)
Lead (Pb)
Nitrous oxides (NO_x)
Ozone (O_3)
Sulfur dioxide (SO_2)
Particulate matter less than 10 microns diameter (PM_{10})

The NAAQS, published under 40 CFR 50, are ambient air quality goals for criteria pollutants established for every state. Primary NAAQS are established to protect the public health, especially the most air pollution sensitive sectors of our population: infants and geriatric citizens. Therefore, Primary NAAQS are established at very low concentrations compared to OSHA PELs, ACGIH TLVs, NIOSH RELs, or AIHA WEELs, which are all designed to protect the most vigorous and healthy sector of our population: the working population. Secondary NAAQS are established to protect the environment. Attainment requirements for primary NAAQS are given in Table 13.7.

Table 13.7
Attainment of NAAQS

Primary NAAQS	Non-Attainment Status	Compliance Dates
CO	Moderate AQMDs	12/31/1995
	Serious AQMDs	12/31/2000
PM$_{10}$	Moderate AQMDs	12/31/1994
	Serious AQMDs	12/31/2001
O$_3$	Marginal AQMDs	11/15/1993
	Moderate AQMDs	11/15/1996
	Serious AQMDs	11/15/1999
	Severe AQMDs	11/15/2005
	Extreme AQMDs	11/15/2010
NO$_x$	Any	Pre-approved SIP
SO$_2$	Any	Pre-approved SIP
Pb	Any	Pre-approved SIP

Scientific terms that often confuse novices, *primary* and *secondary* pollutants, are not regulatory terms. Primary pollutants are generated at the source and emitted directly into the atmosphere, while secondary pollutants are the products of photochemical reactions in the atmosphere.

1. *Carbon monoxide*

Carbon monoxide (CO) is a colorless, odorless, non-irritating gas generated by incomplete combustion of carbon-based fuels. When CO is inhaled, it combines with the iron compounds in red blood cells to form carboxyhemoglobin, and thus prevents the blood cell from collecting an oxygen molecule, which is needed by the tissues and organs. CO has a half-life of 4 hours in human blood, but this can be reduced to 1 hour (or 90 minutes, depending on the individual) if pure oxygen is administered by face mask. At low levels of CO exposure the oxygen depletion in blood may not be noticed. At higher levels (10–20% blood COHb), headaches and fatigue are symptomatic. Around 40% blood COHb causes mental confusion and loss of coordination. Convulsions begin at about the 50–60% level of blood COHb. At very high levels (above 70% blood COHb), death occurs from lack of oxygen delivery to vital organs. As low as 9% blood COHb, myocardial hypoxia alters the heart's electrical activity as well as its contractile force, and ventricular fibrillation is initiated. Levels of COHb as low as 7% have been noted to decrease attention span and learning capacity.

Atmospheric CO is buffered by free radicals found in the environment:

$$CO + OH^- \rightarrow CO_2 + H^+ \qquad (13.1)$$

Some CO pollution is mitigated naturally. In air pollution exposures, however, the concentration of carbon monoxide that may affect an infant or an elderly, emphysemic patient must be considered. The fraction of blood saturated with CO is dependent on the ratio of the partial pressures of CO and O_2, and this relationship is called the Haldane equation:

$$\frac{COHb}{Hb} = M(p_{CO}/p_{O_2}) \hspace{2cm} (13.2)$$

Where:

COHb = Carboxyhemoglobin
Hb   = Hemoglobin
K   = Dissociation constant dependent on blood pH and temperature, 210 at normal pH and 98.6°F
p   = Partial pressure of CO or O_2

CO concentrations around 1000 ppm (0.1% volume or 1160 mg/m³) may be fatal within hours. At 1% volume and greater (>10,000 ppm or >11,600 mg/m³), CO is lethal within a few minutes. The effects of chronic CO exposure is indicated in so-called heavy smokers, who typically experience headache, dizziness, and dyspnea (shortness of breath).

2. Nitrogen oxides

Nitrogen oxides (NO_x) are produced from burning fuels. The six forms of NO_x are: nitrous oxide (N_2O), nitric oxide (NO), nitrogen dioxide (NO_2), nitrogen trioxide (N_2O_3), nitrogen tetroxide (N_2O_4), and nitrogen pentoxide (N_2O_5). NO_x reacts with VOCs to form photochemical smog. NO_x is also a precursor for acid rain, reacting with water to form nitric acid. Nitrogen dioxide causes generalized *chlorosis* in plants. Compared to other irritant gases such as sulfur dioxide, NO_x vapors are less water soluble and are, therefore, lower respiratory tract irritants (or whole lung irritants). Because the reaction with water proceeds more slowly to make acid, NO_x penetrates deep into the lungs and does more damage. However, initially, in the upper respiratory passages, NO_x is less irritating and, therefore, exposed people do not feel as if they are being harmed. Once the irritation starts in the lower respiratory passages, the lungs fill with a fluid, causing a pneumonia-like condition which can cause death by drowning, or, at the very least, scarring of tissue.

The most significant pollutant of the NO_x species is NO_2, which becomes involved in a photolytic cycle to produce ozone in the troposphere.

$$NO_2 + \mu v \rightarrow NO + O$$

$$O_2 + O \rightarrow O_3 \hspace{2cm} (13.3)$$

$$NO + O_3 \rightarrow NO_2 + O_2$$

While there is a conservation of nitrogen and oxygen species on paper, a net accumulation of ozone occurs because unburned hydrocarbons and other anthropogenic free radicals in the troposphere react with NO to block the forward reaction.

3. Ozone

Ozone is a molecular variety of oxygen containing three oxygen atoms instead of two. Lightning is the primary natural source of ozone in the troposphere, although the oxidation of ammonia and soil bacteria also emit ozone. Anthropogenic ozone sources include combustion processes and spark-producing equipment. Secondary anthropogenic ozone is produced by reactions of hydrocarbons or VOCs with nitrogen and oxygen species in the troposphere. In the stratosphere, a thin layer of ozone provides an effective barrier, protecting us from harmful effects of the sun's ultraviolet energy. However, in the troposphere, ozone is a pollutant, producing smog when it reacts with compounds of incomplete combustion and volatile organics from paints, solvents, and sprayed-on chemicals. Olefins, such as ethylene and propylene, are the most common hydrocarbons in the atmosphere and are the most reactive, as shown in Table 13.8.

Table 13.8
Relative Reactivity of Hydrocarbons

Hydrocarbon Species	Relative Re-activity
Olefins	1.4
Ethylene	1.0
Aromatics	0.9–3.0
Alkanes (paraffins)	~0
Acetylene	~0

Ozone, a *phytotoxic* pollutant (poisonous to plants), causes a green plant pathology called *chlorosis*, a breakdown of chlorophyll. In humans, ozone causes respiratory and eye irritation. Ozone can lead to coughing, chest pain, dyspnea, edema, chronic bronchitis, bronchiolitis, emphysematous fibrosis, and septal fibrosis. The upper respiratory tract may become dehydrated upon ozone exposure, bronchial irritation may lead to some of the problems already mentioned, but headache, fatigue, and visual impairment may also be a result of exposure. Accelerated aging has also been linked to ozone, as well as immune system impairment and alteration of proteins and amino acids in the body in biochemical processes (ozonization and ozonolysis of unsaturated fatty acids).

Photochemical smog is produced when hydrocarbon free radicals (R) react with the nitrogen and oxygen species examined above. A whole series of reactions takes place over a period of time in sunlight, leading to photochemical smog:

$$OH^- + RH \rightarrow R^- + H_2O$$
$$R^- + O_2 \rightarrow ROO^-$$
$$ROO^- + NO \rightarrow RO^- + NO_2$$
$$RO^- + O_2 \rightarrow R'CHO + HO_2 \qquad (13.4)$$
$$HO_2 + NO \rightarrow NO_2 + OH^-$$
$$ROO^- + O_2 \rightarrow RO^- + O_3$$
$$ROO^- + NO_2 + O_2 \rightarrow R''CO_3NO_2$$
$$R''CO_3^- + NO_2 \rightarrow R''CO_3NO_2$$

The latter reactions involve one of the chief components of photo-chemical smog, peroxyacyl nitrate (PAN). In fact, Table 13.9 lists a typical composition of smog:

Table 13.9
Typical Components of Photochemical Smog

Photochemical Compound	Composition %
Ozone (O_3)	$\leq 90\%$
NO_x (mostly NO_2)	$\leq 10\%$
Peroxyacyl nitrate (PAN)	$\leq 0.6\%$
Free radical O_2, aldehydes, ketones, alkyl nitrates	Trace

4. Sulfur dioxide

Sulfur dioxide (SO_2) is mostly produced by burning high-sulfur coal in power plants. Industrial processes that produce SO_2 are paper mills, metal smelters, and sulfuric acid production. SO_2, which is very soluble in water, is an important precursor for acid rain. The unstable sulfur trioxide (SO_3) is also regulated as equivalent sulfur dioxide, but the sulfate, SO_4, is regulated as particulate matter.

Gaseous sulfur oxides can irritate lungs and nasal passages and compli-cate, if not cause, respiratory diseases such as asthma. Exposure to SO_2 concentrations far greater than found in pollution conditions are required to effect any serious damage to humans or animals. Acid gases also cause *blotchy leaf damage* on plants by reducing chlorophyll concentration in the leaf. The dilute sulfuric acid, formed by dissolving sulfur oxides in rain-water, is very destructive to concrete structures as well.

5. Particulate matter

Particulate matter (PM) includes dust, soot, and other tiny, solid materi-als released into the atmosphere by *biogenic* or *anthropogenic* sources, in-cluding sulfates as mentioned above. Biogenic sources of PM include emis-sions such as pollens from trees, shrubs, and grasses, but bacteria, protistans, and phytoplankton also emit PM. Also, the decay of organisms produces PM. Observe closely the next time you rake leaves that have already begun to decay. Diesel fuel is a large source of anthropogenic PM in cities. Municipal incinerators are another major source. Mixing and application of fertilizers are a major rural source of anthropogenic PM, as is field and slash burning. A major suburban PM problem is burning of wood in fireplaces and woodstoves. According to the EPA, wintertime air pollution from wood smoke has become so bad in some localized areas of the country that the local governments have been forced to curtail the use of woodstoves and fireplaces during certain weather conditions. Many industrial operations produce significant amounts of PM, as do mining and mineral beneficiation processes.

Early EPA documents and reference books may refer to total suspended particulate (TSP), but this term has fallen into disuse in favor of PM and PM_{10}. PM, especially PM_{10} (PM finer than 10 microns aerodynamic diameter), can cause eye, nose, and throat irritation and more serious lung ailments. The nasal hair collects most PM from 5 to 10 microns in diameter. PM larger than 0.5 microns do not diffuse in the inhaled breath but settle out in the nasal passages in the mucus. From 0.5 to 0.1 microns, minimal diffusion occurs, allowing the PM to be carried further into the bronchial passages by incoming air. The smaller the particle, the more rapidly and completely diffusion occurs. Considerable diffusion occurs with particles that are 0.01 microns and less in diameter and these smallest diameter PM typically find their way into the alveoli. Accumulated PM in the airways increases the resistance to airflow and increases the flow of mucus. Continual insult by PM such as pollen produces a persistent hypersecretion of mucus and enlarges the mucus-secreting glands. If the PM is a sulfate or nitrate aerosol or smoke, paralysis of the cilia occurs and cleansing of the upper airways ceases or, at least, is impaired.

Table 13.10 gives a summary of the sources of the criteria pollutants and Table 13.11 summarizes their deleterious effects on humanity and the environment.

C. Air Quality Management Districts

Air Quality Management Districts (AQMD) have been established to help states/regions attain compliance with NAAQS requirements. For instance, the South Coast Air Quality Management District is the designation for the Los Angeles air basin, the worst district for nonattainment of ozone, PM, CO, and NO_x NAAQS in the country. Larger subdivisions, called Air Quality Control Regions (AQCR), may be composed of two or more AQMD and have the primary responsibility to meet primary and secondary NAAQS.

Each AQCR is required to monitor ambient air and determine if it is in compliance with the NAAQS. Regions which meet NAAQS requirements for criteria pollutants are called *attainment* areas. Regions which do not meet NAAQS for a criteria pollutant are called *nonattainment* areas for that pollutant. States do not have to limit air emissions of criteria pollutants in attainment areas unless ambient air quality will be affected (see PSD discussion below). That is why it is easier to get air permits in some states than in others. Nonattainment areas are classified as *marginal, moderate, serious, severe*, or *extreme*, depending on how polluted the air is. Table 13.12 summarizes the number of nonattainment areas in the country by pollutant.

Table 13.10
Criteria Pollutant Source Summary

Pollutant	Anthropogenic	Natural
Particulate matter	• Burning of wood • Burning fossil fuel • Industrial plants • Plowing • Burning off fields • Unpaved roads	• Volcanoes • Forest fires • Storms
Lead	• Leaded gasoline • Spray painting • Smelters • Battery manufacture	
Carbon monoxide	• Burning fossil fuel	• Smoldering fires
Nitrogen oxide	• Burning fossil fuel (especially cars)	• Soil bacteria
Ozone	• Reaction of VOC/NOx • VOC from burning fuel • Solvent, paint, glue • Other organics	• Lightning
Sulfur oxide	• Burning of coal/oil • Paper plants • Metal processes	

Table 13.11
Deleterious Effects of Criteria Air Pollutants

Pollutant	Health Effect	Environmental Effect	Property Damage
Particulate matter	• Nose/throat irritation • Lung damage • Bronchitis • Early death	• Reduced visibility (haze)	• Soil and discolor • Structures • Clothing • Furniture
Lead	• CNS damage • Animal carcinogen • Digestive problems	• Wildlife damage	• None
Carbon monoxide	• Headache • Anoxic hypoxia • Death	• None	• None
Nitrogen dioxide	• Respiratory illnesses	• Acid rain • Leaf damage • Lake damage • Reduced visibility	• Stone • Buildings • Statues • Monuments
Ozone	• Dyspnea • Asthma • Stuffy nose • Eye irritation • Immune deficiency • Aging of lung tissue	• Plant damage • Tree damage • Reduced visibility (smog)	• Rubber damage • Fabric damage
VOCs (ozone precursor)	• Some cause cancer • Other serious health effects	• Plant damage	• None
Sulfur dioxide	• Breathing problems • Permanent lung damage	• Acid rain • Tree damage • Lake damage • Leduce visibility	• Stone • Buildings • Statues • Monuments

Marginal nonattainment areas require Reasonably Available Control Technologies (RACT–sometimes referred to as Reasonably Available Control Measures, RACM) for industries emitting the nonattained pollutant. PM controls include dry scrubbers, cyclones, and baghouses for power plants and other industries, as well as restrictions on sources such as wood stoves, agricultural burning, and dust stirred up from fields and roads from normal use and construction. RACT for ozone control means removal or destruction of VOCs from tail gases. Carbon adsorbers and catalytic converters are two control measures typically encountered in industry. The acid gases are typically scrubbed in wet scrubbers or absorber towers.

Under the Amendments of 1990, cities that currently do not meet primary NAAQS must attain them by deadlines set in the law. Most had 6 years, but Los Angeles had 20 years. State programs are to complement EPA efforts on behalf of these cities.

Table 13.12a
Nonattainment Area Summary–Ozone

Extreme Ozone Nonattainment Area - 1
LA (South Coast Air Basin)

Severe Ozone Nonattainment Areas - 9
Chicago-Gary-Lake Co., IL-IN; Houston-Galveston-Brazoria, TX; Milwaukee-Racine, WI; New York-Northern New Jersey-Long Is., NY-NJ-CT; Southeast Desert, CA; Baltimore, MD; Philadelphia-Wilmington-Trenton, PA-NJ-DE-MD; San Diego, CA; Ventura, CA

Serious Ozone Nonattainment Areas - 12
Atlanta, GA; Baton Rouge, LA; Beaumont-Port Arthur, TX; Boston-Lawrence-Worcester, MA-NH; El Paso, TX; Greater CT; Portsmouth-Dover-Rochester, NH; RI; Metro Sacramento, CA; San Joaquin Valley, CA; Western MA; Washington, D.C.-MD-VA

Moderate Ozone Nonattainment Areas - 33
Atlantic City, NJ; Charleston, WV; Charlotte-Gastonia, NC; Cincinnati-Hamilton, OH; Cleveland-Akron-Lorain, OH; Dallas-Ft. Worth, TX; Dayton-Springfield, OH; Detroit-Ann Arbor, MI; Grand Rapids, MI; Greensboro-Winston Salem-High Point, NC; Huntington-Ashland, WV-KY; Kewaunee Co., WI; Knox & Lincoln Co., ME; Lewiston-Auburn, ME; Louisville, KY-IN; Manitowoc Co., WI; Miami-Ft. Lauderdale-W. Palm Beach, FL; Monterey Bay, CA; Muskegon, MI; Nashville, TN; Parkersburg, WV; Phoenix, AZ; Pittsburgh-Beaver Valley, PA; Portland, ME; Raleigh-Durham, NC; Reading, PA; Richmond, VA; Salt Lake City, UT; San Francisco Bay Area; Santa Barbara-Santa Maria-Lompoc, CA; Sheboygan, WI; St. Louis, MO-IL; Toledo, OH

Marginal Ozone Nonattainment Areas - 41
Albany-Schenectady-Troy, NY; Allentown-Bethlehem-Easton, PA; Altoona, PA; Birmingham, AL; Buffalo-Niagara Falls, NY; Canton, OH; Columbus, OH; Door Co., WI; Edmonson Co., KY; Erie, PA; Essex Co., NY; Evansville, IN; Greenbrier Co., WV; Hancock & Waldo Co., ME; Harrisburg-Lebanon-Carlisle, PA; Indianapolis, IN; Jefferson Co., NY; Jersey Co. IL; Johnstown, PA; Kent & Queen Anne's Co., MD; Knoxville, TN; Lake Charles, LA; Lancaster, PA; Lexington-Fayette, KY; Manchester, NH; Memphis, TN; Norfolk-Virginia Beach-Newport News, VA; Owensboro, KY; Paducah, KY; Portland-Vancouver, OR-WA; Poughkeepsie, NY; Reno, NV; Scranton-Wilkes-Barre, PA; Seattle-Tacoma, WA; Smyth Co., VA; South Bend-Elkhart, IN; Sussex Co. ,DE; Tampa-St Petersburg-Clearwater, FL; Walworth Co., WI; York, PA; Youngstown-Warren-Sharon, OH-PA

---

1. *Ozone transport regions*

Ozone transport regions are areas where interstate transport of air pollutants from one or more states contributes significantly to the nonattainment of the ozone NAAQS in one or more states. The CAAA establishes one such region by law, which includes Connecticut, Delaware, Maine, Maryland, Massachusetts, New Hampshire, New Jersey, New York, Pennsylvania, Rhode Island, Vermont, and the District of Columbia. Within the ozone transport area, volatile organic chemical (VOC) reduction targets are established and maintained. Here, for instance, a stationary source that potentially emits 50 TPY or more VOCs is classified as a major source and is subject to the requirements of major sources in Moderate Nonattainment areas. Similarly, NO_x reduction targets are set for major sources in the ozone transport region.

Table 13.12b
Nonattainment Area Summary - Carbon Monoxide

Serious Carbon Monoxide Nonattainment Area - 1
LA (South Coast Air Basin)

Moderate Carbon Monoxide Nonattainment Areas - 41
Albuquerque, NM; Anchorage, AK; Baltimore, MD; Boston, MA; Camden-Philadelphia, PA; Chico, CA; Cleveland, OH; Colorado Springs, CO; Denver-Boulder, CO; Duluth, MN; El Paso, TX; Fairbanks-North Star, AK; Fort Collins, CO; Fresno, CA; Grant's Pass, OR; Hartford-New Britain-Middletown, CT; Klamath Falls, OR; Lake Tahoe-South Shore, CA; Las Vegas, NV; Longmont, CO; Medford, OR; Memphis, TN; Minneapolis-St. Paul, MN; Missoula, MT; Modesto, CA; New York-Long Is., NY-CT-NJ; Ogden, UT; Phoenix, AZ; Portland-Vancouver, OR-WA; Provo-Orem, UT; Raleigh-Durham, NC; Reno, NV; Sacramento, CA; San Diego, CA; San Francisco-Oakland-San Jose, CA; Seattle-Tacoma, WA; Spokane, WA; Stockton, CA; Syracuse, NY; Washington, D.C.-MD-VA; Winston-Salem, NC

Table 13.12c
Nonattainment Area Summary - Particulate Matter

Serious Particulate Matter Nonattainment Areas - 5
Coachella Valley, CA; Las Vegas, NV; Owens Valley, CA; San Joaquin Valley, CA; South Coast Air Basin (LA)

Moderate Particulate Matter Nonattainment Areas - 41
Ada Co., ID; Allegheny Co.-Liberty-Lincoln-Glassport boroughs-Clairton, PA; Anchorage, AK; Archuleta Co., CO; Adams-Denver-Boulder Co., CO; Arostook Co., ME; Bannock-Power Co., ID; Bonner-Shoshone Co., ID; Brooke Co., WV; Cook-LaSalle Co., IL; Cuyahoga Co., OH; Dona Ana Co., NM; El Paso Co., TX; Flathead-Lincoln-Lake-Missoula Co., MT; Fremont Co., CO; Inyo-San Bernardino-Kern-Mono-Stanislaus-Madera Co., CA; Jackson-Josephine-Klamath Co., OR; Jefferson Co., OH; Juneau, AK; King-Pierce-Thurston Co., WA; Lake Co., IN; Lane Co., OR; Madison Co., IL; Olmstead Co., MN; Pitkin Co., CO; Prowers Co., CO; Ramsey Co., MN; Riverside (east)-San Bernardino (part) Co., CA; Rosebud Co., MT; Salt Lake-Utah Co., UT; San Miguel Co., CO; Santa Cruz-Pima-Maricopa-Pinal-Gila Co., AZ; Sheridan Co., WY; Silver Bow Co., MT; Spokane Co., WA; Union Co., OR; Walla Walla Co., WA; Washoe Co., NV; Wayne Co., MI; Yakima Co., WA; Yuma, AZ

Cities failing for ozone are ranked from marginal to extreme, as seen in Table 13.12a. The more severe cases are expected to institute more rigorous control programs for VOC sources. States may have to initiate or upgrade inspection and maintenance programs. Vapor recovery may have to be installed at gasoline filling stations in ozone nonattainment areas in order to reduce hydrocarbon emissions (which are photochemically reactive and produce ozone) from even the smallest sources. Major stationary sources of NO_x (fossil fuel burners) in the ozone nonattainment areas also have to reduce emissions in order to bring ambient ozone levels into compliance.

Areas failing for carbon monoxide are ranked from moderate to serious. States may have to initiate or upgrade by enhanced enforcement vehicle inspection and maintenance programs and also adopt transportation controls for some cities in these areas. Improved inspection and maintenance machines are used to measure the pollution released by a car when it is being driven versus idling at a test station. Congress and the EPA expect emission inspection and maintenance programs to pay off in reduction of air pollution from cars. These areas are also having to require the car manufacturers to provide cars that burn fuel cleaner. Other measures are high occupancy vehicle (hov) lanes on freeways and surcharges on parking fees for one-occupant vehicles.

Areas failing to attain NAAQS for PM, specifically PM-10, are ranked moderate. The states involved will have to force industry to implement RACT, such as particulate scrubbers, cyclones, baghouses, electrostatic precipitators, and other industrial control devices, and also curtail residential use of wood stoves and fire places.

The primary NAAQS are established to protect public health. Secondary NAAQS have the goal of protecting the public welfare by protecting materials and vegetation. Each state is responsible for setting the source standards which will allow the state to comply with the NAAQS. The state is required to develop a State Implementation Plan (SIP) (see Table 13.13) consisting of regulations that establish methods for attaining NAAQS. The EPA is the approval authority for SIPs. When preparing an application for a permit, whether for a new one or for renewal of an existing permit, make sure you understand the air quality goals of your state in its SIP.

D. New Source Performance Standards (NSPS)

New Source Performance Standards (NSPS) are established by the EPA in 40 CFR 60. Under Section 111 of CAAA, industrial facilities are subject to standards for their particular industrial category, if the facility is constructed or modified after the date on which EPA proposed the NSPS. These standards are technology forcing and based on best demonstrated technology systems. NSPS reflect economically achievable controls and apply in all states evenly, in order to discourage industries from shopping for sites with less stringent control technology requirements.

Table 13.13
State Implementation Plans

Regulations Establishing Methods for Attaining NAAQS:

- A description of air quality
- Inventory of sources
- Compliance schedule for attaining NAAQS
- Description of state permitting process
- Monitoring and reporting requirements for permittees
- Enforcement procedures

Each industrial category assigned an NSPS is given concentration limits for various criteria pollutants and monitoring and reporting requirements. Continuous or periodical monitoring is specified. In-line continuous monitoring, which is quite expensive, is the preferred method but not always feasible. Reporting requirements include anticipated startup date, actual startup date, physical or operational changes which may increase emissions, monitoring results, date for conducting opacity observations, malfunctions, and notification of inoperative monitoring devices, with estimated date of repair.

E. Permits

Permits, until recently, have been a state function. Each state has two types of permits to control sources: permits to construct and operating permits. States typically base permit requirements on the information listed in Table 13.14.

Permits may be issued for an entire facility as a stationary source or for an individual source within the facility. Either way, the estimation of emission rates (amount and type) of each pollutant must be determined by one of the methods listed in Table 13.15.

Table 13.14
Permit Information Requirements

Amount of each pollutant
Type of pollutant
Air pollution control equipment

Table 13.15
Methods for Estimation of Emission Rates

Continuous emission monitoring system
Emission factors
Mass balance
Stack test

1. *Emission estimation*

Only the emission factors and mass balance methods are available for estimating emissions for plants in the design phase. Emission factors are real averages related to the quantity of pollutants released to the atmosphere by the activity associated with the release of that pollutant. An emission factor is an emission rate expressed as the weight of pollutant divided by a unit production or raw material consumption rate, such as pounds of PM emitted per ton of coal combusted. Using emission factors published by the EPA for various processes permits the estimation of emissions from similar sources. The emission factor may be an average of all available data of acceptable quality, but generally the influence of process parameters such as temperature, pressure, humidity, raw material quality, reactant concentrations, and others are not considered. Therefore, the emissions calculated by emission factors, while based on real-world data, are theoretical since many conditions could vary from those that produced the original data and the calculated results are likely to differ from the estimated facility's actual emissions. In some cases, such as in the estimation of VOC emissions from storage tanks, emissions are determined by applying empirical formulas that relate emissions to variables such as tank diameter, bulk liquid temperature, and wind velocity. Emission factors computed thusly yield more precise estimates than do factors derived from broad statistical averages. Multiply a production or consumption rate by the appropriate emission factor and the result is the emission rate (usually in tons per year).

Sample Calculation: The combustion of distillate oil (D.O.) in an industrial boiler produces carbon monoxide, among other emissions. Generally,

CO emissions (kg/day) = emission factor x kg D.O./day

According to AP-42, the CO emission factor for industrial boilers is 0.6 kg/1000 L D.O. If our industrial boiler will burn 150,000 L D.O./day, how much CO will it emit?

CO = 0.6 kg/1000 L D.O. X 150,000 L D.O./day = 90 kg CO/day

Where emission factors are used to predict emissions from new or proposed sources, review the latest process literature and technology to determine whether such sources are likely to have different emission characteristics from those of typical sources averaged in AP-42.

Mass balances are a process engineering technique which can arrive at an emission estimate mathematically. A process flow sheet listing each stream, components, temperatures, pressures, and flow rates is required.

This information can be retrieved from plant files, from the process designer's files or from the vendor of a particular piece of equipment.

2. Permits to construct

A *permit to construct* allows construction and installation of a new source to proceed. The permit is not issued until the application is thoroughly reviewed, the source has been modeled for impact on ambient air quality, and, in many cases, the public has had time to comment. Construction may not proceed until the permit is in hand. The professional engineer (PE) who certified the construction application must determine, near the completion of construction, that the emissions source(s) has (have) been installed in accordance with approved plans and send a certification to the permitting authority. Once the certification of construction is received and a verification inspection by the permitting authority is satisfied, the authority will issue a Permit to Operate. Performance testing is required within 60 days after achieving maximum production rate, but not later than 180 days after startup. These requirements may vary in some permit jurisdictions.

Tests, at any rate, must be conducted in accordance with approved methods. Many states have been so backlogged on operating permit applications that they never get around to issuing them. In other states it may take from 6 to 18 months but the permit will eventually get issued.

3. Title V operating permits

The 1990 Amendments strengthen the ability of the EPA and the states to enforce air pollution standards by requiring that all air pollution control obligations of a single source be covered by one 5-year operating permit. The states have set up EPA-approved programs to administer the Federal Title V Operating Permits. Stationary sources pay annual fees to the states to provide the funds to operate these programs in an effective and timely manner. Now, one permit for each facility lists all its air sources and permits can theoretically be issued in a more timely fashion.

The CAAA establishes an application shield which did not exist at the federal level before. If an existing facility submits an application in good faith, which is still being processed when new regulations governing those specific sources become effective, then the facility can continue operating as if the public had commented and a permit had been issued. However, facilities that do not submit applications in a timely manner or that contribute to the delay in permit issuance are not shielded and would be in violation of the law, if they continue to operate. Another CAAA concept is the permit shield, which amounts to: compliance with your public reviewed permit is in compliance with the law. However, if a source by itself, or a group of sources, contributes to an imminent and substantial endangerment to public health or welfare, the EPA may force the source(s) to cease the emissions or curtail them, regardless of permit.

CAAA requires that the achievement of SIP be taken into account upon permit renewal. Also, if a facility fails to consistently comply with its permit, the permit issuance and public participation process can be reopened. If a good-faith effort was not made in submitting information or if a change in the source process occurs, the permit may also be reopened. Anytime it is reopened, the shield may be negated and the public is invited back to the negotiating table.

F. Prevention of Significant Deterioration

Prevention of Significant Deterioration (PSD). The goals of the PSD program are to 1) ensure that economic growth occurs in harmony with the preservation of existing clean air areas, 2) protect the public health and welfare from any adverse effect that might occur, even at air pollution concentrations better than the NAAQS, and 3) preserve, protect, and enhance the air quality in areas of special natural recreational, scenic, or historic value, such as national parks and wildlife areas. Requirements for PSD review are pollutant specific for new *major stationary sources* and *major modifications* to major stationary sources before construction beings in an attainment area. The PSD process is part of a broader process called a New Source Review (NSR), the goal of which is to preserve the attainment of the NAAQS. Three categories of PSD areas are summarized in Table 13.16 with their implications.

These designations have been confused with nonattainment areas. All three PSD area categories are in attainment by definition but have different goals for maintaining their attainment status. In attainment areas, PSD is used for NSR; while in nonattainment, areas a nonattainment area (NAA) permit is used. A single source may need a PSD review for one pollutant but an NAA review for another.

Table 13.16
PSD Areas

Class I - Attainment Areas to be Preserved as Pristine Areas
Such as national parks and some international border areas: only minor emissions increases allowed
Class II - Other Attainment Areas
Most of the rest of U.S. attainment areas Emissions increases can accommodate industrial growth
Class III - Attainment Areas so Clean as to be Pristine
No areas have been designated Class III to date Larger emissions increases could theoretically be tolerated

1. *Stationary sources*

A *major stationary source* belongs to any of 28 categories (Table 13.17) and has the potential to emit 100 tons per year (TPY) or more of any pollutant subject to regulation, or any other source that has the potential to emit 250 TPY or more.

Table 13.17
PSD Source Categories

1. Fossil fuel-fired steam electric plants, >250 MBTU/hr
2. Coal cleaning plants with thermal dryers
3. Kraft pulp mills
4. Portland cement plants
5. Primary zinc smelters
6. Iron and steel mills
7. Primary aluminum ore reduction plants
8. Primary copper smelters
9. Municipal incinerators, >250 TPD refuse
10. Hydrofluoric acid plants
11. Sulfuric acid plants
12. Nitric acid plants
13. Petroleum refineries
14. Lime plants
15. Phosphate rock processing plants
16. Coke oven batteries
17. Sulfur recovery plants
18. Carbon black plants (furnace type)
19. Primary lead smelters
20. Fuel conversion plants (coal-to-gas)
21. Sintering plants
22. Secondary metal production plants
23. Chemical process plants
24. Fossil fuel boilers (or combinations) >250 MBTU/hr
25. Petroleum storage/transfer
26. Taconite ore processing plants
27. Glass fiber processing plants
28. Charcoal production plants

A *stationary source* includes all emitting activities for an industrial grouping that are located on contiguous or adjacent properties and under common control. A *major modification* is a physical change or a change in method of operation that results in a contemporaneous significant net emissions increase of a regulated pollutant. A net emissions increase is the sum of the increase in *actual* emissions from a *particular* modification or new source, and any other increases and decreases in actual emissions that are contemporaneous with the particular change and has a positive value when the following equation is used:

Net Emissions Change
EQUALS
Emissions *increases* associated with the proposed modification
MINUS
Source-wide creditable contemporaneous emissions *decreases*
PLUS
Source-wide creditable contemporaneous emissions *increases*

Significant emission rate changes are summarized in Table 13.18.

Table 13.18
Significant Emission Rate Changes

Pollutant	Allowable Emissions (TPY)
Carbon monoxide	100
Nitrogen oxides	40
Sulfur dioxide	40
PM - total	25
PM_{10}	15
Ozone (rate given for VOCs)	40
Lead	0.6
Asbestos	0.007
Beryllium	0.0004
Mercury	0.1
Vinyl chloride	1
Fluorides	3
Sulfuric acid mist	7
Hydrogen sulfide	10
TRS (total reduced sulfur compounds-includes hydrogen sulfide)	10
Benzene	Any
Arsenic	Any
Radionuclides	Any
Radon-222	Any
Polonium-210	Any
CFC-11, -12, -112, -114, -115	Any
Halon 1211, 1301, 2402	Any

To be *contemporaneous* with a proposed new source, changes in actual emissions must have occurred after January 6, 1976, and within the period starting 5 years before the date construction is expected to commence and ending when the emission increase from the modification occurs.

Sample Calculation: Your company plans to construct a new nitric acid unit, to be called Unit #7, and your management wants to know if it can avoid a costly PSD review. The plant is already a major NO_2 source and Unit #7 constitutes a modification to the source. Unit #7 will have the potential to emit 90 TPY of NO_2 when operational. PSD significant emissions increase level for NO_2 is 40 TPY. The history of existing nitric acid units are shown in Example Table 13.1.

Example Table 13.1

Emission Unit #	History
1, 2, & 5	Built in 1975 and still operational
3 & 4	Built in 1975; operations ceased 01/91
6	PSD permitted; construction commenced 01/91 and was operational on 01/92
7	New modification; construction scheduled to commence 01/95 and unit is expected to be operational 01/97

Units #1 and #2 are permitted to emit 150 TPY SO_2 each and are not to be physically modified, but the planned modification will remove a bottleneck that will increase the operational capacity of these acid plants.

From company records, you compile actual emissions that are shown in Example Table 13.2 for the years 1988 through 1994.

Example Table 13.2–Actual Emissions (TPY)

Year	Unit #1	Unit #2	Unit #3	Unit #4	Unit #5	Unit #6
1988	80	135	70	95	60	0
1989	85	130	85	85	70	0
1990	90	150	75	90	75	0
1991	120	100	0	0	80	0
1992	125	95	0	0	85	85
1993	115	85	0	0	75	80
1994	100	100	0	0	70	75

The contemporaneous period for this modification extends from 01/90, 5 years prior to 01/95 when the modification is scheduled to commence, until 01/97, when the plant is supposed to be operational. Preliminary talks with the permitting authority reveal that the new unit will be able to get a permit restricting NO_2 emissions to 90 TPY, pending a satisfactory PSD review. Therefore, the potential to emit for Unit #7 is 90 TPY and taken by itself would trigger a PSD review.

If you restrict operations of units #1 and #2 to 42 weeks per year, will you be able to "net out" of a PSD review?

Units #1 and #2 will not be modified but their emissions will increase, so this increase must be included as part of net increase for the proposed modification.

Unit change = new allowable - old actual
Hours of operation = 42 x 7 x 24 = 7056
Unit #1 change = 150 (7056/8760) - (115 + 100)/2 = +13.3 TPY
Unit #2 change = 150 (7056/8760) - (85 + 100)/2 = +28.3 TPY

The shutdown and decommissioning of Units #3 and #4 occurred during the contemporaneous period. The retirement of these units was not relied upon for the previous PSD permit, so the change is creditable for netting increases for this modification.

Unit #3 change = 0 - (85 + 75)/2 = -80 TPY
Unit #4 change = 0 - (85 + 90)/2 = -87.5 TPY

Unit #5 will not be affected by the modification and its emissions will not change.

Unit #6 was allowed under PSD and was considered when issuing a PSD permit that is still in effect; therefore, its increase is not creditable for this modification and is not used here.

Net change = #1 + #2 + #3 + #4 + #7
= +13.3+28.3-80-87.5+90 = -35.9 TPY

A PSD review will not be required. Your management team, upon learning of this, asks you to explore the possibility of leaving Unit #4 as a standby unit. This amounts to not accepting federally enforceable limits (retirement) on this plant, but management still wants to avoid PSD review.

Alternative net change = #1 + #2 + #3 + #7
= +13.3+28.3-80+90 = +51.6 TPD
Inform your management that the alternative plan will require a PSD analysis.

Table 13.19 shows what an applicant must do to obtain a PSD permit.

Table 13.19
PSD Requirements

1. Apply best available control technology (BACT)

2. Conduct ambient air quality analysis

3. Analyze impact on soil, vegetation, and visibility

4. Not adversely affect a Class I area

5. Survive adequate public participation

Table 13.20
Some Technologies Available for BACT Options

Generally Available Options
Existing controls for Source Category
Existing controls for similar Source Category
Innovative controls
CO BACT Options
Catalytic oxidation
Good combustion operating/maintenance practices
Thermal oxidation
NO$_x$ BACT Options
Flue gas recirculation
Low NO$_x$ burners
Selective catalytic reduction
Selective noncatalytic reduction
Water/steam injection into stack gas
PM BACT Options
Centrifugal filtration (cyclone)
Electrostatic precipitation (wet or dry)
Fabric filtration (baghouse)
Venturi scrubbing
Wet scrubbing (tower)
Information Resources for BACT
Air & Waste Management Association Engineering Manual
EPA BACT Clearinghouse Bulletin Board
Equipment vendors
Technical journals
Trade journals

2. *Best available control technology (BACT)*

BACT analysis, conducted on a case-by-case basis, considers the energy, environmental, and economic impacts of achieving the maximum degree of emissions reduction achievable for the construction proposal. The BACT emissions limitations must meet any standard of performance that is applicable to the source as specified in 40 CFR 60 or 61. BACT varies from industry to industry and no pre-approved lists exist. Table 13.20 lists some (not all) technologies available that are commonly examined for potential BACT. The appropriate control technology is literally determined on a case-by-case basis for new or modified sources in an attainment area. For instance, in southern California–remember the South Coast AQMD is not in attainment but modified sources in serious, severe, and extreme nonattainment areas must meet either RACT or BACT–BACT for petroleum refineries includes the use of magnetically driven seal-less pumps to reduce fugitive emissions from leaking seals. Fugitives from open top storage tanks are reduced by the use of additives to lower the surface tension of the stored liquid. Tank covers are another type of BACT. Floating media, such as ping pong balls, can be used to cover the surface of a volatile liquid in order to satisfy BACT in some places. The EPA is required to issue technical guidance on what constitutes BACT, but this information is flexible as the source may demonstrate energy and economic impact as well as the environmental impact. In brief, BACT is typically negotiable, so start with the Regional EPA office and the state or AQMD air office for guidance before finalizing your plans for a new or modified source.

The ambient air quality analysis, often but not always conducted by the permitting authority for the applicant, must show that the new emissions will not violate NAAQS or an applicable PSD increment. Proposed emission increases may not impair visibility, or affect soils or vegetation adversely. The indirect effects of commercial, industrial, residential, or other development in connection with the proposed source must also be considered. If the emissions will have an impact on a Class I area, the Federal Land Manager and the manager directly responsible for the lands must be notified. The public must also have a chance to review and comment on the project.

3. Potential to emit

Potential to emit is the maximum capacity of a stationary source to emit a pollutant under its physical and operational design. Physical or operational limitations on a source's capacity to emit a pollutant are not considered unless the limitation or its effect on emissions is federally enforceable. For instance, even if an emitting process is only operated for a limited amount of time, you must calculate the emissions as if they occur continually (8760 hours/year), unless you agree to accept a permit condition limiting hours of operation. That means your company can be penalized by administrative procedure (without trial) if you are found to be operating at other than allowed times.

Sample Calculation: A manufacturing process operates intermittently for 3500 hours per year and produces 100 TPY PM-10. What is the source's potential to emit? No controls have been implemented to reduce PM. What are maximum hourly emissions?

The potential to emit is determined as follows:

100 TPY x (8760 hr/yr, 3500 hr/yr) = 250 TPY PM-10

Average emissions are:

100 TPY x 2000 lb/ton, 3500 hr/yr = 57 lb/hr (round to 60)
The same answer is obtained by (250 x 2000/8760).

But, what about maximum emissions? Startups, shutdowns, and other process swings and upsets can cause emissions to exceed the average, plus the definition of average is that 50% of the sum of the averaged values is greater than the average; so if you report the average only, and that number is used to establish your hourly allowable emission rate, you will be out of compliance at least half the time! Therefore, you must either establish the maximum hourly rate by measurement (emission testing), or by vendor information, or by historical data (operating logs or EPA emission factors), or by simply guessing. A guess with some scientific basis uses 2.57 times the average (99% confidence level, statistically). You may still be out of compliance sometimes, which is now trackable for those under continuous emission monitoring mandates.

Table 13.21 lists acceptable emission limitations for potential to emit calculations. For major sources, fugitive emissions, if quantifiable, are considered in permit application analyses. Fugitive emissions are those that are generated by some means other than combustion or evaporation. Common examples of fugitive sources are shown in Table 13.22. Sometimes, fugitives do not exit the plant through a stack, but this is not always the case, and should not be considered as defining fugitives, as a capture hood may easily be used in many cases.

Table 13.21
Potentially Acceptable Limitations on Potential to Emit[*]

1. Installation and operation of required air pollution control equipment at prescribed efficiencies.
2. Restrictions on design capacity limitations [under certain circumstances].
3. Restricted hours of operation.
4. Restrictions on types or amount of materials processed, combusted, or stored.

Limitations must be federally enforceable

Table 13.22
Some Fugitive Emission Sources

Acid Gases
Fuming acid plants
Particulate Matter
Basic oxygen furnace
Coal piles
Machining of metals
Metallurgical processes
Quarries
Road dust
VOCs
Leaky valves and flanges, rubber, or Plastic extrusion tanks

G. Stratospheric Ozone Depletion

In the stratosphere (9–31 miles above the Earth's surface), the photolysis of diatomic oxygen molecules produces ozone (triatomic oxygen molecules). A thin layer of ozone in the stratosphere protects the life on Earth from receiving harmful amounts of ultraviolet radiation. Researchers at the University of California at Irvine pointed out in 1978 that this ozone layer was being jeopardized by anthropogenic organo-halogen compounds such as chlorofluorocarbons (CFCs). Photodissociation of CFCs in the stratosphere produces chlorine radicals and chlorate molecules. Finally, the dissociation of N_2O produces NO. Each of these radicals further reacts with ozone to produce fully oxidized species, consuming ozone.

Each year, NASA compiles ozone concentration data for the stratosphere and reports to the international scientific community. A genuine problem is universally recognized although causes and solutions are not universally agreed upon. The international community has responded by agreeing to the conditions of the Montreal Protocol of 1987 and the London Accords a year later. A timetable for phasing out the offensive anthropogenic chemicals was devised and agreed to. In 1990, after NASA reported an unusual amount of ozone depletion over the northern hemisphere, President Bush issued an abbreviated schedule of phaseout for U.S. industry. In Table 13.23, the U.S. (expedited) timetable for the phaseout of CFCs is shown. A *Safe Alternatives Policy* mandated by Title VI of the CAAA tasks EPA to provide information about alternatives to the regulated community.

Table 13.23
U.S. Timetable for Phaseout of
Ozone-Depleting Substances (ODS)

Class I ODS (Halons only) - January 1, 1994
Halons (a generic branch of Class I ODSs) are fire extinguishment agents, but are among the most potent ODS.
Class I ODS (remainder) - January 1, 1996
Includes CFC solvents, aerosols, and foaming agents, carbon tetrachloride, and methyl chloroform (1,1,1-trichloroethane).
Class II ODS (potent HCFCs) - January 1, 2003
HCFCs with most severe ozone-depleting effects, currently used as substitutes for CFCs.
Class III ODS (less potent HCFCs) - January 1, 2030
All other HCFCs.

H. Hazardous Air Pollutants

The EPA also directly establishes regulations to control the emissions of certain *hazardous air pollutants* (called *air toxics*) and several categories of new industrial sources. The CAAA requires the reduction of 189 air toxics within 10 years. Air toxics are typically carcinogens, mutagens, reproductive toxins, neurotoxins, or are suspected of being one of these. Generally, ambient standards do not work well for managing air quality with respect to these pollutants, so they are regulated separately, sometimes as individual toxic compounds, sometimes as generic groups.

One of the chief differences between criteria pollutants and HAPs is that the criteria pollutants have well-established threshold doses, whereas the high emotional interest in HAPs is based on their having no threshold, or, at least, no verified threshold. HAPs are also more likely to bioaccumulate. Although the lungs are the major route of entry for HAPs, many target organs are affected by HAPs as a class. To understand the prevalence of these pollutants in our society, one merely has to examine the EPA Annual Toxics Release Inventory to see that they are emitted by a variety of industrial categories and represent some of the heftiest emitters.

1. *Maximum achievable control technology*

The EPA is required to develop technology-based standards for listed source categories. These standards, called *MACT* for *Maximum Achievable Control Technology*, include process changes, substitution of raw materials, equipment design specifications, designated work practices, as well as emission capture and control. MACT for new major sources is at least as stringent as the emission levels achieved in practice from a major source in the same category. For existing major sources, MACT means either 1) the average emissions controls achieved by best performing 12% of similar sources, or 2) the average emissions controls achieved by the five best-performing sources in the category, if the category has fewer than 30 sources. Small sources, called *area sources*, are either required to implement MACT for their category or *generally available control technology* (GACT).

Table 13.24a
NESHAP Categories

November 15, 1992

Commercial dry cleaners - dry to dry machines
(perchloroethylene)
Commercial dry cleaners - transfer machines
(perchloroethylene)
Industrial dry cleaning - dry to dry machines
(perchloroethylene)
Industrial dry cleaning - transfer machines
(perchloroethylene)
Synthetic organic chemical manufacturing

December 31, 1992
Coke oven

The EPA is required to publish a list of the new source categories and issue MACT for the control of air toxic emissions from these sources. The new standards will include the control of fugitive emissions, which have been largely unregulated in the past. A timetable for compliance will be included as part of each standard.

Table 13.24 lists the categories of industries subject to NESHAPs. Companies that implement partial controls, before the deadlines set for MACT for their category, can receive extensions for full implementation. This is called the "credit for early reduction" program.

Health-based standards aim at preventing human health effects in the exposed or potentially exposed population. An epidemiologic model is used in conjunction with typical dispersion models to predict whether physiological effects will be experienced by exposed individuals.

Table 13.24b
NESHAP Categories for November 15, 1994

Acrylonitrile-butadiene-styrene production
Aerospace industries
Asbestos processing
Butyl rubber production
Chromic acid anodizing
Commercial sterilization
Decorative chromium electroplating
Epichlorohydrin elastomers production
Epoxy resins production
Ethylene-propylene elastomers
Gasoline distribution (stage 1)
Halogenated solvent cleaners
Hard chromium electroplating
Hypalon (TM) production
Industrial process cooling towers
Magnetic tapes
Methyl methacrylate-acrylonitrile-butadiene-styrene
Methyl methacrylate-butadiene-styrene terpolymers
Neoprene production
Nitrile-butadiene rubber production
Non-nylon polyamide production
Petroleum refineries (all others)
Polybutadiene rubber production
Polyethylene terephthalate production
Polystyrene production
Polysulfide rubber production
Printing/publishing
Secondary lead smelting
Shipbuilding/ship repair
Solid waste treatment, storage, disposal facilities (TSDF)
Styrene-acrylonitrile production
Styrene-butadiene rubber and latex production
Wood furniture

Table 13.24c
NESHAP Categories for November 15, 1995

Publicly owned treatment works (POTW)

Table 13.24d
NESHAP Categories for November 15, 1997

Acetal resins production	Hydrazine production	Polyvinyl acetate emulsions production
Acrylic fibers/modacrylic fibers	Hydrochloric acid production	Polyvinyl alcohol production
Aerosol can-filling facilities	Hydrogen cyanide production	Polyvinyl butyral production
Amino resins production	Hydrogen fluoride production	Portland cement manufacturing
Autos and light duty trucks	Iron and steel manufacturing	Primary aluminum production
Benzyltrimethylammonium chloride production	Iron foundries	Primary copper smelting
Butadiene dimers production	Mineral wool production	Primary lead smelting
Carboxymethylcellulose production	Municipal landfills	Pulp and paper production
Cellophane production	Nylon 6 production	Rayon production
Chelating agents production	Oil and natural gas production	Reinforced plastic composites production
Chlorine production	Paper and other webs	Rubber chemicals production
Chromium chemicals manufacturing	Petroleum refineries (cat-crackers; cat-reformers; sulfur)	Secondary aluminum production
Chromium refractories production	Pharmaceutical production	Semiconductor manufacturing
Cyanuric chloride production	Phenolic resins production	Sewage sludge incineration
Electric arc furnaces - nonstainless steel	Phosphate fertilizer plants	Sodium cyanide production
Electric arc furnaces - stainless steel	Phosphoric acid manufacturing	Stationary internal combustion engines
Ferroalloys production	Photographic chemical manufacturing	Stationary turbines
Flexible polyurethane foam production	Polycarbonates production	Steel foundries
	Polyesters production	Steel pickling - HCl process
	Polyether polyol production	Wood treatment
	Polymethyl methacrylate resins production	Wool fiberglass manufacturing

Table 13.24e
NESHAP Categories for November 15, 2000

Alkyd resins production	Large appliances
Alumina processing	Lead acid battery manufacturing
Ammonium sulfate-caprolactam byprod-	Lime manufacturing
uct plants	Maleic anhydride copolymers production
Antimony oxides manufacturing	Metal cans
Asphalt/coal tar application	Metal coils
Asphalt concrete manufacturing	Metal furniture
Asphalt processing	Metal parts and products, miscellaneous
Asphalt roofing manufacturing	Methylcellulose production
Baker's yeast manufacture	OBPA/1,3-diisocyanate production
Boat manufacturing	Organic liquids distribution (nongasoline)
Butadiene-furfural cotrimer	Paints, coatings, adhesives
Captafol production	Paint stripper users
Captan production	Phthalate plasticizers production
Carbonyl sulfide production	Plastics parts and products
Cellulose ethers production	Plywood/particle board manufacturing
Cellulose food casing manufacture	Polyvinyl chloride and copolymers
Chlorinated paraffins production	Primary magnesium refining
4-Chloro-2-methylphenoxyacetic acid	Printing, coating, and dyeing of fabrics
production	Process heaters
Chloroneb production	Quarternary ammonium compounds pro-
Chlorothalonil production	duction
Clay products manufacturing	Rocket engine test firing
Coke byproduct plants	Site remediation
2,4-D salts and esters production	Sodium pentachlorophenate production
Dacthal (TM) production	Spandex production
4,6-Dinitro-o-cresol production	Symmetrical tetrachloropyridine production
Dodecanedioic acid production	Tire production
Dry cleaning (petroleum)	Tordon (TM) acid production
Engine test facilities	Uranium hexafluoride production
Ethylidene norbornene production	Vegetable oil production
Explosives production	
Flat wood paneling	
Fume silica production	
Hazardous waste incineration	
Industrial boilers	
Industrial/commercial boilers	

I. Acid Rain

The Amendments of 1990 also made provisions for control of acid deposition (also called acid rain or acid precipitation). That forest, crop, and city alike are being adversely affected by acidic rain is undisputed. Sulfur dioxide, one of the acid precursors, damages buildings, statues, and other concrete or rock structures by corrosion. Many times, the problem is international in scope with heavy emitters of acid precursors on one side of a border and the affected areas on the other side (U.S.–Canada; U.S.–Mexico; Scandinavia–Germany). What is not clear is nature's own role, as scientists have found evidence of acidic pond sludges at a depth that suggests the same sort of acid rain fell in the 600 to 700 A.D. era. Nevertheless, the present damage is obvious and the sources are well understood. Therefore, Congress acted to control acid rain in 1990. The U.S. response to acid deposition, per CAAA, is to place emission limits on fossil-fuel burning electric power generators which are by far the category of acid emitters with the largest volume of emissions. A two-phased, market-based system, which is expected to reduce emissions by 50%, has been devised for control of sulfur dioxide emissions, the major pollutant, from power plants. By the year 2000, total annual emissions of sulfur dioxide are to be capped at 8.9 million tons, a reduction of 10 million tons from 1980 levels. Power plants will be given emission allowances based on fixed emission rates set by law, based on previous fossil-fuel use. Each allowance is worth 1 ton of sulfur dioxide smoke stack emissions. Allowances are set below the current level of releases in order to force emission reductions. Utilities will pay penalties if emissions exceed the allowances held, but these may include unused allowances obtained legally from other power plants. Unused allowances may be banked or sold. Bonus allowances will be given to power companies for installing *clean coal technology*, using renewable energy sources (solar, wind power, ocean thermal currents, tidal currents, geothermal), or by encouraging energy consumers to conserve energy.

In Phase I, large, high-emission utilities, located mainly in the eastern and midwestern states, must achieve reductions by 1995. Phase II, which commences on January 1, 2000, imposes limits on smaller, cleaner utilities as well as tightens limits on Phase I plants. Oxides of nitrogen emissions must also be reduced by the utilities; but instead of emissions allowances similar to those just explained for SO_2, the EPA will set performance standards for NO_x.

All acid rain sources must install continuous emission monitoring devices to track and assure compliance with standards and allowances. Penalties will be stiff for those facilities that exceed allowances for SO_2 or emission standards for NO_x. Acid rain provisions of the CAAA are to be incorporated into the new federal operating permits.

III. CONTROL TECHNIQUES

After the best efforts at waste minimization and pollution prevention within the typical industrial plant; some residual remains that can add no value to the product, a waste stream; and, where this residual is borne by a compressible fluid stream, air pollution needs to be controlled. The industrial hygienist can assist the plant engineering team by collecting information about the gaseous effluents of the plant. Figure 13.1 shows an industrial hygienist measuring pollutant discharge concentrations and control apparatus effectiveness for rooftop exhaust stacks. Typical information collected by the industrial hygienist is included in Table 13.25.

Figure 13.1 Industrial hygienist measures stack emissions using a portable infrared ambient air sampling instrument (Courtesy: *Foxboro Environmental*, East Bridgewater, MA).

Table 13.25
Air Pollution Control Data

1. Physical properties of the effluent contaminants. Size, shape, density, and size spectrum.
2. Chemical properties of the effluent contaminants. Chemical composition, corrosiveness, humidity.
3. Process factors: volumetric flow rate, velocity, temperature, and pressure.
4. Facility factors: equipment size, plant layout, materials of construction, and safety requirements.
5. Operational factors: maintenance costs, utility costs, and disposal costs.

A. Control of Particulate Matter

Particulate matter (PM) is removed from air streams by either air filtration or by dust collection. Small amounts of PM are removed by filtration while dust collectors remove PM from heavily-laden, large volume streams. In order to properly select dust removal equipment, the criteria listed in Table 13.26 must be provided to the designer.

Table 13.26
Criteria for Dust Collector Selection

Concentration of particulate in stream
Particle size analysis of particulate
Degree of particulate removal required
Temperature, pressure, and flow rate of gas stream
Characterization of contamination
Utility requirements
Preferred method of disposal of collected particulate

1. *Gravity settling chamber*

The oldest, least efficient type of dust collection is the *gravity settling chamber* consisting of a large cross-sectional area. As an air stream enters the large chamber from a small-diameter duct, its velocity is reduced, allowing the inertia of the dust particles to cause them to settle due to gravity. The collection efficiency of a gravity chamber is expressed as a function of the terminal velocity of the particles ranging from 40 to 50 microns in aerodynamic diameter. Smaller particles are collected with very low efficiency.

$$\eta = \frac{100 \, U_{infinity} \, A_h}{Q} \tag{13.5}$$

Where:

   h = Collection efficiency, percent by weight
   U_α = Terminal velocity of settling particles, ft/min
   A_h = Horizontal area of gravity chamber, ft^2
   Q = Volumetric flow rate of gas stream, ft^3/min

If the linear flow of the gas stream exceeds about 10 fps, the particles can be re-entrained by the gas.

Operating Tip:
To improve the collection efficiency of a gravity chamber, increase the effective horizontal area or reduce the volumetric flow rate of the gas stream.

2. Cyclone

The mechanism of separation by a *cyclone* is centrifugal force, which depends on the difference in the density of the dust particle and the air that entrains it. As the path of entrained PM is changed, inertia separates it mechanically from the air stream. Once the PM collides with a wall, it drops by gravity, much as a gravity separator works, to a collection drum at the bottom of the cyclone. Cyclonic dust collectors are generally inexpensive, low maintenance, and create the lowest pressure drops of any control devices available. The particle size collected by cyclone separation is:

$$D_{pc} = \sqrt{\frac{9\mu b}{2\pi \, N_e \, V_i \, (\rho_p - \rho_g)}} \qquad (13.6)$$

Where:

D_{pc} = Diameter of particle collected at 50% efficiency
m = Gas viscosity, lb/ft-sec
b = Cyclone inlet diameter, ft
N_e = Number of turns within cyclone, 5 assumed
V_i = Inlet gas velocity, ft/sec
r_p = Particle density, lb/ft^3
r_g = Gas density, lb/ft^3

3. Impingement separators

Impingement separators, a variation on the gravity separator, use a baffle in a plenum chamber to remove PM by halting its movement with respect to the air stream. The efficiency of impingement is related to the percentage of particles that strike the stationary object, which is obtained graphically by plotting a dimensionless separation number, N_S, against efficiency.

$$N_S = \frac{D_p^2 \, V \, \rho_p}{18\mu \, D_b} \qquad (13.7)$$

Where:

D_p = Particle diameter, ft
V = Relative velocity gas to target, ft/sec
r_p = Particle density, lb/ft^3
m = Gas viscosity, lb/ft-sec
D_b = Target diameter, ft

4. Packed towers

Packed towers are deep beds of granular or porous materials in cylindrical towers. The packing changes the direction and velocity of the gas stream, allowing both particle inertia and impaction to remove entrained particles. The disadvantage of using a packed tower for PM is that the dust will eventually plug the packing. Extraordinary maintenance effort is typically required to prevent this from happening. Also, packed tower efficiency is sensitive to changes in volumetric flow. The mechanisms of removal of PM by packed towers are 1) impaction of particles on the bed material as the air stream enters the tower, 2) direct interception as remaining PM tries to travel through the bed with the air stream, and 3) diffusion as the air stream expands and loses velocity in fresh packing material which is relatively clean with respect to PM.

5. Electrostatic precipitators

Impingement separators typically require more electrical power than *electrostatic precipitators* (ESP) or fabric filters. However, ESPs can be used at higher air stream temperatures than baghouses can. Hot, moist gases, such as those emitting from combustion processes, are handled more effectively in ESPs or scrubbers than in baghouses (BH) because the filters quickly plug and the bags degrade fast in the heat and moisture. ESPs can drop efficiency quickly from 98% to around 75% without careful management. This drop represents around 1200% increase in emissions. ESP management includes periodic removal of the collected PM, called rapping the ESP. Some units must be shut down for mechanical cleaning of the plates. Sticky PM reduces collection plate efficiency. ESP efficiency is calculated as follows:

$$E = 100 - 100\ [exp(-AE_oE_pa)/(V2\pi\mu)] \qquad (13.8)$$

Where:

    E = Percent efficiency
    A = Surface area of collector electrodes, ft^2
    V = Volumetric flow rate, ft^3/min
    E_o = Charging field, volts/ft
    E_p = Collecting field, volts/ft
    a = Particle radius, ft
    m = Gas viscosity

6. Baghouses

Fabric filters or *baghouses* (BH), on the other hand, typically become more efficient at removing dry particulate as time passes. Open the door of a baghouse and you find why these filters got their name. Inside are a number of cloth bags held rigid by the flow of the gas stream, which passes either from inside out or from outside in through the fabric. As dust grains plug the fabric pores, the increased resistance of the filter more effectively removes PM from the air stream, until it becomes so plugged as to begin to hinder operations. However, it is the coating of particulate on the fabric (called the cake), not the fabric itself (which is merely a structure for the particulate coating), which does the filtering. Generally, the cake resistance, called *cake pressure drop*, increases in proportion to time and is independent of air flow except at the very beginning or very end of the fabric filter's life. Cake pressure drop expressed as a function of time and gas velocity is:

$$\Delta P_{cake}(t) = KCtV^2 \qquad (13.9)$$

Where:

   K = Constant
   C = Dust concentration in air stream
   t = Time
   V = Superficial velocity

and:

$$V = Q / A \qquad (13.10)$$

   Q = Volumetric flow rate
   A = Filter face area

Sample Calculation: If a baghouse collects an air stream at 1200 cfm with a dust loading of 10 grains per cubic foot, how long would it take to reach maximum permissible pressure drop at 20 grains per cubic foot loading? Assume that the initial resistance in both cases is 1" w.g. and that it takes 6 hours to reach the maximum permissible pressure drop of 4" w.g. under the original conditions.

    At twice the loading, the baghouse maximum permissible pressure drop will be reached in half the time: 3 hours. Remember, cake pressure drop increases in direct proportion to time, not gas flow. Since a given thickness of the fabric coating produces a given pressure drop, if the grain loading doubles and everything else is constant, the same pressure drop will be achieved in half the time.

    The pressure drop of fabric filtration is the sum of the resistances of the cloth and the filter cake, as per this relationship:

$$\Delta P_t = \Delta P_f - \Delta P_i = K_2 L_t V^2 \qquad (13.11)$$

Where:

   DP_t = Pressure drop at time t, lb_f/ft^2 (due to dust cake)

DP_f = Total filter resistance at time t, lb_f/ft^2
Dp_i = Initial resistance of clean filter, lb_f/ft^2
K_2 = Proportionality constant, $lb_f\ sec^2/lb_m\ ft$
L_t = PM concentration in carrier gas, lb_m/ft^3
t   = Accumulated time since cleaning, sec
V   = Superficial filtering velocity, ft/sec

This expression can be used along with a pressure drop measurement when the filter is clean and when the fabric is totally plugged, to predict time until the next cleaning or the impact of various PM concentrations.

B. Control of Acid Gases/Vapors
1. *Packed Towers*

Packed towers are excellent scrubbers for acid gases and vapors because of their compact designs, ability to receive relatively hot contaminated stream, and constant pressure drop. Absorbing acid gas in water or weak acid is a very effective process. Processes that rely on the solubility of a gas in water, or aqueous solution, are governed by *Henry's law*:

$$P_A = H\,X_A \tag{13.12}$$

Where:
P_A = Partial pressure of gas A
H   = Henry's Law constant
X_A = Mole fraction of gas A dissolved in the liquid

Generally, the contaminated gas is fed to the bottom of the tower and the fresh water or weak acid is fed at the top and allowed to drop down through the bed or plates.

Operating Tip:

To prevent flooding in a packed tower, increase the surface area or decrease the air flow rate.

C. Control of Toxic Gases/Vapors

1. *Recirculation*

Recirculation of scrubbed air is not recommended because wet scrubber effluents are saturated with moisture; few environments can tolerate this amount of moisture. Cleaned air can be safely recirculated when the contaminants being removed are nuisance hazards where the health risk is minimal.

2. *Source reduction*

Source reduction and pollution prevention can be as effective or more effective than control measures. Not emitting the air pollutant in the first place is far superior to cleaning the waste gas stream before it enters the atmosphere. Therefore, when planning a new source, modifying an existing source, or looking for compliance assurance, carefully examine the possibilities for source reduction or pollution prevention. You are referred to reference books on these topics for more information.

3. *Stack height and air pollution dilution*

Tall stacks are used for diluting air pollution in the vastness of the atmosphere. Generally, once the polluted gas stream gets above the inversion layer, it is mixed with such a large volume of air that the concentration is no longer a threat to human health or the environment. Dispersion, therefore, may be used as control, though this does not hold true for the acid gases, ODS, or HAP/VOC. Dispersion of a plume of stack gas is enhanced by the highest possible height of the stack. *Effective stack height* (h_e) is the vertical distance that a plume rises over a stack (called *plume rise* or h_r) plus the physical height of the stack (h'):

$$h_e = h_r + h' \qquad\qquad (13.13)$$

While the effective height is needed in order to complete modeling runs to check impact on ambient concentrations in the vicinity of the discharge, the plant engineers need to know how high to build the stack in the first place. Generally, stack gas exit velocities must be at least 60 feet per second (fps) in order to control plume rise. Natural draft produces exit velocities around 10 fps, so blowers are needed to achieve 60+ fps. Natural draft is produced by the buoyant lift of warm air and mathematically is a function of exit gas temperature, air temperature, stack height, and wind velocity over the top of the stack. The stack height creates two gradients that drive the stack gas flow: atmospheric pressure and air density differences at the top and bottom of the stack.

Neglecting wind velocity, the theoretical draft at the base of a stack of height h' is:

$$D_r = (h'/5.2)(\rho_{ca} - \rho_{HG}) \qquad (13.14)$$

Where, for h' in feet:

   D_r = Theoretical draft, inches of water

   ρ_{ca} = Density of cold air outside stack (0.0743 lb/ft³ at 60°F and 1 atm

   ρ_{HG} = Density of hot gas inside stack, lb/ft³

The net effective draft (D'$_r$) is determined by subtracting the effects of kinetic energy (KE) and friction losses (F'), both in inches of water, from the theoretical draft:

$$D_{r'} = D_r - KE - F' \qquad (13.15)$$

and where:

$$KE = (12\ \rho_{HG}/\rho_{H_2O})/(v^2/2g_c) \qquad (13.16)$$

Where:

   ρ_{H_2O} = Density of water, lb/ft³

   v = Stack exit velocity, fps

   g_c = Gravitational constant, 32.174 ft lb$_m$/sec²lb$_f$

Friction losses are calculated as:

$$F' = (12\ \rho_{HG}/\rho_{H_2O})/F \qquad (13.17)$$

For those turbulent or viscous flow systems where fluid density, viscosity, and linear velocity are constant (assume they are), the Fanning friction factor (F) is determined by:

$$F = 4f(v^2/2g_c)(L/D) \qquad (13.18)$$

The length (L) is the sum of the stack height, the duct length, and additional equivalent lengths for bends, enlargements, constrictions, and other turbulence creating structures in the duct. Often, the value of 50D is added to the sum of stack height and duct length to approximate. The friction factor (*f*) is a function of the Reynold's number (N_{Re}) and relative roughness of the duct or stack interior surface. Experimental data for *f* have been tabulated in engineering handbooks. For exit velocities of 60 fps or more, the stack gas flow will be turbulent, meaning the N_{Re} will be greater than 1,000,000; and for smooth stacks, the friction factor is estimated to be:

$$f = (0.04)/N_{re}^{0.16} \qquad (13.19)$$

Normally, stack gases lose 1°F for every foot of stack height. Sometimes, the temperature difference from stack entrance to exit can be sufficient to require the calculation of draft and stack height to be accomplished by computer in order to make the friction and kinetic energy calculations over small intervals using average temperatures for each interval. If entrance or exit constrictions or enlargements exist, the friction losses must be adjusted. Air pollution control devices are also sources of pressure drop (increased draft requirement) in tall stacks. A common practice is to add $3v^2/2g_c$ to the calculated value of F in order to account for these additional losses. The practice is to build the stack 190 times the D_r' in feet. Table 13.27 gives typical stack gas and related values for a combustion type of off-gas.

Table 13.27
Typical Gas Values for Tall Stacks

Stack Gas Constant Parameter	Value
Atmospheric pressure (sea level), in. Hg	29.92
Atmospheric (cold air) temperature, °F	62
Exit velocity, ft/sec (minimum)	60
Friction coefficient (f), rough concrete	0.016
Hot gas density @ 0°F/1 atm, lb/ft³	0.09
Hot gas temperature, °F	400 - 500

Some people equate air pollution with what they see exiting from a stack. Actually, visible clouds from stacks are independent of control device efficiency, meaning the visibility has little to do with pollution. Generally, visible clouds are 90% moisture.

The Gaussian dispersion equation is a generalized mathematical model giving ground level concentration for point sources in which the air pollution is dispersed along direction of wind:

$$C(x, y, z) = \frac{Q}{\pi \, \upsilon_x \sigma_y \sigma_z} \exp^{-\frac{y^2}{2\sigma_y^2} + \frac{h^2}{2\sigma_z^2}} \qquad (13.20)$$

Where:

    C   = Concentration at a point (x,y,z) from the stack

    Q   = Emission rate

    u_x   = Average downwind velocity

s_y = Lateral (crosswind) dispersion coefficient

s_z = Vertical dispersion coefficient

h   = Release height

IV. PLUME BEHAVIOR

Plume behavior includes coning, fanning, trapping, looping, lofting, and fumigation. When the atmosphere is neutral or slightly stable (dry air temperature decreases with altitude less than or equal to 1.7°F/1000 ft), the plume appears as a nearly perfect cone extending horizontally from the top of the stack. This is called coning and the plume grows in all directions the further away from the stack it gets, promoting good dilution of the stack gas. Looping occurs when the dry lapse rate is greater than 1.7°F/1000 ft (extremely unstable) in which the plume looks as if it is weaving up and down through the atmosphere as it gets further away from the stack. This is the best air mixing condition short of violent weather. In the case of an inversion (a warm temperature layer caps the local atmosphere), the plume has no tendency to move up or down, so it fans out (fanning) in a thin, flat layer in the direction in which the wind meanders, looking much like a fan from the stack. When the stack gas is released just above a mild inversion, the plume is in a lofting condition in which it grows more in the upward direction than earthward. Fumigation happens when the inversion is above the stack and the plume grows more in the earthward direction than upward. The plume is essentially reflected off the inversion and fumigates the ground level. A similar condition occurs in trapping in which the inversion over the stack has a weak lapse rate between it and the ground and a cone shaped plume is directed earthward.

A. Adiabatic Lapse Rate

Adiabatic lapse rate is a 5.5°F drop in temperature of moist air for each 1000 feet of elevation rise. A positive lapse rate means the air temperature is increasing with increasing altitude. Normally, under adiabatic conditions, the lapse rate is negative, but when the lapse rate is positive, an inversion exists and very little mixing of pollutants and air takes place. Therefore, pollutants tend to accumulate near the ground. Understanding plume conditions, lapse rates, and inversions allows you to instantly interpret how well stack gas is being mixed and dispersed.

V. AIR POLLUTION/EMISSION SAMPLING/MONITORING TECHNIQUES

A. Ambient Monitoring

While the naked eye can detect some gross pollution, especially PM, it is not a reliable monitoring device. *Smoke readers* can approximate the amount of pollution in a stack plume by estimating its opacity.

1. *Emission sampling and monitoring*

In order to get a representative sample of particulate matter from a stack, the sampling must ordinarily be performed *isokinetically*, meaning that the velocity of the air drawn into the probe, is equal to the velocity of the stack. If the velocity of the stack is greater than the velocity of the probe an excess of small particles will be collected by the probe because the air stream diverges around the probe tip. Particles that are unable to follow the stream line of the diverging air stream, the smaller particles, continue straight into the probe. However, isokinetic sampling is usually not necessary when the bulk of particles are 5 microns or greater in diameter. Figure 13.2 demonstrates the collection of duct velocities to determine appropriate sample collection velocities to ensure isokinetic sampling.

Sample Calculation: If a probe is 0.199 inches in diameter and the stack velocity is 3500 fpm, what probe flow rate is required to achieve *isokinetic sampling?*

Isokinetic sampling means that the same linear velocity moves through the probe as the stack.

$$\text{cfm} = 3500 \text{ fpm} \times (0.199 \text{ in}/12 \text{ in/ft})^2 \times \pi/4 = 0.756 \text{ cfm}$$

The number of velocity points required by EPA Method 1 ranges from 8 to 25 per diameter on each of two perpendicular diameters. The duct diameter and the proximity of upstream and downstream obstructions, which cause turbulence, must be known in order to select the specified number of points according to EPA instructions.

The EPA Method, "Determination of Stack Gas Velocity and Volumetric Flow Rate," does not require an actual sample. However, the stack diameter, velocity head, and barometric pressure are required in order to compute the actual flow rate.

In-stack sampling for particulates is typically preferred in order to avoid problems of aerosol deposition in the sampling lines. Nevertheless, in-stack systems may require heating to avoid condensation from supersaturated effluent streams and condensation on cold sampler surfaces.

Figure 13.2 Measuring duct velocities to determine appropriate sample collection velocities for stack sampling (Courtesy: *TSI Incorporated*, St. Paul, MN).

Orsat analysis is a field measurement of CO_2, CO, O_2, and H_2 in flue gas. Moisture cannot be determined since the method uses absorption of a water solution to quantify the four gases named.

VI. OTHER SIGNIFICANT LEGISLATION

Before the CAAA, handling of hazardous waste was strictly a concern of RCRA (the Resource Conservation and Recovery Act). The CAAA directed the EPA to investigate and regulate any air emissions from RCRA solid waste management units (SWMUs). The EPA has treated SWMUs as a source category and has developed appropriate standards under NSPS and NESHAPS. Emissions from CWA (Clean Water Act) wastewater treatment plants were another overlooked category. Again, the EPA now has appropriate air emissions standards under NSPS and NESHAPS for wastewater plants.

Global warming may be the pollution controversy for the twenty-first century. The greenhouse effect is a beneficial condition that maintains an average temperature on the Earth conducive to supporting life as we know it. Certain gases in the atmosphere such as carbon dioxide act as a temperature trap to retain some of the solar energy that is absorbed by the earth during daylight hours. No one ever suspected that the vast bulk of these greenhouse gases could ever be affected by the human race. Anthropogenic sources of greenhouse gases, however, are beginning to modify the average temperature by retaining more solar energy than heretofore captured. Whether or not the phenomena observed are true indicators of a climate change is a controversial issue among scientists and policy makers. Climate change, if real, would have beneficial as well as detrimental effects on the human race. While some food-producing areas of the planet would dry up and become desert or become covered by a rising ocean due to melting at polar caps, other areas that are not food producing now would become ideal for agriculture. Virtually all atmospheric oxygen is produced by photosynthesis and the overall coverage of the earth by green plants might not change much. Industrial hygienists should keep abreast of developments, as they might play a role in measurement of atmospheric gases and other data collection.

A. Air Pollution and the Shrinking Planet

Diesel engines produce more nitrogen oxides and PM but less carbon monoxide than gasoline engines. At highway speeds, diesels primarily produce nitrogen oxides as pollutants, as shown in Table 13.28.

Table 13.28
Typical Diesel Emissions

Pollutant	lb/ton #2 DO	Ambient Standard (mg/m^3)
Carbon monoxide	15	10,000
Hydrocarbon	15+	160
Nitrogen oxide, NOx	49	100
Particulate matter	34	75
Sulfur dioxide	10	80

Accidental chemical release. Another way that industrial hygienists are involved in air pollution is in the modeling of accidental chemical releases. Earlier, the Gaussian plume dispersion equation was discussed. This model can be used for accidental release by providing an appropriate source term (puff, or batch release, as opposed to continuous release). The EPA has several release models and several private engineering firms also have some excellent models. The industrial hygienist is in a position to provide knowledge about the health effects of the pollutants as well as understand source modeling and dispersion modeling, making communication with plant engineers and operation staff more cooperative and harmonious.

REVIEW QUESTIONS

1. The criteria pollutants include:

    a. Lead, particulate, chromium, volatile organics
    b. Lead, particulate, sulfur dioxide, volatile organics
    c. Lead, particulate, sulfur dioxide, ozone
    d. Lead, particulate, sulfur trioxide, ozone

2. A stationary source is:

    a. A parked vehicle
    b. A piece of process equipment that never moves
    c. An automobile or airplane
    d. An industrial plant or surface mine

3. The law called the Clean Air Act is:

    a. A specific law passed by Congress in 1963
    b. A specific law passed by Congress in 1971
    c. A body of laws passed over a 40-year period
    d. A body of laws passed between 1967 and 1975

4. Each stack at a new industrial plant must meet:

    a. National Ambient Air Quality Standards
    b. New Source Review
    c. Prevention of Significant Deterioration review
    d. Stratospheric Ozone Protection Standards

5. Nonattainment areas are classified as:

    a. Borderline, Moderated, Serious, Severe, Extreme
    b. Marginal, Moderate, Serious, Severe, Extreme
    c. Marginal, Moderate, Serial, Several, Extreme
    d. Marginal, Moderate, Serial, Severe, Enhanced

6. A situation that may force the source to cease or curtail emissions, regardless of permit conditions is:

    a. Operating beyond the life of the permit
    b. Failure of a control device
    c. An imminent and substantial endangerment to public health or welfare
    d. Loss of operating logs in a fire

7. A permit shield may be negated when:

    a. A good-faith effort was not made in submitting the application
    b. The source fails to consistently comply with the permit
    c. A process change is made
    d. Any of the above conditions apply

8. A primary pollutant:

    a.   Is emitted directly from a source
    b.   Comes from a primary source
    c.   Is produced by a chemical reaction with primary ozone
    d.   Can do more serious harm than secondary pollutants

9. Which of the following is a secondary pollutant?

    a.   Lead
    b.   Ozone
    c.   Volatile organic chemicals
    d.   Particulate matter

10. A source operates 16 hours per day, 5 days per week and emits an average 1.0 lb/hr PM for 50 weeks every year. The maximum emission rate is 3.0 lb/hr. What is the potential to emit for the source?

    a. 2.00 TPY   **b.** 4.38 TPY **c.** 6.00 TPY   **d.** 13.14 TPY

11. If the source described in Question 10 belongs to a stationary source category that requires a certain scrubber with 97% PM removal efficiency and the plant agrees to limiting operations to sixteen hours daily in the Operating Permit, what is the potential to emit?

    a. 0.262 TPY **b.**   0.060 TPY **c.** 0.087 TPY **d.**   0.180 TPY

12. The source mentioned in Question 10 is:

    a.   A major mobile source
    b.   A major stationary source
    c.   A minor area source
    d.   Not regulated

13. Which of the following situations is a major source?

    a.   A natural gas compressor station with potential to emit 200 TPY NO_2
    b.   A coal-burning electric plant rated at 200 MMBTU/hr and potential to emit 200 TPY NO_2
    c.   A chemical plant with potential to emit 140 TPY NO_2
    d.   A petroleum storage yard with a 299,000 barrel capacity

14. Which of the following situations does not require a New Source Review?

    a.   A natural gas compressor station with potential to emit 200 TPY NO_2 proposing a 300 TPY net increase
    b.   A natural gas compressor station with potential to emit 300 TPY NO_2 proposing a 35 TPY net increase
    c.   A chemical processing plant with potential to emit 200 TPY NO_2 proposing a 50 TPY increase
    d.   A chemical processing plant with potential to emit 80 TPY NO_2 proposing a 120 TPY increase

15. A sulfuric acid plant converts sulfur dioxide (SO_2) into sulfur trioxide (SO_3) with 97.5% efficiency and produces 1000 mg of 100% sulfuric acid (H_2SO_4) per day. You are completing an air permit application for SO_2 emissions without benefit of stack data. According to AP-42, the emission factor for sulfur dioxide from a sulfuric acid plant is:

E.F. (kg SO_2/mg 100% H_2SO_4) = 682 - [(6.82)(% SO_2 to SO_3 conversion)]

Estimate the SO_2 emissions from this plant in kg/day.

   a. 17,000 kg/day   b.   17 kg/day   c.   675,351 kg/day   d.   675 kg/day

ANSWERS

1.	c.	9.	b.	
2.	d.	10.	d.	3 lb/hr x 8760 hr/yr , 2000 lb/T = 13.14 TPY
3.	c.	11.	a.	3 lb/hr x 0.03 X 16 hr/d x 7 d/w X 52 w/yr , 2000 lb/T = 0.262 TPY
4.	b.	12.	d.	
5.	b.	13.	c.	
6.	c.	14.	b.	
7.	d.	15.	a.	% SO_2 to SO_3 conversion = 97.5
8.	a.			(6.82)(97.5) = 665

EF = 682-665 = 17 kg SO_2/mg 100% H_2SO_4
Emissions = 17 kg/mg X 1,000 mg/day = 17,000 kg SO_2/day

References

Brownell, F. W., Clean Air Act, Chap. 6, in *Environmental Law Handbook*, Arbuckle, J. G., et al, Eds. Government Institutes, Inc., 1993,

Cockerham, L. G. and Shane, B. S., Eds., *Basic Environmental Toxicology*, Boca Raton, FL, CRC Press/Lewis Publishers, 1994.

Finlayson-Pitts, B. J. and Pitts, Jr., J. N., *Atmospheric Chemistry: Fundamentals and Techniques*, New York: John Wiley & Sons, 1986.

Griffin, R. D. *Principles of Air Quality Management*, Boca Raton, FL: CRC Press/Lewis Publishers, 1994.

Landis, W. G. and Yu, M. H., *Introduction to Environmental Toxicology: Impacts of Chemicals upon Ecological Systems*, Boca Raton, FL: CRC Press/Lewis Publishers, 1995.

Lipfert, F. W., *Air Pollution and Community Health: A Critical Review and Data Sourcebook*, New York: Van Nostrand Reinhold, 1994.

Little, James W. Air quality, in *Environmental Science and Technology Handbook*, Ayers, et al., Eds., Rockville, MD: Government Institutes, Inc., 1994. p. 125.

McCollough, G. T., et al., Control of industrial stack emissions, *The Industrial Environment - Its Evaluation and Control*. U.S. Department of Health, Education, and Welfare, Public Health Service, Center for Disease Control, National Institute for Occupational Safety and Health, 1973, Chap. 43.

Patrick, D. R., *Toxic Air Pollution Handbook*. New York: Van Nostrand Reinhold, 1994.

Philp, R. B., *Environmental Hazards & Human Health*. Boca Raton, FL: CRC Press/Lewis Publishers, 1995.

Scott, R. M., *Introduction to Industrial Hygiene*. Boca Raton, FL: CRC Press/Lewis Publishers, 1995.

Sheriff, R. H., Air pollution, *Industrial Hygiene Study Guide*. 4th ed., New Jersey Section American Industrial Hygiene Association, 1989.

U.S. EPA. AP-42. 4th ed., Supplement F. *Compilation of Air Pollutant Emission Factors, Volume I: Stationary Point and Area Sources*, Research Triangle Park, NC: Office of Air and Radiation, July 1993.

U.S. EPA. *New Source Review Workshop Manual: Prevention of Significant Deterioration and Nonttainment Area Permitting*, Research Triangle Park, NC: Office of Air Quality Planning and Standards, October 1990.

U.S. EPA. EPA 400-K-93-001, *The Plain English Guide to the Clean Air Act*. Washington, D.C.: Office of Air and Radiation (ANR-443), April 1993.

Chapter 14

INDUSTRIAL HYGIENE MANAGEMENT

Anthony Joseph, PhD

I. CORPORATE POLICY

A. Introduction

Industrial hygiene corporate policy must support and mirror the corporate mission statement in order for it to be successful and meaningful. Most corporate mission statements declare that the primary responsibility of management is to maximize return on investment. According to Birkner, value creation is at the heart of the business process and must become a central issue for health and safety professionals. Petersen says that a policy should do three things, namely:

- Affirm a long-range purpose
- Commit management at all levels to reaffirm and reinforce this purpose in daily decisions
- Suggest the scope that can be left to the discretion and decision by lower-level management

Levy and others say that a policy should address the following:

- A commitment to provide the greatest possible safety to all employees and ensure that all facilities and processes are designed with this objective
- A requirement that all occupational injuries and accidents be reported, and corrective actions taken to assure that similar incidents do not occur
- Clear explanations to all employees of potetnial exposures to hazards in the workplace, and the establishment of training programs to inform employees of how to reduce and minimize potential risks
- Regularly scheduled system safety analyses of all processes and workstations to identify potential hazards so that corrective actions can be taken before accidents occur

- Disciplinary procedures for employees who engage in unsafe behavior and for supervisors who encourage, or permit, unsafe activities

Health policies will vary from one organization to another. However, a good corporate health policy should include the following items:

- A mission or philosophy statement
- A list of measurable objectives
- Scope of the policy
- Responsibility, authority and accountability
- Supports involvement and commitment

The corporate policy must be generic, covering the activities of the company. The operational or departmental policy must be specific and address detailed issues as they relate to successful implementation of the corporate policy.

B. Mission or Philosophical Statements

The company's mission or philosophical statement normally sets the tone for policy statements. As an example, a management philosophy that all accidents can be prevented is in harmony with the statement that "workers will be held responsible and accountable for accidents."

C. Objectives and Scope

The scope must identify the activities covered in the policy. Examples of activities may include: prevention of injuries to visitors; prevention of damage to property by fire; or prevention of water pollution. For each activity included in the scope, there should be at least one corresponding objective associated with it. The objectives must be achievable, measurable, justifiable, consistent, ethical and legal, and understandable. As an example, "One of our objectives for this year is to reduce the number of lost time accidents to zero." All terms used in the objective must have common meaning to all people involved in achieving the objective. Note carefully, the objectives must be consistent with the company's mission statement and other objectives. An example of an inconsistent objective in a production-oriented company is to state that "Our number one priority is to ensure a healthy workplace." Clearly this objective is meaningless, since the mission of the company is to optimize profit. A suggested alternative to such an objective statement is "Our number one priority is to optimize production in a healthy workplace."

D. Responsibility, Authority, and Accountability

Ideally, one person should be responsible for each activity identified in the scope. The duty of this person is to ensure that the objectives associated with the activity are achieved. For example, the purchasing manager can be assigned the responsibility, authority, and accountability for purchasing materials, equipment, and machinery that meet acceptable safety standards. An important component of the responsibility, authority, and accountability must be the *empowerment* of the person to command and determine courses of action if acceptable standards are not met. The limitations of the duties under the three elements of responsibility, authority, and accountability must also be clarified. For example, when purchasing protective gloves, the required specifications must be provided by the health and safety professional, not the purchasing manager. The culture of some companies or organizations of self-protection, self-preservation, and liability demands that a chain of responsibility, authority, and accountability be included in the policy. Provisions to ensure the effectiveness and usefulness of the policy must be consistent with the implementation culture of the company. Figure 14.1 shows an industrial hygienist working with management to assign accountability for a new process scheduled for installation in a plant.

Figure 14.1 Industrial hygiene professionals team with management to assign health & safety accountability for a new process being installed at a plant (Courtesy: *Mine Safety Appliances Company*, Pittsburgh, PA).

E. Support, Involvement, and Commitment

The policy should be written, publicized, and signed by the company's top management. The names of the persons associated with the development of the policy can help in the acceptance of the policy. The language of the policy should encourage involvement and participation of the workforce, foster teamwork, and ensure commitment. The policy should seek contribution, support, involvement, and commitment to reaffirm and reinforce the objectives of the policy in daily decisions.

F. Summary

The *acid test* of a good corporate health policy is to know that the mission or philosophy of the company is reflected in the policy, contains a list of measurable objectives, clearly defined scope, responsibility, authority, accountability; and promotes support, involvement, and commitment of all employees. The policy will be effective once these elements are addressed.

II. COORDINATING INDUSTRIAL HYGIENE PROGRAMS - MEDICAL, ENVIRONMENTAL, SAFETY, AND LEGAL

Integrating the information from all departments connected to the successful carrying out of industrial hygiene (IH) programs is a prerequisite. Conventional departments or services connected to industrial hygiene programs are medical, environmental, safety, engineering, finance, personnel, and legal. An understanding of the roles of each of these departments or services in advancing a healthy workplace, and knowing what is a good industrial hygiene program, are paramount to the integration and coordinating processes. It is suggested that a large portion of the role of the IH department should be devoted to integrating information from departments and coordinating working committees. The engineering department, for example, can assure that machinery, equipment, and the facilities are designed or modified according to health standards. The medical department can certify that all injuries and illnesses are properly treated, recorded, and investigated. In addition, the medical personnel can help in identifying injury trends and disorders. The legal department can verify that advertisement and services are legally acceptable, and the rights of workers are not infringed. The environmental department can look beyond the workplace to confirm that the health risk to the public and the environment are marginal, and environmental standards are in compliance. Departments that should be included in the integration are normally determined by the size and structure of the organization. Despite the size and structure of the organization, medical services should play a role in industrial hygiene programs. The medical department or service can help identify patterns of employee injuries, disorders, and complaints, which can provide early detection of potentially hazardous operations and processes. Examples of subtle and difficult injuries to detect are those caused by cumulative trauma.

An IH program must identify, or provide the means to recognize associated health hazards, provide controls measures, promote worker involvement, and show management commitment. The most effective industrial hygiene programs today result from team effort involving interactions among many groups within the organization. Coordinating industrial hygiene programs should be the responsibility of the health and safety manager/professional, or the chair of the health and safety committee. Experience has shown that assigning the responsibility and authority for administering programs to a plant-wide committee is more effective. The composition of the plant-wide committee should include representation from all affected departments. This committee can serves as the company's safety committee responsible for establishing policy, setting objectives, and overseeing activities of subcommittees.

It can be argued that all the departments in an organization are essential to the successful execution of IH programs. However, specific departments are directly connected. The contributions from these departments should be integrated and coordinated. As indicated in the previous section, the culture of the organization will influence the coordinating process. In a proactive organization totally committed to health and safety, a healthy workplace will be the goal of the workers and management. Therefore, when a worker is assigned to a new task, the worker will be educated regarding specific health hazards associated with the new task. Clearly, coordinating IH programs in this setting will require representation from all the departments.

III. INDUSTRIAL HYGIENE SURVEYS

One of the responsibilities of the industrial hygienist is to accurately determine employee exposure to environmental stressors. The ultimate success of performing this responsibility rests upon the validity and reliability of the measurements collected during a survey. Industrial hygiene field surveys or exposure assessments are intended to help identify and quantify health stressors in the workplace. It can be motivated by legal compliance, complaint investigation, insurance request, management concerns, or health and safety activities of the company. Whatever the reason for the survey, several factors must be determined, such as the purpose of the survey and areas to be surveyed. If sampling, monitoring, and analyses are to be part of the survey, then the pertinent factors mentioned in previous chapters of this text must be addressed. Factors such as: Which employee or employees are at maximum risk? Locations to be sampled? Sampling size? What should be the sample duration? and When should samples be taken? Field surveys will help clarify or answer most of these factors.

An industrial hygiene survey should follow the flow of the materials into the plant from cradle to grave, including the various processes involved in the operation. An initial walk-through survey of the plant is advisable if initial familiarization of the operation is needed, or a baseline of conditions is required. A checklist, a sketch of the layout of the plant, and a flow diagram of the operation are useful tools for aiding in the decision process of identifying areas or operations for intensive inspection.

A survey should be structured to capture the environmental stressors, the effectiveness of control measures in use, and how workers perform their jobs. In surveying how workers do their job tasks, it would be beneficial to evaluate production levels and variations, exposure sources, and job functions. Interviewing and observing workers can reveal information regarding hazard evaluation and adequacy of controls. Workers' participation should be an important component of an IH survey. Each worker is an expert in his or her job, and should be actively involved in identifying health hazards. Also, if modifications are necessary to reduce a hazard, worker acceptance of the new equipment, tools, work methods, or personal protection equipment is essential for the successful implementation of the change. Therefore, workers should actively participate in IH surveys. The maintenance department and contracted workers can also play a critical role in the success of an IH survey. The extent to which workers and contractors are included in an IH survey depends on the purpose and objective for performing the survey. In short, an IH survey should systematically identify health stressors, effectiveness of control measures, and health risks associated with job operations.

IV. RECORDKEEPING AND REPORTING

Recordkeeping and reporting workplace accidents and injuries are important activities for an organization. OSHA requires that employers with more than 10 employees must record all injuries and diseases if they result in death, hospital admittance of three or more workers, one or more lost workdays, restriction of work or motion, loss of consciousness, transfer to another job, or medical treatment other than first aid. The data recorded in the OSHA 200 Log provides a broad estimate of work-related illness and injuries. The data is required to be confidential; therefore, in-depth investigations using the log is not possible.

OSHA Standard 29 CFR 1910.20, Access to Employee Exposure and Medical Records, requires that employee exposure records be preserved for at least 30 years. This information must be readily available to employees and their representative(s). Collected data should be reliable and valid to enable prediction and trend analysis.

Other than injuries and illnesses, data is also collected in surveys. Complete and detailed records must be kept on sampling procedures, sampling conditions, and sample results. The hygienist must document that sampling was conducted according to accepted professional standards. Records should include the identity of the equipment and collection devices used, the calibration procedures and results, the identity of the analytical laboratory and related laboratory reports, and the air-sampling calculations.

The conditions under which the sampling was conducted should also be carefully documented to ensure the integrity and usefulness of the results. Anything that might help interpret or explain the final air sample result should be recorded. For example, in a production welding operation, the record should contain the name and location of the welder, the material being welded, the welding rods used, the number of pieces welded, the personal protective equipment used, and the use and location of local exhaust ventilation.

Many industrial hygiene programs have developed air sampling forms to ensure that all the necessary information is collected. OSHA has developed an air-sampling worksheet for its Industrial Hygiene Compliance Officers.

Background data, such as laboratory reports and field notes, need only be retained for 1 year since information on the sampling method, the analytical and mathematical methods, and summary of other background information is retained for the required 30 years.

V. QUALITY ASSURANCE AND QUALITY CONTROL PROGRAMS

A. Introduction

Most errors in industrial hygiene evaluation result from incorrect sampling. As a result, quality assurance and quality control programs must be incorporated into the sampling procedure. Errors can occur in four broad categories. They are:

- Sampling performance of the collector
- Sample representation
- Documentation of the sampling events
- Protection of the sample

B. Sampling Performance of the Collector

All samples must always be considered hazardous to the health of the person performing the sampling. The samples can be a physical, biological, or chemical hazard that can result in injury or adverse reaction while sampling. Therefore, the first step in assuring quality sampling is the safety of the collector. The collector must be properly trained and qualified to conduct the sampling using standardized protocol.

C. Sampling Representation

The number of samples, size, equipment, time, and space used are critical in assuring quality of the sample. The equipment must be maintained in reliable working order. It should be properly calibrated. Quality checks should be done occasionally to ensure consistency in the measuring instruments. The use of proper measuring procedures, established standards, and accredited laboratories for analysis will help guarantee quality assurance.

D. Documentation

Industrial hygiene sampling has the potential to lead to legal contests. As a result, each sampling event must be legally defensible. Appropriate documentation and records should be maintained for all stages of the sampling exercise. This includes data on site conditions before, during, and after the field investigation. Keeping a sampling log book should be maintained by the collector. Information such as time and date of the sampling event, names of persons collecting the samples, duration and flow rates of sampling equipment, list of sampling instruments or equipment, calibration protocols, results of field tests performed, lists of samples obtained, and weather and site conditions. Sample collection techniques were discussed in Section III, *Industrial Hygiene Surveys*.

The most important part of the documentation in legal contests is the chain-of-custody. This documentation must be able to clearly establish who collected the sample, how the sample was collected, handling procedures employed, where and when the sample was collected, and other critical sample components. It is advisable to treat the log as a legal document. Therefore, signatures, dates, and times should be included for individuals who handle samples from collection to ultimate analysis. Other information that could be included in the log are the analytical protocols, observations on site conditions at the time of sampling, precautions related to transporting the sample to the laboratory, verification of the analytical procedures used by the laboratory, and responsible persons at each stage of the sampling. Proper labeling of the samples is very important. The information provided on each label must agree with the specific sample and the information in the log.

The evaluation of the results should be documented. This should include: a list of the limitations of the results, deviations from standardized sampling procedures, major sources of errors, precautions, QC, and any other factors that can influence the validity and reliability of the results.

E. Sample Protection

Precautions should be taken to ensure that samples are not contaminated or degraded. The required containers for collecting the samples and the number of blanks and duplicates should be established and followed. Special attention must be placed on handling, storing, transporting, and shipping of samples. In recent times the security of samples is also a major consideration.

F. Other Considerations

One of the most important goals of any industrial hygiene program is to accurately determine employee exposure to environmental stressors. The ultimate success of accurately determining exposure hinges upon the validity and reliability of the measurements and analysis. A comprehensive quality assurance program that can track and cross-reference the sample at every stage is essential.

VI. ETHICS

A. Introduction

This section covers a brief discussion on the ethical dimensions and moral principles in industrial hygiene management. Industrial hygienists are thought to be objective; however, this is not quite true. There are many instances when their activities are influenced by individual, social, political, and cultural values. Such values may influence who is sampled, how the data is collected, analytical techniques employed, laboratory selection, data interpretation, and even how the result is presented. Often, disagreements result because of lack of clearly defined policies and practices, and not ethical dimensions. Sometimes, statements (or lack of statements) by the policymakers reflect values or the industrial hygienist's uncertainty. Therefore, the lack of clearly delineated policies and practices can contribute to influence the approach selected by the industrial hygienist. The policies and practices inherent in established regulations, codes of practice, and procedures should be used to guide decisions. This will reduce uncertainties or value-ridden policies and practices. When uncertainties persist, it may be prudent to wait until there is more evidence, or a conservative policy is adopted. Ethical analyses can make unique contributions to decision-making and action.

B. Selected Issues

Autonomy, involvement, knowledge, allegiance, and interest are issues constantly confronting the industrial hygienist in the performance of duty. In making decisions to control health hazards, the tendency is to be autonomous. The decision should be implemented with respect and consideration for the persons that will be affected. An excellent public relations strategy is to involve workers, labor representatives, and the community when developing corporate health and safety policies. The ethical obligations and responsibilities of these groups should be leveraged and nurtured.

Freedom and non-interference are tightly guarded and highly cherished *rights* beyond privacy, confidentiality, and fairness. Personal interest and goals differ significantly. These rights can invoke ethical and moral issues. This is especially obvious when an unpopular decision has to be made.

Ethical issues should be a test of the allegiance to the company and independence as a professional. The knowledge and training received help to identify the industrial hygienist as morally and ethically responsible to the workers, the company, and the community.

In examining these issues, it is important to understand the very real and personal consequences demanded of health professionals. Their actions can either enhance their own reputation, status, and esteem, or incur the wrath and distrust of their employer, workers, and peers. In short, their decisions can affect their family and social life, financial standing, employability, and respectability.

C. Codes of Ethics

Codes of ethics are written by professional organizations to provide guidance to difficult ethical issues. For example, the objectives of the code of ethics for the practice of industrial hygiene are stated as: "to provide standards of ethical conduct for industrial hygienists as they practice their profession and exercise their primary mission, to protect the health and well-being of working people and the public from chemical, microbiological, and physical health hazards present at, or emanating from, the workplace." Although some of these codes are ambivalent, they are very helpful in guiding the actions of the professional. However, ethical codes cannot solve moral and ethical dilemmas that arise daily. Ultimately, the professionals have to decide.

VII. EPIDEMIOLOGY

A. Introduction

Epidemiology is concerned with identifying subgroups of a population at high risk for a particular disease, providing evidence for causal associations, estimating dose-response relationships, and determining the effectiveness of exposure control measures. Applying the information generated from epidemiological studies can be beneficial in managing the health of workers in the workplace. These studies can assist in the identification and prevention of specific work-related health problems. Workers are constantly exposed to changing health or environmental stresses. Therefore, the health-risk relationship between any group of workers and their work environment is dynamic. This dynamic relationship is further complicated by non-workplace health stresses. The net effect can be anxiety and stress for the industrial hygienist who is responsible for pinpointing work injuries and protecting workers' health.

Three applications of epidemiological studies are relevant in managing the health of workers and assessing workplace injury. They are:

- Surveillance of already recognized occupational disease or injury
- Systematic study of the relationship between health effects and known or suspected workplace hazards
- Evaluation of the effectiveness of controls or interventions in the reduction of the injury

These three applications are built on measures of responses and not exposure. This is not typical for the industrial hygienist. Exposure measurements are what we can measure. This is the product of intensity and duration. A dose is the amount of a substance delivered to the organs or tissues where the effects are manifested. Epidemiological health responses are either discrete, such as an illness or disease (re: occupational asthma), or combined, such as a biological parameter (re: urinary hippuric acid level). The study of dose or exposure is significant if the strength of the association between exposure and response is weak. Exposure estimates or studies are more relevant to acute effects. Exposure measurements do not consider clinical manifestations, latency period, or chronic exposure. Evaluation of a biological index, such as toxic agents in blood, urine, or exhaled air, can provide a more accurate value of exposure. A biological index of exposure is a measure of the body burden via all routes of entry. The important lesson from this brief discussion is that a wide range of methods to estimate both current and past exposure exist. Each method has advantages and disadvantages. Having knowledge of the limitations and assumptions inherent in each method is important. An accurate measure of exposure is equally important as the measurement of health outcomes in arriving at an unbiased and precise estimate of an exposure-response relationship.

B. Selected Measures in Epidemiology

The single most important parameter in *epidemiology* is 'to define the population at risk.' This is critical for the interpretation of the *measures*. Rates and risks are the two most frequently used epidemiologic measures. Defined crudely, a *rate* is the frequency of a disease per unit size of the population being studied, and a *risk* is a comparison between rates. The simplest rate is the *prevalence rate* that is the number of cases at a single point in time. This can be mathematically expressed as:

$$\text{Prevalence Rate} = \frac{\text{Number of Cases}}{\text{Total Population}} \qquad (14.1)$$

The incidence rate is the ratio of number of new cases during a specific period to the population at risk during that same time, mathematically expressed as:

$$\text{Prevalence Rate} = \frac{\text{Number of New Cases}}{\text{Total Population}} \qquad (14.2)$$

These two rates are known as *crude rates* and are employed when the populations under study are not homogeneous.

Rates can provide valuable information. However, care must be exercised in comparing rates. The distribution of the population should be standardized or adjusted. An example of rates is detailed in Table 14.1. The rates of the community are considered crude rates.

Table 14.1
Age Effects on Incidences of Viral Hepatitis

Area	# of Cases	Population at Risk	Incidence Rates
Community A	1200	60000	0.2000
Population < 50 years	900	10000	0.0090
Population ≥ 50 years	300	50000	0.0060
Community B	800	80000	0.010
Population < 50 years	600	12000	0.0500
Population ≥ 50 years	200	68000	0.0029

The two most frequently used risk measures are relative and attributable. *Relative risk* is the ratio of rates and *attributable risk* is the difference in rates. The relative risk is a comparison of two rates, exposed to unexposed populations. This is a measure of the potency of the hazard. The attributable risk measures the size of the injury burden in the population under study. Relative risks are presented in epidemiological studies as the preferred measure of association between an exposure and a disease outcome, while attributable risk measures are preferred for setting control priorities.

Epidemiological studies can be categorized into three broad groups; specifically, cross-sectional, case-control, and cohort. The *cross-sectional* design is characterized by collection of both exposure and non-exposure information at one point in time, while *cohort studies* measure the incidence over time. Cross-sectional cohort studies have many advantages over longitudinal studies. Cross-functional studies enable the examination of disease morbidity or measures of physiological function. Because the subjects are alive, information can also be collected for non-occupational risk factors. Also, these studies require less time to complete since both disease prevalence and exposure data are collected simultaneously.

Two major limitations in epidemiological studies are:

- They are less appropriate for investigating causal relationships because they are based on prevalent rather than incident data.
- They are based on actively employed workers, and do not include those who have left or retired before the

study. The absence of this latter category may result in an underestimate in risk since these workers may have left because of an occupational impairment.

Longitudinal cohort studies focus on exposure and project ahead to an outcome or disease incidence. Thus, several outcomes can be studied in the same population. Longitudinal cohort studies can occur completely in the past, which is *retrospective*, or both in the past and future, which is a *prospective study*. The major difference in the two approaches lies with the data collection. Data collected on exposure retrospectively are dependent on the quality of records, while prospective depends on the specific study plan. In addition, retrospective studies normally rely on outcomes recorded for other purposes. The inclusion of the entire population in cohort studies has been the major advantage over cross-sectional studies. However, most cohort studies suffer from long-term follow-up.

Case-control studies start by identifying disease and eliciting information about exposures. The case-control design is effective when the disease is rare or when multiple possible exposures are being explored in the history of a specific disease. The major advantage of the case-control study is the relative simplicity and reduced cost. In addition, this type of study is well suited for the assessment of several different exposures. Of all the epidemiological studies, case-control is most susceptible to biases.

The interpretation of epidemiological studies depends on the strength of the association, supporting evidence for causality, and the validity of the observed association. In studies of discrete health outcomes, the relative risks represent the measure of the strength of a given association. The attributable risk is the preferred measure for the association of the other two types of studies. Additional evidence that can assist in the interpretation of the studies is provided by toxicological studies consisting of dose-response relationships. Monson developed a guide to evaluate epidemiological studies consisting of a set of questions addressing the collection, analysis, and interpretation of data. Most epidemiological studies cannot be replicated because the population is constantly changing. As a result, conclusions can differ for similar studies.

VIII. CASE STUDY

A. Introduction

A construction company in a community of 35,000 residents plans to expand its operations to include recycling of lead and copper pipes. Although this is the major employer in the community, some residents are opposing the expansion for health reasons. To counteract the protest, management hired a trained industrial hygienist for the plant. The company is already involved in sand-blasting, chemical treatment of materials, spray painting, and galvanizing. The safety department consists of a safety manager and a clerk. The existing safety manager was promoted 1 year ago from a position as a machine shop supervisor. He has no formal training in safety. Within the first 6 months of the industrial hygienist been hired, he did the following:

- Developed a corporate policy
- Performed field surveys and developed inspection checklists
- Established reporting and recording systems
- Reviewed epidemiological studies related to the construction industry

He involved the community and workers as much as possible in his activities.

The corporate policy he developed was accepted by the company. It read as follows:

"The corporate policy of this company is that every employee is entitled to work under the safest possible conditions for the construction industry. As a result, every credible attempt will be undertaken to prevent accidents, protect property, and preserve health. The company will maintain a safe and healthy workplace. It will provide safe working equipment and necessary personal protective equipment. In case of an injury, the best available first aid and medical service will be provided. It is management's belief that all accidents can be prevented by taking common-sense precautions. All accidents must be reported and recorded. The collected data will help direct our efforts to develop prevention, intervention strategies, and controls.

Coordinating the health and safety program is Mr. Thomas Sam, vice-president. The overall effectiveness of the health and safety program is his responsibility. He is also responsible for presiding at the weekly safety meetings of employees to discuss health and safety concerns and issues. A written record of these meetings will be maintained. The supervisor or foreman is responsible for the performance of all jobs according to the health and safety work rules set forth in the employees safety manual. In addition, the supervisor or foreman will conduct health and safety inspections daily and submit a written report to the plant manager. The industrial hygienist is responsible for coordinating the implementation of health programs, and the safety manager is responsible for the implementation of safety programs. These two professionals are also responsible for passing on health and safety information to all supervisors, foremen, and contractors.

All work-related incidents must be recorded by the supervisor or foreman on the company's injury form. This form must be submitted to the safety manager within 1 hour of the incident for investigation. The health and safety department will make recommendations to the foreman or supervisor to reduce the risk of a similar incident recurring.

If all of us do our part always, we will create a safe and healthy work-place. In so doing, we will improve our safety record, reduce our insurance costs, and be more competitive in our industry, thus helping to safeguard our jobs."

These activities initiated by the industrial hygienist improved the company's relationship with the community, increased production, reduced injury rates, and lowered medical costs. This, in essence, fulfilled the prime mission of the company: to maximize profit.

REVIEW QUESTIONS

1. You are hired as a health & safety professional at a family-owned art shop in operation for the past 20 years, employing 80 artists and 26 workers. The shop is involved in acrylic, oil-base, watercolors, sand, and lithographic painting and art work. The company is planning to expand, seeking to employ 60 more employees. There currently are no safety and health policies or programs. The county the plant resides in recently introduced an ordinance requiring businesses to include appropriate health and safety measures when applying for approval to expand. As the newly hired health and safety manager, suggest a health policy statement.

2. Describe the role of the industrial hygienist in managing and promoting health programs in a large industrial company with medical staff, a legal department, an environmental department, and a health and wellness department.

3. The ultimate success of an industrial hygiene sampling process depends on the reliability and validity of the measurements collected in the workplace. Several factors must be determined when deciding to conduct an employee exposure survey. List four critical considerations.

4. List three strategies that can aid in the performance of conducting an industrial hygiene survey.

5. What sampling records are recommended for maintenance?

6. Based on the following data, calculate the following rates and risks: **(a)** exposed disease rate; **(b)** non-exposed disease rate; **(c)** relative risk; **(d)** attributable risk.

Disease	Exposure	
	Present	Absent
Present	800	200
Absent	400	100

7. List 10 questions that can assist the industrial hygienist in evaluating the collection of data for epidemiologic assessments.

8. The outcome of many legal contests involving industrial hygiene sampling activities were determined on the grounds of failure to follow accepted protocols. Discuss the most important element in a quality control/quality assurance program to maintain sampling *truth*.

ANSWERS

1. The health policy statement should include, but not be limited to: (a) a philosophical statement as it relates to the thinking of management; (b) objectives in measurable terms; and (c) both line and staff roles and responsibilities, authorities, and accountabilities. Clearly, one of the objectives of the company should be to comply with the local ordinance. As indicated in this chapter, the "acid test" of an effective corporate health policy is one that integrates the overall mission or philosophy of the company, and is embodied by measurable objectives. The scope is clearly defined, while responsibility, authority, and accountability are clearly delineated. The policy also is designed to have the necessary support, involvement, and commitment of employees at all levels.

2. The role an industrial hygienist can play in managing and promoting health programs includes: (a) coordination of health programs; (b) integration of information from all departments having a stake in the successful implementation of health programs; (c) ensuring that health hazards are properly controlled; and (d) providing training and compliance assistance.

3. (a) Which employees are at risk and should be monitored? (b) When should samples be collected? (c) What should be the sample duration? (d) Where should samples be collected and under what conditions or circumstances? (e) How many samples should be collected? (f) What laboratory should be selected to analyze samples? (g) How should samples be handled?

4. (a) Follow the flow of materials; (b) conduct a pre-project walkthrough; (c) use a checklist; (d) interview and observe the work tasks being performed; (e) seek and elicit worker involvement and participation; (f) systematically identify stressors.

5. Detailed and complete records should be kept for: (a) sample protocols employed; (b) field conditions; (c) sample results; (d) sample collection equipment and calibration procedures; (e) laboratory analyzing samples and any generated analytical reports; (f) any calculations employed to determine exposure; (g) other data that can assist in the interpretation of results.

6. (a) exposed disease rate = 800/1200 = 0.67
   (b) non-exposed disease rate = 200/300 = 0.67
   (c) relative risk = [(800/1200)/(200/300)] =1
   (d) attributable risk = (800/1200 - 200/300) = 0

7. **(a)** What were the objectives of the study? **(b)** What was the association? **(c)** What was the primary outcome of interest? **(d)** Was it accurately measured? **(e)** What was the primary exposure of interest? **(f)** Was this accurately measured? **(g)** What type of study was conducted? **(h)** What was the study base? **(i)** Was the subject selection based on outcome or the exposure of interest? **(j)** Could the selection have differed with respect to other factors of interest? **(k)** Were these likely to have introduced potential biases? **(l)** Was subject assignment to exposure or disease categories accurate? **(m)** Were possible mis-assignments equal for all groups?

8. The most important element of a quality control/quality assurance program to establish truth and data reproducibility is documentation. Appropriate documentation and records should be kept for all phases of a sampling exercise, from start to ultimate delivery of a report interpreting findings. Information to document includes: field conditions during all sample activities; equipment/calibration; chain-of-custody forms; field observations; protocols employed; types of samples collected; analytical techniques employed; sample handling, labeling, and transport protocols utilized; and laboratory QA/QC procedures.

References

Birkner, L. and Birkner, R. K., Creating value with health and safety, *Occupational Hazards*, April 1997, 57-58.

Bisesi, M. S. and Kohn, J. P., *Industrial Hygiene Evaluation Methods*, Boca Raton, FL: CRC Press, 1995.

Hernberg, S., *Introduction to Occupational Epidemiology*, Lewis Publishers, Inc., Chelsea, MI, 1992.

Levy, B. S. and Wegman, D. H., Eds., *Occupational Health: Recognizing and Preventing Work-Related Disease*, 3rd ed., Little, Brown, Boston, 1995.

Monson, R. R., *Occupational Epidemiology*. 2nd ed., CRC Press/Lewis Publishers, Boca Raton, FL, 1989.

Olsen, J., et al., *Searching for Causes of Work-Related Diseases: An Introduction to Epidemiology at the Work Site*, Oxford Medical Publications, Oxford, 1991.

Petersen, D., *Techniques of Safety Management: A Systems Approach*, 3rd ed., Aloray, New York, 1989.

Plog, B., et al., Eds., *Fundamentals of Industrial Hygiene*, 4th ed., National Safety Council, Itasca, IL, 1996.

Rothman, K. J., *Modern Epidemiology*, Little, Brown, Boston, 1986.

Chapter 15

CONSTRUCTION ISSUES

Martin B. Stern, MPH, CIH

I. OVERVIEW

Construction traditionally has been an industry plagued by safety-related concerns and issues. Injuries from falls, strains and sprains, and injuries associated with heavy equipment and machinery have predominated in this industry. However, construction workers are routinely exposed to adverse health agents during the performance of their jobs. Exposure can result from welding operations, preparation of structures for painting, coating operations, demolition of facilities and structures, or excavation of soils contaminated with hazardous wastes and substances. There are numerous construction activities and tasks that can result in worker exposure. This chapter will present an overview of the potential industrial hygiene ramifications associated with today's construction work.

Figure 15.1 Exposure potential during grinding (Courtesy: *Mine Safety Appliances Company*, Pittsburgh, PA).

1-56670-197-X/99/$0.00=$.50
© 1999 by CRC Press LLC

There are a variety of exposure events that can be encountered during the performance of construction work, regardless of whether the operation is related to new structures and facilities, demolition of inactive structures and facilities, or the clean-up of contamination from previous industrial or construction activities. Figure 15.1 demonstrates exposure potential during a grinding operation.

"Routine" construction activities and operations present potential exposure concerns for an industrial hygienist. Welding can expose workers to nonionizing radiation, painting can present volatile organic compound exposure, while installation of fiberglass can create dermal and respiratory irritation. As detailed in Chapter 12, Ergonomics, Kohn and Winterberger refer to operations present during construction activities that potentially impact construction workers from an ergonomic standpoint. These activities can include the workers having poor static positions, repetitive motions, awkward postures, overhead work, and exposure to vibrating tools or equipment which can result in the construction worker experiencing a repetitive trauma disorder.

The mitigation of problems affecting the external environment, such as the remediation and clean-up of abandoned hazardous waste sites or response and control of emergency releases of hazardous substances into the environment, pose a potential exposure risk to the public and require to be addressed, as well. Workers at these sites can routinely be exposed to a myriad of hazardous substances, depending upon the activities that were undertaken at the site previously. Contaminants encountered at these sites can include compounds that are acutely hazardous, carcinogenic, teratogenic, pyrophoric, and quite often, combinations that can produce a variety of toxicological end points.

There are also construction activities that are designed to *remove* potential contamination from structures or their components. In the past, asbestos and lead were routinely applied to, or incorporated into building structures and/or components. As these materials age and become damaged, they create exposure potential, and often require abatement interaction, such as removal or encapsulation. OSHA, the U.S. EPA, and most states have stringent requirements related to the management of damaged asbestos-containing materials.

These contaminants have been removed from bridges, tunnels, towers, internal and external building components, obsolete cable containing lead, and asbestos-related demolition activities. Quite often, the intention of removal activities is to protect occupants covered under the guise of OSHA's General Industry Standard from exposure to lead or asbestos. Unfortunately, the workers that are involved in the removal of lead and asbestos have potential exposure risks that can far exceed the potential exposures experienced by occupants residing in the facilities where the asbestos or lead is being removed. These projects require the use of respiratory protection, chemical-resistant clothing, mechanical ventilation, and other control mechanisms that impose additional constraints on the workers performing the removal operations.

The industrial hygienist is presented with a host of many challenges when seeking to recognize, evaluate, and control exposures in the construction work environment. This chapter will help to identify industrial hygiene applications encountered in the construction industry, and present potential mechanisms to protect the construction work population from potential exposures.

II. CONSTRUCTION INDUSTRIAL HYGIENE CHALLENGES & ISSUES

A. Workforce Considerations

There is a substantive difference when one considers the demographics associated with the workforce comprising the construction industry versus the work population that comprises the manufacturing, or general industry, sector. The general industry workforce normally consists of a fixed, stratified work population. For the most part, in a manufacturing facility, the workers are a sustained element of the facility where they work. Generally, they remain with an employer over an extended period, so their presence is expected, and can be monitored and protected properly. As well, the operations that are performed in a manufacturing plant are similar to that of the worker: consistent, relatively unchanging, and somewhat predictable. The worker can readily identify with processes that can result in potential exposure. The industrial hygienist in this role can effectively manage and ensure appropriate employee training, medical surveillance, proper use of personal protective equipment, and maintain and evaluate potential exposure pathways in the workplace.

In contrast, construction work is considered transient, both in terms of the workers and the work sites. Depending on the type of construction project, the workforce can change on a daily basis. In addition to the ever-changing workforce, the work site can change dramatically from day to day, as can the potential hazards that exist on a given construction site. How can the industrial hygienist adequately ensure that the health of the workforce is sufficiently protected? The industrial hygiene professional or site safety and health officer in the construction industry has to have an intimate comprehension of the workplace and of the construction activities being undertaken in order to protect the health (and safety) of the site workers. The potential exposures can range from simple asphyxiants to compounds that are known carcinogens, as well as movements and actions involving materials handling and repetitive tasks.

In order to protect the workforce properly, the industrial hygienist has to be aware of, and comprehend all job tasks and functions that will be performed at the work site. Understanding these tasks and functions will enable the industrial hygienist to effectively evaluate all potential exposure pathways that may be encountered and design appropriate protective equipment, engineering solutions, and administrative control measures.

For demolition projects involving buildings, it is prudent for the industrial hygienist to concern him or herself with potential exposure to building-related contaminants such as asbestos, lead, hazardous operations that took place in the facility such as laboratories, manufacturing processes, or disposal operations; exposure to rodents and their droppings; exposure to elevated noise, in addition to safety-related issues such as fall hazards, struck by/against, lacerations, etc.

The industrial hygienist, in addition to having a thorough understanding of the work tasks and operations that are being performed, must also have an intimate knowledge of the work site and all of its subtle (and not so subtle) characteristics. Comprehension of both the work operations and the work site enables the industrial hygienist in the construction arena to properly protect the work population.

B. Facility/Structure Construction

A construction employee working on a project involving the construction of a building or the erection of a structure has substantially different exposure pathways, depending upon the given stage or construction phase being undertaken. Each phase of construction may present differing exposure threats that will have to be independently considered and addressed. The first phase of activities conducted on a construction project involving the erection of a structure or building typically involves surface landscape preparation and excavation. This can include the removal and clearing of unwanted or unneeded vegetation and removal of unwanted site debris, such as discarded waste or trash, leveling of surface soils, and excavation and removal of piled or unwanted soils. Each of these activities can result in potential exposures for site workers.

1. Construction site hazards

Activities involving the clearing of brush and vegetation can result in potential construction worker exposure to insects and plant allergens. Stings from bees, wasps, and hornets, fire ants in certain areas of the United States, spiders, and other insects can result in minor to severe reactions to workers that are bitten or stung. Workers with allergies or who are hypersensitive, can experience pronounced effects, and sometimes death, after encounters with insects while clearing brush from a site.

Similar responses can be expected from workers exposed to plant toxins such as poison ivy, oak, or sumac. These plants are sensitizers, and each successive exposure will result in a more pronounced effect. Exposures to insects and plants can easily be controlled if the worker is provided with an appropriate, over-the-counter insect repellent and a disposable tyvek coverall, and appropriate hygienic techniques are employed at the completion of the workday.

Depending on the site, there may also be electrical hazards, confined space issues, man-holes, sewers, and other hazards present that will require careful evaluation and planning to properly protect site workers.

2. Determination of past uses

Excavation operations can result in worker exposure to contaminants released during the disturbance of soils. A Phase I Environmental Assessment is a tool that can be utilized to assess current and past site uses, including: hazardous material/waste handling practices; presence of underground or above-ground storage tanks; asbestos presence; and other potential issues and practices that could have contributed contamination. This data can be helpful in designing appropriate personal protective equipment (PPE) strategies, based upon the potential contaminants present. As well, Phase I information can be used to help determine if air monitoring and sampling will be necessary during site activities.

For example, if a site was utilized previously as a landfill, there is unlimited potential as to what can be encountered from a contaminant standpoint during excavation activities. Prior to the establishment of the Resource Conservation and Recovery Act (RCRA) in 1976, there was limited control over what could or could not be disposed of in a landfill. As a result, there are industrial and hazardous wastes, contaminated materials drums and barrels of unknown contents, and a host of other materials that present exposure potential to today's construction workers.

The industrial hygienist should ensure that any historical records that are available are carefully reviewed to determine materials that may have been disposed or utilized at the site. If historical data indicates that potential contamination may exist at the site, soil characterization may be warranted. This is accomplished to assess if surface and subsurface soils are contaminated, to what extent, and identify compounds and their given concentrations. Contaminants such as heavy metals, polycyclic aromatic compounds, or other soil-adhering contaminants could be potentially present. Proper sample collection techniques, including appropriate containers and handling protocols, must be employed for soil sample collection. The given protocols will be prescribed based upon the material suspected of being present and the location where the sample is being collected. In the event that sampling is warranted, one should refer to state environmental protection agency and federal U.S. EPA requirements to determine the appropriate sample collection methodologies to employ. The U.S. EPA's protocols deal with environmental media such as air, water, soil, etc., as compared to personal and area air sampling protocols administered by OSHA, AIHA, NIOSH, and the ACGIH.

The laboratory selected to perform the analysis should be able to provide counsel as to appropriate techniques for sample collection for the media you desire to assess and should maintain any required analytical certification to perform the analyses requested.

The sample results will help to determine the appropriate PPE that should be utilized during site activities and any air sampling/monitoring that should be conducted during site excavation work. Depending on the types of materials used or disposed at the location, personal air sampling and/or air monitoring may be required to evaluate any of the following parameters: volatile organic compounds monitoring, and monitoring for combustible gases generated from the decay of organic materials and detritus, hydrogen sulfide, or other contaminants. The industrial hygienist should prepare an air sampling and monitoring strategy after a careful review of historical site information is completed. On many construction locations, air sampling and monitoring may not necessary, nor PPE for that matter (beyond *construction essentials*, specifically: a hard hat, safety eyewear, and steel-toed boots). If historical site information and soil characterization results indicate the potential presence of contaminants, air sampling and monitoring should be conducted during the performance of any work operation where contaminants may become airborne to ensure that the prescribed PPE is appropriate and serve to continually measure airborne contaminant level to prevent overexposure to site workers. For example, if elevated volatile organic compound concentrations are detected based on analytical results, then appropriate personal protective clothing may include an air purifying respirator with organic vapor cartridges, a disposable tyvek coverall, boot covers over steel-toed boots, nitrile gloves over canvas gloves, safety eyewear, and a hard hat. The contaminant and its corresponding soil concentration will assist in the determination of an appropriate PPE strategy during the performance of excavation activities. Figure 15.2 demonstrates air monitoring during site excavation activities.

3. Site worker interpersonal "challenges"

The industrial hygienist assigned a construction/excavation health & safety role should be aware that the workers will go out of their way to express the reasons and why the personal protective equipment is not essential, necessary, or anything that they need to wear at work. Some of the reasons that may be given will include: "PPE has never been worn performing the job previously," or "why do I have to listen to a punk kid who has never worked hard and knows nothing about construction, tell me what I need to wear on the job?" The worker's diatribe can be filled with expletives. It is very important to communicate effectively with the workers and gain their respect. Sharing information about previous experiences ("war stories") will help generate worker trust.

Prescribed PPE should be assigned based on its ability to protect the worker from the hazards present, but the PPE should also be comfortable, aesthetically pleasing, and not interfere with the work operation being performed. If a worker is assigned PPE that does not meet these criteria, it will be very difficult to get the worker to don the given PPE. The industrial hygienist must be prepared to meet resistance when prescribing and assigning PPE. Gaining the support and trust of the workforce will have to be earned.

The industrial hygienist must win the respect and approval of the workers to achieve desired objectives. This is part of the role the industrial hygienist must play in order to protect the health of the construction worker.

Figure 15.2 Monitoring soils during excavation (Courtesy: *Mine Safety Appliances Company*, Pittsburgh, PA).

4. Sources of exposure
a. Noise

Upon completion of excavation activities, pilings may be driven, depending on the type of structure that is being erected. Piles are installed to provide structural support. This operation generates excessive noise levels that can exceed OSHA impact noise criteria of 140 decibels, on the A-weighting (dBA). As well, there is potential for workers to exceed OSHA Permissible Noise Limits as listed in Table 15.1. OSHA mandates the establishment of a Hearing Conservation Program for workers exposed to noise at or above 85 dBA as an 8-hour time-weighted average. Components of a Hearing Conservation Program include requirements related to: training, written program, monitoring, audiometric testing, and hearing protection. It should be mentioned that the ACGIH criteria for noise is based on a permissible noise limit of 85 dBA as an 8-hour time-weighted average, which is less than the OSHA 8-hour PNL.

Any individuals working in the general vicinity of pile installation should be properly protected with appropriate hearing protection. Noise monitoring will help determine what hearing protection will be needed. Depending on the measured noise intensity levels, multiple hearing protectors may be required, such as ear muffs in conjunction with ear plugs. The workers will be required to be educated on proper utilization and selection of hearing protectors, donning procedures, and hearing physiology.

Table 15.1
OSHA Permissible Noise Limits, 29 CFR 1926.52

Duration (Hours)	Permissible Noise Limit (dB)
8	90
6	92
4	95
3	97
2	100
1.5	102
1	105
½	110
¼	115

b. *Concrete*

After pile installation is completed, concrete will be poured and the steel infrastructure for the facility will be erected. If concrete is blended/mixed on-site, the workers can potentially be exposed to crystalline silica. Exposure monitoring should be performed periodically to evaluate if airborne concentrations are below the applicable OSHA PEL for silica as detailed in Table 15.2, which is defined in 29 CFR 1926.55 (OSHA Construction Standard) criteria. Workers may require the use of an appropriate air purifying respirator (APR) equipped with high-efficiency particulate air (HEPA) cartridges. APR utilization will be determined based on airborne concentration levels exceeding values listed in Table 15.2. Regardless of whether the cement mix is preblended in cement trucks or blended on-site, contact can result in dermatitis, since the material often has an acidic pH (less than 7.0). Appropriate gloves, such as neoprene or polyvinyl alcohol gloves, should be worn over canvas work gloves to minimize any dermal contact. Protective eyewear should be utilized when the mix is transferred from the mixer to the pour locations, as material may be splashed to prevent ocular contact with particulate matter generated during the mixing and blending operation.

c. *Welding*

Rebar, which is used to reinforce concrete, will require hot cutting and welding, as will the erecting of the steel infrastructure. These activities can result in worker exposure to fumes, ultraviolet light, ozone, and heat associated with welding activities. The type of welding being performed will indicate the potential exposures present. If there is insufficient natural air movement at the location of the weld, where feasible and practical, fans can be installed as a quasi-engineering control to remove fumes and other welding-generated contaminants from the workers' breathing zone. It should be anticipated that during welding operations, workers will require the use of a welding helmet that contains a UV-resistant lens piece. Several manufacturers have designed welding helmets to also serve as air purifying respirators equipped with chemical cartridges to capture fumes. Spark-resistant clothing should be employed as well to protect critical body parts of the workers from sparks, heat, and hot metals generated during the welding activities.

Table 15.2
Silica PEL, OSHA Standard 29CFR1926.55

Silica Form	PEL (mmpcf)[1]
Crystalline	
Quartz (respirable)	$250/(\% SiO_2 + 5)$ or $10\ mg/m^3/(\% SiO_2 + 2)$
Quartz (total dust)	$30\ mg/m^3/(\% SiO_2 + 2)$
Cristobalite and tridymite - use ½ the value calculated from the count or mass formulae for quarts	
Noncrystalline	
Amorphous, including diatoma-ceous earth	20
Silicates (less then 1% silica)	20
Silicates, Portland cement	20
Silicates, soapstone	20
Silicates, talc (non-asbestiform)	20
Silicates, talc (fibrous), use asbestos limit	–
Graphite (natural)	15

1. mppcf = Millions of particles per cubic foot, converted to million particles per cubic centimeter by multiplying 35.3 x mppcf.

After concrete operations are completed, the superstructure will be erected. As the structure is erected, there will be a myriad of metal-working tasks that will need to be accomplished. These activities will include welding operations, as well as activities involving the use of heavy machinery and hand tools. Fumes generated from welding operations will have to be sufficiently controlled. Metal-working activities can result in excessive noise exposure requiring the use of appropriate hearing protection. Noise monitoring will help determine if hearing protection will reduce employee noise exposure below the OSHA Permissible Noise Limits as detailed previously in Table 15.1.

d. *Fire proofing*
A fire-proofing agent will be applied to steel beams and decking to increase the overall fire rating for the materials. During the application process, workers should be protected with a half-face APR with HEPA cartridges, and appropriate gloves with canvas work gloves to minimize dermal contact. Since the material is generally spray-applied, it may be prudent to equip workers with disposable tyvek coveralls. The material applied as the fire-proofing agent can produce localized irritation upon contact, these materials are composed of man-made fibrous material such as mineral wool.

e. Trades activities

As a facility structure starts to "take shape," plumbing fixtures, electrical components, and fixtures are added. During these activities, workers can potentially be exposed to welding fumes, particulate matter, solvents associated with adhesives, compounding materials, and cleaning agents. As wallboard, or sheetrock, is added to a structure, workers can be exposed to particulate matter generated during the cutting, installation, and sanding of the sheetrock. Appropriate dermal protection and safety eyewear should be donned, as this dust can be irritating and result in contact dermatitis. Figure 15.3 shows an employee with a disposable respirator (dust mask) protecting him from dusts generated during sanding.

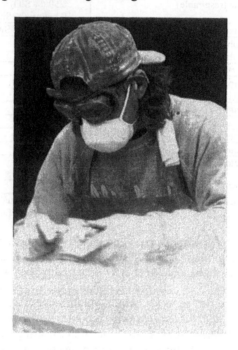

Figure 15.3 Exposure during sanding   (Courtesy: *Mine Safety Appliances Company*, Pittsburgh, PA).

f. Wallboard/sheetrock

After the wallboard has been installed, it will be coated either with paint or with a wallpaper covering. These products generally contain a volatile organic constituent and appropriate ventilation or respiratory protection should be employed. A variety of volatile organic compounds (VOCs) can be present in the paints, wallpaper adhesives, and surface preparation products. Consideration should be given to use low-VOC products. If general ventilation is sufficient to maintain VOC concentrations at acceptable concentrations, respiratory protection will not be necessary. Air monitoring with a direct reading instrument (DRI) will help to identify airborne contaminant concentrations to assess existing respiratory protection is sufficient.

g. *Carpet/floor tile installation*

Carpet and floor tiles are utilized as a decorative touch to cover flooring and improve overall structure aesthetics. The adhesive utilized to affix these materials to the substrate often contains VOCs. Appropriate dermal protection will be necessary, and respiratory protection may be required if airborne concentrations become elevated. Mechanical ventilation can be used to reduce airborne levels during application and installation. As with coating operations, monitoring with an applicable DRI will serve to assess ambient concentrations.

h. *Forklift trucks*

Diesel- or gasoline-powered forklifts can generate exhaust emissions consisting of particulate matter, carbon monoxide, oxides of sulfur, and oxides of nitrogen. Proper ventilation is particularly critical when the units are utilized in areas with limited or no natural (or mechanical) ventilation, as will often be the case when moving materials inside a facility. In order to reduce airborne contaminant concentrations, ancillary ventilation units may be necessary. Whenever possible, electric or propane forklifts should be utilized for any work activities that need to be conducted in enclosed areas, as dangerous confined space situations could arise.

C. Conclusion

The construction activities described illustrate a broad, simplified overview of some of the industrial hygiene issues that can be encountered while performing construction-related activities and operations. It should not be misconstrued as a complete overview, merely to direct that industrial hygiene implications exist throughout all phases of construction. The industrial hygienist must have an intimate understanding of the work tasks that are going to be performed and any media that will be impacted during the performance and completion of the construction effort. The industrial hygienist must be cognizant of the fact that the workforce, as well as the workplace, rapidly evolve and change as a given project progresses. At every step and phase of a construction project there are different stressors that can adversely impact worker health. Appropriate control mechanisms must be properly planned for and addressed prior to the commencement of a given project phase. Though briefly mentioned, the industrial hygienist most likely will also have responsibility for safety issues as well, so proper planning during the design phases of a project will focus on potential areas that may detract from worker safety and health.

III. HAZARDOUS WASTE OPERATIONS

A. Background

The hazardous waste and remediation industry has evolved from its inception in the late 1970s, early 1980s, to become a multibillion dollar industry. This industry was created to address the cleanup and remediation of uncontrolled hazardous waste sites that had served as industrial waste disposal sites throughout the first three-quarters of the 20th century. In order to address these cleanup activities, the Comprehensive Environmental Cleanup and Liability Responsibility Act (CERCLA) was passed in 1980. CERCLA, or Superfund as it commonly became known to be called, because of the funding mechanisms that were established, created criteria for the cleanup and remediation of sites contaminated with improperly disposed hazardous waste.

B. Love Canal: The Evolution of CERCLA

CERCLA became a necessity with the discovery of an abandoned hazardous waste disposal site in Niagara Falls, New York: the *infamous Love Canal Site*. This site served as an industrial dumping ground for over 10 years before the property was donated for the cost of $1.00 to the Niagara Falls Board of Education. The Board of Education erected an elementary grammar school and playground on a site that contained over 19,000 tons of various hazardous wastes. There were over 80 different contaminants disposed at the site, including strong acids, bases, and over a dozen known or suspected carcinogens. Concern was raised by community residents when inordinate numbers of illnesses and cancer incidents were thought to be recurring in the community. As investigations were conducted, it was learned that the school was built on an industrial waste dumping ground. Actions at the Love Canal site cost millions of dollars to address, including monies expended to relocate displaced homeowners, remedial investigations to characterize the scope and magnitude of site contamination, feasibility studies to determine the most prudent cleanup approaches, and cleanup operations.

C. "Typical" CERCLA Sites

Throughout the United States there are similar horror stories depicting instances where industrial waste disposal sites were implicated in adverse health events. These exposure events contributed to the passage of CERCLA, as abandoned hazardous waste sites pose potential exposure threat to public health. Contained in Table 15.3 is an overview of a number of Massachusetts sites on the U.S. EPA Region I CERCLA National Priorities List targeted for cleanup operations, and the contamination that needs to be addressed at each site. As depicted in the table, there is great deal of variability in the contaminant streams present at these sites. In order to mitigate this exposure threat, the contamination has to be remediated. Personnel involved in the cleanup of abandoned hazardous waste sites are at a much greater exposure risk than the general public who they are striving to protect.

Table 15.3
Examples of Massachusetts U.S. EPA Region 1:
National Priorities List Sites

Site Name & Location	Brief Description of Past Site Activities	Description of Waste Streams Present, or Contaminants Suspected to be Present
Otis Air National Guard/Camp Falmouth, MA	Site was used as a sanitary landfill and firefighting training area	Soil and groundwater contaminated with trichloroethane, tetrachloroethylene, and dichloroethylene
Atlas Tack Corp, Fairhaven, MA	Former manufacturing location of cut and wire tacks, steel nails, and similar items	Groundwater is contaminated with cyanide and toluene; on-site soil is contaminated w/toluene, ethyl benzene, heavy metals, pesticides, PCBs, and polycyclic aromatic hydrocarbons
Baird & McGuire, Holbrook, MA	McGuire was a chemical mixing & batching company from 1912 - 1983. Later activities included mixing, packaging, storing, and distributing pesticides, disinfectants, soaps, floor waxes, and solvents	Groundwater is contaminated w/pesticides and organic/inorganic chemicals; Cochato River sediments contain significant levels of VOCs, arsenic, and pesticides (especially DDT and chlordane); site soils contain VOCs, polycyclic aromatic hydrocarbons, pesticides, dioxin, and heavy metals
Blackburn and Union Privileges, Walpole, MA	These areas were part of 10 distinct water privileges. Miscellaneous manufacturing and processing operations accomplished at the site, including snuff, iron, nails, cotton, wool asbestos heavy metal processing and brake	Site soils, sediments, and groundwater are contaminated w/asbestos, lead, arsenic, nickel, VOCs, and non-VOCs
Charles-George Reclamation and Trust Landfill, between Tyngsborough, MA and Nashua, NH	During its active site operation, site was licensed to accept household, industrial, and hazardous wastes, including drummed and bulk chemicals	Groundwater contamination includes: benzene, tetrahydrofuran, arsenic, and 2-butanone; sediment-contaminants include benzo(a)pyrene
Fort Devens, Fort Devens, MA	Fort serves as an Army training site for active military personnel	Monitoring well contamination includes arsenic, cadmium, chromium, lead, iron, and magnesium; sediments contain arsenic, cadmium, chromium, lead, and mercury contamination; surface soils contain polycyclic aromatic hydrocarbon and petroleum hydrocarbon contamination

Table 15.3 (cont'd)

Site Name & Location	Brief Description of Past Site Activities	Description of Waste Streams Present, or Contaminants Suspected to be Present
Hanscom Field/ Hanscom Air Force Base, Towns of Bedford, Concord, Lexington, and Lincoln, MA	Former U.S. Air Force base/air field, currently is operated by the Massachusetts Port Authority as a civilian airport	Contamination identified in groundwater and soils includes chlorinated and aromatic solvents, jet fuel, and other petroleum-related hydrocarbons
Haverhill Municipal Landfill, Haverhill, MA	Site has served as an industrial and municipal landfill beginning in the late 1930s	Soils are contaminated with benzoanthracene, dibenzofuran, and VOCs. Liquids present on site have been found to contain chromium and arsenic. Drums have been found to contain VOCs, including xylene and toluene; groundwater is contaminated w/VOCs, arsenic, lead, mercury, manganese, and chromium
Hocomonco Pond, Westborough, MA	Site has been used for the following operations: wood-treating, involving the saturation of wood-products w/creosote; asphalt mining; and dry cement distributed in bulk	Soils, sediments, and groundwater are contaminated w/creosotes, carcinogens, arsenic, chromium, and other heavy metal compounds
W.R. Grace & Co (Acton Plant), Acton & Concord, MA	Site was former location of American Cyanamid and Dewey & Almy Chemical Company; these companies manufactured sealant products for rubber containers, latex products, plasticizers, resins, and other products; operations at W.R. Grace include the production of materials used to process concrete, make sealing compounds, latex products, and paper & plastic separators	Groundwater is contaminated w/VOCs and lead, arsenic, chromium, nickel, and other heavy metals; cadmium contamination is contained in site sediments. Soil and sludges are contaminated with arsenic and VOCs, including vinyl chloride, ethyl benzene, benzene, 1,1-dichloroethylene, and bis(2-ethylhexyl)phthalate

D. Industrial Hygiene Responsibilities at Hazardous Waste Sites

1. *Development of site-specific safety and health plan*

The unknown variability of the contaminant stream potentially present at hazardous waste sites raises a great challenge to the industrial hygiene professional. Though these responsibilities are similar to problems and issues encountered by a construction industrial hygienist, there is a substantive difference. The contaminants encountered at hazardous waste sites have far greater hazard and toxicity potential than contaminants typically encountered at construction sites. Hazardous waste cleanup projects can involve unknown materials, mixed waste streams, incompatible materials, pyrophoric compounds, flammable and combustible agents, and a host of other hazardous compounds. A detailed site-specific health and safety plan (SSHASP) must be developed prior to the commencement of any site activities at a hazardous waste site. The SSHASP, a requirement of the OSHA's *Hazardous Waste Operations Standards* covered in 29 CFR 1910.120 (General Industry) and 29 CFR 1926.65, is a tool employed to assess all potential safety and health risks that may be encountered during the performance of site activities, and develop a coordinating control strategy to adequately protect site workers.

The SSHASP requires that the industrial hygienist or site safety & health officer, have, at a minimum, a thorough understanding of contaminant streams and pathways that are potentially present at the given site. This knowledge is essential to determine appropriate PPE, air monitoring strategies, potential hazardous atmospheres that may be encountered, and other issues of critical importance. As well, the SSHASP will list the potential hazards associated with all site work activities. This is a critical component, as the potential hazards associated with activities such as drum handling or drilling are far greater than operations that do not involve intrusive activities or the direct handling of potentially contaminated materials. Table 15.4 details essential components that the SSHASP addresses.

2. Assignment of personal protective equipment (PPE)

Proper PPE assignment is essential to ensure that site workers are sufficiently protected from exposure to potential site contaminants. PPE is selected based on site-specific criteria that can be impacted by any or all of the following: airborne contaminant concentrations, hazard potential of site contaminants, presence of oxygen-deficient environments, on-site confined spaces, whether or not a *known* (or unknown) contaminant is present, etc. Proper assignment of PPE during the performance of site work activities is essential to minimize and prevent potential exposure to site workers. Due to the uncontrollable nature of hazardous waste sites, engineering controls are generally not practical or feasible, so proper PPE plays an important role in exposure control. PPE *levels* were developed to provide guidance as to when PPE should be upgraded or downgraded during site work activities. Table 15.5 contains an overview of PPE levels for hazardous waste site work, as well as detailing when a given PPE level should be employed. Figure 15.4 demonstrates and example of a PPE ensemble utilized for hazardous waste site work.

Table 15.4
Components of a Site-Specific Health & Safety Plan (SSHASP)

Component	Description
Hazard analysis	Assessment of safety & health hazards associated with all potential site activities and operations
Training requirements	Specific training necessary to conduct site activities, including: use of PPE; hazards potentially present on-site; initial hazardous waste operations training; use of air monitoring equipment; medical surveillance requirements; and work practices to minimize exposure
PPE	Detail PPE to be used during all site work tasks and operations
Medical surveillance requirements	Detail medical surveillance requirements required for site operations
Air monitoring requirements	Detail of the types of air monitoring that will be performed to prevent exposure and identify potentially oxygen-deficient or flammable atmospheres
Site control measures	Delineation of work zones, specifically: the hot zone (area where contamination exists); decontamination zone (area to enter hot zone and remove contamination after exiting hot zone); and control zone (area to suit up, stage equipment, etc.–no contamination exists in this area)
Decontamination	Protocols to remove site contamination from personnel and equipment prior to entering control zone
Emergency response	Procedures to contact emergency response personnel in the event of a site emergency
Confined space entry	Procedures to enter confined spaces potentially present on-site
Spill containment	Detail procedures to control the inadvertent release of contaminants during site operations

The use of PPE during hazardous waste site work can have severe physiologic effects on workers. Hazardous waste site workers routinely experience heat-related ailments because of the PPE that is utilized and the types of activities that are undertaken. A detailed program should be in place to evaluate thermal ambient parameters and worker physiologic responses. Chapter 10, Thermal Hazards, contains a thorough overview of the requirements necessary to establish an effective heat stress program.

Figure 15.4 Level B PPE during waste site work (Courtesy: *Mine Safety Appliances Company*, Pittburgh, PA).

Air monitoring with direct reading instrumentation for volatile organic compounds should be conducted periodically (or as conditions change) with a photoionization detector or an organic vapor analyzer to ensure the level of PPE prescribed for the given site task is sufficient to adequately protect site workers from exposure. Details of monitoring frequency, instrumentation, and calibration protocols should be clearly delineated in the site-specific health and safety plan.

Table 15.5
PPE Level Selection Criteria for Hazardous Waste Site Work

PPE Level	Description of PPE Elements Potentially Included	When PPE Level Should be Employed
D	Coveralls; work gloves; boots/shoes steel toe and shank; chemical-resistant boot coverings (if necessary); safety glasses or chemical splash goggles; hard hat	(1) Atmosphere contains no known hazard; and (2) Work functions preclude splashes, immersion, or the potential for unexpected inhalation of or contact w/hazardous levels of any chemicals
C	Full-facepiece or half-facepiece air purifying respirators; hooded chemical-resistant clothing; chemical-resistant outer boot covering, boots w/ steel toe and shank; inner and outer chemical-resistant glove; hard hat	(1) Atmospheric contaminants, liquid splashes, or other direct contact will not be absorbed through the exposed skin or create an adverse effect on contact; (2) Type of air contaminants have been identified, airborne concentrations are known, and an air-purifying respirator is an acceptable means to reduce contaminant exposure; and all criteria for the use of an air-purifying respirator are met
B	Positive-pressure, full-facepiece self-contained breathing apparatus (SCBA), or positive pressure supplied air respirator w/escape SCBA; hooded chemical-resistant clothing; inner and outer chemical-resistant glove; chemical-resistant boot covering over steel toe and shank boots; hard hat	(1) Types and concentrations of airborne contaminants have been identified and require a high level of respiratory protection, but *less* skin protection; (2) There is potential for oxygen deficiency (less than 19.5%); (3) There are airborne vapors or gases present as detected by direct-reading instrumentation, though they are not expected to pose skin hazards
A	Positive-pressure, full face-piece SCBA; totally encapsulating, chemical-resistant suit; inner and outer chemical-resistant gloves; chemical-resistant, steel toe and shank boots; hard hat; disposable protective suit, gloves, and boots	(1) Hazardous substance has been identified and requires highest level of protection for skin, eyes, and respiratory system based on measured or potential airborne contaminant concentrations; (2) Substances present or suspected to be present are known to present a high hazard to the skin; (3) Operations conducted in poorly ventilated areas, and assessments have not been conducted to downgrade to a different PPE Level

3. *Delivery of appropriate training*

It is imperative that general site workers such as equipment operators, laborers, and site supervisors who have exposure potential need to receive appropriate training, as hazardous waste site workers can be exposed to a myriad of contaminants, depending upon what was disposed at the given site. Workers can encounter oxygen-deficient environments and flammable/explosive environments as well. An effective training program will ensure that workers are properly educated on the potential safety and health hazards present at these work sites. According to the OSHA HazWoper Standards, a minimum of 40 hours of off-site training covering the following elements must be presented to individuals performing general site work activities:

- Names of personnel and alternates responsible for site safety and health
- Safety, health and other hazards present at hazardous waste sites
- Use of personal protective equipment
- Work practices by which the employee can minimize risks from hazards
- Safe use of engineering controls and equipment on the site
- Medical surveillance requirements, including recognition of symptoms and signs which might indicate overexposure to hazards
- Components of site safety and health plans.

In addition to 40-hours of classroom-type training, site field workers are also required to have a minimum of 3 days of actual, supervised field experience under the direct oversight of a trained, experienced supervisor. The intention of this requirement is to ensure that information delivered to site workers during the 40-hour training class has been retained by the student, and to evaluate/supervise workers during actual conditions.

Hazardous waste site workers who occasionally access site for specialized activities, such as sample collection, land surveying operations, geophysical surveying, and other similar activities where overexposure to an applicable OSHA Permissible Exposure Limit is unlikely, are only required to receive 24 hours of training instruction. One day of supervised on-site activity is also required under the supervision of a trained, qualified individual.

E. Emergency Response Operations

In addition to abandoned hazardous waste sites, the HazWoper Standards also address emergency cleanup operations. These types of situations can include: transportation mishaps; accidental releases of hazardous chemicals at manufacturing locations; release of hazardous materials during handling and transfer operations; and other unexpected events that require and necessitate the need for prompt attention. Incidental releases of hazardous substances where the material can be absorbed, neutralized, or controlled at the time of release by individuals in the immediate work area are not covered as an OSHA emergency response operation under the guise of the HazWoper Standards, but not addressed or covered under the guise of either the general industry or construction Hazard Communication Standards. Individuals involved in teams responding to emergency response operations must be properly educated on the inherent hazards associated with any and all situations that they may be expected to encounter and the potential hazards associated with each type of situation.

For example, a *HazMat Response Team* receives an emergency call from a state police hotline that there is an overturned tanker on a highway spilling liquid. Based on a call that has provided limited information, the emergency response team will have to mobilize and prepare for any type of potential waste stream, from a nonhazardous material, to a compound that is extremely toxic, presenting an immediate health threat for all site response personnel. When the HazMat response team arrives at the site where the tanker truck has overturned, team members try to collect as much information as possible to potentially identify the material being released from the overturned tanker. Information can be obtained from the driver of the tanker if he/she has not sustained a debilitating or serious injury, DOT placarding contained on the exterior of the truck, shipping papers, MSDSs, or other sources of information that may identify the material being released. The more information that can be collected by the HazMat team, the more effectively they can prescribe PPE to suit up and control the release episode. Background information must be collected quickly, as material continues to be released from the tanker as the HazMat team collects data. Figure 15.5 shows workers responding to a tanker car emergency.

After the Hazmat team has determined what the potential waste stream is, work zones are established. The Control Zone is positioned upwind from the overturned tanker. The Decontamination Zone serves to prevent the migration of contamination from the Hot Zone into the Control Zone. The Hot Zone (or Exclusion Zone) is the area where the contamination exists. This zone should be located downwind from the Control Zone to prevent contamination migration. After appropriate PPE has been donned, workers will enter the Hot Zone. Upon entering the Hot Zone, workers will perform operations that will include: air monitoring/surveillance, determining the approximate amount of material that was released, identifying if potential ignition sources are present if dealing with flammables; however, the number one responsibility will be to control the release of the material from the tanker using shut-off valves, sorbent boom, socks, or pillows, diking around the tanker, etc. This is accomplished to prevent further environmental contamination and minimize additional adverse environmental effects.

Depending on the toxicity of the material that has been released, workers may have to exit the Hot Zone to change into new PPE, add an air cylinder, or replenish lost fluids. Once PPE has been replaced and the release has been *controlled*, site workers clean up any other contamination that has impacted soil or surface water. Materials used to clean up contamination are disposed of in appropriate salvage drums. This example presents an overview of some of the operations that are undertaken in response to the emergency release of hazardous contaminants.

Figure 15.5 Workers in Level A PPE responding to a tanker car emergency (Courtesy: Mine Safety Appliances Company, Pittsburgh, PA).

Hazardous waste operations present great challenges to industrial hygienists and site safety specialists. There are numerous variables that have to be considered to properly protect the worker and prevent additional environmental damage while cleaning up existing contamination. Worker physiological response to site stressors and PPE must be carefully thought out and planned for in order to properly prepare for site operations. Careful planning and preparation will serve to prevent potential site hazards from adversely impacting site workers as well as the general public.

IV. ASBESTOS

Asbestos is one of the most widely regulated workplace contaminants in the United States because of its demonstrated carcinogenic potential, having regulatory coverage by OSHA, the U.S. EPA, and a myriad of state and local requirements that include licensing, disposal, notification, and other elements.

Asbestos has been linked to cancers of the lung, pleural linings of the lung and peritoneum, and occasionally, to the gastrointestinal tract. Studies have indicated that there is a synergistic effect exhibited when an individual is exposed to asbestos and is a tobacco smoker. As depicted in Table 15.6, if a worker is exposed to asbestos and smokes a pack of cigarettes per day, there is 50 times greater chance the individual will contract cancer of the lung. This value increases to 100 times when a two pack-a-day smoker is exposed to asbestos. Asbestos is the only known causative agent to produce mesothelioma, a rare cancer type, in humans. In addition to its carcinogenic potential, asbestos also produces a debilitating lung disease, *asbestosis*. Asbestosis is a disease form consisting of diffuse interstitial fibrosis which results in substantially reduced lung capacity and pulmonary function.

Table 15.6
Synergistic Effects of Asbestos and Tobacco Smoke on Lung Cancer

# of Packs of Cigarettes Smoked per Day	Exposure to Asbestos	Increased Risk of Lung Cancer
Zero	None	N/A
Zero	Yes	5 Times
One Pack	None	10 Times
One Pack	Yes	50 Times
Two Packs	Yes	100 Times

A. Asbestos Hazard Emergency Response Act

Asbestos's role as a carcinogen led to the establishment of the Asbestos Hazard Emergency Response Act (*AHERA*), 40CFR763, in 1984, a program designed to prevent asbestos exposure to children and develop a mechanism to manage asbestos in schools. AHERA's coverage is applicable to all elementary and secondary schools, grades K-12, regardless of whether the school is public, private, or parochial. Table 15.7 contains a summary of AHERA's key provisions. AHERA's provisions established a management infrastructure to assess, inspect, evaluate, document, educate, and inform all individuals having a need to know of the asbestos hazards present in the school. It was felt that the establishment of a unified, systematic approach to manage all issues related to asbestos, from inspection and assessment to abatement actions, would prevent unnecessary exposures in schools.

A fundamental provision of AHERA's coverage was the establishment of a process mandating inspections and re-inspections of the primary and secondary schools. The purpose of the inspections is to assess the location, condition, and exposure potential of asbestos-containing building materials (ACBM) based upon the ACBM's *friability*. According to AHERA, a material is considered friable if it is capable of being reduced to powder or dust with hand or mechanical pressure. AHERA categorized ACBM in seven hazard assessment categories, based on potential to create exposure events. The AHERA Standard recognized that certain asbestos types had greater potential to create exposure episodes based on the given materials ability to release fibers that can become airborne. This fiber release capability is related to the material's friability. Table 15.8 depicts the AHERA assessment categories. The type of ACBM will dictate the minimum number of samples that will have to be collected to verify whether a *homogeneous* sampling area can be called non-ACM.

Table 15.7
Key AHERA Provisions

AHERA Issue	Description
Inspection	Established process to conduct initial inspections to assess condition, with regularly scheduled re-inspections
Management Plan	Written operations & maintenance (O&M) program detailing asbestos location, condition, and applicable response action necessary in the event of condition change, damage, accidental fiber release, routine operations & maintenance, etc.
Certification/Training Requirements	Established certification program requiring specific training requirements for individuals performing activities in schools
Analytical/Laboratory Protocols	Procedures to address bulk sample/analytical collection techniques based on surface area/linear footage, and type of material
Notification	Established procedures to notify school employees and community members of asbestos response actions occurring in schools

In conjunction with the AHERA inspection process, each school was required to develop and maintain a written *Management Plan* indicating the locations of ACM and the applicable condition assessment; critical response actions to prepare for incidents involving releases of ACM, and other issues necessary for a school to effectively manage asbestos-containing building materials in schools.

Table 15.8
AHERA Assessment Categories

Assessment Categories
• Damaged/significantly damaged thermal system insulation (TSI)
• Damaged friable surfacing ACM
• Significantly damaged friable surfacing ACM
• Damaged or significantly damaged friable miscellaneous ACM
• ACBM with potential for damage
• ACBM with potential for significant damage
• Any remaining friable ACBM or friable suspected ACBM

B. Asbestos Response Actions

Asbestos response actions have created significant challenges for the industrial hygienist. While the overall purpose of an asbestos response action is to potentially reduce overall exposure to the occupants of a facility, the physical action of *abating* asbestos creates substantial exposure for the workers involved in the abatement process. The three primary types of asbestos abatement actions can be classified as either: 1) removal; 2) encapsulation; and 3) enclosure. Each abatement process is designed to address ACM to prevent the release of asbestos fibers into ambient air. Abatement activities are conducted using *wet methods*. Asbestos being abated should be constantly saturated with an *amended water solution*. Amended water consists of approximately 15 milliliters of surfactant-wetting agent to 1 liter of water. The use of wet methods during an abatement operation assists in minimizing airborne fiber concentrations in the regulated area.

1. *Removal*

Each asbestos abatement process addresses asbestos control by employing a different approach. The *removal* of asbestos involves physically removing the asbestos-containing materials from the substrate that they are insulating or covering. This process may entail scraping, cutting, manual handling of ACM, and other activities that can potentially result in airborne fiber release. It is the only asbestos abatement operation that actually results in the elimination of the asbestos from the substrate, and subsequently, the facility. For small sections of pipe insulation requiring an abatement action, glove-bags can be employed to contain the asbestos.

2. *Encapsulation*

Encapsulation involves the application of an appropriate encapsulant which serves to seal in the asbestos to prevent fibers from being released. The encapsulating material typically is an adhesive-mastic material that physically binds the asbestos fibers. Latex-based paints have been demonstrated to be effective as encapsulants, especially for thermal system insulation (TSI) on pipe and elbow materials. The asbestos remains in place, though the threat of fiber release has been controlled because the asbestos has been effectively sealed. Encapsulation can be compared to sealing a pile of cooking flour with glue. Prior to applying the glue, the flour can easily become airborne. After application of the glue, the flour is sealed and cannot become airborne. One potential drawback associated with encapsulation is that the asbestos still remains within the facility, and if the encapsulated section of asbestos is damaged, hit, abraded, etc., there is the risk of fiber release, and thus exposure.

3. *Enclosure*

Enclosure involves the establishment of a physical, airtight barrier between the asbestos and the ambient environment. An airtight wall or barrier is constructed to prevent the asbestos from becoming airborne. An example may be a sheetrock enclosure constructed around a TSI riser in a classroom. By enclosing the asbestos insulation, the classroom occupants are protected from potential fiber release episodes because the asbestos is physically enclosed within the sheetrock structure. Other response actions are designed from an operations and maintenance standpoint to address the minor repair of ACM that is damaged. These repair operations may involve removal, encapsulation, enclosure, or combinations of each. All asbestos response actions can result in potential exposure to workers. Similar to encapsulation since the asbestos is not removed from the facility, the potential for damage and subsequent fiber release remains.

C. Personal Protective Equipment for Abatement Actions

Workers who perform asbestos abatement activities should be equipped with an appropriate respirator that will be selected based on the given operations that are going to be performed, previous historical air sample results, and other related concerns. At a minimum, a half-mask air purifying respirator equipped with HEPA cartridges. If air sampling indicates elevated airborne asbestos concentrations, workers may have to upgrade respiratory protection to a more effective respirator, as detailed in Table 15.9. All details must be captured and documented in a written respiratory protection program.

Table 15.9
Respirator Selection Criteria

Conditions of Use	Respirator to Utilize
Not exceeding 1 f/cc (10 X OSHA PEL of 0.1 f/cc)	Half-mask APR with HEPA cartridges, other than a disposable respirator
Not exceeding 5 f/cc (50 X OSHA PEL)	Full face piece APR with HEPA cartridges
Not exceeding 10 f/cc (100 X OSHA PEL)	Any powered APR equipped with HEPA cartridges, or any supplied air respirator
Not exceeding 100 f/cc (1000 X OSHA PEL)	Full face piece supplied air respirator operated in pressure demand mode
Greater than 100 f/cc or unknown concentration (greater than 1000 X OSHA PEL)	Full face piece supplied air respirator operated in pressure demand mode equipped with an auxiliary positive pressure SCBA

Source: OSHA Standard 29 CFR 1926.1101.

Disposable hooded, full-body tyvek coveralls are worn during removal operations, with appropriate boot coverings, as well as latex gloves. Hard hats, canvas work gloves, boots with steel toe and shank should also be utilized during site activities.

Since asbestos abatement actions are labor intensive, work is generally accomplished indoors where air-conditioning has been shut off and PPE is essential to complete project, workers have an increased risk of heat-related injuries. A heat stress program should be implemented to ensure that workers are sufficiently protected from heat-related injuries. The program should include provisions for physiologic monitoring, liquid replenishment, rest breaks, addition of cooling vests, fans, and other components to protect and prevent heat stress. Refer to Chapter 10 for more detailed coverage of heat exposure topics.

D. Asbestos Work Zone Establishment
1. *Maintenance of the regulated area*

Abatement operations are accomplished with containment to ensure that there is segregation between the area where the asbestos is being abated, the *regulated area*, and the area where asbestos is not present. The regulated area is maintained under negative pressure using a quasi-local exhaust ventilation technique, *negative air machines*. The purpose of these units is to ensure that pressures within the containment are significantly less than ambient pressure outside the containment area. Static pressures should range from 0.02 to 0.10 inches of water gauge to maintain negative pressure within the regulated area.

Pressure measurements can be collected with a manometer or a pressure gauge (refer to Chapter 6 for greater detail on measurement techniques). If static pressure is demonstrated to be reduced, work should be stopped and the negative air machine checked and static pressure values corrected. Prior to commencing any work activities, static pressures should be reevaluated. In addition to ensuring proper static pressures are maintained, negative air machines are equipped with HEPA filtration at the exhaust ports so that containment area contamination is not exhausted to the outside. Air should be exhausted outside of the facility where the abatement project is being undertaken. In order to ensure proper air exchange within the regulated area, there should be no less than four (4) air changes per hour (ACH). The following equation can be employed to determine the appropriate number of negative air machines:

$$\# NAM = \frac{ACH \times V}{NAM\ Field\ Eff \times 60} \tag{15.1}$$

Where:

NAM	**= Negative air machine**
ACH	**= Air changes per hour, for asbestos abatement, use 4**
V	**= Room volume, ft^3**
NAM Field Eff	**= Design volumetric flow rate, ft^3/min (CFM) of NAM**

Sample Calculation: An abatement project to remove damaged asbestos-containing boiler and pipe insulation is to be undertaken in a boiler room that has the following dimensions: Length = 80 ft; Height = 20 ft; and Width = 60 ft. The negative air machines (NAMs) to be used for the project are designed with a volumetric flow rate of 2000 CFM. Assuming that all of the space in the room will be included in the establishment of the containment area, how many NAMs will be necessary to maintain 4 ACH?:

$$\# NAM = \frac{ACH \times V}{NAM \; Field \; Eff \times 60}$$

$$\# NAM = \frac{4 \times [(80 \; ft) \; (60 \; ft) \; (20)]}{2000 \; ft^3/min \times 60}$$

$\# NAM = 3.2$ (round up to next #), so 4

To maintain 4 ACH, there will be 4 NAMs needed.

Polyethylene sheeting with a thickness of at least 6 mils is used to construct the containment barrier of the regulated area. Typically, the polyethylene is affixed to a plywood frame. *Critical barriers* are established around air supply and return ducts and grilles, electrical equipment, telecommunications equipment, communications equipment, and other key apparati that could be damaged if contacted with water.

2. Decontamination

A staged decontamination chamber is constructed to provide controlled access and egress into the abatement work area. The purpose of the staged decontamination chamber is to systematically remove any contamination obtained in the regulated area. Upon exiting the regulated area, workers enter an *equipment drop room*, followed by a *shower/decontamination room*, then a *clean change room*. Any gross contamination or debris on workers or equipment is removed prior to exiting the regulated area. Equipment utilized in the containment area that is reuseable, is dropped in the equipment drop area, as it will be decontaminated for future use. Reusable equipment may include: specialized tools/power tools; portable generators; scaffolding; negative air machines; etc. This area is equipped with appropriate bags and labels to dispose of non-reusable, contaminated PPE (tyvek coveralls, boot covers, gloves, etc.). Workers exit the equipment room and the shower area. The shower room is in place to decontaminate workers. All shower water is contained and containerized, as it is assumed to be contaminated with asbestos. After showering, workers proceed to the clean change room to don street clothes and exit the work site.

Upon completion of an asbestos abatement operation, the work area is thoroughly cleaned of any residual dust or debris utilizing HEPA vacuums and/or wet wiping techniques. All exposed surfaces are cleaned to remove any residual asbestos potentially present.

E. Air Sampling

1. Personal sampling

Air sampling is conducted during all abatement activities related to evaluate airborne asbestos concentrations. Mixed cellulose ester fiber filter cassettes with 50 millimeter length cowl, having a 25 millimeter diameter are used to collect airborne asbestos samples. Personal sampling is conducted to assess exposure to asbestos during work operations. The current

OSHA PEL for asbestos is 0.1 fibers/cubic centimeter (f/cc) as an 8-hr TWA; the Excursion Limit is 1 f/cc. The recommended flow rate for the collection of personal samples is 0.5 liters per minute (LPM) to 2.5 LPM.

2. *Area sampling*

Area samples are collected from multiple locations, including: in the regulated area; at the exhaust ports of the negative air machines; outside of the decontamination chamber; and general ambient samples collected at strategic locations outside of the containment area. For area samples, the recommended flow range is 0.5 LPM to 16 LPM. The higher upper flow rate value enables industrial hygienists to collect large sample volumes in less time. Area samples are routinely analyzed microscopically on-site so that results are available *real-time*. Each type of area sampling has a specific function when conducted during abatement operations. Area samples collected within the containment area during abatement work to assess if wet methods are sufficient to control airborne asbestos fibers, or if greater amended water needs to be applied. As well, this data can be used to determine if respiratory protection is acceptable or if respiratory protection should be upgraded. Area samples collected at the exhaust ports of the negative air machines are analyzed to ensure that the HEPA filtration is functioning properly and has not become overloaded, causing fiber breakthrough. Appropriate HEPA filtration changes will be made if necessary based on results. Samples collected at strategic locations outside of the containment area are collected to ensure the containment's integrity. Corrective actions are taken to repair any breach of containment if needed, based on results. Area samples are also collected at the exit point of the decontamination chamber to ensure that employees and equipment have been properly decontaminated so that asbestos fibers are not carried through the decontamination chamber into the clean environment. The collection of personal and area samples during abatement project work is essential to protect site workers, the containment environment, and the ambient environment outside of the containment area.

Area sampling is also conducted in the containment area at the completion of the project, which is known as *clearance sampling*. Clearance sampling is conducted using aggressive techniques, including the use of leaf blowers to ensure that there is no asbestos still requiring action. If clearance samples come up "clean," the contractor can commence breakdown of the containment area and the project can be considered successfully completed.

F. OSHA Requirements

In addition to coverage by the AHERA Standard for response actions in primary or secondary schools, all abatement activities are covered under the guise of the OSHA Asbestos Construction Standard, 29 CFR 1926.1101, *Asbestos*. The OSHA Standard has established requirements related to the abatement of asbestos to ensure that workers are properly protected during work operations. Table 15.10 provides an overview of some of the key OSHA Construction Standard provisions. The objective of the OSHA Standard is protect workers involved in the abatement of asbestos, as well as establishing detailed criteria to ensure that adjacent spaces are not contaminated during the completion of the given abatement project. There are four work activity *classes* that are covered in the OSHS Asbestos Construction Standard. Class I activities involve the removal of TSI and surfacing ACM or Presumed ACM (PACM). Class II asbestos work activities are defined as those involving the removal of ACM or PACM that is not TSI or surfacing material. Class III work activities include repair and maintenance operations involving ACM or PACM, and Class IV work involves maintenance and custodial activities where employees may come in contact with ACM or PACM, but will not disturb the materials. Class III and IV activities involve work operations for employees covered under the guise of OSHA's General Industry Asbestos Standard, so there is some overlap between the two Standard's coverages.

The OSHA Standard presents an infrastructure for construction and general industry employers to address and manage a myriad of asbestos-related exposure concerns, while providing guidance as to the procedures to employ to address abatement and response actions. Asbestos will continue to present exposure threats to workers going forward, so industrial hygienists will need to be knowledgeable of asbestos-containing materials and their exposure risk.

V. OTHER CONSTRUCTION HEALTH-RELATED HAZARDS

Thus far, this chapter has presented examples of specific construction-related activities and the potential health hazards encountered during the performance and completion of various construction projects. However, there are numerous other activities that can result in worker exposure. These activities can include exposures resulting from abrasive blasting operations; roofing and road resurfacing; structural demolition; and bridge and concrete repairs. There are other construction operations, as well, that will result in potential worker exposure to potential stressors. The following section will provide examples of contaminants that can be encountered in the construction industry and the applications where the toxins can be expected to be present. Each contaminant presents exposure potential that can result in serious health implications. Proper control strategies will have to be implemented to protect workers from exposure. It will be the role of the industrial hygienist/site safety & health officer to adequately prescribe appropriate control intervention strategies to prevent exposure to the workplace.

Table 15.10

Overview of Components of OSHA Standard 29 CFR 1926.1101

Standard Element	Scope
Permissible Exposure Limit (PEL) Criteria	A PEL of 0.1 f/cc was established, as an 8-hr TWA and an Excursion Limit of 1.0 f/cc were established
Regulated Areas	Establishment of barriers in *regulated areas* where abatement work is to be undertaken (regulated area refers to an area where Class I, II, III, or IV work activities are to be undertaken and exposure is likely to exceed the PEL and/or the Excursion Limit)
Hygienic Requirements	Prohibits smoking, eating, or drinking in regulated areas
Exposure Assessment	Performance of an initial exposure assessment before conducting a new operation that could potentially result in asbestos exposure, including employee notification requirements
Asbestos Handling Procedures	Ensure that wet methods are employed when addressing/handling asbestos during abatement operations
Prohibitions	During asbestos removal operations the following activities are prohibited: (1) use of high-speed abrasive disc saws; (2) use of compressed air; (3) dry sweeping; (4) employee rotation as a means of reducing employee exposure to asbestos
Work Practice Protocols	Specific procedures on how to address asbestos, depending on the Class of work that is being performed, I, II, III, or IV
Competent Persons	A competent person shall be assigned to ensure proper asbestos hazard recognition/assessment and select appropriate control strategies. This individual will be able to: (1) identify workplace asbestos hazards and select appropriate control strategies; (2) obtain appropriate AHERA-type training based on the Category of work to be evaluated
Training	Specific training requirements are mandated based upon the Category of asbestos work to be performed
PPE Requirements	Establishment of required PPE to be employed during the performance of activities that potentially impact asbestos
Decontamination	Mandates procedures to remove asbestos contamination from workers and equipment prior to exiting the job site
Landlord Requirements	Stipulates that landlords must inform tenants/building occupants of activities that impact/disturb asbestos
Appropriate Signage	Provide appropriate danger signage in areas in which employees can be reasonably expected to enter which contain ACM or presumed ACM (PACM)
Medical Surveillance	Requirements for medical surveillance programs which include baseline exams, annual surveillance, and post-employment evaluations for any employee

Table 15.11 contains an overview of contaminants that are routinely involved in construction exposure incidents, as well as the typical applications/work operations where the contaminant is encountered. As depicted in the table, construction workers have potential exposure to a variety of chemical toxins that produce outcomes ranging from irritation to carcinogenesis. Depending on the given construction operation, there may be potential exposure to multiple contaminants. The site safety officer/industrial

hygienist must design control strategies that protect workers from all potential contaminants that may be present at a construction site.

Table 15.11
Contaminants Potentially Present at Construction Sites

Contaminant	Construction Operation/Task
Silica	Abrasive blasting; concrete work; (sawing, cutting, drilling, etc.); concrete/masonry demolition; dry sweeping/air blowing concrete, rock or sand dust:
Diisocyanates	Painting w/isocyanate- or polyurethane-containing paints; application of polyurethane water-proofing membranes; installation of roofing foam
Lead	Welding; renovation/demolition activities; lead-coated structure repair; plumbing work on lead lines; lead-paint abatement
Metal fumes	Welding, cutting, brazing
Carbon monoxide	Use of gasoline-powered equipment
Xylenes; toluene	Paint/coating operations; degreasing/solvent applications
Formaldehyde	Installation/application of formaldehyde-containing pressed wood products
Fiberglass	Fiberglass installation, including: piping, attic, ducts, acoustical insulation
Exposure to hot/cold environments	Any outdoor construction activities; work in indoor environments with limited/no thermal (hot or cold) conditioning or control
Epoxies	Use of caulks, adhesives; flooring installation
Ethers	Paint/coating operations; certain solvent applications
Asbestos	Demolition projects; constructions; renovations

VI. CONCLUSION

Construction workers have a variety of contaminants that can potentially be encountered during the performance of job activities. Many construction activities have work tasks that can result in exposure to multiple contaminants. This chapter presented an overview of some typical construction-related activities that can result in worker exposure. The objective was to focus reader attention to the fact the construction workers can have exposure potential, from the beginning phases of the construction of a facility, to the ultimate demolition of the structure if applicable. Past industrial waste disposal practices resulting in abandoned hazardous waste sites have created exposure during the cleanup and remediation of the sites to protect the public from exposure. As well, protecting the public from the exposures associated with lead and asbestos in facilities and on facility components has evolved into a construction industry segment designed at the abatement of the hazards associated with these materials. Other construction activities have also been touched upon that can result in potential exposures.

If the industrial hygienist or site safety & health officer is to truly protect the health (and safety) of the construction workforce, the individual will have to have an intimate knowledge of the workers, the tasks to be performed and the equipment utilized, and the site where the given work will be accomplished. This knowledge will have to include information regarding past site historical activities to ensure that previous site activities will not contribute to the worker's overall exposure potential. Each job/project will need to be addressed as an individual event, and planned and prepared for accordingly. Without this approach, the construction workforce will not be properly protected from the hazards potentially present on the sites.

VII. CASE STUDY

A construction project is underway to repair a damaged multilevel parking garage. The concrete has cracked in numerous locations, many expansion joints are in need of repair, and the concrete has sustained extensive water damage, as it was never coated with an appropriate water-proofing membrane. The parking garage is a four-level deck. The bottom three levels are below-ground. The top level is open on all sides, while the bottom three levels are completely enclosed. Automotive exhaust is removed from the bottom three levels with a ducted exhaust ventilation system. The upper level utilizes natural air movement to exhaust any generated automotive emissions.

Concrete and expansion joint repair is completed without incident; however from the commencement of the water-proofing phase of the project, there have been numerous problems that have resulted in three separate incidents of worker illness involving multiple workers. The Corporate Safety Officer from CMC Construction responded to the first job shutdown but could not positively identify the source of worker illness. Workers were experiencing respiratory discomfort/difficulty breathing. A number of workers commented that they could not catch their breath. Three workers had developed dermal irritation on their wrists, face, and necks. A review of the MSDS for the water-proofing coating revealed that the product was a polyurethane-based material, with toluene diisocyanate (TDI) as a key component. The MSDS referenced that mechanical ventilation was needed to maintain airborne contaminant levels below the applicable OSHA PELs or ACGIH TLVs. The ventilation system contained in the parking deck should suffice to remove airborne contaminant levels, though the Site Safety Officer never bothered to determine if the system was functioning as per design criteria requirements. All workers were utilizing appropriate chemical-resistant clothing, though no respiratory protection was employed. The MSDS referenced that if airborne concentrations exceeded the applicable OSHA PEL, then supplied air respiratory protection would be necessary, though respiratory protection would not be necessary if proper ventilation was in place.

When the Site Safety Officer received a call from the Project Engineer that the job was shut down a second time, an industrial hygiene consulting firm was contacted to determine what was going on and offer recommendations for corrective action. Each time work was stopped, coating operations were being accomplished in one of the below-ground parking decks. Upon review of the MSDS for the coating product, the Consultant Industrial Hygienist requested to speak with the Site Engineer to review air change rates for the ventilation system. Design specifications detailed that the system was designed for 4.0 air changes per hour (ACH). Based on the amount of product being applied and the volume of the deck, the industrial hygiene consultant determined that 4.0 ACH would be more than ample to maintain airborne TDI concentrations below the PEL and TLV, respectively. In order to ensure that the system was operating at 4.0 ACH, the Industrial Hygiene Consultant evaluated the ventilation effectiveness of the system. Upon review of the exhaust grilles for the ventilation system on the three below-ground decks, the Industrial Hygiene Consultant determined that the exhaust ventilation system was completely inoperable or totally shut down, as no exhaust ventilation volumes were measured.

The Industrial Hygiene Consultant learned through discussions with a building mechanic and a watch engineer that the system had been inoperable for several years. Requests had been submitted to upgrade the system, but monies were never approved because of the perceived low priority.

In order to correct the ventilation system's deficiencies, it was explained to the Industrial Hygiene Consultant that it would cost approximately $3500 to repair the units and get them up and running again. Other options included the utilization of five negative air machines at a rental cost of $200 per machine per week. The below-ground coating work was expected to take another 3 weeks to complete, assuming no additional problems. The total rental cost for the negative air machines would be $3000. The last option available was to equip all workers with supplied air respirators. The total cost to equip workers with this control strategy would be approximately $1750 for the project's duration, which would include all required training, medical surveillance, and written program development. Each option will result in potential down time (3 to 5 days), though the first option, repair of the system, will result in a down time of 2 to 3 weeks. It is anticipated that there is an additional 2 to 3 weeks of coating that still needs to accomplished on the top deck, however. Regardless of the option selected, the Industrial Hygiene Consultant has recommended that air sampling be conducted, workers employ supplied air respiratory protection until results indicate that it is unnecessary, and an evaluation be conducted to determine that workers properly utilize chemical-resistant clothing. The latter recommendation must be conducted prior to restarting coating operations on the top deck. If you were the Industrial Hygiene Consultant, what would your recommendations include and what do you think the consultant determines?

Upon review of the donning procedures for chemical-resistant clothing, it was learned that the workers who experienced dermal irritation of the wrist, neck, and face did not have PPE properly covering these body surface areas. Apparently, the gloves were rolled back to create a "cool" look; unfortunately, this left several inches of exposed skin that was coming into contact with the coating during the application process. The same workers also left the hoods of their tyvek coveralls down, so sections of their necks and face were exposed. The consultant delivered a tailgate session on how to properly don PPE for all workers so that they would adequately protect themselves when using the waterproofing membrane. The Industrial Hygiene Consultant recommended that the owner of the parking garage repair the ventilation system, as the system would need to be functioning to remove automotive exhaust when the garage is back in full operation and function. The garage owner agreed when he compared the costs of the three recommended alternatives that repair of the system was the only logical solution, even though the cost was the highest. It was a "no-brainer."

Upon repair of the system, the Industrial Hygiene Consultant determined that the system was operating at approximately 3.85 ACH. Air monitoring with workers previously trained in supplied air respiratory protection revealed that airborne concentrations were nondetectable, so the Industrial Hygiene Consultant's work was completed (after receiving a hefty check from CMC Contracting).

REVIEW QUESTIONS

1. Which hazard can be encountered at a construction site?

    a. Asbestos   **b.** Lead   **c.** Thermal hazards   **d.** All of the above

2. During the removal of asbestos, if the airborne concentration fiber of 4.0 f/cc in the containment area, the appropriate respirator would be:

    Half-face APR   **b.** Full-face APR   **c.** Full-face PAPR   **d.** SCBA

3. When working on a project involving the erection of a new structure, an industrial hygienist, or site safety & health officer, must be prepared to encounter the following contaminant during the application of paint:

    a. Acid aerosols   **b.** VOCs   **c.** Particulate matter   **d.** Silica

4. On a construction site, the use and operation of heavy machinery can result in potential overexposure to:

    a. Volatile organic compounds   **b.** Noise   **c.** Thermal hazards   **d.** Asbestos

5. Application of polyurethane-based waterproofing membrane can result in exposure to toluene diisocyanate. This compound can cause the following health-related effect(s) from overexposure:

    a. Chemical asthma   **b.** Respiratory discomfort   **c.** Dermal irritation   **d.** All of the above

6. Hazardous waste work requires the establishment of work zones. The zone expected to contain the most hazardous environments is the:

    a. Exclusion Zone   **b.** Decontamination Zone   **c.** Control Zone   **d.** Contaminant Zone

7. Level A PPE is designed to afford the maximal level of protection related to:

    a. Respiratory protection   **b.** Face/head   **c.** Dermal   **d.** All of the above

8. During an asbestos abatement project, how many air changes per hour is recommended as a minimum?

    a. 2.0 ACH   **b.** 3.0 ACH   **c.** 4.0 ACH   **d.** 5.0 ACH

9. When analyzing asbestos air samples, the following microscopic technique *is not* an appropriate analytical technique:

    a. TEM   **b.** PLM   **c.** PCM   **d.** SEM

10. Which regulation addresses asbestos?

    a. NESHAPs   **b.** AHERA   **c.** TSCA   **d.** All of the above

11. Given a containment area volume of 100 ft x 100 ft x 10 ft, and assume the negative air machine removes/exhausts air at rate of 2000 CFM. How many negative air machines (NAMs) are necessary to maintain an acceptable number of air changes per hour?

    a. 2   **b.** 3   **c.** 4   **d.** 5

12. The asbestos abatement process involving establishing an impermeable or impenetrable barrier between the damaged asbestos and the ambient environment is called:

    a. Encapsulation   **b.** Removal   **c.** Dissociation   **d.** Enclosure

Answers

1.   d.
2.   b.
3.   b.
4.   b.
5.   d.
6.   a.
7.   c.
8.   c.
9.   b.
10. d.
11. c.

$$\# \text{ of NAM} = \frac{\text{ACH x V}}{\text{NAM Field Eff x 60}}$$
$$\frac{4 \text{ x (100 ft) (100 ft) (10 ft)}}{2000 \text{ CFM x 60}}$$
$$= 3.33 \text{ or 4 NAMs}$$

12. d.

References

American Conference of Governmental Industrial Hygienists, 1994-*1995 Threshold Limit Values for Chemical Substances and Physical Agents and Biological Exposure Indices*, ACGIH, Cincinnati, 1994.

Belard, J., Beeckman, D. C., Hause, M. G., Wassel, J. T., and Stanevich, R. T., Hazardous waste abatement simulation in three controlled environments, *Professional Safety*, Vol. 41, No. 6, 1996, 33-36.

Bender, J. R., Konzen, J. L., and Devitt, G. E., *Occupational Exposure, Toxic Properties, and Work Practice Guidelines for Fiber Glass*, AIHA Publications, Fairfax, VA, 1991.

Bimonte, A., Making do just won't do when it comes to safety, *Professional Safety*, Vol. 41, No. 6, 1996, 8.

Chiras, D. D., *Environmental Science, A Framework for Decision Making*, 2nd ed., Benjamin/Cummings Publishing Company, Inc., Menlo Park, CA, 1988.

Clayton, G. D. and Clayton, F. E., Eds., *Patty's Industrial Hygiene and Toxicology, Volume 2A: Toxicology*, 3rd ed., John Wiley & Sons, New York, 1981.

Clayton, G. D. and Clayton, F. E., Eds., *Patty's Industrial Hygiene and Toxicology, Volume 2B: Toxicology*, 3rd ed., John Wiley & Sons, New York, 1981.

Coluccio, V. M., Ed., *Lead-Based Paint Hazards Assessment and Management*, Van Nostrand Reinhold, New York, 1994.

Confer, R. G., *Workplace Health Promotion Industrial Hygiene Program Guide*, Lewis Publishers, Boca Raton, FL, 1994.

Exposure Evaluation Division, Office of Toxic Substances, Office of Pesticides and Toxic Substances, U.S. Environmental Protection Agency, Guidance for Controlling Asbestos-Containing Materials in Buildings 1985 Edition, EPA 560/5-85-024, 1985.

Findley, M. E. and Timmons, T. N., Team safety in construction tapping into underground knowledge, *Professional Safety*, Vol. 40, No. 7, 1995, 23-25.

Harrington, J. M., Ed., *Recent Advances in Occupational Health*, Number Two, Churchill Livingstone, New York, 1984.

Hill, D. C., Waste remediation issues & technologies for the future, *Professional Safety*, Vol. 41, No. 3, 1996, 28-31.

Hortsman, S. W. and O'Keefe, T. P., Influence of ventilation on irradiance from a welding operation, *Professional Safety*, Vol. 41, No. 4, 1996, 45-46.

http://www.epa.gov/region01/remed/sfsites/atlastak.html; *Atlas Tack Corp., Massachusetts*, EPA Region 1 NPL Sites New England.

http://www.epa.gov/region01/remed/sfsites/baird.html; *Baird & McGuire Massachusetts*, EPA Region 1 NPL Sites New England.

http://www.epa.gov/region01/remed/sfsites/blackbrn.html; *Blackburn and Union Privileges Massachusetts*, EPA Region 1 NPL Sites New England.

http://www.epa.gov/region01/remed/sfsites/charlgeo.html; *Charles-George Reclamation Trust Landfill Massachusetts*, EPA Region 1 NPL Sites New England.

http://www.epa.gov/region01/remed/sfsites/fortdev.html; *Fort Devens Massachusetts*, EPA Region 1 NPL Sites New England.

http://www.epa.gov/region01/remed/sfsites/hanscom.html; *Hanscom Field/Hanscom Air Force Base Massachusetts*, EPA Region 1 NPL Sites New England.

http://www.epa.gov/region01/remed/sfsites/haverhill.html; *Haverhill Municipal Landfill Massachusetts*, EPA Region 1 NPL Sites New England.

http://www.epa.gov/region01/remed/sfsites/hocomonc.html; *Hocomonco Pond Massachusetts*, EPA Region 1 NPL Sites New England.

http://www.epa.gov/region01/remed/sfsites/index.html#MA; *EPA Region 1 - NPL Site Index: Massachusetts*.

http://www.epa.gov/region01/remed/sfsites/otisair.html; *Otis Air National Guard Camp Edwards Massachusetts*, EPA Region 1 NPL Sites New England.

http://www.epa.gov/region01/remed/sfsites/wrgrace.html; *W.R. Grace & Co., Inc. (Acton Plant) Massachusetts*, EPA Region 1 NPL Sites New England.

Kominsky, J., Freyberg, Gerber, R., Centifonti, G., and Brownlee, J., Evaluation of the implementation of asbestos operations and maintenance programs in New Jersey schools, *AIHA Journal*, Vol. 57, No. 11, 1996.

Marsicano, L., Getting lost in the rubble–is industrial hygiene receiving the attention it deserves in the construction health and safety arena?, *The Synergist*, Vol 6, No. 5., 1995, 18-19.

Mossman, B. T., Bignon, J., Corn, M., Seaton, A., and Gee, J. B. L., Asbestos: scientific developments and implications for public policy, *Science*, Vol. 247, January 19, 1990, 294-301.

Nwaelele, O. D., Prudent owners take proactive approach, *Professional Safety*, Vol. 41, No. 4, 1996, 27-29.

Office of Pesticides and Toxic Substances, U.S. Environmental Protection Agency, *Managing Asbestos in Place A Building Owner's Guide to Operations and Maintenance Programs for Asbestos-Containing Materials*, 20T-2003, July 1990.

Ross, K., New glovebag regulations allow safer, cheaper asbestos removal, *Occupational Health & Safety*, Vol. 65, No. 9, 1996 ,52-55.

Turk, J. and Turk, A., *Environmental Science*, 3rd ed., Saunders College Publishing, Philadelphia, 1983.

U.S. Department of Health and Human Services, Public Health Service, Centers for Disease Control, National Institute for Occupational Safety and Health, *An Evaluation of Glove bag Containment in Asbestos Removal*, NIOSH Publication No. 90-119, Oct 1990.

U.S. Department of Health and Human Services, Centers for Disease Control and Prevention, National Institute for Occupational Safety & Health, Preventing asthma and death from diisocyanate exposure, *NIOSH Alert*, No. 96-111, 1996.

U.S. Department of Health and Human Services, Centers for Disease Control and Prevention, National Institute for Occupational Safety & Health, Preventing silicosis and death in construction workers, *NIOSH Alert*, No. 96-112, 1996.

U.S. Department of Labor, Occupational Safety & Health Administration, *Title 29 Code of Federal Regulations, Part 1926.65, Appendix E - Training Curriculum Guidelines - Non-mandatory*, Feb 13, 1996.

U.S. Department of Labor, Occupational Safety & Health Administration, *Title 29 Code of Federal Regulations, Part 1926.65, Appendix B - General Description of the Levels of Protection and Protective Gear - Non-mandatory*, Aug 22, 1994.

U.S. Department of Labor, Occupational Safety & Health Administration, *Title 29 Code of Federal Regulations, Part 1910.120, Hazardous waste operations and emergency response*, Mar 7, 1996.

U.S. Department of Labor, Occupational Safety & Health Administration, *Title 29 Code of Federal Regulations, Part 1926.55, Appendix A - Gases, vapors fumes, dusts, and mists*, Oct 17, 1986.

U.S. Department of Labor, Occupational Safety & Health Administration, *Title 29 Code of Federal Regulations, Part 1926.1101, Asbestos*, as detailed in 59 Federal Register (FR) 40964, Aug 10, 1994; 60FR9624, Feb. 21, 1995; 60FR33343, Jun 28, 1995; 60FR33972, Jun 29, 1995; 60FR36043, Jul13, 1995; 60FR50411, Sep. 29, 1995; 61FR5507, Feb 13, 1996; 61FR43454, Aug 23, 1996.

U.S. Department of Labor, Occupational Safety & Health Administration, *Title 29 Code of Federal Regulations, Part 1926.1101, Appendix F-Work practices and engineering controls for Class I asbestos operations - non-mandatory*, as detailed in 59 Federal Register (FR) 40964, Aug 10, 1994; 60FR33972, June 29, 1995.

U.S. Department of Labor, Occupational Safety & Health Administration, *Title 29 Code of Federal Regulations, Part 1926.1101, Appendix H-Substance technical information for asbestos - non-mandatory*, as detailed in 59 Federal Register (FR) 40964, Aug 10, 1994; 60FR33972, June 29, 1995.

U.S. Environmental Protection Agency, *Title 40 Code of Federal Regulations, Part 763, Asbestos Hazard Emergency Response Act*, Mar 7, 1996.

Chapter 16

Evolving Industrial Hygiene Issues

Martin B. Stern, MPH, CIH

I. OVERVIEW

In the 1990s, industrial hygienists have been introduced to a plethora of issues that have challenged traditional approaches to problem-solving, and forced the development of innovative techniques and protocols to protect the health of the working population. Global workers in the 1990s can potentially be exposed to microbial agents, including tuberculosis; construction site contaminants; and exposure to a myriad of hazardous waste site contaminants. The industrial hygienist of today must also be versed in total quality management and environment, health, and safety management systems such as the International Standards Organization 9000 and 14000 Series Standarda, and the Occupational Safety and Health Administration's (OSHA's) Voluntary Protection Program, as well as utilizing performance metrics to assess program effectiveness, or ineffectiveness, on an on-going basis.

These issues have remanded the hygienist to come up with new personal protective equipment approaches, engineering controls, and administrative controls, while modifying traditional approaches to exposure assessment. The industrial hygienist must also demonstrate an intimate understanding of management principles and philosophies, as management and business are the primary drivers that are determinants if an industrial hygiene and environmental health and safety program will be adequately supported and funded.

A practicing industrial hygienist in the 1990s is required to be versed and knowledgeable on issues that have evolved over the last 20 or 30 years. These issues are reflective of how the United States' industrial population has matured from manufacturing to service industries. As well, the environmental movement has facilitated the development of new industries and consulting enterprises aligned with the protection of the environment. The transition away from heavy manufacturing has required that the industrial hygienist think differently in the resolution of worker exposure issues.

How has the industrial hygienist evolved to address these issues as they have become so prevalent in the 1990s? With regard to these diversified issues, the industrial hygienist has had to adjust the traditional approaches that have been the staple of industrial hygiene for the past 50 or so years, and adapt accordingly. This chapter will focus on a handful of issues that present challenges to the 1990s industrial hygienist and will continue to be of concern as we proceed to the 21st century.

Figure 16.1 Conducting an indoor air quality survey in an administrative office building (Courtesy: Foxboro Environmental, East Bridgewater, MA).

II. INDOOR AIR QUALITY

A. General Indoor Air Quality Considerations

Indoor air quality issues in commercial and administrative office buildings have evolved to present unique challenges to the industrial hygienist, as viewed in Figure 16.1, which identifies an industrial hygienist performing an indoor air quality survey in a "typical administrative office location." Utilizing traditional sampling approaches and correlating the resulting sample results to an existing OSHA Permissible Exposure Limit (PEL) Standard, could potentially lead an industrial hygienist without knowledge of indoor air quality scenarios to erroneously conclude that a problem does not exist. An industrial hygienist without knowledge of indoor air quality issues could conclude this type of decision, even if there were documented incidents of worker illness. A conclusion of "no problem" would cause the affected building occupants, in addition to suffering from a real or perceived illness, to also question the industrial hygienist's qualifications, capabilities, and integrity. Their problem would not be resolved, despite the fact that the industrial hygienist had concluded that a problem "did not exist."

An industrial hygienist familiar with the assessment and resolution of indoor air quality (IAQ) events would understand that correlating air sampling data obtained during an IAQ survey to OSHA PELs is inappropriate. The industrial hygienist would know that the workers in an administrative or commercial office building are exposed to a myriad of contaminants, generally at levels three to five orders of magnitude below the corresponding OSHA PEL. A contaminant at a concentration approaching an OSHA PEL could be indicative of a serious problem in commercial and administrative facilities. The industrial hygiene professional who addresses these types of issues is versed in ventilation and the American Society of Heating, Refrigerating, and Air-Conditioning Engineers, Inc. (ASHRAE) requirements.

1. ASHRAE Standard 62 requirements

Regarding ASHRAE requirements, these Standards have been drafted to provide guidance to maintain acceptable indoor air quality. ASHRAE Standard 62-1989, entitled *Ventilation for Acceptable Indoor Air Quality*, contains guidance which recommends that air handling equipment provide a minimum of 20 cubic feet per minute (CFM) of outdoor air (OA) per person in typical administrative office space. The 20 CFM requirement is designed to maintain airborne carbon dioxide (CO_2) levels below 1000 ppm to remove occupant-generated odors and bioeffluents. CO_2 is a byproduct of human respiration, and levels above 1000 ppm can indicate that an HVAC system is not properly supplying (or removing) air from the occupied spaces. When insufficient air supply or removal occurs in a facility, occupant and facility odors can be become more prevalent. The 1000 ppm value was not meant to serve as a "health-related" standard, only as an indicator that there is a potential air delivery or return problem occurring with the HVAC system.

ASHRAE 62-1989 also recommends that smoking rooms be equipped to provide a minimum of 60 CFM of OA per person, reception areas should provide occupants with a minimum of 15 CFM per person, while conference rooms should be designed to provide occupants with a minimum of 20 CFM per person. The Standard also contains minimum OA ventilation rates for telecommunications facilities, restaurants, theaters, bathrooms, and other facilities. The industrial hygienist should be familiar with computational and instrumental techniques to ensure that adequate outdoor air is provided to meet occupant demands:

$$ACH = \frac{Q \; C}{V} \qquad\qquad (16.1)$$

Where:

ACH = Air changes per hour
Q   = Volumetric air flow rate into the space, CFM
C   = Constant, 60
V   = Volume of space, ft³

This equation can assist industrial hygienists in determining if an occupied space is being supplied with adequate quantities of outdoor air. Typically, the ACH value will range from 0.5 ACH to 2.0 ACH. In administrative and commercial office, these values may be as low as 0.1 ACH.

ASHRAE 62-1989 contains recommendations for acceptable occupancy loads per square foot area, acceptable airborne pollutant concentrations, relative humidity ranges, control of outdoor air which is contaminated, and other provisions related to maintaining acceptable ventilation and indoor air quality in administrative and commercial space. During 1996, ASHRAE Standard 62-1989 was undergoing substantial revision. Based on preliminary evaluation of some of the key criteria of the proposed changes, it appears that the revised Standard will offer recommendations to ensure that sources of contaminants are properly addressed during the design phase of HVAC activities. Contained in Table 16.1 is an overview of key components proposed in ASHRAE Standard 62. ASHRAE is placing a great deal of responsibility to address IAQ measures on design engineers to ensure that ventilation rates account for all pollutant sources. The draft Standard's coverage of how to assess contaminant concentrations and sources is uncertain though. As was the case in Standard 62-1989, ASHRAE references multiple sources and data to derive acceptable contaminant concentrations in the indoor environment. The revised Standard also references *acceptability criteria* to ensure that ambient air meets the comfort requirements of the occupants of the space. The issue of acceptability criteria is somewhat controversial, though, as perception will be a primary gauge in developing proper ventilation rates to remove odors and contaminant concentrations. This may create particular difficulty when addressing hypersensitive and immuno-compromised individuals. These individuals, potentially single members of an entire work location, may experience discomfort and health-related effects at concentrations below values referenced in the Standard's revision.

The revised Standard 62 aggressively addresses design criteria to prevent the accumulation of moisture on facility structural components inside, as this can lead to the growth and amplification of pathogenic organisms. The revised Standard also contains information to ensure that preventive maintenance routines are performed to identify and mitigate potential sources of microbiological contamination. As the revised ASHRAE Standard 62 evolves, it will be interesting to see its overall impact on the health and comfort of administrative and office workers.

Table 16.1
Key Components of ASHRAE Standard 62-1989R, Public Review Draft

Component	Description of Criteria
Revised scope and coverage	Draft changes take into consideration chemical, physical, and biological contaminants, as well as moisture and temperature that can potentially impact human health and air quality comfort perception
Revised purpose	To ensure that acceptable IAQ is derived based on multiple criteria, including: ventilation rates; contaminant source management, and control mechanisms (air cleaning)
Revised definition of "acceptable IAQ"	Refers to IAQ that presents no dissatisfaction to a majority of a space's occupants, and there is likely to be no readily identified contaminants present at concentrations that pose significant health risk
Design ventilation rates	Multiple procedures are referenced to establish acceptable ventilation rates: **(1)** *Simple systems procedure* - outdoor air and supply air rates are determined based on floor area (simple systems only); **(2)** *Prescriptive procedure* - outdoor air rates are based on occupancy and floor area, while supply air rates are floor area dependent; **(3)** *Analytical procedure* - design criteria is based on contaminant limits and maximum health and comfort requirements
Construction & system start-up	Procedures developed to prevent contamination of adjacent spaces during construction activities, and establishment of post-construction star-up procedures, including: **(1)** filtration; **(2)** construction area isolation; **(3)** post-construction *purge*; **(4)** system inspection; **(5)** generation of documentation indicating activities to be performed
Operations & maintenance	Recommendations included to ensure that there is a responsible party identified to ensure preventive maintenance operations are performed as required in a timely manner, including: filter maintenance; condensate pans; dampers; intakes; etc.

Source: *ASHRAE Standard 62-1989R, Ventilation for Acceptable Indoor Air Quality Public Review Draft.*

2. ASHRAE Standard 55 requirements

ASHRAE Standard 55-1992, entitled *Thermal Conditions for Human Occupancy*, offers recommendations for acceptable temperature ranges to maintain occupant comfort in administrative and commercial facilities. The Standard takes into consideration seasonal variability when establishing acceptable ambient temperature ranges. Temperature ranges were established and corrected for clothing expected to be worn during each given season. Acceptable thermal comfort ranges have been generated for winter months and summer months. During summer months, ASHRAE 55-1992 recommends that indoor temperature ranges be maintained in the 73° to 79°F range, while the Standard recommends that temperature ranges in winter months be maintained between 68° and 75°F. ASHRAE Standard 55 also employs the use of psychrometrics to assist in the development of acceptability criteria. These temperature ranges were established with the intention of appeasing the comfort of 90% of a facility's occupants.

3. *IAQ survey techniques*

IAQ surveys involve the use of instrumental techniques and approaches that can be somewhat different than sampling protocols typically employed when evaluating worker exposure in the industrial sector. Usually, IAQ investigations are conducted with the use of direct-reading instruments that are portable, durable, and easy to use. Rarely do IAQ surveys employ integrated sampling techniques. There are other issues that can impact indoor air quality as well: temperature, relative humidity, particulate matter, and a host of other elements are often implicated in IAQ episodes. The important issue to remember is that the industrial hygienist has to approach a potential indoor air quality problem using techniques and approaches that are substantially different from traditional approaches, including having a thorough understanding of HVAC systems and their components, knowledge of how to read blueprints, and how to correlate data obtained in IAQ investigations into meaningful results. Table 16.2 contains an overview of instruments often utilized in IAQ surveys, as well as the parameters they evaluate. Each of the referenced parameters can be linked as an IAQ contaminant based on the presence of these compounds in numerous IAQ studies in the past.

Table 16.2
Examples of Instruments Used for IAQ Surveys

Parameter Evaluated	Instruments Potentially Employed
Carbon dioxide	Infrared CO_2 instrument
Carbon monoxide	Electrochemical sensor
Organic compounds	Passive dosimeter badge, pocket photoionization detectors, infrared ambient air samplers
Temperature/relative humidity	Thermohygrometer
Formaldehyde	Passive dosimeter badges
Ozone	Infrared or electrochemical sensor
Particulate matter	Direct reading respirable dust monitor
Bioaerosols	Cascade impactor

The parameters referenced in Table 16.2 are routinely involved in IAQ events. Selection of the appropriate instrument is essential to ensure the desired results will be obtained. Figures 16.2 and 16.3 contain examples of IAQ instrument currently involved in IAQ surveys today.

The key is to be able to associate the given IAQ contaminant with a typical source, as depicted in Table 16.3. Table 16.3 also contains potential mitigation strategies. An IAQ investigator must be prepared to conduct a thorough, detailed assessment to identify the given source (or sources) of IAQ contamination. There are many other parameters that can be implicated in IAQ events, but typically the parameters that are referenced in Table 16.3 are most often implicated in events involving IAQ.

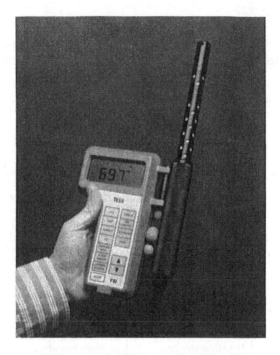

Figure 16.2 Directing indoor air quality instrument for carbon dioxide, carbon monoxide, temperature, and relative humidity (Courtesy: *TSI Incorporated*, St. Paul, MN).

Figure 16.3   Infrared, direct-reading ambient air sampling instrument (Courtesy: *Foxboro Environmental*, East Bridgewater, MA).

Table 16.3
Contaminant Pathways and Mitigation Strategies

Contaminant	Potential Sources	Mitigation Strategies
Carbon dioxide	Problems with HVAC system distribution and removal	Determine potential HVAC system disconnects and repair accordingly
Carbon monoxide	Re-entrainment of incomplete combustion sources into outdoor intakes	Move (or raise) exhaust stack to alternative location
Organic compounds	Construction activities, adhesives, chemical compounds utilized in the facility	Identify sources; if chemicals: ensure that handling is conducted per material safety data sheets (MSDSs), including adhering to ventilation requirements; segregate construction activities from adjacent space, and ensure that air supply and returns grilles are blocked
Formaldehyde	Off-gassing of newly installed carpeting, partitions, or pressed wood products	Increase ventilation to areas where new furnishings are installed, elevating temperatures in areas of newly installed furnishings ("bake-out") helps to expedite off-gas process
Microbiologicals	Leaking, water-damaged furnishings and structural materials, or pooled HVAC condensate	Remediate/remove microbial contaminated materials and disinfect nondisposable structural components with an appropriate biocidal agent
Particulate matter	Construction activities, improper/non-existent particulate filter maintenance, entrainment of outdoor particulate sources	Ensure air handling units are equipped with appropriate particulate filter and change-out is accomplished on a routine, regularly scheduled basis; If "blow-through" particulate matter is identified on the surface of supply diffusers or return grilles, investigate prospect of interior duct cleaning; segregate construction activities from adjacent space and ensure that air supply and returns grilles are blocked
Ozone	Printers, photo-reproduction equipment, or other electrical equipment sources	Identify sources of ozone and ensure appropriate ventilation control exists, either local exhaust or dilution ventilation
Polycyclic aromatic hydrocarbons	Roofing or road resurfacing operations	Ensure that outdoor air intakes are set at minimum settings while roofing (rooftop intakes) or road-resurfacing (ground-level intakes)
(4)-Phenyl cyclohexene	Off-gassing of newly installed carpeting	Recommend prior to installing new carpet, carpet to be installed is baked-out by manufacturer, and upon installation, utilize maximum outdoor ventilation rates, preferably 100% outdoor air
Fibers	Breakdown of interior acoustical duct insulation, release of fibers from carpet or office furnishings	Wet-wipe and/or HEPA vacuum surfaces potentially contaminated with fibers

Each IAQ parameter can have multiple sources/events that can lead to elevated concentrations of the contaminant, and subsequently impact occupant health or comfort. For example, carbon dioxide measurements exceeding 1000 ppm are indicative of a potential problem with the supply (or removal) of air to/from the occupied spaces, or the design ventilation rates are insufficient to meet the needs of the current space's occupancy loading. The industrial hygienist would seek to evaluate *why* the carbon dioxide value exceeded 1000 ppm, and determine *what* within the HVAC or its components was not operating properly. The industrial hygienist may discover any of the following deficiencies to be present:

- Short-circuited duct work

- Outdoor air dampers completely shut

- Return plenums not equipped with penetrations to return "waste" air back to air handling unit for exhaust to outside or recirculation

- Supply diffusers or return grilles blocked

- Variable air volume (VAV) box(es) not working properly

- System not designed to meet current occupancy loading

- Diffusers/return grilles not properly placed

- Office configuration prevent proper mixing of supply, or exhaust of "waste" air

These elements represent a sampling of some of the issues that may lead to airborne carbon dioxide concentrations in excess of 1000 ppm. Another potential problem that occurs when carbon dioxide values are excessive is related to the system's inability remove occupant- and building-generated pollutants and odors. If a system is not removing the waste products of the respiration process, it is unlikely that other pollutant loads are being effectively purged either. The key is that the industrial hygienist has to be cognizant that there are potentially multiple contaminant sources and pathways involved in IAQ episodes and events.

B. Microbial Contamination

An IAQ contaminant that merits separate discussion is an event and episode involving exposure to microbial contamination. Microbiological and bioaerosol issues have been put to the forefront in the industrial hygiene and health sectors due to an exposure event in administrative and commercial office buildings, hospitals, hotels, schools, and cruise ships. Some of the events have resulted in fatalities. Bioaerosol contamination can adversely impact the health of individuals who work in administrative and commercial office buildings, hospitals, schools, and hotels. It is recognized that there are other professions where exposure to microbiological agents is a predominant exposure consequence, as depicted in Table 16.4.

Table 16.4
Biological Agents and the Potential Industries Impacted

Biological Agent	Potential Occupation Impacted
Anthrax	Agricultural workers
Brucellosis	Veterinarians, farm workers
Campylobacteriosis	Veterinarians, packing plant workers, poultry processors, livestock producers
Chlamydia	Poultry workers
Cytomegalovirus	Daycare workers, nursery workers
Erysipelothrix infections	Veterinarians, fish handlers
Giardiasis	Day care workers
Leptospirosis	Animal husbandry handlers
Q Fever	Veterinarians, dairy workers
Shigellosis	Mental hospital workers, hospital workers, daycare workers
Sporotrichosis	Farmers, gardeners, horticulturists

Source: *Patty's Industrial Hygiene and Toxicology, Fourth Edition, Vol. 1A.*

This section's effort will focus primarily on activities associated with exposure in commercial and administrative buildings. Workers from these types of locations are generally unsuspecting participants of bioaerosol episodes. Most commercial and administrative facilities have ambient environmental conditions and components that can serve as optimum reservoirs for the growth, amplification, and dissemination of microbial agents.

Microbial contamination in administrative and commercial facilities can result from a number of different events. These events can be characterized as maintenance-related occurrences, episodic/catastrophic events, and/or from improper/inadequate engineering designs. Each of these occurrences contributes significantly to outbreaks involving pathogenic organisms. An industrial hygienist must be somewhat familiar with each pathway in order to properly assess incidents involving microbial agents and provide recommendations as to how the contamination can be effectively mitigated and addressed.

1. Bacterial contaminants

Microbial organisms typically implicated in indoor air quality episodes can be classified into three major categories: bacterial, fungal, or protozoan agents. Typical bacterial organisms implicated in microbial episodes include *Clostridium, Streptococcus, Salmonella, Myobacterium, Pseudonomas, Acinetobacter, Bacillus,* and Thermophilic Actinomycetes. Typically, bacterial agents cause fever and malaise, changes in white blood cell counts, respiratory illness, shock, and in certain concentration levels, even death. Individuals with preexisting allergies or lung conditions and ailments, such as asthma, have been shown to have decreased lung function when exposed to elevated airborne concentrations of bacterial organisms. There has been difficulty in assessing what constitutes "elevated" bioaerosol levels. The American Conference of Governmental Industrial Hygienist's, in its work *Guidelines for the Assessments of Microorganisms in Indoor Air,* has recommended that airborne bacterial concentrations in excess of 1000 colony-forming units per cubic meter (1000 CFUs/m³) is indicative of a contamination. There has been support for considering elevated airborne bioaerosol levels when indoor concentrations exceed outdoor concentrations by 200 CFUs/m³ for *like* microbial species.

2. Fungal agents

Fungal agents routinely identified in IAQ episodes include *Cladosporium, Penicillium, Aspergillus, Alternaria,* and *Mucor*. Fungi are consistently implicated as allergenic agents, with fever and asthmatic responses being commonly viewed symptomatic responses. Fungal agents can also cause hypersensitivity pneumonitis as well. There has not been a clear correlation between airborne fungal concentration and the development of illness; however, the airborne assessment concentration expected to result in increased incidence of illness should be similar to the approach employed for the assessment of bacterial agents.

3. Protozoa

Protozoans that are implicated in indoor air quality episodes include amoebae and ciliates. Protozoans can cause infections. Individuals who are immunocompromised are particularly susceptible to the effects of protozoans when they enter the bloodstream. Severe eye infections are routinely associated with exposure to protozoan species.

4. HVAC maintenance concerns

Improper and ineffective maintenance is one of the primary causes of microbiological contamination in administrative and commercial office facilities. When preventive maintenance programs are insufficient or lacking, often the result is airborne bioaerosol contamination. Typical heating, ventilation, and air-conditioning system components that require routine, scheduled maintenance include: outdoor air intakes; particulate filters; heating and cooling coils; air delivery ductwork, especially if it is lined with a sound dampening agent such as fibrous glass; variable air volume boxes; air supply diffusers; and air supply and return grilles.

As outdoor air is brought into a heating, ventilation, and air-conditioning systems, it passes through an intake which may or may not contain screens designed to prevent the infiltration of gross contamination such as leaves, bird droppings, rodents, or other larger sources of contamination from infiltrating the air delivery equipment. If the screens and outdoor air intakes are not periodically cleaned and freed of any identified debris, the resulting buildup of organic matter can serve as a nutrient source for a host of pathogenic organisms. Organisms that utilize organic matter located at an outdoor air intake or screen can be easily entrained in the air delivery equipment and disseminated to occupied spaces.

Upon entering the air handling equipment, the outdoor air is mixed with air that has been returned from the occupied spaces and passed over a particulate filter bank. Generally, these filters have a low to moderate capture efficiency. Routine, scheduled filter maintenance is an essential component of a preventive maintenance program. Particulate matter that builds up on a filter can "break through" the filter and be disseminated through the air delivery system. Particulate matter often is laden with pathogenic organisms which utilize the organic constituency of the particulate matter as a primary nutrient source. The filter type will determine when changing is necessary. Most manufacturers recommend changing filters based upon a given pressure drop across the face of the filter, as measured with a magnehellic pressure gauge. In most instances, trade professionals will opt for changing filters based on a scheduled time frame, or based on visible evidence of particulate buildup on the filter. Periodically, the frame which houses the filters should be cleaned and removed of any particulate buildup. This should be accomplished approximately twice a year, as this debris can be heavily contaminated with microbial agents.

After the air passes through the particulate filter bank, it is conditioned, either heated or cooled, by heating and cooling coils. Mechanical cooling generates a great deal of moisture from the cooling coils. The moisture is captured in a condensate pan. This moisture must be freely draining, or accumulated moisture can serve as a reservoir for microbial organisms. As well, if generated condensate does not readily flow from the cooling coil drip pan, it can be distributed through the air delivery system and elevate the moisture content in air, making conditions more favorable to sustain and support microbial growth and amplification. Any pooled condensate should be treated with an appropriate biocidal agent. A regular preventive maintenance program should seek to inspect and ensure that the cooling and heating coils are free of any suspect microbial debris and condensation. Identified suspect contaminant debris should be cleaned with an appropriate disinfecting agent accordingly. Evaluation and inspection of the cooling coils are especially critical during months when facilities utilize air conditioning.

Upon traversing the air handling equipment, air is delivered via ductwork which may contain interior acoustical lining. Routinely, this material is fibrous glass, and if there is an organic nutrient source present, such as particulate matter and suitable thermal environmental conditions, this material serves as an optimal growth media for fungal and bacterial agents. Though generally not included as an inspected element in a preventive maintenance program, interior insulation should be inspected for the presence of suspect microbial contamination approximately every 3 to 4 months. If the inspection identifies the presence of microbial contamination, mitigation is accomplished in a manner comparable to the removal of asbestos. Containment areas are established and maintained under negative air to prevent cross-contamination of microbial agents. Workers involved in the remediation of contaminated duct insulation don air-purifying respirators with high-efficiency particulate air (HEPA) cartridges, disposable tyvek coveralls, and a canvas work glove which is placed over a latex glove. During remedial activities, air sampling is conducted outside of the containment area to ensure that bioaerosol contamination is not migrating outside of the established work zone. Improper removal techniques can result in the contamination of sections of facilities free of airborne microorganisms.

In a variable air volume system, the main air supply ductwork then distributes air to duct branches, where air is then distributed from variable air volume boxes to duct branches which supply air to the occupied spaces. Variable air volume boxes are controlled by thermostats in the occupied spaces. These boxes often are lined with acoustical insulation which can become contaminated with microbial agents if not maintained properly. Inspection of the variable air volume boxes for the presence of contamination should be accomplished two to three times a year. The insulation should be removed under containment and negative air in the event contamination is identified.

After delivery to the occupied space, air is returned through grilles to an open plenum, or ducts which return the air to the air handling unit. Some of the return air is exhausted to the outside, while some of the air is mixed with outdoor air for redistribution. It is understood that there are numerous system configurations and variations, but the discussion of the above system was included to provide an overview of components that could potentially could be involved in microbial episodes.

The improper design of the components of heating, ventilation, and air-conditioning equipment can contribute to microbial contamination in administrative and commercial buildings. Subtle design discrepancies can contribute to the growth and amplification of microbial species. For example, cooling coil condensate pans that are designed for the drainage of the condensate generated from the coils, if improperly pitched, will not drain sufficiently. Pathogenic organisms can utilize the non-draining condensate as a growth reservoir, and when an appropriate nutrient source is present, they can amplify and be disseminated through the air handling equipment and be delivered to the occupied space. Table 16.5 contains an overview of typical design deficiencies that can result in microbial contamination. Though the components are not be to be an inclusive listing, it helps to focus the industrial hygienist's attention on HVAC system design flaws that have been identified in situations involving bioaerosol contamination. Professionals should be aware that a thorough and detailed assessment of the HVAC systems and its components will be necessary to evaluate all potential sources of microbial contamination.

Episodic events can contribute to incidents involving microbial contamination. Examples of episodic events that can result in contamination include: leaking roofs, walls, and facility structural components; release of fire sprinkler system water; leaking pipes and water-containing vessels; damaged heating, ventilation, and air-conditioning system elements; and other episodes that can contribute moisture to any facility components. Facility components potentially involved in these episodes are: carpeting, furnishings, wallboard, ceiling tiles, or other porous construction-related materials. When a furnishing or porous construction material has been contaminated with microbial organisms, generally the material cannot be disinfected and must be disposed. Attempts to disinfect porous items contaminated with microorganisms often prove fruitless, as subsequent growth and amplification quite often recur.

Table 16. 5
HVAC Design Deficiencies That Can Lead to
Microbial Contamination

HVAC System Component	Potential Design Deficiency/Problem	Resulting Affect
Acoustical duct insulation	Employed in system which utilizes humidification or typically has elevated outdoor relative humidity values	Damp acoustical insulation can serve as an optimal growth medium for microbial agents
Coil drip pan	Not equipped with properly sized drain	Condensate/moisture accumulates and serves as reservoir for microbial agents
Cooling coil drip pan	Not properly pitched	Condensate/moisture accumulates and serves as reservoir for microbial agents
Fan	Placed too close to cooling coils	Condensate blow-through passes into air delivery ducts, elevating relative humidity value
Humidification system	Recirculated water used	Recirculated water can serve as reservoir for host of pathogenic organism
Outdoor air intakes	Placed in close proximity to cooling tower or other moisture source	Aerosol/moisture potentially containing microbial contamination is entrained in air delivery equipment and distributed through system
Outdoor air intakes	Insufficient outdoor air quantities	Airborne concentrations of pathogenic organisms can become elevated in occupied spaces, since adequate dilution ventilation is not provided
Particulate filters	Insufficient particulate removal efficiency/capability	Particulate matter, potentially containing biogenic materials passes into air delivery system

When an episode occurs which results in the release and accumulation of moisture and water, it is imperative that the source of the release or leak be addressed and repaired immediately prior to conducting any remedial cleanup strategies. Once the release or leak has been controlled, water-damaged furnishings that are damp should be dried with a wet-dry vacuum to remove any residual moisture. Any furnishings or porous construction materials suspected to be contaminated with microbial agents should be removed and replaced. The removal and replacement of contaminated materials should be conducted utilizing similar methods as employed for asbestos abatement projects. The removal of facility components contaminated with microbial agents should be conducted utilizing the establishment of polyethylene containment around the area to be remediated, in conjunction with the use of negative air machines with HEPA filtration to prevent the contamination of "clean areas." Appropriate disinfecting agents should be utilized to remove any residual organisms potentially present after contaminated materials have been removed. All contaminated materials should be double-wrapped with polyethylene to prevent the release and dissemination of bioaerosols during the disposal process.

Airborne microbial contamination can be easily addressed in the workplace if facility management and industrial hygiene staff are prepared to promptly address situations involving releases of moisture/condensate. A regular on-going preventive maintenance program, coupled with thermal parameter control, will help to ensure the prevention of microbial contamination.

5. *Legionella pneumophila*

Legionella pneumophila is a ubiquitous organism that thrives in natural water bodies such as ponds, lakes, streams, and man-made warm-water sources, such as cooling towers, evaporative condensers, hot water tanks, and other similar systems and components. Legionella pneumophila, which has 34 identified serotypes, has been positively identified as a causative agent in numerous fatal outbreaks. The Legionella pneumophila bacterium is a causative agent of a respiratory disorder and febrile illness, including Legionnaires' disease and Pontiac fever. Legionnaires' disease has an exceedingly high overall case-fatality ratio, though individuals exposed to the organism can successfully be treated with antibiotics. The incubation period for the disease is 2 to 10 days. Immunocompromised individuals have greater susceptibility of contracting the disease as compared to "healthy" individuals.

The most famous Legionella outbreak occurred in 1976 at a hotel in Philadelphia, Pennsylvania, during an American Legion Convention. Of the 221 people afflicted with the illness, 34 people died; 221 people came down with symptoms typical of exposure to the organism; 72 of these individuals were not attendees of the conference, but had passed in the general vicinity of the hotel. The outbreak was traced to cooling towers heavily contaminated with the organism. The cooling towers were located directly adjacent to an outdoor air intake that served as one of the hotel's primary sources of outdoor air. Aerosolized Legionella droplet nuclei were entrained in the hotel's air handling/delivery system and distributed throughout the hotel. Unsuspecting American Legion conference attendees never realized they were being exposed until it was too late.

Cooling towers and humidification systems are routinely involved in Legionella outbreaks, as the water temperature contained within these systems can serve as an optimum reservoir for the Legionella bacterium to thrive. Typically, Legionella pneumophila bacteria amplify in reservoirs with temperatures ranging from 75°F to 110°F. The optimal temperature range for the growth and amplification of the organism is from 95°F to 99°F. The bacterium generally cannot amplify at temperatures below 68°F and is killed when temperatures exceed 131°F. Legionella pneumophila can exist in pHs as low as 2.0. Building systems which have been implicated in Legionella outbreaks have included: hot water tanks; cooling towers; condensate which has pooled in air delivery ducts; stagnant water which has accumulated in a building from leaks or damaged components; evaporative condensers/cooling coils condensate; or variable air volume boxes which are dysfunctional and condensate has accumulated or pooled.

Prevention and control of the Legionella pneumophila bacterium can be addressed by temperature maintenance and by employing the use of appropriate biocidal agents. For systems such as hot water storage tanks that have been implicated in Legionella outbreaks, the water temperature in the tanks was below 131°F. Quite often, water temperature is maintained at lower temperature values because of considerations associated with energy conservation and monetary savings purposes or because of faulty and dysfunctional equipment. Whatever the reason, when temperature falls within the desired optimal range, the potential for an incident involving the organism increase dramatically. If the temperature is maintained above 131°F, the bacterium cannot flourish and exist.

Preventive treatment and mitigation of Legionella pneumophila can be accomplished utilizing appropriate biocidal compounds. A chlorine solution is an effective tool to combat the Legionella bacterium. For a system with an existing contamination problem, the ratio of chlorine added to the system is based upon the concentration of the organism in the systems. All components and parts which come in contact with the contaminated water should be cleaned with the chlorine solution and a metal brush or bristled scrub brush. Workers assigned the task of decontaminating systems contaminated with the organism should don appropriate personal protective equipment which, at a minimum, should include an air-purifying respirator with HEPA cartridges, a disposable tyvek coverall, and canvas work gloves placed over a latex glove. In addition to chlorine, biocidal compounds such as potassium n-methyldithiocarbamate, disodium cyanothioidocarbonate, and bis-tributyltin oxide with dimethylbenzyl ammonium chlorides have been shown to be effective at treating waters sources contaminated with Legionella.

To prevent future occurrences of Legionella from impacting and contaminating heating, ventilating, and conditioning components, a routine preventive maintenance program should be implemented that includes application of an appropriate biocidal compound at regularly scheduled intervals. The extent of the Legionella problem will dictate the frequency of application of the biocidal compound selected.

In order to effectively prevent the amplification of the Legionella pneumophila bacterium in warm-water systems, it is essential that a routine preventive maintenance program is in place to ensure that visible algal growth is not identified on the substrate of water system components. Coupling visible inspection with routine cleaning and disinfection with an appropriate chlorine-based agent or comparable alternative will help to reduce the likelihood that an exposure event will occur. As well, facility structural components and mechanical system components should be routinely investigated to ensure that condensate is not pooling. Where potential design problems exist that are conducive to outbreak episodes, routine inspection and biocidal treatments can help to ensure that Legionella organism counts will be maintained at concentration levels that present minimal to no health risk. The overall key is for the industrial hygienist to partner with the facility management representative to work together to ensure that sources are properly rectified and corrected in a timely manner. By accomplishing this, the potential exposure threats associated with the organism can be managed in an effective manner.

C. Tuberculosis

Occupational exposure to tuberculosis has been catapulted to the forefront over the last several years. Workers in professions where daily interactions with potentially infected individuals include: health care, corrections officers/workers, drug treatment center employees, homeless shelters, long-term care facilities for the elderly, or other similar job activities. Individuals who work in nondescript jobs such as heating, ventilation, and air-conditioning mechanics, who perform maintenance on systems in high-risk facilities, also are at risk of exposure to TB. OSHA has indicated that exposure to TB resulted in 12 confirmed occupational deaths between 1993 and 1996. Several hundred other workers were involved in exposure incidents resulting in infection to TB. (Note: OSHA has drafted a Standard to address the TB issue as this publication was going to print).

The bacterium *Myobacterium tuberculosis* is the organism responsible for causing the tuberculosis infection. Individuals exposed to the tuberculosis bacterium generally develop a pneumonia-like condition. Symptoms of TB exposure include productive cough, coughing up blood, weight loss, appetite loss, lethargy, weakness, night sweats, and fever. The disease is spread from person to person via airborne transmission from exposed/infected individuals. Infectious droplet nuclei are typically transmitted in the sputum of infected individuals during bouts of sneezing or coughing. Individuals in the vicinity of infected individuals are at an increased risk of exposure to the infection if not properly protected. The handling of materials that have been contacted by an infected individual have not been shown to be a source of TB transmission. These materials can include clothing, cooking/eating utensils, reading materials, games, books, and other items.

Envision a crowded homeless shelter, or an emergency room in a hospital. There are people located all over the place, some are sitting, some are standing, many are pacing, or meandering the room. The temperature in the room is very warm, in excess of 75°F. Some people are sweating, though they are stilling wearing their coats. There are people who are coughing, sneezing, sick. Some of the individuals in the rooms are infected with the tuberculosis bacterium. Those that are infected are coughing periodically. The contents of their cough contain droplet nuclei contaminated with the myobactierum tuberculosis. The bacterium is thus disseminated through the ambient environment. The aerosolized bacterium can be inhaled into the lungs of unsuspecting individuals. The workers may or may not be familiar with the tuberculosis bacteria, or its mode of transmission. As a result, some of the workers are protected, but some have no protection, and thus have potential for unnecessary exposure. The tuberculosis bacterium is stable when it is airborne and suspended in droplets, remaining viable.

As the above hypothetical example depicted, the potential for exposure to the tuberculosis bacteria is a definitive problem to workers in these professions, as well as other professions where there is day-to-day interaction with members of society potentially contaminated with the bacterium. The message, however, is very clear: exposure potential is a real possibility for these work populations. According to the Center for Disease Control in Atlanta, Georgia, in 1990, there were 25,000 cases of TB reported in the U.S. In 1991, there were over 26,000 reported TB cases in the U.S. by the CDC. However, in 1994 in U.S. correctional facilities, there were over 24,000 cases of TB reported alone. These numbers should continue to rise in the high-risk professions. There is also a threat of antibiotic-resistant strains of TB. Adequate protection must be provided for the workers. An issue which creates additional concern is that droplet nuclei can remain airborne and be distributed by typical air movement associated with an air delivery system.

1. *Worker protection*

In order to adequately protect workers who are involved in professions that have a high risk for interaction or contact with individuals infected with the bacterium, a preventive program must be developed and implemented. Workers must be trained and educated on the hazards of TB, personal protective equipment measures, mechanisms to minimize the transmission of the bacterium, methods to identify high-risk populations, and mechanisms to ensure adequate engineering controls exist to minimize exposure potential. A written exposure control program should be developed detailing all procedures essential to minimize and control exposure to TB.

Individuals who have contracted, or are suspected to be infected with the TB bacterium, must be isolated from nonexposed individuals. These individuals should be segregated in a TB isolation room. A TB isolation room should be equipped with a ventilation system that is independent from the rest of the facility's air delivery system. Air flow into a TB isolation room should be maintained under negative pressure to prevent aerosolized TB from escaping into uncontaminated ambient air space. A protocol should be in place to routinely evaluate the directionality of the air flow to ensure the room is maintained under negative pressure. OSHA has established protocols to assess and ensure that TB isolation rooms are maintained under negative pressure by utilizing smoke tubes or other air movement directional indicators. It is essential that all individuals expected to have TB, or who have TB, are isolated in the TB isolation room as soon as it is known that they present an exposure risk. These individuals should be isolated until they are rendered noninfectious or the diagnosis of TB is effectively ruled out. Workers who have to gain entry into an isolation room should be equipped with an air purifying respirator with high-efficiency particulate air (HEPA) cartridges.

2. Engineering control

Engineering controls are necessary to minimize and prevent transmission of airborne concentration levels of the TB bacterium. There are four primary ventilation techniques that can be employed to control airborne infectious nuclei. Local exhaust ventilation can be utilized to remove airborne TB. This ventilation technique can be employed in situations where there are, and will be, individuals present who are infected, or are expected to be infected, with tuberculosis. The second ventilation technique involves the use of directional air flow to control airborne droplet nuclei close to its source of generation. This technique utilizes airflow to direct contaminated air away from workers. Effective general ventilation (dilution ventilation) should be used to provide an adequate number of air changes. ASHRAE recommends that areas likely to be contaminated with airborne infectious TB be equipped with an air delivery system that can accommodate a minimum of six air changes per hour, though ideally there are greater than or equal to twelve air changes per hour provided. The system should be equipped with high-efficiency particulate air (HEPA) filtration and the use of an appropriate in-line ultraviolet germicidal irradiation (UVGI) source to eliminate any residual airborne TB bacteria. The use of HEPA filtration in conjunction with UVGI will help to prevent airborne infectious droplet nuclei from contaminating additional spaces.

OSHA estimates that there are over 5.6 million workers who have an exposure risk to come in contact with TB-infected individuals. As a result, OSHA developed and established criteria for employers of high-risk professions to prevent and control potential worker exposure to TB, as well as establishing inspection criteria to assess facilities where TB exposures can be expected. In response to a perceived workplace crisis, OSHA established guidelines to which all high-risk facilities should adhere, as depicted in Table 16.6. Components range from medical surveillance to engineering controls. OSHA will cite employers who fail to adhere to the guidelines under the guise of the *General Duty Clause*, § 5(a)(1).

Table 16.6

OSHA Guidelines to Protect Workers from Exposure to TB

Inspection/Program Element	Description of Employer's Requirement
Early TB identification	Protocol must be established to identify individuals with active TB
Medical surveillance program	Offering of initial exams and TB skin tests; Periodic evaluations; Reassessments to monitor changes
Case management program	Establish and implement training program to communicate hazards associated with TB '
Engineering controls	Establishment of TB isolation rooms, booths, and hoods maintained under negative pressure, while employing HEPA filtration and direct exhaust to outside
Respiratory protection	Comply with all elements of 29 CFR 1910.134, *Respiratory protection*
Proper signage	As per 29 CFR 1910.145, *Specifications for accident prevention signs and tags*, appropriate warning/hazard recognition signage should be posted outside TB isolation rooms
OSHA 200 Logs	Positive TB skin tests and contraction of TB are reportable incidents/illnesses

Source: OSHA, 1996

3. OSHA inspections

OSHA has indicated that an employer's obligation to protect its workers are detailed in the OSHAct of 1970, regardless of whether or not a health and safety regulatory standard exists. The program that OSHA expects employers to utilize has been extracted from CDC requirements and recommendations for control and prevention of TB exposure. Between October 1, 1993, and March 31, 1995, OSHA conducted over 260 inspections involving potential employee exposure to TB. Approximately 43% of the inspections were in hospitals ($n=113$), 13% were in long-term care facilities ($n=35$), 11% were in correctional facilities ($n=29$), and the remaining 32% of the OSHA investigations were conducted in miscellaneous workplaces, such as clinics, homeless shelters, or drug treatment centers ($n=83$)

Controlling and preventing exposure to TB will continue to challenge industrial hygienists well into the 21st century, as cases continue to be identified. The industrial hygienist will be partnering with medical professionals, epidemiologists, engineers, public health professionals, as well as with the management and workforce from institutions routinely implicated in TB outbreaks to control and prevent additional exposure.

III. HEALTH AND SAFETY MANAGEMENT SYSTEMS

In addition to typical industrial hygiene and health-related issues, the industrial hygienist of today must be versed in environmental and safety elements, as well as management philosophies and techniques. The utilization and establishment of management systems have been effective mechanisms to help the industrial hygienist bridge the gap required to address traditional problems, environmental and safety issues, and management. There are numerous management systems that have shown promise in the environmental, health, and safety (EH&S) arena to assist in achieving the desired objectives.

A. OSHA Voluntary Protection Program

The OSHA Voluntary Protection Program (VPP) has been an active OSHA program since its inception in 1982. The purpose of the OSHA VPP is to foster excellence in industry with occupational safety and health requirements. Companies that apply for and receive acceptance in the OSHA VPP are required to participate in an intensive review of all of their major safety and health compliance programs to ensure that written programs are in place, employees have been trained and are familiar with training course elements, environmental monitoring has been accomplished as required, and accident and injury statistics are below the applicable Standard Industrial Code (SIC) average for the given industry in question. If a company meets the rigorous demands of the OSHA VPP inspection and assessment, the company is granted Merit Status, or Star Status, which is the highest status that can be achieved. Companies that attain Merit or Star status receive exemptions from routine scheduled OSHA inspections.

The OSHA VPP program incorporates the effective management of occupational safety and health by integrating reduction of accident and injury rates, joint labor-management safety committees which foster teamwork, compliance with all applicable OSHA Standards, and establish written programs and processes to ensure the continued performance of health and safety criteria, in addition to consistently meeting and exceeding compliance requirements. Each of these components forms the overall framework of a *management system* which addresses compliance, promotes continuous improvement, and fosters teamwork to protect worker safety and health and promote a safe and healthy workplace.

The OSHA VPP has been a staple of success in industry, as participants are gauged upon their on-going success with worker safety and health as compared to their industrial counterparts. Though the OSHA VPP is primarily intended to be a health & safety-based "management system," the key elements and components of the OSHA VPP process can be easily extracted and expanded to include coverage for a facility's environmental programs. The development of sound written programs, ensuring effective training delivery and participation, fostering teamwork between management and labor, and the other components detailed in the OSHA VPP process can assist a facility to systematically improve and address environment, health, and safety compliance and performance through an effective management system process.

B. International Standards Organization Standards

The International Standards Organization (ISO) has established management system standards to ensure that quality protocols and principles are incorporated into manufacturing processes and production procedures. By involving representatives from many nations, the ISO Standards are often utilized as the minimum criteria recognized for the sale and transfer of products and services in the global market. The United States is represented on the ISO Standards Committee by the American National Standards Institute (ANSI). ISO developed the 9000 Standard Series to provide a benchmark quality management system and quality assurance for organizations to follow. The overall objective of the ISO 9000 Series is to develop documented, written procedures and protocols, establish a process of internal assessments/audits, and incorporate a third-party certification review to evaluate the overall functional effectiveness of an organization's implementation of ISO 9000 components.

ISO developed an Environmental Management System, the ISO 14000 Series, which is applicable to an organization's management of environmental concerns, in a coordinated, documented management system. ISO 14000 seeks to provide organizations with a management system framework to integrate key environmental provisions into business and management activities on a daily basis. The ISO 14000 Series is framed in a manner similar to the ISO 9000 Series Standards. Table 16.7 contains a summary of the key requirements and provisions of the ISO 14000 Series Standard. The overall objective of the Standard is for participants to a establish a quality-based management system to document key operational protocols, while establishing clear-cut definitions of the roles and responsibilities for functions and individuals involved. Written procedures formulate the framework for Standard adherence.

C. Chemical Manufacturer's Association–Responsible Care®

The Chemical Manufacturer's Association (CMA) has established a management system process entitled *Responsible Care®* for companies involved in the manufacture of chemicals. The Responsible Care® program establishes key EH&S criteria that manufacturers of chemicals who participate in the program have to effectively address. The Responsible Care® program transcends into areas that expand beyond the typical scope of EH&S issues, including: product stewardship, participation with regulatory standard development, provision of assistance to other participating CMA companies, self-evaluation/audit requirements, as well as coverage of traditional EH&S-related issues.

Table 16.7
Key Components of ISO 14000 Series Standard

ISO 14000 Series Component	Overview of Component's Requirements
Environmental Policy	Develop an environmental policy that includes commitment to improvement and pollution prevention; commitment to comply with regulations; provides a framework for environmental objective setting and review; is documented, implemented, and communicated to all employees; and is made available to the public.
Planning	Includes procedures to ensure that environmental concerns are properly planned for and addressed to prevent significant environmental impacts through the use of legal strategies and objective and goal setting. Written programs are to be established for all relevant environmental functional areas detailing requirements.
Implementation & operation	ISO 14001 addresses the establishment of clearly defined roles & responsibilities. Sufficient resources will be allocated to ensure environmental program effectiveness. Top management commitment will be necessitated to support environmental policy requirements. Key environmental training and awareness needs will be identified and addressed
Checking & corrective action	Protocols will be developed and implemented to evaluate adherence to environmental standards and good management practices. Procedures will be developed and implemented to ensure correction of any identified deficiencies.
Management review	Periodic reviews will be conducted by an organization's top management to ensure the management system is adequately addressing environmental concerns, compliance, policy, and implementation needs and requirements.

The CMA Responsible Care® program has established *Codes of Management Practices* which consist of key EH&S functional areas requiring that chemical manufacturers develop and implement management system components. Table 16.8 contains an overview of the functional areas included in the Responsible Care® Codes of Management Practice, as well as each Code's primary objective. Each Code represents a quasi-management system in a critical EH&S area. As well, each code has independent objectives and requirements. Participating CMA members are expected to conduct self-evaluations, or self-audits, for each of the primary CMA Codes. Checklists have been developed to simplify the process and focus participating companies on critical areas of concern.

Table 16.8

Chemical Manufacturer's Association Responsible Care®
Codes of Management Practices

Applicable Code	Code's Objective
Employee Health & Safety	Safeguard and improve employee safety and health
Pollution Prevention	Encourage participating companies to reduce and minimize waste
Distribution	Reduce/minimize risks associated with the shipment/transport and storage of chemicals
Product Stewardship	To include EH&S issues in the all cycles of the life of a product
Process Safety	Prevent fires, explosions, and releases of chemicals
Community Awareness and Emergency Response	Establish procedures to prepare the community for catastrophic events

The CMA's *Employment Health & Safety* Management Practice contains provisions which enable participants to develop sound management protocols related to worker safety and health. This Practice requires that participants elicit management commitment to support health and safety initiatives. This is accomplished through the dedication of sufficient resources, both monetary and people in scope. This CMA Practice seeks to ensure that participating companies are staffed with adequate numbers of health and safety professionals to provide technical guidance and counsel in the workplace.

D. The AIHA's Occupational Health & Safety Management System

It is felt by many that the same *Total Quality Management* (TQM) principles employed that serve as the framework for ISO 9000 and draft ISO 14000 Series Standards, have direct application to an occupational safety and health management system standard. The American Industrial Hygiene Association's (AIHA's) Occupational Health & Safety Management System appears to be a document that could be extracted into an ISO-supported Standard, as the AIHA's management system is designed to address requirements related to health and safety.

The AIHA took initiative and established its own recommended approach for developing and implementing a health & safety management system. The AIHA management system document is titled: *Occupational Health & Safety Management System: An AIHA Guidance Document*. The components of the AIHA management system are designed to improve overall performance with health and safety concerns that could potentially impact the employee in the workplace. The approach is designed to ensure that companies think critically about these issues in their respective workplaces and develop a detailed, documented approach and strategy. By employing the components of the AIHA system, serious attention is given to the assessment of the hazards in the workplace and how they can be mitigated, in a calculated, well-thought written plan.

Table 16.9
Summary of AIHA Management System Criteria

Component	Description/Objective
Health & safety policy	States organizational health & safety objectives, and helps to ensure top management commitment
Responsibility and authority	Ensures that all members of an organization accountable for health & safety issues understand their roles and responsibilities, while having accountability for their actions
Resources	Seeks to ensure that organizations are sufficiently funded to adequately address health & safety issues
Written documentation	Develop a manual that documents all components that organization will employ to adhere to the management system requirements
Procedures	Establish operational procedures detailing the recognition, evaluation, and control of health & safety hazards in the workplace
Planning	Periodically, an assessment should be conducted to evaluate deficiencies (or potential deficiencies that could likely be expected) and establish corrective intervention strategies
Compliance and conformance review	Define and document procedures to ensure a process exists to stay abreast of pending/current legislation/regulation and develop appropriate organizational responses accordingly
Goals & objectives	Establish health & safety objectives and verify success of the objectives through key metrics such as accident/injury rates, and illness rates, etc.
Design control	Establish and maintain procedures to ensure that work stations are designed in a manner to control potential hazards, both health & safety in scope
Document & data control	Develop and implement procedure to ensure management of all elements related to health & safety documentation
Purchasing	Establish procedures to evaluate potential toxicological impact associated with product purchased for the organization, as well as adequately ensuring contractor adherence to health & safety stipulations
Communication systems	Develop and implement procedures to communicate critical health & safety requirements to employees and contractors in a timely manner to prevent and minimize hazards and potential exposures
Inspection and evaluation	Establish protocol to routinely inspect/audit the workplace for health & safety deficiencies
Corrective & preventive action	Establish procedures to implement corrective actions and procedures
Training	Ensure that employees are provided with appropriate training to educate them of the hazards of the workplace, while adhering to all regulatory requirements related to training

The AIHA management system stipulates imperative requirements to which the employer must adhere. A summary of some of the key elements of the AIHA management system are contained in Table 16.9. Other components not covered in Table 16.9 include: hazard identification and traceability; process control; receiving, inspection, and record keeping; test equipment protocols; inspection and evaluation status; control of nonconforming processes or devices; handling, storage, and packaging of hazardous materials; control of records; internal management system audits; operations and maintenance; and statistical techniques. Readers are encouraged to review the AIHA health & safety management system in detail to assess all requirements.

It would seem prudent to combine the components of the ISO 14000 Standard and the AIHA management system to form an EH&S management system that can serve as an inclusive management system for an organization. The application of such a management system is of particular value, as there is so much overlap when addressing environment and health & safety issues in industry today. The CMA approach appears to currently meet the EH&S management system needs of chemical manufacturers, though components of the CMA program can be easily applied to any manufacturing operation's criteria. Organizations that participate in a TQM/management system's approach to coordinate and manage EH&S programs will incorporate effective planning, implementation, strategic objective setting, and continual evaluation of program effectiveness to ensure success. In the long run, these programs will help to minimize compliance woes, improve accident/injury rates, and reduce costs associated with negative EH&S actions.

IV. METRICS

In order to effectively manage an industrial hygiene program, an industrial hygienist must have an operational understanding and working knowledge of how to incorporate metrics into the given programs that are managed by the industrial hygiene professional. *Metrics* are measures of performance to evaluate if an objective or goal is successfully attained. If properly addressed, it can enable the industrial hygienist to translate technical environment, health, and safety data and information into business language and terminology and effectively communicate the criticality of the industrial hygiene profession to today's business leaders. In the past, industrial hygienists have been unable to demonstrate value-added propositions as they presented their findings and outcomes in technical language. Though this language is easily understood within the confines of the industrial hygiene community, to those who have the responsibility to run businesses and corporations, this language may as well be foreign. Metrics can be the tool, if properly used, to transition the technical industrial hygiene language into business terminology.

Metrics are employed to ensure that a desired endpoint has been successfully achieved. For example, if a facility has problems related to repetitive strain injuries, specifically for carpal tunnel syndrome, metrics can be utilized to assess if an intervention strategy is successful. Potential intervention strategies may include delivery of training; process modification; workstation adjustment; or adjustment of worker positioning. Metrics are employed to evaluate if the control mechanism is obtaining the desired outcome. Sample metrics could potentially be as simple as evaluating if lost workday case rate incidences associated with repetitive strain injuries (RSIs) show a statistically significant decrease, or possibly determining if workers' compensation expenditures related to carpal tunnel illnesses are reduced after the intervention strategy is delivered, as compared to workers' compensation results prior to the delivery of any of the previously mentioned intervention strategies.

A. Establishment of a Metrics Program

One of the first steps prior to establishing a metrics program is to prioritize which health and safety issues will be targeted for intervention. This can be determined by evaluating accident/injury statistics, assessing workers' compensation/disability expenditures, audit/self-audit result summations, and regulatory enforcement actions. There are other data sources that may be available which can identify potential areas that may require performance improvement as well. Any data that helps to focus on target problem areas should be evaluated.

Table 16.10
Prioritization of EHS Risks Encountered in the Workplace

Potential Components to Determine Risk Priority	Potential Rating Scheme/Rationale
Regulatory Significance	The greater the regulatory significance for noncompliance, the higher the rating
Risk to Employee Health & Safety	The greater potential for serious employee injury or health impairment, the more weighting the issue receives
Ease of Developing Intervention Strategy	If a program will require extensive time and effort to develop the intervention strategy, it receives a low rating
Cost of Issue to Close	If the issue requires extensive revenue to address, the issue receives a low scoring
Revenue Expended on Issue	If there has been extensive revenue expended on the problem, the issue receives a higher rating
Number of Times Event/Incident Occurred	The greater the frequency of the incident, the higher the rating
Previous Regulatory Citations	Incidents that have resulted in citations/penalties in the past receive higher priority
Audit Results that Identify Compliance Gaps	Identified compliance gaps from audits receive highest priority/scoring as compared to favorable audit findings

In order to assess which areas should be addressed, risk management techniques need to be employed. Table 16.10 contains potential components and ratings scenarios that can be used to assist in the establishment of priority. When determining which health and safety issues to address for intervention, any, or all, of the potential rating approaches identified in Table 16.10 can be utilized. The key is for the industrial hygienist to prioritize the myriad of issues that will be present, and determine the most critical issues to address based on data, statistics, and potential impact to workers and the business.

B. Use of Metrics

After prioritizing which elements will be addressed, metrics will be employed to evaluate the success of intervention strategies used to address identified problems. For example, if an organization has identified that there is a potential problem with slips, trips, and falls, as indicated in elevated lost workday case rates, an investigation should be conducted to determine potential causes. These could include: wet working services; spilled debris, either liquid or solid, not being expeditiously cleaned; workers rushing to get tasks accomplished and moving too quickly through the work space; proper signage not present indicating that work tasks are being accomplished in egress point, and other similar activities. An intervention strategy will need to be developed to educate the workforce on the actions (or lack of actions) that can potentially lead to injuries associated with slips, trips, or falls. This program may include the use of formal classroom training, brief safety-break educational awareness sessions, handouts, hands-on demonstrations of events typically involved in slip/trip/fall events in the organization, or other mechanisms that convey the criticality of slip/trip/fall events to the organization. Metrics are employed to measure if the selected intervention strategies are effective. Potential metrics that may be utilized by an industrial hygienist or other EH&S professional may include assessing if overall lost workday rates or lost workday case rates improve after the intervention strategies are addressed or the conductance of an assessment to investigate if overall worker's compensation cost expenditures are reduced after the delivery of the prescribed intervention strategy. Regardless of the mechanism employed, by addressing a technical EH&S response action in terms of metrics, and translating the collected information into business language, the management of the organization will comprehend the results' meanings because the information has been presented in business language. Instead of being perceived as an overhead cost to an organization, health and safety, and environmental issues as well, can be perceived as constantly adding value to the organization or business.

V. CONCLUSION

Modern-day industrial hygienists will continue to have to recognize, evaluate, and control exposures and hazards that were not addressed by their counterparts in previous years. In addition to being technically competent and versed in the full gamut of traditional industrial hygiene *rubrics*, today's industrial hygienist must also have knowledge of safety and environmental issues, business philosophies, and evolving technical issues such as IAQ assessment and mitigation or tuberculosis control. Issues that may emerge to challenge the industrial hygienist in the near future may include control and mitigation of exposures associated with hantavirus, a vector-borne disease which is transmitted from the droppings, sputum, urine, and bites of infected rodents. Hantavirus exposure can also result from the handling of any equipment or materials that have been contaminated. Hantavirus gains entry into the body through broken skin, through the eyes, or potentially via the ingestion of contaminated materials. Workers most likely to be exposed to hantavirus are those who potentially work in the presence of the deer mouse, the cotton mouse, pinon mice, brush mice, and the western chipmunk, the most common carriers. These workers can include sanitation workers, telecommunications workers, and other workers potentially exposed to rodents or rodents droppings during the performance of their job responsibilities. Hantavirus produces a febrile respiratory illness that has resulted in numerous fatalities. Other challenges may be manifested from operations involving the refractory ceramic fiber and mineral wool, which has been utilized as a replacement insulating material for asbestos. Data has been presented that suggests a potential carcinogenicity of this material. Risk and exposure assessment techniques currently in wide use in the environmental field to assess the hazard potential of contamination impacting hazardous waste site contamination are considered in the industrial hygiene field as effective tools to evaluate multivariate exposures impacting the workforce. In the future, these tools may also be used in the day-to-day functioning of the industrial hygienist.

The key is for the industrial hygienist to prepare and meet the challenges presented in the workplace. Ultimately, the individuals that reap the rewards of effective industrial hygiene program management are the workers, as unnecessary exposures are controlled and mitigated. By keeping the work population healthy, the business also reaps benefits, as workers' compensation expenditures are reduced, the need for replacement labor is minimized, and productivity increases, or is not impacted negatively. As the clock heads toward the 21st century, the industrial hygiene profession will meet the new century, prepared for the challenges that await.

VI. CASE STUDY

A. IAQ Consultation

An industrial hygiene consulting firm with *no* IAQ survey experience, No-Clue Consulting, is contacted by a facility manager to investigate a potential indoor air quality problem in a building. Unfortunately, the consultant fails to identify the direct source of the problem, although air monitoring results for carbon dioxide and temperature are indicative of a potential problem. In order to gain closure on the issue, the facility manager has contacted you to address the situation because of extensive experience investigating and correcting indoor air quality concerns in the past.

Table 16.11
IAQ Air Monitoring Data by No-Clue Consulting Firm

Location	Time	Carbon Dioxide (ppm)	Temperature (°F)	Relative Humidity (%)
Christina's office	0900	550	69.5	42
Marty's office	0910	525	70	41
Chad's office	0920	550	69	40
Christina's office	1040	825	70.5	41
Marty's office	1050	850	70	4
Chad's office	1100	850	70	43
Christina's office	1300	950	73	43
Marty's office	1310	950	73	43
Chad's office	1320	975	73.5	40
Outside	1325	325	39	19
Christina's office	1440	1200	74.5	38
Marty's office	1445	1275	74.5	37
Chad's office	1450	1250	75.5	38
Christina's office	1555	1455	77	35
Marty's office	1600	1500	78	35
Chad's office	1610	1475	77.5	34
Christina's office	1710	1350	76	37
Marty's office	1720	1350	76	36
Chad's office	1725	1325	76.5	35

Upon review of the air monitoring data collected by the No-Clue industrial hygienist, you recognize that the carbon dioxide concentrations and temperature measurements increase linearly during the course of the day, and begin to decrease at the end of the workday (refer to Table 16.11 for a summary of No-Clue's air monitoring results). The data contained in Table 16.11 indicate that carbon dioxide and temperature measurements peak at 4:00 p.m. (1600) in Marty's office as carbon dioxide was measured to be 1500 ppm, and temperature reached 78°F. Measurements collected from Christina's office and Chad's office were also elevated.

In the report prepared by No-Clue, you learn that the air handling unit and its components that supply air to the area of concern were in good operating condition and involved in an on-going, documented, preventive maintenance program. One issue that surprises you, though, is that the No-Clue investigator did not evaluate the air supply or return system. The report referenced "time constraints" as the reason. Interviews with Marty, Chad, and Christina revealed that they felt the air was *stuffy*, *hot*, and *heavy*, especially in the latter part of the day. What could potentially be causing the problem?

You scratch your head and request a copy of an HVAC blueprint from the facility manager. Upon review of the blueprint, you see that the air supply system is a variable air volume (VAV) system, as you see VAV boxes on the prints. The area of concern is supplied by VAV box # 18-SW. You suspect that there may be ductwork blocked or disconnected from the VAV box. When you look above the hung ceiling, you see that all ductwork is attached, you do not see any disconnected or blocked ductwork, and everything appears to be in place. You go back to the blueprints to review the return system. It is an open-plenum system that does not appear to have any special components. You see that there are four fire-rated penetrations equipped with fire dampers designed to close during an emergency fire event as identified in blueprints. The penetrations serve as the only openings in the floor to decking sheetrock that exist in the open plenum space to allow "waste" air to be returned to the air handling unit for redistribution or exhaust to the outside. During nonemergency situations, those dampers are fixed in the open position.

Upon inspection, it is learned that none of the four fire dampers were ever installed, so return air has no way of effectively being removed from the occupied space. You recommend to the facility manager that the dampers be added as per the design requirements, which the facility manager agrees to address. After the four dampers have been installed, you are called back for subsequent air monitoring, which demonstrates that your recommendation was correct, as carbon dioxide and temperature values were well within the recommended ASHRAE guidelines.

REVIEW QUESTIONS

1. Of the contaminants listed below, which can be involved in episodes involving poor IAQ?

    a. Carbon dioxide   **b.** Sulfur dioxide   **c.** Microorganisms   **d.** All of the above

2. An effective approach to evaluate potential IAQ-related problems is to develop an air monitoring strategy that is coupled with the following activity(s):

    a. Assessment of HVAC system components and preventive maintenance procedures
      b. Evaluation of potential sources near outdoor air intakes
      c. Bulk samples of ceiling tiles and floor tiles
      d. Both a & b

3. The organism responsible for recent TB outbreaks in health care settings, drug treatment centers, and corrections facilities is:

    a. Bacillus tuburculinia
    b. Tubercullin pneumophila
    c. Myobacterium tuberculosis
    d. Tuberculosis cladosporium

4. Which of the listed elements is essential to mitigate and control exposure to Legionella pneumophila?

    a. Use of supplied air respirators
    b. On-going treatment program with an appropriate biocidal agent
    c. Routine wipe sampling with mixed cellulose ester fiber filters
    d. Manual scraping of identified contamination with an appropriate polyethylene-based brush

5. Exhaust from rooms where individuals with known exposure to TB reside should be treated prior to recirculation/redistribution with following control apparatus:

    a. HEPA filtration   **b.** Activated carbon   **c.** UVGI   **d.** Both a & c

6. What management system has been specifically designed to serve as a health and safety management system?

    a. ISO 9000 Series
    b. ISO 14000 Series
    c. CMA's Management Practice
    d. AIHA's Management System

7. Microbial contamination resulting from the Legionella pneumophila bacterium can be controlled if water temperatures of water-based systems are maintained above:

    a. 100°F   **b.** 125°F   **c.** 135°F   **d.** 150°F

8. Management systems are based on the following principles:

    a. TQM   **b.** CTD   **c.** RSI   **d.** ISO

9.   Metrics can be used as a tool to assess if:

    a. Desired intervention strategies are successful
    b. Accident/injuries result from a given cause
    c. Industrial hygiene samples are collected using random sampling techniques
    d. Air contaminants are attributed to a specific process operation

10. When addressing potential exposure incidents involving microorganisms, the following source(s) will routinely be involved in exposure episodes:

    a. Facility roofing materials
    b. Exterior HVAC sheet metal
    c. Cooling coils
    d. Both a & b

11. When determining appropriate data to assess potential environmental, health and safety issues that require intervention, it is felt that which of the following data elements should be evaluated?

    a. Lost workday case rates
    b. Identified deficient audit items
    c. Previous regulatory citations
    d. All of the above

12. When developing a strategy to investigate an IAQ episode to investigate complaints related to stuffiness and dead air, what is the *most important issue* that should be considered?

    a. Assurance that exhaust from reproduction machines is properly ventilated
    b. Number of workers impacted
    c. Proper function of the HVAC system and its components
    d. Location of OA intakes

13. All of the following elements represent components that can be utilized to establish a risk-based approach to problem resolution/correction except:

    a. Use of data related to the number of workers at the site
    b. Workers' compensation expenditures
    c. Top causes of lost workday case incidents
    d. Use of motor vehicle accident data

14. To ensure that porous materials such as fabric partitions or carpeting contaminated with microbial contamination are properly addressed, the *most prudent* approach is:

    a. Disinfect with an appropriate biocidal agent
    b. Dispose of the material using HEPA-filtered negative air containment
    c. Dry the material remove visible material with a vacuum
    d. Wash the materials with a soap/water/chlorine solution

ANSWERS

1.	d.		8.	a.
2.	d.		9.	a.
3.	c.		10.	c.
4.	b.		11.	d.
5.	d.		12.	c.
6.	d.		13.	a.
7.	c.		14.	b.

References

American Conference of Governmental Industrial Hygienists (ACGIH), *Guidelines for the Assessment of Bioaerosols in the Indoor Environment*, ACGIH, Cincinnati, 1989.

American Industrial Hygiene Association, Graphically speaking-OSHA TB inspection, *The Synergist*, Vol. 6, No. 9, 1995, 7.

American Industrial Hygiene Association, *Occupational Health & Safety Management System: An AIHA Guidance Document*, AIHA Publications, Fairfax, VA, 1996.

American Society of Heating Refrigerating & Air-Conditioning Engineers, Inc.(ASHRAE), *BSR/ASHRAE Standard 62-1989R, Ventilation for Acceptable Indoor Air Quality Public Review Draft*, ASHRAE, Atlanta, 1996.

American Society of Heating Refrigerating & Air-Conditioning Engineers, Inc. (ASHRAE), *Standard 62-1989, Ventilation for Acceptable Indoor Air Quality*, ASHRAE, Atlanta, 1989.

American Society of Heating Refrigerating & Air-Conditioning Engineers, Inc. (ASHRAE), *ANSI/ASHRAE Standard 55-1992, Thermal Conditions for Human Occupancy*, ASHARE, Atlanta, 1992.

Barbaree, J. M., Controlling Legionella in cooling towers factors affecting the transmission of Legionella from aerosol-emitting equipment to people are described, *ASHRAE Journal*, June 1991.

Blair, C. H., Legionella makes a comeback, *Occupational Health & Safety*, Vol. 65, No. 9, 1996, 67-69.

Burge, H. A., *Bioaerosols*, CRC Press, Inc., Boca Raton, FL, 1995.

Burton, D. J., ASHRAE IAQ standard 62 proposed revisions, *Occupational Health & Safety*, Vol. 65, No. 4, 1996, 28.

Chemical Manufacturers Association, Employee health and safety code of management practices, Responsible Care A Public Commitment, Chemical Manufacturer's Association, January 1992.

Clayton, G. D. and Clayton, F. E., Eds., *Patty's Industrial Hygiene and Toxicology*, 4th ed. Vol. 1, Part A; John Wiley & Sons, New York, 1991.

Dobos, R. T., Tuberculosis: an occupational illness revisited, *Professional Safety*, Vol. 40, No. 1995, 39-41.

Eckhardt, R., Introducing quality principles into safety regulatory strategies, *Professional Safety*, Vol. 40, No. 5, 1995, 34-36.

Figura, S. Z., Globalizing OH&S: is ISO the answer? *Occupational Hazards*: Vol. 85, No. 5, 1996, 51-53.

Hansen, D. J., *The Work Environment, Volume One Occupational Health Fundamentals*, Lewis Publishers, Boca Raton, FL, 1991.

Hansen, M. A., International standardization of safety management systems: Is there a need?, *Professional Safety*, Vol. 41, No. 7, 1996, 56-58.

Herbert, D. A., Safety as a competitive edge how you can set objectives that improve business performance, *Industrial Safety & Hygiene News*, Vol. 30, No. 10, 1996, 36-37.

http://www.cmahq.com/cmaprograms/rc/elements, Elements of responsible care, *Codes of Management Practices*, 4/8/98.

Industrial Safety & Hygiene News-Heads Up, Getting serious about finances, *Industrial Safety & Hygiene News*, Vol. 30, No. 10, 1996, 24.

International Standards Organization, *Environmental Management Systems-General Guidelines on Principles, Systems, and Supporting Techniques (ISO/TC 207/SC1)*, February 1995.

International Standards Organization, *Environmental Management Systems-Specification with Guidance for Use (ISO/TC 207/SC1)*, February 1995.

Kay, J. G., Keller, G. E., and Miller, J.F., *Indoor Air Pollution Radon Bioaerosols & VOC's*, Lewis Publishers, Chelsea, MI, 1991.

Kerbel, W., Indoor air quality taking a second look at blueprints, *Occupational Health & Safety*, Vol. 64, No. 1, 1995, 47-49, 72.

Markiewicz, D., Managing Risks, Cost-justifying safety and health, *Industrial Safety & Hygiene News*, Vol. 30, No. 10, 1996, 22.

McGinley, M. A, ISO management models: panacea or paradigm? *The Synergist*, Vol. 7, No. 9, 1996, 30-31.

McQuiston, F. C. and Parker, J. D., *Heating Ventilating, and Air Conditioning Analysis and Design*, 4th ed., John Wiley & Sons, Inc., New York, 1994.

PathCon Laboratories, *Technical Bulletin 2.4-A Suggested Air Sampling Strategy for Microorganisms in Office Settings*, Pathogen Control Associates, Norcross, GA, Not Dated.

Rautiala, S., Reponen, T., Hyvärinen, A., Nevalainen, A., Husman, T.,Vehviläinen, A., and Kalliokoski, P. Exposure to airborne microbes during the repair of moldy buildings, *AIHA Journal*, Vol. 57, No. 3, 1996, 279-284.

Redinger, C. F. and Levine, S.P., *New Frontiers in Occupational Health & Safety: A Management Systems Approach and the ISO Model*, AIHA Publications, Fairfax, VA, 1996.

Shands, L and Turner, R., Occupational health implications of hantavirus pulmonary syndrome, *The Synergist*, Vol. 6, No. 11, 1995, 18-19.

Smith, R. B., OSHA's TB rulemaking, *Occupational Health & Safety*, Vol. 64, No. 4, 1995, 48-51.

Swartz, G., OSHA's Voluntary Protection Program, *Professional Safety*, Vol. 40, No. 1, 1995, 21-23.

U.S. Department of Health and Human Services, Centers for Disease Control, Essential components of a tuberculosis prevention and control program, *and* Screening for tuberculosis and tuberculosis infection in high-risk populations, *Morbidity and Mortality Weekly Report*, Vol. 44, No. RR-11, 1995.

U.S. Department of Health and Human Services, Centers for Disease Control and Prevention, *TB/HIV Connection: What Health Care Workers Should Know*, U.S. Department of Health and Human Services, Centers for Disease Control and Prevention, Atlanta, 1995.

U.S. Department of Health and Human Services, Centers for Disease Control and Prevention, Prevention and control of tuberculosis in facilities providing long-term care to the elderly, *Morbidity and Mortality Weekly Report*, Vol. 39, No. RR-10, 1990.

U.S. Department of Health and Human Services, Centers for Disease Control and Prevention, *Controlling TB in Correctional Facilities*, U.S. Department of Health and Human Services, Centers for Disease Control and Prevention, Atlanta, 1995.

U.S. Department of Labor, Occupational Safety & Health Administration, Office of Compliance Assistance, *OSHA Instruction CPL 2.106, Enforcement Procedures and Scheduling for Occupational Exposure to Tuberculosis*, February 9, 1996.

U.S. Department of Labor, Occupational Safety & Health Administration, *OSHA Enforcement Guidelines for Occupational Exposure to Tuberculosis*, May 7, 1992.

Weinstein, M.B., Total quality management approach to safety management, *Professional Safety*, Vol. 41, No. 7, 1996, 18-22.

INDEX

A

Milton Keynes UK
Ingram Content Group UK Ltd.
UKHW021937071024
449327UK00022B/1838